Nanostructure Science and Technology

More information about this series at http://www.springer.com/series/6331

Kenneth I. Ozoemena • Shaowei Chen
Editors

Nanomaterials in Advanced Batteries and Supercapacitors

Springer

Editors
Kenneth I. Ozoemena
Council for Scientific and Industrial Research
Pretoria, South Africa

Shaowei Chen
University of California
Santa Cruz, CA, USA

ISSN 1571-5744 ISSN 2197-7976 (electronic)
Nanostructure Science and Technology
ISBN 978-3-319-26080-8 ISBN 978-3-319-26082-2 (eBook)
DOI 10.1007/978-3-319-26082-2

Library of Congress Control Number: 2016944233

Printed on acid-free paper

This Springer imprint is published by Springer Nature
The registered company is Springer International Publishing AG Switzerland

Preface

This decade has witnessed rapid development in the global quest for clean, sustainable energy and an outburst in new technologies, such as smart phones and electric vehicles. These developments have stimulated intense research interest in advanced electrical energy storage systems. Every market analysis continues to predict that advanced batteries, most notably lithium-ion batteries, and supercapacitors (also known as ultracapacitors) will dominate electrical energy storage technologies for a plethora of applications, ranging from portable electronics to next-generation environmentally friendly technologies such as electric vehicles and smart grids. Batteries and supercapacitors are complementary electrochemical energy storage systems; the former are characterized by high-energy density, while the latter are known for high-power density. Supercapacitors are most suited for applications that require energy pulses during short periods of time, such as in emergency doors, escalators, regenerative braking energy recovery systems in vehicles and metro-rails, and "stop-start" applications in modern cars. In addition to their applications in electric vehicles and portable electronics, batteries and supercapacitors are proving extremely useful in utility-scale energy storage, offering services such as: (i) price arbitrage (i.e., storing 'cheap' electricity during the off-peak periods when the cost of electricity generation is usually low and using it during the expensive peak times), (ii) industrial peak-shaving or demand charge reduction (i.e., using stored energy during peak periods in order to avoid penalties for breach of contractual peak demand), (iii) balancing power or frequency regulation (i.e., compensating for excess electricity generation and utilization from the grid), (iv) island and off-grid storage (i.e., to augment the electricity generated from the variable renewable energy sources such as solar and wind), (v) transmission and distribution (T&D) upgrade deferral (i.e., in a situation where the existing grid's capacity is barely enough to meet the required peak demand in a given area or to store the peak power supply from distributed variable renewable energy sources), (vi) voltage control/support (i.e., mainly to improve power quality and local grid congestion), and (vii) security of electricity supply (i.e., mainly as power backup or to avoid the socioeconomic problems arising from load shedding).

Despite the growing interests in research and technological applications of batteries and supercapacitors, these energy storage systems still fall short of some critical requirements, such as energy density, power density, cycle life, and safety. Nanostructuring or nanoscale engineering of electrode materials has emerged as one of the most elegant research and development strategies to improve the performance of batteries and supercapacitors and revolutionize their applications.

Nanomaterials in Advanced Batteries and Supercapacitors is unique in that it provides an authoritative source of information on the use of nanomaterials to enhance the electrochemical performance of existing electrode materials for lithium-ion batteries, magnesium-ion batteries, and supercapacitors. The book covers the state-of-the-art design, preparation, and engineering of nanoscale functional materials as effective electrodes for advanced batteries and supercapacitors, as well as perspectives and challenges in future research. Contributing authors are world experts in the field and carefully chosen to ensure an in-depth coverage of the various topics related to advanced battery and supercapacitor systems. The 15 chapters cover the critical components of the energy storage systems that are the cathode, anode and separators.

Chapters 1 and 2 describe the developments in nanostructured cathode materials for the development of high-performance lithium-ion batteries (LIBs). Chapter 1 deals with the three most important structures (spinel, layered, and olivines), while Chap. 2 focuses on the manganese-based orthosilicates.

Chapters 3, 4, and 5 describe the developments in nanostructured anode materials for LIBs, ranging from metal oxides and lithium alloys to the titanates.

Chapter 6 is the only chapter dedicated to magnesium batteries, which is an excellent example of the emerging multivalent battery systems. The interest in magnesium electrochemistry is due to its ability to deliver much higher volumetric energy density (3833 mAh cm^{-3}) than Li (2061 mAh cm^{-3}). A high volumetric energy density is more desirable for mobile devices than for stationary energy storage.

Chapters 7, 8, 9, 10, and 11 deal with various aspects of supercapacitor electrode systems, including metal oxides (notably the low-cost manganese oxides), carbon nanostructures, and suspension electrodes for flowable large-scale energy storage.

Chapters 12 and 13 discuss separators and solid polymer electrolytes used in LIBs and supercapacitors.

Chapters 14 and 15 focus on the computational/mathematical modeling and simulation of electrode materials as essential value additions to LIBs and supercapacitors, enriching our understanding of the energetics, reaction mechanism of electrode processes, and providing useful insights critical to achieving science-based rational design of better electrode materials.

Enlarging on the theme of electrochemical energy, this book has a companion volume organized by the same editors entitled *Fuel Cell Catalysis Based on Engineered Nanomaterials*. We sincerely thank the authors and reviewers, without whose support these books would have been impossible. We hope that the readers will be greatly enriched by the contents of this work. Enjoy reading!

Pretoria, South Africa
CA, USA

Kenneth I. Ozoemena
Shaowei Chen

Contents

Contributors

Timothy S. Arthur Toyota Research Institute of North America, Ann Arbor, MI, USA

Craig E. Banks College of Chemistry and Chemical Engineering, Central South University, Changsha, People's Republic of China; Faculty of Science and Engineering, School of Chemistry and the Environment, Division of Chemistry and Environmental Science, Manchester Metropolitan University, Lancs, UK

Paul M. Ejikeme Department of Chemistry, University of Pretoria, Pretoria, South Africa; Department of Pure and Industrial Chemistry, University of Nigeria, Nsukka, Nigeria

Yury Gogotsi Department of Material Science and Engineering, A.J. Drexel Nanomaterials Institute, Drexel University, Philadelphia, PA, USA

Rosalind J. Gummow College of Science, Technology and Engineering, James Cook University, Townsville, QLD, Australia

Kelsey B. Hatzell Department of Material Science and Engineering, A.J. Drexel Nanomaterials Institute, Drexel University, Philadelphia, PA, USA; Lawrence Berkeley National Laboratory, Berkeley, CA, USA

Renzong Hu School of Materials Science and Engineering, Guangdong Provincial Key Laboratory of Advanced Energy Storage Materials, South China University of Technology, Guangzhou, People's Republic of China

Innocent S. Ike Materials for Energy Research Group (MERG), University of the Witwatersrand, Johannesburg, South Africa; Department of Science and Technology/National Research Foundation, DST/NRF Centre of Excellence in Strong Materials (CoE-SM), University of the Witwatersrand, Johannesburg, South Africa; School of Chemical and Metallurgical Engineering, University of the Witwatersrand, Johannesburg, South Africa; Department of Chemical Engineering, Federal University of Technology, Owerri, Imo State, Nigeria

Sunny Iyuke School of Chemical and Metallurgical Engineering, University of the Witwatersrand, Johannesburg, South Africa

Xiaobo Ji College of Chemistry and Chemical Engineering, Central South University, Changsha, People's Republic of China

Jin Jia New Energy Research Institute, School of Environment and Energy, Guangzhou Higher Education Mega Center, South China University of Technology, Guangzhou, People's Republic of China

Mesfin Kebede Energy Materials Unit, Materials Science and Manufacturing, Council for Scientific & Industrial Research (CSIR), Pretoria, South Africa

Guang-Yin Liu Institute of Metal Research, Chinese Academy of Science, Shenyang, People's Republic of China; Nanyang Normal University, Nanyang, China

Y. Liu College of Energy and Institute for Electrochemical Energy Storage, Nanjing Tech University, Nanjing, China; College of Chemistry and Chemical Engineering, Jiangxi Normal University, Nanchang, China

Xiaojun Liu New Energy Research Institute, School of Environment and Energy, Guangzhou Higher Education Mega Center, South China University of Technology, Guangzhou, People's Republic of China

Hong-Ze Luo Institute of Metal Research, Chinese Academy of Science, Shenyang, People's Republic of China; Council for Scientific and Industrial Research, Pretoria, South Africa

Hongze Luo Materials Science and Manufacturing, Council for Scientific and Industrial Research (CSIR), Pretoria, South Africa

Katlego Makgopa Department of Chemistry, University of Pretoria, Pretoria, South Africa

Rapela Regina Maphanga Materials Modelling Centre, University of Limpopo, Sovenga, South Africa

Nomthandazo Mutangwa Materials Science and Manufacturing, Council for Scientific and Industrial Research (CSIR), Pretoria, South Africa

Phuti E. Ngoepe Materials Modelling Centre, University of Limpopo, Sovenga, South Africa

Patrick T. Nonjola Material Science and Manufacturing, Council for Scientific and Industrial Research (CSIR), Pretoria, South Africa

Kenneth I. Ozoemena Energy Materials Unit, Materials Science and Manufacturing, Council for Scientific & Industrial Research (CSIR), Pretoria, South Africa; Department of Chemistry, University of Pretoria, Pretoria, South Africa; School of Chemistry, University of the Witwatersrand, Johannesburg, South Africa

Kumar Raju Energy Materials, Materials Science and Manufacturing, Council for Scientific and Industrial Research (CSIR), Pretoria, South Africa; Department of Energy, University of Madras, Chennai, Tamil Nadu, India

Nikhilendra Singh Toyota Research Institute of North America, Ann Arbor, MI, USA

Samuel A. Suthanthiraraj Department of Energy, University of Madras, Chennai, Tamil Nadu, India

X.W. Wang College of Energy and Institute for Electrochemical Energy Storage, Nanjing Tech University, Nanjing, China

Lei Wen Shenyang National Laboratory for Materials Science, Institute of Metal Research, Chinese Academy of Science, Shenyang, People's Republic of China

Z.B. Wen College of Chemistry and Chemical Engineer, Jiangxi Normal University, Nanchang, China

Y.P. Wu College of Energy and Institute for Electrochemical Energy Storage, Nanjing Tech University, Nanjing, China; New Energy and Materials Laboratory (NEML), Department of Chemistry & Shanghai Key Laboratory of Molecular Catalysis and Innovative Materials, Fudan University, Shanghai, China

Yuping Wu Department of Chemistry, New Energy and Materials Laboratory, Fudan University, Shanghai, People's Republic of China

Zhibin Wu College of Chemistry and Chemical Engineering, Central South University, Changsha, People's Republic of China

F. Yu College of Chemistry and Chemical Engineer, Jiangxi Normal University, Nanchang, China

Hai-Tao Zheng Institute of Metal Research, Chinese Academy of Science, Shenyang, People's Republic of China; Council for Scientific and Industrial Research, Pretoria, South Africa

Haitao Zheng Energy Materials, Materials Science and Manufacturing, Council for Scientific and Industrial Research (CSIR), Pretoria, South Africa

Kai Zhou New Energy Research Institute, School of Environment and Energy, Guangzhou Higher Education Mega Center, South China University of Technology, Guangzhou, People's Republic of China

Weijia Zhou New Energy Research Institute, School of Environment and Energy, Guangzhou Higher Education Mega Center, South China University of Technology, Guangzhou, People's Republic of China

Min Zhu School of Materials Science and Engineering, Guangdong Provincial Key Laboratory of Advanced Energy Storage Materials, South China University of Technology, Guangzhou, People's Republic of China

Y.S. Zhu College of Energy and Institute for Electrochemical Energy Storage, Nanjing Tech University, Nanjing, China

Yirong Zhu College of Chemistry and Chemical Engineering, Central South University, Changsha, People's Republic of China

About the Editors

Kenneth I. Ozoemena, PhD, is a Chief Research Scientist at the Council for Scientific and Industrial Research (CSIR), Pretoria, South Africa. His current research focuses on functional nanomaterials for the development of electrochemical energy systems (notably, advanced batteries, supercapacitors, and fuel cells).

Shaowei Chen, PhD, is Professor of Chemistry at the University of California at Santa Crux, USA. His current research focuses on novel functional nanomaterials and their potential applications in nanoelectronics, chemical/biological sensing, photovoltaics, fuel cells, and supercapacitors.

Chapter 1
Next-Generation Nanostructured Lithium-Ion Cathode Materials: Critical Challenges for New Directions in R&D

Kenneth I. Ozoemena and Mesfin Kebede

1.1 Introduction

Rechargeable lithium-ion battery (LIB) was introduced to the world by John Bannister Goodenough in 1980, during his career as the Head of the Inorganic Chemistry Laboratory at Oxford University. Goodenough's brainchild was the layered lithium cobalt oxide (Li_xCoO_2) cathode material, which was the first lithium-ion battery cathode material to become commercialised. Since the commercialisation of Li_xCoO_2 in 1991 by Sony [1], LIBs have continued to capture global attention. It should be encouraging to researchers that Goodenough, who invented this technology at the age of 57, is still active and regularly publishes in the field of LIB technology. This decade has witnessed an indescribable outburst of technologies utilising LIBs as the main power sources. Some of these technologies that utilise LIBs as power sources include the portable electronics and consumer products (such as laptop computers, smart phones, tablets and digital cameras), electric vehicles (e.g. Tesla Model S, BMW i8, Nissan Leaf and Chevrolet Volt, to mention but a few) and home storage (notably, Tesla's Powerwall® that was launched on 30 April 2015 but sold out barely a week after because of the huge market demand). The market of mobile electronics, including cell phones, tablets,

K.I. Ozoemena (✉)
Energy Materials Unit, Materials Science and Manufacturing, Council for Scientific & Industrial Research (CSIR), Pretoria 0001, South Africa

Department of Chemistry, University of Pretoria, Pretoria 0002, South Africa

School of Chemistry, University of the Witwatersrand, Johannesburg 2050, South Africa
e-mail: kozoemena@csir.co.za

M. Kebede
Energy Materials Unit, Materials Science and Manufacturing, Council for Scientific & Industrial Research (CSIR), Pretoria 0001, South Africa

K.I. Ozoemena, S. Chen (eds.), *Nanomaterials in Advanced Batteries and Supercapacitors*, Nanostructure Science and Technology,
DOI 10.1007/978-3-319-26082-2_1

digital cameras and laptop computers, has increased very rapidly in the past decade. In addition to their applications in electric vehicles and portable electronics, it is now common knowledge that LIBs have become very useful in grid/utility-scale energy storage, offering critical services such as

1. *Security of electricity supply* (i.e. as power back-up and to avoid the socio-economic problems of load shedding),
2. *Price arbitrage* (i.e. storing 'cheap' electricity with LIBs during the off-peak periods when the cost of electricity is usually cheap and using it during the expensive peak times),
3. *Industrial peak shaving* or *demand charge reduction* (i.e. using stored electricity during peak periods in order to avoid penalties for breach of contractual peak demand),
4. *Balancing power* or *frequency regulation* (i.e. compensating for excess electricity being generated in the grid and/or utilised from the grid),
5. *Island and off-grid storage* (i.e. to augment the variable renewable energy sources such as solar and wind power used by people who live or work in remote areas where there is no access to the national grid),
6. *Transmission and distribution (T&D) upgrade deferral* (i.e. in a situation where the existing grid's capacity is barely enough to meet the required peak demand in a given area or to store the peak power supply from distributed variable renewable energy sources) and
7. *Voltage control/support* (i.e. to enhance power quality and local grid congestion).

However, despite these strides made by LIBs, the LIB technology has rightly been criticised for its slow pace of development. Indeed, the rapid development of portable electronics and the urgent need for grid stabilisation/application and renewable energy utilisation (wind and solar storage) for a greener and sustainable environment have continued to put even more pressure on the development of more reliable battery technologies.

The extensive interest on LIBs stems from their extraordinary inherent features such as high energy density, high capacity, long cycle lifetime, lightweight and environmental friendliness. The need to achieve higher energy densities in LIBs is one of the main drivers for the development of next-generation LIB technologies. Interestingly, energy density (i.e. amount of electrical energy per unit mass or volume) is intrinsically linked with the values of cell's voltage and its capacity (i.e. E_{cell} (wh g^{-1}) = Q (mAh g^{-1}) $\times$ V (Volt)). On the other hand, the capacity of any battery cell is determined by the electrode with the lower capacity (i.e. cathode), meaning that the overall energy density of any LIB is limited by the capacity of the cathode material. Little wonder, therefore, the identification of high-voltage, high-capacity, low-cost, lightweight and safe cathode materials has been predicted to be the main driver for the development of next-generation, high-performance LIB technologies [2–4].

Over the years, several types of LIB cathode materials have been proposed, but progress on commercialisation has been generally slow. Today, the three most popular LIB cathode materials are derived from the layered, spinel and olivine structures (Fig. 1.1) [5] such as the lithium cobalt oxide layer ($LiCoO_2$ (LCO) which is the first cathode material to be commercialised in 1991 by Sony), lithium nickel cobalt manganese oxide layer ($LiNi_{1/3}Co_{1/3}Mn_{1/3}O_2$, (NCM)), lithium nickel cobalt aluminium oxide layer ($LiNi_{0.8}Co_{0.15}Al_{0.05}O_2$, (NCA)), lithium- and manganese-rich manganese nickel cobalt oxide (e.g. $Li_{1.2}Mn_{0.54}Ni_{0.13}Co_{0.13}O_2$), lithium manganese oxide spinel ($LiMn_2O_4$ (LMO)), lithium manganese nickel oxide spinel ($LiMn_{1.5}Ni_{0.5}O_4$ (LMNO)) and lithium iron phosphate olivine ($LiFePO_4$ (LFP)).

Table 1.1 presents the main features of these LIB cathode materials [6]. With the exception of LMNC and LMNCA, all the cathode materials presented in Table 1.1 are already commercially available. It is evident that these commercially available

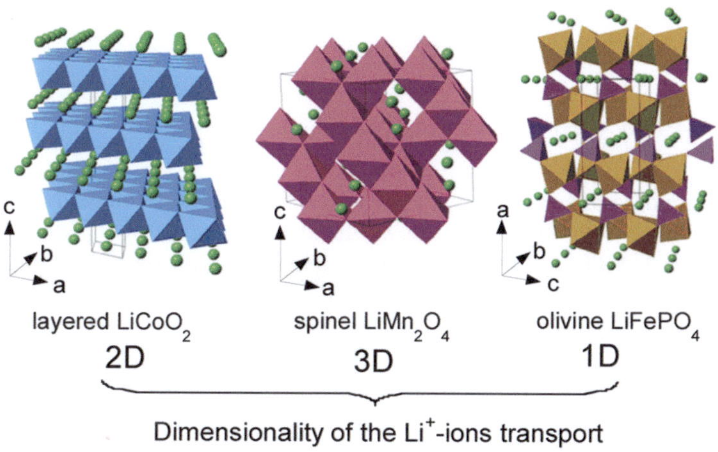

Fig. 1.1 Crystal structures of the three lithium-ion battery cathodes in which the Li-ions move through the 2D (*layered*), 3D (*spinel*) and 1D (*olivine*) frameworks (Reprinted with permission from Ref. [5], copyright 2014, Elsevier)

Table 1.1 Electrochemical characteristics of the lithium-ion battery cathode materials

Cathode material	Structure	Average voltage, V (vs. Li/Li^+)	Specific capacity (mAh/g)	Specific energy (Wh/kg)
$LiCoO_2$ (LCO)	Layered	3.9	140	546
$LiNi_{1/3}Co_{1/3}Mn_{1/3}O_2$ (NMC)	Layered	3.8	160–170	610–650
$LiNi_{0.8}Co_{0.15}Al_{0.05}O_2$ (NCA)	Layered	3.8	180–200	680–760
$Li_{1.2}Mn_{0.54}Ni_{0.13}Co_{0.13}O_2$ (LNMC)	Layered	4.5	>250	>1125
$Li_{1.2}Mn_{0.52}Ni_{0.13}Co_{0.13}Al_{0.02}O_2$ (LNMCA)	Layered	4.5	>250	>1125
$LiMn_2O_4$ (LMO)	Spinel	4.1	100–120	410–492
$LiMn_{1.5}Ni_{0.5}O_4$ (LMNO)	Spinel	4.8	120–148	576–710
$LiFePO_4$ (LFP)	Olivine	3.45	150–170	518–587

cathode materials have almost attained their capacity and energy limits (100–200 mAh g^{-1}; 410–760 Wh kg^{-1}). Besides, their applications are still fraught with several challenges such as poor cyclability and poor rate capability. Thus, further developments are needed to enhance the performance of these existing cathode materials and also to develop new ones to obtain high-performance LIBs that will meet the ever-increasing demand of new and ever-emerging technologies.

Recent studies and reviews [2–12] have continued to show that nanostructured cathode materials provide new opportunities for enhancing the performance criteria (including capacity, rate capability, cyclability and safety) of existing LIBs and developing new ones. This chapter describes research trends on nanostructured LIB cathode materials aimed at enhancing the performance of next-generation LIBs. We focus on the LCO, NMC, LMO, NCA, the high-voltage lithium manganese nickel oxide spinel ($LiNi_{0.5}Mn_{1.5}O_4$, LMNO) and high-energy lithium- and manganese-rich HENMC such as the LMNC and LMNCA.

1.2 Layered Compounds ($LiMO_2$)

The structure of layered compound $LiMO_2$ is a distorted rock salt (α -$NaFeO_2$-type) crystal structure [13]. Figure 1.2 shows the structure of $LiMO_2$ [14]. Upon deep lithiation, the $LiMO_2$ will undergo a phase transformation from the rhombohedral space group R3m ('O3' phase) to the monoclinic space group C_2/m ('O1' phase), leading to a decrease of symmetry. The MO_2 slabs and Li layers are stacked alternatively while the oxygen anions form a cubic-close-packed array along the *C* axis while the cations (Li^+ and M^{3+} ions) alternately occupy the close-packed (111) interstitial octahedral-site planes. It is easy for the two-dimensional Li+ ion diffusion to occur in the space between <MO_2> layers since the edge-shared octahedral <LiO_6> arrangement permits the movement of ion from one (vacant) octahedral-site plane to another via a tetrahedral site. Also, the edge-shared octahedral <MO_6> is able to maintain a direct <M–M> interaction, thus increasing the electronic conductivity.

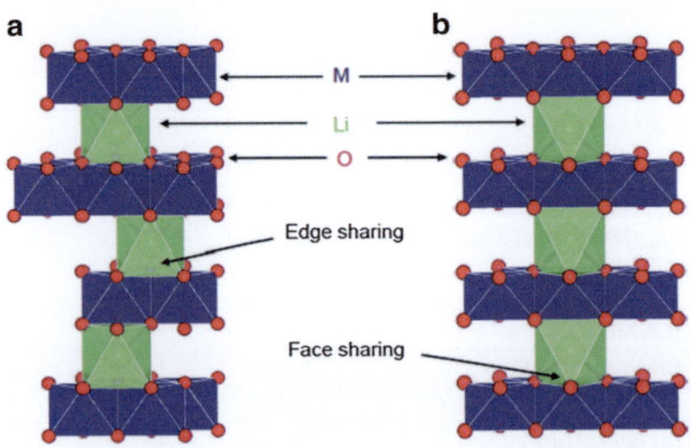

Fig. 1.2 Schematic representation of a layered $LiMO_2$ structure: (**a**) 'O3' phase, rhombohedral system (space group R3m) ('O3' phase), and (**b**) 'O1' phase, monoclinic space group C2/m ('O1' phase) (Reprinted with permission from Ref. [14], Copyright 2014, Royal Society of Chemistry)

The layered $LiCoO_2$ is the most popular LIB cathode material. Unfortunately, due to its intrinsic structural instability, it can only deliver about 140 mAh g^{-1} capacity (i.e. half of its theoretical capacity of 274 mAh g^{-1}, meaning that only half of the Li^+ is practically accessible for extraction/insertion). The toxic and expensive cobalt ion in the cathode material has also made $LiCoO_2$-based LIB environmentally undesirable and expensive. Thus, research on layered $LiCoO_2$ has dramatically shifted to the replacement of the Co ions, fully or partially, with the more abundant and environmentally benign transition metal ions (notably the Mn and Ni) to form layered materials such as the $LiNi_{0.5}Mn_{0.5}O_2$ (LMN) and the Li–Mn–Ni–Co–O layered compound (commonly referred to as the 'NMC' materials).

1.2.1 $LiNi_{0.5}Mn_{0.5}O_2$

The layered material $LiNi_{0.5}Mn_{0.5}O_2$ (LMN), regarded as a solid solution of $LiMnO_2$ and $LiNiO_2$, was first reported by Ohzuku and Makimura in 2001 [15]. The material shows good electrochemistry with a capacity of about 200 mAh g^{-1} and charge–discharge voltages appearing around 3.6–4.3 V (with Ni^{2+}/Ni^{4+} redox couple) [16]. Unlike the classical layered material that consists of pure MO_2 slab and Li layer, the LMN contains about 8–10 % of its Ni ions found in the Li layer. The existence of this unusual Li–Ni interlayer mixing impedes the free movement of Li-ion, thus negatively affecting rate capability. In fact, Li diffusion coefficient in $LiNi_{0.5}Mn_{0.5}O_2$ is about one order of magnitude lower than that in the parent $LiCoO_2$ [17]. Thus, the main research challenge has been to eliminate the Ni in the Li layers of the structure to improve the rate capability. To achieve this, Kang et al. in 2006 [18] introduced a new synthesis strategy involving 'ion exchange' that was able to reduce the Ni ion content to about 4 %, significantly improving the rate capability of the $LiNi_{0.5}Mn_{0.5}O_2$.

1.2.2 $LiNi_{1/3}Co_{1/3}Mn_{1/3}O_2$ (NCM)

The layered material, $LiNi_{1/3}Co_{1/3}Mn_{1/3}O_2$ (NMC or NMC-333), is regarded as a solid solution of $LiCoO_2$, $LiMnO_2$ and $LiNiO_2$ and was first reported by Ohzuku and Makimura in 2001 [19]. It displays similar electrochemistry as the $LiNi_{0.5}Mn_{0.5}O_2$ with reversible capacity of about 200 mAh g^{-1} but a slightly wider voltage window (3.6–4.7 V) than the $LiNi_{0.5}Mn_{0.5}O_2$. The observed redox couples are due to Ni^{2+}/Ni^{4+} and Co^{3+}/Co^{4+}. The introduction of the NMC was quite encouraging as the presence of the additional Co ions in the structure was able to significantly reduce the Li–Ni interlayer mixing to 1–6 % [20–25]. The NMC is characterised with some unique advantages such as high specific capacity, superior thermal stability (due to the presence of the inactive Mn^{4+} that stabilises the structural framework), low cost and less toxicity (due to low amount of Co ion in the material compared to the LMN). Considering the phase transition depicted in Fig. 1.2, it must be cautioned that electrochemical cycling of layered material is performed within the potential window where the 'O3'/'O1' phase transition is eliminated (i.e. with an upper cut-off potential $\leq$4.4 V) to obtain high cycle reversibility and capacity retention. Within the right potential window, NMC-333 is structurally stable and up to 0.7 mol Li^+ per formula unit could be extracted providing 10–20 mAh g^{-1} extra storage capability compared to the parent LCO [14].

To improve the performance of this cathode material, several preparation strategies have been reported, including solid-state reaction [26], sol–gel [27], co-precipitation methods [28] and microwave-assisted synthesis [29].

1.2.3 $LiNi_{0.8}Co_{0.15}Al_{0.05}O_2$ (NCA)

The NCA is a nickel-rich lithium-ion cathode material commercialised and used by Tesla. One of the major drawbacks of this cathode material is safety. Zaghib et al. [30] have shown that the thermal stability of NCA is very poor compared to other cathode materials such as the LMO and LFP (Fig. 1.3). From the DSC spectra in Fig. 1.3, spinel and olivine cathodes have delayed onset temperature by at least 70 °C compared to layered NCA cathode material. The exothermic reaction enthalpies decrease as NCA (−941 J g^{-1}) > LMO (−439 J g^{-1}) > LFP (−250 J g^{-1}), confirming that NCA is thermally unsafe and completes its reaction much earlier than the onset temperature of LMO and LFP.

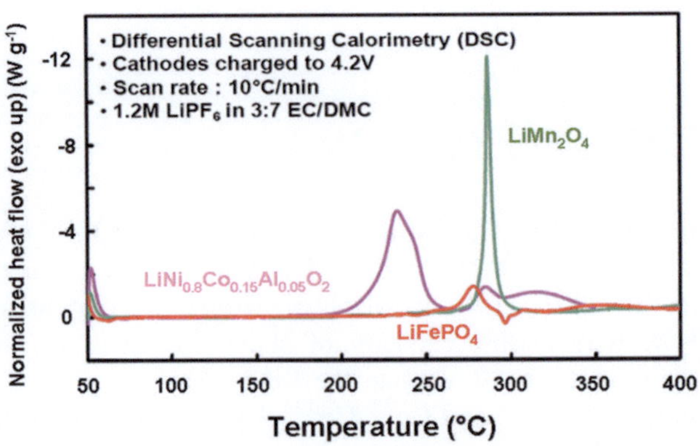

Fig. 1.3 DSC spectra of overcharged layered, spinel and olivine cathodes with traces of 1.2 mol L^{-1} $LiPF_6$ in EC-EMC (3:7) electrolyte at 10 °C min^{-1} (Reprinted with permission from Ref. [30], copyright 2014, Elsevier)

1.2.4 xLi_2MnO_3 • $(1-x)LiMO_2$ (M = Transition Metals)

High-energy lithium- and manganese-rich layered cathode materials (HENCMs) have continued to draw global attention because of their high specific capacities (>280 mAh g^{-1} with 3.6 V or larger operating voltages when these materials are charged to over 4.6 V at room temperature) and specific energy densities (>1000 Wh kg^{-1}) at a lower cost compared to the $LiCoO_2$. The HENCMs (Fig. 1.4) are regarded as 'composite mixture' of monoclinic Li_2MnO_3 structure (space group: C2/m) and rhombohedral $LiMO_2$ structure (space group: R3m) and generally written in two-component notation [7], xLi_2MnO_3 • $(1-x)LiMO_2$ (M = Mn, Co, Ni, Cr, Fe, etc.). For example, $0.5Li_2MnO_3$•$0.5LiMn_{1/3}Ni_{1/3}Co_{1/3}O_2$ is the same as the $Li_{1.20}Mn_{0.54}Ni_{0.13}Co_{0.13}O_2$ (HENCM, Fig. 1.5) and the $0.5Li_2MnO_3$•$0.5LiMn_{0.42}Ni_{0.42}Co_{0.16}O_2$ as $Li[Li_{0.2}Mn_{0.567}Ni_{0.166}Co_{0.067}]O_2$.

Thackeray and co-workers (then working at the CSIR, South Africa) first introduced the HENMCs in 1991 and the two-component notation [31, 32]. These researchers had synthesised layered $Li_{2-x}MnO_{3-x/2}$ $(0 < x < 2)$ by chemically leaching Li_2O from the rock salt phase Li_2MnO_3 $(Li_2O \cdot MnO_2)$ with acid at

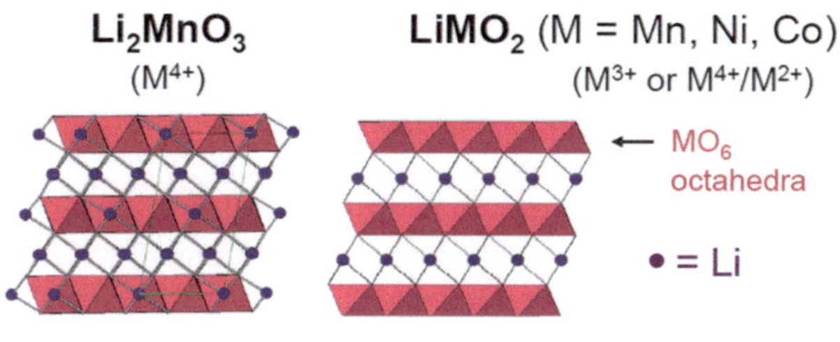

Fig. 1.4 Crystal structure of the (**a**) monoclinic Li_2MnO_3 structure (space group: C_2/m) and (**b**) rhombohedral $LiMO_2$ structure (space group: R3m, M = Ni, Co, Mn, Fe, Cr, etc.) (*Source*: Presentation by Dr Michael Thackeray, Argonne National Laboratory, USA. http://energy.gov/sites/prod/files/2014/07/f17/es049_thackeray_2014_p.pdf)

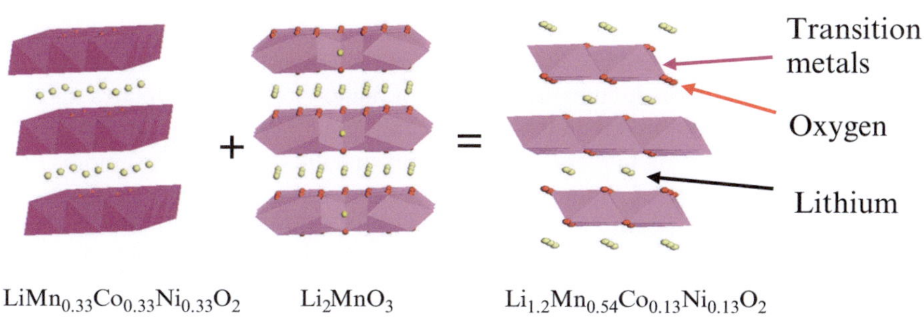

Fig. 1.5 The HENMC ($Li_{1.20}Mn_{0.54}Ni_{0.13}Co_{0.13}O_2$) is considered as the solid solution of $LiMn_{0.33}Co_{0.33}Ni_{0.33}O_2$ and Li_2MnO_3 (The crystal structures were generated using Materials Studio 5.0 software)

25 °C, which upon relithiation in an electrochemical cell, obtained the layered $Li_{1.09}Mn_{0.91}O_2$ or $0.2Li_2MnO_3 \cdot 0.8LiMnO_2$. The $0.2Li_2MnO_3 \cdot 0.8LiMnO_2$ was found to be structurally more stable than the pure $LiMnO_2$ during electrochemical cycling, and then the xLi_2MnO_3 • (1-x)$LiMO_2$ material concept was first introduced. In fact, the xLi_2MnO_3 • (1-x)$LiMO_2$ became even more attractive when in 1999, Kalyani et al. [33] reported that the monoclinic Li_2MnO_3 material could be electrochemically activated by charging the Li/Li_2MnO_3 cell to 4.5 V.

Despite the various advantages of the HENMCs, three key challenges have continued to conspire against their applications. These challenges are:

(i) Low initial coulombic efficiency: This is associated with the large irreversible capacity that occurs during the first charge cycle due to the removal of Li_2O from the $LiMO_3$. This initial irreversible reaction arising from the first charge plateau above 4.4 V vs. Li/Li + can result in 'mass ionised lithium and the formation of a solid-electrolyte interface (SEI) layer, thus reducing the energy density of the lithium-ion battery and presenting a serious trade-off in lithium-ion battery design' [7]. In other words, such removal of the Li_2O from the cathode is capable of physically damaging the surface of the electrode during cycling and then increasing the cell impedance.

To improve the initial coulombic efficiency, Thackeray and co-workers have used acid-treatment methods with NH_3 and HNO_3 [34]. Other methods include the surface coating with nanostructured Al_2O_3, $AlPO_4$ or RuO_2 [35] as well as ruthenium doping (i.e. substituting part of the Mn with Ru) [36]. The best performance of 86 % initial coulombic efficiency was achieved with 284 mAh/g discharge capacity at room temperature and ruthenium substitution of 5 mol %. The performance achieved by ruthenium substitution may be related to the decreased content of the Li_2MnO_3 component.

(ii) Slow rate of Li-ion transport (poor rate capability): The speed at which lithium-ion can be extracted and inserted into the electrode is mainly controlled by the inactive Li_2MO_3 component inside the HENCM. Although the Li_2MO_3 can be activated during the first charge process, the LMO_2 component still exhibit poor rate of lithium-ion diffusion [7]. Thus, to enhance the electrochemical kinetics of Li-ion, research efforts are directed towards the use of low amount of the $LiMO_3$ in the structure, shortening the diffusion pathway of Li-ion and eliminating interface reaction barrier in the structure. These efforts include the preparation of nanowires or nanoplates of the HENCMs, coating with insulating materials such as (Al_2O_3 and $AlPO_4$), decoration with nanofibers, doping with metals, etc. For example, Ma et al. [9] reported that nanofiber-decorated $Li_{1.2}Mn_{0.54}Ni_{0.13}Co_{0.13}O_2$ exhibited rate capability of ~120 mAh/g at 5 C compared to ~70 mAh/g of the pristine particle at 5 C; Ozoemena and co-workers [10] showed that aluminium-doped material ($Li_{1.2}Mn_{0.52}Ni_{0.13}Co_{0.13}Al_{0.02}O_2$) showed rate capability of 75 mAh/g at 5C compared to the undoped counterpart of 35 mAh/g at 5 C; $Li[Li_{0.15}Ni_{0.25}Mn_{0.6}]O_2$ nanowires showed rate capability of 95 % at 4C (=1200 mA/g); crystal habit-tuned nanoplate materials of Li

$[Li_{0.17}Ni_{0.25}Mn_{0.58}]O_2$ with enhanced (010) nanoplates showed high-rate performance [37]. Manthiram and co-workers [35] have modified the surface of the HENCMs with insulating materials (Al_2O_3 and $AlPO_4$) and reported enhanced rate capability compared to the unmodified counterparts, further indicating that crystal grain and particle surface modification are crucial to enhancing the rate capability of these layered materials.

(iii) Voltage decay during cycling (rapid capacity fade): Rapid capacity fade has been attributed to phase transformation from the layered structure into the cubic spinel-like structure [38]. In fact, according to Reed et al. [39], the transition from layered $Li_{0.5}MO_2$ (delithiation) into the ideal cubic spinel phase $(Li)_{8a}[M_2]_{16d}O_4$ simply requires moving just one-fourth of the transition metal ion from the octahedral sites (3b sites) of the M planes into the empty octahedral sites (3a sites) of the lithium planes, which becomes 16d positions of spinel without changing the framework of closed-paced oxygen arrays. Thus, it is now believed that the rapid capacity fade observed in HENCMs involve both internal and surface phase transition to a cubic spinel-like phase by the migration of transition metal ions.

The main strategy used by researchers to prevent voltage decay in HENCMs is surface coating with materials such as oxides, fluoride, phosphates and mild acid treatment. Some of the popular coating materials include Al_2O_3, CeO_2, ZrO_2, SiO_2, ZnO, $AlPO_4$ and $LiNiPO_4$. However, these surface-coating treatments have hardly completely overcome the voltage decay, confirming that phase transitions from layered structure into cubic spinel-like structure occur on the surface of the particle as well as inside the particle bulk. Thus, there is an urgent need to explore how best to minimise or completely eliminate the changes occurring within structure, and one of the means of doing this could be found in the doping of the HENCMs. Ozoemena and co-workers [10] recently reported that aluminium doping (i.e. aluminium substitution for manganese) coupled with microwave irradiation could help stabilise the structure and improve cyclability.

1.3 Spinel Compounds

The two most popular spinel materials for LIBs are the LMO and LMNO (Fig. 1.6) [40].

1.3.1 $LiMn_2O_4$ (LMO)

In 1983, Thackeray and his co-workers at the CSIR South Africa first reported spinel $LiMn_2O_4$ (LMO) as the cathode for lithium-ion battery [41–43]. In a recent review by Michael Thackeray reminiscing the '20 Golden Years of Battery R&D at CSIR (1974–1994)' [44], he said 'It must be particularly rewarding to CSIR that the

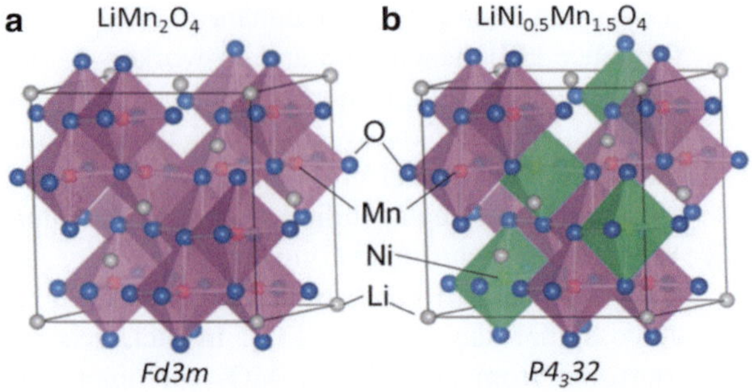

Fig. 1.6 Crystal structure of (**a**) $LiMn_2O_4$ (LMO) and (**b**) $LiM_{1.5}Ni_{0.5}O_4$ (LMNO) (Reprinted with permission from Ref. [40]. Copyright 2013 Royal Society of Chemistry)

first commercially-available lithium-battery powered electric cars, the Nissan Leaf and General Motor's Chevy Volt, due to be released to the general public in 2010, both operate with lithium-manganese-oxide spinel as a major cathode component in the batteries'.

As illustrated in Fig. 1.6a, $LiMn_2O_4$ adopts the spinel structure with space group: Fd-3 m [40]. The conventional unit cell of spinel $LiMn_2O_4$ has 56 atoms of which 8 lithium (Li) atoms occupy 8a tetrahedral sites, 16 manganese (Mn) atoms occupy 16d octahedral sites and 32 oxygen atoms are placed at 32e sites forming a cage [45] as displayed in Fig. 1.6a. The oxygen framework of $LiMn_2O_4$ is the same as that of layered $LiMnO_2$ structure. Mn cations still occupy the octahedral site, but 1/4 of them are located in the Li layer, leaving 1/4 of the sites in the transition metal layer vacant. Li-ions occupy the tetrahedral sites in Li layer that share faces with the empty octahedral sites in the transition metal layer. The structure is based on a three-dimensional Mn_2O_4 host and the vacancies in transition metal layer ensure the three-dimensional Li diffusion pathways. The interstitial spaces in the Mn_2O_4 framework can be viewed as a network of 8a tetrahedral sites, which can function as pathways for transporting lithium ions in the structure.

Generally, $LiMn_2O_4$ (LMO) spinel is a very attractive material for the positive (cathode) electrode in rechargeable lithium-ion batteries because of its several benefits such as low cost, high abundance, low toxicity and environmental compatibility compared to the layered oxides such as $LiCoO_2$ and $LiNiO_2$ [46, 47]. The theoretical capacity of LMO is 148 mA h g^{-1} and its operating potential is 4 V. Moreover, the spinel $LiMn_2O_4$ can deliver a high energy density comparable to popular $LiCoO_2$ as it has high voltage and high specific capacity which is only 10 % less than that of $LiCoO_2$. Thus, it has drawn a lot of attention for its use as a 4 V cathode in Li− ion batteries.

Despite the various advantages of LMO, this cathode material suffers from severe capacity fading during storage and electrochemical cycling at elevated temperatures above 55 ° C [41–43]. The two main causes considered as sources for the capacity fading are:

(i) Slow dissolution of Mn^{2+} into the electrolyte generated by the disproportional reaction due to the presence of trace acids, especially HF: $2Mn^{3+}_{(solid)} \rightarrow Mn^{4+}_{(solid)} + Mn^{2+}_{(solution)}$ (with Mn^{2+} going into the solution) [48–50] in the 4 V region.
(ii) Jahn–Teller distortion [48–50] in the 3 V region [3], which is the generation of new phases during cycling and the related micro-strains. Here, there is the transformation of the crystal symmetry from cubic to tetragonal structure, and it is this structural transition that deteriorates its cycle life and is said to occur when the average manganese valence number (nMn) is equal or less than 3.5 [51, 52]. The stress generated by this phenomenon leads to cracking of particles and loss of electric contact upon cycling. The manganese ions in LMO are believed to exist as 50 % Mn^{3+} and 50 % Mn^{4+} (i.e. nMn = 3.5+). High content of Mn^{3+} ions causes capacity fading and leads to the dissolution of the cathode material into the electrolyte.

The conventional strategies for improving the cycling performance of LMO include:

(i) Making the spinel structure lithium-rich (Li-excess) [53–55]
(ii) Doping with cations or substituting a small amount of trivalent Mn ions (Mn^{3+}) with other metal ions [45, 56, 57] such as those of Mg, Al, Zn, Ti, Cr, Fe, Co, Ni, Cu, Nd and La [45, 53, 58–66], the dopant ions being assumed to occupy the 16d octahedral sites of Mn ions in the spinel lattice and stabilise the spinel structure from lattice distortion [45, 67, 68]
(iii) Coating with metal oxides such as Al_2O_3, B_2O_3, ZnO, CoO, MgO, ZrO_2, CeO_2, La_2O_3 and carbon that were chosen as a coating layer for the surface of spinel $LiMn_2O_4$ and demonstrated improved properties in comparison to bare $LiMn_2O_4$ [69–79]

Ozoemena and co-workers recently showed that effect of microwave irradiation on the performance of pristine and Al-doped LMO [12]. The authors clearly proved that microwave-assisted synthesis improved the performance of the $LiAl_xMn_{2-x}O_4$ ($x = 0$ and 0.3) (Fig. 1.7). The authors investigated the effects of microwave irradiation on the products obtained at the pre-annealing (LMO-ma and LMOA-ma) and post-annealing (LMO-am and LMOA-am) steps. When compared with the commercial sample (LMO-comm) and those that did not undergo any microwave irradiation (LMO-a and LMOA-a), it became evident that microwave irradiation is a potentially useful strategy to optimise the lattice parameter, average manganese valence, particle size and morphology, reversibility of the deintercalation/intercalation processes and capacity retention upon continuous cycling.

Kebede et al. [80] recently synthesised spinel LMO with small nickel content substitution and reported 99 % capacity retention at 0.2C rate as it is shown in Fig. 1.8.

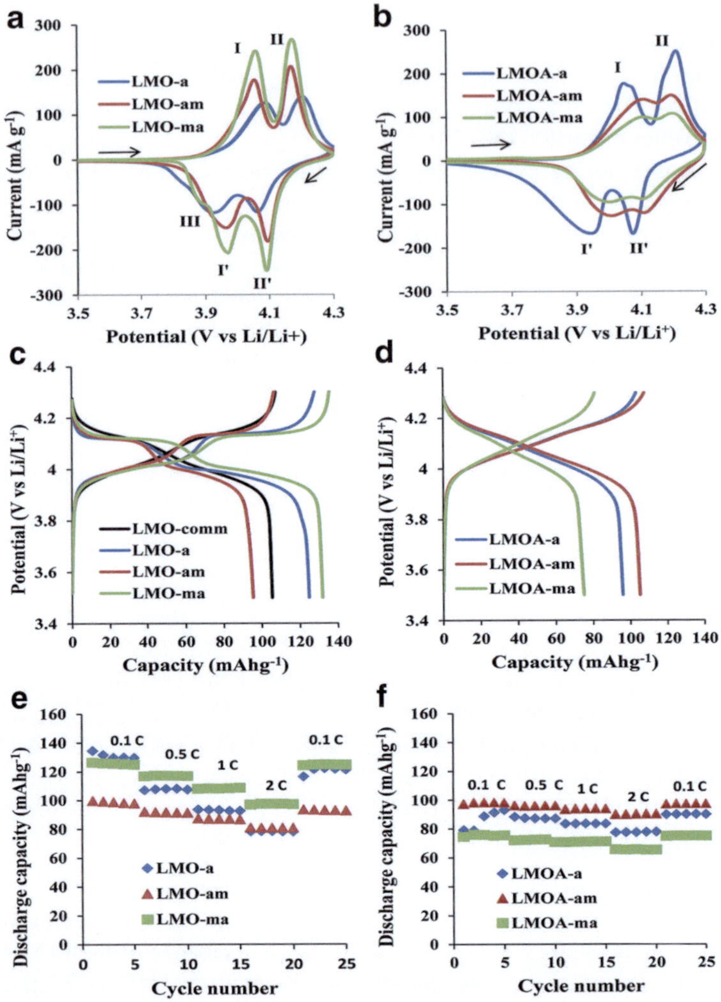

Fig. 1.7 Cyclic voltammograms of LMO (**a**) and LMOA (**b**) coin cells at 0.1 mV s^{-1}; galvanostatic charge–discharge of LMO (**c**) and LMOA (**d**) coin cells at 0.1 C; plots of discharge capacity vs. cycle number for the LMO (**e**) and LMOA (**f**) coin cells at different current densities (0.1–2 C) between 3.5 and 4.3 V range (Reproduced from ref. 12. Copyright 2015, Royal Society of Chemistry)

1.3.2 *High-Voltage $LiMn_{1.5}Ni_{0.5}O_4$ (LMNO)*

High-voltage $LiMn_{1.5}Ni_{0.5}O_4$ (LMNO) is regarded as one of the most promising next-generation lithium-ion battery cathode materials because of its high energy density (*ca.* 4.8 V vs. Li/Li^+; 672 Wh kg^{-1}), good cycling stability and fast three-dimensional Li^+ diffusion channels which makes it attractive for deployment in

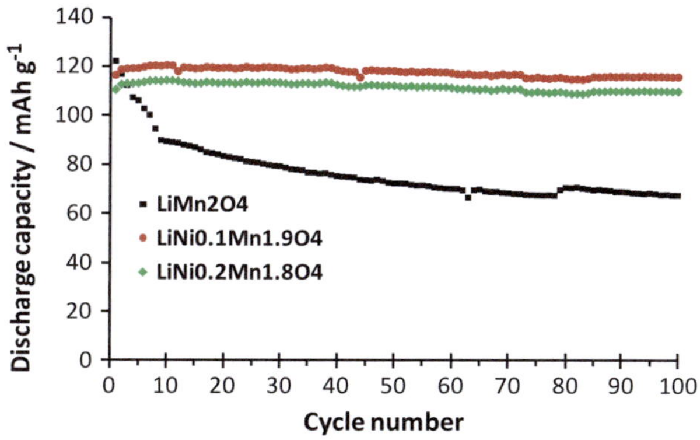

Fig. 1.8 The discharge capacity vs. cycle number for $LiNi_xMn_{2-x}O_4$ ($x = 0$, 0.1, 0.2,) at 0.2 C rate (Reproduced from Ref. [80]. Copyright 2014, Elsevier)

various applications ranging from portable electronics to stationary/utility. In terms of safety, Oh et al. demonstrated that LMNO possess satisfactory thermal stability of up to 250 °C [81].

LMNO adopts the same spinel structure as the $LiMn_2O_4$ wherein the Ni ions occupy the sites where Mn ions originally exist. LMNO has two different crystal structures: (i) the stoichiometric ordered structure, composed only of Mn^{4+} and (ii) the non-stoichiometric disordered, face-centred cubic structure (*Fd3m*), composed of Mn^{3+} and Mn^{4+}. In the ordered structure LiNi0.5Mn1.5O4 (primitive simple cubic structure ($P4_332$), Li-ions occupy the 8c sites, Ni ions occupy 4b sites, and Mn ions occupy 12d sites, while the O atoms reside in the 8c and 24e sites. The non-stoichiometric, disordered structure ($LiNi_{0.5}Mn_{1.5}O_{4-\delta}$) is a face-centred cubic structure (*Fd3m*), and Ni and Mn ions are randomly distributed in 16d sites, while Li and O occupy 8a tetrahedral sites and 32e sites, respectively. In the ordered structure, LiNi0.5Mn1.5O4, the valence of Ni ions is 2+ pushing all Mn ions to Mn4+. Unlike the LiMn2O4 spinel, the redox couple of LiNi0.5Mn1.5O4 is switched from Mn3+/Mn4+ to Ni2+/Ni4, lifting the voltage from 4.1 V to 4.7 V. Such high discharge voltage enlarges the energy density of the LMNO with the potential to be easily coupled with anode materials that possess better safety and relatively higher voltage (e.g. $Li_4Ti_5O_{12}$). The preparation of either of the LMNO structures is determined by the annealing temperatures; ordered spinel at T = 700 °C while disordered spinel at T > 700 °C [82], (Eq. 1.1) [83–85]:

$$LiMn_{1.5}Ni_{0.5}O_4 \leftrightarrow \alpha Li_xNi_{1-x}O + \beta LiMn_{1.5+y}Ni_{0.5-y}O_{4-x} + \gamma O_2(T > 700\,°C) \quad (1.1)$$

The electrochemical performance of this spinel as a cathode material for lithium-ion battery is intricately linked to the (i) presence of Mn^{3+} ions and/or the degree of disorder, (ii) doping/substitution with cations and (iii) presence of $Li_yNi_{1-y}O$ impurity. These factors explain why it still remains a challenge to correlate synthesis, structure and performance of LMNO spinel material. Some of the challenges that must be overcome to allow the commercialisation of LMNO include:

(i) *Difficulty to obtain pure LMNO:* It is somewhat difficult to synthesise phase-pure LiNi0.5Mn1.5O4 with controlled morphology because of undesirable particle growth during sintering process at high temperatures. The formation of the disordered LMNO is usually accompanied with the $Li_xNi_{1\text{-}y}O$ phase (impurity) that lowers the obtainable capacity.

(ii) *The ability to control the degree of disorder:* Mn^{4+} is redox-inactive while Mn $^{3+}$ is redox-active (with a signature peak plateau around 4 V) and improves the conductivity of the LMNO. If every Mn ion in the LMNO spinel is maintained at the 4+ state, we can only obtain the poor-performing ordered LMNO. There is need to control the concentration of Mn^{3+} in the spinel to avoid the danger of capacity fading due to the generation of soluble Mn^{2+} in the electrolyte via disproportionation reactions.

(iii) *Enhanced rate capability and ionic conductivity:* The conductivity of the disordered LMNO is enhanced with small amount of Mn^{3+} and thus exhibits better rate capability than its ordered counterpart. The diffusion coefficient of lithium-ion in LMNO ranges between 10^{-10} and 10^{-16} $cm^2\ s^{-1}$ depending on the composition and morphologies [86–88].

(iv) *Electrolyte decomposition:* The high voltage of the LMNO operates beyond the upper voltage limit (4.5 V) of the current electrolytes (i.e. LiPF6 and carbonates) used in lithium-ion batteries. This problem, unfortunately, results to the decomposition of the electrolyte and the formation of unstable SEI on composite cathode side during electrochemical cycling. It is important to direct some efforts towards finding new electrolytes that will tolerate upper voltage limits and exhibit better chemical stability with LMNO.

Several strategies are currently being investigated to mitigate these challenges, for example, doping a small amount of metal ions into LMNO forming 'bi-doped' spinel. The dopants (e.g. Fe, Cr and Ti) are said to have the potential to reduce the impurity phases and enhance the cation disorder [89–93]. Also, Co and Cu may improve electronic conductivity and/or lithium diffusion coefficient [94, 95]. Wang et al. [96] showed that by doping with a trace amount of ruthenium, LMNO showed marginal capacity fade after 500 cycles at 10 C. Although doping has the potential to improve the performance of the LMNO, some researchers have reported [89, 95, 96] that some dopants such as Fe, Cu, Al and Mg tend to occupy the tetrahedral sites and become inactive ions, thereby reducing the capacity and also blocking the lithium diffusion pathways.

Apart from doping, particle size modification (i.e. nanostructuring) is believed to be an effective strategy to enhance rate capability by shortening the path length of Li diffusion and enhance conductivity. Interestingly, the bulk properties of nanosized particles are generally similar to micro-sized particles, albeit the surface areas of the nanostructures may increase undesirable surface reactions. Also, most nanostructures do not possess the expected packing density and thus do not provide the needed volumetric energy density (Wh/L).

The amount of Mn^{3+} ions is considered quite critical in determining the electrochemical performances of the spinel. To buttress the importance of this issue,

Xiao et al. [97] showed that the concentration of Mn^{3+} ions in the LMNO spinel lattice is controlled by combining post-synthesis annealing and partially substituting Ni^{2+} with Cr^{2+}. In summarising their findings, the authors concluded that the 'comparison of the electrochemical performances of spinels with different Mn^{3+} contents demonstrates that careful control of the amount of Mn^{3+} ions and, thus, the disordered phase, is the key for synthesis of high performance spinel and provides valuable clues for understanding the structure–property relationships in energy materials'. Motivated by this statement, Ozoemena and co-workers [11] recently introduced a microwave-assisted synthesis strategy that proved the ability of microwave irradiation to tune the concentration of Mn^{3+} for improved performance of the disordered LMNO. Thus, today, there are five reported methods for controlling the Mn^{3+} content in LMNO: (i) careful control of the cooling rate after high temperature calcination [98]; (ii) partial substitution of Ni and/or Mn with elements such as Co, Al, Ti, Fe, Cr, Ru or Mg [93, 94, 97, 99, 100]; (iii) combined post-synthesis annealing and partial substitution of Ni with Cr [97]; (iv) long-hour acidic treatment [101]; and (v) microwave-assisted synthesis [11].

Figure 1.9 shows the XPS results of the pristine LMNO samples obtained at 700 and 800 °C without microwave assistance (LMNO-700 and LMNO-800) and with microwave assistance (LMNO-700-mic and LMNO-800-mic). Clearly, from

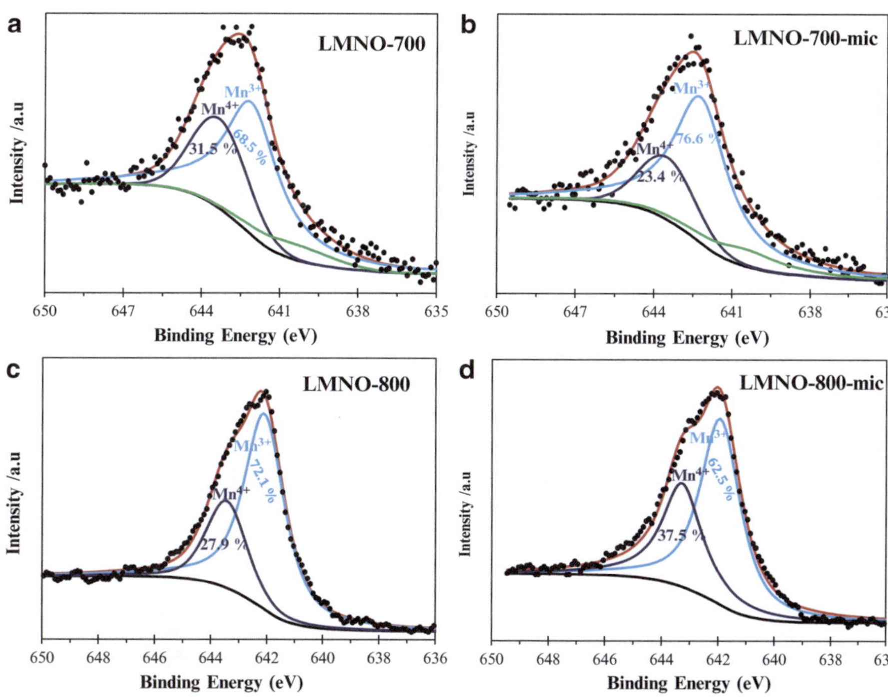

Fig. 1.9 X-ray photoelectron spectra of the Mn $2p_{3/2}$ peak of (**a**) LMNO-700, (**b**) LMNO-700-mic, (**c**) LMNO-800 and (**d**) LMNO-800-mic (Reproduced from Ref. [11]. Copyright 2013, American Chemical Society)

Fig. 1.9 (as also summarised in Table 1.2), microwave irradiation is able to tune the concentration of Mn^{3+} and Mn^{4+} in LMNO samples. This was in agreement with their XRD data that predicted a slight downward adjustment of the Mn^{3+} content of the disordered sample.

Figures 1.10a, c compare the 1st and 25th discharge curves of the LMNO samples. Unlike the ordered spinel materials (obtained at 700 °C), the disordered samples (obtained at 800 °C) showed well-defined plateaus at 4 V due to the Mn^{3+}/Mn^{4+} redox couple, signature of 'disordered' spinel. In addition, the Ni^{2+}/Ni^{3+} and

Table 1.2 Mn $2p_{3/2}$ peak positions, cation distribution and average manganese valence number

Spinel powdered sample	Binding energy position (eV)		Cation distribution			Average Mn valence
	Mn^{4+}	Mn^{3+}	Mn^{4+} (%)	Mn^{3+} (%)	Mn^{3+}/Mn^{4+}	Mn^{n+}
LMNO-700	643.2	642.1	31.5	68.5	2.17	3.315
LMNO-700-mic	643.4	642.2	23.4	76.6	3.27	3.234
LMNO-800	643.4	642.1	27.9	72.1	2.58	3.279
LMNO-800-mic	643.3	641.9	37.5	62.5	1.67	3.375

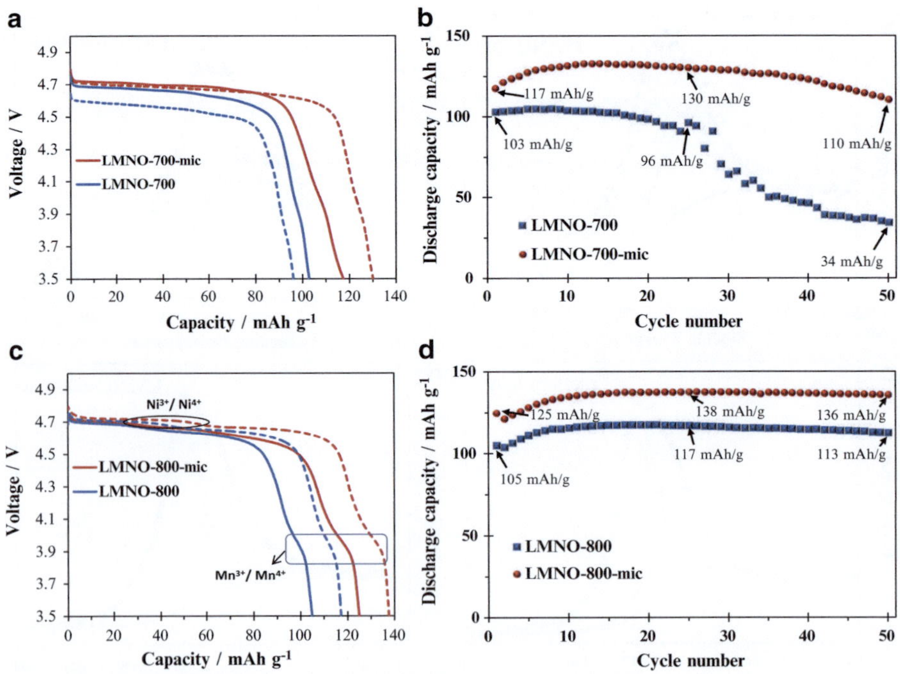

Fig. 1.10 The discharge curves of (**a**) LMNO and LMNO-mic and (**c**) LMNO and LMNO-800 at 0.1 C, 1st cycle (*solid line*) and 25th cycle (*broken line*). Capacity retention ability (cyclability) of (**b**) LMNO-700/LMNO-700-mic and (**d**) LMN-800/LMN-800-mic at 0.1C (Reproduced from Ref. [11]. Copyright 2013, American Chemical Society)

Ni^{3+}/Ni^{4+} redox couples of the disordered spinel are activated upon cycling, suggesting some structural changes induced by electrochemical cycling. The increase in capacity upon cycling for the disordered sample was attributed to the wetting of the electrodes with the electrolytes prior to stabilisation of the redox process. Figs. 1.10b, d compare the cycle stability of the spinel samples, proving that microwave irradiation improved the capacity retention of both the ordered and disordered spinels. It was also shown that the LMNO-800-mic gave better rate capability than the pristine spinel materials, indicative of a shorter path length for lithium-ion diffusion in the former than the latter.

1.4 Olivine Compounds ($LiMPO_4$)

The orthorhombic structure of the olivine, $LiMPO_4$ (abbreviated as LMP, where M = Fe, Mn, Ni, Co or combinations of thereof), is shown in Fig. 1.11. The structure consists of PO_4 framework strongly bonded in three-dimensional layers of M cations in octahedral coordination [102]. It is this strong 3D bonding framework that is solely responsible for the strong thermal stability that characterises the olivines. It has been shown by both theory [103, 104] and experiments [105] that the guest Li-ion moves in this framework along the [010] channels in the open space between the PO_4 units. The Li-ion moves inside the channel like a slalom racer, swerving and nonlinear, but in a 1D trajectory. The major problem with this property is that any defect or impurity in the structure can block a channel and prevent the diffusion of the Li-ions to all the sites in a channel. Thus, the Li-ion conductivity is extremely sensitive to any defect or impurity in the olivine structure. Today, unlike in the past, pure LMPs can be prepared for industrial application.

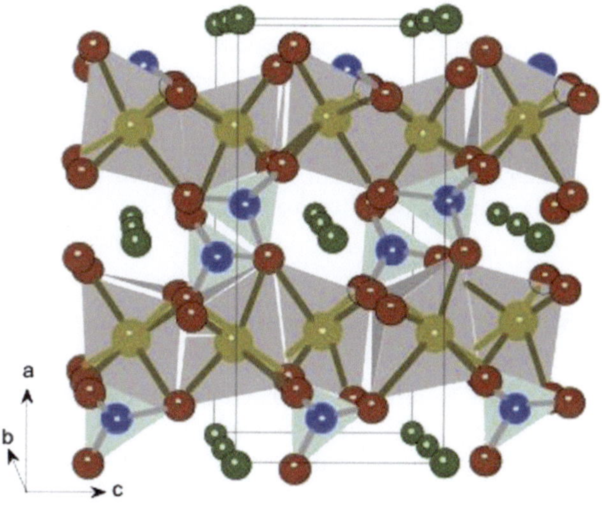

Fig. 1.11 Scheme of the olivine structure of $LiMPO_4$ (M = Fe, Mn, Ni, Co) where *green* is Li, *light brown* is M, *blue* is P, *red* is O (Reprinted with permission from Ref. [102]. Copyright 2013, Elsevier)

The most famous member of the LMP family is the $LiFePO_4$ (LFP, Fig. 1.1) which has been successfully developed for several commercial applications because of its low cost, non-toxicity, high safety and excellent cycling performance. LFP was first reported by John Goodenough and co-workers in 1997 [106]. The LFP has emerged as the new generation of rechargeable LIB for several applications such as computers, power tools, portable electronics, hybrid electric vehicles and large-scale power storage. One of the excellent properties of the LFP is its safety which stems from the strong P–O bonds in the framework that prevent the release of oxygen and thus the possible explosion of the LFP-based batteries. In fact, it is these strong P–O bonds that explain why LFP-based batteries do not really require complicated and expensive battery monitoring systems (BMS) as the other LIBs. Despite that the LFP exhibits excellent cycling performance and possesses admirable advantages, this cathode material still suffers from low energy density due to low voltage (3.45 V) and poor rate capability due to the 1D ionic and poor intrinsic electronic conductivity. To mitigate these challenges, several strategies are being explored, including (i) coating or surface modifications with particles with good electron-conducting properties such as conductive carbons, (ii) nanosizing with the aim of shortening the Li-ion diffusion path in the solid state as well as decreasing the anti-site defect to increase the Li-ion conductivity and (iii) doping (both divalent and aliovalent dopings) either Li or Fe sites with other metal ions such as Co, Mn and Ni to improve voltage.

The next-generation polyanions are likely to be found in the high-voltage LMPs such as the $LiCoPO_4$ (4.8 V) [107], $LiNiPO_4$ (5.1 V) [108] and $Li_3V_2PO_4$ (>4.6 V) [102, 109]. However, the utilisation of these potential cathode materials is hampered by their inherent low electronic/ionic conductivity, poor rate capability, cycling stability and compatible electrolytes. Thus, the new research direction is focussed on the development of LiMnPO4 (4.1 V) [106, 110] and doping $LiFePO_4$ (3.45 V) with Mn, Co or Ni in Fe site (i.e. 'divalent doping') [111–114] to improve the electrochemical performance at optimal voltage.

The huge success of the LFP has spurred growing interest on the silicate compounds, Li_2MSiO_4 (M = Mn, Fe, Co and Ni). Despite the growing attention on the Li_2MSiO_4 complexes and their high theoretical capacity arising from the ability for two lithium ions to be reversibly extracted from the host structure, the properties of these cathode materials have yet to satisfy the requirements for commercialisation. Recent development on the Li_2MSiO_4 cathode is described elsewhere in this book by Prof Rosalind Rossouw.

1.5 Summary and Future Prospects

Lithium-ion batteries (LIB) have emerged as a key driving power for several technological applications ranging from mobile electronics to electric vehicles and stationary storage/utility applications. Of the three major components of the LIBs (i.e. cathode, anode and electrolyte), the cathode material is usually the most expensive and the determining factor of the overall energy density of the battery. Thus, to keep pace with the ever-emerging technologies that require LIBs as power sources, the need for an intense research focus on new and improved LIB cathode materials cannot be overemphasised. In this chapter, we have reviewed the breakthroughs in the last decades on the three successful structural groups of LIB cathode materials (layered oxides, spinel oxides and olivine compounds) and provided the readers an understanding of the critical scientific challenges faced by these cathode materials.

Different application imposes different requirements on the LIBs (i.e. high energy density, high power density, long cycle life, safety and low cost). None of the available cathode materials meets (or will meet) the stringent requirements for different applications, thus the search for new LIB cathode materials will continue for many years to come. Interestingly, however, it has become evident that the performance of the LIB cathode materials can be improved by what we may describe as the 'quadruple combination processes' *(QCP)*, which are (i) *nanostructuring*, (ii) *doping with cations and anions*, (iii) *coating/surface modification* and (iv) *microwave irradiation*. While nanostructuring is generally aimed at shortening the diffusion path length for Li-ion, the other three processes serve to stabilise the structure and enhance the electrochemical performance. Thus, when nanostructuring is combined with one or all of the other processes, it is possible to overcome kinetic and electronic limitations, thus allowing LIB cathode materials that would once have been thought of as unsuitable for electrodes to be revisited.

References

1. Nagaura T, Tozawa K (1990) Lithium ion rechargeable battery. Prog Batter Solar Cells 9:209
2. Song H, Lee KT, Kim MG, Nazar LF, Cho J (2010) Recent progress in nanostructured cathode materials for lithium secondary batteries. Adv Funct Mater 20:3818–3834
3. Xu B, Qian D, Wang Z, Meng YS (2012) Recent progress in cathode materials research for advanced lithium ion batteries. Mater Sci Eng R Rep 73:51–65
4. Kraytsberg A, Ein-Eli Y (2012) Higher, stronger, better... . A review of 5 volt cathode materials for advanced Lithium-Ion batteries. Adv Energy Mater 2:922–939
5. Julien CM, Mauger A, Zaghib K, Groult H (2014) Comparative issues of cathode materials for Li-ion batteries. Inorganics 2:132–154
6. Brodd RJ (2012) Batteries for sustainability: selected entries from the encyclopedia of sustainability science and technology. Springer, New York
7. Yu H, Zhou H (2013) High-energy cathode materials (Li_2MnO_3–$LiMO_2$) for lithium-ion batteries. J Phys Chem Lett 4:1268–1280
8. Xu X, Lee S, Jeong S, Kim Y, Cho J (2013) Recent progress on nanostructured 4V cathode materials for Li-ion batteries for mobile electronics. Mater Today 16:487–495
9. Ma D, Zhang P, Li Y, Ren X (2015) $Li_{1.2}Mn_{0.54}Ni_{0.13}Co_{0.13}O_2$-encapsulated carbon nanofiber network cathodes with improved stability and rate capability for Li-ion batteries. Sci Rep 5:11257
10. Jafta CJ, Raju K, Mathe MK, Manyala N, Ozoemena KI (2015) Microwave irradiation controls the manganese oxidation states of nanostructured (Li

[$Li_{0.2}Mn_{0.52}Ni_{0.13}Co_{0.13}Al_{0.02}]O_2$) layered cathode materials for high-performance lithium ion batteries. J Electrochem Soc 162:A768–A773
11. Jafta CJ, Mathe MK, Manyala N, Roos WD, Ozoemena KI (2013) Microwave-assisted synthesis of high-voltage nanostructured $LiMn_{1.5}Ni_{0.5}O_4$ spinel: tuning the Mn3 content and electrochemical performance. ACS Appl Mater Interfaces 5:7592–7598
12. Nkosi FP, Jafta CJ, Kebede M, le Roux L, Mathe MK, Ozoemena KI (2015) Microwave-assisted optimization of the manganese redox states for enhanced capacity and capacity retention of $LiAl_xMn_{2-x}O_4$ ($x = 0$ and 0.3) spinel materials. RSC Adv 5:32256–32262
13. Ohzuku T, Brodd RJ (2007) An overview of positive-electrode materials for advanced lithium-ion batteries. J Power Sources 174:449–456
14. Robert R, Villevieille C, Novák P (2014) Enhancement of the high potential specific charge in layered electrode materials for lithium-ion batteries. J Mater Chem A 2:8589–8598
15. Ohzuku T, Makimura Y (2001) Layered lithium insertion material of $LiNi_{1/2}Mn_{1/2}O_2$: a possible alternative to $LiCoO_2$ for advanced lithium-ion batteries. Chem Lett 744–745
16. Yoon W, Grey CP, Balasubramanian M, Yang X, McBreen J (2003) In situ X-ray absorption spectroscopic study on $LiNi_{0.5}Mn_{0.5}O_2$ cathode material during electrochemical cycling. Chem Mater 15:3161–3169
17. Whittingham MS (1976) Electrical energy storage and intercalation chemistry. Science 192:1126–1127
18. Kang K, Meng YS, Breger J, Grey CP, Ceder G (2006) Electrodes with high power and high capacity for rechargeable lithium batteries. Science 311:977–980
19. Ohzuku T, Makimura Y (2001) Layered lithium insertion material of $LiCo_{1/3}Ni_{1/3}Mn_{1/3}O_2$ for lithium-ion batteries. Chem Lett 642–643
20. Lee M, Kang Y, Myung S, Sun Y (2004) Synthetic optimization of $Li[Ni_{1/3}Co_{1/3}Mn_{1/3}]O_2$ via co-precipitation. Electrochim Acta 50:939–948
21. Whitfield P, Davidson I, Cranswick L, Swainson I, Stephens P (2005) Investigation of possible superstructure and cation disorder in the lithium battery cathode material $LiMn_{1/3}Ni_{1/3}Co_{1/3}O_2$ using neutron and anomalous dispersion powder diffraction. Solid State Ion 176:463–471
22. Yabuuchi N, Koyama Y, Nakayama N, Ohzuku T (2005) Solid-state chemistry and electrochemistry of $LiCo_{1/3}Ni_{1/3}Mn_{1/3}O_2$ for advanced lithium-ion batteries II. Preparation and characterization. J Electrochem Soc 152:A1434–A1440
23. Kim J, Chung H (2004) The first cycle characteristics of $Li[Ni_{1/3}Co_{1/3}Mn_{1/3}]O_2$ charged up to 4.7 V. Electrochim Acta 49:937–944
24. Li D, Muta T, Zhang L, Yoshio M, Noguchi H (2004) Effect of synthesis method on the electrochemical performance of $LiNi_{1/3}Mn_{1/3}Co_{1/3}O_2$. J Power Sources 132:150–155
25. Choi J, Manthiram A (2005) Role of chemical and structural stabilities on the electrochemical properties of layered $LiNi_{1/3}Mn_{1/3}Co_{1/3}O_2$ cathodes. J Electrochem Soc 152:A1714–A1718
26. Jiang X, Sha Y, Cai R, Shao Z (2015) The solid-state chelation synthesis of $LiNi_{1/3}Co_{1/3}Mn_{1/3}O_2$ as a cathode material for lithium-ion batteries. J Mater Chem A 3:10536–10544
27. Santhanam R, Rambabu B (2009) Improved high rate cycling of Li-rich $Li_{1.10}Ni_{1/3}Co_{1/3}Mn_{1/3}O_2$ cathode for lithium batteries. Int J Electrochem Sci 4:1770–1778
28. Bai Y, Li Y, Wu C, Lu J, Li H et al (2015) Lithium-rich nanoscale Li1. 2Mn0. 54Ni0. 13Co0. 13O2 cathode material prepared by co-precipitation combined freeze drying (CP–FD) for lithium-ion batteries. Energy Technol 3:843–850
29. Hsieh C, Mo C, Chen Y, Chung Y (2013) Chemical-wet synthesis and electrochemistry of $LiNi_{1/3}Co_{1/3}Mn_{1/3}O_2$ cathode materials for Li-ion batteries. Electrochim Acta 106:525–533
30. Zaghib K, Dubé J, Dallaire A, Galoustov K, Guerfi A et al (2012) Enhanced thermal safety and high power performance of carbon-coated $LiFePO_4$ olivine cathode for Li-ion batteries. J Power Sources 219:36–44
31. Rossouw M, Thackeray M (1991) Lithium manganese oxides from Li_2MnO_3 for rechargeable lithium battery applications. Mater Res Bull 26:463–473
32. Rossouw M, Liles D, Thackeray M (1993) Synthesis and structural characterization of a novel layered lithium manganese oxide, $Li_{0.36}Mn_{0.91}O_2$, and its lithiated derivative, $Li_{1.09}Mn_{0.91}O_2$. J Solid State Chem 104:464–466

33. Kalyani P, Chitra S, Mohan T, Gopukumar S (1999) Lithium metal rechargeable cells using Li_2MnO_3 as the positive electrode. J Power Sources 80:103–106
34. Kang S, Johnson C, Vaughey J, Amine K, Thackeray M (2006) The effects of acid treatment on the electrochemical properties of $0.5Li_2MnO_3°$ $0.5LiNi_{0.44}Co_{0.25}Mn_{0.31}O_2$ electrodes in lithium cells. J Electrochem Soc 153:A1186–A1192
35. Manthiram A (2011) Materials challenges and opportunities of lithium ion batteries. J Phys Chem Lett 2:176–184
36. Yu H, Zhou H (2012) Initial coulombic efficiency improvement of the $Li_{1.2}Mn_{0.567}Ni_{0.166}Co_{0.067}O_2$ lithium-rich material by ruthenium substitution for manganese. J Mater Chem 22:15507–15510
37. Wei G, Lu X, Ke F, Huang L, Li J et al (2010) Crystal habit-tuned nanoplate material of Li $[Li_{1/3-2x/3}Ni_xMn_{2/3-x/3}]O_2$ for high-rate performance lithium-ion batteries. Adv Mater 22:4364–4367
38. Yu H, Kim H, Wang Y, He P, Asakura D et al (2012) High-energy 'composite' layered manganese-rich cathode materials via controlling Li_2MnO_3 phase activation for lithium-ion batteries. Phys Chem Chem Phys 14:6584–6595
39. Reed J, Ceder G, Van Der Ven A (2001) Layered-to-spinel phase transition in Li_xMnO_2. Electrochem Solid-State Lett 4:A78–A81
40. Mukai K, Ikedo Y, Kamazawa K, Brewer JH, Ansaldo EJ et al (2013) The gradient distribution of ni ions in cation-disordered $Li[Ni_{1/2}Mn_{3/2}]O_4$ clarified by muon-spin rotation and relaxation (μSR). Rsc Adv 3:11634–11639
41. Thackeray M, David W, Bruce P, Goodenough J (1983) Lithium insertion into manganese spinels. Mater Res Bull 18:461–472
42. Goodenough J, Thackeray M, David W, Bruce P (1984) Lithium insertion/extraction reactions with manganese oxides. Rev Chim Mineral 21:435–455
43. David W, Thackeray M, Bruce P, Goodenough J (1984) Lithium insertion into β MnO_2 and the rutile-spinel transformation. Mater Res Bull 19:99–106
44. Thackeray M (2011) Twenty golden years of battery R&D at CSIR, 1974–1994. S Afr J Chem 64:61–66, Review article
45. Kebede MA, Phasha MJ, Kunjuzwa N, le Roux LJ, Mkhonto D et al (2014) Structural and electrochemical properties of aluminium doped $LiMn_2O_4$ cathode materials for Li battery: experimental and ab initio calculations. Sustain Energy Technol Assess 5:44–49
46. Arico AS, Bruce P, Scrosati B, Tarascon J, Van Schalkwijk W (2005) Nanostructured materials for advanced energy conversion and storage devices. Nat Mater 4:366–377
47. Şahan H, Göktepe H, Patat Ş, Ülgen A (2011) Improvement of the electrochemical performance of $LiMn_2O_4$ cathode active material by lithium borosilicate (LBS) surface coating for lithium-ion batteries. J Alloys Compd 509:4235–4241
48. Aurbach D, Levi M, Gamulski K, Markovsky B, Salitra G et al (1999) Capacity fading of $Li_xMn_2O_4$ spinel electrodes studied by XRD and electroanalytical techniques. J Power Sources 81:472–479
49. Xia Y, Zhou Y, Yoshio M (1997) Capacity fading on cycling of 4 V $Li/LiMn_2O_4$ cells. J Electrochem Soc 144:2593–2600
50. Shin Y, Manthiram A (2004) Factors influencing the capacity fade of spinel lithium manganese oxides. J Electrochem Soc 151:A204–A208
51. Martinez S, Sobrados I, Tonti D, Amarilla J, Sanz J (2014) Chemical vs. electrochemical extraction of lithium from the Li-excess $Li_{1.10}Mn_{1.90}O_4$ spinel followed by NMR and DRX techniques. Phys Chem Chem Phys 16:3282–3291
52. Tong Q, Yang Y, Shi J, Yan J, Zheng L (2007) Synthesis and storage performance of the doped $LiMn_2O_4$ spinel. J Electrochem Soc 154:A656–A667
53. Gummow R, De Kock A, Thackeray M (1994) Improved capacity retention in rechargeable 4 V lithium/lithium-manganese oxide (spinel) cells. Solid State Ion 69:59–67
54. Reddy M, Raju MS, Sharma N, Quan P, Nowshad SH et al (2011) Preparation of $Li_{1.03}Mn_{1.97}O4$ and $Li_{1.06}Mn_{1.94}O_4$ by the polymer precursor method and X-ray, neutron diffraction and electrochemical studies. J Electrochem Soc 158:A1231–A1236

55. Jiao F, Bao J, Hill AH, Bruce PG (2008) Synthesis of ordered mesoporous Li–Mn–O spinel as a positive electrode for rechargeable lithium batteries. Angew Chem Int Ed 47:9711–9716
56. Shin Y, Manthiram A (2003) Influence of the lattice parameter difference between the two cubic phases formed in the 4 V region on the capacity fading of spinel manganese oxides. Chem Mater 15:2954–2961
57. Choi W, Manthiram A (2007) Influence of fluorine on the electrochemical performance of spinel $LiMn_{2-y-z}Li_yZn_zO_{4-\eta}F_\eta$ cathodes. J Electrochem Soc 154:A614–A618
58. Lee K, Myung S, Bang HJ, Chung S, Sun Y (2007) Co-precipitation synthesis of spherical $Li_{1.05}M_{0.05}Mn_{1.9}O_4$ (M = Ni, Mg, Al) spinel and its application for lithium secondary battery cathode. Electrochim Acta 52:5201–5206
59. Lee YJ, Park S, Eng C, Parise JB, Grey CP (2002) Cation ordering and electrochemical properties of the cathode materials $LiZn_xMn_{2-x}O_4$, $0 < x \leq 0.5$: a 6Li magic-angle spinning NMR spectroscopy and diffraction study. Chem Mater 14:194–205
60. Hernan L, Morales J, Sanchez L, Santos J (1999) Use of Li–M–Mn–O [M = Co, Cr, Ti] spinels prepared by a sol–gel method as cathodes in high-voltage lithium batteries. Solid State Ion 118:179–185
61. Kim J, Myung S, Yoon C, Kang S, Sun Y (2004) Comparative study of $LiNi_{0.5}Mn_{1.5}O_{4-\delta}$ and $LiNi_{0.5}Mn_{1.5}O_4$ cathodes having two crystallographic structures: Fd3m and P4332. Chem Mater 16:906–914
62. Guohua L, Ikuta H, Uchida T, Wakihara M (1996) The spinel phases $LiM_yMn_{2-y}O_4$ (M = Co, Cr, Ni) as the cathode for rechargeable lithium batteries. J Electrochem Soc 143:178–182
63. Molenda J, Marzec J, Świerczek K, Ojczyk W, Ziemnicki M et al (2004) The effect of 3d substitutions in the manganese sublattice on the charge transport mechanism and electrochemical properties of manganese spinel. Solid State Ion 171:215–227
64. Ohzuku T, Takeda S, Iwanaga M (1999) Solid-state redox potentials for $Li[Me_{1/2}Mn_{3/2}]O_4$ (Me: 3d-transition metal) having spinel-framework structures: a series of 5 volt materials for advanced lithium-ion batteries. J Power Sources 81:90–94
65. Singhal R, Das SR, Tomar MS, Ovideo O, Nieto S et al (2007) Synthesis and characterization of nd doped $LiMn_2O_4$ cathode for Li-ion rechargeable batteries. J Power Sources 164:857–861
66. Tu J, Zhao X, Zhuang D, Cao G, Zhu T, Tu J (2006) Studies of cycleability of $LiMn_2O_4$ and $LiLa_{0.01}Mn_{1.99}O_4$ as cathode materials for Li-ion battery. Phys B Condens Matter 382:129–134
67. Yang S, Jia J, Ding L, Zhang M (2003) Studies of structure and cycleability of $LiMn_2O_4$ and $LiNd_{0.01}Mn_{1.99}O_4$ as cathode for Li-ion batteries. Electrochim Acta 48:569–573
68. Hwang B, Santhanam R, Liu D, Tsai Y (2001) Effect of al-substitution on the stability of $LiMn_2O_4$ spinel, synthesized by citric acid sol–gel method. J Power Sources 102:326–331
69. Lim S, Cho J (2008) PVP-assisted ZrO_2 coating on $LiMn_2O_4$ spinel cathode nanoparticles prepared by MnO_2 nanowire templates. Electrochem Commun 10:1478–1481
70. Cho J (2008) VO_x-coated $LiMn_2O_4$ nanorod clusters for lithium battery cathode materials. J Mater Chem 18:2257–2261
71. Arumugam D, Kalaignan GP (2010) Synthesis and electrochemical characterization of nano-CeO_2-coated nanostructure $LiMn_2O_4$ cathode materials for rechargeable lithium batteries. Electrochim Acta 55:8709–8716
72. Arumugam D, Kalaignan GP (2010) Synthesis and electrochemical characterizations of nano-La_2O_3-coated nanostructure $LiMn_2O_4$ cathode materials for rechargeable lithium batteries. Mater Res Bull 45:1825–1831
73. Luan X, Guan D, Wang Y (2012) Enhancing high-rate and elevated-temperature performances of nano-sized and micron-sized $LiMn_2O_4$ in lithium-ion batteries with ultrathin surface coatings. J Nanosci Nanotechnol 12:7113–7120
74. Zhao S, Fan X, Deng Y, Nan C (2012) Structure and electrochemical performance of single-crystal $Li_{1.05}Ni_{0.1}Mn_{1.9}O_{3.98}F_{0.02}$ coated by Li–La–Ti–O solid electrolyte. Electrochim Acta 65:7–12

75. Amatucci G, Blyr A, Sigala C, Alfonse P, Tarascon J (1997) Surface treatments of $Li_{1+x}Mn_{2-x}O_4$ spinels for improved elevated temperature performance. Solid State Ion 104:13–25
76. Cho J, Kim T, Kim YJ, Park B (2001) Complete blocking of Mn^{3+} ion dissolution from a $LiMn_2O_4$ spinel intercalation compound by Co_3O_4 coating. Chem Commun 1074–1075
77. Han J, Myung S, Sun Y (2006) Improved electrochemical cycling behavior of ZnO-coated $Li_{1.05}Al_{0.1}Mn_{1.85}O_{3.95}F_{0.05}$ spinel at 55 C. J Electrochem Soc 153:A1290–A1295
78. Li C, Zhang H, Fu L, Liu H, Wu Y et al (2006) Cathode materials modified by surface coating for lithium ion batteries. Electrochim Acta 51:3872–3883
79. Fergus JW (2010) Recent developments in cathode materials for lithium ion batteries. J Power Sources 195:939–954
80. Kebede MA, Kunjuzwa N, Jafta CJ, Mathe MK, Ozoemena KI (2014) Solution-combustion synthesized nickel-substituted spinel cathode materials ($LiNi_xMn_{2-x}O_4$; $0 \leq x \leq 0.2$) for lithium ion battery: enhancing energy storage, capacity retention, and lithium ion transport. Electrochim Acta 128:172–177
81. Oh S, Park S, Kim J, Bae YC, Sun Y (2006) Improvement of electrochemical properties of LiNi0.5Mn1.5O4 spinel material by fluorine substitution. J Power Sources 157:464–470
82. Kunduraci M, Al-Sharab JF, Amatucci GG (2006) High-power nanostructured $LiMn_{2-x}Ni_xO_4$ high-voltage lithium-ion battery electrode materials: electrochemical impact of electronic conductivity and morphology. Chem Mater 18:3585–3592
83. Oh SH, Chung KY, Jeon SH, Kim CS, Cho WI, Cho BW (2009) Structural and electrochemical investigations on the LiNi0.5−xMn1.5−yMxyO4 (M = Cr, Al, Zr) compound for 5V cathode material. J Alloys Compd 469:244–250
84. Wang L, Li H, Huang X, Baudrin E (2011) A comparative study of fd-3m and P4332 "$LiNi_{0.5}Mn_{1.5}O_4$". Solid State Ion 193:32–38
85. Kunduraci M, Amatucci G (2006) Synthesis and characterization of nanostructured 4.7 V $Li_xMn_{1.5}Ni_{0.5}O_4$ spinels for high-power lithium-ion batteries. J Electrochem Soc 153: A1345–A1352
86. Kunduraci M, Amatucci G (2008) The effect of particle size and morphology on the rate capability of 4.7 V $LiMn_{1.5+\delta}Ni_{0.5-\delta}O_4$ spinel lithium-ion battery cathodes. Electrochim Acta 53:4193–4199
87. Xia H, Meng Y, Lu L, Ceder G (2007) Electrochemical properties of nonstoichiometric $LiNi_{0.5}Mn_{1.5}O_{4-\delta}$ thin-film electrodes prepared by pulsed laser deposition. J Electrochem Soc 154:A737–A743
88. Yang J, Xu JJ (2006) Synthesis and characterization of carbon-coated lithium transition metal phosphates $LiMPO_4$ (M = Fe, Mn, Co, Ni) prepared via a nonaqueous sol–gel route. J Electrochem Soc 153:A716–A723
89. Fey GT, Lu C, Kumar TP (2003) Preparation and electrochemical properties of high-voltage cathode materials, LiMyNi0.5−yMn1.5O4 (M = Fe, Cu, Al, Mg; y = 0.0–0.4). J Power Sources 115:332–345
90. Arunkumar T, Manthiram A (2005) Influence of chromium doping on the electrochemical performance of the 5V spinel cathode $LiMn_{1.5}Ni_{0.5}O_4$. Electrochim Acta 50:5568–5572
91. Aklalouch M, Amarilla JM, Rojas RM, Saadoune I, Rojo JM (2008) Chromium doping as a new approach to improve the cycling performance at high temperature of 5V LiNi0.5Mn1.5O4-based positive electrode. J Power Sources 185:501–511
92. Li D, Ito A, Kobayakawa K, Noguchi H, Sato Y (2006) Structural and electrochemical characteristics of $LiNi_{0.5-x}Co_{2x}Mn_{1.5-x}O_4$ prepared by spray drying process and post-annealing in O_2. J Power Sources 161:1241–1246
93. Liu J, Manthiram A (2009) Understanding the improved electrochemical performances of Fe-substituted 5 V spinel cathode $LiMn_{1.5}Ni_{0.5}O_4$. J Phys Chem C 113:15073–15079
94. Ito A, Li D, Lee Y, Kobayakawa K, Sato Y (2008) Influence of co substitution for ni and mn on the structural and electrochemical characteristics of LiNi0.5Mn1.5O4. J Power Sources 185:1429–1433

95. Yang M, Xu B, Cheng J, Pan C, Hwang B, Meng YS (2011) Electronic, structural, and electrochemical properties of $LiNi_xCu_yMn_{2-x-y}O_4$ ($0 < x < 0.5$, $0 < y < 0.5$) high-voltage spinel materials. Chem Mater 23:2832–2841
96. Wang H, Xia H, Lai MO, Lu L (2009) Enhancements of rate capability and cyclic performance of spinel $LiNi_{0.5}Mn_{1.5}O_4$ by trace Ru-doping. Electrochem Commun 11:1539–1542
97. Xiao J, Chen X, Sushko PV, Sushko ML, Kovarik L et al (2012) High-performance $LiNi_{0.5}Mn_{1.5}O_4$ spinel controlled by Mn3 concentration and site disorder. Adv Mater 24:2109–2116
98. Zheng J, Xiao J, Yu X, Kovarik L, Gu M et al (2012) Enhanced li ion transport in $LiNi_{0.5}Mn_{1.5}O_4$ through control of site disorder. Phys Chem Chem Phys 14:13515–13521
99. Locati C, Lafont U, Simonin L, Ooms F, Kelder E (2007) Mg-doped LiNi0.5Mn1.5O4 spinel for cathode materials. J Power Sources 174:847–851
100. Reddy M, Manoharan SS, John J, Singh B, Rao GS, Chowdari B (2009) Synthesis, characterization, and electrochemical cycling behavior of the Ru-doped spinel, $Li[Mn_{2-x}Ru_x]O_4$ ($x = 0$, 0.1, and 0.25). J Electrochem Soc 156:A652–A660
101. Park JS, Roh KC, Lee J, Song K, Kim Y, Kang Y (2013) Structurally stabilized LiNi0.5Mn1.5O4 with enhanced electrochemical properties through nitric acid treatment. J Power Sources 230:138–142
102. Zaghib K, Guerfi A, Hovington P, Vijh A, Trudeau M et al (2013) Review and analysis of nanostructured olivine-based lithium recheargeable batteries: status and trends. J Power Sources 232:357–369
103. Islam MS, Driscoll DJ, Fisher CA, Slater PR (2005) Atomic-scale investigation of defects, dopants, and lithium transport in the $LiFePO_4$ olivine-type battery material. Chem Mater 17:5085–5092
104. Nishimura S, Kobayashi G, Ohoyama K, Kanno R, Yashima M, Yamada A (2008) Experimental visualization of lithium diffusion in $LixFePO_4$. Nat Mater 7:707–711
105. Yamada A, Koizumi H, Nishimura S, Sonoyama N, Kanno R et al (2006) Room-temperature miscibility gap in $LixFePO_4$. Nat Mater 5:357–360
106. Padhi A_K, Nanjundaswamy K, Goodenough J (1997) Phospho-olivines as positive-electrode materials for rechargeable lithium batteries. J Electrochem Soc 144:1188–1194
107. Wolfenstine J, Allen J (2004) $LiNiPO_4$–$LiCoPO_4$ solid solutions as cathodes. J Power Sources 136:150–153
108. Wolfenstine J, Allen J (2005) Ni^{3+}/Ni^{2+} redox potential in $LiNiPO_4$. J Power Sources 142:389–390
109. Huang H, Yin S, Kerr T, Taylor N, Nazar LF (2002) Nanostructured composites: a high capacity, fast rate $Li_3V_2(PO_4)_3$/carbon cathode for rechargeable lithium batteries. Adv Mater 14:1525–1528
110. Yamada A, Chung S (2001) Crystal chemistry of the olivine-type $Li(Mn_yFe_{1-y})PO_4$ and $(Mn_yFe_{1-y})PO_4$ as possible 4 V cathode materials for lithium batteries. J Electrochem Soc 148:A960–A967
111. Li C, Hua N, Wang C, Kang X, Wumair T, Han Y (2011) Effect of mn 2 -doping in $LiFePO_4$ and the low temperature electrochemical performances. J Alloys Compd 509:1897–1900
112. Liu Q, Liu Z, Xiao G, Liao S (2013) Enhancement of capacity at high charge/discharge rate and cyclic stability of $LiFePO_4$/C by nickel doping. Ionics 19:445–450
113. Čech O, Thomas JE, Visintin A, Sedlarikova M, Vondrák J, Moreno S (2012) Cobalt doped $LiFePO_4$/C composite material for li-ion cathodes. ECS Trans 40:93–98
114. Mandal S, Saha SK (2012) Ni/graphene/Ni nanostructures for spintronic applications. Nanoscale 4:986–990

Chapter 2
Li_2MnSiO_4 Nanostructured Cathodes for Rechargeable Lithium-Ion Batteries

Rosalind June Gummow

2.1 Introduction

Rechargeable lithium-ion battery technology dominates the portable electronics market following the rapid growth in demand since the introduction of the first, commercial lithium-ion battery by Sony in 1990 [1]. Motivated by concerns for global warming and environmental degradation, researchers have focused on extending the use of rechargeable lithium-ion batteries to large-scale applications. Lithium-ion batteries are attractive candidates for these applications due to their high energy density, high efficiency, and long cycle life. These applications include use as power supplies for low emission hybrid and plug-in electric vehicles and as standby storage to mitigate the unavoidable intermittency of renewable energy technologies like solar and wind power [2, 3]. These extended, large-scale applications place increasingly stringent demands on battery performance and demand the development and optimization of new battery chemistries to meet the challenges for commercial acceptance [3, 4]. In large-scale applications, factors such as safety, toxicity, cost, and abundance of raw materials become highly significant. In addition, high charge and discharge rate requirements mean that nanostructured materials are essential to meet performance targets.

The original, commercial secondary lithium-ion batteries were based on a lithium transition metal oxide cathode (e.g., $LiCoO_2$) and a carbon anode [1]. These two electrodes are insertion materials capable of accommodating lithium ions within their lattice, with very little change to the overall structure. During the charge cycle, lithium ions are extracted from the cathode and at the same time the

R.J. Gummow (✉)
College of Science, Technology and Engineering, James Cook University,
James Cook Drive, Douglas, Townsville, QLD 4811, Australia
e-mail: rosalind.gummow@jcu.edu.au

K.I. Ozoemena, S. Chen (eds.), *Nanomaterials in Advanced Batteries and Supercapacitors*, Nanostructure Science and Technology,
DOI 10.1007/978-3-319-26082-2_2

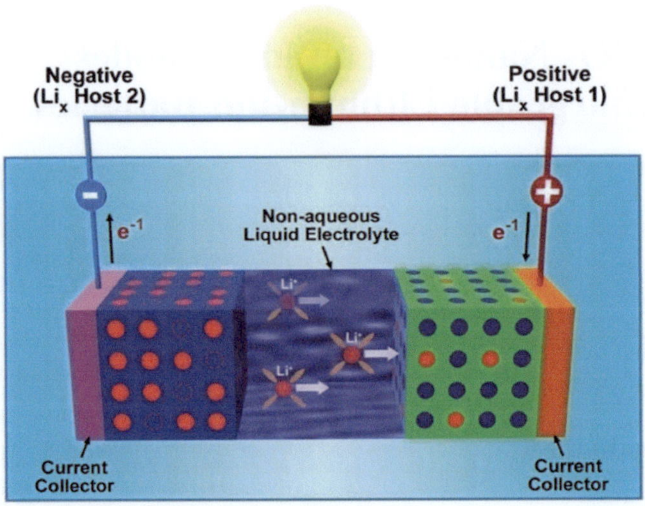

Fig. 2.1 Schematic of a traditional Li-ion battery cell in which, during discharge, Li ions migrate through the electrolyte and electrons flow through the external circuit, both moving from the anode (negative) to the cathode (positive) (Reprinted with permission from Yang Z, Zhang J, Kintner-Meyer MCW, et al. (2011) Chemical Reviews 111:3577–3613. Copyright 2011, American Chemical Society)

transition metal is oxidized. The lithium ions are transported by the electrolyte to the anode where they combine with electrons from the external circuit to form lithium atoms within the carbon matrix. During discharge this process is reversed; lithium atoms at the anode are ionized and lithium ions move through the electrolyte to the anode while electrons move through the external circuit. At the cathode, the lithium ions move into the crystal lattice and the transition metal cations are simultaneously reduced by electrons from the external circuit to maintain charge neutrality. The movement of the lithium ions from anode to cathode and back is the basis of operation of the cells and the origin of the term lithium-ion battery. A schematic diagram of a lithium-ion cell is shown in Fig. 2.1.

Over the years, modifications to the cell chemistry have included changes to the anode, cathode, and electrolyte compositions. The driving force for these innovations has been to increase the electrochemical cell capacity, improve safety and reliability, reduce cost, and increase the reversibility of lithium-ion insertion and extraction with repeated cycling. The performance of electrochemical cells is determined by the crystallographic structure, chemical composition, morphology, and microstructure of electrode materials. In addition, for high rate applications, the optimal nanostructure and architecture of electrodes are essential to provide rapid charge transfer across the electrode/electrolyte interface and fast transfer of lithium ions and electrons within the active material [5]. Significant advances have been made in improving the capacity of anode materials for lithium-ion batteries [6, 7], but finding suitable high-capacity cathode materials to match the anode performance remains a challenge.

Initially, the selection of suitable cathodes was confined to those materials that had a high intrinsic electronic conductivity, for example, the lithium transition metal oxides. Safety concerns and the desire to move away from relatively scarce, toxic, and expensive transition metals, like cobalt, prompted a broadening of the search to include polyanionic compounds like the lithium transition metal phosphates and silicates [8]. The phosphates and silicates were attractive as the strong covalent bonds in the polyanionic units reduce the likelihood of oxygen evolution in the charged state, compared to the oxides [9]. Oxygen evolution poses a hazard when cells are overcharged resulting in undesirable thermal runaway. Polyanionic compounds also offered the possibility of using nontoxic, abundant metals like Fe and Mn to replace the Co and Ni commonly used in lithium transition metal oxide cathodes. Along with their promise of improved safety, however, comes the disadvantage of their low intrinsic electronic conductivity. These cathode materials, when produced as micron-sized particles, show high polarization when charged at high current rates, resulting in very low capacities. It is believed that this polarization is due to both the poor electronic conductivity of the materials resulting in poor charge transfer and the low lithium-ion diffusion rates within the materials. These limitations have driven researchers to explore nanostructuring cathode materials both to increase the surface area to maximize charge transfer and to reduce the diffusion length to minimize the effect of the low lithium diffusion rates [10].

Lithium iron phosphate ($LiFePO_4$), first proposed by Padhi and Goodenough [9] in the 1990s, is the most commercially successful polyanionic cathode to date. $LiFePO_4$ has excellent rechargeability and consists of abundant, nontoxic materials and has proved very attractive for a range of commercial applications. $LiFePO_4$ discharges at ~3.4 V vs. a Li/Li^+ anode with a capacity of 170 $mAhg^{-1}$ (95 % of theoretical capacity). Nanoparticles of $LiFePO_4$ coated with a layer of conductive carbon perform well even at high current rates [11, 12].

The success of $LiFePO_4$ prompted renewed interest in other polyanionic compounds as lithium-ion battery cathodes [13]. In particular, the lithium transition metal orthosilicates Li_2MSiO_4, where M = Mn, Fe, or Ni have attracted the attention of lithium ion battery researchers [14]. In this chapter the potential of Li_2MnSiO_4 as a secondary lithium-ion battery cathode is explored with particular emphasis on the effect of nanostructuring on electrochemical performance. In this chapter, we will first discuss the challenges associated with the use of Li_2MnSiO_4 cathodes, secondly we will consider the advantages of nanostructuring, and thirdly we will review the published data on nanostructured Li_2MnSiO_4 cathodes.

2.2 The Attraction of Li_2MnSiO_4 as a Lithium-Ion Battery Cathode

Among the lithium transition metal orthosilicates Li_2MSiO_4 (where M = Mn, Co, or Ni), Li_2MnSiO_4 offers the greatest potential as a high-voltage, high-energy-density cathode material. Firstly, it is composed of safe, nontoxic, and abundant elements, offering the possibility for the formation of low-cost cathodes. Secondly, if both lithium ions could be extracted reversibly from Li_2MnSiO_4, the theoretical capacity would be a very attractive 330 $mAhg^{-1}$ (roughly double the capacity of $LiFePO_4$ electrodes), with the cathode operating with both the $Mn^{2+/3+}$ and Mn $^{3+/4+}$ electrochemical couples [15, 16]. Of the three Li_2MSiO_4 cathodes, Li_2FeSiO_4 shows the most stable capacity with cycling but at a relatively low average voltage of ~2.5 V which lowers the energy density of the cathode. Recent studies have indicated that more than one Li ion can be reversibly extracted from Li_2FeSiO_4 although this remains controversial and further studies are needed to confirm this conclusively [13, 17, 18]. Systematic studies of Li_2CoSiO_4 by Bruce et al. [19] showed low capacities and poor reversibility with cycling, and this has been confirmed by other investigators [20]. Li_2MnSiO_4 is the most likely of the three compounds to exhibit reversible extraction of more than one lithium ion per formula unit as the Mn^{4+} state is more stable than Fe^{4+} or Co $^{4+}$ and is predicted to be accessible within the voltage stability range of common electrolytes [14].

2.3 Challenges of Lithium Manganese Orthosilicate (Li_2MnSiO_4) Cathodes

2.3.1 Multiple Structural Forms

All the lithium transition metal silicates, including Li_2MnSiO_4, can exist in a wide variety of structural forms. There are at least four known forms of Li_2MnSiO_4 at ambient pressure [15, 21–23]. In all cases the Li, Mn, and Si cations are in tetrahedral coordination in a distorted hexagonally close-packed oxygen array, but the arrangement of the tetrahedra differs in each case (Fig. 2.2) [24].

In practice, the multitude of structural forms makes the synthesis of well-defined, single-phase products challenging. The most widely studied forms are the orthorhombic polymorphs ($Pmn2_1$ and *Pmnb*) which differ only slightly in thermodynamic stability [25]. The monoclinic $P2_1/n$ polymorph is a high temperature form typically prepared at 900 °C, although it can be synthesized at lower temperatures with the partial substitution of Mn with Mg [22, 26]. The *Pn* polymorph has only been prepared by ion exchange from the sodium analogue [23]. The analysis of the cathodes is complicated by the fact that the materials frequently contain multiple phases of Li_2MnSiO_4 as well as secondary impurity phases like MnO and Li_2SiO_3. This complexity makes the production of well-defined and reproducible samples difficult, and small changes in the synthesis conditions can result in very different phase compositions with resulting differences in electrochemical performance.

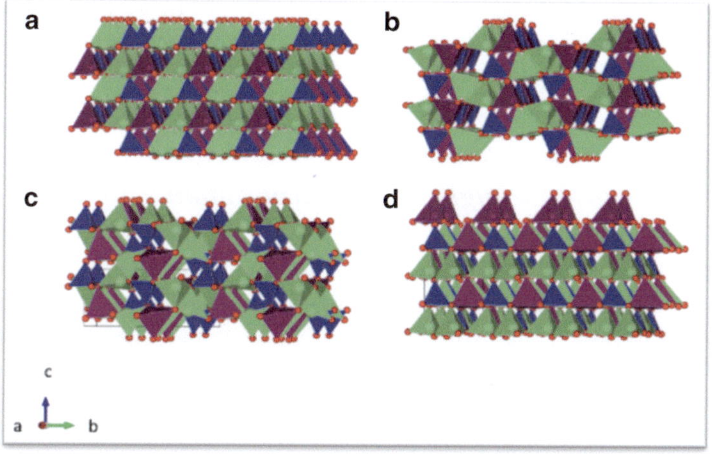

Fig. 2.2 Crystal structures of the four known ambient pressure polymorphs of Li_2MnSiO_4 (**a**) *Pmn*2_1, (**b**) *Pmnb*, (**c**) *P*2_1/*n*, and (**d**) *Pn*. Li tetrahedra are shown in *green*, Mn tetrahedra in *purple*, and Si tetrahedra in *blue*; *red spheres* represent oxygen atoms (Reprinted from Gummow and He [24], copyright (2014), with permission from Elsevier)

2.3.2 Low Lithium-Ion Diffusion Rates

The lithium-ion diffusion pathways in the different structural forms of Li_2MnSiO_4 are distinctly different. The orthorhombic forms *Pmn*2_1 and *Pmnb* (Fig. 2.2a and b) have two-dimensional pathways for lithium-ion diffusion, while the monoclinic forms (*P*2_1/*n* and *Pn*) have lithium positions connected in three dimensions (Fig. 2.2c and d). Calculations have shown that the lithium-ion diffusion pathways in these materials are curved and anisotropic and that all are poor conductors of lithium ions [27, 28]. This finding suggests that lithium diffusion in Li_2MnSiO_4 electrodes may be limiting in high rate applications.

2.3.3 Low Electronic Conductivity

In Li_2MnSiO_4, MnO_4 tetrahedra are surrounded by insulating SiO_4 tetrahedra resulting in very low intrinsic conductivity. The conductivity of Li_2MnSiO_4 at room temperature is of the order of 10^{-16} Scm^{-1} but increases a 100-fold at 60 °C [29, 30]. Unfortunately, this is the case for all structural forms of Li_2MnSiO_4, and this is confirmed by calculations of the band gap for different structural forms [28]. This low conductivity means that the material is effectively electrochemically inactive when particles are micron sized. Without nanostructuring, Li_2MnSiO_4 is completely ineffective as a cathode material but nanostructuring and carbon coating dramatically increase the electrochemical performance and offer hope that the potential of this material may be realized in practice. For Li_2MnSiO_4, as with other polyanionic materials with low intrinsic electronic and ionic conductivity, nanostructuring is essential to achieve reasonable electrochemical performance.

2.3.4 Volumetric Changes and Amorphization

Density functional theory calculations indicated that the extraction of both lithium ions from Li_2MnSiO_4 should be feasible within the electrolyte stability window of common electrolytes [25]. However, density functional theory simulations by Kokalj et al. [30] also indicated that there is a dramatic volume change associated with extraction of more than 1 Li per formula unit in Li_2MnSiO_4 ($-17\,\%$ for $Li_{0.5}MnSiO_4$ and $-27\,\%$ for $MnSiO_4$). It appears that such large volumetric changes cannot be accommodated within the crystallographic structure and lead to structural collapse and amorphization when lithium is extracted [29, 31].

2.4 Advantages of Nanostructuring Li_2MnSiO_4 Cathodes

2.4.1 Carbon Coating

Carbon is an excellent electronic conductor. When nanoparticles of a poorly conducting electrode material are coated with a layer of carbon, this increases the conductivity of the composite electrode. The carbon coating acts as a distributed current collector and provides a continuous pathway for rapid electronic conduction throughout the nanostructured electrode. In this way, the contact resistance between adjacent particles is dramatically reduced [32]. Thin carbon coatings are also generally permeable to Li ions and therefore do not inhibit the diffusion of lithium into the active material. Reduction of the internal resistance of the composite cathode results in a reduction in polarization at high current densities and therefore in increased high-rate capacities. Excess carbon in composite electrodes decreases the overall energy density of the electrode, and therefore the amount of carbon should be kept to a minimum.

2.4.2 Maximizing the Surface Area

In nanosize particles the surface-to-volume ratio is maximized. This implies that there is a greater area for interaction between the electrolyte and the electrode particles. This results in reduced polarization and greater utilization of the electrode material. However, the increased surface area for interaction can potentially result in problems with nanostructured cathode materials if they are unstable in contact with the electrolyte, leading to increased degradation of the electrode surface. In this case, carbon coating has the advantage of reducing the interaction between the active material and the electrolyte and effectively stabilizing the interface.

2.4.3 Reducing the Lithium-Ion Diffusion Length

Due to the low lithium-ion diffusion rate in Li_2MnSiO_4 discussed in Sect. 3.2, nanostructuring is important to reduce the lithium-ion diffusion length, i.e., the length that Li ions need to diffuse from the surface of the individual particles into the bulk. This becomes increasingly critical at high current rates. Nanostructuring reduces polarization at high current rates and enables higher capacities to be obtained from cathodes before the voltage cutoff is reached.

2.5 Synthesis and Electrochemistry of Nanostructured Li_2MnSiO_4

2.5.1 Pechini Sol–Gel Synthesis

From the outset, it was recognized that minimizing particle size and incorporating a carbon coating were critical to achieving optimized electrochemical performance in Li_2MnSiO_4 cathodes. Pechini sol–gel synthesis was adopted by Dominko [15] and co-workers in their pioneering work to prepare 70 nm particles of the $Pmn2_1$ form. In their initial report, although there were islands of carbon between particles, there was no discrete carbon coating. The results of electrochemical tests were disappointing, yielding reversible capacities corresponding to only 0.6 Li ions on the first discharge cycle followed by a rapid capacity fade on subsequent cycles [15] (Fig. 2.3a). Refinements to the synthesis technique in later studies reduced the particle size to 20–50 nm (Fig. 2.3b and c), with a uniform distribution of ~5 wt% of carbon, but this failed to improve the room temperature electrochemical cycling performance compared to the earlier studies [16]. Electrochemical cycling experiments at 60 °C showed a plateau near 4 V corresponding to the extraction of 1.5 Li ions but this capacity was not recovered on discharge. It was not clear if this observed charge capacity corresponded to Li extraction or to irreversible side reactions with the electrolyte at high voltage. Even after ball milling to reduce the particle size and improve the distribution of carbon, there was only a slight reduction in polarization. All samples consistently showed a rapid loss of electrochemical capacity with repeated charge–discharge cycling. Even after adopting the same nanostructuring tactics that had proved so successful with $LiFePO_4$, the

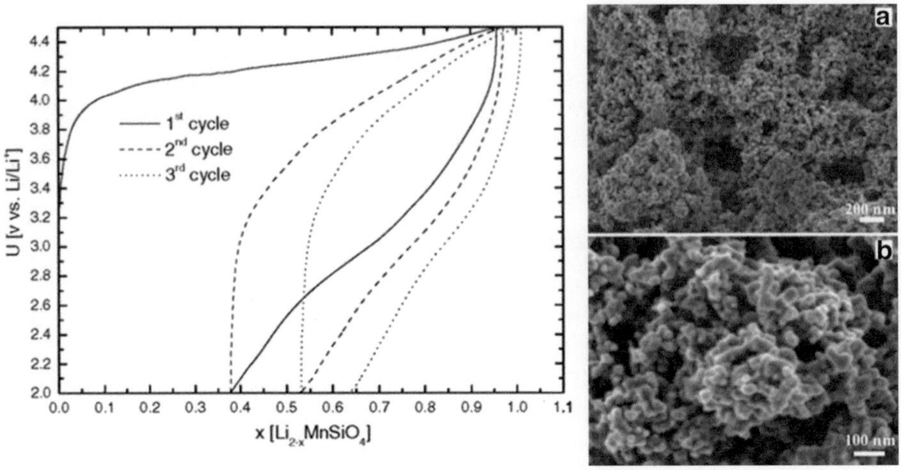

Fig. 2.3 (**a**) Voltage profiles of Li_2MnSiO_4 cycled at room temperature and at a rate of C/30 (Reprinted from Dominko et al. [15], copyright (2006), with permission from Elsevier). SEM micrographs of (**b**) Li_2MnSiO_4 microstructure and (c) Li_2MnSiO_4 particles (Reprinted from Dominko et al. [16], copyright (2007), with permission from Elsevier)

authors failed to achieve a dramatic improvement in the performance of Li_2MnSiO_4 electrodes.

To understand the persistent loss of capacity with cycling, *ex situ* x-ray diffraction studies of partially charged cathodes were conducted. The results revealed a gradual loss of x-ray peak intensity with lithium extraction on charge and no recovery of the peaks when the cell was subsequently discharged (Fig. 2.4). This indicated that the Li_2MnSiO_4 structure underwent irreversible change when lithium was extracted, and these results were confirmed by Li et al. [31] and by further in situ studies by Dominko et al. [29]. The experimentally observed structural collapse (Fig. 2.4) was attributed to the fact that the crystal lattice was unable to accommodate the large volume changes associated with the extraction of more than one lithium ion per formula unit [30].

2.5.2 Alternative Sol–Gel Routes

Refinements to the sol–gel synthesis route were attempted. Deng et al. [33] used a modified citric acid-assisted route with tetraethyl orthosilicate (TEOS) as the silicon source but, although 10.5 wt% carbon was successfully incorporated in the product, the primary particle size was ~200 nm. The electrochemical cycling results for this product were similar to those found in earlier studies. Belharouak et al. [34] also used a modified sol–gel route with $Si(CH_3COO)_4$ as the silicon source but as-prepared samples had primary particles in the 200–300 nm range. As-prepared samples were effectively electrochemically inactive (Fig. 2.5a).

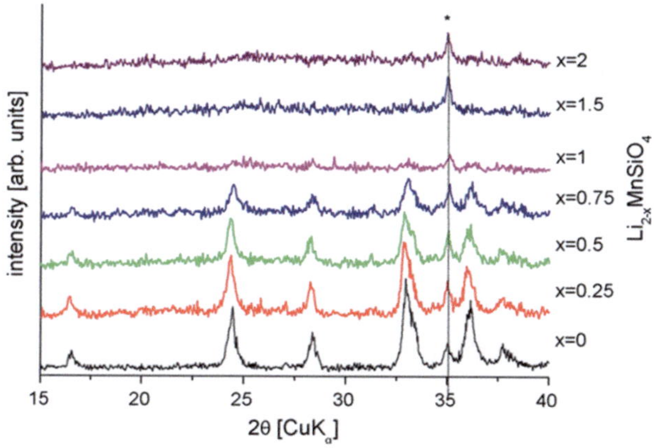

Fig. 2.4 X-ray diffraction patterns of compositions obtained with electrochemical oxidation of Li_2MnSiO_4-based composites. The value of x corresponds to the expected chemical composition of the electrode that is based on the amount of charge passed through the cell. The remaining diffraction peak marked with asterisk denotes MnO impurity (Reprinted from Dominko [29], copyright (2007), with permission from Elsevier)

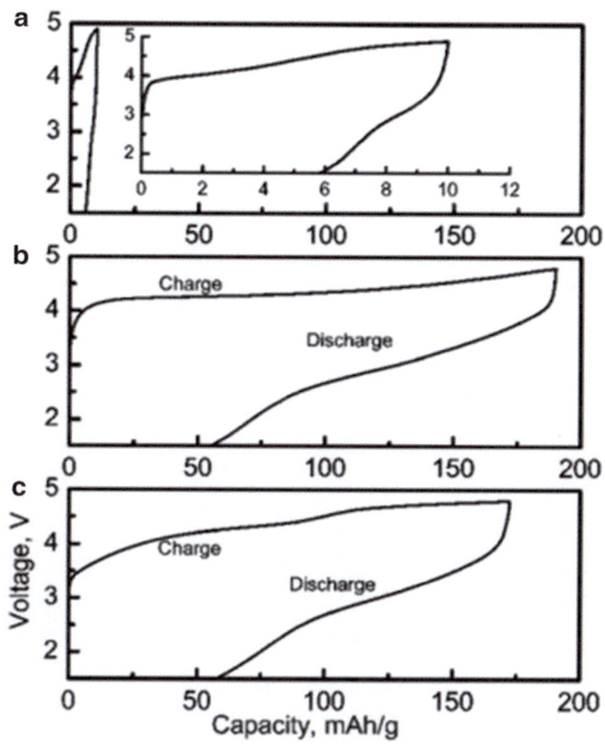

Fig. 2.5 Voltage profiles of (**a**) as-prepared Li_2MnSiO_4, (**b**) carbon-coated Li_2MnSiO_4, and (**c**) high-energy ball-milled Li_2MnSiO_4 (Reprinted with permission from Belharouak et al. [34], copyright (2009) American Chemical Society)

However, dramatic improvement in electrochemical capacity was demonstrated after ball milling to reduce the particle size and coating the particles with a conductive layer of carbon to reduce the charge-transfer resistance and the inter-particle contact resistance (Fig. 2.5b and c). However, even after milling, the particles were agglomerated and the electrochemical capacity remained limited to approximately that equivalent to the reversible extraction, and insertion of one Li ion per formula unit and the capacity retention with cycling was poor.

2.5.3 Solution Synthesis

Li et al. [31] adopted a solution synthesis route using Li and Mn acetates with TEOS in a water–ethanol solution refluxed at 80 °C for 24 h. Sucrose was added as a carbon source and the product was calcined at 600 °C in flowing N_2. The final particle size of 20–30 nm was again comparable to that obtained with Pechini synthesis and the structural form of the product was the $Pmn2_1$ polymorph (Fig. 2.6d). The initial discharge capacity obtained with this material in electro-chemical cycling tests was 210 $mAhg^{-1}$ at a current rate of 5 mAg^{-1}. This was the first report of a discharge capacity corresponding to insertion of more than 1 Li ion per formula unit (Fig. 2.6c) possibly due to the well-defined carbon coating on the surface of the primary particles shown in the TEM image in Fig. 2.6e. Once again the loss of capacity with repeated cycling was confirmed to be due to a loss of crystallinity with lithium extraction (Fig. 2.6b).

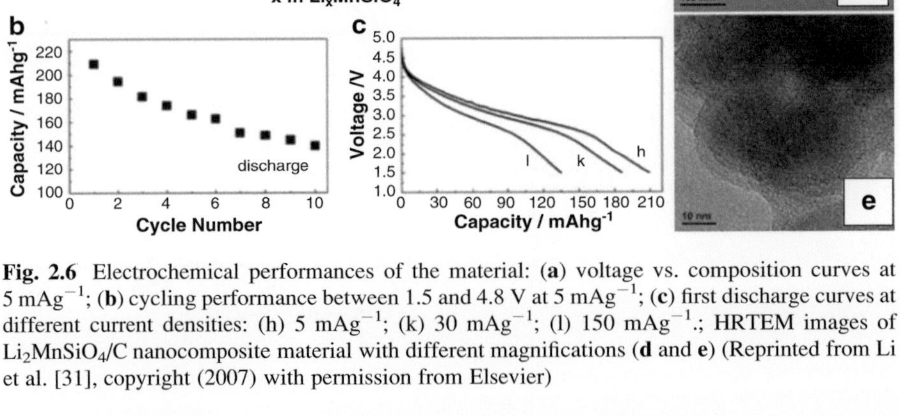

Fig. 2.6 Electrochemical performances of the material: (**a**) voltage vs. composition curves at 5 mAg^{-1}; (**b**) cycling performance between 1.5 and 4.8 V at 5 mAg^{-1}; (**c**) first discharge curves at different current densities: (h) 5 mAg^{-1}; (k) 30 mAg^{-1}; (l) 150 mAg^{-1}.; HRTEM images of Li_2MnSiO_4/C nanocomposite material with different magnifications (**d** and **e**) (Reprinted from Li et al. [31], copyright (2007) with permission from Elsevier)

2.5.4 Polyol Synthesis

Liu et al. [35] adopted a polyol synthesis route for the production of Li_2MnSiO_4. Lithium and manganese acetates and tetraethyl silicate were refluxed in ethylene glycol at 196 °C for 16 h. The product was dried at 150 °C and milled with sucrose to provide a carbon source. After calcination at 600 °C in an inert atmosphere, 30 nm particles of the $Pmn2_1$ polymorph of Li_2MnSiO_4 were formed. The electrochemical cycling performance of this material was comparable to that reported in earlier studies, and the rapid drop in capacity with cycling was again confirmed to be due to a loss of crystallinity when lithium was extracted.

2.5.5 Hydrothermal Synthesis

Sol–gel synthesis techniques and variations thereof failed to result in any dramatic breakthroughs in the performance of Li_2MnSiO_4. Hydrothermal synthesis, in which the precursors are reacted in a sealed autoclave at moderate temperatures but elevated pressures, was introduced as an alternative. This method was initially utilized very successfully to synthesize individual polymorphs of Li_2MnSiO_4 for crystallographic structural investigations [36]. However, no electrochemical evaluation of these materials was performed.

Manthiram et al. [37] used microwave-assisted hydrothermal synthesis to prepare Li_2MnSiO_4 at 30 b and 300 °C. The most significant advantage of this technique is that the synthesis time was reduced to only 25 min resulting in energy saving. After calcination at 650 °C in Ar, the sample consisted of agglomerated 20 nm primary particles with a well-defined layer of carbon on the surface of the particles (Fig. 2.7a and b). Despite high discharge capacities of over 200 $mAhg^{-1}$ on the first two discharge cycles at room temperature (Fig. 2.7c), these materials rapidly lost capacity with repeated charge–discharge cycling. The Jahn–Teller distortion of the manganese cations and dissolution of manganese in the electrolyte

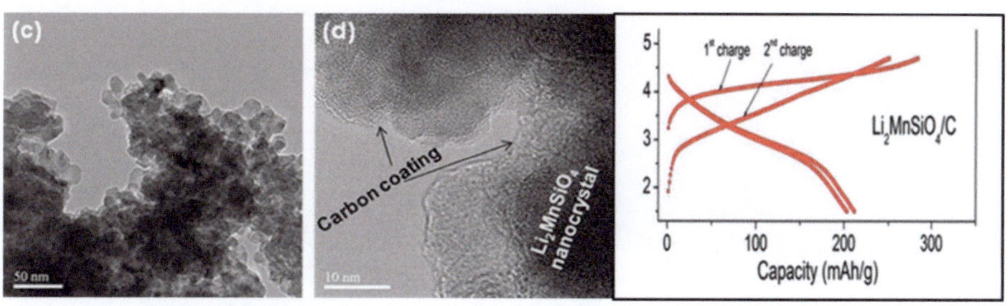

Fig. 2.7 (**a**) TEM and (**b**) HRTEM images of the Li_2MnSiO_4/C nanocomposite. (**c**) Charge–discharge profiles recorded at C/20 rate and room temperature of Li_2MnSiO_4/C (Reprinted with permission Manthiram [37]. Copyright (2010) American Chemical Society)

was given as an explanation for the capacity decline with cycling. The initial discharge capacity at elevated temperature (55 °C) was 250 $mAhg^{-1}$ but the capacity decreased even more rapidly with cycling at this temperature probably due to increased interaction with the electrolyte at the elevated temperature.

2.5.6 *Molten Carbonate Flux Synthesis*

Kojima et al. [38] used a $(Li_{0.435}Na_{0.315}K_{0.25})_2CO_3$ flux with Li_2SiO_3 and either $MnC_2O_4.2H_2O$ or $Mn(OH)_2$ precursors. Reactions were carried out in an atmosphere of CO_2/H_2 100:3 v/v. This method enabled the production of single-phase products of the $Pmn2_1$ form at 500–650 °C, significantly reducing the synthesis temperature compared to conventional sol–gel routes. A key finding was that the choice of manganese precursor dramatically affected the morphology of the final Li_2MnSiO_4 product and this, in turn, resulted in widely different electrochemical performances. The samples with flake-like morphology prepared with $Mn(OH)_2$ precursors (Fig. 2.8b and d) gave higher initial capacities and better capacity retention compared to the more spherical particles prepared with oxalates (Fig. 2.8a and c). However, in both cases, capacity loss with cycling was significant with only 55 % capacity retention after 20 cycles for the better performing samples prepared from $Mn(OH)_2$.

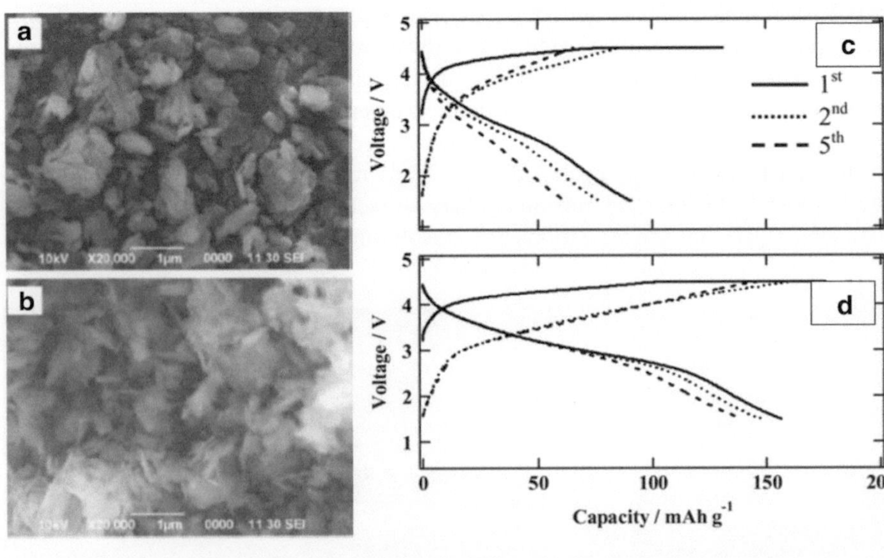

Fig. 2.8 SEM images and charge–discharge curves of the Li_2MnSiO_4 synthesized from $MnC_2O_4.2H_2O$ (LMS-MCO) (**a** and **c**) and manganese hydroxide (LMS-MOH) (**b** and **d**) at 30 °C with a current density of 10 mAg^{-1} in the voltage range of 1.5–4.5 V (Reproduced from Kojima et al. [38], by permission of The Electrochemical Society)

2.5.7 *Supercritical Solvothermal Synthesis*

2.5.7.1 Monodisperse Li_2MnSiO_4 Particles

Supercritical solvothermal synthesis utilizes very high-pressure, moderate temperatures and has the advantage that synthesis times can be reduced to the order of minutes. The requirement of specialized equipment for this type of synthesis would, however, make it expensive and difficult to apply for large-scale manufacture despite the reduced reaction times. This technique was used by Kempaiah et al. [39] to synthesize monodisperse 5–20 nm particles of Li_2MnSiO_4 (Fig. 2.9a–d). The material was found by x-ray diffraction to be a well-defined $Pmn2_1$ polymorph with no obvious impurities. The heat-treated, conductive polymer PEDOT was used for the first time to form a conductive coating on the surface of the particles (Fig. 2.9d). Although excellent capacities were reported for the initial discharge cycle, approaching 300 mAhg^{-1}, there was a rapid drop in discharge capacity after the first cycle at room temperature, followed by a gradual decline in subsequent cycles (Fig. 2.10a and c). The capacities reported for cycling at 40 °C were above 250 mAhg^{-1} for the first 20 cycles (Fig. 2.10b and d). It should

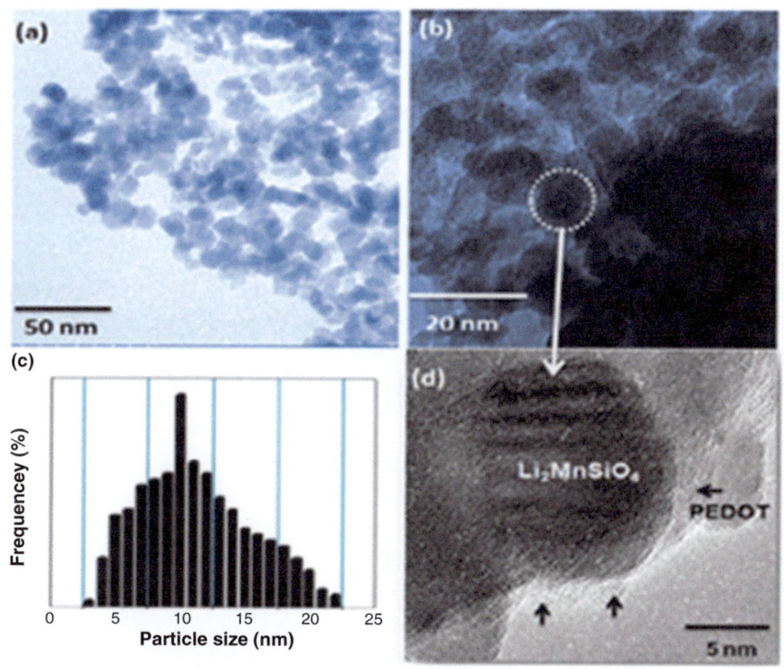

Fig 2.9 (**a**) and (**b**) TEM and HRTEM images of as-synthesized Li_2MnSiO_4 and PEDOT/Li_2MnSiO_4 nanoparticles, (**c**) particle size distribution and (**d**) HRTEM image showing PEDOT-coated Li_2MnSiO_4 nanoparticles (Reproduced from Kempaiah et al. [39] with permission of The Royal Society of Chemistry)

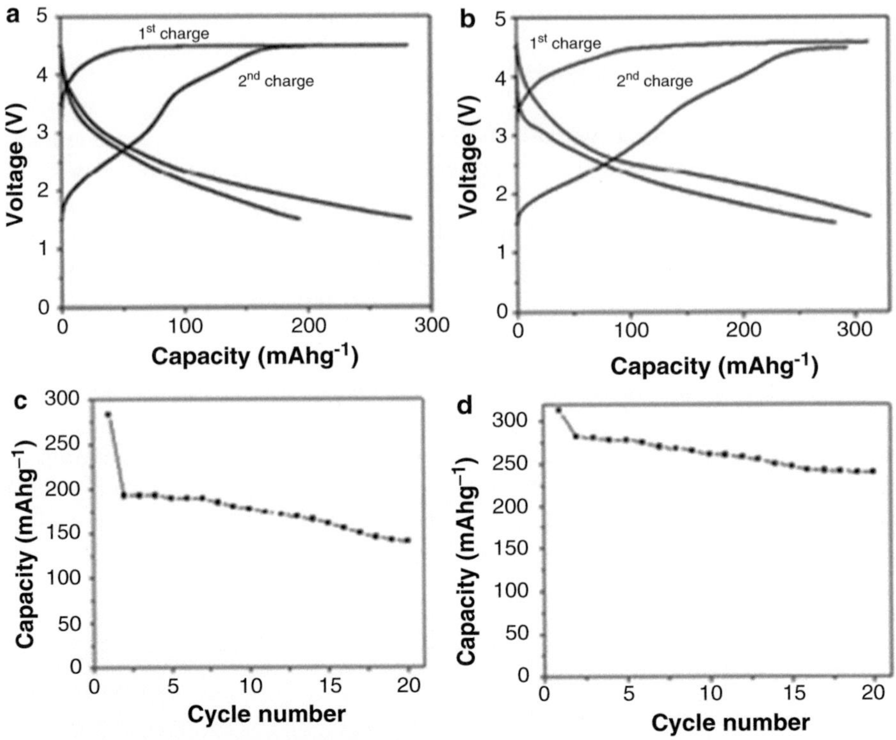

Fig. 2.10 Charge–discharge profiles (**a** and **b**) and cyclic performance (**c** and **d**) of PEDOT/Li_2MnSiO_4 nanoparticles recorded at C/20 in room temperature (**a** and **c**) and at 40 °C (**b** and **d**), respectively (Reproduced from Kempaiah et al. [39] with permission of The Royal Society of Chemistry)

be noted that, although the capacities were high, much of the discharge capacity was recorded at low voltages (below 2.5 V) with only 180 $mAhg^{-1}$ above 2.5 V for the first discharge cycle at 40 °C (Fig. 2.10b); this would reduce the overall energy density of the cathode.

2.5.7.2 Nanosheet Morphology

The success and wide-ranging applications for graphene have prompted great interest in nanosheet morphology [40] The formation of nanosheets only a few nanometers in thickness and with lateral dimensions in the micron range has led to the emergence of a new class of materials with novel properties and a very high specific surface area. Nanosheets of a wide range of materials have been synthesized including polyanionic cathodes like $LiFePO_4$, $LiMnPO_4$, and $LiCoPO_4$ which show improved high rate performance and cycle stability compared to bulk materials [41]. Rangappa et al. [42] applied the supercritical fluid technique to form

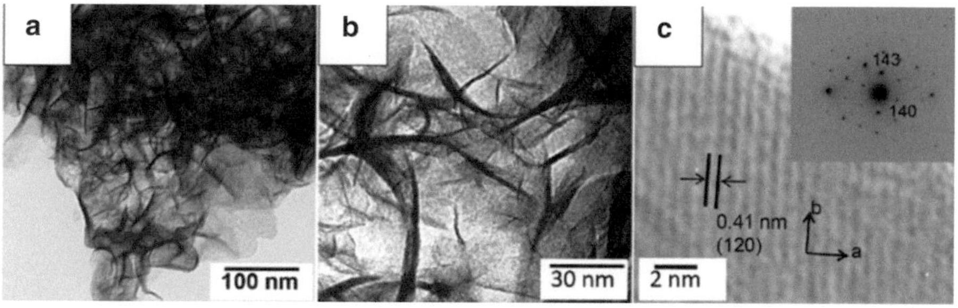

Fig. 2.11 TEM images of the as-prepared samples. (**a** and **b**) Li_2MnSiO_4 nanosheets; (**c**) HRTEM images of Li_2MnSiO_4 nanosheets with SAED patterns in the inset (Reprinted with permission from Rangappa et al. ([42]). Copyright (2012) American Chemical Society)

nanosheets of Li_2MnSiO_4 with a thickness of only several atoms and lateral dimensions of 100–300 nm (Fig. 2.11a–c). Nanosheet morphology has the advantage of dramatically reducing the lithium ion diffusion length in the material and also potentially accommodating the volumetric changes associated with lithium extraction and insertion. However, the complex synthetic pathway made it difficult to control the phase purity of the product, and x-ray analysis results revealed the presence of peaks of unidentified impurities. The electrochemical results showed high discharge capacities above 300 $mAhg^{-1}$ for the first 20 cycles at 45 °C and a current rate of 0.02 C, followed by a reported dramatic loss of capacity (Fig. 2.12). This loss of capacity was probably due to structural collapse of the Li_2MnSiO_4 as x-ray diffraction of cathodes after cycling showed no remaining peaks of the initial Li_2MnSiO_4 phase. This indicates that the nanosheet morphology was not successful in overcoming the structural instability associated with lithium insertion and extraction in Li_2MnSiO_4. However, the nanosheet morphology was successful in reducing the polarization of the cathodes resulting in a gently sloping discharge curve and high capacities. Unfortunately no cycling data was given for higher current rates as it is in this regime that the benefits of this morphology are likely to be most significant.

2.5.7.3 The Use of Surfactants to Create Complex Nanostructures

Deveraju et al. [43] performed synthesis reactions using the supercritical fluid technique and incorporating surfactants in the synthesis. By varying the amount and nature of the surfactant, the reaction time, and temperature, they formed samples with a wide range of particle sizes and morphologies, including monodisperse nanoparticles and complex, hierarchical structures (Fig. 2.13). Carbon coating was achieved by wet milling with PEDOT and carbon black followed by heat treatment at 450 °C in a reducing atmosphere. The complex synthesis resulted in somewhat impure products in which it was difficult to control the phase content. No detailed structural analysis was performed on the products. The electrochemical

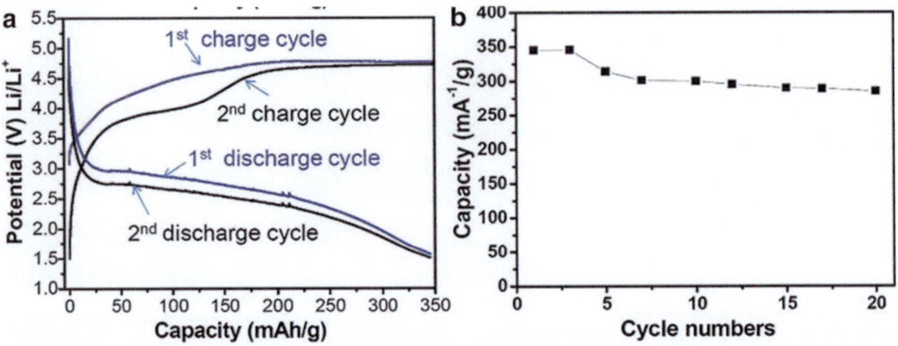

Fig. 2.12 Charge and discharge profile of first and second cycles. (**a**) Li_2MnSiO_4 samples measured at 45 °C temperatures at 0.02 C rates. (**b**) The cyclic performance of Li_2MnSiO_4 samples (Reprinted with permission from Rangappa et al. [42]. Copyright (2012) American Chemical Society)

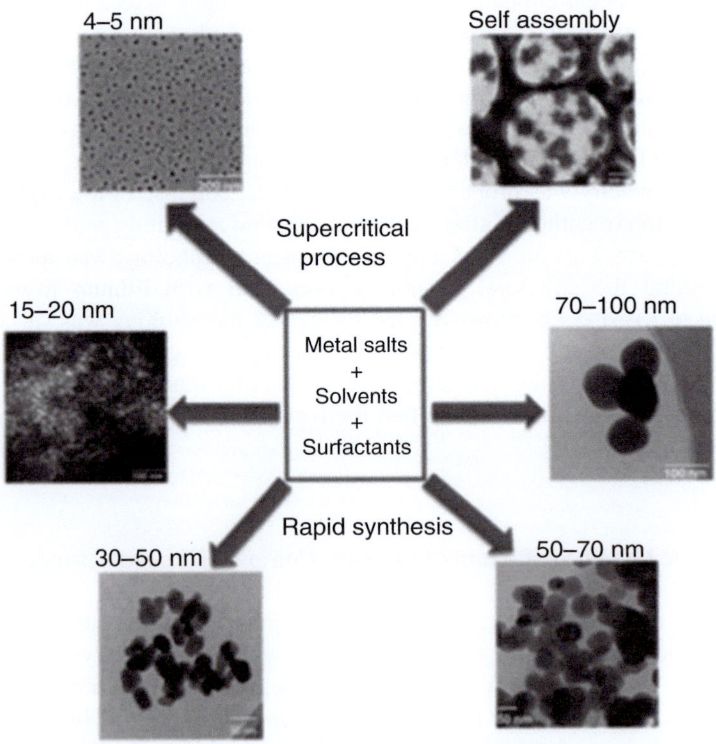

Fig. 2.13 Schematic of nanocrystalline Li_2MnSiO_4 positive electrode materials obtained via a supercritical fluid process and rapid reaction times (4–30 min). The process enables the synthesis of particles with a diameter of between 4 and 100 nm (Reprinted from Devaraju et al. [43] with permission of the Royal Society of Chemistry)

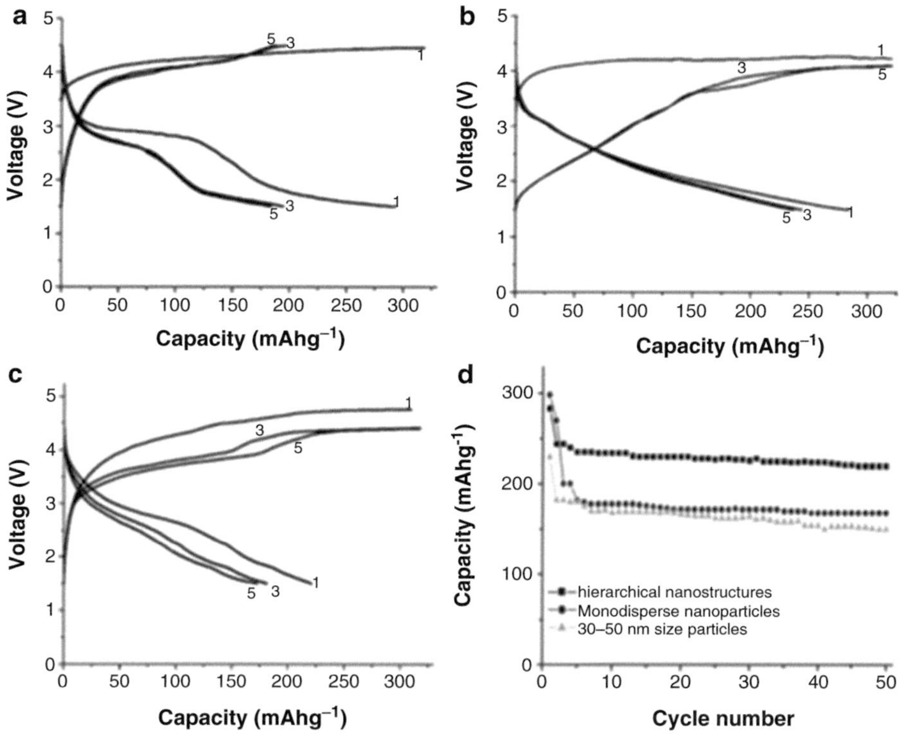

Fig. 2.14 Charge–discharge characteristics of Li_2MnSiO_4-positive electrode materials and cycle performance measured at room temperature with 0.05 C. (**a**) Charge–discharge curves of monodisperse nanoparticles showing typical two-step discharge profiles; (**b**) charge–discharge curves of hierarchical nanostructures showing a gentle slope-like discharge profile and nearly two lithium ion capacity for the first cycle and more than one lithium capacity for the 3rd and 5th cycles; (**c**) charge–discharge curves of 30–50 nm sized particles, with more than one lithium capacity for the 1st and 3rd cycles and nearly one lithium ion capacity for the 5th cycle; and (**d**) the cycle performance. The hierarchical nanostructures showed excellent cycling performance with a stable capacity up to 50 cycles; monodisperse particles showed a decrease in cycle performance after a few cycles due to detachment from the carbon source; 30–50 nm sized particles showed decent cycle performance (Reprinted from Devaraju et al. [43] with permission of the Royal Society of Chemistry)

performance of several of these materials is shown in Fig. 2.14. The highest capacities were reported for the complex hierarchical structures (Fig. 2.14b) although the average voltage was lower than that of the nanoparticle electrodes due to the absence of the voltage plateau region (Fig. 2.14a). Excellent capacity retention with cycling was observed but it should be noted that the composite cathodes contained about 40 w% carbon. This high carbon content would reduce the energy density of the cathodes. Good cycling stability with high carbon loading was earlier reported by Aravindan et al. [44].

2.5.8 *Using Carbon Supports*

2.5.8.1 Reduced Graphene Oxide Networks

Reduced graphene oxide (RGO) has recently attracted interest as a cathode material for lithium-ion batteries, and it has been demonstrated that Faradaic reactions occur between lithium and functional groups on the graphene oxide surface [45]. Zhao et al. [46] reported the use of carbon coating and a reduced graphene oxide network to support homogenously distributed nanoparticles of Li_2MnSiO_4 (Fig. 2.15). The x-ray diffraction data indicate the presence of the $Pmn2_1$ polymorph of Li_2MnSiO_4 with a minor MnO impurity. The electrochemical capacity at low current rates (0.05 C) was reported to be 290 $mAhg^{-1}$, and, even at the 1C rate, the material delivered a capacity of 170 $mAhg^{-1}$ after 700 cycles based on the mass of the Li_2MnSiO_4 (Fig. 2.16a and b). The total carbon content in the composite was calculated to be 26 % after synthesis, and a further 20 % of Ketjenblack was added to form the cathodes for electrochemical testing. At least some of the

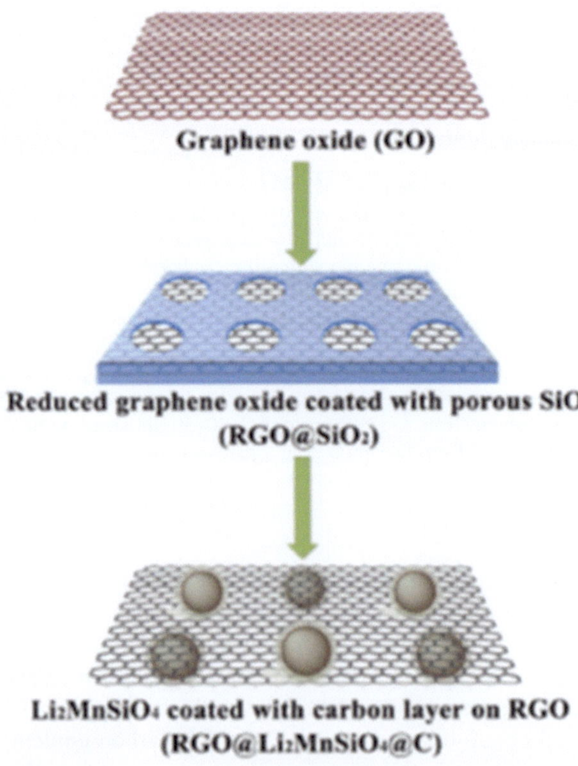

Fig. 2.15 Schematic synthesis of the RGO@Li_2MnSiO_4@C composite (Reprinted from Zhao et al. [46] with permission of the Royal Society of Chemistry)

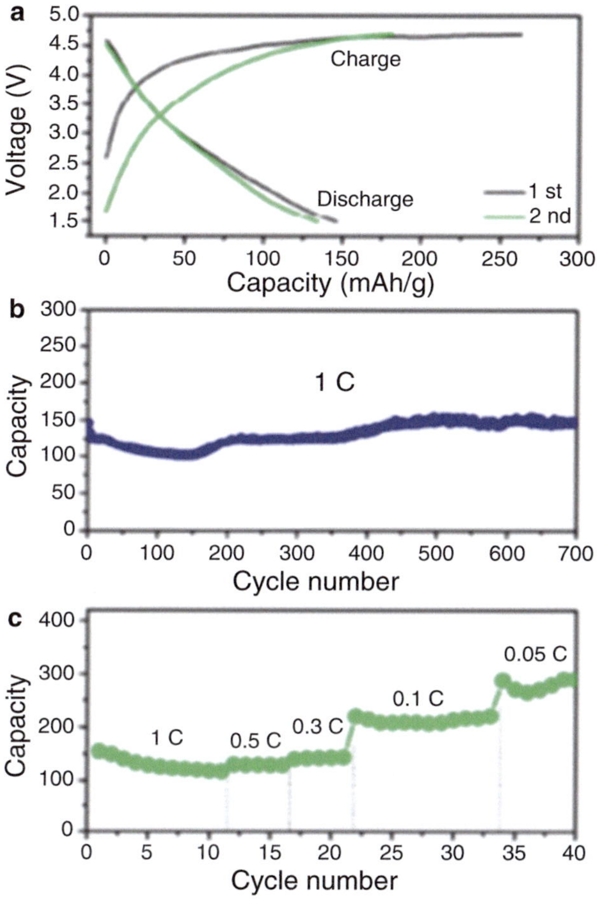

Fig. 2.16 (**a**) The first and second charge–discharge curves of the RGO@Li2MnSiO4@C composite at a current density of 1C between 1.5 and 4.7 V. (**b**) Discharge capacities vs. cycle number of the RGO@Li_2MnSiO_4@C composite at 1C. (**c**) Rate 5 performance of the RGO@Li_2MnSiO_4@C composite under various current densities (1C = 166 mA g-1) (Reprinted from Zhao et al. [46] with permission of the Royal Society of Chemistry)

observed capacity is likely due to the electrochemical activity of the RGO network itself, which is not taken into consideration in the capacity calculations [45]. The composite cathode had increased conductivity due to the presence of the conductive RGO, and the authors claimed that the RGO network and carbon coating prevented Mn dissolution into the electrolyte resulting in a stable electrochemical performance over 700 cycles in contrast to the findings of Rangappa et al. [42] who reported a dramatic capacity loss after 20 cycles. The high carbon loading is again a common factor in the cathodes that show good cycle stability over a large number of cycles, but would seriously reduce the energy density of the cathodes in practice.

2.5.8.2 Electrospinning to Form Li_2MnSiO_4/Carbon Nanofiber Cathodes

Electrospinning is a well-known technique that has been applied to produce fibers of many materials including silicon electrodes consisting of silicon nanoparticles encapsulated in a carbon sheath [47]. Earlier attempts to produce Cr-doped Li_2MnSiO_4 nanofibers resulted in material with a large MnO impurity [48]. The combined effect of Cr doping and nanofiber morphology showed an improved capacity and capacity retention with cycling compared to Li_2MnSiO_4 nanoparticles but the capacity faded with repeated cycling. Park et al. [49] used electrospinning to produce composite Li_2MnSiO_4/C nanofibers that consisted of nanoparticles of single-phase Li_2MnSiO_4 embedded in a carbon matrix (Fig. 2.17). The nanocrystals of the active phase and the carbon layer were synthesized simultaneously to

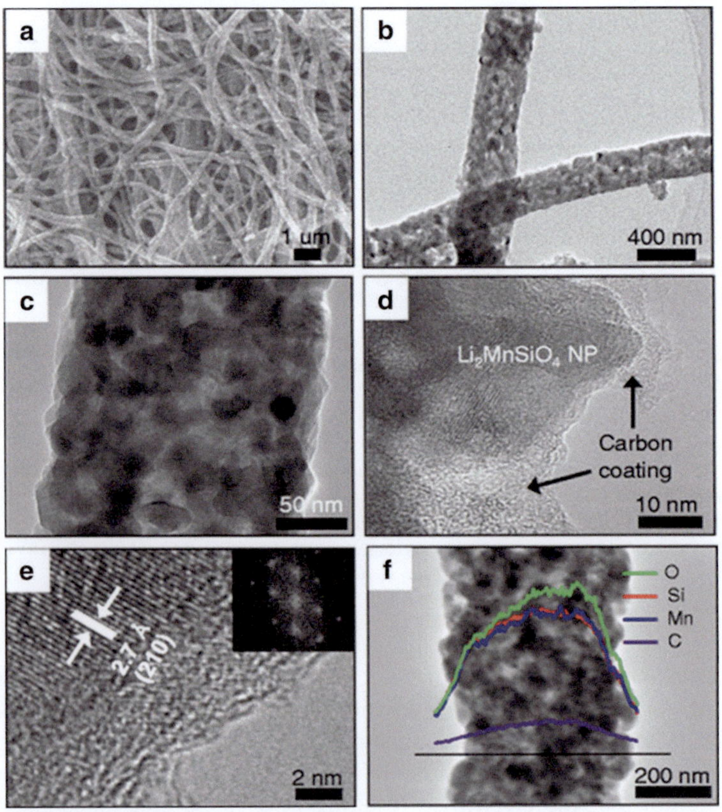

Fig. 2.17 (**a**) SEM image of as-spun Li_2MnSiO_4/C nanofibers, (**b**) low-magnification TEM image, (**c**) STEM image, (**d**) high-magnification TEM image, (**e**) HRTEM image, and (**f**) EDX line scan elemental mapping image for Li_2MnSiO_4/C nanofibers. The insert in e is the SAED pattern of Li_2MnSiO_4 (Reprinted from Park et al. [49], Fig. 2.1, copyright (2014). With kind permission from Springer Science and Business Media)

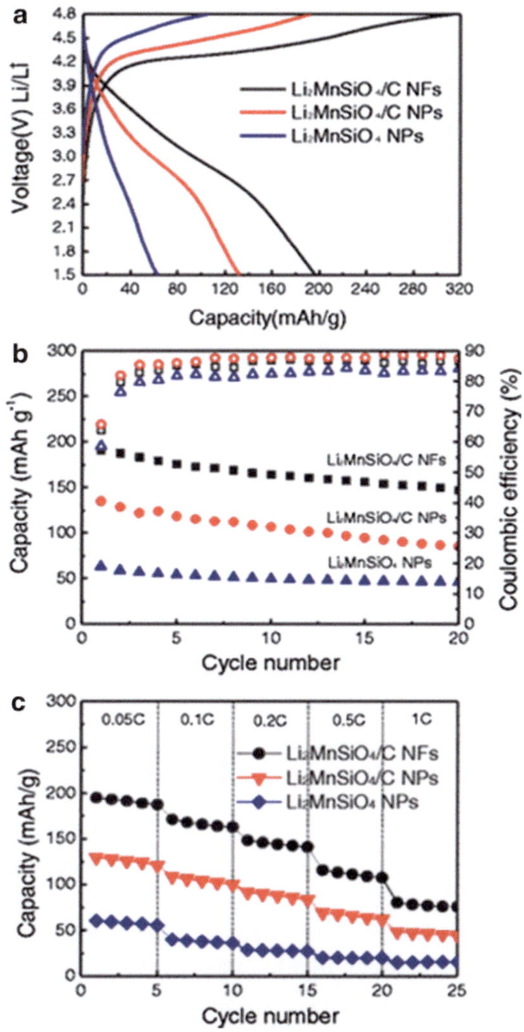

Fig. 2.18 The electrchemical properties of Li_2MnSiO_4 NPs, Li_2MnSiO_4/C NPs, and Li_2MnSiO_4/C NFs in the potential window of 1.5 ~ 4.8 V at a rate of 0.05 C at room temperature. (**a**) The first charge–discharge profile, (**b**) cycle performance, and (**c**) rate capability at various C rates (Reprinted from Park et al. [49], Fig. 2.1, copyright (2014). With kind permission from Springer Science and Business Media)

produce the composite product. The one-dimensional nanofibers had a diameter of approximately 400 nm, and Li_2MnSiO_4 nanoparticles of ~30 nm diameter were confined within the fibers. TGA analysis revealed that there was only ~6 wt% carbon in the nanofibers. The electrochemical performance of these composite nanofibers was shown to be superior to that of nanoparticles of Li_2MnSiO_4 as well as conventional carbon-coated nanoparticles with a first cycle discharge capacity of 197 mAg^{-1} at room temperature at a current rate of 0.05C (1C = 330 mAg^{-1}) (Fig. 2.18a). Capacity retention over 20 cycles was reported to be 77 %, comparable to that of the nanoparticulate samples tested, indicating that the amorphization commonly reported for these materials was not inhibited by the nanofiber morphology (Fig. 2.18b). The one-dimensional nanofiber morphology was shown by impedance spectroscopy to improve the electron transport and rate capability of the Li_2MnSiO_4 nanofiber electrodes compared to electrodes consisting of active material with conventional carbon coating (Fig. 2.18c). However, the electrochemical capacity and capacity retention with cycling of the Li_2MnSiO_4 nanofiber electrodes was inferior to that reported for the RGO-supported material discussed in 5.8.1 although this may be as a consequence of the lower total carbon content in the electrode (~26 %) compared to ~46 % for the RGO composites [46].

2.5.9 *Macroporous and Mesoporous Structures*

Ordered porous materials with both macro- and mesopores have been synthesized and applied widely in materials chemistry. The use of ordered porous morphology for the active material in lithium ion battery electrodes has attracted much attention from researchers [50–55]. This method of nanostructuring cathodes has recently also been applied to polyanionic cathodes like $LiFePO_4$ [55, 56]. Porous cathodes have several advantages including the easy penetration of the electrolyte into the bulk of the material, the reduction of the diffusion distance in the thin walls of the pores, and the accommodation of volumetric changes in the material during lithium insertion and extraction [51]. Introducing porosity into cathodes unavoidably decreases the volumetric energy density of the material but, if the gain in performance is significant, then the actual practical energy density may still be improved [52].

2.5.9.1 Mesoporous, Carbon-Supported Li_2MnSiO_4

Kawase et al. [57] attempted to deposit Li_2MnSiO_4 nanoparticles within the mesopores of an ordered mesoporous carbon matrix as an alternative to a distributed carbon coating on the surface of nanoparticles. The carbon matrix would then function to enhance the conductivity of the composite cathode. However, x-ray diffraction analysis showed that these samples contained major impurities of MnO and that the peaks assigned to Li_2MnSiO_4 were poorly defined.

2.5.9.2 Mesoporous Cathodes from a Mesoporous Silica Template

Gummow and He [58] used a SBA-15 silica template and an emulsion synthesis technique to produce a Mn-deficient lithium manganese silicate product that retained the mesoporous symmetry of the SBA-15 precursor (Fig. 2.19). Cycling results for this material showed higher capacities and better capacity retention with cycling compared to nonporous material with the same composition (Fig. 2.20). This report showed the retention of the mesoporous symmetry with cycling although the crystallinity of the lithium manganese silicate was lost during the

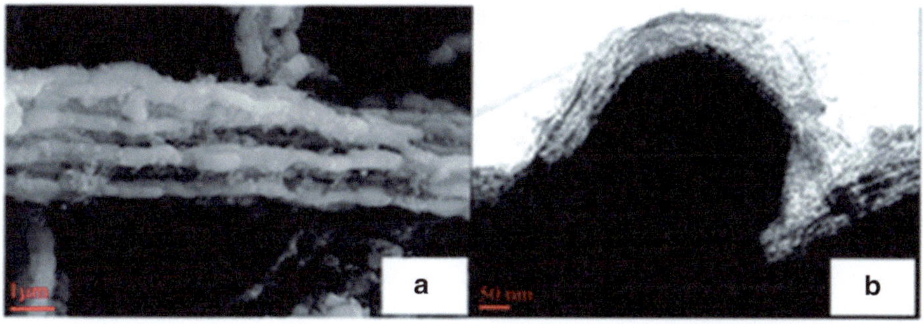

Fig. 2.19 Images of the mesoporous Mn-deficient lithium manganese silicate prepared from an SBA-15 mesoporous SiO_2 template (**a**) SEM image and (**b**) TEM image (Reproduced from Gummow and He [58] with permission of The Royal Society of Chemistry)

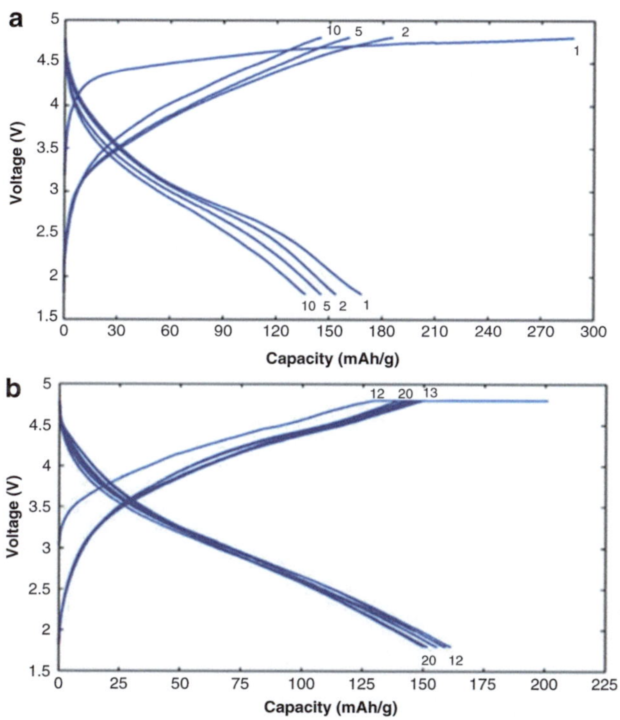

Fig. 2.20 Galvanostatic cycling curves of half cells containing mesoporous cathodes: (**a**) cycles 1–10 CC cycling and (**b**) cycles 12–20 CC–CV cycling at room temperature and 20 mAg^{-1} (Reproduced from Gummow and He [58] with permission of The Royal Society of Chemistry)

cycling process. This indicated that the amorphization previously reported for this material was not prevented by the mesoporous morphology. However, despite this loss of crystallinity, the sample cycled with excellent reversibility in the constant current-constant voltage regime.

2.5.9.3 Hierarchical Macroporous and Microporous Li_2MnSiO_4

He and Manthiram [59] reported the use of hierarchical macro- and mesoporous Li_2MnSiO_4 cathodes produced using a hard template of close-packed 430 nm diameter poly(methyl methacrylate) (PMMA) spheres. The PMMA template was impregnated with the precursors for Li_2MnSiO_4 as well as a resin as carbon source. After decomposing the PMMA by heating in an inert atmosphere, a replicated, ordered macroporous Li_2MnSiO_4 product was formed with average pores of 200–400 nm (Fig. 2.21a–c). The pore walls were composed of 20–40 nm nanocrystals of Li_2MnSiO_4 with additional disordered mesopores between the nanocrystals in the walls. Decomposition of the resin also provided a carbon coating on the surface of the nanocrystals to improve electrical conductivity with ~20 wt.% residual carbon in the composite cathodes. In electrochemical testing, the maximum discharge capacity reached was 200 mAhg^{-1} on the 3rd discharge cycled at C/10 (16 mAg^{-1}) at 45 °C. At higher current rates, C/2 and C, the highest recorded capacities were 160 and 119 mAhg^{-1}, respectively. Capacity retention was good with about 112 % of the initial capacity reported after 40 cycles at 1 C (95 mAhg^{-1}) with a total carbon content in electrodes of ~30 %. Higher capacities

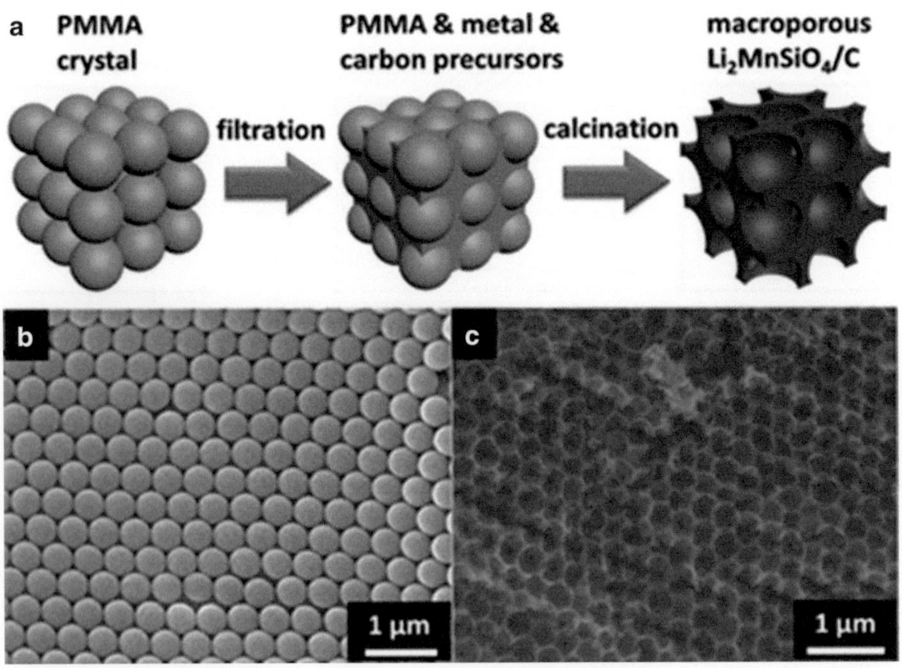

Fig. 2.21 (**a**) Schematic representation of the preparation process of the hierarchical ordered macroporous Li_2MnSiO_4/C with PMMA as the hard template; SEM images of the (**b**) PMMA spheres and (**c**) the resulting Li_2MnSiO_4/C replica (Reprinted from He and Manthiram [59]. Copyright (2014) WILEY-VCH Verlag GmbH & Co. KGaA, Weinheim with permission)

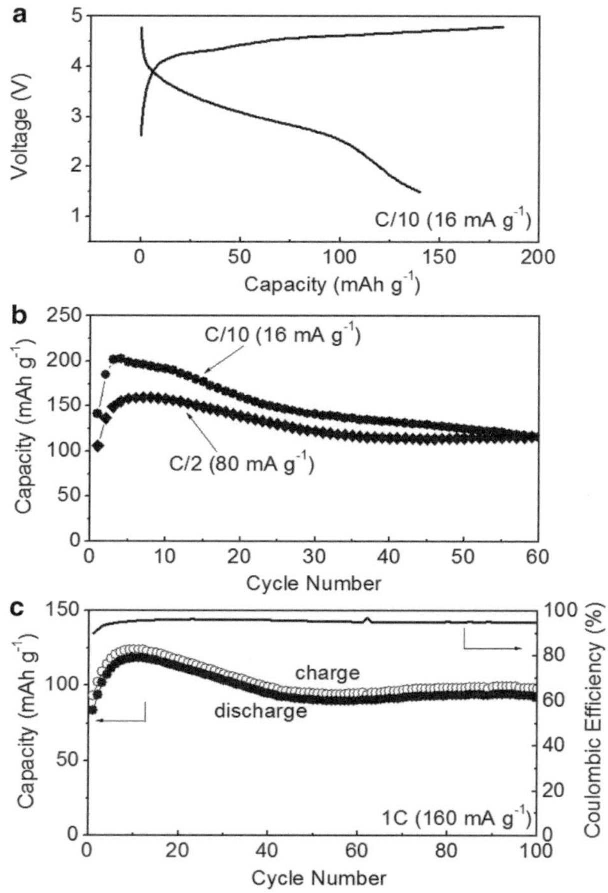

Fig. 2.22 Electrochemical performance of the Li 2 MnSiO 4/C cathodes at 45 °C: (**a**) initial charge–discharge profile at C/10, (**b**) cycling performance at C/10 and C/2, (**c**) cycling performance and Coulombic efficiency over 100 cycles at 1 C rate (Reprinted from He and Manthiram [59]. Copyright 2014 WILEY-VCH Verlag GmbH & Co. KGaA, Weinheim with permission)

were recorded for cells cycled at 55 °C but the capacity retention with cycling was reduced and found to be very sensitive to the voltage range that was chosen (Fig. 2.22).

2.6 Spray Pyrolysis

Spray pyrolysis is a commonly used technique to produce ceramic powders with a well-defined composition and a narrow particle size distribution [60]. Spray pyrolysis, combined with high-energy ball milling, has recently been applied to form Li_2MnSiO_4/C composites using glucose as the carbon source [61–63]. Shao and Taniguchi found the best performance for composites produced with a 0.1 mol l^{-1} glucose solution to give 12 wt% residual carbon in the composite and 15.2 wt% C in the electrodes with primarily particles in the 40 nm range. The 1st cycle discharge capacity was approximately 170 $mAhg^{-1}$ at a current rate of 0.1 C with cycling in the range 1.5–4.8 V. However, even with this optimized composition, the cathodes showed poor capacity retention with cycling at room temperature with a residual capacity of only ~100 mAg^{-1} after 20 cycles. Moriya et al. [61] adopted a very similar strategy except that in their synthesis the glucose was added into the mixture prior to the pyrolysis step. In this way, the glucose was carbonized at the same time as the formation of the Li_2MnSiO_4 powder. The residual carbon in the composite was also determined to be 12 wt% in close agreement to the study of Shao et al. [63] but the composite electrodes consisted of 20 mg of active material and 15 mg of teflonized acetylene black (TAB) resulting in a much higher carbon loading in the electrodes than in the case of Shao et al. The particle size was also found to be only 10–20 nm and particles had an amorphous carbon coating. Moriya et al. [61] reported a stable discharge capacity of ~200 $mAhg^{-1}$ at 0.1 C and 25 °C for 60 cycles but it should be noted that the lower voltage cutoff was 1 V and the capacity above 1.5 V was only approximately 150 $mAhg^{-1}$. The authors also found it necessary to replace the anode and anode current collector after every 20 cycles because of degradation during cycling. This further serves to confirm the findings of other investigators [44, 46] that stable cycling of Li_2MnSiO_4 cathodes is possible with high carbon loading. Moriya et al. [62] carried out a detailed structural analysis of cycled cathodes of their spray-pyrolyzed Li_2MnSiO_4 material. The material could be indexed as the $Pmn2_1$ orthorhombic polymorph of Li_2MnSiO_4. In contrast to other structural studies, Moriya et al. found that although the long-range order of the sample was lost on charge, when lithium was subsequently reinserted on discharge, the peaks corresponding to Li_2MnSiO_4 re-appeared (Fig. 2.23). Pair distribution function (PDF) analysis indicated that the local structural order was disrupted when lithium was extracted due to local distortions of the MnO tetrahedra to accommodate the smaller Mn^{3+} and Mn^{4+} cations generated on charge. Importantly, the structural order was restored when lithium was reinserted, as evidenced by the re-appearance of the diffraction peaks.

2.7 Conclusion

After more than 15 years of research effort by lithium-ion battery researchers, there is now a good understanding of many of the structural and electrochemical properties of Li_2MnSiO_4. The wide range of synthesis methods and nanostructuring strategies that have been explored shows the persistence and determination of researchers to find the optimum methods to get the most out of this material in terms of electrochemical performance. Sophisticated nanostructures, e.g., nanosheets, nanofibers, and complex hierarchical porous structures, have been formed and have delivered dramatic improvements in high rate performance due to the increased surface area for interaction with the electrolyte and the reduced diffusion length for lithium ions. Recent structural studies by Moriya et al. [62]

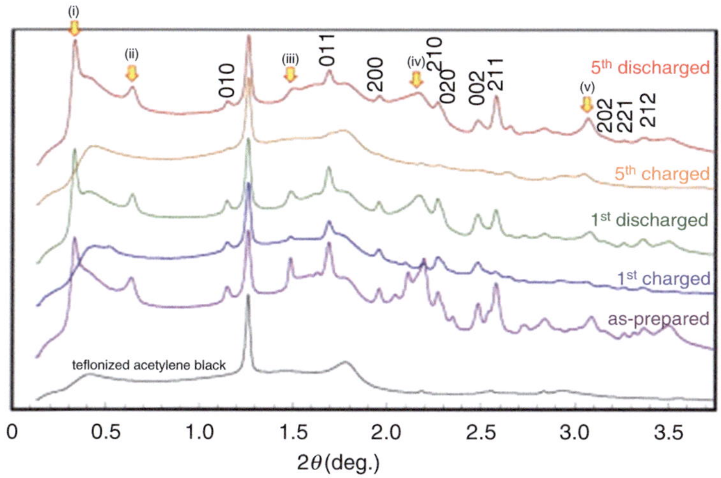

Fig. 2.23 Synchrotron-based HE-XRD patterns of the as-prepared, 1st-charged, 1st-discharged, 5th-charged, and 5th-discharged samples and the teflonized acetylene black. The carbon-hybridized Li_2MnSiO_4 samples were charged or discharged in the delithiation of 1.25 Li/formula unit. The hkl indices refer to the underlining β-Li_2MnSiO_4 structure (Reprinted from Moriya et al. [62]. Copyright 2014, with permission from Elsevier)

have also indicated that nanostructuring, coupled with high carbon loading, can allow the structure to reversibly accommodate the distortions to the MnO_4 tetrahedra that inevitably occur with Li extraction.

References

1. Nagaura T, Tozawa K (1990) Prog Batter Solar Cells 9:209
2. Armand M, Tarascon JM (2008) Building better batteries. Nature 451:652–657
3. Thackeray MM, Wolverton C, Isaacs ED (2012) Electrical energy storage for transportation-approaching the limits of, and going beyond, lithium-ion batteries. Energy Environ Sci 5:7854–7863
4. Xu T, Wang W, Gordin ML et al (2010) Lithium-ion batteries for stationary energy storage. JOM-US 62:24–30
5. Arico AS, Bruce P, Scrosati B et al (2005) Nanostructured materials for advanced energy conversion and storage devices. Nat Mater 4:366–377
6. Song T, Xia J, Lee J-H et al (2010) Arrays of sealed silicon nanotubes as anodes for lithium ion batteries. Nano Lett 10:1710–1716
7. Wang Y, Zeng HC, Lee JY (2006) Highly reversible lithium storage in porous SnO_2 nanotubes with coaxially grown carbon nanotube overlayers. Adv Mater 18:645–649
8. Armand M (2002) World Patent WO02/27823
9. Padhi AK, Nanjundaswamy KS, Goodenough JB (1997) Phospho-olivines as positive-electrode materials for rechargeable lithium batteries. J Electrochem Soc 144:1188–1194
10. Nazar LF, Goward G, Leroux F et al (2001) Nanostructured materials for energy storage. Int J Inorg Mater 3:191–200

11. Ravet N, Chouinard Y, Magnan JF et al (2001) Electroactivity of natural and synthetic triphylite. J Power Sources 97–98:503–507
12. Ellis BL, Lee KT, Nazar LF (2010) Positive electrode materials for Li-ion and Li-batteries. Chem Mater 22:691–714
13. Gong Z, Yang Y (2011) Recent advances in the research of polyanion-type cathode materials for Li-ion batteries. Energy Environ Sci 4:3223–3242
14. Islam MS, Dominko R, Masquelier C et al (2011) Silicate cathodes for lithium batteries: alternatives to phosphates? J Mater Chem 21:9811–9818
15. Dominko R, Bele M, Gaberscek M et al (2006) Structure and electrochemical performance of Li_2MnSiO_4 and Li_2FeSiO_4 as potential Li-battery cathode materials. Electrochem Commun 8:217–222
16. Dominko R, Bele M, Kokalj A et al (2007) Li_2MnSiO_4 as a potential Li-battery cathode material. J Power Sources 174:457–461
17. Lv D, Bai J, Zhang P et al (2013) Understanding the high capacity of $Li_2FeSiO4$: in situ XRD/XANES study combined with first principles calculations. Chem Mater 25:2014–2020
18. Dominko R, Sirisopanaporn C, Masquelier C et al (2010) On the origin of the electrochemical capacity of Li $_2Fe_{0.8}Mn_{0.2}SiO_4$. J Electrochem Soc 157:A1309
19. Bruce PG, Lyness C, Delobel B et al (2007) The lithium intercalation compound Li_2CoSiO_4 and its behaviour as a positive electrode for lithium batteries. Chem Commun 46:4890–4892
20. Gong ZL, Li YX, Yang Y (2007) Synthesis and electrochemical performance of Li_2CoSiO_4 as cathode material for lithium ion batteries. J Power Sources 174:524–527
21. Gummow RJ, Sharma N, Peterson VK et al (2012) Crystal chemistry of the Pmnb polymorph of $Li_2MnSiO4$. J Solid State Chem 188C:32–37
22. Politaev VV, Petrenko AA, Nalbandyan VB et al (2007) Crystal structure, phase relations and electrochemical properties of monoclinic Li_2MnSiO_4. J Solid State Chem 180:1045–1050
23. Duncan H, Kondamreddy A, Mercier PHJ et al (2011) Novel Pn polymorph for Li_2MnSiO_4 and its electrochemical activity as a cathode material in Li-ion batteries. Chem Mater 23:5446–5456
24. Gummow RJ, He Y (2014) Recent progress in the development of Li_2MnSiO_4 cathode materials. J Power Sources 253:315–331
25. Arroyo-deDompablo ME, Dominko R, Gallardo-Amores JM et al (2008) On the energetic stability and electrochemistry of Li_2MnSiO_4 polymorphs. Chem Mater 20:5574–5584
26. Gummow RJ, Sharma N, Peterson VK et al (2012) Synthesis, structure, and electrochemical performance of magnesium-substituted lithium manganese orthosilicate cathode materials for lithium-ion batteries. J Power Sources 197:231–237
27. Kuganathan N, Islam MS (2009) Li_2MnSiO_4 lithium battery material: atomic-scale study of defects, lithium mobility, and trivalent dopants. Chem Mater 21:5196–5202
28. Fisher CAJ, Kuganathan N, Islam MS (2013) Defect chemistry and lithium-ion migration in polymorphs of the cathode material Li_2MnSiO_4. J Mater Chem A 1:4207–4214
29. Dominko R (2008) Li_2MSiO_4 (M = Fe and/or Mn) cathode materials. J Power Sources 184:462–468
30. Kokalj A, Dominko R, Mali G et al (2007) Beyond one-electron reaction in Li cathode materials: designing $Li_2Mn_xFe_{1-x}SiO_4$. Chem Mater 19:3633–3640
31. Li YX, Gong ZL, Yang Y (2007) Synthesis and characterization of Li_2MnSiO_4/C nanocomposite cathode material for lithium ion batteries. J Power Sources 174:528–532
32. Wang Y, Li H, He P et al (2010) Nano active materials for lithium-ion batteries. Nanoscale 2:1294–1305
33. Deng C, Zhang S, Fu BL et al (2010) Characterization of $Li_2MnSiO4$ and Li_2FeSiO_4 cathode materials synthesized via a citric acid assisted sol–gel method. Mater Chem Phys 120:14–17
34. Belharouak I, Abouimrane A, Amine K (2009) Structural and electrochemical characterization of Li_2MnSiO_4 cathode material. J Phys Chem C 113:20733–20737
35. Liu WG, Xu YH, Yang R (2010) Synthesis and electrochemical properties of Li_2MnSiO_4/C nanoparticles via polyol process. Rare Met 29:511–514

36. Mali G, Meden A, Dominko R (2010) Li-6 MAS NMR spectroscopy and first-principles calculations as a combined tool for the investigation of Li_2MnSiO_4 polymorphs. Chem Commun 46:3306–3308
37. Manthiram A, Muraliganth T, Stroukoff KR (2010) Microwave-solvothermal synthesis of nanostructured Li_2MSiO_4/C (M = Mn and Fe) cathodes for lithium-ion batteries. Chem Mater 22:5754–5761
38. Kojima A, Kojima T, Tabuchi M et al (2012) Synthesis of Li_2MnSiO_4 cathode material using molten carbonate flux method with high capacity and initial efficiency. J Electrochem Soc 159: A532–A537
39. Kempaiah DM, Rangappa D, Honma I (2012) Controlled synthesis of nanocrystalline Li_2MnSiO_4 particles for high capacity cathode application in lithium-ion batteries. Chem Commun 48:2698–2700
40. Geim AK (2009) Graphene: status and prospects. Science 324:1530–1534
41. Rui X, Zhao X, Lu Z et al (2013) Olivine-Type nanosheets for lithium ion battery cathodes. ACS Nano 7:5637–5646
42. Rangappa D, Murukanahally KD, Tomai T et al (2012) Ultrathin nanosheets of Li_2MSiO_4 (M = Fe, Mn) as high-capacity Li-ion battery electrode. Nano Lett 12:1146–1151
43. Devaraju MK, Tomai T, Unemoto A et al (2013) Novel processing of lithium manganese silicate nanomaterials for Li-ion battery applications. RSC Adv 3:608–615
44. Aravindan V, Karthikeyan K, Kang KS et al (2011) Influence of carbon towards improved lithium storage properties of Li_2MnSiO_4 cathodes. J Mater Chem 21:2470–2475
45. Ha SH, Jeong YS, Lee YJ (2013) Free standing reduced graphene oxide film cathodes for lithium ion batteries. ACS Appl Mater Interfaces 5:12295–12303
46. Zhao Y, Wu C, Li J et al (2013) Long cycling life of Li_2MnSiO_4 lithium battery cathodes under the double protection from carbon coating and graphene network. J Mater Chem A 1:3856–3859
47. Hwang TH, Lee YM, Kong B-S et al (2011) Electrospun core–shell fibers for robust silicon nanoparticle-based lithium ion battery anodes. Nano Lett 12:802–807
48. Zhang S, Lin Z, Ji L et al (2012) Cr-doped Li_2MnSiO_4/carbon composite nanofibers as high-energy cathodes for Li-ion batteries. J Mater Chem 22:14661–14666
49. Park H, Song T, Tripathi R et al (2014) Li_2MnSiO_4/carbon nanofiber cathodes for Li-ion batteries. Ionics 20:1351–1359
50. Jiao F, Bruce PG (2007) Mesoporous crystalline β-MnO2—a reversible positive electrode for rechargeable lithium batteries. Adv Mater 19:657–660
51. Jiang C, Hosono E, Zhou H (2006) Nanomaterials for lithium ion batteries. Nano Today 1:28–33
52. Bruce PG (2008) Energy storage beyond the horizon: rechargeable lithium batteries. Solid State Ion 179:752–760
53. Jiao F, Bao J, Hill AH et al (2008) Synthesis of ordered mesoporous Li–Mn–O spinel as a positive electrode for rechargeable lithium batteries. Angew Chem Int Ed 47:9711–9716
54. Yan HW, Sokolov S, Lytle JC et al (2003) Colloidal-crystal-templated synthesis of ordered macroporous electrode materials for lithium secondary batteries. J Electrochem Soc 150: A1102–A1107
55. Doherty CM, Caruso RA, Smarsly BM et al (2009) Colloidal crystal templating to produce hierarchically porous $LiFePO_4$ electrode materials for high power lithium ion batteries. Chem Mater 21:2895–2903
56. Vu A, Stein A (2011) Multiconstituent synthesis of $LiFePO_4$/C composites with hierarchical porosity as cathode materials for lithium ion batteries. Chem Mater 23:3237–3245
57. Kawase T, Yoshitake H (2012) Cathodes comprising Li_2MnSiO_4 nanoparticles dispersed in the mesoporous carbon frameworks, CMK-3 and CMK-8. Microporous Mesoporous Mater 155:99–105
58. Gummow RJ, He Y (2014) Mesoporous manganese-deficient lithium manganese silicate cathodes for lithium-ion batteries. RSC Adv 4:11580–11584

59. He G, Manthiram A (2014) Nanostructured Li_2MnSiO_4/C cathodes with hierarchical macro-/mesoporosity for lithium-ion batteries. Adv Funct Mater 24:5277–5283
60. Messing GL, Zhang S-C, Jayanthi GV (1993) Ceramic powder synthesis by spray pyrolysis. J Am Ceram Soc 76:2707–2726
61. Moriya M, Miyahara M, Hokazono M et al (2014) Synthesis of hybrid Li_2MnSiO_4 nanoparticles with carbon for cathode materials with stable charge/discharge cycles. J Electrochem Soc 161:A97–A101
62. Moriya M, Miyahara M, Hokazono M et al (2014) High-energy X-ray powder diffraction and atomic-pair distribution-function studies of charged/discharged structures in carbon-hybridized Li_2MnSiO_4 nanoparticles as a cathode material for lithium ion batteries. J Power Sources 263:7–12
63. Shao B, Taniguchi I (2014) Synthesis of Li_2MnSiO_4/C nanocomposites for lithium battery cathode employing sucrose as carbon source. Electrochim Acta 128:156–162

Chapter 3
Metal Oxides and Lithium Alloys as Anode Materials for Lithium-Ion Batteries

Mesfin Kebede, Haitao Zheng, and Kenneth I. Ozoemena

3.1 Introduction

The rechargeable lithium–ion batteries (LIBs) use lithium cobalt oxide ($LiCoO_2$) as the cathode, graphite (C) as the anode material, and a nonaqueous Li-ion-conducting medium as the electrolyte. On discharging, the Li-ions are deintercalated from anode graphite, pass across the electrolyte, and are intercalated between the $LiCoO_2$ layers in the cathode. Charging reverses this process. The electrons pass through the external circuit to run a certain load such as cell phone, laptop, etc., as it is illustrated in Fig. 3.1. The electrochemical reactions which occur in a typical LIB using $LiCoO_2$ cathode and graphite anode are as follows:

At the anode:

$$6C + xLi^+ + xe^- \leftrightarrow Li_xC_6 \tag{3.1}$$

At the cathode:

$$LiCoO_2 \leftrightarrow Li_{1-x}CoO_2 + xLi^+ + xe^- \tag{3.2}$$

M. Kebede (✉) • H. Zheng
Energy Materials Unit, Materials Science and Manufacturing, Council for Scientific & Industrial Research (CSIR), Pretoria 0001, South Africa
e-mail: mkebede@csir.co.za

K.I. Ozoemena
Energy Materials Unit, Materials Science and Manufacturing, Council for Scientific & Industrial Research (CSIR), Pretoria 0001, South Africa

Department of Chemistry, University of Pretoria, Pretoria 0002, South Africa

School of Chemistry, University of the Witwatersrand, Johannesburg 2050, South Africa

K.I. Ozoemena, S. Chen (eds.), *Nanomaterials in Advanced Batteries and Supercapacitors*, Nanostructure Science and Technology,
DOI 10.1007/978-3-319-26082-2_3

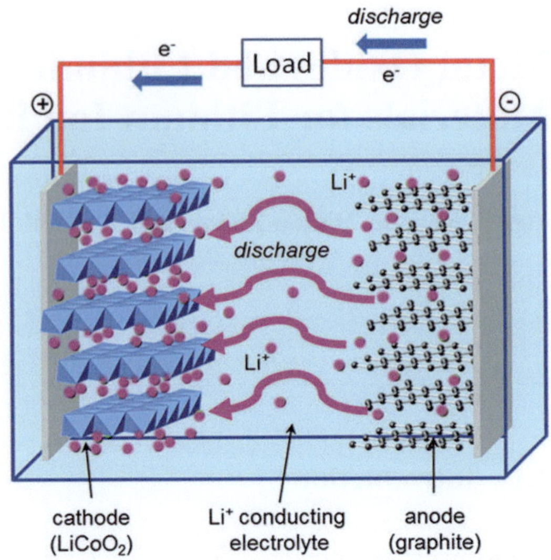

Fig. 3.1 Schematic of lithium intercalation/deintercalation mechanism for lithium-ion battery

The potential difference between cathode and anode is called the cell potential or electromotive force (emf) of the lithium–ion battery. The cell potential of the lithium–ion battery is defined as

$$E^{\mathrm{o}}\mathrm{cell} = \mathrm{E^o(cathode)} - \mathrm{E^o(anode)} \tag{3.3}$$

$$E^{\mathrm{o}}\mathrm{cell} = \mathrm{E^o\ reduction} - \mathrm{E^o oxidation} \tag{3.4}$$

The anode materials for LIBs are categorized into three main groups based on the electrochemical reaction mechanism during charge–discharge process [1]. The first group of the materials, carbon-based materials (such as graphite, graphene), and titanium-based materials (such as TiO_2, $Li_4Ti_5O_{12}$) follow the lithium intercalation/deintercalation reaction mechanism. When these materials are used as anode for lithium–ion battery, the lithium–ions enter and leave without destroying their crystal structure. As an intercalation function anode material, the compound needs to meet some important conditions. First, the compound must be crystalline, and there must be empty sites in its crystal lattice which creates conducive condition for Li intercalation–deintercalation. Second, the host compound must contain a transition metal or rare-earth metal which can exhibit one or more stable valence states (e.g., Ti^{4+} to Ti^{3+}) to allow for efficient cycling process.

The second group of anode materials operates by adopting the alloy/dealloy reaction mechanism and includes elements like Si and Sb and metals like Sn, In, and Cd. These materials can give rise to Li storage and cycling behavior by virtue of alloying–dealloying reactions at low potentials ($V \leq 1$ V vs Li metal) and hence can be considered as prospective anodes for LIBs. Oxides of these metals are also equally valuable and being used in the form of simple and complex oxides. Actually, in the process the oxides will first reduce to their respective metals by Li metal under electrochemical conditions, followed by Li alloy formation. For example, in the case of tin oxides, during the first discharge reaction with Li, SnO or SnO_2 undergoes crystal structure destruction (amorphization) and formation of nano-Sn metal dispersed in amorphous Li_2O as indicated in Eqs. 3.5 and 3.6, followed by alloy formation ($Li_{4.4}Sn$) as expressed in Eq. 3.7.

$$SnO + 2Li^+ + 2e^- \rightarrow Sn + Li_2O \tag{3.5}$$

$$SnO_2 + 4Li^+ + 4e^- \rightarrow Sn + 2Li_2O \tag{3.6}$$

$$Sn + 4.4Li \leftrightarrow Li_{4.4}Sn \tag{3.7}$$

Equations 3.5 and 3.6 are irreversible reactions, which correspond to the irreversible capacity loss during the first discharge–charge cycle, whereas Eq. 3.7 contributes to the reversible capacity. Thus, the theoretical reversible capacities of SnO and SnO_2 are 875 and 782 mA h g^{-1}, respectively. Generally, anodes working by alloying mechanism have high theoretical capacity, but they suffer rapid capacity fading due to exaggerated volume change during cycling process. The usual approach to solve this problem is nanosizing the active materials and coating with carbon.

The third category of anode materials functions by the principle of conversion (redox) reaction; these are transition metal (TM) oxides such as Mn_xO_y, NiO, Fe_xO_y, CuO, Cu_2O, MoO_2, etc. The reaction mechanism involved during Li storage and cycling is different from those of intercalation/deintercalation and alloying/dealloying. It basically involves the formation and decomposition of Li_2O along with the reduction and oxidation of transition metal nanoparticles. The total electrochemical reaction of the process can be expressed as

$$TMO + 2Li^+ + 2e^- \leftrightarrow TM + Li_2O \tag{3.8}$$

where TM = Mn, Fe, Co, Ni, Cu. Normally, the compound Li_2O is electrochemically inert, but it can be activated during electrochemical cycling. Li_2O helps to improve the capacity retention to some extent by buffering the volume expansion.

So far, rechargeable Li-ion batteries (LIBs) have not only been successfully applied as the most efficient energy storage technology for a wide range of consumer electronic devices like cellular phones, laptops, and digital cameras, but their usage has also stretched into powering of electric vehicles and storage of renewable energies (such as wind and solar) [2, 3]. At present, there is a huge attempt to achieve effective energy storage systems which are the key for the exploitation of alternative energy and thus for the replacement of fossil fuels and traditional energy sources. In this regard, rechargeable LIBs play a promising and significant role owing to their high gravimetric and volumetric energy, high power density, long life cycle, and low self-discharge property [4–6]. However, it should be noted that to utilize LIBs in hybrid electric vehicles (HEV), plug-in hybrid electric vehicles (PHEV), and pure electric vehicles (PEV), increased energy density and power density (2–5 times of 150 Wh/kg) are required compared to the ones currently used for portable electronics which can only offer between 120 and 150 Wh/kg [7]. The good news is in their recent achievement in 2014, Hitachi announced the LIB technology which can offer 335 Wh/kg by using high-capacity silicon anode instead of graphite. Obviously, in order to increase the energy density of lithium-ion batteries, it is crucial to use high-voltage cathode active materials such as $LiNi_{0.5}Mn_{1.5}O_4$ (provides about 5 Vvs Li metal) or high-capacity anode materials such as silicon or tin. One of the main challenges to the design and use of high-voltage cathode in LIBs is electrolyte decomposition that occurs at potentials more than 4.2 V vs Li/Li^+ [8–10]. This challenge explains why the research for anode materials with high capacity is getting broad attention and considered as the promising path to develop LIBs for vehicle applications.

3.2 Lithium Intercalation/Deintercalation Reaction-Based Anode Materials

Carbon-based materials such as graphitic anode materials are the most familiar and commercialized anode materials which operate by adopting the intercalation–deintercalation reaction mechanism. The other most recognized types of anode compounds which work with the same reaction mechanism are lithium titanate ($Li_4Ti_5O_{12}$) and titanium dioxide (TiO_2). In this section we focus on the titanium-based anode materials.

3.2.1 Titanium-Based Oxides

Titanium-based oxides basically comprise spinel lithium titanate $Li_4Ti_5O_{12}$ (LTO) and titanium dioxide (TiO_2) along with its different polymorphs. Titanium-based oxides have attracted tremendous attention as next-generation anode materials. Similar to the commercialized graphite anode materials, the titanium-based oxides operate with the intercalation–deintercalation reaction mechanism [11, 12]. They are currently considered as potential graphite substitutes; actually LTO is already commercially available as possible replacement or alternative to graphite as negative electrode of lithium–ion batteries [13, 14].

The strongest benefit of using titanium-based compounds compared to graphite is that they operate at high voltage around 1.5 V which characteristically mitigates the issue of the formation of the solid electrolyte interface (SEI) and the development of dendrites which occurs at graphite. Besides, titanium-based oxides are inexpensive and less toxic with excellent safety performance and exhibit insignificant volume change with zero strain during lithium intercalation and deintercalation; consequently they provide excellent calendar life [15]. However, the main drawback with titanium-based anodes is that their theoretical capacities are relatively low and they have low electronic conductivity. As a result of low electronic conductivity, titanium oxides have poor rate capability.

The structure, morphology, and size of titanium-based oxide anode materials critically determine the electrochemical performance and the lithium insertion/removal capacity of the material. Particularly, nanostructured titanium oxides are preferred to achieve better capacity, longer life cycle, and higher rate capability

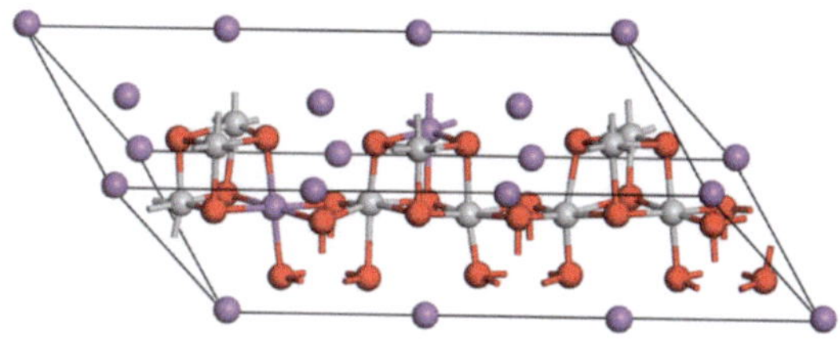

Fig. 3.2 Crystal structure of $Li_4Ti_5O_{12}$ (LTO). *Pink spheres*, Li; *gray spheres*, Ti; *red spheres*, O (Reproduced with permission from Ref. [17]. Copyright 2012 Elsevier)

than the bulk materials [1, 13, 16]. The typical electrochemical lithium storage process of intercalation-type anode materials can be described by the following reaction:

$$MO_x + ye^- + yLi^+ \leftrightarrow Li_yMO_x \tag{3.9}$$

where M represents a metal.

3.2.1.1 Spinel $Li_4Ti_5O_{12}$ (LTO)

The crystal structure of spinel $Li_4Ti_5O_{12}$ is shown in Fig. 3.2, and its phase transforms to rock salt $Li_7Ti_5O_{12}$ as it goes through the lithiation process [17]. $Li_4Ti_5O_{12}$ shows excellent lithium–ion reversibility at high equilibrium potential of around 1.5 V vs Li/Li^+ which makes it promising new power sources for hybrid electric vehicles (HEVs) due to their long life cycle and high energy density [17, 18]. Some of the shortcomings of the spinel $Li_4Ti_5O_{12}$ anode material are its low theoretical capacity of 175 mA h g^{-1} and poor electronic and lithium ionic conductivities which then limit its high-rate capability. To enhance the conductivity of LTO, three different approaches have been implemented so far. The first is surface modification such as coating with carbon to increase the conductivity of LTO. The second is doping with cations or anions and incorporation of second phase with high electronic conductivity to improve the conductivities of LTO. The third approach is to improve the Li-ion diffusion in LTO by downsizing the LTO to the nanoscale.

The surface modification of spinel $Li_4Ti_5O_{12}$ improves the cyclability of the spinel $Li_4Ti_5O_{12}$ materials since it decreases the surface area to retard the side reactions between the electrode and the electrolyte. It is also feasible to improve the conductivity of LTO by coating it with conductive materials. Some of the materials as LTO coatings are Ag, Cu, Fe, carbon, SnO_2, etc. Coating of LTO using the high-conducting metal silver (Ag) would considerably increase the electronic conductivity and decrease polarization of electrode, thereby improving its high-rate discharge capacity and cycling stability [19]. Huang et al. [20] reported Ag-coated $Li_4Ti_5O_{12}$ synthesized by solid-state reaction and also studied the effect of the Ag content on the electrochemical performance and electronic conductivities of LTO.

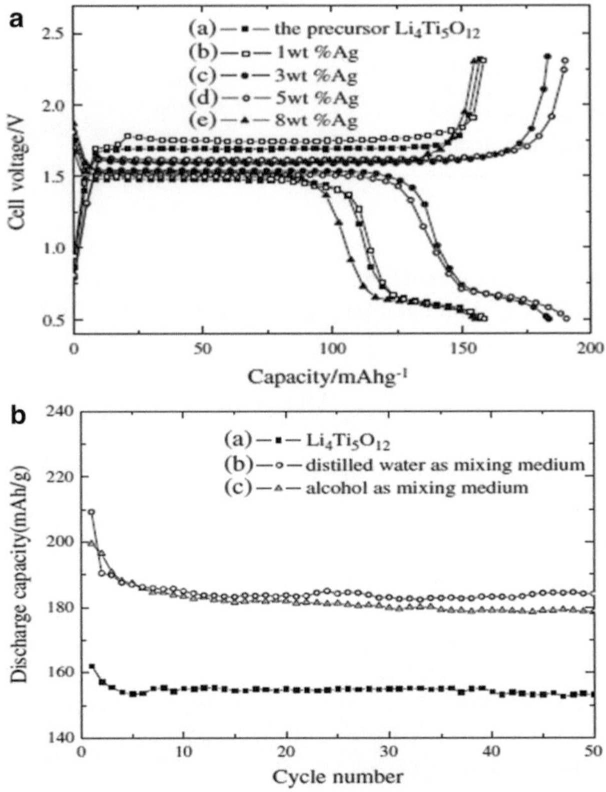

Fig. 3.3 (**a**) The second charge–discharge curves for the $Li_4Ti_5O_{12}$/Ag composites with different Ag content. (**b**) Cycling performance for the precursor $Li_4Ti_5O_{12}$ and the $Li_4Ti_5O_{12}$/Ag composites. The coin cells were cycled at 2 C rate between 2.3 and 0.5 V (Reproduced with permission from Ref. [20]. Copyright 2006 Elsevier)

Their result confirmed that coating with Ag greatly enhanced the electronic conductivity of $Li_4Ti_5O_{12}$ anode materials. $Li_4Ti_5O_{12}$/Ag composite remarkably improved reversible capacity and the cycling stability over the pristine $Li_4Ti_5O_{12}$, especially at high discharge rates [20]. As it is shown in Fig. 3.3, the first discharge capacities were respectively 162.2, 163.2, 206.5, 209.3, and 175 mA h g^{-1} for the pristine $Li_4Ti_5O_{12}$ and $Li_4Ti_5O_{12}$/Ag composites with the Ag contents of 1, 3, 5, and 8 wt% at 2 C rate. It was found that Ag additive greatly improved the first discharge capacity of $Li_4Ti_5O_{12}$ materials at 1.5 V potential. The discharge capacities after 50 cycles were 153, 155.4, 180.2, 184.2, and 152.3 mA h g^{-1} with the corresponding capacity degradation of 2.67 %, 2.14 %, 1.91 %, 4.55 %, and 1.81 % compared to their corresponding second cycle, respectively, for the $Li_4Ti_5O_{12}$/Ag composites with the Ag contents of 0, 1, 3, 5, and 8 wt%. It was obvious that the content of Ag additive played a remarkable role in enhancing the electrochemical performance of the composites, specially 3 wt% of Ag, which gave the best first-cycle capacity and superior capacity retention.

Copper, Cu, can be considered as a good replacement to Ag due to its high electronic conductivity and low cost. He et al. [21] presented a novel approach involving the coating of the lithium titanate electrodes with Cu prepared by electroplating. In their work the electronic conductivity of the composites was enhanced due to the surface modification of Cu and then significantly improved the surface intercalation reaction and reduced cell polarization, thereby improving the high-rate performance and cycling stability of the composite. The next low-cost, abundant, and conductive metal is iron, Fe. Wang et al. [22] have shown enhanced specific capacity and cyclic performance of LTO by surface modification with Fe. They used solid-state reaction method to prepare $Li_4Ti_5O_{12}$/Fe composite and evaluated the electrochemical performance of pristine $Li_4Ti_5O_{12}$, $Li_4Ti_5O_{12}$/Fe, and $Li_4Ti_5O_{12}$ reground. The specific capacity of $Li_4Ti_5O_{12}$/Fe is 123.4 mA h g^{-1} at 5C rate, much higher than 59.5 mA h g^{-1} of the pristine material LTO. The Fe-modified $Li_4Ti_5O_{12}$ anode also showed superior cycling stability with an initial discharge capacity of 141 mA h g^{-1} at 2C rate and still maintained 96.7 % of its initial capacity after 300 cycles, while the capacity of LTO is 89.6 mA h g^{-1} and capacity retention is 90.8 %.

The next effective approach to improve the poor rate capability and electronic conductivity of $Li_4Ti_5O_{12}$ is doping with aliovalent metal ions. Although coating with carbon or a metallic conducting layer is an efficient way to improve its conductance, the method does not improve the lattice electronic conductivity or lithium–ion diffusion coefficient within the crystal. Doping with various ions enhances the electronic conductivity of $Li_4Ti_5O_{12}$ materials; this process transforms some of Ti^{4+} into Ti^{3+} to generate mixed Ti^{3+}/Ti^{4+} and thus increases the concentration of electrons [18, 23]. Substitution of a small quantity of Li^+, Ti^{4+}, or O^{2-} by other metallic ions can greatly improve the kinetics of materials in terms of capacity delivery, cycle stability, and especially rate capability. Thus, one of the effective approaches to improve the performance of LTO is doping it with aliovalent metal ions such as Mg, V, Mn, Fe, Cr, Ta, Ni, Al, Zr, and Sn at the Li, Ti, and O sites [24]. For instance, Huang et al. reported that the Al^{3+} doping at the Li site and the Ti site of $Li_4Ti_5O_{12}$ increases the reversible capacity and cycling stability considerably compared with pristine $Li_4Ti_5O_{12}$ [25]. Li et al. synthesized Zr-doped $Li_4Ti_5O_{12}$ using solid-state reaction and reported that Zr^{4+} doping at the Ti site of $Li_4Ti_5O_{12}$ increased the cycling stability of $Li_4Ti_5O_{12}$ at high C rate [26].

Apart from surface coating with conductive material and doping aliovalent metal ions, nanostructured $Li_4Ti_5O_{12}$ [27, 28] has appeared to be interesting because its large surface area increases the electrolyte/electrode contact area, which leads to a decrease of the current density per unit surface area and an increase in the charge/discharge rate [29]. Also, the nanostructure provides a shorter path for lithium–ion and electron transport and results in improved kinetic performance. Shen et al. [30] succeeded in growing LTO nanowires on titanium foil and further showed an improvement in the conductivity of LTO nanowires by introducing Ti^{+3} ions through hydrogenation. Their samples exhibited excellent rate capability and delivered a discharge capacity of 173 mA h g^{-1} at 0.2C rate which is almost equal to the theoretical value with good life cycle. Their samples still deliver a discharge

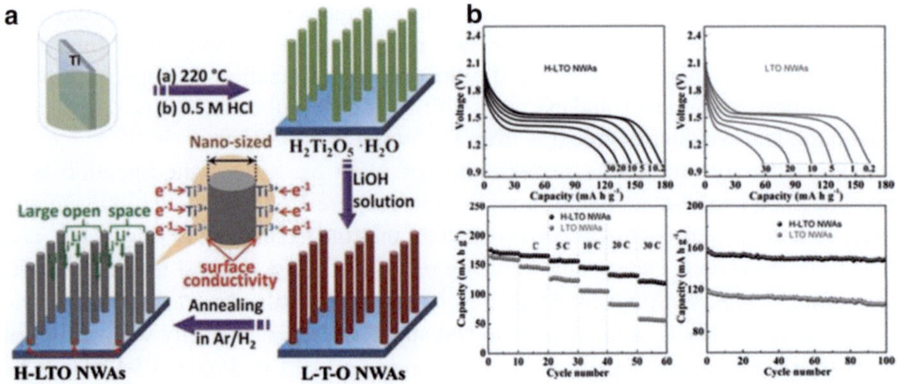

Fig. 3.4 (**a**) Schematic representation of the fabrication of hydrogenated LTO (H-LTO). (**b**) Electrochemical performance of H-LTO and LTO nanowires (Reproduced with permission from Ref. [30]. Copyright 2012 John Wiley and Sons)

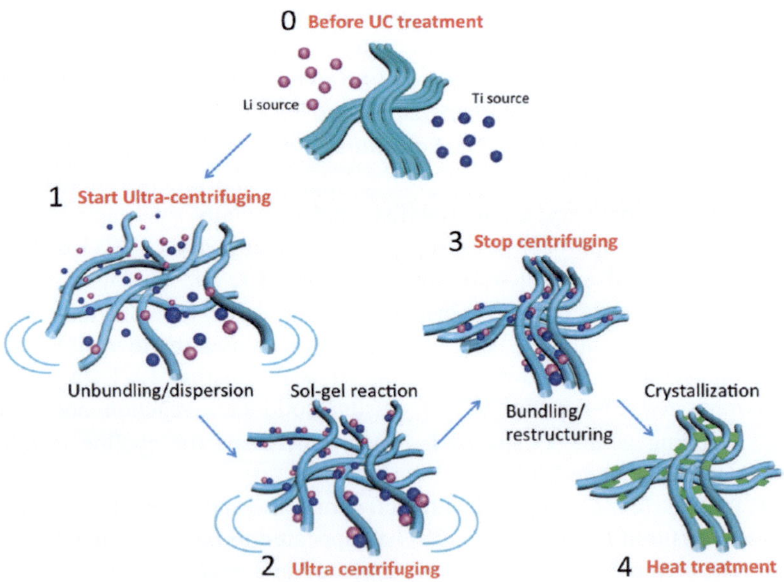

Fig. 3.5 Concept of ultracentrifuging (UC) treatment (Reproduced with permission from Ref. [31]. Copyright 2013 American Chemical Society)

capacity of 121 mA h g^{-1} at 30C as it is shown in Fig. 3.4. These superior electrochemical performances are due to the enhancement of the electron conductivity in hydrogenated LTO with respect to the pristine one. Naoi et al. introduced the so-called ultracentrifuging (UC) treatment to prepare high-performance LTO [31] (refer Fig. 3.5).

3.2.1.2 Titanium Dioxide (TiO_2)

Titanium dioxide is found in several polymorphic forms, the most familiar ones being the anatase, rutile, and brookite. TiO_2 is a very attractive material for anode applications in the LIBs due to its low cost, natural abundance, and environmentally friendliness. TiO_2 is considered as anode for lithium–ion battery since it is one of the transition metal oxides that can be reduced (from Ti^{4+} to Ti^{3+}) upon lithium insertion at lower potential. TiO_2 crystal structure creates favorable empty sites for lithium intercalation since its structures are all built upon the connection of TiO_6 octahedra [32, 33]. Although the anatase structure consists of edge-sharing TiO_6 octahedra, the rutile and the brookite frameworks exhibit both corner- and edge-sharing configurations. Besides, TiO_2 exhibits excellent safety and stability characteristics at the operative potential of 1.5 V vs Li/Li^+ that makes it a promising next-generation anode material for HEV applications just like spinel LTO. The theoretical capacity of TiO_2 is $330\ \mathrm{mA\ h\ g^{-1}}$ almost double of LTO ($175\ \mathrm{mA\ h\ g^{-1}}$), if it is able to host 1 mol of lithium per 1 mol of TiO_2 with $LiTiO_2$ stoichiometry. TiO_2 anatase has been considered as the most electrochemically active form, whereas there are many reports on the anode applications of TiO_2 rutile and brookites. However, the practical attainable capacity of TiO_2 has been reported to be only half of its theoretical value due to the blocking of further Li-ion insertion of TiO_2 resulting from the strong repulsive force between Li-ions. This behavior has reportedly limited the application and development of TiO_2 as anode materials for LIBs. The lithium intercalation/deintercalation process in titania typically depends on its crystallinity, particle size, structure, and surface area.

TiO_2 anatase

TiO_2 anatase adopts a tetragonal structure, with lattice parameters of $a = b = 3.776$ Å and $c = 9.486$ Å as shown in Fig. 3.6. TiO_2 is well recognized as a lithium insertion host material owing to its stacking of zigzag chain crystal structure consisting of highly distorted edge-sharing TiO_6 octahedra [34]. Its crystal structure arrangement forms open channels which is convenient for the insertion/extraction of Li^+ during discharge/charge [29].

Anatase-phase titanium dioxide is an attractive anode material for Li-ion batteries due to its high stable potential plateau of about 1.8 V which makes it suitable to be paired with high-voltage (5 V) cathode materials such as $LiNi_{0.5}Mn_{1.5}O_4$ [35, 36]. However, its shortcoming is the low electrical conductivity which results in poor rate capability and poor cycling performance during the lithiation process. Anatase phase of TiO_2 with a potential of 1.8 V can intercalate and deintercalate 1 lithium per formula unit at room temperature. There has been a thorough investigation attempting the exploitation of anatase-phase mesoporous TiO_2 for

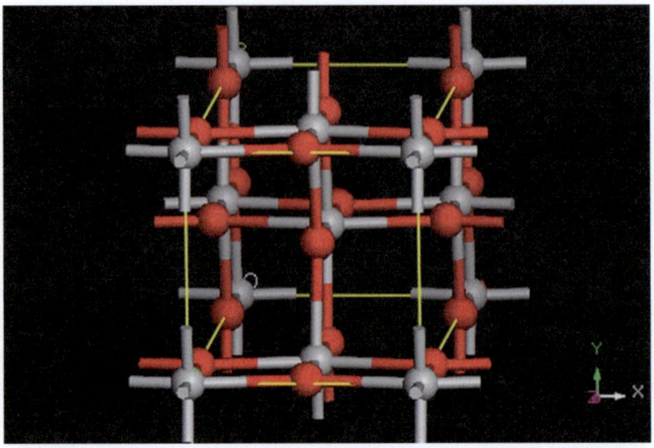

Fig. 3.6 Crystallographic representation of anatase TiO_2. *Gray* and *red spheres* are Ti and O atoms, respectively

high-power lithium battery application. The principal reaction that governs the electrochemical processes of a TiO_2/Li half-cell can be expressed as follows:

$$TiO_2 + x(Li^+ + e^-) \leftrightarrow Li_xTiO_2 \tag{3.10}$$

At first the maximum number of Li^+ that can be inserted is considered to be 0.5, leading to a theoretical capacity of 167.5 mA h g^{-1} [37]. Yang et al. first reported the high-energy (0 0 1) facet of anatase TiO_2; then after that nanocrystals/nanosheets of anatase TiO_2 with large fraction of exposed high-energy (0 0 1) facets have been successfully synthesized by different groups [38–40]. Chen and Lou [34] have investigated the lithium storage capabilities of the anatase TiO_2 nanosheets with exposed high-energy (0 0 1) surfaces. The TiO_2 nanosheets were synthesized through a modified hydrothermal method, and their result shows that these nanosheets exhibit much lower initial irreversible capacity losses compared to other ordinary anatase TiO_2 nanocrystals and excellent capacity retention upon repeated cycling.

TiO_2 rutile

TiO_2 rutile adopts a tetragonal structure, with lattice parameters of $a = b = 4.594$ Å and $c = 2.486$ Å as shown in Fig. 3.7. The rutile polymorph is the stable form of TiO_2. The anatase form transforms to rutile form by heating to temperatures $T > 700$ °C and also by high-energy ball milling.

The rutile structure is made up of TiO_6 octahedra that share edges to form single chains extending along the c-axis giving a (1×1) tunnel structure; the Li-ions can occupy these tunnels during intercalation. TiO_2 rutile has a theoretical capacity of 335 mA h g^{-1} corresponding to the insertion of 1 lithium per mol of TiO_2 according to the following equation:

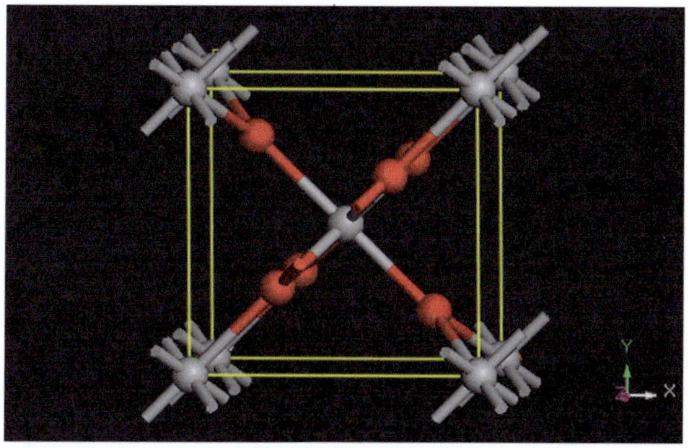

Fig. 3.7 Crystallographic representation of rutile TiO_2. *Gray* and *red spheres* are Ti and O atoms, respectively

$$TiO_2 + x(Li^+ + e^-) \leftrightarrow Li_xTiO_2 \text{ with } x = 1 \tag{3.11}$$

Bulk rutile TiO_2 can only accommodate negligible Li (< 0.1 Li per TiO_2 unit) at room temperature [41]; it only performs a successful Li^+ insertion and extraction at relatively high temperature of 120C in a lithium polymer cell [42]. However, interestingly, the Li reactivity increases with decreasing the particle size. In 2006 Hu et al. [43] reported up to 0.8 mol Li insertion into nanostructured rutile TiO_2 (10 nm × 40 nm) at room temperature as it is shown in Fig. 3.8 while only 0.1–0.25 mol of Li into micrometer-sized rutile.

TiO_2 brookite

TiO_2 brookite adopts an orthorhombic structure, with lattice parameters of $a = 9.166$ Å, $b = 5.436$Å, and $c = 5.135$ Å as shown in Fig. 3.9. Its crystal structure comprises of TiO_6 octahedra which share both edges and corners. The channels observed along the [001] direction in brookite are good for allowing lithium mobility [44].

Pure TiO_2 brookite with orthorhombic structure is recently being synthesized and investigated for its Li electrochemical reactivity and as anode for lithium–ion battery application. It exhibits as good result as its counterpart rutile and anatase structured TiO_2 anode materials. Generally, in the synthesis of TiO_2, brookite phase occurs as a minority phase. In 2007, Reddy et al. synthesized pure brookite by the thermolysis and precipitation of $TiCl_4$ at 100°C [45, 46] and showed for the first time that the intercalation of Li into the nanophase brookite TiO_2 can reach about 0.9 Li per formula unit. The nanosized (10 nm) brookite was tested as a Li-ion battery anode, which delivered around 375 mA h g^{-1} first-cycle discharge capacity and retained reversible capacity of 170 mA h g^{-1} for more than 40 cycles as it is displayed in Fig. 3.10. The specific capacity of brookite varied with the size of the particles, where 20 and 33 nm sized brookite delivered 60 and 35 mA h g^{-1} after

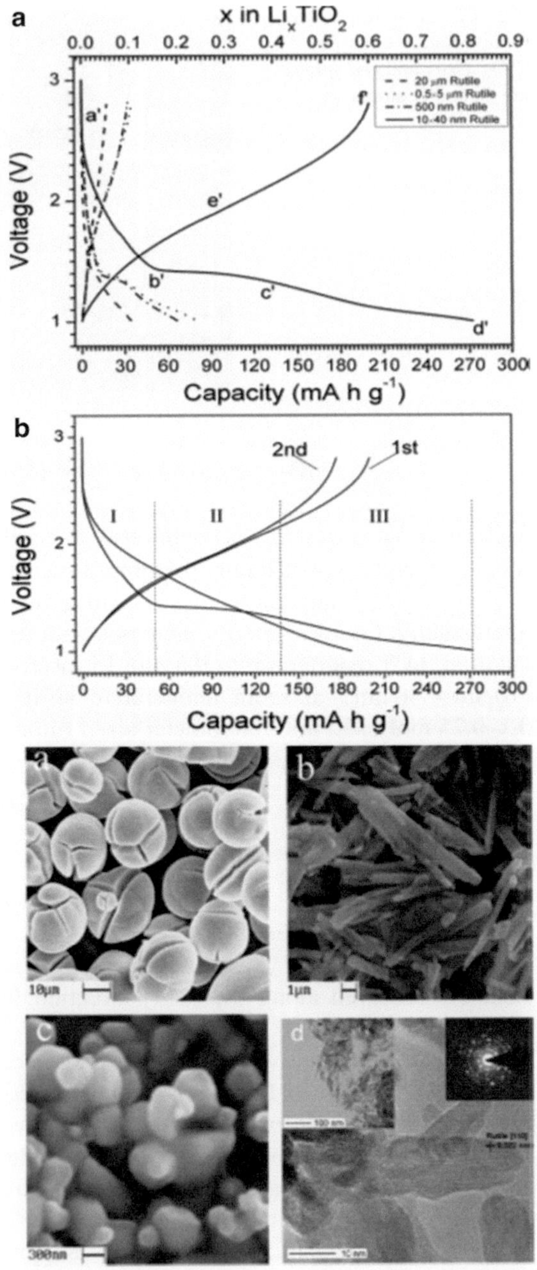

Fig. 3.8 Voltage profiles of nanostructured TiO_2 rutile at a rate of C/20 between 1 and 2.8 V. SEM images of nanocrystalline TiO_2 rutile (Reproduced with permission from Ref. [43]. Copyright 2006 John Wiley and Sons)

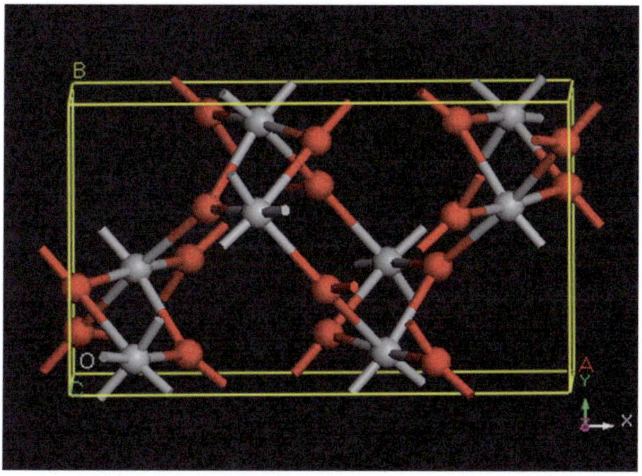

Fig. 3.9 Crystallographic representation of brookite TiO_2. *Gray* and *red spheres* are Ti and O atoms, respectively

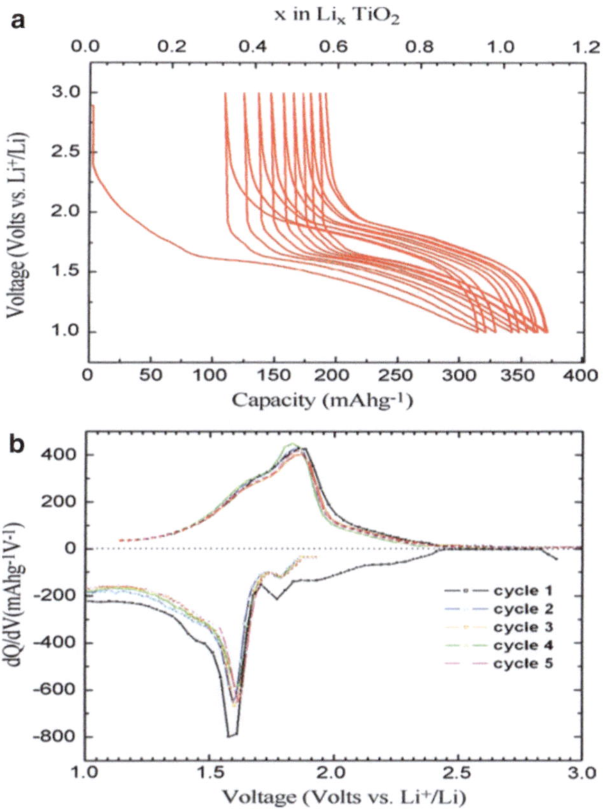

Fig. 3.10 (**a**) Voltage–capacity profiles of brookite TiO_2 in the voltage range 1.0–3.0 V at *C*/10 rate; (**b**) the corresponding differential capacity plot (Reproduced with permission from Ref. [45]. Copyright 2008 Electrochemical Society)

50 cycles, respectively. Also, Lee et al. [47] synthesized brookite by urea precipitation and investigated 10–20 nm sized brookite with multiwalled carbon nanotubes, which delivered 160 mA h g^{-1} over 50 cycles. Li insertion/extraction was observed at 1.7 and 2.02 V, respectively.

3.3 Alloying–Dealloying Reaction-Based Anodes

There are several elements, metals, and metal oxides which perform according to alloying–dealloying reaction mechanism when they are used as anode materials for lithium–ion battery. The materials include Si, Ge, Sb, Zn, Sn, SnO_2, and ZnO, to mention a few. Here we focus on SnO and SnO_2, and Si and Sn would be discussed later in Sect. 3.5. The alloying−dealloying reaction of these materials takes place at low potentials ($\leq$ 1V vs Li metal) and hence they can be considered as prospective anodes for LIBs.

3.3.1 Binary Tin Oxides

Tin oxides, SnO_2 and SnO, have been considered as promising anode materials [48] to replace the presently used carbonaceous materials due to their ability to deliver larger reversible capacities. The chemistry of tin oxides reacting with lithium is believed to proceed in two steps [49]. Initially Li reduces the tin to form small clusters of tin metal, dispersed in Li_2O framework. Then, Li-ions can be reversibly inserted into the metallic tin to form Li–Sn alloy. The alloying of Li directly into bulk tin metal causes a huge volume expansion, producing cracks of the electrode, and rapid loss of capacity. Thus, it is difficult to use for a practical battery. However, in the tin oxide systems, the oxide framework is thought to absorb some of the expansion and to stabilize the insertion process. Tin oxides are also nontoxic and readily available and so would make a good choice for new battery technology.

3.3.1.1 Tin Dioxide, SnO_2

Among the anode materials for lithium–ion batteries, tin dioxide (SnO_2) attracted much attention of the researchers due to its high theoretical capacity, low cost, low toxicity, and high abundance [48, 50, 51]. The theoretical capacity of tin dioxide is as high as 1494 mA h g^{-1}, with a reversible capacity of 782 mA h g^{-1} [52]. The electrochemical reaction between lithium and Sn occurs at relatively low potential of 0.6 V vs Li/Li^+. This will benefit to obtain a high overall cell potential of more than 3 V when SnO_2 anode is coupled with cathode materials such as

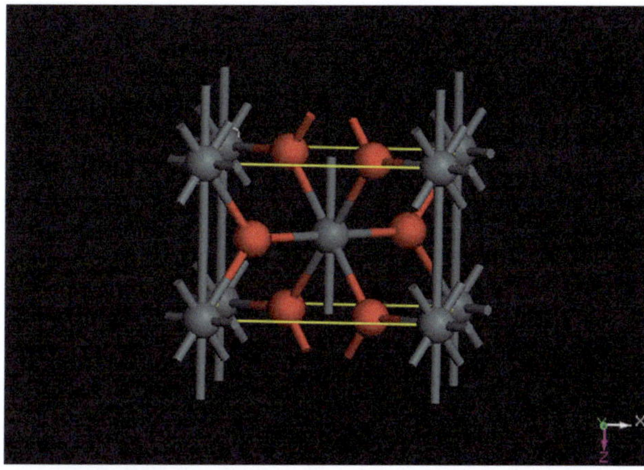

Fig. 3.11 Crystallographic representation of SnO_2. *Gray* and *red spheres* are Sn and O atoms, respectively

$LiNi_{0.5}Mn_{1.5}O_4$. On the other hand, SnO_2 exists in the form of tetragonal crystal structure with lattice constants of a = b = 4.737Å and c = 3.18 Å as shown in Fig. 3.11, and it is easy to synthesize. Idoda et al. [48] in 1997 first reported that SnO_2 anode material is electrochemically active to store lithium in lithium–ion battery. Interestingly, in their paper they assumed that lithium can be reversibly stored in $SnO_2 (SnO_2 + xLi^+ + xe^- \leftrightarrow Li_xSnO_2)$ and follows intercalation–deintercalation mechanism. In the following year it was experimentally proven that SnO_2 is irreversibly reduced to Li_2O and metallic Sn in the initial discharge [53]. This explained the low initial coulombic efficiency of the material, which is one of the major obstacles to its commercial application.

$$SnO_2 + xLi^+ + xe^- \rightarrow Li_2O + Sn \tag{3.12}$$

Courtney and Dahn [49] further proved that the lithium storage in SnO_x is by Li–Sn alloying and dealloying. They proposed that the theoretical lithium storage capacity of Sn by the Li–Sn alloying is 991 mA h g^{-1}, more than twice the theoretical capacity of commercial graphite carbon.

$$Sn + xLi^+ + xe^- \leftrightarrow Li_xSn \; (0 \leq x \leq 4.4) \tag{3.13}$$

Currently both the above reactions in Eqs. 3.12 and 3.13 are well recognized and become the basis for evaluating the electrochemical performances of SnO_2 anodes.

In a SnO_2 -based LIB electrode, as Sn alloys with Li to form Li_xSn $(0 \leq x \leq 4.4)$ during lithium insertion process, there is a huge volume expansion more than 300 %, which results in the disintegration and rapid capacity decrease over extended charge–discharge cycling. This severe capacity fading of SnO_2 is the main challenge to apply it as efficient anode for lithium–ion battery. There are different strategies to overcome this problem such as nanostructuring the active particles and coating with carbon layers to buffer the volume expansion.

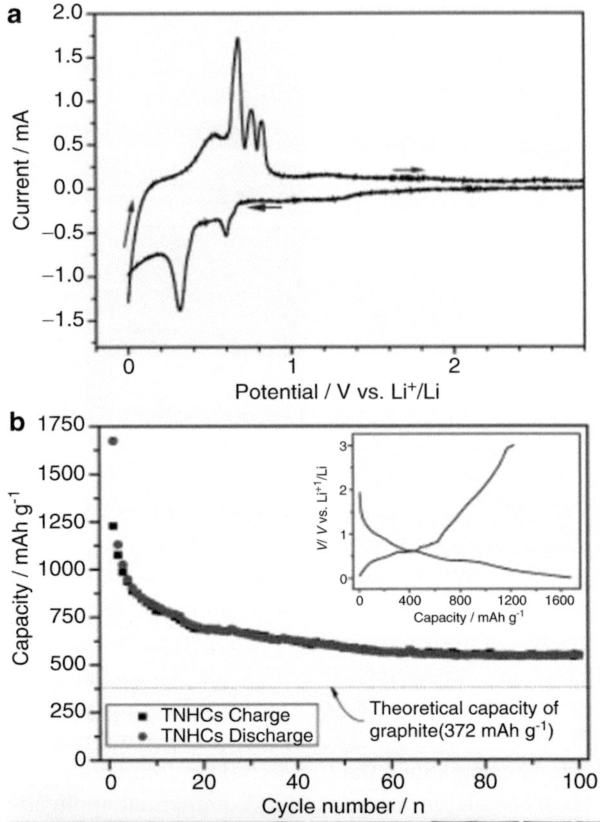

Fig. 3.12 (**a**) The cyclic voltammograms of TNHCs, the scan speed is 0.2 mV s^{-1}; (**b**) The discharge/charge capacity profiles of TNHCs in 5 mV-3 V (vs. Li^+/Li) voltage window and C/5, the inset shows the first discharge/charge profiles of TNHCs cycled at a rate of C/5 (Reproduced with permission from Ref. [67]. Copyright 2008 John Wiley and Sons)

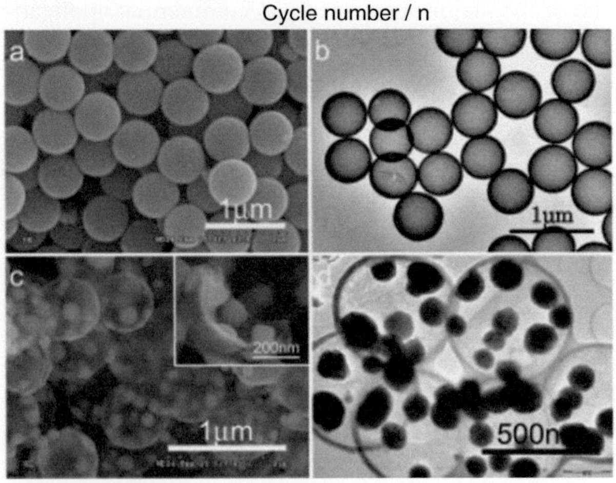

For the nanostructured SnO_2 particles, it has been established that small particles are able to minimize changes in volume as well as the strain in the lithium insertion/de-insertion thereby improving life cycle [54]. With this understanding, numerous nanostructured SnO_2 have been prepared such as SnO_2

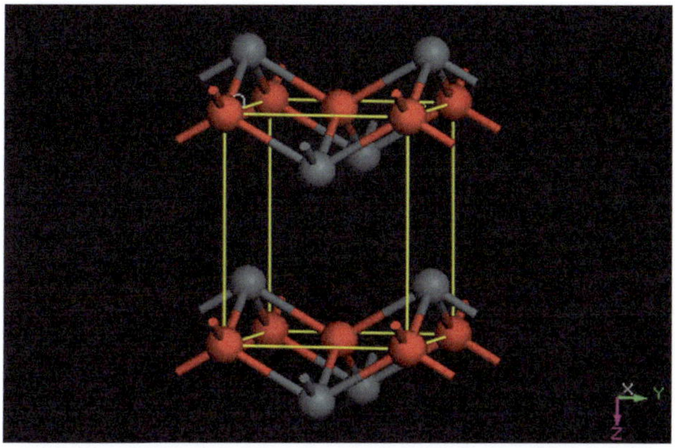

Fig. 3.13 Crystallographic representation of SnO. *Gray* and *red spheres* are Sn and O atoms, respectively

nanorods [55, 56], SnO_2 nanowires [50], SnO_2 nanotubes [57], and SnO_2 hollow nanospheres [58]. Ko et al. [59] proposed a promising synthetic technique, which is self-supported nanostructuring, for the direct growth of one-dimensional SnO_2 nanowires on the current collector. They compared their product SnO_2 nanowire with SnO_2 nanopowder and Sn nanopowder. The SnO_2 nanowire electrode delivered a discharge capacity of 510 mA h g^{-1} at the 50th cycle, which exceeded that of SnO_2 nanopowder and Sn nanopowder electrodes. Furthermore, the SnO_2 nanowire electrode displayed superior rate capabilities with a rechargeable discharge capacity of 600 mA h g^{-1} at 3C (where 1 C = 782 mA h g^{-1}), 530 mA h g^{-1} at 5C, and 440 mA h g^{-1} at 10C.

The other effective technique to buffer the high volume expansion in cycling SnO_2 is by coating it with carbon. It is worth mentioning that Sony has introduced an amorphous composite anode material consisting of tin, cobalt, and carbon to replace graphite [60]. Cobalt and carbon in Sony's tin composite anode serve to make that the material remains amorphous upon continuous lithiation–delithiation [61]; tin is the electrochemically active material that operates by reversibly alloying–dealloying with lithium and offers a theoretical specific capacity of about 990 mA h g^{-1} [62]. The method followed by Sony was successful to buffer the strain due to volume change during the alloying/dealloying process upon cycling. Normally, the volume change leads to a pulverization of the electrode active material, finally resulting in a loss of electrical contact and accordingly in a rapid capacity fading [63, 64]. Moreover, Sony's tin composite is able to suppress the aggregation of tin particles upon continuous lithiation–delithiation which is usually observed for tin oxide-based electrodes [49, 65]. Since cobalt is toxic as well as expensive, other cheap and economical approaches were designed. The method is generally related to the integration of metallic tin in carbonaceous matrices. In this regard successful attempts were reported by the encapsulation of tin nanoparticles in hollow carbon spheres [66, 67], tin–carbon core–shell particles [68], carbon-coated hollow tin nanoparticles [69], tin-filled carbon nanotubes [70], tin nanoparticles encapsulated in nano and submicron-sized carbon particles [71], and very recently tin nanoparticles embedded in micron-sized disordered carbon [60].

For instance, tin nanoparticles were encapsulated in elastic hollow carbon spheres (TNHCs) [67] as shown in Fig. 3.12. The cyclic voltammograms (CVs) of TNHCs show four separated peaks at 0.4–0.8 V on the positive current side which are related with the extraction of Li from Sn and the broad background is related with the extraction from carbon shells. The peaks originate from the phase transitions between Sn, Li_2Sn_5, LiSn, Li_7Sn_3, Li_5Sn_2, $Li_{13}Sn_5$, Li_7Sn_2, and $Li_{4.4}Sn$. The material has specific capacity higher than 800 mA h g^{-1} in the initial ten cycles and higher than 550 mA h g^{-1} after the 100th cycle. It is approximately 50 % higher than the theoretical specific capacity of currently used graphite even after the 100th cycle, which amounts to 66.2 % of its theoretical specific capacity (831 mA h g^{-1}).

One thing is common for all the aforementioned carbon-coated SnO_2 samples; they achieved improvement relative to pristine tin-based lithium–ion anodes. However, the most important thing is there has to be homogeneous dispersion of fine tin nanoparticles incorporated in micron-sized carbon particles in order to secure stable specific capacities exceeding 450 mA h g^{-1} for more than 100 cycles [72, 73].

3.3.1.2 Tin Monoxide, SnO

Tin monoxide, SnO, has a tetragonal structure with lattice parameters of $a = b = 3.796$Å and $c = 4.816$Å as shown in the Fig. 3.13 below and is made up of SnO6 octahedra.

The group of Dahn [74] has investigated the factors which are responsible for Li cyclability of SnO and reported that materials with small particles, small grains, and smooth sloping voltage profiles which do not change with cycle number give the best cycling performance. Courtney and Dahn [65] have shown that bulk SnO can deliver an initial reversible capacity of as high as 825 mA h g^{-1}; the problem is the capacity fades drastically on repeated cycling in the voltage window 0.0–1.3 V vs Li. However, as the voltage window was narrowed to 0.4–1.3 V, the capacity was found to be stable for the first ten cycles. Sakamoto et al. [75] studied how the particle size alone influences the initial and reversible capacities by fixed grain size. They reported on the preliminary Li cycling studies of SnO prepared by precipitation method (surface area 87 m^2g^{-1}) and a commercial sample (surface area 28 m^2 g^{-1}); they showed a reversible capacity of 657 and 590 mA h g^{-1} in the voltage range 0–1.2 V at current of 0.02 mA cm^{-2}.

3.3.2 *Ternary Tin Oxides, M_xSnO_y*

Various nonmetal oxides and metal oxides were added into SnO_2 compound to improve the electrochemical performances, and all ended up in forming ternary oxides [52, 76, 77]. Ternary tin oxides, M_xSnO_y, mostly consist of electrochemically inactive elements. By preparing composites of electrochemically inactive elements and SnO_2, it is possible to improve capacity retention of alloy-forming metal oxide anodes. The electrochemically inactive matrices can buffer the unit cell volume changes that take place during alloying–dealloying reactions, thereby minimizing the capacity fading. The main disadvantage of the mixed tin oxides is that the theoretical reversible capacity will be less than that of the binary tin oxides.

The commonly accepted Li cycling mechanism for ternary tin oxides, M_xSnO_y, is known. Firstly, during discharge reaction with Li metal, the crystal structure breaks down, or amorphization of the lattice of the original mixed oxide occurs, just like the binary tin oxides. This leads to the formation of metal (M) or metal oxides (MO_y) and formation of nano-Sn metal dispersed in amorphous Li_2O. This is followed by alloy formation, $Li_{4.4}Sn$. Studies have shown that in the cases where M is alkali, alkaline earth, or rare-earth metal, only metal oxides (MO or MO_y) are formed during the first discharge. However, when M = Zn, Co, Mn, Ni, or Fe, the respective nanoparticles of M are formed along with Li_2O depending on the magnitude of free energy of formation of respective oxide in comparison to Li_2O.

The most reported ternary tin oxides M_2SnO_4 (M = Zn, Co, Mn, Mg) adopt a spinel structure with Sn in tetrahedral O coordination. In the case of Co_2SnO_4, both tin-based oxides and cobalt-based oxides have attracted a considerable attention because of their potential application in Li-ion batteries [48, 78]. The theoretical capacity of SnO_2 and CoO can reach 1494 and 715mA h g^{-1}, respectively, so that tin- or cobalt-based oxides are likely to be the potential candidates of anode materials for Li-ion batteries [78]. Connor and Irvine [79] reported on the Li cycling behavior of ternary tin oxides, M_2SnO_4 (M = Mg, Mn, Co, Zn), in the range 0.02–1.5 V vs Li at low current rates. Best reversibility is noted for M = Mn and Zn and the worst when M = Mg. They found a correlation between the initial reduction potential of the spinel oxides and enthalpy of formation of the metal oxide, MO. MgO would not be expected to be reduced at these potentials, and so no Mg metal would be formed to alloy with lithium. The Mg, Mn, and Co spinels should behave electrochemically as the SnO_2, with slightly different products:

$$M_2SnO_4 + 4Li^+ + 4e^- \rightarrow 2(MO + Li_2O) + Sn \tag{3.14}$$

$$M_2SnO_4 + 8Li^+ + 8e^- \rightarrow 2Li_2O + Sn + 2M \tag{3.15}$$

$$Sn + xLi^+ + xe^- \leftrightarrow Li_xSn\ (0 < x < 4.4) \tag{3.16}$$

whereas the Zn_2SnO_4 spinel does have quite different properties, as Li readily reduces and alloys with the zinc. Previous studies on the Li cycling behavior of Zn_2SnO_4 showed that upon insertion of Li-ions, its crystal structure collapses and metallic Sn and Zn as well as Li–Sn and Li–Zn alloys are formed [80]. The overall electrochemical reaction mechanism that occurs during insertion of Li can be described by the following equations:

$$8\text{Li} + \text{Zn}_2\text{SnO}_4 \rightarrow 2\text{Zn} + \text{Sn} + 4\text{Li}_2\text{O} \tag{3.17}$$

$$4.4\text{Li} + \text{Sn} \rightarrow \text{Li}_{4.4}\text{Sn} \tag{3.18}$$

$$\text{Li} + \text{Zn} \rightarrow \text{LiZn} \tag{3.19}$$

The other interesting ternary tin oxide considered here is lithium stannate (Li_2SnO_3). Li_2SnO_3 contains SnO_6 octahedra and adopts a monoclinically distorted layer structure, similar to that of Li_2MnO_3. Its formula can be written as $Li(Li_{1/3}Sn_{2/3})O_2$, and it contains LiO_6 octahedra in both the Li layers and the (Li–Sn) layers. Courtney et al. [49] studied the electrochemical role of atoms which are not involved in the alloying–dealloying process. They used in situ XRD to characterize their materials; there were no tin peaks at the end of the initial charge, and this could be the result of the formation of three moles of Li_2O per mole of Li_2SnO_3 compared to two moles per mole of SnO_2, which should slow down the accumulation of tin metal. The presence of the Li_2O matrix in Li_2SnO_3 would enhance the life cycle of the battery compared with pristine SnO_2.

3.4 Conversion (Redox) Reaction-Based Anodes

Many transition metal oxides with rock salt and spinel structure such as CoO, NiO, Fe_3O_4, and Mn_3O_4 follow the conversion accompanying the reduction and oxidation reaction mechanism in the electrochemical lithium storage process. Basically, these metal oxides are first converted to metallic clusters embedded in a Li_2O matrix in the lithiation and then transfer to metal oxides at the subsequent delithiation, which involves reversible reduction and oxidation of metal oxides, coupled with the formation and decomposition of Li_2O.

3.4.1 Transition Metal Oxides with Rock Salt Structure (TMO; TM = Mn, *Fe, Co, Ni, or Cu)*

Transition metal oxide anodes with rock salt structure (TMO; TM = Mn, Fe, Co, Ni, or Cu) will be reviewed in this section. The electrochemical reaction mechanism comprising these compounds, together with lithium, involves the reduction (oxidation) of the transition metal along with the composition (decomposition) of lithium compounds. Anodes based on these compounds exhibit high reversible capacities (ranging $600 - 1000$ mA h g^{-1}) owing to the participation of a high number of electrons in the conversion reactions [1, 81–83]. The appropriate electrochemical conversion reactions can be described as follows:

$$\text{TM}x\text{O}y + 2y\text{Li}^+ + 2y\text{e}^- \leftrightarrow y\text{Li}_2\text{O} + x\text{TM}^0 \tag{3.20}$$

where TM = Fe, Co, Cu, Mn, Ni. Here the reaction mechanism which takes place during Li storage and cycling is different from that of the classical Li intercalation/deintercalation and alloying/dealloying. It typically involves the formation and decomposition of Li_2O along with the reduction and oxidation of metal nanoparticles, respectively. Generally, Li_2O is electrochemically inert, but it can be formed/decomposed during electrochemical cycling due to the electrochemically formed nanoparticles. The first discharge reaction with Li metal involves crystal structure destruction or amorphization of lattice followed by the formation of nanoparticles of metal embedded into the Li_2O matrix. During charge, the reformation of TMO is seen as a consequence of decomposition of Li_2O. Binary oxides, TMO, with TM = Mn, Co, and Ni adopt the cubic rock salt structure.

Cobalt oxide, CoO, has a high theoretical reversible specific capacity of 715 m A h g^{-1} when discharged to 0 V vs Li^+/Li [84–86]. Bulk CoO anode materials experience the large irreversible capacity in the first cycle and suffer severe capacity fading due to large volume expansion upon insertion of lithium–ions and poor electrical conductivity, and that limits their possible application [87, 88]. The group of Tarascon [78, 89] first reported on the use of TMO with TM = Fe, Co, Ni, and Cu as anode materials through the conversion reaction mechanism and demonstrated that transition metal oxides (TMO) (where TM is Co, Ni, Cu, or Fe) can reach electrochemical capacities of more than 700 mA h g^{-1} with 100 % capacity retention for up to 100 cycles by the electrochemically driven size confinement of the metal particles and buffering of the Li_2O matrix [78].

Poizot et al. [78] first reported about the Li storage capability of nanosized NiO through conversion reaction; they achieved an initial discharge capacity of around 600 mA h g^{-1} in the voltage range 0.01–3.0 V at C/5 rate. In their result the discharge capacity slowly degrades to 200 mA h g^{-1} at the end of 50 cycles. Later, Huang et al. [90] synthesized NiO–Ni composite and examined the Li cycling of the composites; they reported a reversible capacity of 500 mA h g^{-1} for NiO and 720 mA h g^{-1} for Ni–NiO composite at the end of the 30th. Mesoporous NiO–Ni composite was synthesized by Hosono et al. [91], and they found a capacity of 695 mA h g^{-1}. Recently, Aravindan et al. [92] produced electrospun NiO nanofibers and reported a capacity of 1280 and 784 mA h g^{-1} for first discharge and charge, respectively. The cell displayed a discharge capacity of 583 mA h g^{-1} (1.62 mol of Li) after 100 cycles which is 75 % of its initial reversible capacity as it is shown in Fig. 3.14.

CuO is an attractive material for anode materials which is easily obtained via conversion reactions. Its attraction stems from its abundance, low cost, ease of synthesis, high theoretical capacity of 674 mA h g^{-1}, and environmental friendliness [93, 94]. The challenge to use CuO as anode materials is its poor cyclability and low reversible capacity at high C rates which is due to its low conductivity [95, 96]. Various nanostructured CuO anode materials with superior performance have been synthesized using different methods. Among them, recently, Sahay

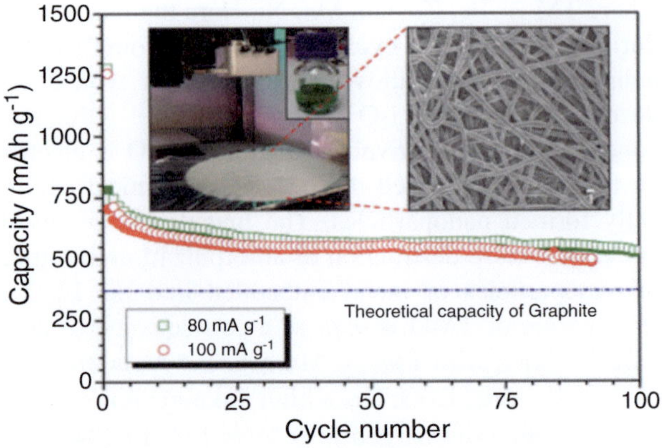

Fig. 3.14 SEM image of electrospun NiO fibers and plot of capacity vs cycle number for Li/NiO fibers where filled and open symbols correspond to the charge and discharge capacity, respectively (Reproduced with permission from Ref. [92]. Copyright 2013 Elsevier)

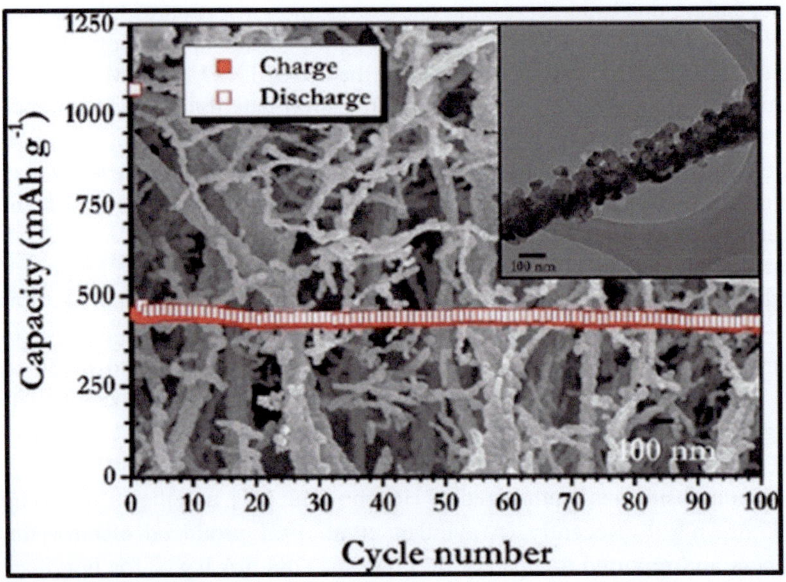

Fig. 3.15 SEM and TEM image and typical charge–discharge curves of CuO nanofiber cycled between 0.005 and 3 V vs. Li (Reproduced with permission from Ref. [97]. Copyright 2012 American Chemical Society)

et al. [97] attempted to improve the cycling stability of CuO by synthesizing one-dimensional CuO nanofibers with interconnected particulates by electrospinning technique and reported very stable reversible capacity of 452 mA h g^{-1} at current density of 100 mA g^{-1} as shown in Fig. 3.15.

The early report by Poizot et al. [83] on the initial results of MnO–Li system was not promising; it was indicated that MnO powder electrode could be discharged and charged under a very small current density (1/300 C). Later the group of Tirado [98] prepared sub-micrometer MnO that delivers an initial reversible capacity of about

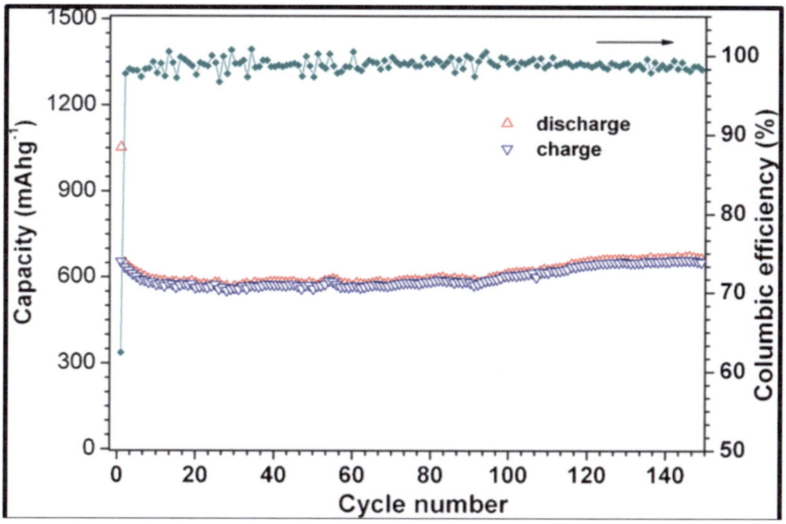

Fig. 3.16 Cyclic performance and coulombic efficiency of the C/MnO-L electrode (Reproduced with permission from Ref. [100]. Copyright 2010 Elsevier)

650 mA h g^{-1}, and it retained only 390 mA h g^{-1} after 25 cycles. MnOx nanoparticles were synthesized by Ji et al. [99] using the mixture of MnO and Mn_3O_4 confined to porous carbon nanofibers, and they examined their Li cyclability. They got a first discharge capacity of 785 mA h g^{-1} in the range 0.01–3.0 V at 50 mA g^{-1}, which stabilized to 600 mA h g^{-1} between 10 and 50 cycles. Recently, the group of Li et al. [100, 101] prepared nanocrystalline MnO thin-film electrode by a pulsed laser deposition (PLD) method and reported the reversible lithium storage capacity of the MnO thin-film electrodes at 0.125C is over 472 mA h g^{-1} and can be retained more than 90 % after 25 cycles. They also used mechanical milling to synthesize carbon-coated MnO and to decrease the particle size of MnO and labeled their sample as C/MnO-L. MnO powder materials show reversible capacity of 650 mA h g^{-1} with average charging voltage of 1.2 V. It can deliver 400mA h g^{-1} at a rate of 400m A g^{-1}. The cyclic performance of MnO is improved significantly after decreasing particle size and coating with a layer of carbon as it is shown in Fig. 3.16. Their result has shown that the kinetic property of the MnO anode is not as poor as the early study by Poizot and co-workers.

The electrochemical performance of FeO as anode for Li-ion battery was first studied by Poizot et al. [83]; they reported an initial capacity of around 750 mA h g^{-1} which fades to around 300 mA h g^{-1} after 50 cycles in voltage range 0.01–3.0 V at 0.2C. There are a few reports on the Li cyclability of the FeO–Li system. Recently, Gao et al. [102] successfully prepared FeO/C composites by a facile carbothermal reduction of the mixtures of nanosized Fe_2O_3 and acetylene black (AB). The FeO/C composite exhibited superior electrochemical performance than the Fe_2O_3/AB mixture for the same AB addition. A specific capacity of 511 mA h g^{-1} is obtained for the FeO/C composite, and a value of only 396 mA h g^{-1} is obtained for the Fe_2O_3/AB mixture. The capacity retention after 50 cycles was more than 96 % for the FeO/C composites with AB additions higher than 50 wt%, while a value of only 70–78 % was obtained for the Fe_2O_3/AB mixtures with the same AB additions.

3.4.2 *Transition Metal Oxides with Spinel Structure (TM₃O₄,* TM = Co, *Fe, or Mn)*

Co_3O_4 exists in the inverse spinel structure as it is displayed in Fig. 3.17 [103]. The group of Tarascon [83] first reported about the electrochemical properties of Co_3O_4 anode for Li-ion battery by conversion reaction, and they found a stable and reversible capacity of 800 – 900 mA h g^{-1} in the voltage range 0.005–3.0 V. The conversion reaction is expressed as $Co_3O_4 + 8Li^+ + 8e^- \leftrightarrow 3Co + 4Li_2O$, which typically means that the material can store and cycle 8 mol of Li per mole of Co_3O_4 with a theoretical capacity of 890 mA h g^{-1}.

With the purpose of improving the cycling stability, many researchers have focused on studying nanostructured Co_3O_4 and reported that the initial charge capacities increased mostly between 800 mA h g^{-1} and 1000 mA h g^{-1}, and the capacity retention ratios were improved (> 85% after 40 or 50 cycles) at low rates such as 0.1 and 0.2 C. Recently, Huang et al. [104] synthesized micro-nanostructured Co_3O_4 powders with cube morphology that achieved high initial discharge capacities (1298 mA h g^{-1} at 0.1 C and 1041 mA h g^{-1} at 1 C) with impressive rate capability and excellent capacity retention (99.3 %, 97.5 %, 99.2 %, and 89.9 % of the first charge capacities after 60 cycles at 0.1 C, 0.2 C, 0.5 C, and 1 C, respectively) as depicted in Fig. 3.18.

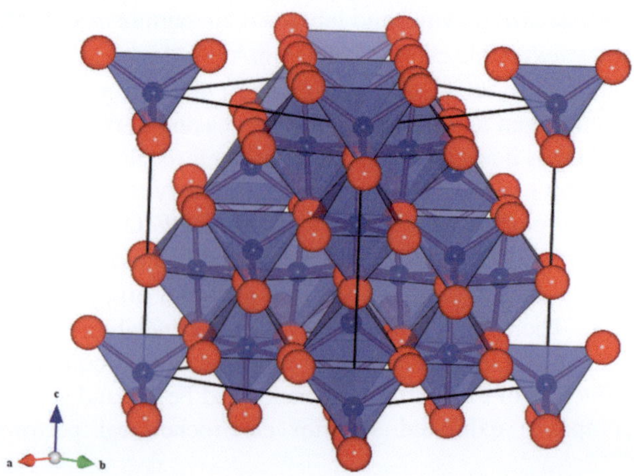

Fig. 3.17 General crystallographic presentation of spinel oxides like Co_3O_4. O in *red*; Co in *blue*. CoO4 tetrahedra and CoO6 octahedra are shown. Outline of the unit cell is also shown (Reproduced with permission from Ref. [103]. Copyright 2013 American Chemical Society)

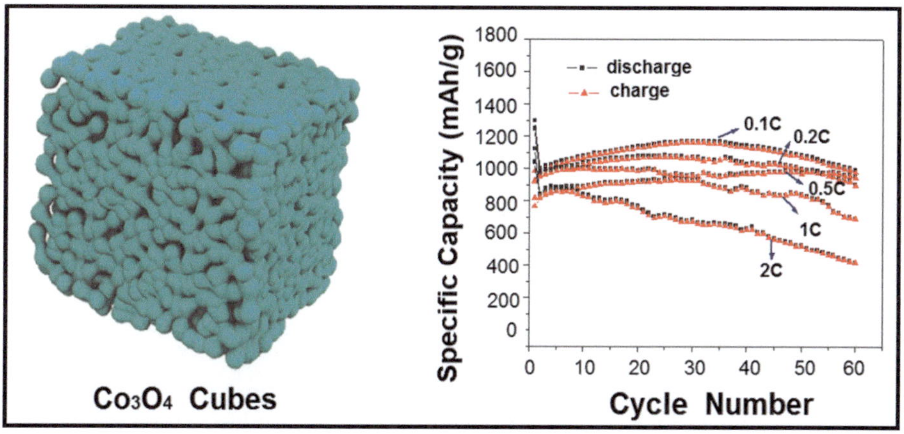

Fig. 3.18 Schematic representation of mesoporous Co_3O_4 cubes with micro-nanoscale structures and cycling performances of Co_3O_4 electrodes at different rates of 0.1 C, 0.2 C, 0.5 C, 1 C, and 2 C (Reproduced with permission from Ref. [104]. Copyright 2014 American Chemical Society)

Magnetite Fe_3O_4 adopts an inverse spinel structure similar to Co_3O_4 (refer to Fig. 3.17).

At present Fe_3O_4 has attracted great research interest as an anode material due to its natural abundance, low cost, environmental friendliness, and high theoretical capacity value (928 mA h g^{-1}). During the charge/discharge process, the reaction mechanism of Li storage in Fe_3O_4 can be described as $Fe_3O_4 + 8Li^+ + 8e^- = 3Fe + 4Li_2O$. The drawbacks of Fe_3O_4 are low electronic conductivity and high volume change during Li insertion–extraction process which leads to pulverization. In order to tackle these problems, two approaches are commonly applied. The first one is to synthesize nanostructured Fe_3O_4, and the second is coating it with carbon. The group of Tarascon [105] used a two-step electrode design consisting of the electrochemically assisted template growth of Cu nanorods onto a current collector followed by electrochemical plating of Fe_3O_4 which showed very good Li cyclability. They observed capacities in the range 800 – 900 mA h g^{-1} when the anode materials were cycled in the voltage range 0.01–3.0 V vs Li at current rates of 0.25 C and 0.5 C. They achieved a very good C rate capability even for a current rate of 8 C. Recently, He et al. [106] synthesized carbon-encapsulated Fe3O4@C@PGC (porous graphitic carbon) nanoparticles, and the electrode exhibits high rate capability (858, 587, and 311 mA h g^{-1} at 5, 10, and 20 C, respectively, 1 C = 1 A g^{-1}) and extremely excellent cycling performance at high rates (only 3.47 % capacity loss after 350 cycles at a high rate of 10 C), which is the best one ever reported for an Fe_3O_4 -based electrode including various nanostructured Fe_3O_4 anode materials and composite electrodes.

Mn_3O_4 is a promising anode material for LIBs with high theoretical capacity of 936 mA h g^{-1} which crystallizes in the inverse spinel structure. Pasero et al. [107]

have compared the Li cycling performances of Mn_3O_4 and $Mn_{2.6}Co_{0.4}O_4$ in the voltage range 0.25–3.5 V at 0.2 C rate. In their result, pristine Mn_3O_4 showed a reversible and constant capacity of 250 mA h g^{-1} up to ten cycles, whereas the Co-doped $Mn_{2.6}Co_{0.4}O_4$ anode material showed a constant capacity of about 400 mA h g^{-1} with the same cycling conditions.

3.5 Lithium Alloys

Since Dey reported the electrochemical alloying of lithium metal with other elements in organic electrolytes, various lithium alloys such as Li–Si and Li–Sn have been investigated as a potential high-capacity anode material for lithium–ion batteries [108–110]. Due to the high-capacity lithium-rich binary alloys, Li–Si and Li–Sn have gained considerable attention among the various lithium alloy elements. Therefore, herein we summarize the electrochemical characteristics of silicon and tin as anode materials for lithium–ion batteries.

3.5.1 Silicon

Silicon has been considered as one of the best candidates to substitute graphite as anode materials for Li-ion batteries because of its high gravimetric capacity of 4200 mA h g^{-1} and relatively low discharge potential (0.4 V versus Li/Li^+) and is also abundant, cheap, and environmentally friendly. The major challenge to apply Si as anode for LIB is its poor cycle performance due to its severe volume change during the discharge/charge reactions [111–113] which leads to its fracture and pulverization. To overcome the limitations of Si, various approaches have been applied such as nano-engineering of Si, the use of active–inactive binary alloys, and Si dispersion in active and/or inactive matrices.

Among nano-engineered Si anode materials, Si nanowire is getting much attention. In 2008, Chan et al. [114] showed that silicon nanowire battery electrodes solve the issues of pulverization and capacity fading as they can accommodate large strain without pulverization, provide good electronic contact and conduction, and display short lithium insertion distances. They achieved the theoretical charge capacity for silicon anodes and maintained a discharge capacity close to 75 % of this maximum, with little fading during cycling. Then after the report of Chan et al, many other groups reported some works using Si nanowire anode materials. For instance, Kang et al. [115] prepared Si nanowire anode and studied the cyclability over 50 cycles; the capacity retention was 1600mA h g^{-1} (i.e., 42 % of its first capacity of 3825mA h g^{-1}) at the 50th cycle while still showing coulomb efficiencies above 90 %. This indicates that the conductivity of the wires may decreases upon cycling or the contact with the current collector is destroyed. Thus, the Si

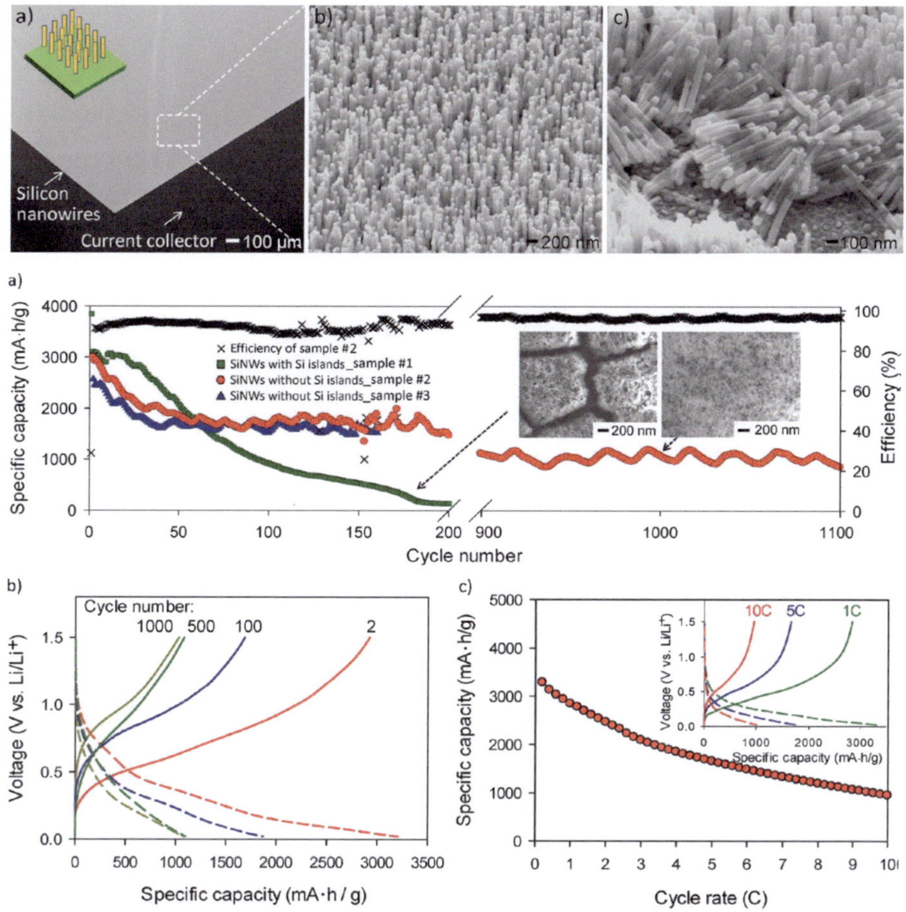

Fig. 3.19 (Top a, b, c) Schematic and SEM images of SiNWs; (Middle) (**a**) Specific capacity (delithiation) measurements of three Si nanowire samples prepared with and without the formation of Si islands on current collectors, (Bottom) (**b**) Charge/discharge profiles of Si nanowires prepared without Si islands for 2nd, 100th, 500th, and 1000th cycles, and (**c**) Discharge capacities vs cycle rate (0.2–10.0C) of Si. Reproduced with permission from ref 120. Copyright 2013 American Chemical Society

nanowires required further treatment to mitigate the fading problem such as covering the wires with amorphous Si particles [114, 116] or mixing Si wires with carbon-based additives to improve the conductivity [117, 118].

Moreover, with regard to nano-engineering of Si, a significant achievement was made by confining active Si at the nanoscale, which effectively minimizes the mechanical strains induced by the volume change of Si [114, 119]. Cho and Picraux [120] recently prepared Si nanowire anodes by anodic aluminum oxide (AAO) template growth to eliminate Si underlayer islands demonstrated promising behavior as an anode material for lithium–ion batteries as it is displayed in Fig. 3.19.

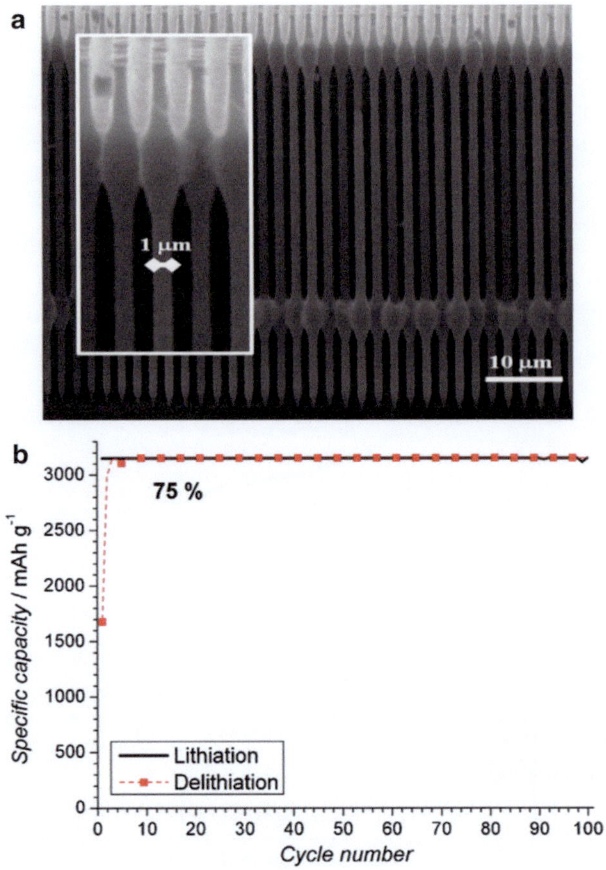

Fig. 3.20 (**a**) Second generation of optimized wires with 1 μm diameter. (**b**) Cycling performance of the optimized second-generation SWAs with C/10 current for the first four cycles and C/2 current for the following ones (Reproduced with permission from Ref. [122]. Copyright 2013 Elsevier)

They achieved significantly enhanced cycling stabilities with retentions of more than 1000 mA h g^{-1} discharge capacity over 1100 cycles for cycling at 0.5 C within the voltage range of 0.02–1.5 V versus Li/Li^+. They clearly demonstrated that eliminating the Si island under Si nanowire further improves the cyclability of the material. In the Si nanowire form without underlayer island, the pulverization of Si can be effectively avoided by relaxing facile strain during cycling.

The group of Föll [121, 122] has recently modified the silicon wire anodes (SWAs) designed to be free-standing silicon microwires embedded at one end in a copper current collector as shown in Fig. 3.20; the silicon microwire anode materials exhibited almost zero capacity fading after 100 cycles using 0.5 C rate. In the first cycle, the irreversible loss of 46 % observed may be due to the formation of an SEI layer on the Si wires. After cycle 2 the capacity remains constant for 100 cycles, with coulomb efficiency of over 99 %. These SWAs exhibited the same large gravimetric capacity as thinner SWAs. Surprisingly, they have good long term stability at relatively fast cycling rates.

To overcome huge volume changes of Si during charge/discharge cycling, the other strategy accommodating the mechanical stress is to synthesize Si-based composites with various buffer phases such as Si/SiOx composite [123] where crystalline Si was confined in amorphous SiOx matrix and improved the performance of the anode material.

3.5.2 *Tin, Sn*

Lithium can form intermetallic phases (Li_xSn) with tin at room temperature if the metal is polarized to a sufficiently negative potential in a Li^+ cation contained liquid organic electrolyte [54]. Similar to silicon, tin can also react with lithium at elevated temperatures to give different intermetallic compounds such as Li_2Sn_5, LiSn, Li_7Sn_3, Li_5Sn_2, $Li1_3Sn_5$, Li_7Sn_2, and $Li_{17}Sn_4$ in electrochemical lithiation process [124]. By producing lithium-rich phase of $Li_{17}Sn_4$, the lithiation process provides a theoretical capacity of 960 mA h g^{-1}. As shown in Fig. 3.21, the plateau potential of tin was 0.4 V versus Li^+/Li in its lithiation process; this in turn benefits in getting a high-voltage full cell by combining it with high-voltage cathode like $LiNi_{0.5}Mn_{1.5}O_4$ [125]. Thus, the tin-based alloys are also attractive anode materials for rechargeable lithium–ion batteries. Although Sn has a lower gravimetric capacity than Si and metallic lithium, Sn is an attractive anode material due to its volumetric capacity of about 2000 mA h cm^{-3}, which is higher than that of even metallic Li and is comparable to Si [126].

Although tin is an attractive anode candidate due to its low cost and high abundance in nature, the application of tin as an anode is still far from commercialization. Sn-based anode materials commonly suffer from severe capacity fading during the charge–discharge cycling of lithium–ions, which mainly results from the pulverization of active materials caused by the huge volume change of tin upon

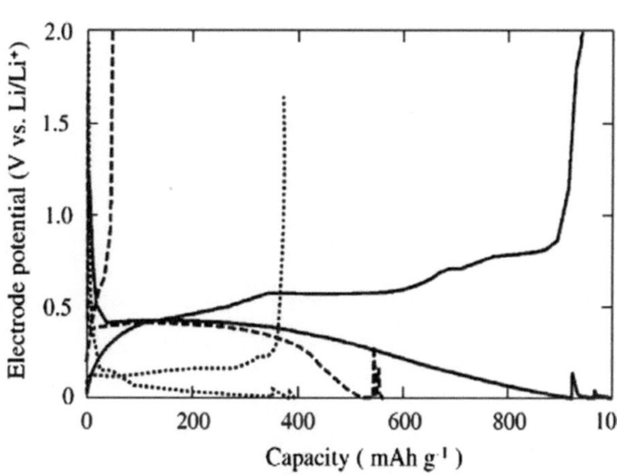

Fig. 3.21 Electrochemical behaviors of various Sn electrodes (*solid line*, electrodeposited Sn; *dashed line*, Sn powder; *dotted line*, graphite) (Reproduced with permission from Ref. [125]. Copyright 2002 Elsevier)

cycling. Therefore, various approaches have been tried to effectively accommodate the volume expansion. Nanostructured electrode materials generally enhance the ability to buffer the stress during the lithiation/delithiation processes. Actually, nano-engineering of Sn-based materials into special nanoarchitectures is an efficient way to improve their electrochemical performance, including specific capacity, rate capability, and cycling stability. Particularly, 1D nanowire or nanotube-based materials are attractive candidates for LIB electrodes since they have the ability to provide efficient 1D pathway for electron transport, allow ease of electrolyte penetration, as well as facilitate strain relaxation. Luo et al. developed a hybrid 1D Sn nanostructure with graphene sheets, where Sn-core/C-sheath coaxial nanocables were directly integrated onto the reduced graphene oxide (RGO) surface by an RGO-mediated growth procedure [127]. The nanocables exhibited specific capacities higher than 760 mAh g^{-1} in the initial ten cycles and more than 630 mAh g^{-1} after the 50th cycle, being much better than those values for the RGO-supported Sn particles (RGO–Sn) as is shown in Fig. 3.22c.

Hollow 1D structure is also effective in improving the lithium storage properties of Sn-based materials similar to 1D nanowires, nanorods, and coaxial nanocables. The electrochemical performance of Sn nanotube ensures stable capacity retention as the local empty space of hollow or porous nanostructured electrode materials can accommodate the large volume change and retain their high capacities. Wu et al. synthesized tubular 1D Sn-based anode materials using porous anodized aluminum oxide (AAO) membrane-templated fabrication methods as shown in Fig. 3.23c [128]. Shafiei and Alpas successfully achieved continuous formation of Sn nanotubes by electrochemical deposition of Sn nanograins on the surface of carbon fiber as depicted in Fig. 3.23a [129]. Their result showed that the Sn-coated carbon fiber had capacity retention of 50 % after 20 cycles as compared to the 23 % for pristine Sn at relatively large current densities.

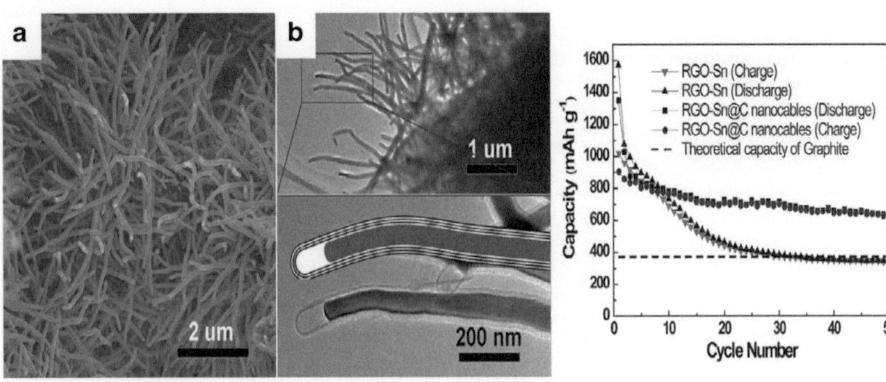

Fig. 3.22 Scanning electron microscopy (SEM) image (**a**) and TEM image (**b**) of the Sn-core/C-sheath nanocables. (**c**) Comparative cycling performance of different samples (Reproduced with permission from Ref. [127]. Copyright 2012 John Wiley and Sons)

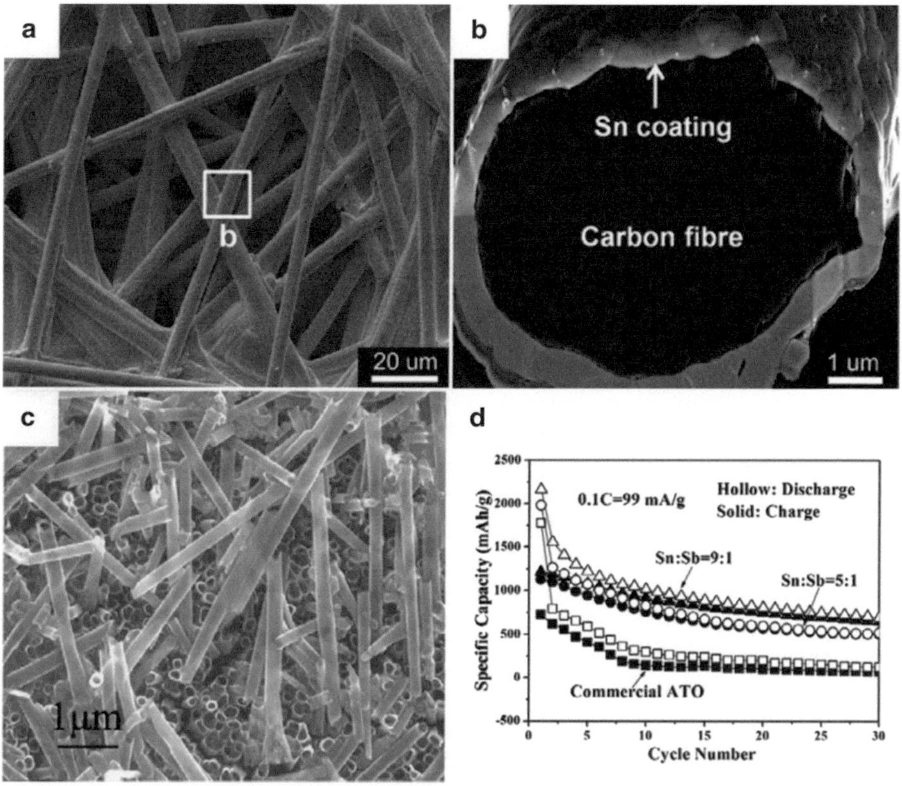

Fig. 3.23 SEM images of (**a**) the Sn-coated carbon fiber electrode and (**b**) a focused ion beam-milled cross section of a Sn-coated carbon fiber. (**c**) SEM image of the antimony-doped SnO2 nanotubes. (**d**) Comparative cycling performance of different samples (Reproduced with permission from Refs. [128] and [129]. Copyright 2011 Elsevier)

3.6 Summary and Future Perspective

In summary, this chapter outlines metal oxides and lithium alloy anode materials as promising and with high prospect to replace graphite anode for lithium-ion battery to achieve both higher energy and higher power density. The anode materials were categorized into three groups according to their reaction mechanism with lithium.

The first group comprises intercalation/deintercalation anodes particularly titanium-based oxides, spinel $Li_4Ti_5O_{12}$ and TiO_2 (anatase, rutile, and brookite phase). Titanium-based anodes are attractive since they operate at high voltage around 1.5 V which typically mitigates the issue of the formation of the solid electrolyte interface (SEI) that occurs during graphite usage and provide excellent cyclability. Specifically, nanostructured titanium-based oxides are preferred to achieve better capacity, longer life cycle, and higher rate capability than the bulk materials.

The second group of anodes operates according to alloying/dealloying reaction mechanism. Alloy-based anode, tin, and silicon are the best candidates to substitute graphite as anode materials for Li-ion batteries. The major challenge to apply Si and Sn as anode for LIB is their poor cycle performance due to its severe volume change during the discharge/charge reactions which leads to fracture and pulverization. To overcome the limitations of Si, nano-engineered modified silicon wire anodes (SWAs) designed to be free-standing silicon microwires embedded at one end in a copper current collector exhibited almost zero capacity fading after 100 cycles using 0.5 C rate. Similar to silicon, nano-engineering of Sn-based materials into special nanoarchitectures is an efficient way to improve their electrochemical performance, including specific capacity, rate capability, and cycling stability.

The third category of anodes works with conversion (redox) reaction mechanism; many transition metal oxides with rock salt and spinel structure such as CoO, NiO, Fe_3O_4, and Mn_3O_4 compounds exhibit high reversible capacities (ranging 600 -1000 mA h g^{-1}) owing to the participation of a high number of electrons in the conversion reactions. The current trend to improve the cycling stability of transition metal oxide anodes is by developing their nanostructured counterpart.

References

1. Goriparti S, Miele E, De Angelis F, Di Fabrizio E, Proietti Zaccaria R, Capiglia C (2014) Review on recent progress of nanostructured anode materials for Li-ion batteries. J Power Sources 257:421–443
2. Dunn B, Kamath H, Tarascon JM (2011) Electrical energy storage for the grid: a battery of choices. Science 334:928–935
3. Scrosati B, Hassoun J, Sun Y (2011) Lithium-ion batteries. A look into the future. Energy Environ Sci 4:3287–3295
4. Marom R, Amalraj SF, Leifer N, Jacob D, Aurbach D (2011) A review of advanced and practical lithium battery materials. J Mater Chem 21:9938–9954
5. Girishkumar G, McCloskey B, Luntz A, Swanson S, Wilcke W (2010) Lithium–air battery: promise and challenges. J Phys Chem Lett 1:2193–2203
6. Goodenough JB, Park K (2013) The li-ion rechargeable battery: a perspective. J Am Chem Soc 135:1167–1176
7. Thackeray MM, Wolverton C, Isaacs ED (2012) Electrical energy storage for transportation—approaching the limits of, and going beyond, lithium-ion batteries. Energy Environ Sci 5:7854–7863
8. Manthiram A (2011) Materials challenges and opportunities of lithium ion batteries. J Phys Chem Lett 2:176–184
9. von Cresce A, Xu K (2011) Electrolyte additive in support of 5 V li ion chemistry. J Electrochem Soc 158:A337–A342
10. Capiglia C, Saito Y, Kageyama H, Mustarelli P, Iwamoto T et al (1999) 7Li and 19F diffusion coefficients and thermal properties of non-aqueous electrolyte solutions for rechargeable lithium batteries. J Power Sources 81:859–862
11. Kavan L, Grätzel M (2002) Facile synthesis of nanocrystalline (spinel) exhibiting fast li insertion. Electrochem Solid-State Lett 5:A39–A42
12. Saravanan K, Ananthanarayanan K, Balaya P (2010) Mesoporous with high packing density for superior lithium storage. Energy Environ Sci 3:939–948

13. Ren Y, Liu Z, Pourpoint F, Armstrong AR, Grey CP, Bruce PG (2012) Nanoparticulate TiO_2 (B): an anode for lithium-ion batteries. Angew Chem 124:2206–2209
14. Du Pasquier A, Huang C, Spitler T (2009) Nano batteries with high power capability and improved cycle-life. J Power Sources 186:508–514
15. He Y, Ning F, Li B, Song Q, Lv W et al (2012) Carbon coating to suppress the reduction decomposition of electrolyte on the electrode. J Power Sources 202:253–261
16. Moretti A, Kim G, Bresser D, Renger K, Paillard E et al (2013) Investigation of different binding agents for nanocrystalline anatase anodes and its application in a novel, green lithium-ion battery. J Power Sources 221:419–426
17. Yi T, Jiang L, Shu J, Yue C, Zhu R, Qiao H (2010) Recent development and application of as anode material of lithium ion battery. J Phys Chem Solid 71:1236–1242
18. Zhu G, Wang Y, Xia Y (2012) Ti-based compounds as anode materials for Li-ion batteries. Energy Environ Sci 5:6652–6667
19. Huang S, Wen Z, Zhu X, Gu Z (2004) Preparation and electrochemical performance of Ag doped. Electrochem Commun 6:1093–1097
20. Huang S, Wen Z, Zhang J, Gu Z, Xu X (2006) composite as electrode materials for lithium-ion battery. Solid State Ion 177:851–855
21. He Z, Wang Z, Cheng L, Li T, Li X et al (2013) Conductive surface modification with copper of as anode materials for Li-ion batteries. Mater Lett 107:273–275
22. Wang B, Cao J, Liu Y (2014) Surface modification of spinel with Fe for lithium ion batteries. Mater Technol 29:124–128
23. Zhang Q, Li X (2013) Recent developments in the doped- anode materials of lithium-ion batteries for improving the rate capability. Int J Electrochem Sci 8:6449–6456
24. Park JS, Baek S, Park Y, Kim JH (2014) Improving the electrochemical properties of Al, Zr co-doped as a lithium-ion battery anode material. J Korean Phys Soc 64:1545–1549
25. Huang S, Wen Z, Gu Z, Zhu X (2005) Preparation and cycling performance of and co-substituted compounds. Electrochim Acta 50:4057–4062
26. Li X, Qu M, Yu Z (2009) Structural and electrochemical performances of as anode material for lithium-ion batteries. J Alloys Compd 487:L12–L17
27. Shen L, Yuan C, Luo H, Zhang X, Yang S, Lu X (2011) In situ synthesis of high-loading – graphene hybrid nanostructures for high rate lithium ion batteries. Nanoscale 3:572–574
28. Capsoni D, Bini M, Massarotti V, Mustarelli P, Ferrari S et al (2009) Cr and Ni doping of : cation distribution and functional properties. J Phys Chem C 113:19664–19671
29. Deng D, Kim MG, Lee JY, Cho J (2009) Green energy storage materials: nanostructured and Sn-based anodes for lithium-ion batteries. Energy Environ Sci 2:818–837
30. Shen L, Uchaker E, Zhang X, Cao G (2012) Hydrogenated nanowire arrays for high rate lithium ion batteries. Adv Mater 24:6502–6506
31. Naoi K, Naoi W, Aoyagi S, Miyamoto J, Kamino T (2012) New generation "nanohybrid supercapacitor". Acc Chem Res 46:1075–1083
32. Guler MO, Cevher O, Cetinkaya T, Tocoglu U, Akbulut H (2013) High capacity anode materials for Li-ion batteries. Energy Convers Manage 72:111–116
33. Zhang H, Li G, An L, Yan T, Gao X, Zhu H (2007) Electrochemical lithium storage of titanate and titania nanotubes and nanorods. J Phys Chem C 111:6143–6148
34. Chen JS, Lou XW (2009) Anatase nanosheet: an ideal host structure for fast and efficient lithium insertion/extraction. Electrochem Commun 11:2332–2335
35. Yang M, Lee Y, Xu B, Powers K, Meng YS (2012) TiO_2 flakes as anode materials for Li-ion-batteries. J Power Sources 207:166–172
36. Ohzuku T, Takeda S, Iwanaga M (1999) Solid-state redox potentials for (Me: 3d-transition metal) having spinel-framework structures: a series of 5 volt materials for advanced lithium-ion batteries. J Power Sources 81:90–94
37. Lou XW, Archer LA (2008) A general route to nonspherical anatase hollow colloids and magnetic multifunctional particles. Adv Mater 20:1853–1858

38. Dai Y, Cobley CM, Zeng J, Sun Y, Xia Y (2009) Synthesis of anatase nanocrystals with exposed {001} facets. Nano Lett 9:2455–2459
39. Han X, Kuang Q, Jin M, Xie Z, Zheng L (2009) Synthesis of titania nanosheets with a high percentage of exposed (001) facets and related photocatalytic properties. J Am Chem Soc 131:3152–3153
40. Wu B, Guo C, Zheng N, Xie Z, Stucky GD (2008) Nonaqueous production of nanostructured anatase with high-energy facets. J Am Chem Soc 130:17563–17567
41. Takai S, Kamata M, Fujine S, Yoneda K, Kanda K, Esaka T (1999) Diffusion coefficient measurement of lithium ion in sintered by means of neutron radiography. Solid State Ion 123:165–172
42. Kavan L, Fattakhova D, Krtil P (1999) Lithium insertion into mesoscopic and single-crystal (rutile) electrodes. J Electrochem Soc 146:1375–1379
43. Hu Y, Kienle L, Guo Y, Maier J (2006) High lithium electroactivity of nanometer-sized rutile. Adv Mater 18:1421–1426
44. Bokhimi X, Morales A, Aguilar M, Toledo-Antonio J, Pedraza F (2001) Local order in titania polymorphs. Int J Hydrog Energy 26:1279–1287
45. Reddy MA, Pralong V, Varadaraju U, Raveau B (2008) Crystallite size constraints on lithium insertion into brookite. Electrochem Solid-State Lett 11:A132–A134
46. Reddy MA, Kishore MS, Pralong V, Varadaraju U, Raveau B (2007) Lithium intercalation into nanocrystalline brookite. Electrochem Solid-State Lett 10:A29–A31
47. Lee D, Park J, Jin Choi K, Choi H, Kim D (2008) Preparation of brookite-type /carbon nanocomposite electrodes for application to Li ion batteries. Eur J Inorg Chem 2008:878–882
48. Idota Y, Kubota T, Matsufuji A, Maekawa Y, Miyasaka T (1997) Tin-based amorphous oxide: a high-capacity lithium-ion-storage material. Science 276:1395–1397
49. Courtney IA, Dahn J (1997) Electrochemical and in situ X-ray diffraction studies of the reaction of lithium with tin oxide composites. J Electrochem Soc 144:2045–2052
50. Park M, Wang G, Kang Y, Wexler D, Dou S, Liu H (2007) Preparation and electrochemical properties of nanowires for application in lithium-ion batteries. Angew Chem 119:764–767
51. Lou XW, Deng D, Lee JY, Archer LA (2008) Preparation of /carbon composite hollow spheres and their lithium storage properties. Chem Mater 20:6562–6566
52. Rong A, Gao X, Li G, Yan T, Zhu H et al (2006) Hydrothermal synthesis of as anode materials for Li-ion battery. J Phys Chem B 110:14754–14760
53. Liu W, Huang X, Wang Z, Li H, Chen L (1998) Studies of stannic oxide as an anode material for lithium-ion batteries. J Electrochem Soc 145:59–62
54. Winter M, Besenhard JO (1999) Electrochemical lithiation of tin and tin-based intermetallics and composites. Electrochim Acta 45:31–50
55. Lei D, Zhang M, Hao Q, Chen L, Li Q et al (2011) Morphology effect on the performances of nanorod arrays as anodes for Li-ion batteries. Mater Lett 65:1154–1156
56. Liu J, Li Y, Huang X, Ding R, Hu Y et al (2009) Direct growth of nanorod array electrodes for lithium-ion batteries. J Mater Chem 19:1859–1864
57. Wang J, Du N, Zhang H, Yu J, Yang D (2011) Large-scale synthesis of nanotube arrays as high-performance anode materials of Li-ion batteries. J Phys Chem C 115:11302–11305
58. Yuan L, Guo Z, Konstantinov K, Liu HK, Dou S (2006) Nano-structured spherical porous anodes for lithium-ion batteries. J Power Sources 159:345–348
59. Ko Y, Kang J, Park J, Lee S, Kim D (2009) Self-supported nanowire electrodes for high-power lithium-ion batteries. Nanotechnology 20:455701
60. Bresser D, Mueller F, Buchholz D, Paillard E, Passerini S (2014) Embedding tin nanoparticles in micron-sized disordered carbon for lithium-and sodium-ion anodes. Electrochim Acta 128:163–171
61. Dahn J, Mar R, Abouzeid A (2006) Combinatorial study of and alloy negative electrode materials for Li-ion batteries. J Electrochem Soc 153:A361–A365
62. Wen CJ, Huggins RA (1981) Thermodynamic study of the lithium-tin system. J Electrochem Soc 128:1181–1187

63. Besenhard J, Hess M, Komenda P (1990) Dimensionally stable Li-alloy electrodes for secondary batteries. Solid State Ion 40:525–529
64. Boukamp B, Lesh G, Huggins R (1981) All-solid lithium electrodes with mixed-conductor matrix. J Electrochem Soc 128:725–729
65. Courtney IA, McKinnon W, Dahn J (1999) On the aggregation of tin in SnO composite glasses caused by the reversible reaction with lithium. J Electrochem Soc 146:59–68
66. Lee KT, Jung YS, Oh SM (2003) Synthesis of tin-encapsulated spherical hollow carbon for anode material in lithium secondary batteries. J Am Chem Soc 125:5652–5653
67. Zhang W, Hu J, Guo Y, Zheng S, Zhong L et al (2008) Tin-nanoparticles encapsulated in elastic hollow carbon spheres for high-performance anode material in lithium-Ion batteries. Adv Mater 20:1160–1165
68. Jung YS, Lee KT, Ryu JH, Im D, Oh SM (2005) Sn-carbon core-shell powder for anode in lithium secondary batteries. J Electrochem Soc 152:A1452–A1457
69. Cui G, Hu Y, Zhi L, Wu D, Lieberwirth I et al (2007) A one-step approach towards carbon-encapsulated hollow tin nanoparticles and their application in lithium batteries. Small 3:2066–2069
70. Prem Kumar T, Ramesh R, Lin Y, Fey GT (2004) Tin-filled carbon nanotubes as insertion anode materials for lithium-ion batteries. Electrochem Commun 6:520–525
71. Deng D, Lee JY (2009) Reversible storage of lithium in a rambutan-like tin–carbon electrode. Angew Chem Int Ed 48:1660–1663
72. Hassoun J, Derrien G, Panero S, Scrosati B (2008) A nanostructured Sn–C composite lithium battery electrode with unique stability and high electrochemical performance. Adv Mater 20:3169–3175
73. Hassoun J, Lee K, Sun Y, Scrosati B (2011) An advanced lithium ion battery based on high performance electrode materials. J Am Chem Soc 133:3139–3143
74. Courtney IA, Dahn J (1997) Key factors controlling the reversibility of the reaction of lithium with SnO2 and Sn2 BPO 6 glass. J Electrochem Soc 144:2943–2948
75. Sakamoto J, Huang C, Surampudi S, Smart M, Wolfenstine J (1998) The effects of particle size on SnO electrode performance in lithium-ion cells. Mater Lett 33:327–329
76. Martos M, Morales J, Sanchez L (2002) Mechanochemical synthesis of anode materials for Li-ion batteries. J Mater Chem 12:2979–2984
77. Choi SH, Kim JS, Yoon YS (2004) Fabrication and characterization of composite anode thin film for lithium ion batteries. Electrochim Acta 50:547–552
78. Poizot P, Laruelle S, Grugeon S, Dupont L, Tarascon J (2000) Nano-sized transition-metal oxides as negative-electrode materials for lithium-ion batteries. Nature 407:496–499
79. Connor P, Irvine J (2001) Novel tin oxide spinel-based anodes for Li-ion batteries. J Power Sources 97:223–225
80. Becker SM, Scheuermann M, Sepelák V, Eichhöfer A, Chen D et al (2011) Electrochemical insertion of lithium in mechanochemically synthesized. Phys Chem Chem Phys 13:19624–19631
81. Ji L, Lin Z, Alcoutlabi M, Zhang X (2011) Recent developments in nanostructured anode materials for rechargeable lithium-ion batteries. Energy Environ Sci 4:2682–2699
82. Li X, Wang C (2013) Engineering nanostructured anodes via electrostatic spray deposition for high performance lithium ion battery application. J Mater Chem A 1:165–182
83. Poizot P, Laruelle S, Grugeon S, Tarascon J (2002) Rationalization of the low-potential reactivity of 3d-metal-based inorganic compounds toward li. J Electrochem Soc 149:A1212–A1217
84. Zhang L, Hu P, Zhao X, Tian R, Zou R, Xia D (2011) Controllable synthesis of core–shell Co@CoO nanocomposites with a superior performance as an anode material for lithium-ion batteries. J Mater Chem 21:18279–18283
85. Yang R, Wang Z, Liu J, Chen L (2004) Nano particles embedded in porous hard carbon spherules as anode material for li-ion batteries. Electrochem Solid-State Lett 7:A496–A499

86. Liu H, Bo S, Cui W, Li F, Wang C, Xia Y (2008) Nano-sized cobalt oxide/mesoporous carbon sphere composites as negative electrode material for lithium-ion batteries. Electrochim Acta 53:6497–6503
87. Lou XW, Deng D, Lee JY, Feng J, Archer LA (2008) Self-supported formation of needlelike Co3O4 nanotubes and their application as lithium-ion battery electrodes. Adv Mater 20:258–262
88. Binotto G, Larcher D, Prakash A, Herrera Urbina R, Hegde M, Tarascon J (2007) Synthesis, characterization, and li-electrochemical performance of highly porous powders. Chem Mater 19:3032–3040
89. Poizot P, Laruelle S, Grugeon S, Dupont L, Tarascon J (2001) Searching for new anode materials for the li-ion technology: time to deviate from the usual path. J Power Sources 97:235–239
90. Huang X, Tu J, Zhang B, Zhang C, Li Y et al (2006) Electrochemical properties of NiO–Ni nanocomposite as anode material for lithium ion batteries. J Power Sources 161:541–544
91. Hosono E, Fujihara S, Honma I, Zhou H (2006) The high power and high energy densities li ion storage device by nanocrystalline and mesoporous Ni/NiO covered structure. Electrochem Commun 8:284–288
92. Aravindan V, Suresh Kumar P, Sundaramurthy J, Ling WC, Ramakrishna S, Madhavi S (2013) Electrospun NiO nanofibers as high performance anode material for li-ion batteries. J Power Sources 227:284–290
93. Hu Y, Huang X, Wang K, Liu J, Jiang J et al (2010) Kirkendall-effect-based growth of dendrite-shaped CuO hollow micro/nanostructures for lithium-ion battery anodes. J Solid State Chem 183:662–667
94. Xiang J, Tu J, Yuan Y, Wang X, Huang X, Zeng Z (2009) Electrochemical investigation on nanoflower-like CuO/Ni composite film as anode for lithium ion batteries. Electrochim Acta 54:1160–1165
95. Morales J, Sanchez L, Martin F, Ramos-Barrado J, Sanchez M (2005) Use of low-temperature nanostructured CuO thin films deposited by spray-pyrolysis in lithium cells. Thin Solid Films 474:133–140
96. Xiang J, Tu J, Huang X, Yang Y (2008) A comparison of anodically grown CuO nanotube film and film as anodes for lithium ion batteries. J Solid State Electrochem 12:941–945
97. Sahay R, Suresh Kumar P, Aravindan V, Sundaramurthy J, Chui Ling W et al (2012) High aspect ratio electrospun CuO nanofibers as anode material for lithium-ion batteries with superior cycleability. J Phys Chem C 116:18087–18092
98. Aragón M, Pérez-Vicente C, Tirado J (2007) Submicronic particles of manganese carbonate prepared in reverse micelles: a new electrode material for lithium-ion batteries. Electrochem Commun 9:1744–1748
99. Ji L, Medford AJ, Zhang X (2009) Porous carbon nanofibers loaded with manganese oxide particles: formation mechanism and electrochemical performance as energy-storage materials. J Mater Chem 19:5593–5601
100. Zhong K, Xia X, Zhang B, Li H, Wang Z, Chen L (2010) MnO powder as anode active materials for lithium ion batteries. J Power Sources 195:3300–3308
101. Yu X, He Y, Sun J, Tang K, Li H et al (2009) Nanocrystalline MnO thin film anode for lithium ion batteries with low overpotential. Electrochem Commun 11:791–794
102. Gao M, Zhou P, Wang P, Wang J, Liang C et al (2013) FeO/C anode materials of high capacity and cycle stability for lithium-ion batteries synthesized by carbothermal reduction. J Alloys Compd 565:97–103
103. Reddy M, Subba Rao G, Chowdari B (2013) Metal oxides and oxysalts as anode materials for li ion batteries. Chem Rev 113:5364–5457
104. Huang G, Xu S, Lu S, Li L, Sun H (2014) Micro-/Nano-structured anode with enhanced rate capability for lithium-ion batteries. ACS Appl Mater Interfaces 6(10):7236–7243
105. Taberna P, Mitra S, Poizot P, Simon P, Tarascon J (2006) High rate capabilities -based Cu nano-architectured electrodes for lithium-ion battery applications. Nat Mater 5:567–573

106. He C, Wu S, Zhao N, Shi C, Liu E, Li J (2013) Carbon-encapsulated nanoparticles as a high-rate lithium ion battery anode material. ACS Nano 7:4459–4469
107. Pasero D, Reeves N, West A (2005) Co-doped: a possible anode material for lithium batteries. J Power Sources 141:156–158
108. Dey A (1971) Electrochemical alloying of lithium in organic electrolytes. J Electrochem Soc 118:1547–1549
109. Wen CJ, Huggins RA (1981) Chemical diffusion in intermediate phases in the lithium-silicon system. J Solid State Chem 37:271–278
110. Saint J, Morcrette M, Larcher D, Laffont L, Beattie S et al (2007) Towards a fundamental understanding of the improved electrochemical performance of silicon–carbon composites. Adv Funct Mater 17:1765–1774
111. Park C, Kim J, Kim H, Sohn H (2010) Li-alloy based anode materials for li secondary batteries. Chem Soc Rev 39:3115–3141
112. Wang JW, He Y, Fan F, Liu XH, Xia S et al (2013) Two-phase electrochemical lithiation in amorphous silicon. Nano Lett 13:709–715
113. Kim H, Lee E, Sun Y (2014) Recent advances in the Si-based nanocomposite materials as high capacity anode materials for lithium ion batteries. Mater Today 17(6):285–297
114. Chan CK, Peng H, Liu G, McIlwrath K, Zhang XF et al (2007) High-performance lithium battery anodes using silicon nanowires. Nat Nanotechnol 3:31–35
115. Kang K, Lee H, Han D, Kim G, Lee D et al (2010) Maximum li storage in Si nanowires for the high capacity three-dimensional Li-ion battery. Appl Phys Lett 96:053110
116. Yang Y, McDowell MT, Jackson A, Cha JJ, Hong SS, Cui Y (2010) New nanostructured/silicon rechargeable battery with high specific energy. Nano Lett 10:1486–1491
117. Ding N, Xu J, Yao Y, Wegner G, Lieberwirth I, Chen C (2009) Improvement of cyclability of si as anode for Li-ion batteries. J Power Sources 192:644–651
118. Chan CK, Patel RN, O'Connell MJ, Korgel BA, Cui Y (2010) Solution-grown silicon nanowires for lithium-ion battery anodes. ACS Nano 4:1443–1450
119. Magasinski A, Dixon P, Hertzberg B, Kvit A, Ayala J, Yushin G (2010) High-performance lithium-ion anodes using a hierarchical bottom-up approach. Nat Mater 9:353–358
120. Cho J, Picraux ST (2013) Enhanced lithium ion battery cycling of silicon nanowire anodes by template growth to eliminate silicon underlayer islands. Nano Lett 13:5740–5747
121. Quiroga-González E, Carstensen J, Föll H (2013) Optimal conditions for fast charging and long cycling stability of silicon microwire anodes for lithium ion batteries, and comparison with the performance of other si anode concepts. Energies 6:5145–5156
122. Quiroga-González E, Carstensen J, Föll H (2013) Good cycling performance of high-density arrays of Si microwires as anodes for li ion batteries. Electrochim Acta 101:93–98
123. Lee J, Lee KT, Cho J, Kim J, Choi N, Park S (2012) Chemical-assisted thermal disproportionation of porous silicon monoxide into silicon-based multicomponent systems. Angew Chem 124:2821–2825
124. Goward G, Taylor N, Souza D, Nazar L (2001) The true crystal structure of Li17M4 (, Sn, Pb)–revised from. J Alloys Compd 329:82–91
125. Tamura N, Ohshita R, Fujimoto M, Fujitani S, Kamino M, Yonezu I (2002) Study on the anode behavior of Sn and Sn–Cu alloy thin-film electrodes. J Power Sources 107:48–55
126. Tamura N, Ohshita R, Fujimoto M, Kamino M, Fujitani S (2003) Advanced structures in electrodeposited tin base negative electrodes for lithium secondary batteries. J Electrochem Soc 150:A679–A683
127. Luo B, Wang B, Liang M, Ning J, Li X, Zhi L (2012) Reduced graphene oxide-mediated growth of uniform tin-core/carbon-sheath coaxial nanocables with enhanced lithium ion storage properties. Adv Mater 24:1405–1409
128. Wu FD, Wu M, Wang Y (2011) Antimony-doped tin oxide nanotubes for high capacity lithium storage. Electrochem Commun 13:433–436
129. Shafiei M, Alpas AT (2011) Electrochemical performance of a tin-coated carbon fibre electrode for rechargeable lithium-ion batteries. J Power Sources 196:7771–7778

Chapter 4
Sn-Based Alloy Anode Materials for Lithium-Ion Batteries: Preparation, Multi-scale Structure, and Performance

Renzong Hu and Min Zhu

4.1 Introduction

The explosive demand for upcoming electronic vehicles and hybrid electronic vehicles (EV/HEV) as well as smart grids requires the large and reliable energy storage devices, outputting high capacity, superior power, and long life span. Among different kinds of candidates, Li-ion batteries have attracted most attention and are widely used as small power sources in different fields owing to its high capacity. However, the commonly used graphite anodes have been restricted in the capacity of the cell per unit weight (theoretically 372mAh/g vs. LiC_6) and/or per unit volume due to its low density, as well as the battery safety due to the low lithiation potential (<0.2 V vs. Li) of graphite [1]. From these points of view, developing new advanced anode materials, with higher energy density, long life, and improved safety, is of great importance for the next-generation high-performance Li-ion batteries.

It has been known that Li could form intermetallic phases (Li_xM) with numerous metals M (M = Sn, Mg, Al, Sb, Ge, Si, etc.) at room temperature, which offer much higher Li storage capacities than lithiated graphite (LiC_6) [2]. Figure 4.1 compares the specific capacity (mAh/g) and charge density (mAh/cm^3) as well as lithiation potential (V vs. Li) of different metal anodes with the commercial LMO ($LiCoO_2$, $LiFePO_4$, etc.) cathodes, which indicates the much higher energy density of the Li-alloy anodes. In addition, it also shows that many transition metal oxides (TMO), such as Fe_2O_3 and Co_3O_4, can also store Li by conversion reaction and deliver high capacity of more than 600mAh/g. And, thus, replacing graphite anodes

R. Hu • M. Zhu (✉)
School of Materials Science and Engineering, Guangdong Provincial Key Laboratory of Advanced Energy Storage Materials, South China University of Technology, Guangzhou 510640, P.R. China
e-mail: memzhu@scut.edu.cn

K.I. Ozoemena, S. Chen (eds.), *Nanomaterials in Advanced Batteries and Supercapacitors*, Nanostructure Science and Technology,
DOI 10.1007/978-3-319-26082-2_4

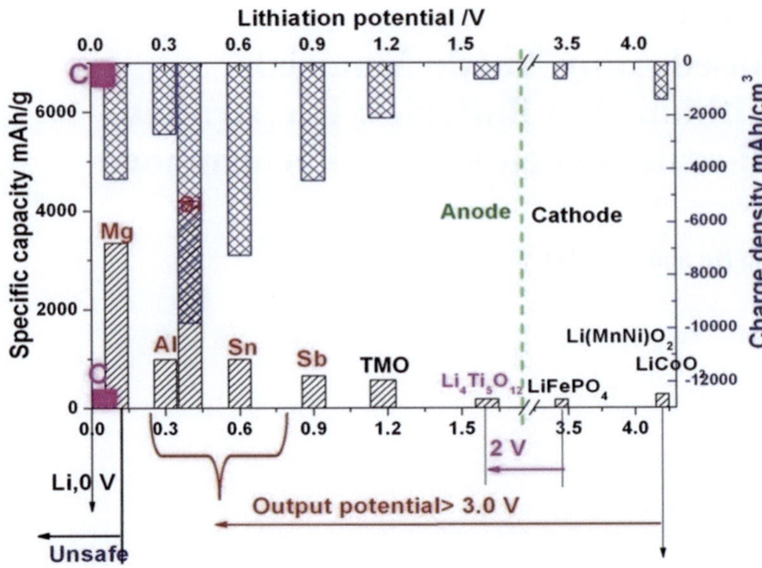

Fig. 4.1 Specific capacity, lithiation potential of different kinds of electrode materials

with metal-based anodes is an attractive route to achieve higher performance in next generation of Li-ion batteries [3–5].

Among those potential anode materials mentioned above, significant efforts have been devoted to Sn-based materials in recent years [6–8]. Sn-based anodes not only have a high theoretical capacity but also have moderate operation potential to avert the safety concern of Li deposition and co-insertion of solvents to active materials as is the case with graphite anodes. It also has excellent electronic conductivity greater than Si, Ge, TMO, and so on. These advantages make Sn one of the most promising alternative anode materials [9] and have attracted tremendous research. In 2005, the commercial Sn–Co–C anode was successfully launched by Sony's in "Nexelion" batteries [10].

However, the capacity and cycle performance of the Sn-based anodes should be further improved to meet wide application in Li-ion batteries. Thus, the quest for high-capacity materials, microstructure design, and new fabrication methods has been widely concerned. Great effort has been made in the aspects of nanostructuring, alloying, and so on [5, 11]. Many works have focused on designing and preparing multi-scaled structures in various Sn-based multiphase materials (Sn–M alloys, SnO_2, Sn–C, etc.), which revealed the critical influences of multi-scale structure, such as multilayers, core–shell, porous, and nanocomposite, on Li storage behaviors especially cycle performance of Sn-based anodes. In this chapter, we review the research advances mainly concerning Sn-based alloy and Sn–C composite anodes made in the last decade. Although SnO_2 is also an important type among those Sn-based anode materials, we do not review herein the research progress of it because the lithiation and delithiation mechanism of SnO_2 was different from those of the Sn-based metal and intermetallics [12, 13].

4.2 Structure Variation and Properties of Li–Sn Alloys

It has been proved by both ab initio calculations and experiment work that, at operating potential range of 1.0–0V vs. Li, Sn could electrochemically react with Li to form different Li-rich alloy phases, such as Li_7Sn_2, Li_7Sn_3, $Li_{13}Sn_5$, and $Li_{22}Sn_5$, with definite potential plateau in discharge–charge curves for each two-phase reaction [14], as shown in Fig. 4.2. However, the Li insertion/extraction potentials of the Li_xSn were found to be dependent on the microstructure of electrode as well as discharge–charge rates. Furthermore, voltage hysteresis and capacity gap would exist between discharge and recharge. These should be due to volume stress/strain induced by the Li–Sn compound formations because the hosting matrix Sn has dramatical structure/volume variation during the Li-alloying process [15].

As shown in Fig. 4.3a, β-Sn has a tetragonal structure (space group: $I4_1/amd$, no. 141), with lattice constants of $a = 0.5831$ nm and $c = 0.3182$ nm. When Li insert

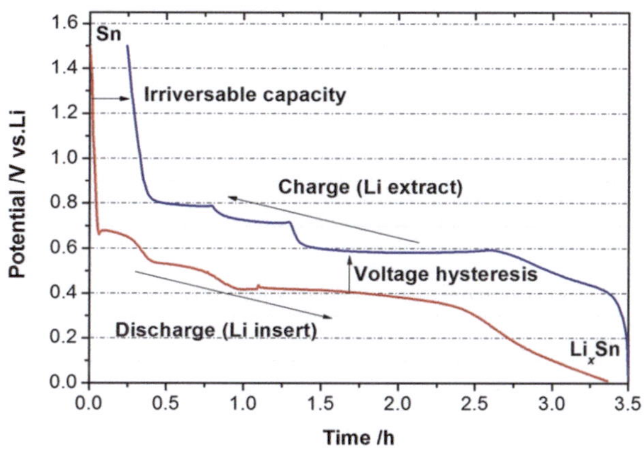

Fig. 4.2 Typical discharge–charge curves of the pure Sn electrode

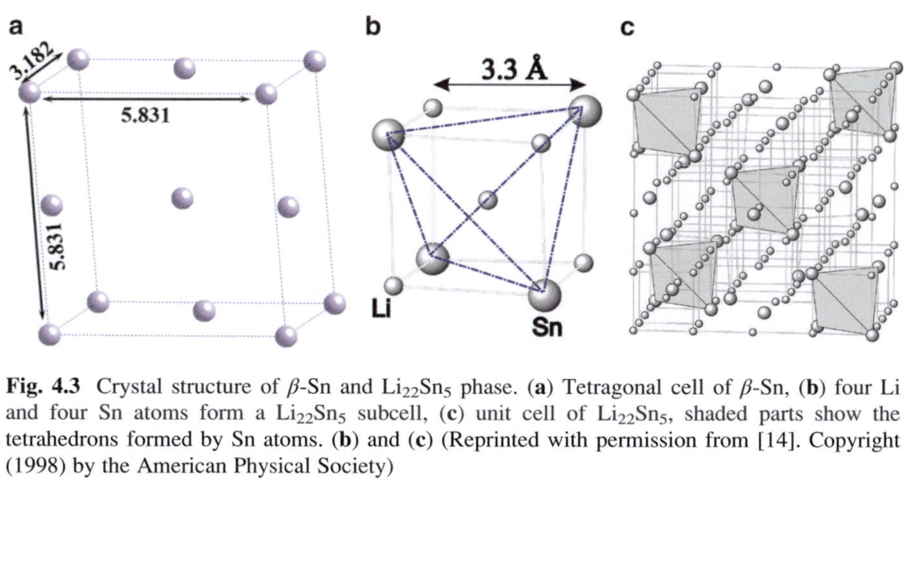

Fig. 4.3 Crystal structure of β-Sn and $Li_{22}Sn_5$ phase. (**a**) Tetragonal cell of β-Sn, (**b**) four Li and four Sn atoms form a $Li_{22}Sn_5$ subcell, (**c**) unit cell of $Li_{22}Sn_5$, shaded parts show the tetrahedrons formed by Sn atoms. (**b**) and (**c**) (Reprinted with permission from [14]. Copyright (1998) by the American Physical Society)

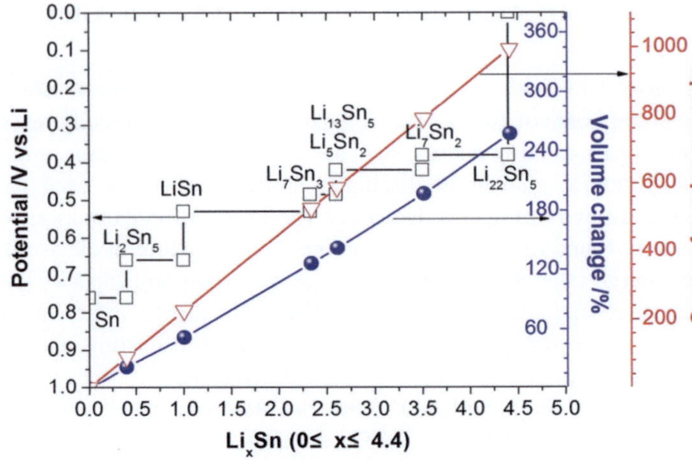

Fig. 4.4 The theoretical formation potentials, specific capacities, and volume changes of Li_xSn phases (Reprinted with permission from Ref. [8]. Copyright (2012) by Springer)

into Sn to reach a stoichiometric ratio of $Li_{22}Sn_5$, the unit cell of Sn will convert to be a *bcc* subcell with a lattice constant $a = 0.3296$ nm shown in Fig. 4.3b. The Sn and Li atoms are arranged on this *bcc* network to give a $6 \times 6 \times 6$ superstructure with $a = 1.978$ nm. All the Sn atoms in $Li_{22}Sn_5$ occupy the corner of the tetrahedron, while Li atom stores at the center of the Sn tetrahedron as shown in Fig. 4.3c. The theoretical specific electrochemical capacity of pure Sn anode is 994 mAh/g (vs. $Li_{22}Sn_5$), and its packing density is 75.46 mol/L which is close to that of Li metal (76.36 mol/L). Thus, the specific volume capacity of Sn anode amounts to 7262 mAh/cm^3, which is almost nine times higher than that of graphite (Fig. 4.1).

However, as shown in Fig. 4.4, Li storage in Sn by the formation of various Li–Sn alloys is accompanied by large volume changes, which is up to 256 % in $Li_{22}Sn_5$. Furthermore, each Li_xSn ($0 \leq x \leq 4.4$) intermetallic compound forms at a definite potential, suggesting that the volume of these Li_xSn phases would be mutation rather than gradual change and consequently induce large internal stress. In addition, by first principle calculations, Shenoy et al. [16] demonstrated that the bulk, shear, and Young's moduli of isotropic Li_xSn alloys decreased monotonically with increasing Li concentration, resulting in elastic softening. This is very different from the intercalation compounds of graphite, which shows small volume changes of less than 10 %.

Owing to the large volume expansion associated with the $Sn \rightarrow Li_xSn$ transformation together with the brittleness of Li–Sn intermetallic compounds, the Sn-based anodes usually suffer from inferior cyclability. This is caused by easy pulverization by large Li-driven volume variation during the charge and discharge reactions, which leads to the loss of electric contact between particles and between particles and current collectors. In addition, the increasing of exposed fresh surface increases the formation of solid electrolyte interface film, which also causes capacity loss [17]. These materials' failure leads to the degradation of the anode. This problem has been the major challenge in the practical application of Sn-based anodes. Great efforts are being made worldwide to find solutions to overcome this limitation.

4.3 Pure Sn Anodes and Electrode–Electrolyte Interface Properties

4.3.1 Sn Particle/Rod/Wire Anodes

It has been proved that large absolute volume changes could be avoided when the size of the metallic host particles was kept small. Besenhard et al. [18] reported the importance of morphology effects by showing the cycling data of coarse (about 5 μm) and fine particulate (less than 1 μm) Sn. They found that, in the case of coarse Sn, the formation of large cracks and the peeling of Sn from the substrate was much sever and appeared much earlier during cycling than in the case of fine tin, which resulted in poorer cyclability in coarse Sn anode. Similar results could be also found in Sn nanorod electrode. By an AAO template-assisted growth method, Fu et al. [19] fabricated an electrode with Sn nanorods with diameter of about 250 nm uniformly distributed on copper substrate. The Sn nanorod electrode exhibited better capacity retention and rate capability compared to the Sn planar electrode. However, the capacity of the Sn nanorod electrode also began to decay after ten cycles.

It should be pointed out that, due to its low melt point being 231.7 °C, the recrystallization temperature of Sn is below 25 °C and the nanosized Sn has high activity to grow up or coalesce even at room temperature. This is probably the reason that there was very little research reported to anode behavior of pure Sn nanoparticles and nanowires [20, 21]. Thus, most of the Sn nanoparticles and nanowires were combined with different type of carbon (C) to form Sn–C nanocomposite electrodes which will be discussed later in Sect. 5. In striking contrast, pure Sn thin films, which are two-dimensional free-binder electrodes, have been widely studied [8].

4.3.2 Pure Sn Thin Film Anodes

A lot of attempts have been made on the preparation and electrochemical performances of pure Sn thin film anodes, and those films are generally Sn thin layer coating on current collector. Most of the pure Sn thin film anodes have been prepared by electrodeposition [17, 22–28], magnetron sputtering [29–31], electron beam deposition (EBD) [32], or vacuum evaporation [33]. Morimoto et al. [22] prepared a Sn thin film anode with a rough surface on a Cu foil substrate. It had a high initial discharge capacity of 860 mAh/g but a poor cyclability, with a relatively large initial irreversible capacity loss, amounting to about 23 % of the total capacity. In contrast, the Sn thin film with grain size of about 1 μm, which was prepared by a pulse electrodeposition, could yield a higher initial discharge–charge efficiency of 93.0 % [24].

In order to understand the origin of the large irreversible capacity in the electroplated Sn film anodes, the morphology changes of the Sn film surface during lithiation and delithiation were investigated by using in situ AFM. It was revealed that surface roughening by lithium insertion and extraction occurred to a significant extent in the first and second cycles. Inaba et al. [17] considered that this surface roughening destroyed the existing surface film and exposed a large area of active surface and therefore was the reason for the appearance of large irreversible capacities in the initial cycles. Also with the aid of in situ AFM observations, Beaulieu et al. [29] confirmed that, for sputtered crystalline Sn thin film anodes, decomposition of the surface film and consumption of the electrolyte resulted from pulverization of the electrode due to inhomogeneous volume expansion of Sn and that this was a key factor for the poor cycle performance of the electrodes. The above results indicated that the interface properties of electrodes have a great impact on the capacity loss of Sn thin film anodes.

4.3.3 Electrode–Electrolyte Interface of Sn Film Anodes

Due to absence of conductive and binder additives (C, PVDF, etc.), in pure Sn film electrodes, the electrode–electrolyte interface is actually a Sn–electrolyte interface. The interface properties of Sn electrode in Li cells are associated with the formation processes and features of solid electrolyte interphase (SEI) films at the Sn–electrolyte interfaces, as well as Sn corrosion or dissolution which however was neglected [34]. Especially, as illustrated in Fig. 4.5, due to cracks generation and morphology

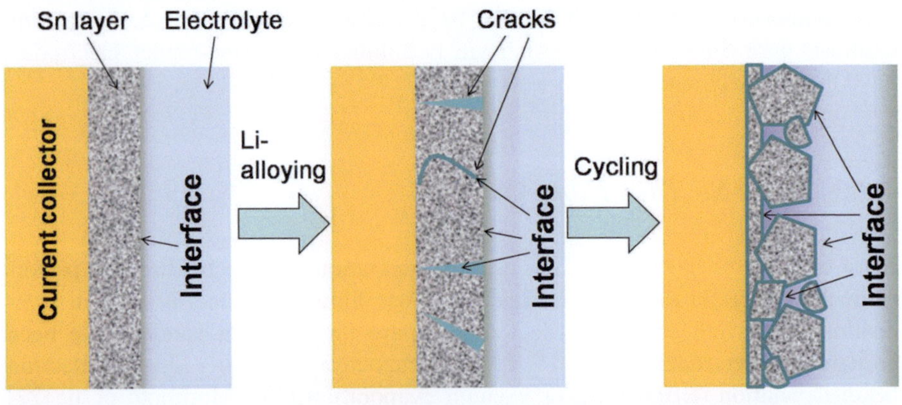

Fig. 4.5 Sn–electrolyte interface variation of Sn film electrode during lithiation/delithiation cycling

change of the Sn layer, the interface/interphases between Sn and electrolyte should be changed during lithiation and delithiation cycling.

And thus, the formation of SEI films is mainly dependent on the properties and surface morphology of the electrode materials, the components of the electrolyte, and the operating conditions of the battery. The salient properties of SEI films include their stability, impedance, Li^+-diffusion kinetics, and ability to prevent decomposition of the electrolyte [35]. These features determine the lithiation/delithiation dynamics, the stability of the electrode–electrolyte interfaces, and the cyclability and self-discharge of a battery. It was initially assumed that SEI films were only formed on graphite materials. However, it was subsequently found that SEI films are common in Sn-based and other alloy-based anode materials and that they have a great influence on the performances of electrodes.

By means of impedance spectroscopy, Hassoun et al. [23] investigated the characteristics of an SEI film formed on the surface of electrodeposited Sn film anodes cycled in Li/Li-ion cells. They found that the SEI film was formed by electrolyte decomposition catalyzed by the Sn surface and that its characteristics depended upon the cycling rate. When the cycling rate was high, there was limited SEI film formation on the surface of the Sn film, and thus, a high reversible capacity was retained. The completed cell, with Sn film anode and $LiNi_{0.5}Mn_{1.5}O_4$ cathode, further confirmed that the formation and decomposition of SEI layers on the Sn film anodes did indeed influence the responses of the batteries, including their cyclability and rate capability. Furthermore, in an amorphous Sn film anode, Chiu et al. [30] found that a relatively thick SEI film formed on the surface of the electrode upon cycling in the range 2.5–0.05 V vs. Li/Li^+, which was effective in suppressing the pulverization of nanosized Sn particles. In contrast, formation of the SEI film was prevented when the surface of the Sn film anode was covered by a smooth SnO_2 layer, which also enhanced the cycle performance of the amorphous Sn thin film anode. These results suggested that the two cases, that is, with a thick SEI film and without SEI formation, would help to improve the cycle performance of Sn film anodes. However, it is also needed to be pointed out that conductivity and Li^+ diffusion kinetics should be worse if the thickness of SEI layer is too high.

Nevertheless, it was generally found that the pure Sn film electrodes were hard to full lithiation even discharge to 0 V vs. Li/Li^+, leading to lower initial capacity than the theoretical capacity of Sn. This might be ascribed to the microstructure and Li^+-diffusion kinetics in the pure Sn film anodes, as well as the properties of the Sn–electrolyte interfaces. Moreover, the volume change of Sn during Li insertion is very large. Thus, even though pure Sn thin film anodes obtained by different fabrication methods possess different microstructures, such as microsized crystal, nanosized crystal, and amorphous, porous, and multilayered structure, none of these could effectively relieve the volume expansion of the pure Sn film anodes. Thus, as shown in Fig. 4.6, pulverization of the active Sn materials was unavoidable during discharge–charge cycling, which resulted in pulverized Sn fragments becoming electrically isolated from the current collectors and consequent capacity loss [32]. To overcome those problems, enhancing the structure stability of the host active materials was proposed, while more improvement had been achieved by using

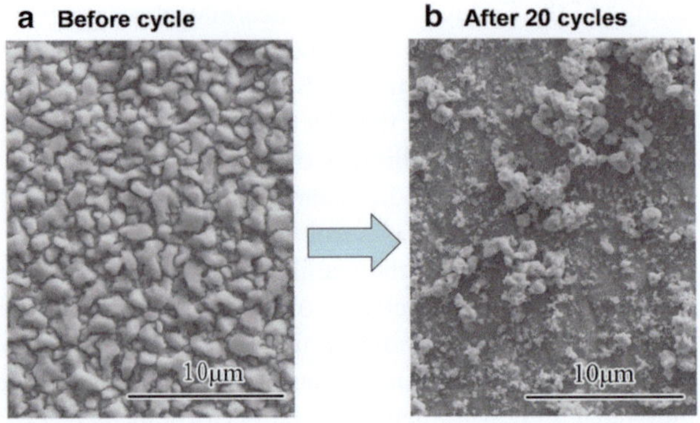

Fig. 4.6 SEM images of electron beam deposited Sn thin film on Cu substrate: (**a**) as-deposited, (**b**) after discharge–charge 20 cycles (Reprinted from Ref. [32] with permission for Elsevier)

multiple phases to accommodate the excessive volume changes of Li–Sn alloying. Accordingly, metallic component, oxide, and carbonaceous materials were selected as the added-phase matrix to support the active Sn.

4.4 Sn–M-Based Alloy Anodes

With respect to the multiphase or multi-element Sn-based systems, Sn–M (M refers to metals) intermetallic compounds and composites attracted much attention in early time. Where M is an active or inactive (not react with Li) metallic component/alloy, which serves as a matrix buffering the volume change of the active Sn phase during Li insertion/extraction, leading to less cracking in the Sn–M anodes and hence improved cycle performance.

4.4.1 *Structure Transition for Lithiation of Cu_6Sn_5*

Among Sn-based alloy anodes, Sn–Cu alloys had been widely studied and regarded as the alternative anode materials to graphite in Li-ion batteries. Cu–Sn alloys were easy to synthesis and the composition for Sn–Cu-based anode generally designed to be corresponding to Cu_6Sn_5 and Cu_3Sn intermetallic compounds. Nevertheless, Cu_3Sn was less concerned as Li-ion anode due to its lower reversible capacity (381mAh/g) [36]. Cu_6Sn_5 has a NiAs-type structure (space group C2/c, $a = 1.1022$ nm, $b = 0.7282$ nm, $c = 0.9827$ nm, $\beta = 98.84°$), in which some of the Cu atoms occupy interstitial bipyramidal sites. Especially, from the [100] projection of Cu_6Sn_5 shown in Fig. 4.7a, the Sn atom layers were seen to be restricted among the Cu atom layers, suggesting the good buffering effect of inactive Cu matrix for active Sn. The Cu_6Sn_5 was evaluated to have a theoretical specific

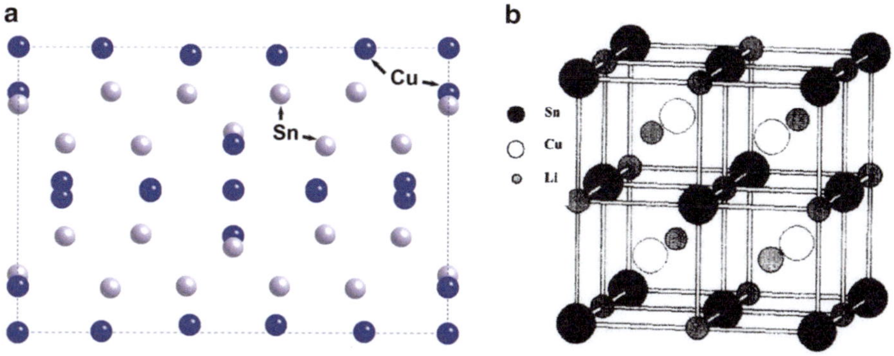

Fig. 4.7 (**a**) Crystal structure of Cu_6Sn_5 phase viewed from [100] projection, (**b**) unit cell of Li_2CuSn (Reproduced from [38] with permission from Elsevier Science)

capacity of 604.8 mAh/g, corresponding to formation of Cu and $Li_{4.4}Sn$ mainly at lithiation potential of 0.4 V vs. Li [37].

The lithiation/delithiation mechanism of Cu_6Sn_5 had been carefully investigated by using in situ methods such as XRD, TEM, etc. It has been confirmed that the lithiation of Cu_6Sn_5 involved a two-step reaction. Firstly, Li^+ inserts into the Cu_6Sn_5 structure to form a topotactic Li_2CuSn phase. As shown in Fig. 4.7b, in the cubic Li_2CuSn structure (space group F-43 m), the Sn atoms and one-half of the Li atoms form two interlinked face-centered cubic arrays. The Cu atoms and remaining Li atoms occupy interstitial sites located at the center of cubes defined by 4Sn and 4Li atoms of the arrays. As Li^+ further insert to the Li_2CuSn, being the second step of reaction, it was decomposed to Cu- and Li-rich phase $Li_{3.5}Sn$ rather than $Li_{4.4}Sn$ [39]. While Li is extracting, the two-step reactions could reverse to restore Cu_6Sn_5. The $Cu_6Sn_5 \rightarrow Li_2CuSn$ transformation leads to a volume change of 140 %, which was much lower than that of $Sn \rightarrow Li_{22}Sn_5$. Furthermore, the in situ formation of Cu phase could well serve as matrix to protect the Sn. Thus, lithiation/delithiation reactions of Cu_6Sn_5 presented good reversibility.

4.4.2 Sn–Cu Alloy Powder Anodes

Many Sn–Cu alloy powders were prepared by mechanical alloying (MA) or ball milling, which had high yield of sample; however the materials' morphology is usually hard to control. When Thackeray et al. [40] first investigated the Cu_6Sn_5 anodes, they prepared them by reaction of Cu and Sn metals under argon at 400°C and then milled to a powder (~400 mesh).

Xia et al. [36] compared the electrochemical performance of Cu_6Sn_5 powders obtained by ball milling, gas atomizing, and melt-spinning techniques. They found that the performances of the compounds were critically dependent on their morphology. The milled Cu_6Sn_5 alloy, consisted of < 1 μm thick flake powders, delivered the best battery performance with a reversible capacity of 200 mAh/g over 50 cycles. Assisted by annealing, they also prepared a series of Ni doped $Ni_xCu_{6-x}Sn5$ ($x=0$, 0.5, 1, 2, 4) alloys, which showed the similar structure with that of Cu_6Sn_5 and Ni_3Sn_2 [41]. With proper amount of Ni doping, the alloy, $Ni_2Cu_4Sn_5$, showed better cycle performance but reversible capacity was low and in the order of 200 mAh/g. In contrast, a higher reversible capacity of 400 mAh/g could be delivered in a Cu_6Sn_5 alloy milled for 110 h, with grain size of 5–10 nm [42]. Dunlap et al. [42] obtained a $Sn_{27}Cu_{33}C_{40}$ composite with ultrafine (*ca.*16 nm) Cu_6Sn_5 dispersed in amorphous carbon also by milling for many hours, and a reversible capacity of about 400 mAh/g was yielded. However, its cyclability was inferior due to the aggregation of the nanosized active materials.

In addition to MA, nanostructured Cu_6Sn_5 alloy and/or their carbon hybrids, which were characterized by multiple-scale, core–shell and porous structure, and so on, were usually fabricated by wet chemical methods [43–46]. A spherical shape Cu–Sn alloy powder with fine size (<0.8 μm) was directly prepared by spray pyrolysis [47] in which Cu and Sn components were well dispersed inside the submicron-sized alloy powders, and delivered capacities ranged from 485 to 313 mAh/g among initial 20 cycles. With a carbothermal reduction using polymer-coated CuO and SnO_2 as precursors, Xia et al.[48] synthesized a core–shell, carbon-coated, nanoscale Cu_6Sn_5 alloy. It delivered a higher reversible capacity of 420mAh/g with capacity retention of 80 % after 50 cycles. The improvement in the cycling ability could be attributed to the fact that the carbon shell prevents aggregation and pulverization of nanosized Cu_6Sn_5 particles during cycling. Nevertheless, the capacity of those Cu_6Sn_5 intermetallic anodes is not very high because the amount of heavy Cu is rather high. However, using transmetalation reactions [49], a core–shell-structured Sn@Cu nanoalloy, with Sn metal cores 8.65 nm in size and outer Cu layer 1.35 nm in size, was synthesized. This alloy demonstrated initial reversible capacities of 720 mAh/g, which was much higher than those of the Cu_6Sn_5 intermetallic anode and Cu_6Sn_5/C hybrids, and good cyclability. This result suggested that less Cu composition combined with careful nanostructure design could lead to superior performance in Sn–Cu alloy powder anodes.

Figure 4.8 summarized the cyclic performance of part of Sn–Cu anode reported in literature. It could be seen that most of the Sn–Cu alloy powders delivered large amount of discharge capacities in the initial cycles, which were higher than the theoretical capacity of Cu_6Sn_5 intermetallic. However, the reversible capacities of the Cu_6Sn_5 generally decreased as increaseing the discharge-charge cycles among the full potential ranges (2.0-0 V). In terms of alloy composition, excess Cu (and other inactive elements) ratio relative to Cu_6Sn_5 shows improved cyclability at the expense of capacity, whereas excess of Sn (and other active elements) ratio results in poor cyclability [36, 56]. In addition, great irreversible capacities were usually caused by the formation of a SEI layer on the surface of nanoscale alloy powders. The contact resistance between nanometer particles results in the poor electric conductivity, and the match of particle size and conductive agent content has a great impact on the electrochemical performance of the nanometer Sn–Cu alloy powder anodes [57]. In contrast, the binder-free Sn–Cu thin film anodes could avoid the above drawbacks, which had also been widely studies.

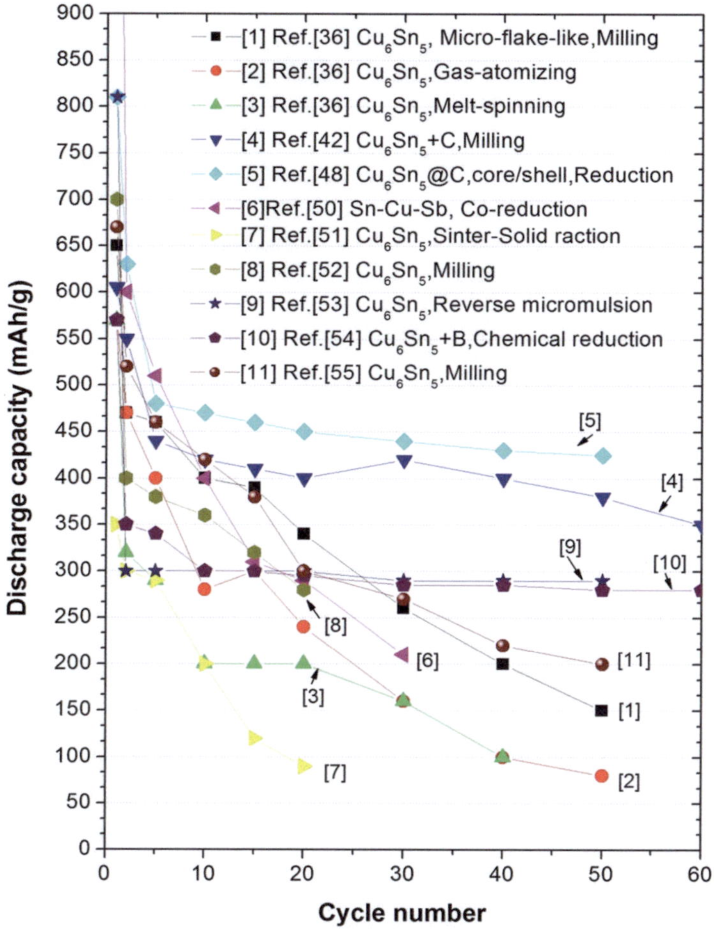

Fig. 4.8 Reversible capacities of some selected Sn–Cu alloy powder anodes versus cycle number among potential range of 0–1.5 V vs. Li/Li^+. Figure constructed from data found in the following Refs [36, 42, 48, 50–55]

4.4.3 *Sn–Cu Thin Film Anodes*

Most of Sn–Cu thin film anodes were prepared by electrodeposition (electroplate). Beattie et al. [56, 58] fabricated Sn–Cu alloy films with varying Cu–Sn stoichiometries by single-bath pulsed electrodeposition. It was nature to found that, as the Cu content in the film increased, specific capacity was sacrificed in favor of capacity retention. For example, a film with a Cu–Sn atomic ratio of 3.83 yielded a specific capacity near 200 mAh/g between 0 and 1.6 V and retained 80 % capacity after 40 cycles. Pu et al. [26] prepared a Cu/Sn/Cu multilayer film anode by deposition of Sn on a Cu foil substrate and then plating the surface of the deposited Sn with a

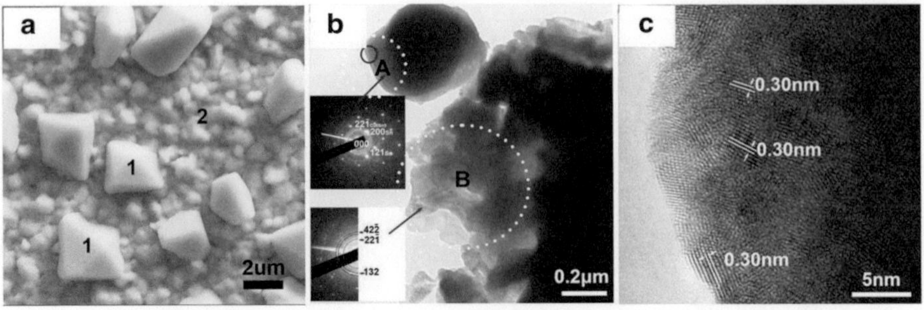

Fig. 4.9 (**a**) SEM image of a Sn/Cu_6Sn_5 composite thin film; (**b**) TEM image; (**c**) HRTEM image of the area encircled by the *solid line* in zone A shown in (**b**) (Reprinted from Ref. [8] with permission from Elsevier)

protective Cu layer. The Cu and Sn in the multilayer film were partially alloyed to form Cu_6Sn_5 and Cu_3Sn after annealing. In this case, the annealed Cu/Sn/Cu multilayer film anode was not cracked after 50 cycles, and it displayed an initial coulomb efficiency reaching 95 % and good cyclability. These results were similar to those reported by Morimoto et al. [22] and Tamura et al. [25].

We directly deposited a Sn/Cu_6Sn_5 composite thin film on a Cu foil substrate by EBD [59], which had a structure of polyhedral microsized Sn particles uniformly dispersed in the Cu_6Sn_5 matrix as shown in Fig. 4.9a. The microsized Sn particles (2–5 μm) were single crystals, while the Cu_6Sn_5 matrix was nanocrystalline, and there was also a nanocrystalline Cu_6Sn_5 layer with thickness of 30 nm covering the surface of these microsized Sn particles, as indicated in Fig. 4.9b and c. The composite thin film layer had a thickness of about 1 μm and presented strong bonding with the Cu substrate. This unique Sn/Cu_6Sn_5 composite thin film anode showed a higher Li^+-diffusion rate and discharge capacity, as well as better cyclability, than those of the Sn–Cu and Cu_3Sn/Cu_6Sn_5 thin film also obtained by EBD [32]. A reversible capacity of more than 400 mAh/g, with high initial coulombic efficiency, was maintained for up to 70 cycles when cycling in potential window of 0.1–1.25 V vs. Li.

In order to further enhance the rapid transport of Li^+ in Cu_6Sn_5 alloy thin film anodes, Liu et al. [60] fabricated Cu_6Sn_5 alloy films with three-dimensional (3D) foam structure by electrochemical deposition process. This 3D Cu_6Sn_5 film anode delivered a reversible capacity of about 400 mAh/g up to 30 cycles between 0 and 1.5 V at a current rate of 0.5 mA/cm^2 (1 C). Furthermore, this anode exhibited superior rate capability, which may be attributed primarily to its unique porous structure and the large surface area for rapid mass transport and rapid surface reactions. Using the electrodeposition process in conjunction with porous Cu foam or porous composite membrane templates, other researchers also prepared various 3D porous Sn–Cu alloy film anodes [61–64], which also displayed greatly enhanced cycle performances and high-rate capabilities. Nevertheless, there was usually a relative low volumetric capacity in the 3D porous hybrids due to large-sized pores and Cu skeletons

It should be pointed that stable reversible capacities of Sn–Cu alloy thin films were generally much lower than the theoretical capacity of Sn and Cu_6Sn_5 phases, although much better cycle performance could be realized in many of them. In other word, the high specific capacities were obviously sacrificed for the enhanced capacity retention.

4.4.4 Sn–Ni Alloy Anodes

Similar to Sn-Cu alloys, Sn and Ni could also alloy to form several intermetallic compounds as illustrated in phase diagram of Sn-Ni. The Ni_3Sn_2alloy has the Ni_2In-type structure as same as that of Cu_6Sn_5 which has relative high capacity of *ca.* 500 mAh/g.[65] Ehinon et al. [66] investigated the Ni_3Sn_4anode by Mössbauer spectroscopy, and confirmed the existence of a two-step reaction, in which the Li-Sn alloying process occurred in the second stage of discharge to form Li_7Sn_2. This mechanism was reversible and allowed restoring the Ni_3Sn_4 phase at the end of the cycle. Lee et al.[67] reported that microcrystalline Ni_3Sn_4 showed a low affinity to Li. However, in case of nanocrystalline Ni_3Sn_4, Li^+ reversibly reacted with Sn atoms at the grain boundaries, and no capacity fading was observed after prolonged cycling. The Ni_3Sn_2 film anodes, prepared by electron-beam deposited on Cu foil at room temperature, also exhibited excellent cyclability over 500 cycles, and did not undergo any crystallographic phase change during cycling. [68]

Osaka et al. [69] prepared Sn–Ni alloy films with different compositions by electrodeposition. They found that a $Sn_{62}Ni_{38}$ alloy was mainly composed of a Ni_3Sn_4 nanophase, which was identified as the key phase for high capacity and long cycle life, retaining 650 mAh/g after 70 cycles between 0.01 and 1.2 V. In contrast, $Sn_{54}Ni_{46}$ alloy, in which only a Sn–Ni metastable phase was identified, showed only low capacity [70]. Crosnier et al. [71] also tested the Sn–Ni alloy anodes prepared by electrodeposition at different current densities. In the thin film deposited at a high current density (20 A/cm^2), the active materials had a multiphase structure with a small particle size (less than 1 μm) and high porosity, giving rise to a high reversible capacity and favorable cycle performance. In contrast, however, the thin film obtained at low current density 0.1 A/cm^2 was composed only of a Ni_3Sn_2 phase with a large particle size (about 5 μm) and high density and showed low reversible capacity. On this basis, it was surmised that the enhanced cycle performance of the Sn–Ni thin film deposited at high current density may be attributed to the smaller particle size, greater porosity, and multiphase characteristics of the active materials in the film.

Accordingly, Hassoun et al. [72] designed and prepared a high-rate, long life Sn–Ni nanostructure electrode by electrodeposition on a nanoarchitectured Cu substrate as shown in Fig. 4.10a. In that electrodes, the Ni_3Sn_4 nanoparticles of about 50 nm in size were uniformly deposited on the surface of the Cu nanorods without any coalescence between them, which is shown in Fig. 4.10b. This nanostructured

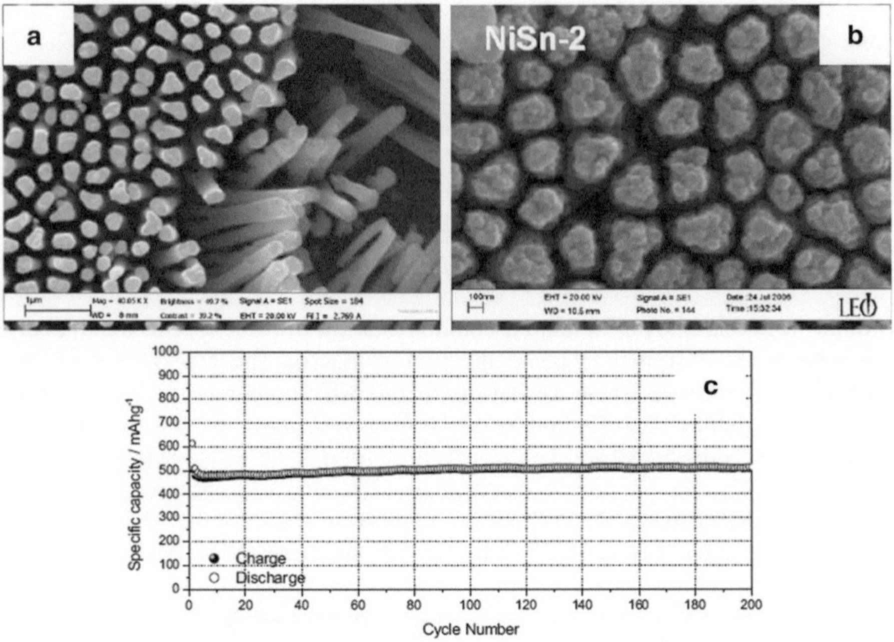

Fig. 4.10 (**a**) SEM image of top views of the Cu nanorod current collector used as a support for the electrodeposition of Sn–Ni samples. (**b**) SEM of top view of Sn–Ni electrode. (**c**) Capacity delivered upon cycling run between 0.02 and 1.5 V at a rate of ca. 0.8C (Reproduced from Ref. [72] with permission of WILEY-VCH Verlag GmbH &Co.)

Ni_3Sn_4 electrodes could deliver a high stable capacity of 500 mAh/g for 200 cycles as shown in Fig. 4.10c. Nevertheless, these kinds of electrodes involved quite complex preparation processes due to use and removal of the AAO templates.

4.4.5 Sn–NiTi Thin Film Anodes

Focusing on simultaneously accommodating the volume strain and stress due to Li–Sn alloying, we combined the Sn with NiTi shape memory alloys. Figure 4.11 illustrated the interactions between stress-induced martensitic transformation and superelasticity of NiTi and the volume effect of Sn and Li–Sn, which demonstrates that the Sn anodes could be effectively buffered by the shape memory effect of NiTi alloys.

To realize the abovementioned strategy, the following preparation process was attempted. Firstly, Sn–TiNi composite thin films were prepared by one-step co-sputtering of Sn and NiTi alloy. We found that, as shown in Fig. 4.12a, the composite films with relative low Sn content (20.2, 25.0 wt.%) had a microstructure of microsized Sn particles uniformly dispersed in porous amorphous NiTi (*a*-NiTi)

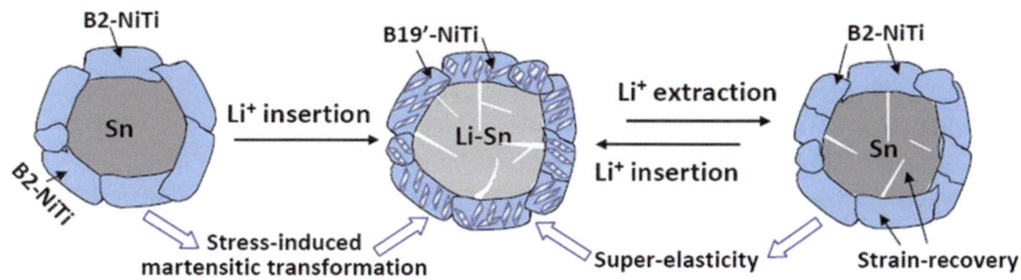

Fig. 4.11 Schematic of morphological changes occurring in Sn–NiTi electrode during charge–discharge cycling. Martensitic transformation ($B2 \rightarrow B19'$) is induced by the stress due to volume expansion of Li insertion into Sn, and this accommodates the stress and strain of Sn. As a result, cracking and pulverizing of the Sn phase is inhibited

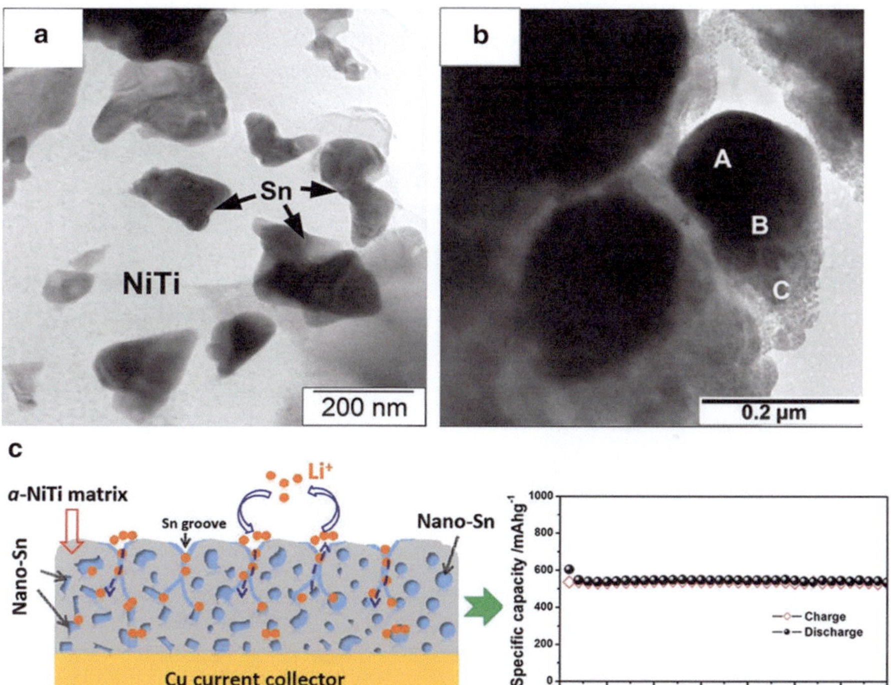

Fig. 4.12 Plane-view transmission electron microscopy (TEM) images of the (**a**) 25 wt.% Sn–NiTi, (**b**) 39 wt.% Sn–NiTi, (**c**) cycle performance of 25 wt.% Sn–NiTi anode (Reprinted from Ref. [73] with permission from The Royal Society of Chemistry)

matrix. In contrast, in the film with a higher Sn content (39.0 wt.%), as can be seen in Fig. 4.12b, a part of the Sn reacted with NiTi to form Ni_3Sn_4 intermetallic phase surrounding the Sn [75]. The inactive NiTi covered and protected most of the surface of Sn in these composite films and resulted in quite low initial irreversible capacity losses and greatly enhanced cycle performances. The 25 wt.% Sn–NiTi electrode delivered a stable capacity of 520 mAh/g at 1C after 40 cycles, as shown in Fig. 4.12c, in relative to the amount of Sn [73].

However, there was no shape memory effect in the *a*-NiTi matrix, and the Sn tended to alloy with Ni to form less active Sn–Ni intermetallics during annealing the above composite electrodes. The improvement is probably owing to the protection of major amount of inactive *a*-NiTi matrix to minor amount of active Sn phase but not superelasticity of NiTi. In addition, the capacity is low if the amount of matrix is included. Thus, we turned to prepare sandwich structured *B2*-NiTi/Sn/*a*-TiNi (referred as *B2*/Sn/*a*; *B2* is the parent phase of NiTi) thin film on stainless steel substrate by stepwise sputtering [74]. The preparation steps for *B2*/Sn/a thin film are as follows: first, a NiTi layer was deposited on a stainless steel substrate, and then it was annealed to form *B2*-NiTi phases. Subsequently, a pure Sn layer and an amorphous NiTi layer were then deposited sequentially on the B2-NiTi layer.

Figure 4.13a shows the cross-section TEM image for sandwich structure of the *B2*/Sn/*a* thin film, in which nanocrystals and microsized single crystals of Sn are present in Sn layer which is illustrated in Fig. 4.13b. The *B2*/Sn/*a* thin film anode

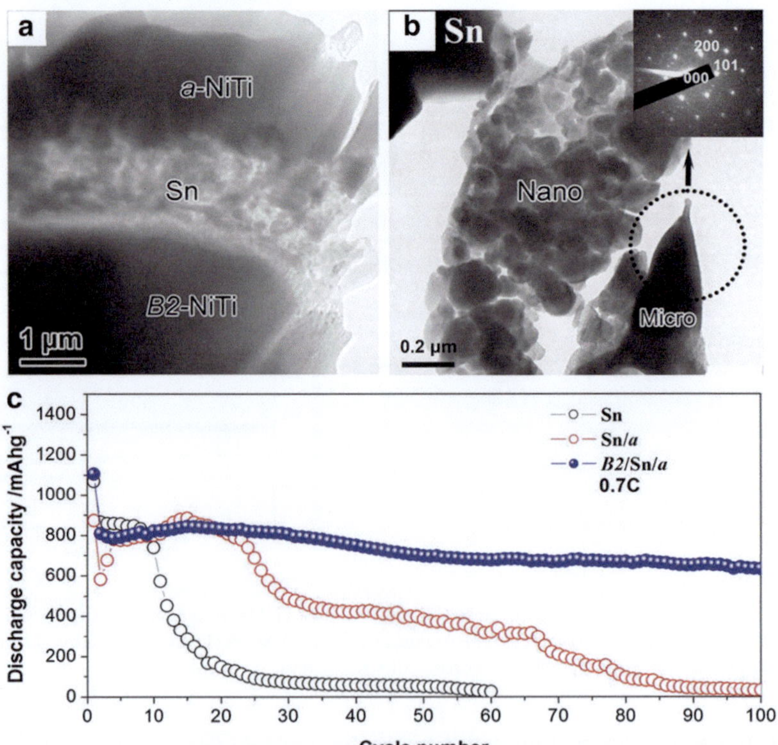

Fig. 4.13 (**a**) Cross-section view TEM image of the *B2*/Sn/*a* thin film; (**b**) TEM image of the Sn layer, the inset SAED pattern is from a microsize single crystal Sn particle; (**c**) comparison of cycle performance for the pure Sn, Sn/*a*, and *B2*/Sn/*a* thin film electrodes at a constant rate of 0.7 C between 0.01 and 1.5 V vs. Li (Reprinted from Ref. [74] with permission from Elsevier)

has good cyclability and high-rate capability, which were much better than those of the pure Sn, Sn/*a*, and Sn–NiTi electrodes without the supporting of *B2*-NiTi matrix (Fig. 4.13c).This was attributed to the superelasticity of *B2*-NiTi layer. Based on the results achieved by ex situ XRD, TEM, and optical microscopy observations [74, 76], we found that martensite (*B19′* phase) was induced in the *B2*-NiTi layer by the stress generated from the volume expansion during Li insertion into the Sn layer in the *B2*/Sn/*a* thin film electrode, and this martensite transforms back to the *B2* phase when the stress is released during the Li extraction. Later on, Cui et al. [77] also confirmed this phenomenon by using in situ TEM observations. The reverse transformation from martensite to parent phase results in crack closure and contraction of the volume of the Sn phase, thus leading to enhanced cyclability. This finding demonstrated a new concept which provides another way to improve the performance of LIBs based on metal-based Li insertion electrodes.

4.4.6 Sn–(Co, Fe, Mo) Alloy Anodes

Sn–Co alloys have also attracted widespread interest, and most of the work focused on designing nanocrystalline and amorphous structure in the electrodes of this system. Tamura et al. [78, 79] electrodeposited several Sn–Co film anodes with different microstructures, such as micro-column, micro-island, nanocrystalline, and amorphous and studied their structures and electrochemical properties. They found that the nanocrystalline film displayed the best anode performance. However, Whittingham et al. [80] found that the nanosized amorphous Sn–Co material showed enhanced cycling behavior over the crystalline Sn–Co analog. The amorphous Sn–Co reacted with Li, showing the typical single-phase behavior by formation of Li_xSnCo, maintained stable capacity over more than 30 cycles.

In addition, large amounts of work have been made on Sn–Co–C composites since the commercial launch of the Sony "Nexelion" batteries. Although the definite composition of the new commercial battery was not yet totally disclosed, it had been reported that the anode material was formed by a Sn–Co–C ternary alloy [10]. However, in this material, the reciprocal and combined role of Co and C might be more complex than that of a mere buffering action, which would be further discussed in the Sn–C section later.

With respect to the Sn–Fe system, attention mainly focused on the nanoparticle $FeSn_2$ intermetallic compound and the Sn_2Fe-C nanocomposite. Tu et al. [81] reported that the $FeSn_2$ intermetallic compound with size of 30–70nm synthesized by solvothermal method showed lower initial discharge capacity and better cyclability than that prepared by chemical reduction process with size of about 80 nm due to the better crystallinity. The two nanoscale $FeSn_2$ powders both delivered a high reversible discharge capacity of about 500 mAh/g. Tirado et al. [82] successfully prepared nanoparticles of crystalline $FeSn_2$ (20 nm in size) below 200 °C. They found that the superparamagnetic Fe nanoparticles, formed in the discharged electrode, avoided the growth of the Sn particles and improved the electrochemical performance. Further, the cycle performance of Sn_2Fe could be much enhanced when combined with the carbon and with the addition of Cu and Al_2O_3 during milling [83, 84], which helps to uniformly disperse the active materials and improve the mechanical stability of the composites.

Bonakdarpour et al. [86] investigated the viability of employing $Mo_{1-x}Sn_x$ thin films anodes, which were prepared by combinatorial sputter deposition. In situ XRD analysis of $Li/Mo_{1-x}Sn_x$ cells with $x < 0.36$ showed no evidence for the formation of new phases during charge–discharge over 75 cycles. However, in films with a higher Sn content of $x > 0.4$, aggregation of Sn clusters and the formation of Li–Sn phase were apparent after the first cycle. Surprisingly, it was found that the inclusion of a small amount of oxygen in the nanostructured Mo–Sn films seemed to prevent Sn aggregation and improved the cycling performance markedly [87].

Sn can easily react with transition metals (Cu, Ni, Co, Fe, Mn, etc.) to form various Sn_xM_y intermetallic compounds. However, in some of these intermetallic phases, such as Cu_3Sn, Ni_3Sn_4, and Co_3Sn_2, the binding energy of Sn–M is higher than that of Li–Sn, and as a consequence some of these, Sn_xM_y intermetallic phases have a relatively low affinity for Li or are even inactive toward Li. Hence, the reversible capacity will decrease if the amounts of these inactive intermetallics in anodes increase. Many studies, especially Dahn's work as shown in Fig. 4.14, suggested that amorphous or nanostructured alloys give the best capacity retention [85]. They had demonstrated that the phase structure of the Sn–M (Ti, V, Cr, Co) alloy anodes was highly dependent on the components and preparation conditions and that these in turn influenced the capacities and cycle performances. Each of these sputtered $Sn_{1-x}M_x$ systems showed an amorphous composition range where

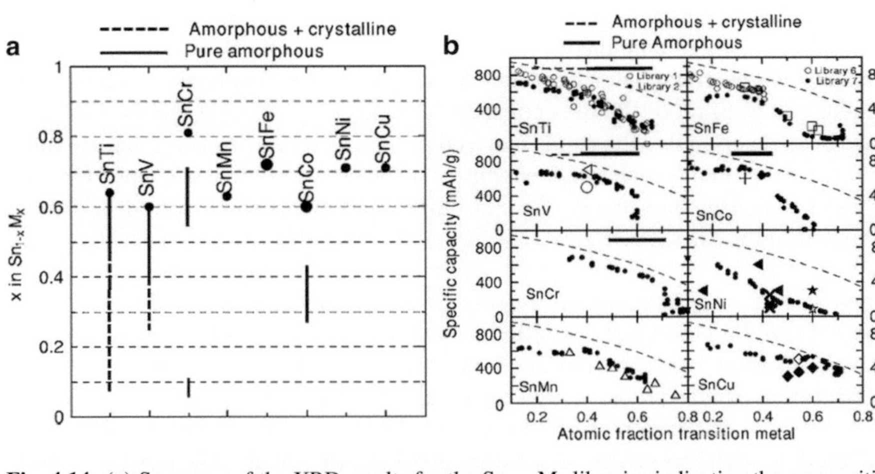

Fig. 4.14 (**a**) Summary of the XRD results for the $Sn_{1-x}M_x$ libraries indicating the composition ranges of the amorphous or nanostructured phases. (**b**) Specific capacity of $Sn_{1-x}M_x$ vs. atomic fraction transition metal for all M (small *circular symbols*) (Derived from Ref. [85] with permission of The Electrochemical Society, Inc.)

the specific capacity for Li decreased with M content. Among these, Sn–Co alloy system could show a good balance between cycling stability and capacity and were thus expected to offer a promising alternative for high-performance anodes.

4.4.7 Sn–Active Element (Sb, Si, Mg, Al) Composite Anodes

The abovementioned Sn-based alloy anodes all contained inactive metallic elements and/or intermetallic phases, which inevitably lowered the specific capacities of electrodes. Therefore, combining Sn with other active elements, such as Sb, Si, Al, and Mg, to form active compounds and/or composites in anodes, has been an important choice because the specific capacity loss could be reduced and/or avoided.

Sn–Sb alloys had attracted researchers' interest and have been widely investigated in earlier research. Sn–Sb intermetallic possesses a rhombohedral structure, in which Sn and Sb atoms alternatively arrange along the *c* direction. It would transfer to a coexistence of Li_3Sb and Li_2Sn after full lithiation, while Sn–Sb could be fully recovered when lithium is extracted [6]. It is due to that both the Sn and Sb have relative low hardness, most of the Sn–Sb alloys, especially the Sn/Sn–Sb composite anodes were not directly prepared by mechanical alloying of the pure Sn and Sb powders. Sn/Sn–Sb composite powders were usually precipitated with $NaBH_4$ from aqueous solutions of the respective chloride salts, mostly in the presence of complexants such as citrates, in order to control the particle sizes properly [88–91].

Sn–Si alloys should deliver high capacity due to the ultrahigh specific capacity of Si. However, what would happen in Sn–Si anodes in view of their volume change? Xiao et al. [92] developed a new class of Si–Sn composites with unique phase-separated nanostructure, where the amorphous Si nanoparticles are precipitated out and embedded within the Sn matrix. They found that the mechanical degradation could be mitigated by preventing the nucleation and propagation of microcracks during lithiation. Beaulieu et al. [93] sputtered a series of amorphous $Si_{1-x}Sn_x (0 < x < 0.5)$ alloy thin films. Among them, a $Si_{0.66}Sn_{0.34}$ thin film, which possessed a short-range ordered diamond structure, showed better conductivity, higher capacity, and superior cyclability than pure Sn or pure Si thin film anodes. The enhanced cyclability was mainly attributed to the absence of the two-phase reaction and therefore a uniform volume expansion of the amorphous Sn–Si film [29]. However, there was obvious charge–discharge potential hysteresis, which was caused by the energy dissipated during the changes in the local atomic environment around the host atoms [94].

Sohn et al. [95] had prepared an Mg_2Sn anode by mechanochemical process. Ex situ XRD and differential capacity plots showed that Li inserted into the Mg_2Sn lattice first followed by alloying with Sn while reaction between Li and Mg was not observed. The electrochemical reactions of Li with Mg_2Sn were further investigated by in situ Mössbauer spectroscopy [96]. It was confirmed that the lithiation initially results in transformation of Mg_2Sn part into Li_xMg_2Sn alloy ($x < 0.5$), and however accompanying with further lithiation, Mg was extruded from the structure with formation of Li_2MgSn ternary alloy.

We explored the immiscible Sn–Al alloys, in which the Al can also deliver a theoretical capacity of 990 mAh/g, as Li-ion anode materials, but has very poor Li-ion diffusion kinetics [97]. Sn–Al thin films prepared by EBD had complex structures. As shown in Fig. 4.15a and b, Sn phases homogeneously dispersed in the Al matrix, in which the Sn phases acted as diffusion channels to enhance the Li^+-diffusion kinetics. Thus, the cycle performance of Sn–Al thin film anodes is much better than those of pure Sn or pure Al thin film anodes [98, 99]. We found that the composition of Sn–Al thin film anodes obviously influenced their capacities and cycle performances. Al-x wt.% Sn ($40 \leq x \leq 60$) thin film electrodes showed a good balance between cycling ability, fast Li^+ diffusion, and acceptable capacity [100]. In particular, the Al-40 wt.% Sn film anode had a unique multi-scale composite structure with faceted large single-crystalline Sn particles and Sn nanocrystallites dispersed in the Al matrix (see Fig. 4.15c, d), and its stable reversible capacity was about 600 mAh/g [99]. However, the capacity and cycle performance of these Sn–Al thin film anodes was still poor due to the slow diffusion of Li^+ in the Al matrix and needs substantial further improvement. Thus, the microstructures of the Sn–Al thin film anodes need more optimal design to greatly facilitate the transport of Li^+ within them, and adding other elements may be also a useful route.

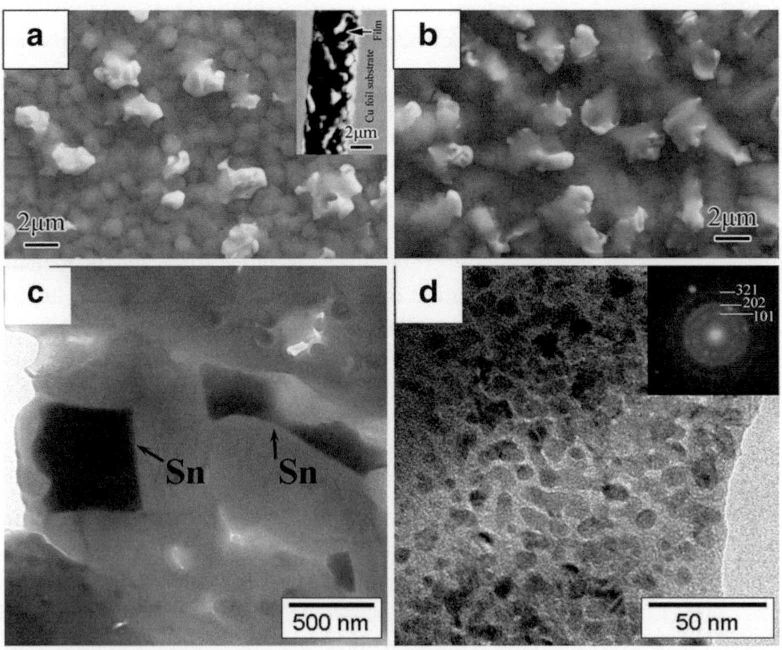

Fig. 4.15 SEM image of (**a**) Al-33 wt.% Sn and (**b**) Al-40 wt.% Sn thin film; (**c**) TEM image of Al-40 wt.% Sn and (**d**) high-resolution TEM image of the Sn nanoparticles in the Al matrix in (**c**) (Reprinted from Refs. [99, 100] with permission from Elsevier and American Chemical Society)

Based on the abovementioned research, nanostructured Sn and Sn–M alloys could yield less absolute volume change during Li-alloying. However, the aggregation of nanosized Sn and Li_xSn during cycle must be effectively suppressed by carbon matrix to meet a good cyclability. Thus, various Sn–C nanostructures were springing up recent years.

4.5 Sn–C Nanocomposite Anodes

Almost all the abovementioned Sn and Sn–M alloys had been combined with different kinds of carbon materials, to form thin film and/or powder electrodes. These Sn-based carbonaceous anodes, in which the active Sn phase mostly had nanoparticle [101–105], core–shell [41, 48, 50, 106], 3D porous structure [102, 107–109], and so on, indeed displayed much enhanced electrochemical performance. Herein, a small amount of materials was selected and discussed.

There were many Sn–C composite thin films, which had been studied as binder-free electrodes. Kostecki et al. [110] prepared a Sn–C nanocomposite by microwave plasma chemical vapor deposition. This Sn–C thin film had a unique structure of thin layers of three-dimensional graphitic carbon decorated with uniformly distributed Sn nanoparticles, which produced reversible capacities of 423 and 297 mAh/g at *C*/25(ca.17 mA/g) and 5 C rates between 0 and 1.1 V, respectively. These performances were much better than those of the sputtered Sn–C thin film anodes [31]. We fabricated a multi-scale Sn–C–Ni composite composed of mainly microsized core–shell particles [111]. As shown in Fig. 4.16, the cores were Sn single crystals and the shells were amorphous sp^2 carbon containing dispersed

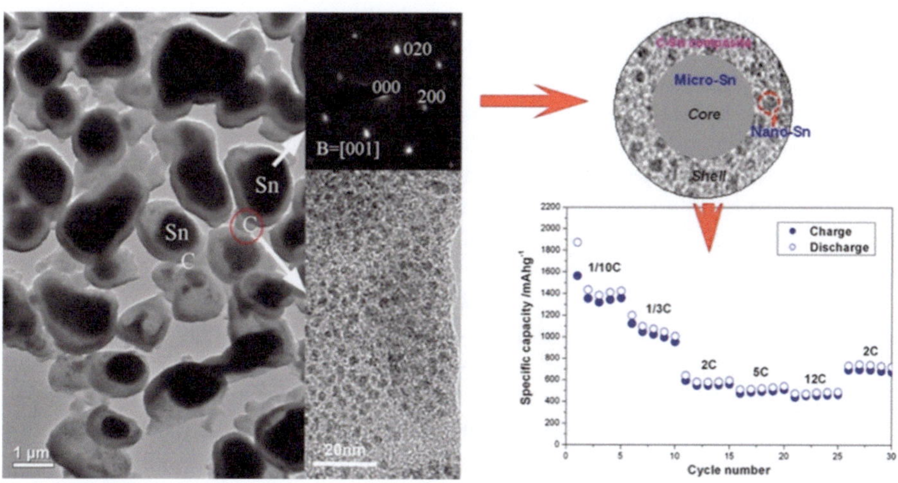

Fig. 4.16 Multi-scale core–shell structure and C rate performance of a Sn–C–Ni composite thin film prepared by electron beam deposition (Reprinted from Ref. [111] with permission from The Royal Society of Chemistry)

nanosized Sn and Ni particles. Both the Sn and the carbonaceous shells reacted with Li and contributed to the high initial capacities of 1872 mAh/g at 1/10C (ca.10 μA/cm^2) and 472 mAh/g at 12 C between 0.01 and 1.5 V. That initial capacity even exceeds the theoretical capacity of the materials constituted of the electrode, which should be mainly due to that a lot of Li-ions reacted with oxides (Ni–O and/or Ti–O) on the surface of the *B2*-NiTi layer in film and increased the total initial discharge capacity of the Sn–NiTi composite films.

Meanwhile, various kinds of transition metals (M) were also added to form Sn–M–C composites by many researchers. For the Sn–M–C thin film anodes, Dahn and coworker have done outstanding contribution [112–116]. By using combinatorial and high-throughput materials science methods, which was mainly based on the co-sputtering system, they investigated Sn–M (Co, Fe, Mn, V, Cu)–C thin film libraries extending over hundred different compositions. It had been found that adding carbon to the amorphous Sn–M binaries, to make ternaries, causes the precipitation of crystalline Sn in the cases of Ti or V which could form stable carbides with carbon. In contrast, the Co did not form the carbide phase with carbon and lead to the wide range of Sn–Co–C composites, which consisted of Sn–Co alloy grains surrounded by a carbon matrix. The initial amorphous atomic arrangements, in certain compositions ($Sn_{0.42}Co_{0.34}C_{0.24}$), were stable over at least 27 charge–discharge cycles of 600 mAh/g [117].

In comparison with thin film fabrications, Sn–C and Sn–M–C nanocomposites prepared by milling and some chemical methods would lead to higher yield of materials and enhanced performance. For example, the abovementioned Sn–Co–C composites were also prepared by milling and electrochemical characterized by Dahn [116, 119] and Hassoun et al. [120], which delivered stable reversible capacity around 400 mAh/g among 100 cycles.

Sn and graphite mixture had ever been milled more than 100 h to form a Sn–C nanocomposite anodes [121]. However, they had large irreversible capacity due to the large amount of defects in the long-term milled carbon. In order to enhance the milling efficiency, we introduced a new method, namely, dielectric barrier discharge plasma-assisted milling (referred as P-milling), to treat various materials [122, 123]. This milling method displays unique advantage in preparing anode materials such as Sn-based, Si-based, and oxide-based [124, 125]. By short-time P-milling (10 h) in Ar atmosphere, a unique Sn–C nanocomposite was obtained with a multi-scale structure of Sn multi-scale particles dispersed in nanosized graphite matrix, as shown Fig. 4.17a, c, and d. This is in strong contrast with microsized Sn in amorphous carbon obtained by conventional milling (refer as C-milling here) with the same milling duration (Fig. 4.17b) [118]. The P-milled Sn–C nanocomposite exhibited much higher reversible capacity and better cyclability. Furthermore, if air or O_2 atm was used during P-milling of Sn–C mixture, SnO_x ($1 \leq x \leq 2$) could be generated in situ on the Sn surface to form a Sn@SnO_x/C nanocomposites, and the obtained composite displays superior electrochemical performance (500 mAh/gat 0.4 C after 70 cycles) to Sn–C and Sn–SnO_2–C nanocomposites milled under argon plasma [126]. These suggested that

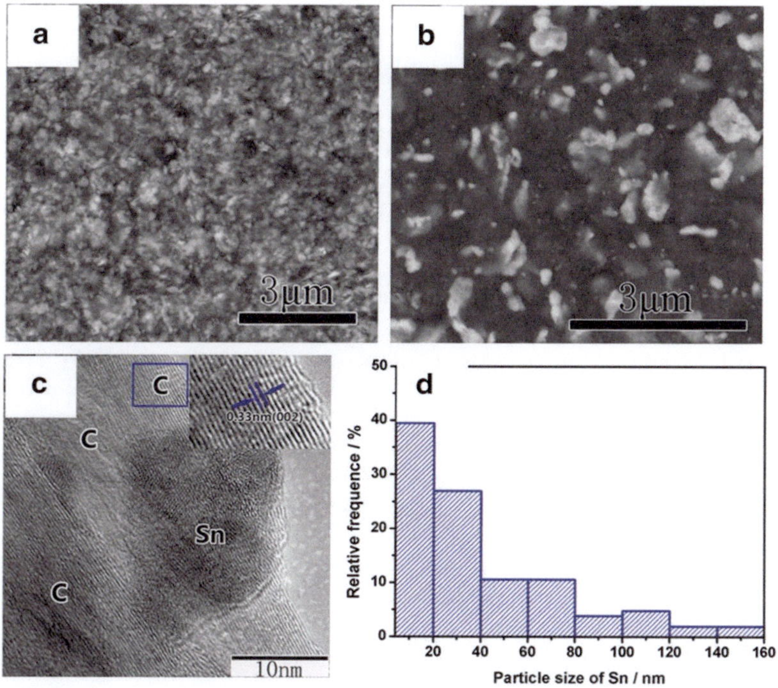

Fig. 4.17 Back-scattering electron SEM images of Sn–C composites. (**a**) P-10 h; (**b**) C-10 h; (**c**) HRTEM images of P-10 h Sn–C composites; (**d**) histograms of Sn particle size distributions of P-10 h (Reprinted from Ref. [118] with permission from The Royal Society of Chemistry)

P-milling is a promising method to prepare Sn-based multiphase nanocomposite anode materials.

With respect to the Sn-based carbonaceous composites, till to now, the best performance should be awarded to the nanostructure Sn–C composites reported by Scrosati and coworkers [101, 127]. This kind of material was achieved by first infiltration of a tin precursor into an organic gel, followed by calcination under argon. During the calcination step, the gel was carbonized and the organic tin precursor was reduced to Sn, formed Sn nanoparticles highly dispersed in the carbon matrix, with a nominal Sn/C composition of 1:1. TEM image given in Fig. 4.18a revealed the uniform dispersion of Sn particles in the carbon matrix, with size of the Sn particles of about 50 nm at the surface but only of few nanometers in the bulk. In a half cells, the Sn-C electrode delivered a stable specific capacity of the order of 500 mAh g–1 at 0.8 C over more than 200 cycles, while it could operate at rates as high as 5 C still delivering a 40 % of its total capacity. These suggest that this Sn-C should also have promising features in full cells.

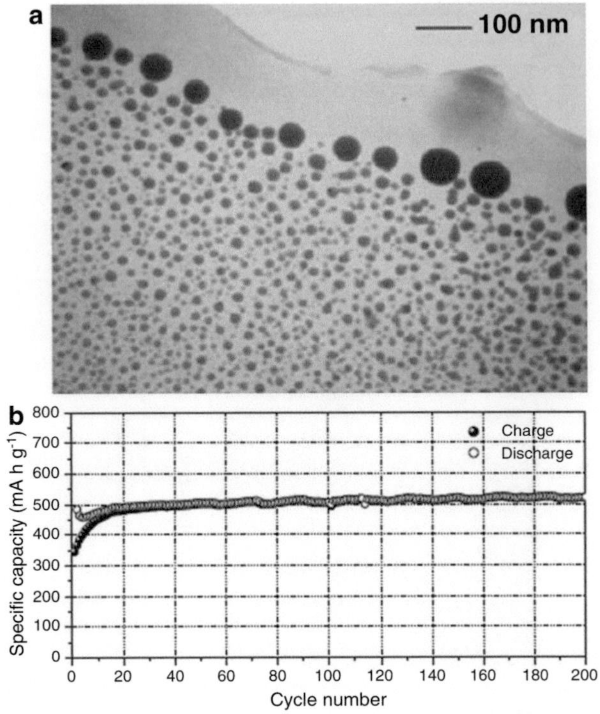

Fig. 4.18 (**a**) TEM image of a Sn–C composite material. (**b**) Cycling response at a 0.8 C rate of a Sn–C composite electrode in an electrochemical cell having lithium metal as a counter electrode and a $LiPF_6$-EC:DMC electrolyte. The measurements were taken at room temperature (Reproduced from Ref. [101] with permission of WILEY-VCH Verlag GmbH &Co.)

4.6 Full Li-ion Batteries with Sn-Based Alloy Material Anodes

It should be pointed that many of the Sn-based negative materials have presented excellent performances in half cells combined with Li counter electrode. But how well could they perform in completed cells? Scrosati and coworkers have made several attempts on full Li-ion batteries by using Sn-based anodes (Sn, Sn–Ni, Sn–Co–C, Sn–C) [9, 23, 118, 126, 127].

A Sn/EC:DMC 1:1 $LiPF_6/LiNi_{0.5}Mn_{1.5}O_4$ Li-ion cell was assembled using the electrodeposited pure Sn film as anode, and its microstructure is shown in Fig. 4.19a [23]. The voltage versus capacity profiles of the cell cycle under optimized conditions, as given in Fig. 4.19b, result as the combination of the expected profile associated with the Li–Sn alloy formation (compared with those in Fig. 4.2) and that related to the $LiNi_{0.5}Mn_{1.5}O_4$ cathode, showing an average operated potential at 3.8 V and stable specific capacity of 100 mAh/g among 20 cycles. These performances were better than those of the Sn–Ni/EC:DMC 1:1 $LiNi_{0.5}Mn_{1.5}O_4$ cell [128].

The electrochemical response of Sn–Co–C composites had also been further investigated in two Li-ion cells, $Sn_{31}Co_{28}C_{41}$/EC:DMC1:1 $LiPF_6/LiNi_{0.5}Mn_{1.5}O_4$ [127] and $Sn_{31}Co_{28}C_{41}$/EC:DMC 1:1 $LiPF_6/LiFePO_4$ [120], which operates around 3.8 and 3.0 V, respectively. As is shown in Fig. 4.20a, the performances of both

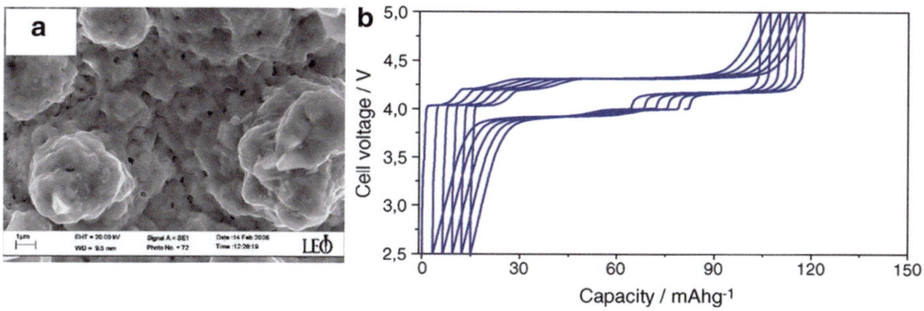

Fig. 4.19 (**a**) SEM image of a typical electrodeposited Sn electrodes prepared by ion, (**b**) voltage vs. specific capacity profiles of a Sn/EC:DMC 1:1 LiPF6/LiNi0.5Mn1.5O4 lithium-ion cell cycled 1 C rate vs. the cathode ($0.2 A/cm^2 g$) (Reproduced from Ref. [23] with permission from Elsevier)

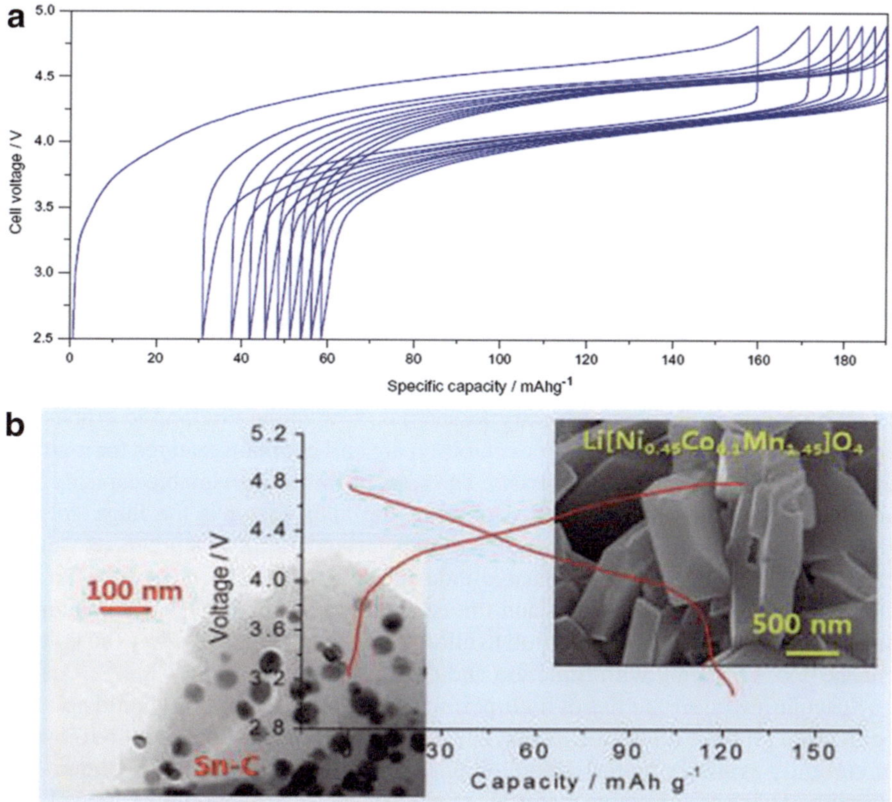

Fig. 4.20 (**a**) Cell voltage vs. specific capacity profile of the SnCoC/EC:DMC 1:1 $LiPF_6$/$LiNi_{0.5}Mn_{1.5}O_4$ lithium-ion cell. The cell is anode to cathode balanced with 1 C rate for the cathode (Reproduced from Ref. [127] with permission of Elsevier). (**b**) Configuration and performance of SnC/$Li[Ni_{0.45}Co_{0.1}Mn_{1.45}]O_4$ Li-ion battery (Reproduced from Ref. [9] with permission of American Chemical Society)

cells were quite good, suggesting that the $Sn_{31}Co_{28}C_{41}$ ternary compound is indeed a promising electrode material for the development of new-generation Li-ion batteries, although some problems still remain on the capacity retention upon prolonged cycling. In contrast, the nanostructure Sn–C invented by Scrosati et al. was systematically evaluated in several completed cells by combining with $LiNi_{0.5}Mn_{1.5}O_4$ [101], $LiFePO_4$ [128], and $Li[Ni_{0.45}Co_{0.1}Mn_{1.45}]O_4$ cathodes [9], respectively, and different types of electrolyte. Among these, the SnC/Li $[Ni_{0.45}Co_{0.1}Mn_{1.45}]O_4$ battery offered excellent performances in terms of cycling life, as shown in Fig. 4.20b, i.e., ca. 100 high-rate cycles; of rate capability, operating at 5C and still keeping more than 85 % of the initial capacity; and of energy density, expected to be in the order of 170 Wh/kg. These effectively demonstrated the practicality of Sn-based anode materials in high-energy Li-ion batteries.

4.7 Summary and Outlook

Sn-based materials have been studied as negative electrodes of Li-ion battery for more than two decades. Lots of material systems were investigated and large amount of publications and patents, which concerns on the preparation, performance, and mechanism of Sn-based electrode materials, have been published. However, until now, there is not any material that could completely replace the commercial graphite anode, as considering together the specific capacity, potential, life span, coulombic efficiency, cost, and so on. Even so, the fact that Sony had updated again its Nexelion technology, which increased capacity from 2.2 Ah to 3.5 Ah in 18650WH1 batteries in 2011, should give hope for the Sn-based anodes.

Large initial irreversible capacity losses, i.e., low initial coulombic efficiency, and fast capacity fading, i.e., poor cyclability, are still common features for many of the reported Sn-based anode materials. The reasons for the irreversible capacity loss and capacity fading are manifold. However, the main factor is the huge volume change and increased surface area of active materials during Li storage, as well as electrode–electrolyte interface change induced by the interaction between Sn and electrolyte. Thus, understanding and tuning the interface between active materials and electrolyte is crucial, in addition to eliminate the problem caused by the volume change accompanying with lithiation and delithiation.

Speculating future studies of high-performance anodes, attention should not only be focused on the exploration of new high-capacity materials and the modification of currently available materials, but emphasis should also be placed on the microstructure design of the active materials of electrodes, in terms of tuning the volume stress/strain of electrodes and maintaining their integrity during cycling. We proposed that manipulating the multiphase and multi-scale structures represents an important strategy for further improving the capacity and cyclability of Sn-based and other high-capacity alloy anodes, together with the development of new materials, new technology, and new mechanism.

4.8 Acknowledgments

This work was supported by the National Science Foundation of China under projects No. 50971060, 51201065, 51231003, and funded by the Guangdong Provincial Key Laboratory of Advanced Energy Storage Materials.

References

1. Tirado JL (2003) Inorganic materials for the negative electrode of lithium-ion batteries: state-of-the-art and future prospects. Mater Sci Eng R Rep 40:103–136
2. Winter M, Besenhard JO, Spahr ME, Novák P (1998) Insertion electrode materials for rechargeable lithium batteries. Adv Mater 10:725–763
3. Arico AS, Bruce P, Scrosati B, Tarascon J-M, van Schalkwijk W (2005) Nanostructured materials for advanced energy conversion and storage devices. Nat Mater 4:366–377
4. Park C-M, Kim J-H, Kim H, Sohn H-J (2010) Li-alloy based anode materials for Li secondary batteries. Chem Soc Rev 39:3115–3141
5. Zhang W-J (2011) A review of the electrochemical performance of alloy anodes for lithium-ion batteries. J Power Sources 196:13–24
6. Winter M, Besenhard JO (1999) Electrochemical lithiation of tin and tin-based intermetallics and composites. Electrochim Acta 45:31–50
7. Chu Daobao LJ, Yuan Ximei, Li Zilong, Wei Xu, Wan Yong (2012) Tin-based alloy anode materials for lithium ion batteries. Prog Chem 24:1466–1476
8. Hu R, Liu H, Zeng M, Liu J, Zhu M (2012) Progress on Sn-based thin-film anode materials for lithium-ion batteries. Chin Sci Bull 57:4119–4130
9. Hassoun J, Lee K-S, Sun Y-K, Scrosati B (2011) An advanced lithium ion battery based on high performance electrode materials. J Am Chem Soc 133:3139–3143
10. Wolfenstine J, Allen JL, Read J, Foster D (2006) Chemistry and structure of Sony's Nexelion Li-ion electrode materials. http://wwwdticmil/dtic/tr/fulltext/u2/a479387pdf
11. Song M-K, Park S, Alamgir FM, Cho J, Liu M (2011) Nanostructured electrodes for lithium-ion and lithium-air batteries: the latest developments, challenges, and perspectives. Mater Sci Eng R Rep 72:203–252
12. Idota Y, Kubota T, Matsufuji A, Maekawa Y, Miyasaka T (1997) Tin-based amorphous oxide: a high-capacity lithium-ion-storage material. Science 276:1395–1397
13. Huang JY, Zhong L, Wang CM, Sullivan JP, Xu W, Zhang LQ et al (2010) In situ observation of the electrochemical lithiation of a single SnO2 nanowire electrode. Science 330:1515–1520
14. Courtney IA, Tse JS, Mao O, Hafner J, Dahn JR (1998) Ab initio calculation of the lithium-tin voltage profile. Phys Rev B 58:15583–15588
15. Hirai K, Ichitsubo T, Uda T, Miyazaki A, Yagi S, Matsubara E (2008) Effects of volume strain due to Li–Sn compound formation on electrode potential in lithium-ion batteries. Acta Mater 56:1539–1545
16. Stournara ME, Guduru PR, Shenoy VB (2012) Elastic behavior of crystalline Li–Sn phases with increasing Li concentration. J Power Sources 208:165–169
17. Inaba M, Uno T, Tasaka A (2005) Irreversible capacity of electrodeposited Sn thin film anode. J Power Sources 146:473–477
18. Yang J, Winter M, Besenhard JO (1996) Small particle size multiphase Li-alloy anodes for lithium-ion batteries. Solid State Ion 90:281–287

19. Jiang D, Tian H, Qiu C, Ma X, Fu Y (2011) Electrodeposition and characterization of assembly of Sn on Cu nanorods for Li-ion microbattery application. J Solid State Electrochem 15:2639–2644
20. Wang C, John Appleby A, Little FE (2001) Electrochemical study on nano-Sn, Li4.4Sn and AlSi0.1 powders used as secondary lithium battery anodes. J Power Sources 93:174–185
21. Hu R, Xin L, Zeng MQ, Liu J, Zhu M (2011) Sn nanowires self-grown on the immiscible Sn-Al alloys and their electrochemical properties as Li-ion battery anode. J Chin Electron Microsc Soc 30:11
22. Morimoto H, Tobishima S-i, Negishi H (2005) Anode behavior of electroplated rough surface Sn thin films for lithium-ion batteries. J Power Sources 146:469–472
23. Hassoun J, Reale P, Panero S (2007) The role of the interface of tin electrodes in lithium cells: an impedance study. J Power Sources 174:321–327
24. Ui K, Kikuchi S, Kadoma Y, Kumagai N, Ito S (2009) Electrochemical characteristics of Sn film prepared by pulse electrodeposition method as negative electrode for lithium secondary batteries. J Power Sources 189:224–229
25. Tamura N, Ohshita R, Fujimoto M, Fujitani S, Kamino M, Yonezu I (2002) Study on the anode behavior of Sn and Sn-Cu alloy thin-film electrodes. J Power Sources 107:48–55
26. Pu W, He X, Ren J, Wan C, Jiang C (2005) Electrodeposition of Sn-Cu alloy anodes for lithium batteries. Electrochim Acta 50:4140–4145
27. Kim R, Nam D, Kwon H (2010) Electrochemical performance of a tin electrodeposit with a multi-layered structure for Li-ion batteries. J Power Sources 195:5067–5070
28. Park J, Eom J, Kwon H (2010) Charge–discharge characteristics of a layered-structure electroplated Cu/Sn anode for Li-ion batteries. Electrochim Acta 55:1825–1828
29. Beaulieu LY, Hatchard TD, Bonakdarpour A, Fleischauer MD, Dahn JR (2003) Reaction of Li with alloy thin films studied by in situ AFM. J Electrochem Soc 150:A1457–A1464
30. Chiu K-F, Lin HC, Lin KM, Lin TY, Shieh DT (2006) The significant role of solid oxide interphase in enhancement of cycling performance of Sn thin-film anodes. J Electrochem Soc 153:A1038–A1042
31. Zhao LZ, Hu SJ, Ru Q, Li WS, Hou XH, Zeng RH et al (2008) Effects of graphite on electrochemical performance of Sn/C composite thin film anodes. J Power Sources 184:481–484
32. Hu RZ, Zhang Y, Zhu M (2008) Microstructure and electrochemical properties of electron-beam deposited Sn-Cu thin film anodes for thin film lithium ion batteries. Electrochim Acta 53:3377–3385
33. Bai HM, Cheng FY, Chen J (2011) Preparation and electrochemical properties of Sn thin film as anode materials for lithium-ion batteries. J Electrochem 17:43–7
34. Qiao R, Lucas IT, Karim A, Syzdek J, Liu X, Chen W et al (2014) Distinct solid-electrolyte-interphases on Sn (100) and (001) electrodes studied by soft x-ray spectroscopy. Adv Mater Interfaces 1:1300115
35. Winter M, Appel WK, Evers B, Hodal T, Möller K-C, Schneider I et al (2001) Studies on the anode/electrolyte interface in lithium ion batteries. Monatshefte fuer Chemie 132:473–486
36. Xia Y, Sakai T, Fujieda T, Wada M, Yoshinaga H (2001) Flake Cu-Sn alloys as negative electrode materials for rechargeable lithium batteries. J Electrochem Soc 148:A471–A481
37. Kepler KD, Vaughey JT, Thackeray MM (1999) Copper-tin anodes for rechargeable lithium batteries: an example of the matrix effect in an intermetallic system. J Power Sources 81–82:383–387
38. Thackeray MM, Vaughey JT, Kahaian AJ, Kepler KD, Benedek R (1999) Intermetallic insertion electrodes derived from NiAs-, Ni2In-, and Li2CuSn-type structures for lithium-ion batteries. Electrochem Commun 1:111–115
39. Choi W, Lee JY, Lim HS (2004) Electrochemical lithiation reactions of Cu6Sn5 and their reaction products. Electrochem Commun 6:816–820

40. Vaughey JT, Kepler KD, Benedek R, Thackeray MM (1999) NiAs- versus zinc-blende-type intermetallic insertion electrodes for lithium batteries: lithium extraction from Li2CuSn. Electrochem Commun 1:517–521
41. Zhang J-j, Zhang Y-m, Zhang X, Xia Y-y (2007) NixCu6 – xSn5 alloys as negative electrode materials for rechargeable lithium batteries. J Power Sources 167:171–177
42. Wang GX, Sun L, Bradhurst DH, Dou SX, Liu HK (2000) Lithium storage properties of nanocrystalline eta-Cu6Sn5 alloys prepared by ball-milling. J Alloys Compd 299:L12–L15
43. Wolfenstine J, Campos S, Foster D, Read J, Behl WK (2002) Nano-scale Cu6Sn5 anodes. J Power Sources 109:230–233
44. Kim DG, Kim H, Sohn HJ, Kang T (2002) Nanosized Sn-Cu-B alloy anode prepared by chemical reduction for secondary lithium batteries. J Power Sources 104:221–225
45. Sarakonsri T, Johnson CS, Hackney SA, Thackeray MM (2006) Solution route synthesis of InSb, Cu6Sn5 and Cu2Sb electrodes for lithium batteries. J Power Sources 153:319–327
46. Chen J, Yang L, Fang S (2012) Hirano S-i. Synthesis of mesoporous Sn–Cu composite for lithium ion batteries. J Power Sources 209:204–208
47. Ju SH, Jang HC, Kang YC (2009) Electrochemical properties of Cu6Sn5 alloy powders directly prepared by spray pyrolysis. J Power Sources 189:163–168
48. Cui W-j, Li F, Liu H-j, Wang C-x, Xia Y-y (2009) Core-shell carbon-coated Cu6Sn5 prepared by in situ polymerization as a high-performance anode material for lithium-ion batteries. J Mater Chem 19:7202–7207
49. Kim MG, Sim S, Cho J (2010) Novel core-shell Sn-Cu anodes for lithium rechargeable batteries prepared by a redox-transmetalation reaction. Adv Mater 22:5154–5158
50. Thorne JS, Dahn JR, Obrovac MN, Dunlap RA (2012) A comparison of sputtered and mechanically milled Cu6Sn5 + C materials for Li-ion battery negative electrodes. J Power Sources 216:139–144
51. Yang R, Huang J, Zhao W, Lai W, Zhang X, Zheng J et al (2010) Bubble assisted synthesis of Sn–Sb–Cu alloy hollow nanostructures and their improved lithium storage properties. J Power Sources 195:6811–6816
52. Cui W, Wang F, Wang J, Liu H, Wang C, Xia Y (2011) A modified carbothermal reduction method for preparation of high-performance nano-scale core/shell Cu6Sn5 alloy anodes in Li-ion batteries. J Power Sources 196:3633–3639
53. Ren J, He X, Wang L, Pu W, Jiang C, Wan C (2007) Nanometer copper–tin alloy anode material for lithium-ion batteries. Electrochim Acta 52:2447–2452
54. Beattie SD, Dahn JR (2003) Single bath, pulsed electrodeposition of copper-tin alloy negative electrodes for lithium-ion batteries. J Electrochem Soc 150:A894–A898
55. Jianguo R, Xiangming H, Changyin J, Chunrong W (2006) Preparation and property of nanometer Cu – Sn alloy anode material for lithium – ion batteries. Acta Metall 42:727–732
56. Beattie SD, Dahn JR (2003) Single-bath electrodeposition of a combinatorial library of binary Cu1-xSnx alloys. J Electrochem Soc 150:C457–C460
57. Hu RZ, Zeng MQ, Zhu M (2009) Cyclic durable high-capacity Sn/Cu6Sn5 composite thin film anodes for lithium ion batteries prepared by electron-beam evaporation deposition. Electrochim Acta 54:2843–2850
58. Shin HC, Liu M (2005) Three-dimensional porous copper–tin alloy electrodes for rechargeable lithium batteries. Adv Funct Mater 15:582–586
59. Jiang T, Zhang S, Qiu X, Zhu W, Chen L (2007) Preparation and characterization of tin-based three-dimensional cellular anode for lithium ion battery. J Power Sources 166:503–508
60. Zhao H, Jiang C, He X, Ren J, Wan C (2007) Advanced structures in electrodeposited tin base anodes for lithium ion batteries. Electrochim Acta 52:7820–7826
61. Ke F-S, Huang L, Cai J-S, Sun S-G (2007) Electroplating synthesis and electrochemical properties of macroporous Sn-Cu alloy electrode for lithium-ion batteries. Electrochim Acta 52:6741–6747
62. Du Z, Zhang S, Jiang T, Bai Z (2010) Preparation and characterization of three-dimensional tin thin-film anode with good cycle performance. Electrochim Acta 55:3537–3541

63. Ehrlich GM, Durand C, Chen X, Hugener TA, Spiess F, Suib SL (2000) Metallic negative electrode materials for rechargeable nonaqueous batteries. J Electrochem Soc 147:886–891
64. Ehinon KKD, Naille S, Dedryvère R, Lippens PE, Jumas JC, Gonbeau D (2008) Ni33Sn4 electrodes for Li-ion batteries: Li-Sn alloying process and electrode/electrolyte interface phenomena. Chem Mater 20:5388–5398
65. Lee H-Y, Jang S-W, Lee S-M, Lee S-J, Baik H-K (2002) Lithium storage properties of nanocrystalline Ni3Sn4 alloys prepared by mechanical alloying. J Power Sources 112:8–12
66. Kim YL, Lee HY, Jang SW, Lee SJ, Baik HK, Yoon YS et al (2003) Nanostructured Ni3Sn2 thin film as anodes for thin film rechargeable lithium batteries. Solid State Ion 160:235–240
67. Mukaibo H, Sumi T, Yokoshima T, Momma T, Osaka T (2003) Electrodeposited Sn-Ni alloy film as a high capacity anode material for lithium-ion secondary batteries. Electrochem Solid-State Lett 6:A218–A220
68. Mukaibo H, Momma T, Osaka T (2005) Changes of electro-deposited Sn-Ni alloy thin film for lithium ion battery anodes during charge discharge cycling. J Power Sources 146:457–463
69. Crosnier O, Brousse T, Devaux X, Fragnaud P, Schleich DM (2001) New anode systems for lithium ion cells. J Power Sources 94:169–174
70. Hassoun J, Panero S, Simon P, Taberna PL, Scrosati B (2007) High-rate, long-life Ni-Sn nanostructured electrodes for lithium-ion batteries. Adv Mater 19:1632–1635
71. Hu R, Liu H, Zeng M, Liu J, Zhu M (2012) Microsized Sn supported by NiTi alloy as a high-performance film anode for Li-ion batteries. J Mater Chem 22:9539–9545
72. Hu R, Zhu M, Wang H, Liu J, Liuzhang O, Zou J (2012) Sn buffered by shape memory effect of NiTi alloys as high-performance anodes for lithium ion batteries. Acta Mater 60:4695–4703
73. Hu R, Liu H, Zeng M, Liu J, Zhu M (2013) Influence of Sn content on microstructure and electrochemical properties of Sn-NiTi film anodes in lithium ion batteries. J Power Sources 244:456–462
74. Hu R, Yang L, Zhu M (2013) Progress on high energy density thin film anode materials for lithium-ion batteries. Chin Sci Bull 58:3140–3156
75. Zhang LQ, Zhang JS, Shao Y, Jiang DQ, Yang F, Guo YP et al (2013) In situ TEM observation of buffering the anode volume change by using NiTi alloy during electrochemical lithiation/delithiation. Nanotechnology 24:325702
76. Tamura N, Kato Y, Mikami A, Kamino M, Matsuta S, Fujitani S (2006) Study on Sn-Co alloy anodes for lithium secondary batteries: I. Amorphous system. J Electrochem Soc 153:A1626–A1632
77. Tamura N, Kato Y, Mikami A, Kamino M, Matsuta S, Fujitani S (2006) Study on Sn-Co alloy electrodes for lithium secondary batteries: II. Nanocomposite system. J Electrochem Soc 153: A2227–A2231
78. Fan Q, Chupas PJ, Whittingham MS (2007) Characterization of amorphous and crystalline tin-cobalt anodes. Electrochem Solid-State Lett 10:A274–A278
79. Zhang CQ, Tu JP, Huang XH, Yuan YF, Wang SF, Mao F (2008) Preparation and electrochemical performances of nanoscale FeSn2 as anode material for lithium ion batteries. J Alloys Compd 457:81–85
80. Nwokeke UG, Alcántara R, Tirado JL, Stoyanova R, Zhecheva E (2011) The electrochemical behavior of low-temperature synthesized FeSn2 nanoparticles as anode materials for Li-ion batteries. J Power Sources 196:6768–6771
81. Lee J-M, Chang W-S, Yu B-C, Kim H, Im D, Doo S-G et al (2010) Enhancement of cyclability using recombination reaction of Cu for Sn2Fe nanocomposite anode for lithium-ion batteries. Electrochem Commun 12:928–932
82. Lee J-M, Jung H, Hwa Y, Kim H, Im D, Doo S-G et al (2010) Improvement of electrochemical behavior of Sn2Fe/C nanocomposite anode with Al2O3 addition for lithium-ion batteries. J Power Sources 195:5044–5048
83. Todd ADW, Mar RE, Dahn JR (2006) Combinatorial study of tin-transition metal alloys as negative electrodes for lithium-ion batteries. J Electrochem Soc 153:A1998–A2005

84. Dahn JR, Turner RL, Mao O, Dunlap RA, George AE, Buckett MM et al (2002) Structure and properties of sequentially sputtered molybdenum-tin films. Thin Solid Films 408:111–122
85. Bonakdarpour A, Hewitt KC, Turner RL, Dahn JR (2004) Electrochemical and in situ XRD studies of the Li reaction with combinatorially sputtered Mo1-xSnx ($0 \leq x \leq 0.50$) thin films. J Electrochem Soc 151:A470–A483
86. Wang K, He X, Ren J, Wang L, Jiang C, Wan C (2006) Preparation of Sn2Sb alloy encapsulated carbon microsphere anode materials for Li-ion batteries by carbothermal reduction of the oxides. Electrochim Acta 52:1221–1225
87. Mukaibo H, Osaka T, Reale P, Panero S, Scrosati B, Wachtler M (2004) Optimized Sn/SnSb lithium storage materials. J Power Sources 132:225–228
88. Simonin L, Lafont U, Kelder EM (2008) SnSb micron-sized particles for Li-ion batteries. J Power Sources 180:859–863
89. Needham SA, Wang GX, Liu HK (2005) Electrochemical performance of SnSb and Sn/SnSb nanosize powders as anode materials in Li-ion cells. J Alloys Compd 400:234–238
90. Xiao X, Wang JS, Liu P, Sachdev AK, Verbrugge MW, Haddad D et al (2012) Phase-separated silicon-tin nanocomposites for high capacity negative electrodes in lithium ion batteries. J Power Sources 214:258–265
91. Beaulieu LY, Hewitt KC, Turner RL, Bonakdarpour A, Abdo AA, Christensen L et al (2003) The electrochemical reaction of Li with amorphous Si-Sn alloys. J Electrochem Soc 150: A149–A156
92. Hatchard TD, Dahn JR (2004) Study of the electrochemical performance of sputtered Si 1-xSnx films. J Electrochem Soc 151:A1628–A1635
93. Kim H, Kim YJ, Kim DG, Sohn HJ, Kang T (2001) Mechanochemical synthesis and electrochemical characteristics of Mg2Sn as an anode material for Li-ion batteries. Solid State Ion 144:41–49
94. Aldon L, Ionica CM, Lippens PE, Larcher D, Tarascon JM, Olivier-Fourcade J et al (2006) In situ 119Sn Mössbauer spectroscopy used to study lithium insertion in c-Mg2Sn. Hyperfine Interact 167:729–732
95. Hamon Y, Brousse T, Jousse F, Topart P, Buvat P, Schleich DM (2001) Aluminum negative electrode in lithium ion batteries. J Power Sources 97–98:185–187
96. Hu RZ, Zhang L, Liu X, Zeng MQ, Zhu M (2008) Investigation of immiscible alloy system of Al-Sn thin films as anodes for lithium ion batteries. Electrochem Commun 10:1109–1112
97. Hu R, Zeng M, Li CYV, Zhu M (2009) Microstructure and electrochemical performance of thin film anodes for lithium ion batteries in immiscible Al-Sn system. J Power Sources 188:268–273
98. Hu R, Shi Q, Wang H, Zeng M, Zhu M (2009) Influences of composition on the electrochemical performance in immiscible Sn-Al thin films as anodes for lithium ion batteries. J Phys Chem C 113:18953–18961
99. Derrien G, Hassoun J, Panero S, Scrosati B (2007) Nanostructured Sn–C composite as an advanced anode material in high-performance lithium-ion batteries. Adv Mater 19:2336–2340
100. Grigoriants I, Soffer A, Salitra G, Aurbach D (2005) Nanoparticles of tin confined in microporous carbon matrices as anode materials for Li batteries. J Power Sources 146:185–189
101. Zhang WM, Hu JS, Guo YG, Zheng SF, Zhong LS, Song WG et al (2008) Tin-nanoparticles encapsulated in elastic hollow carbon spheres for high-performance anode material in lithium-ion batteries. Adv Mater 20:1160–1165
102. Balan L, Schneider R, Ghanbaja J, Willmann P, Billaud D (2006) Electrochemical lithiation of new graphite–nanosized tin particle materials obtained by SnCl2 reduction in organic medium. Electrochim Acta 51:3385–3390
103. Wang B, Luo B, Li X, Zhi L (2012) The dimensionality of Sn anodes in Li-ion batteries. Mater Today 15:544–552

104. Wang Z, Tian W, Liu X, Yang R, Li X (2007) Synthesis and electrochemical performances of amorphous carbon-coated Sn-Sb particles as anode material for lithium-ion batteries. J Solid State Chem 180:3360–3365
105. Yu Y, Gu L, Zhu C, Van Aken PA, Maier J (2009) Tin nanoparticles encapsulated in porous multichannel carbon microtubes: preparation by single-nozzle electrospinning and application as anode material for high-performance Li-based batteries. J Am Chem Soc 131:15984–15985
106. Wang G, Wang B, Wang X, Park J, Dou S, Ahn H et al (2009) Sn/graphene nanocomposite with 3D architecture for enhanced reversible lithium storage in lithium ion batteries. J Mater Chem 19:8378–8384
107. Chen J, Yang L, Fang S, Hirano S-i (2011) Ordered mesoporous Sn-C composite as an anode material for lithium ion batteries. Electrochem Commun 13:848–851
108. Marcinek M, Hardwick LJ, Richardson TJ, Song X, Kostecki R (2007) Microwave plasma chemical vapor deposition of nano-structured Sn/C composite thin-film anodes for Li-ion batteries. J Power Sources 173:965–971
109. Hu R, Liu H, Zeng M, Wang H, Zhu M (2011) Core/shell and multi-scale structures enhance the anode performance of a Sn-C-Ni composite thin film in a lithium ion battery. J Mater Chem 21:4629–4635
110. Todd ADW, Mar RE, Dahn JR (2007) Tin-transition metal-carbon systems for lithium-ion battery negative electrodes. J Electrochem Soc 154:A597–A604
111. Ferguson PP, Martine ML, George AE, Dahn JR (2009) Studies of tin-transition metal-carbon and tin-cobalt-transition metal-carbon negative electrode materials prepared by mechanical attrition. J Power Sources 194:794–800
112. Thorne JS, Sanderson RJ, Dahn JR, Dunlap RA (2010) Combinatorial study of the Sn-Cu-C system for Li-ion battery negative electrode materials. J Electrochem Soc 157:A1085–A1091
113. Al-Maghrabi MA, Thorne JS, Sanderson RJ, Byers JN, Dahn JR, Dunlap RA (2012) A combinatorial study of the Sn-Si-C system for Li-ion battery applications. J Electrochem Soc 159:A711–A719
114. Ferguson PP, Todd ADW, Martine ML, Dahnd JR (2014) Structure and performance of tin-cobalt-carbon alloys prepared by attriting, roller milling and sputtering. J Electrochem Soc 161:A342–A347
115. Dahn JR, Mar RE, Abouzeid A (2006) Combinatorial study of Sn1 − x Co x (0 < x < 0.6) and [Sn0.55Co0.45] 1 − y C y (0 < y < 0.5) alloy negative electrode materials for Li-ion batteries. J Electrochem Soc 153:A361–A365
116. Liu H, Hu R, Zeng M, Liu J, Zhu M (2012) Enhancing the performance of Sn-C nanocomposite as lithium ion anode by discharge plasma assisted milling. J Mater Chem 22:8022–8028
117. Ferguson PP, Todd ADW, Dahn JR (2008) Comparison of mechanically alloyed and sputtered tin-cobalt-carbon as an anode material for lithium-ion batteries. Electrochem Commun 10:25–31
118. Hassoun J, Panero S, Mulas G, Scrosati B (2007) An electrochemical investigation of a Sn-Co-C ternary alloy as a negative electrode in Li-ion batteries. J Power Sources 171:928–931
119. Wang GX, Ahn J-H, Lindsay MJ, Sun L, Bradhurst DH, Dou SX et al (2001) Graphite-Tin composites as anode materials for lithium-ion batteries. J Power Sources 97–98:211–215
120. Zhu M, Dai LY, Gu NS, Cao B, Ouyang LZ (2009) Synergism of mechanical milling and dielectric barrier discharge plasma on the fabrication of nano-powders of pure metals and tungsten carbide. J Alloys Compd 478:624–629
121. Dai LY, Cao B, Zhu M (2006) Comparison on refinement of iron powder by ball milling assisted by different external fields. Acta Metall Sin (Eng Lett) 19:411–417
122. Sun W, Hu R, Liu H, Zeng M, Yang L, Wang H et al (2014) Embedding nano-silicon in graphene nanosheets by plasma assisted milling for high capacity anode materials in lithium ion batteries. J Power Sources 268:610–618

123. Hu R, Sun W, Chen Y, Zeng M, Zhu M (2014) Silicon/graphene based nanocomposite anode: large-scale production and stable high capacity for lithium ion batteries. J Mater Chem A 2:9118–9125
124. Liu H, Hu R, Sun W, Zeng M, Liu J, Yang L et al (2013) Sn@SnOx/C nanocomposites prepared by oxygen plasma-assisted milling as cyclic durable anodes for lithium ion batteries. J Power Sources 242:114–121
125. Hassoun J, Derrien G, Panero S, Scrosati B (2008) A nanostructured Sn–C composite lithium battery electrode with unique stability and high electrochemical performance. Adv Mater 20:3169–3175
126. Hassoun J, Panero S, Scrosati B (2006) Electrodeposited Ni-Sn intermetallic electrodes for advanced lithium ion batteries. J Power Sources 160:1336–1341
127. Hassoun J, Mulas G, Panero S, Scrosati B (2007) Ternary Sn-Co-C Li-ion battery electrode material prepared by high energy ball milling. Electrochem Commun 9:2075–2081
128. Brutti S, Hassoun J, Scrosati B, Lin C-Y, Wu H, Hsieh H-W (2012) A high power Sn–C/C–LiFePO4 lithium ion battery. J Power Sources 217:72–76

Chapter 5
Nanostructured Lithium Titanates ($Li_4Ti_5O_{12}$) for Lithium-Ion Batteries

Lei Wen, Hong-Ze Luo, Guang-Yin Liu, and Hai-Tao Zheng

5.1 Introduction

Lithium-ion batteries (LIBs) have been widely used as power sources for portable electric devices. As one of the most important devices for energy storage, LIBs have many outstanding properties, such as high-energy density, long cycle life, fast charge–discharge ability, and environmental friendliness. In recent years, there are dramatic increased demands in LIBs as a large-scale energy storage and onboard energy storage for electric vehicles (EV), hybrid vehicles (HEV), plug-in hybrid vehicles (PHEV), and electrical energy storage (EES) systems. To meet these requirements, the key factor is to develop electrode materials for LIBs with high safety, high-power density, and long cycle life [1].

Performance of LIBs strongly depends on the properties of cathode and anode materials. Graphite has been commonly used as anode material for commercial LIBs due to its low redox potential close to Li^+/Li, good cycling stability, low cost, and environmental friendliness. However, some shortcomings of graphite inhibit their application in large-scale batteries. For example, graphite's volume expansion/contraction associated with the lithium insertion and extraction process, which results in loss of interparticle electronic contract, consequently, leads to poor

L. Wen (✉)
Shenyang National Laboratory for Materials Science, Institute of Metal Research, Chinese Academy of Sciences, Shenyang 110016, China
e-mail: leiwen@imr.ac.cn

H.-Z. Luo (✉) • H.-T. Zheng
Council for Scientific and Industrial Research, Pretoria 0001, South Africa
e-mail: hluo@csir.co.za

G.-Y. Liu
Nanyang Normal University, Nanyang 473061, China

K.I. Ozoemena, S. Chen (eds.), *Nanomaterials in Advanced Batteries and Supercapacitors*, Nanostructure Science and Technology,
DOI 10.1007/978-3-319-26082-2_5

cycling stability. Another consideration of graphite is their safety, which aroused from lithium dendrite formation due to its low Li intercalation potential at about 0 V (vs. Li^+/Li).

LTO has a lithium intercalation voltage of 1.55 V vs. Li^+/Li with a theoretical capacity of 175 mAhg^{-1}. It is considered as an appealing candidate with its fast Li^+ insertion/de-insertion ability, excellent cycle reversibility, and high safety as well as zero strain during charging/discharging. Since its discovery in 1994 [2], LTO has been investigated as a very promising anode material in large-scale batteries for EVs or renewable energy storage systems. Compared with graphite material, LTO shows a flatter potential plateau and higher operating voltage, which can solve the problem of lithium dendrite deposition entirely. In addition, as the high equilibrium potential of the Ti^{4+}/Ti^{3+} redox couple, it is above the reduction potential of common electrolyte solvents. Therefore, a solid electrolyte interface (SEI) film does not form during the charge and discharge process when LTO as anode material instead of graphite. All of the above advantages make LTO more competitive as a safe anode material for large-scale LIBs [1].

Although LTO has lots of attractive characters as an anode material, two main obstacles prevent its further application for lithium-ion batteries. One is the poor electronic and ionic conductivity of LTO, which leads to poor rate performance [3, 4]. So far, a number of strategies have been done to improve the rate performance of LTO material, including production of nanostructures [5–7], design of unique configurations [8–11], carbon coating [12–14], three-dimensional elements doping at Ti sites [15–17], rare-earth doping [18], and coating LTO with noble metal nanoparticles, oxides, or high conductive phase such as Ag [19], Cu_2O [20], and TiN phase. [3] Among these methods, fabrication of nano-LTO with various structures is believed to be one of the most promising methods to improve its high-rate performance.

Besides the low electronic conductivity, severe gasing during charge–discharge cycles is a critical problem of LTO batteries. Until now, this issue is unsettled but clearly bears heavily on electrode materials, electrolyte, and battery system [21, 22]. In order to further understand how gasing occurs and finds an effective way to remedy it, more investigations and researches are needed for fabrication of LTO materials and gasing mechanisms for LTO batteries.

5.2 Crystal Structure and Lithium Intercalation/ De-intercalation Mechanism of LTO

The $[M_2]O_4$ framework of an $Li[M_2]O_4$ spinel is an attractive host structure for lithium insertion/extraction reactions because it provides a three-dimensional (3D) network of face-sharing tetrahedral and octahedral for lithium-ion diffusion [23]. LTO has face-centered cubic structure, and the space group of LTO is *Fd3m*.

Figure 5.1a shows a spinel structure of LTO, lithium, and titanium atoms randomly distributed on one-half of the octahedral sites (16d sites) and lithium

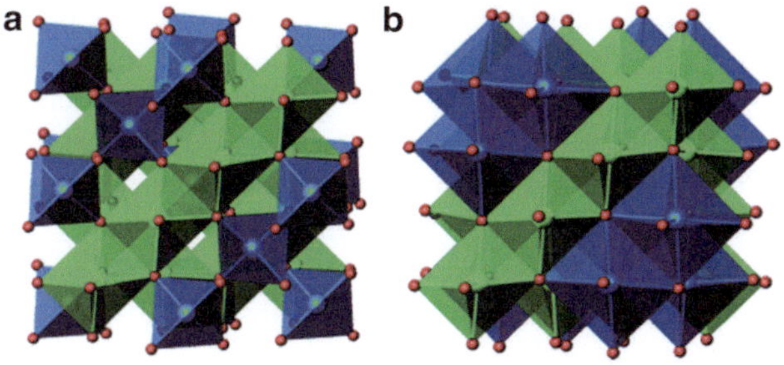

Fig. 5.1 (**a**) LTO spinel structure. *Blue* tetrahedra represent lithium, and *green* octahedra represent disordered lithium and titanium. (**b**) $Li_4Ti_5O_{12}$, rock salt structure. *Blue* octahedra represent lithium, and *green* octahedra represent disordered lithium and titanium (Reprinted with permission from [9] Copyright 2006 American Chemical Society)

atoms filling one-eighth of the tetrahedral sites (8a sites) within the oxygen close-packed lattice (32e sites) [9]. In conventional spinel notation, LTO can be written as $[Li]_{8a}[Li_{1/3}Ti_{5/3}]_{16d}[O_4]_{32e}$. As current is applied and lithium is intercalated into LTO, Ti^{4+} is reduced to Ti^{3+} within the octahedrally coordinated framework, allowing a topotactic transition from the spinel structure of LTO to the rock salt structure of $Li_7Ti_5O_{12}$. Blue octahedra represent lithium, and green octahedra represent disordered lithium and titanium. $Li_7Ti_5O_{12}$ can be written as $[Li_2]_{16c}[Li_{1/3}Ti_{5/3}]O_4$. Phase transformation during lithium intercalation/de-intercalation reaction of LTO materials can be written as follows:

$$[Li]_{8a}\left[Li_{1/3}Ti_{5/3}\right]_{16d}[O_4]_{32e} + Li^+ + e^- \leftrightarrow [Li_2]_{16c}\left[Li_{1/3}Ti_{5/3}\right]_{16d}[O_4]_{32e}$$

The capacity of LTO is determined by the amounts of octahedral sites in materials; 1 mol of LTO can accommodate 3 mol of lithium ions, and the theoretical capacity of LTO is 175 $mAhg^{-1}$. Although two-phase transformation occurs during charge–discharge process, LTO can be considered as a zero-strain material due to the volume change which is only about 0.3 %. from LTO to $Li_7Ti_5O_{12}$, and its lattice parameters increased slightly from 0.836 to 0.837 nm.

5.3 Synthesizing Methods

5.3.1 Solid-State Synthesis

Solid-state reaction method is the most widely used method for the synthesis of polycrystalline materials from a mixture of solid starting reactants. The starting solids do not react together at room temperature over normal time scales, and it is necessary to heat them up to certain temperatures, usually from 1000 to 1500 °C at which the reaction would occur at appreciable rate.

Although various wet-chemical methods can be used for the fabrication of high-purity LTO with high performance, the main disadvantage of these methods is the complicated procedures and low production yield. Unlike wet-chemical methods, economic solid-state reaction using cheap TiO_2 and Li_2CO_3 or LiOH as starting materials is more favorable and environmental friendly for large-scale production. The reaction mechanism is

$$Li_2CO_3 + TiO_2 \rightarrow Li_4Ti_5O_{12}$$

Factors which determine the rate of a solid-state reaction include reaction conditions, structural properties of the reactants, surface area of the solids, their reactivity, and the thermodynamic free energy change associated with the reaction [24]:

(1) Properties of reactants: Anatase TiO_2 has lower crystal density and looser atomic arrangement, which is more favorable for solid-state reaction between TiO_2 and Li_2CO_3. Amatucci et al. [25] synthesized nano-LTO materials by using nanocrystalline anatase TiO_2 and Li_2CO_3 as starting materials. Its high-rate performance is far superior than that of LTO materials obtained by using rutile TiO_2 as starting materials. LiOH and Li_2CO_3 can all be used as lithium source. According to Peramunage et al. [26], under similar conditions, LiOH has lower melting temperature of 450°C. Lower melting temperature allows the products to crystallize and agglomerate into larger particles. On the other hand, the higher melting point of 723°C for Li_2CO_3 enables the reaction to proceed to completion without a significant growth in the particle size of the product.
(2) Mixing methods: Ball milling is one of the most common methods used for mixing starting materials. LTO with small and well-distributed particle size can be prepared via a ball milling-assisted solid-state method [27–30]. High-energy ball milling (HEBM) is a newly developed ball milling process where a powder mixture placed in the ball mill is subjected to high-energy collision from the balls. HEBM can significantly shorten reaction time and lower sintering temperatures. Zaghib et al. [31] compared conventional ball milling and HEBM methods for the fabrication of nanocrystalline LTO phases. It was found that LTO obtained by HEBM has narrower size distribution and finer particles.
(3) Reaction temperature: The maximum fabrication temperature should be lower than 1100°C. Under high temperatures, LTO will decompose into the mixture of Li_2TiO_3 and $Li_2Ti_3O_7$ phases.
(4) Reaction time: Solid-state reaction kinetics is determined by mass diffusion. Under same reaction temperatures, LTO shows higher crystallinity and stability with longer reaction time. But compared with the nano-LTO, micro-LTO has longer diffusion length and thus is unfavorable for high charge–discharge tests.
(5) Atmosphere:Inert atmosphere is unnecessary for the fabrication of LTO with high purity and crystallinity. But under reducing atmosphere, Ti^{4+} can be reduced to Ti^{3+}. The electronic states of oxygen deficient are usually located below conduction bands and even partially overlap with conduction bands when a high density of oxygen vacancies exist, which shows an obvious

advantage of improving the electronic conductivity of the material. The oxygen-deficient LTO shows enhanced lithium-ion insertion/extraction kinetics, especially at high rates. [32]

One main challenge of solid-state reaction is to synthesize high-purity LTO without any impurity. However, the intermittent presence of an unwanted rutile TiO_2 phase in the final LTO has been a significant question for the LTO synthesis [33–35]. It is always believed that the synthetic processes and starting materials were the main reasons for the existence of unwanted rutile TiO_2, including the molar ratio of starting materials, sintering temperature, reaction time, starting material's crystal structure, and particle size [9, 33, 36–38].

In a study on the formation of LTO from anatase TiO_2 and $LiNO_3$, Yuan et al. [36] reported the evolution of rutile TiO_2, which was found to be very difficult to eliminate. The phase transition of TiO_2 from anatase to a more stable rutile phase and the solid-phase reaction between TiO_2 and Li_2CO_3 happened competitively. The mechanism shown in Fig. 5.2 proposed by Yuan et al. [36], which suggested

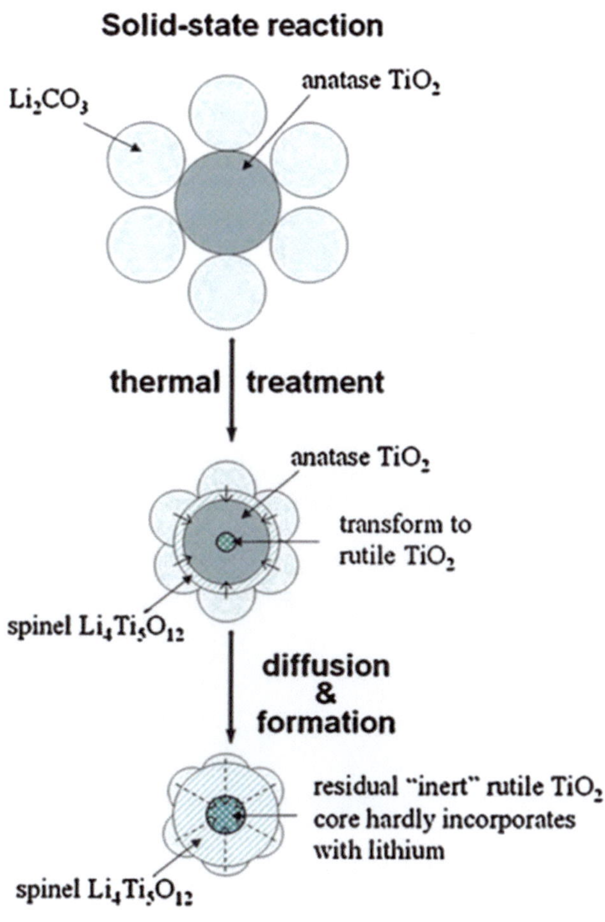

Fig. 5.2 Mechanism illustration of the solid-state reaction for LTO synthesis (Reprinted with permission from [36], Copyright 2010 Elsevier)

that the "inert" rutile TiO_2 transited from anatase TiO_2, was remained in the final product, and it cannot be eliminated even at 750°C for 100 h.

Matsui et al. [39] also observed the presence of rutile TiO_2 in the final LTO powder. Since the existence of this unwanted rutile TiO_2 is detrimental to electrochemical property enhancement, pursuit of LTO materials with high purity and high crystallinity without rutile impurity phase is an essential prerequisite of a successful application.

5.3.2 *Sol–Gel Synthesis*

Sol–gel method is a chemical solution process used to prepare solid materials from small molecules. It has been known since the late 1800s. A sol is a colloidal or molecular suspension in a solvent. A gel is a semirigid mass that forms when the solvent in the sol begins to evaporate. The sol–gel synthesis appears to be a suitable method for the fabrication of nano-LTO materials [40–44]. The sol–gel method can produce high-purity and homogeneous LTO materials with high specific capacity and good cycling and rate performance [40, 45].

Figure 5.3 shows the nano-LTO with a high purity that was synthesized by a modified sol–gel method with ethylenediaminetetraacetic acid and citric acid as a

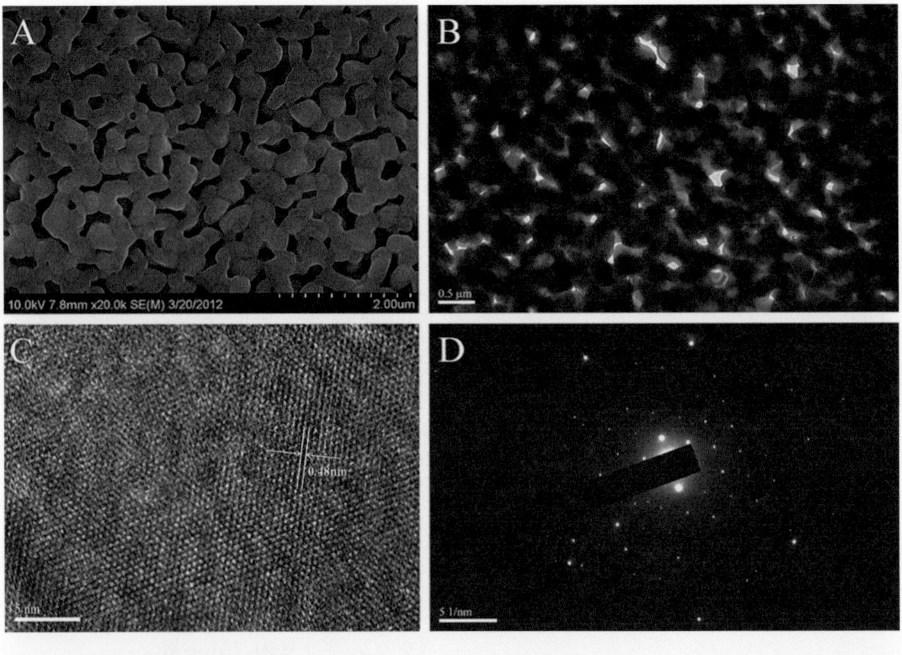

Fig. 5.3 (**a**) SEM, (**b**) TEM, (**c**) HRTEM, and (**d**) SAED images of LTO (Reprinted with permission from [46] Copyright 2013 Elsevier)

bicomponent chelating agent [46]. The nano-LTO shows a network morphology with high dispersion, which reached a capacity of 164 and 108 mAh g^{-1} at 1 and 10 C discharge rate, respectively. It has an excellent cycle performance. The capacity retention ratio is above 97 % after 1000 cycles at 1 C. Shen et al. [40] used sol–gel method to synthesize the nano-LTO material, which showed the initial capacity of 172 mAh g^{-1} in the range 1.0 $\sim$ 2.5 V.

Hao et al. reported several work on synthesis of LTO by the sol–gel method [42, 47–49]. It was found that electrochemical properties of these LTO materials were influenced by the various complex agents, such as acetic acid, citric acid, oxalic acid, and triethanolamine. All samples have an approximately initial discharge capacity $\sim$ 170 mAhg^{-1}. However, LTO synthesized by triethanolamine and citric acid showed discharge capacities of 150 and 137 mAhg^{-1}, while the samples prepared by oxalic acid and acetic acid delivered 133 and 117 mAhg^{-1} after 30 cycles, respectively [47].

Sol–gel process seems to be a viable way to prepare nanocrystalline spinel LTO; however, it is still unsuitable for practical application because of its complicated procedure, low yield, time consuming, and high synthetic cost by using expensive organic reagents [50–52].

5.3.3 Hydro-/Solvothermal Synthesis

Hydrothermal synthesis is a process that utilizes single- or heterogeneous-phase reactions in aqueous media at elevated temperature ($T > 25$ ¤C) and pressure ($P > 100$ KPa) to crystallize inorganic materials directly from the solution. Syntheses are usually conducted at autogenous pressure, which corresponds to the saturated vapor pressure of the solution at the specified temperature [53].

Advantages of the hydrothermal synthesis method include the ability to synthesize materials which are unstable near the melting point and the ability to synthesize large crystals of high quality. Disadvantages of this method are the high cost of equipment and the inability to monitor crystals in the process of their growth. Hydrothermal synthesis can be affected by temperatures and pressures below the critical point for a specific solvent.

The single-crystalline LTO nano-sheets along the (110) facet (as shown in Fig. 5.4) with high reversible capacity and good cycling performance via a simple and low-cost hydrothermal route, using only tetrabutyl titanate, $LiOH \cdot H_2O$ and distilled water, was reported by Li et al. [54]. LTO nano-sheets delivered an initial discharge capacity of 183 mAhg^{-1} together with a discharge capacity of 160 mAhg^{-1} after 100 cycles at 1 C. The discharge capacity could reach up to 120 mAhg^{-1} even after 300 cycles at 10 C.

Shen et al. [55] fabricated LTO nano-wire arrays growing directly on Ti foil and enhance its electronic conductivity by creating Ti^{3+} sites through hydrothermal process. Without using any ancillary materials, the as-synthesized LTO nano-wire

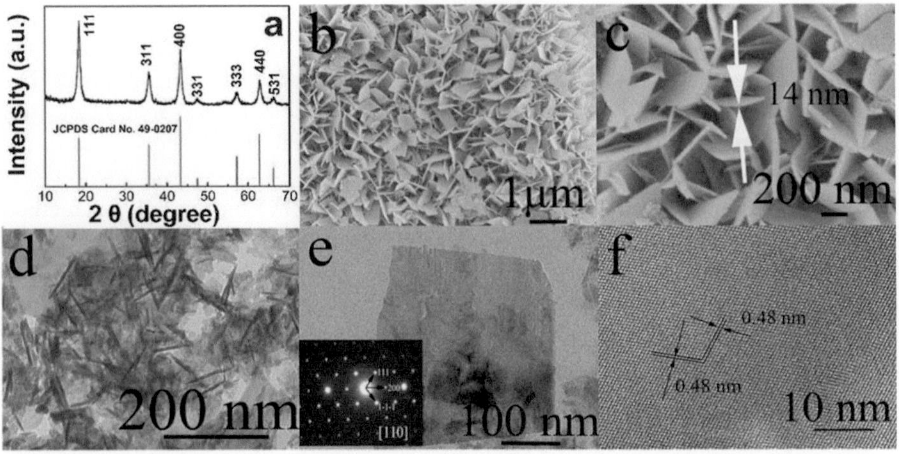

Fig. 5.4 (**a**) XRD pattern of LTO nano-sheets. (**b, c**) SEM, (**d, e**) TEM, and SAEM images (inset in e) of LTO nano-sheets. (**f**) HRTEM image of a LTO nano-sheet (Reprinted with permission from.[54] Copyright 2012 Royal Society of Chemistry)

arrays exhibited an excellent rate capability and a significantly enhanced cycling performance.

The solvothermal method is an alternative method to prepare LTO with small and uniform particles. The difference between the solvothermal and the hydrothermal methods is the solvent used in the preparation process. The solvothermal method uses the organic solvents instead of water as the solvent. Reaction temperature of solvothermal method can be much higher than that in the hydrothermal method when use an organic solvent with high boiling points. Solvothermal method also has better control of the shape distributions and a better crystallinity of the nano-LTO relative to the hydrothermal method.

Tang et al. [10] synthesized a flowerlike LTO by using ethylene glycol solution. The flowerlike LTO exhibited a high reversible capacity and an excellent rate capability of 165.8 $mAhg^{-1}$ at 8 C. The LTO nanotubes/nano-wires with high surface area were synthesized by a low-temperature hydrothermal lithium-ion exchange processing from hydrogen titanate nanotube/nano-wire precursors [56]. The shape and morphology of LTO are controllable by varying hydrogen titanate precursors (nanotube, nano-wire, nano-rod, and nano-belt) from alkaline-hydrothermal approach. The formation temperature of LTO nanotubes/nano-wires is lower than that of bulk materials prepared by solid-state reaction or by sol–gel processing. In hydrothermal syntheses, pH value, concentration of reactants, the pressure, and the solvent have significant effect on the properties on the resultant LTO [10, 56–59].

Hydro-/solvothermal synthesis attracts considerable interest from the theoretical viewpoint as a process for the preparation of nano-materials, but its implementation for large-scale production of LTO is still questionable.

5.3.4 Molten Salt Synthesis

Molten salt synthesis is a simple method to prepare multicomponent oxide, which can provide a liquid reaction environment when serving as reaction media, thus enhancing the crystal formation reaction and lower fabrication temperature [60–62].

Cheng et al. [63] synthesized nano-LTO using LiCl as a high-temperature flux. The synthesized spinel LTO has a discharge capacity over 150 $mAhg^{-1}$ at a current rate of 40 $mAhg^{-1}$ between 1.0 V and 3.0 V. A series of LTO via a composite molten salt method using the mixtures of LiCl and KCl with different molar ratio as the reaction media were reported by Bai et al. [62]. It is found that the melting point of the composite molten salt can effectively influence the formation of particles and leads to different electrochemical performances. A nanocrystalline LTO–TiO_2 duplex phase was synthesized by a molten salt process using an eutectic mixture of $LiNO_3$–LiOH–Li_2O_2 at 400 $\sim$ 500°C. [6] The results indicated that pure LTO phase with very small crystal size can be partially converted to anatase TiO_2 as a secondary phase with increasing heat treatment temperature. Electrochemical testing demonstrates that LTO–TiO_2 with a duplex crystallite size (fine ($<$ 10 nm) and coarse ($>$ 20 nm)) has a stable high charge–discharge capacity with excellent rate capability.

Guo et al. [64] synthesized nano-LTO doped with Fe^{3+} doping in LiCl–KCl molten salts. Doped Fe^{3+} was homogenously distributed in the crystal lattice of LTO and slightly increased the lattice parameters. The addition of 0.1 molar ratio Fe_2O_3 to LTO reduced the average particle size of LTO from 1 μm to about 200 nm. The prepared $Li_4Ti_{4.8}Fe_{0.2}O_{12}$ delivered a capacity of 173.7 $mAhg^{-1}$ at 0.2 C and a capacity of 103.4 $mAhg^{-1}$ at 10 C, much larger than the value of pure LTO (28.7 mAh g^{-1}).

5.3.5 Combustion Synthesis

Combustion synthesis is a well-known method for the preparation of nanocrystalline oxides [65–67]. The method, involving redox reaction between oxidizer and a fuel, is simple, instantaneous, and cost-effective. The method yields nanoscale material almost instantaneously and can be used to prepare a variety of complex oxides in nanocrystalline form [5, 68, 69].

Yuan et al. [70] reported the synthesis of nano-LTO powders by combustion synthesis with combustion reaction initiated at $<$ 250°C. Pure phase and well-crystallized nano-LTO were obtained after a calcination process. The LTO calcined at 700°C exhibited the best and high electrochemical performance, which reached a capacity of 125 mAh g^{-1} at 10 C discharge rate with fairly stable cycling performance. Prakash et al. [5] prepared a nanocrystalline LTO in cubic spinel phase by a

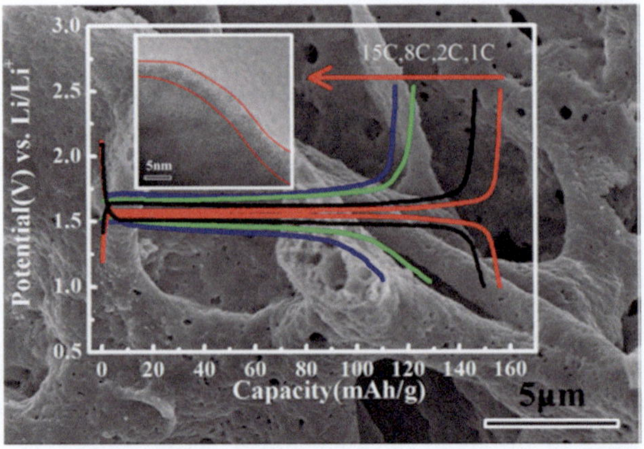

Fig. 5.5 Nitrogen-modified LTO prepared via an one-step combustion process (Reprinted with permission from [71] Copyright 2014 American Chemical Society)

single-step solution–combustion method in less than 1 min. The LTOs were agglomerated with crystallites of varying sizes between 20 and 50 nm plus a three-dimensional interconnected porous network. The LTO delivered a capacity value close to the theoretical value of 175 mAh g^{-1} at C/2 rate. Excellent performance of LTO was ascribed to the nanostructure and highly porous morphology, which provide short diffusion paths for Li diffusion.

The synthesis of nitrogen-modified nano-LTO via a one-step combustion process was reported by Li et al. [71] Fig. 5.5 shows that the nitrogen-modified LTO exhibited high-rate capabilities and high stabilities for lithium storage. Excellent electrochemical performance of the nitrogen-modified LTO was attributed to the nanoparticles with a porous structure and nitrogen modification, which enabled faster ion transport and better electron conductivity. This surface modification is relatively simple and inexpensive and may be a helpful approach to solving the problems of poor electronic conductivity and rate performance.

5.3.6 Spray Drying and Spray Pyrolysis Synthesis

Spray drying process is a continuous and single-step preparation method for producing homogeneous and multicomponent powders from a liquid or slurry by rapid drying with hot gas. Spray pyrolysis is similar to spray drying, except that solute is decomposed at high temperatures in spray pyrolysis and usually the process temperature is higher [72].

Yoshikawa et al. [73] synthesized LTO materials by using a spray drying method, followed by high-temperature calcination for large-scale production. It was found that there was no impurity phase, such as $r-TiO_2$, Li_2TiO_3, and Li_2CO_3 in the final product when the spray drying precursors at the Li/Ti ratio of 0.860. These pure-phase materials showed a higher discharge capacity of $\sim$ 164 $mAhg^{-1}$

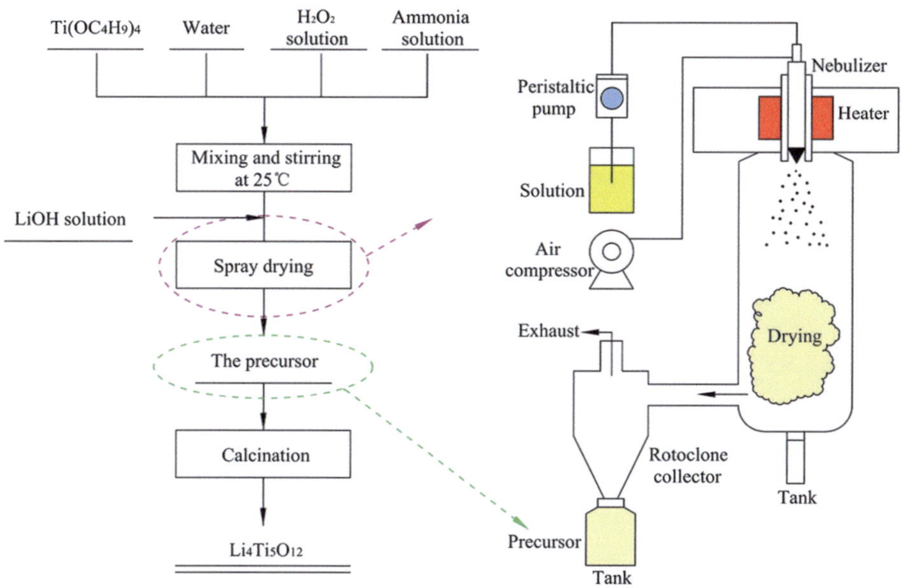

Fig. 5.6 The experimental procedure and spray drying apparatus (Reprinted with permission from [74] Copyright 2012 Elsevier)

in the voltage of 1.2 – 3.0 V, with an excellent cyclic performance and superior rate performance.

Figure 5.6 shows a simple spray drying procedure for synthesizing a spherical-shaped and nano-micron LTO. Reprinted with permission from [74] copyright 2012 Elsevier.

The spherical LTO with a narrow particle size distribution of an average diameter of 0.5 ~ 1 μm and no aggregation was successfully synthesized via a spray pyrolysis method without further treatment as shown in Fig. 5.7 [75]. The purity and electrochemical properties of LTO powders are strongly affected by the synthesis temperature.

The yolk–shell-structured LTO can be prepared by spray pyrolysis by Yang et al. [76]. Du et al. [77] synthesized a nanocrystalline LTO with in situ incorporation of carbon and Ti^{3+} by an in situ spray pyrolysis, producing crystal sizes ranging from 10 to 30 nm. The LTO delivered a discharge capacity of 145.8 $mAhg^{-1}$ at 10 C for up to 500 cycles. In the full battery tests with $LiCo_{0.16}Mn_{1.84}O_4$ as cathode, it also showed excellent capacity and cycling stability.

Compared with the solid-state and sol–gel methods, spray drying and spray pyrolysis are simpler and faster methods. In addition, the particle size distribution is typically narrow and controllable from the micrometer to sub-micrometer order; the purities of the products are high and the compositions of powders are easily controllable [78].

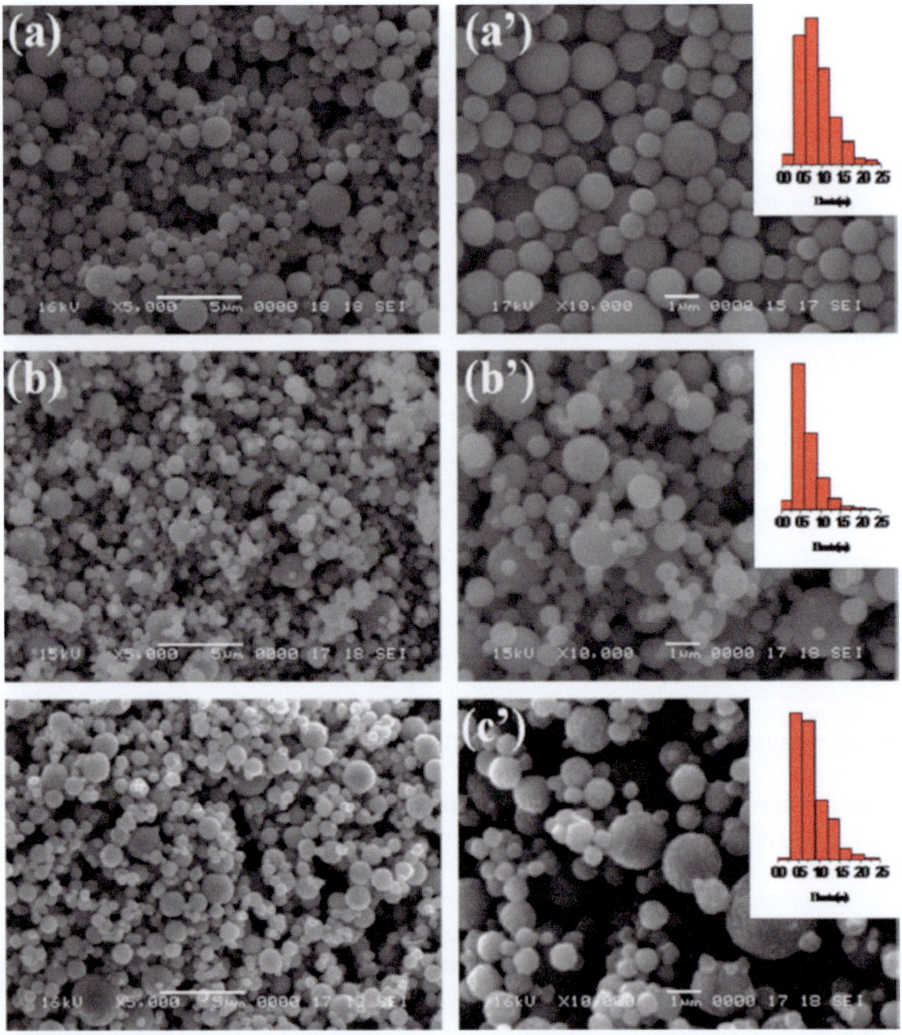

Fig. 5.7 SEM images of LTO synthesized at (**a**) and (**a'**) 800°C; (**b**) and (**b'**) 900°C; (**c**) and (**c'**) 1000°C (Reprinted from [75] with permission from Elsevier)

5.3.7 Microwave Synthesis

Recently, the microwave method has attracted a great deal of attention to synthesize LTO due to its advantages such as low synthetic temperature, short reaction time, and high synthesis rate [38, 79–82].

Li et al. [38] synthesized the sphere-like LTO nanoparticles (40 $\sim$ 50 nm) via a microwave method using Li_2CO_3 and TiO_2 as reactants. The sphere-like nano-LTO

Fig. 5.8 SEM (**a, c**) and TEM (**b, d**) images of LTO synthesized at 130°C (**a, b**) and 170°C (**c, d**). SAED pattern in the inset of (**b**) shows reflection rings corresponding to LTO [81]

has a stable voltage plateau, high discharge capacity, and an excellent cycle performance. Liu et al. [81] reported that nano-flowerlike (Fig. 5.8a, c) and nanoparticle LTO (Fig. 5.8b, d) was synthesized by a microwave-assisted hydrothermal method following heat treatment. The nano-flowerlike LTO (Fig. 5.8a, c) exhibited a layered structure of high specific surface area and provided good reversibility and cycling performance. The nano-flowerlike LTO delivered a specific capacity of 176.7 mAh g^{-1} during the first cycle and remained a capacity of 138.4 $mAhg^{-1}$ after 100 cycles. The better performance of nano-flowerlike LTO is attributed to the larger specific surface area and shorter Li^+ diffusion path.

The microwave synthesis is straightforward, fast, clean, and energy-efficient method that probably promotes this technology to commercial application [79].

5.3.8 Other Methods

Wu et al. [83] synthesized nano-LTO via electro-spinning in a solution containing lithium acetate, titanium tetra(IV)-isopropoxide, polyvinyl acetate, and acetic acid in N,N-dimethylformamide, subsequently followed a calcination. The prepared nano-LTO showed good rate capability and cycling stability. Yin et al. [84] successfully synthesized LTO by a rheological phase method using lithium acetate dihydrate and tetra-n-butyl titanate as the raw materials. The prepared LTO showed excellent high-rate capability and cycling stability due to the high purity and the small particle sizes with narrow particle size distribution.

Liu et al. [85] reported a facile approach for synthesizing nanocrystalline LTO by microemulsion at ambient temperature followed by annealing treatment at 600° C for 2 h. The prepared LTO nanoparticles were in the sizes of 30 $\sim$ 50 nm with high crystallinity, which displays excellent charge–discharge rate capability and cycling stability. Nugroho et al. [34, 86–89] prepared high phase-pure LTO using supercritical fluid synthesis method. The nano-LTO with a high crystallinity can be produced in a very short reaction time (15 min).

As mentioned above, different methods have been investigated for synthesizing LTO in order to improve its electrochemical properties. It can be found that the synthesis methods influence the electrochemical properties of LTO. However, most of the methods are only suitable for small amount production of LTO. So far, the solid-state reaction is still a suitable method to prepare LTO materials for practical mass production.

5.4 Nanostructural Design LTO

Nanostructured LTOs have demonstrated a great improvement on the electrochemical performance, such as specific capacity, rate performance, and cycle life.

Neutron diffraction has proved increased Li-ion capacity in nano-$Li_{4+x}Ti_5O_{12}$ was due to simultaneous occupation of both tetrahedral (8a) and octahedral (16c) sites. It is suggested that the simultaneous 8a and 16c occupation is more easily accommodated in the near-surface region of the particle compared to the bulk. Borghols et al. [90] emphasized that a surface effect plays an important role in nanoscale intercalation compounds. On the other hand, the small size of LTO particle leads to a larger capacity but also causes a lower reversible capacity. The capacities in different sizes of LTOs were shown in Fig. 5.9.

In this section, we summarize the present status of nanoscale LTOs with different morphologies and properties as well as their electrochemical performance.

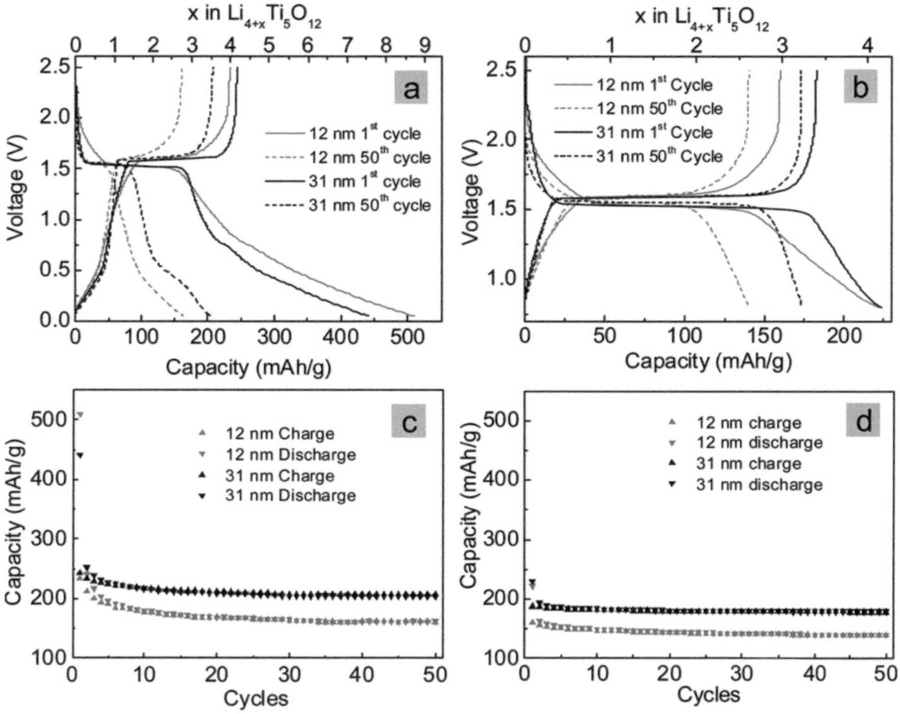

Fig. 5.9 Performances of $Li_{4+x}Ti_5O_{12}$ in different size reported [90]

5.4.1 Zero-Dimensional Structure

The zero-dimensional nano-materials have nano-dimensions in all three directions. Most of these nanoparticles are spherical in shape and the diameter of these particles is in the range of 1 ~ 100 nm. Cube and polygon shapes are also found for this kind of nano-materials.

Kim et al. [91] prepared nano-LTO via a flame spray pyrolysis process. The particle size of the precursor powders was 23 nm. The discharge capacity of the prepared nano-LTO was 157 mAh g^{-1} and the capacity retention is 98 % over 50 cycles. Venkateswarlu et al. [92] prepared nano-LTO with cubic structure and average size around 39 nm by a sol–gel process. The LTO electrode maintains the cubic spinel structure over cycling which results in good cycle performance and the capacity retention. It was worthy to note that the discharging capacity and the charging capacity obtained for the nano-LTO electrode in the potential range of 0.01–1.75 V were 255 and 242 mAh g^{-1}, respectively. Zhang et al. [93] reported a facile approach for synthesizing nanocrystalline LTO via sol–gel process by employing a nonionic surfactant (EO)20(PO)70(EO)20 tri-block copolymer (pluronic P123) as the chelating agent. The prepared pure-phase LTO (~ 100 nm)

Fig. 5.10 Morphology of nanocrystalline LTO: (**a**) representative scanning electron micrographs (SEM) image of LTO (the inset shows HRTEM image); (**b**) high-resolution transmission electron microscopy (HRTEM) image of LTO (the inset shows the SAED pattern for LTO) (Reprinted with permission from [93] Copyright 2011 Elsevier)

with high crystallinity (Fig. 5.10) exhibited high-rate capacity of 108 mAh g^{-1} at 40 C, excellent cycling stability, and retaining over 97 % of its initial capacity after 250 cycles at varying rates.

5.4.2 *One-Dimensional Structure*

For one-dimensional nanostructures, its one-dimension is over the range of the nanometer. These structures include nano-wires, nano-rods, and nanotubes. These materials are several micrometers in length, but with diameter of only a few nanometers. These one-dimensional nanostructures have large surface-to-volume ratio that stimulates contact material with electrolyte efficiently and provides efficient electron transport pathways and facile strain relaxation during charge and discharge of battery [94].

LTO nano-rods were prepared by hydrothermal [95–97], hydrothermal ion exchange [98, 99], and molten salt method. Moreover, the performance of LTO nano-rods was improved by the further surface coating [97].

Kim et al. [100] prepared LTO nano-wires with a diameter of 150 nm. The first discharge capacity was 165 mAh g^{-1} at a 0.1 C rate and 93 % capacity was retained even at 10 C rate. These results were associated with more active surface in the nano-wires which facilitate fast lithium diffusion due to short pathways. Li et al. [56] prepared LTO nanotubes and nano-wires with high surface by a low-temperature hydrothermal lithium-ion exchange method. The shape and morphology of LTO were controllable by varying the hydrogen titanate precursors (nanotube, nano-wire, nano-rod, and nano-belt) from the alkaline-hydrothermal approach. The well-reversible cyclic voltammetric results of both electrodes indicated enhancement of electrochemical kinetics for lithium insertion. Lee et al. [58] synthesized LTO nanotubes using an alkali-hydrothermal reaction with heat

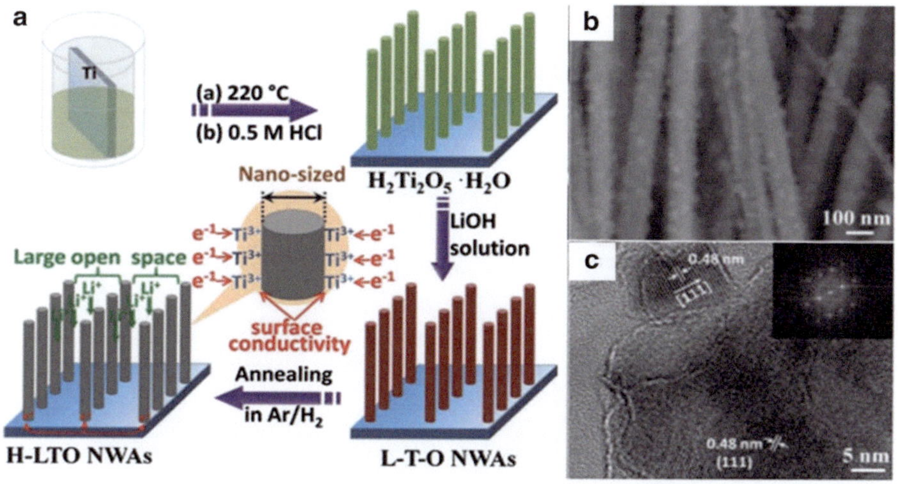

Fig. 5.11 (**a**) Schematic illustration for the fabrication of H-LTO NWAs (nano-wire arrays), (**b**) cross-sectional SEM images of H-LTO NWAs annealing at 600 ° C, (**c**) TEM images of H-LTO NWAs annealing at 600 ° C (Reprinted with permission from [55] Copyright 2012 Wiley)

treatment through a simple structural rearrangement. The LTO nanotubes exhibited a reversible capacity of 156 mAh g^{-1} at a rate of 0.1 C. Shen et al. [55] synthesized LTO nano-wire arrays (NWAs), as shown in Fig. 5.11. Without any ancillary materials, the synthesized LTO NWAs manifested an excellent rate capability and an enhanced cycling performance.

Jo et al. [101] synthesized LTO nano-fibers through electro-spinning. The LTO nano-fibers have displayed an outstanding electrochemical performances with significantly enhanced discharging/charging properties, even at high rates of 10 C, which demonstrated that the nano-fiber offers an attractive architecture for enhancing kinetics.

5.4.3 Two-Dimensional Structure

In this type of nano-materials, two dimensions are over the nanometer range included different kind of nano-films such as coatings, thin-film multilayers, and nano-sheets as well as nano-walls. The area of the two-dimensional nano-materials can be large (several square micrometer) but with the nanoscale range of thickness.

LTO nano-sheets stacked by ultrathin nano-flakes are derived from the interlayer splitting and exfoliation of the layered orthorhombic precursors [102]. The ultrathin nano-flakes were interconnected with thickness of 3 nm. The LTO nano-sheets with a stacked structure delivered an initial discharge capacity of 175.9 mAh g^{-1} and a discharge capacity of 166.8 mAh g^{-1} after 100 cycles at 0.5 C. The discharge capacity still remains 100.2 mAh g^{-1} even at 20 C. It was concluded that ultrathin

Fig. 5.12 Cross-sectional (**a**) and top-view (**b**) SEM images of LTO nano-sheet arrays standing on Ti foil. (**c**) TEM images of large LTO nano-sheets scratched from Ti foil (Reproduced from [103] with permission of The Royal Society of Chemistry)

nano-flakes probably promote the kinetic property of lithium ions and electron diffusion through offering more transporting channels, which is favorable for high-rate performance [102].

The well-aligned LTO nano-sheet arrays were fabricated directly on conductive Ti foil by hydrothermal growth in LiOH solution, as shown in Fig. 5.12 [103]. When used as a binder-free anode, the self-supported LTO nano-sheet arrays standing on Ti foil exhibited an excellent rate capability, a reversible capacity of 163 and 78 mAh g^{-1} at 20 C and 200 C, respectively, and an outstanding cycling performance with the specific capacity of 124 $mAhg^{-1}$ even after 3000 cycles at 50 C. Furthermore, a flexible lithium-ion battery using the LTO nano-sheet arrays as the anode was fully recharged within 30 s and has lighted a light-emitting diode (LED).

Zhao et al. [104] prepared LTO thin-film anode with high discharge capacity and excellent cycle stability by using ink-jet printing technique. It was found that the average thickness of 10-layer LTO film was about 1.7 $\sim$ 1.8 μm, and the active material LTO in the thin film was nanoscale about 50~300 nm. The prepared LTO thin film exhibited a high discharge capacity of about 174 mAh g^{-1} and the discharge capacity in the 300th cycle retained 88 % of the initial discharge capacity.

5.4.4 Three-Dimensional Structure

All dimensions of three-dimensional materials are over the nanoscale range. Three-dimensional materials can be fabricated by assembling individual blocks with the nanoscale (1 ~ 100 nm).

The flowerlike LTO consisting of nano-sheets (as shown in Fig. 5.13a, b) was synthesized by a hydrothermal process in glycol solution and following calcination [10].

This structured LTO exhibited a high reversible capacity and an excellent rate capability of 165.8 $mAhg^{-1}$ at 8 C, indicating potential application for lithium-ion batteries with high-rate performance and high capacity. They also synthesized mesoporous LTO microspheres (Fig. 5.13c, d) by template-free hydrothermal process in ethanol–water mixed solution with heat treatment. The LTO mesoporous spheres exhibited high-rate performance of 114 $mAhg^{-1}$ at 30 C and good capacity retention of 125 $mAhg^{-1}$ after 200 cycles at 20 C. [11]

Fig. 5.13 SEM (**a**) and TEM (**b**) images of the as-prepared product at different magnifications Reprinted from [10] with permission from Elsevier. (**c**) Spinel LTO spheres and (**d**) TEM of mesoporous LTO spheres (Reproduced from [11] with permission of The Royal Society of Chemistry)

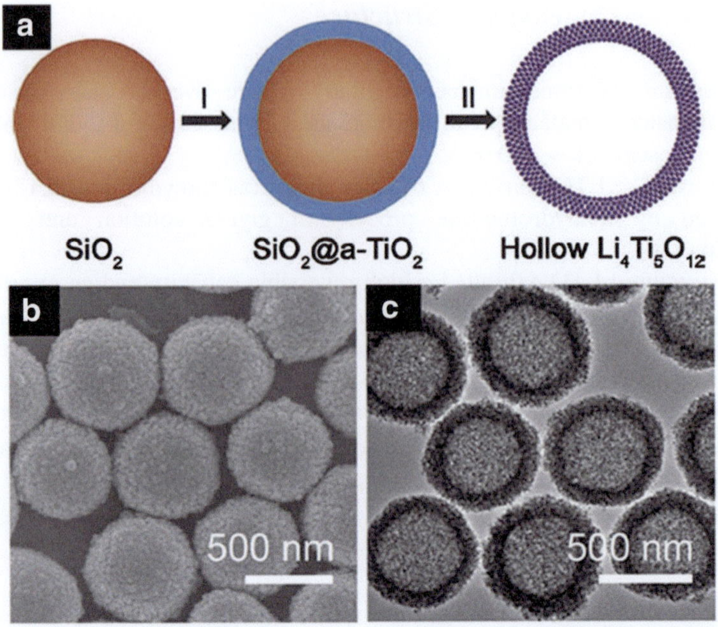

Fig. 5.14 (**a**) Schematic illustration of the formation of mesoporous LTO hollow spheres through a templating approach, FESEM (**b**) and TEM (**c**) images of LTO hollow spheres (Reproduced from [105] with permission of Wiley)

Currently, works on nanoscale electrode materials mainly concentrated on low-dimensional materials, such as nanoparticles, nanotubes, and nano-films. It is found that morphologies have significant effects on the electrochemical properties of electrode materials. Among these methods, hierarchical nano-/micro-structured materials show superior performance. The nano-/micro-hierarchical material has the advantages of both nano- and micro-materials. It not only has a large specific surface area and short lithium diffusion path length but also has higher thermodynamic stability.

LTO hollow spheres with mesoporous shells of tunable thickness were synthesized by a facile template method as shown in Fig. 5.14 [105]. Benefiting from the unique structural features, the prepared LTO hollow structures exhibited remarkable rate capability up to 20 C and stable long-term capacity retention for over 300 cycles.

Chen et al. [106] prepared LTO microspheres by assembling nano-sheets via a hydrothermal process and subsequent thermal treatment. The prepared LTO showed an intriguing and unique sawtooth-like morphology (Fig. 5.15). It exhibited an exceptional high-rate performance and excellent cycling stability, as high as 132 mAh g^{-1} after 200 cycles at 10,000 mAg^{-1}.

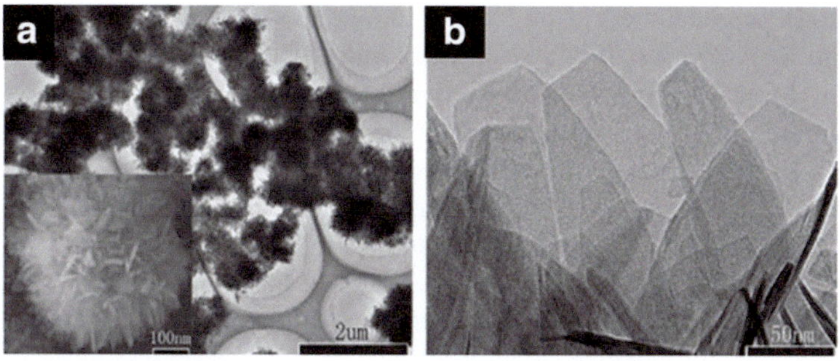

Fig. 5.15 (**a**) SEM and (**b**) TEM micrographs of as-prepared layered hydrous LTO (Reprinted from [106] with permission from Elsevier)

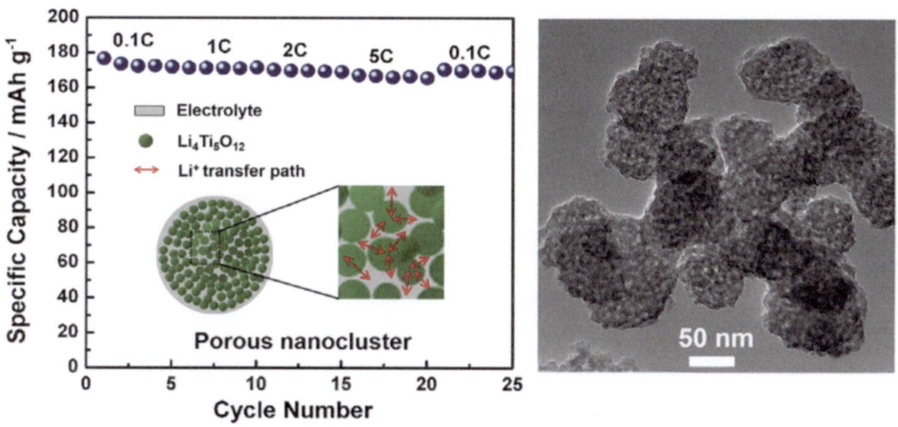

Fig. 5.16 Performance and TEM of porous LTO nano-clusters (Reprinted from [110] with permission from Elsevier)

5.4.5 Porous LTO Materials

Recently, porous electrode materials have attracted more attentions because of their large contact surface area with the electrolyte and the possibility of forming a 3D mixed conducting network in which mesoporous allows both Li^+ and e^- to migrate rapidly, leading to a superior rate performance [9, 107–109].

The porous LTO nano-clusters with high surface were synthesized by a facile solution-based method followed by low-temperature calcination, as shown in Fig. 5.16 [110]. The porous LTO nano-clusters delivered a high initial capacity of $173\ mAhg^{-1}$ at 0.5 C and a high retention of the capacity $162\ mAhg^{-1}$ after 60 cycles.

Nugroho et al. [88] prepared the mesoporous LTO by a facile template-free route in a very short reaction time. Nanoscale, primary LTO particles ($5 \sim 10$ nm) are loosely aggregated and form micron-sized, secondary mesoporous spheres (0.2 ~ 2.0 mm) with a pore size of 2~10 nm. The discharge capacity after 400 cycles at 1 C was 134.9 $mAhg^{-1}$, and the discharge capacity at 10 C was 108.5 $mAhg^{-1}$.

The hierarchically porous LTO microspheres were synthesized via hydrothermal treatment of commercial anatase TiO_2 powder in LiOH solution and a following calcination without employing any surfactants or templates, as shown in Fig. 5.17 [111]. The nano-/micro-materials exhibited rich hierarchical pores and a specific surface area of 57.5 m^2g^{-1}, which delivered good columbic efficiency, ultrahigh-rate capability at high rates, and excellent capacity retention over 200 cycles.

The LTO thin-film electrodes with well-defined mesoporous morphology were prepared by a facile polymer-directed sol–gel route, as shown in Fig. 5.18 [112]. Mesoporous LTO thin-film electrodes exhibited enhanced lithium-ion storage capabilities at short charging times but also are able to maintain stable cycling performance at rates as high as 64 C. It was believed that the unique combination of nanoscale porosity with electrical conductivity of the same order as that of ordinary bulk nanocrystalline LTO is responsible for the facile lithium intercalation observed in these thin-film materials [112].

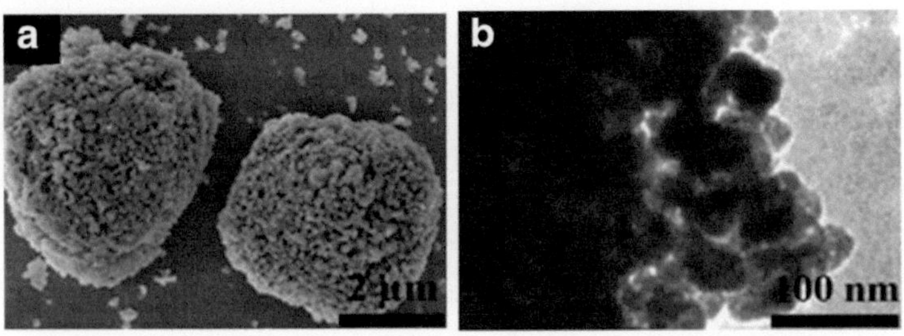

Fig. 5.17 (**a**) SEM of spherical LTO materials, (**b**) TEM image of hierarchical porous LTO microspheres (Reproduced from [111] with permission of The Royal Society of Chemistry)

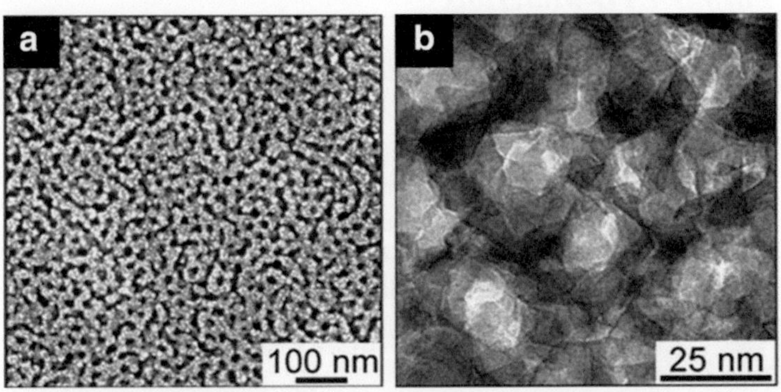

Fig. 5.18 (**a**) SEM and (**b**) TEM images of mesoporous LTO materials (Reproduced with permission from.[112] Copyright 2011 American Chemical Society)

5.4.6 Nano-Carbon Composite Materials and Carbon Coating Materials

Carbon coating and forming LTO/C composite materials have been considered as an effective way to improve rate performance of LTO. LTO can be coated with different carbon forms, such as amorphous carbon, carbon nanotubes, and graphene in order to enhance its electronic conductivity [4, 22, 89, 96, 107, 113–119]. It was found that carbon coating is an effective method to improve the electrochemical performance of LTO anode for lithium-ion batteries in terms of capacity, rate capability, and cyclability.

Using a thermal vapor decomposition method [12], LTO was coated with a uniform nano-thickness graphitized carbon. The coated LTO at 800°C has a 5 nm thick carbon layer and an electrical conductivity of 2.05 S cm^{-1}, which is much higher than that of pristine materials. For simple carbon-mixed materials, the surface of LTO materials contacts electrolyte directly. So, electrochemical process only occurs at the contact point between active materials and carbon conductive. While for carbon-coated LTO materials, lithium intercalation/de-intercalation occurs at the outer carbon layer.

Li et al. [4] synthesized LTO/C with lump morphology by a hydrothermal process followed by low-temperature heat treatment. The synthesized LTO/C delivered a super rate performance with a capacity of 151 and 136 $mAhg^{-1}$ at a charge and discharge rate of 10 and 20 C, respectively. Zhao et al. [107] reported the porous LTO coated with a very thin and uniform N-doped carbon coating layer from ionic liquids, as shown in Fig. 5.19. The material exhibited superior rate

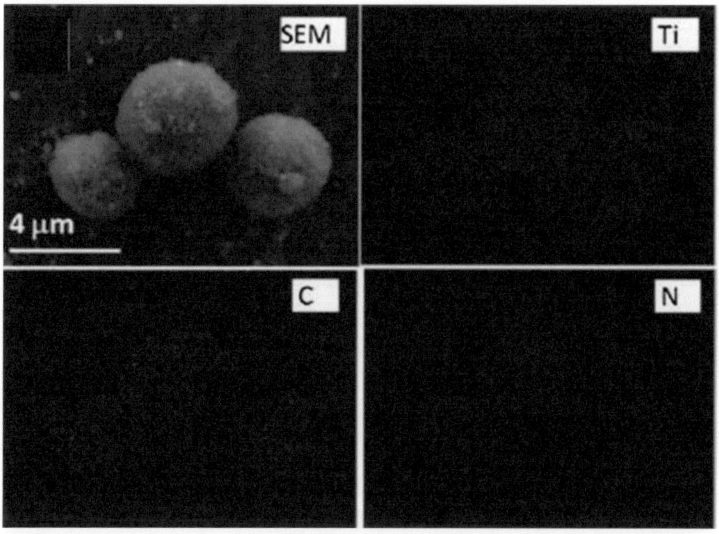

Fig. 5.19 SEM of N-doped carbon coating LTO materials and EIS mapping of Ti, C, and N elements (Reproduced from [107] with permission from Wiley)

capability and excellent cycling performance versus Li. The initial capacity was 150 $mAhg^{-1}$ at 2 C, which decreased to 124 $mAhg^{-1}$ after 2200 cycles at 2 C, with capacity retention of 83 %.

A simple and scalable process was reported to fabricate LTO nano-sheets with N-doped carbon coating [120]. Using this material without any current collector, they developed a flexible film which can be directly used as a flexible electrode for lithium-ion batteries. This flexible electrode shows an excellent rate capability and cycling performance due to the thin carbon coating and porous nano-sheet structures, which produce a highly conductive pathway for electrons and fast transport channels for lithium ions.

Carbon nanotubes (CNTs) and graphene are representatives of one-dimensional and two-dimensional carbon materials. CNTs and graphene materials are expected to be applied in lithium-ion batteries due to their unique structural, mechanical, and electrical properties. CNT or graphene composite materials can effectively improve the specific capacity, cycle life, and high-rate performance.

An in situ growth of LTO on multi-walled carbon nanotubes (MWNT) via a sol–gel method combined with a hydrothermal process and a short post-annealing of morphology-controlled coaxial nano-cables was reported by Shen et al. [121] as shown in Fig. 5.20. Such MWNT@LTO coaxial nano-cables consist of highly conducting MWNT cores and nanoscale LTO sheaths. The discharge capacity in the first cycle of the MWNT@LTO nano-cables was 122 $mAhg^{-1}$ at a rate of 1 C with 5.6 % capacity loss after 100 cycles, and its initial capacity was an order of magnitude higher than that of bulk LTO.

Due to its outstanding mechanical and electronic properties, graphene may have potential applications in LIBs. Figure 5.21 shows a hybrid of nano-LTOs was well-dispersed on electrical conductive graphene nano-sheets [1]. The electron transport was significantly improved by forming a conductive graphene network throughout the insulating LTO nanoparticles. With 5 wt. % graphene added, the hybrid materials delivered a specific capacity of 122 mAh g^{-1} at a very high charge–discharge rate of 30 C and exhibited an excellent cycling performance, with the first discharge capacity of 132.2 mAh g^{-1} and less than 6 % discharge capacity loss over 300 cycles at 20 C. Based on its outstanding electrochemical performance and acceptable initial columbic efficiency, the nano-LTO/graphene composites might be a promising anode material for high-rate lithium-ion batteries.

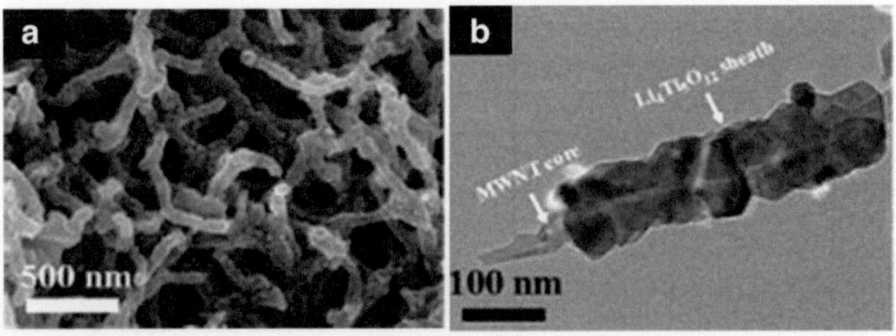

Fig. 5.20 SEM morphologies of TiO_2 and MWNT@LTO nano-cables (Reproduced from [121] with permission of The Royal Society of Chemistry)

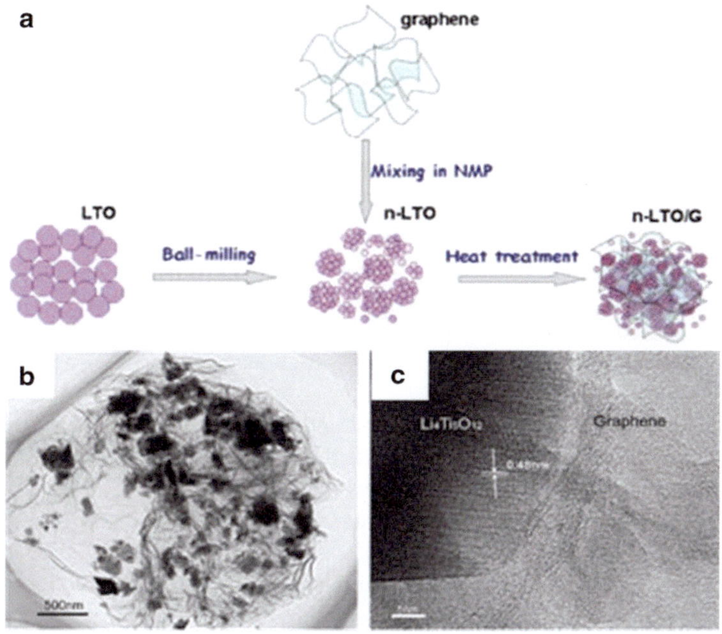

Fig. 5.21 (**a**) Schematic of the preparation of the hybrid of n-LTO and graphene, (**b**) and (**c**) TEM image of n-LTO/G hybrid materials at different magnifications (Reproduced from [1] with permission from Elsevier)

5.4.7 Nano-/Micro-composite Materials

Nanocrystalline material can greatly improve the rate performance, which is mainly due to the smaller particle size. Nano-material has shorter diffusion length, which can significantly improve the dynamic properties and lithium diffusion coefficients. High surface area and porosity of the materials can also enhance the lithium diffusion. On the other hand, nano-materials may have some new lithium storage mechanism, which can also enhance the specific capacity of materials. For example, Poizot et al. [122] investigated the electrochemical performance of nano-transition metal oxide MO (Co, Ni, Cu, and Fe). Its reversible capacity is in the range of $600 \sim 800\ \mathrm{mAhg^{-1}}$. It also has stable cycle performance and high-rate performance. They found that the mechanism of Li reactivity differs from the classical Li insertion/de-insertion or Li-alloying processes and involves the formation and decomposition of Li_2O, accompanying the reduction and oxidation of metal nanoparticles (in the range $1 \sim 5$ nm), respectively.

Although nano-LTO material can obviously improve the electrochemical performance, it still has some serious problems to overcome for practical applications. Major problems are as follows:

(1) Agglomeration: Nano-materials have very large surface area and surface energy, and it is prone to aggregate and agglomerate. During fabrication of electrode, it is difficult to mix them with conductive additives (carbon black) and polymer binder evenly. This uneven mixture results in higher resistance and capacity fading of cells. On the other hand, agglomeration of nano-materials can also deteriorate electrochemical performance during cycling.
(2) High surface energy. Highly reactivity of nano-materials can cause side reactions and electrolyte decomposition. Compared with conventional micro-materials, nano-materials have narrower voltage window. During charge–discharge process, SEI film on the surface of nano-materials becomes thicker, which can increase lithium loss and safety concern.
(3) Nano-materials have lower tap density, which leads to lower volumetric energy density for practical LIBs.
(4) Nano-materials usually need high cost and complicated synthetic procedures.
(5) In most commercial battery systems, the separator has pores with approximately 1 μm diameter. If nanoparticles dissociate from its original electrode, it may cross the separator and agglomerate on opposite electrode, resulting in capacity loss [9].

In order to overcome these problems of nano-LTO materials, the synthesis of nano-/micro-hybrid composites might be an effective way. Nano-/micro-composites can be obtained by assembling nano-LTO primary particles on secondary microspherical particles. This material combined superior electrochemical performance from nano-materials with higher tap density of micro-materials.

5.5 Gasing of LTO-Based Batteries

5.5.1 Performances of LIBs Using LTO Anode Materials and Batteries' Design

Safety is a key concern for LIBs, especially for a high-power energy storage system. At present, there are two considerations on safety of LIBs with graphite anode. Firstly, lithium-ion intercalation into carbon/graphite happened near lithium potential; secondly, there is no end-of-charge indicator in the voltage profile that can signal the onset of catastrophic oxygen evolution form the cathode [123]. Since the safety requirements for LIBs in HEV/PHEV/EV, many LIBs manufacturers have been considering to replace the graphite/carbon negative electrode by LTO that shows 1.5 V potential. Table 5.1 shows the working voltage of LTO-based LIBs coupled with different cathode materials. For example, $LiCoO_2$//LTO (2.4 V), [123] LMO/LTO (2.6 V) [124], $LiNi_{1/3}Co_{1/3}Mn_{1/3}$//LTO (2.5 V) [125], $LiFePO_4$// LTO (1.9 V) [126], and $LiNi_{0.5}Mn_{1.5}O_4$//LTO (3.2 V) [127]. Due to 3D crystal structure, LTO-based LIBs usually show excellent cyclic performance.

Table 5.1 Some typical LTO-based LIBs systems and their electrochemical performance

Cathode materials	Electrolyte	Cyclic performance	Rate performance	Ref.
$LiCoO_2$	1.0 M $LiPF_6$/EC:DMC 1:1 v/v	117,000 cycles under pulse test	No data	[123]
$LiMn_2O_4$	1.2 M $LiPF_6$/EC:EMC 3:7 w/w	55°C, 160 cycles 93 %	No data	[124]
$LiMn_2O_4$	1.0 M $LiPF_6$/EC:DEC: EMC 1:1:1 v/v	300 cycles 89.2 %	10 C/1 C = 89.5 %	[128]
$LiNi_{1/3}Co_{1/3}Mn_{1/3}O_2$	1.2 M $LiPF_6$/EC:EMC 3:7 w/w	55°C, 200 cycles 96.5 %	No data	[129]
$LiNi_{1/3}Co_{1/3}Mn_{1/3}O_2$	1.0 M $LiPF_6$/EC: EMC:DMC 1:1:1 w/w/	25°C, 5000 cycles 89 % 55°C, 720 cycles 80 %	20 C/1 C = 94 %	[130, 131]
$LiNi_{0.5}Mn_{1.5}O_4$	1.0 M $LiPF_6$/EC:EMC 3:7 w/w	1100 cycles, 83 %	No data	[24]
$Li_2Co_{0.4}Fe_{0.4}Mn_{3.2}O_8$	1.0 M $LiPF_6$/PC	10 cycles, 90 %	No data	[132]

EC, ethylene carbonate; *DMC*, dimethyl carbonate; *PC*, propylene carbonate; *EMC*, ethyl methyl carbonate; *DEC*, diethyl carbonate

The accelerated rate calorimetry test on 18650-batteries showed no thermal runaway, explosion, or fire. Relevant results demonstrate that LTO/LMO batteries could be one of the safest Li-ion battery systems [124].

The study on electrochemical performance of the LTO//$LiNi_{1/3}Co_{1/3}Mn_{1/3}O_2$ batteries was carried out by Lu et al. [129]. The batteries showed very good cyclic performances at room and high temperatures. After 30 cycles at 0.1 C, the capacity of LTO//$LiNi_{1/3}Co_{1/3}Mn_{1/3}O_2$ batteries at room temperatures is the same with the capacity at 55°C. It still has stable capacity after 200 cycles. At 55°C, the capacity retention of LTO//$LiNi_{1/3}Co_{1/3}Mn_{1/3}O_2$ and MCMB//$LiNi_{1/3}Co_{1/3}Mn_{1/3}O_2$ batteries is 96.5 % and 83.5 %, respectively. Ohzuku T et al. [24] assembled a 3 V lithium-ion cell using high-voltage $LiNi_{0.5}Mn_{1.5}O_4$ cathode and LTO as anode. This cell showed a quite flat operating voltage of 3.2 V with excellent cyclic performance. Accelerated cycle tests indicated that 83 % of the initial capacity can be achieved even after 1100 cycles. A 3 V battery was assembled through LTO anode combined with doped high-voltage spinel cathode, $Li_2Co_{0.4}Fe_{0.4}Mn_{3.2}O_8$ material [132]. Although somewhat affected by the instability of the electrolyte in the charge region, this design still demonstrates the feasibility of this innovative concept.

When graphite was used as the anode, generally, excess graphite was used to ensure the safety of LIBs due to its very flat voltage plateau (100 mV vs. Li/Li^+). Usually the capacity ratio of anode/cathode is about 1.05 $\sim$ 1.10 for this purpose. Figure 5.22 illustrates the charge and discharge curves of Li/graphite and Li/$LiCoO_2$ batteries.

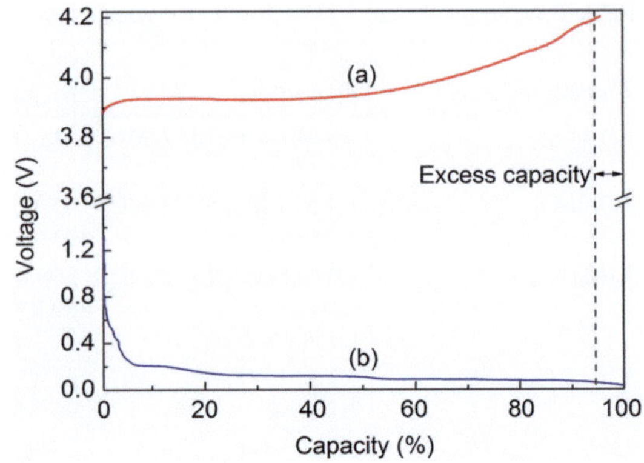

Fig. 5.22 Charge and discharge curves of (**a**) Li/$LiCoO_2$ and (**b**) Li/graphite

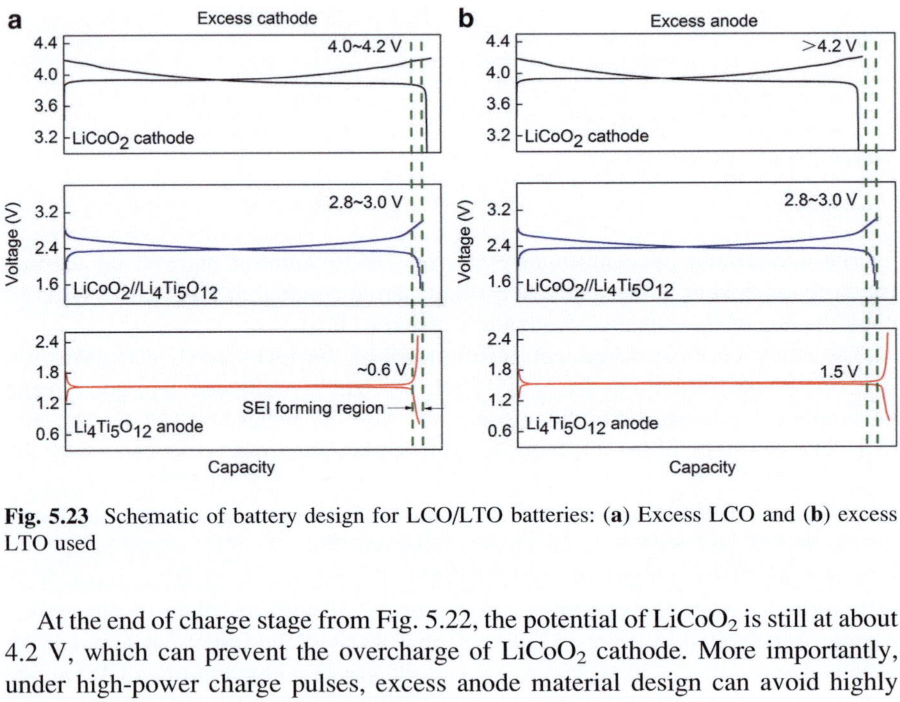

Fig. 5.23 Schematic of battery design for LCO/LTO batteries: (**a**) Excess LCO and (**b**) excess LTO used

At the end of charge stage from Fig. 5.22, the potential of $LiCoO_2$ is still at about 4.2 V, which can prevent the overcharge of $LiCoO_2$ cathode. More importantly, under high-power charge pulses, excess anode material design can avoid highly reactive lithium deposited on the surface of the electrode materials.

However, for LTO-based LIBs, excess cathode (such as $LiCoO_2$, $LiMn_2O_4$, $LiFePO_4$, $LiNi_{1/3}Mn_{1/3}Co_{1/3}O_2$, etc.) design should be used. It is mainly because of the voltage change of LTO materials at the end of discharge stage.

Figure 5.23 shows the design principles of LTO battery. As shown in Fig. 5.23a, the battery capacity was determined by the LTO anode (i.e., excess LCO cathode materials were used to design the battery and capacity ratio of LCO/LTO which is 1.05). Thus, for this excess cathode-designed batteries, at the upper limit of 3.6 V cutoff voltage, the potential of LCO cathode still remains at 3.8 V versus Li/Li^+,

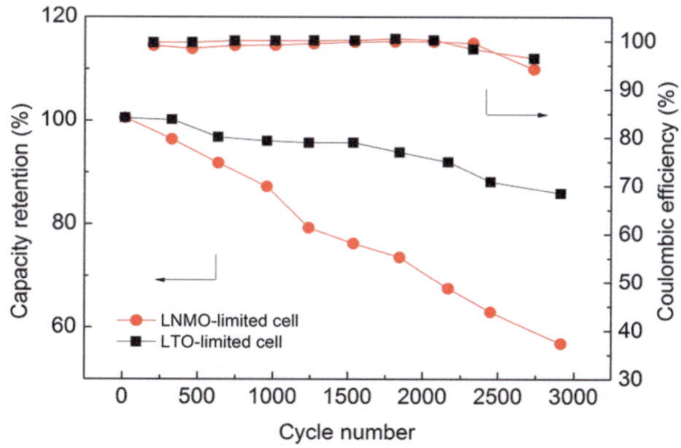

Fig. 5.24 Cycling performance of LTO-based LIBs with different battery design (Reprinted from [133] with permission from Elsevier)

and the potential of the LTO drops down from 1.5 to 0.6 ~ 0.8 V. The voltage window between 0.6 ~ 0.8 V corresponds the SEI forming region, which is shown in Fig. 5.23a. This design can prevent the overcharge of LCO and thus effectively enhance the cyclic life of LTO-based LIBs.

While the battery capacity is limited by the LCO cathode, as shown in Fig. 5.23b, under the same conditions, the potential of the LTO anode still remains at 1.5 V but that of the LCO cathode rises to above 4.4 V where many adverse side reactions, such as overcharge and electrolyte decomposition at higher potential, could occur between the electrolyte and the strong oxidative cathode (Fig. 5.23b). This design can cause the overcharge of the cathode material and severe capacity fading for LTO-based battery systems [133].

The effect of the capacity matchup between cathode and anode in the $LiNi_{0.5}Mn_{1.5}O_4$/LTO cell system on cycling property, choice of electrolyte, high voltage, and overcharge tolerances was investigated and detailed by Xiang et al. [133]. Excess cathode design shows superior cycling performance and less limitation of electrolyte choice than the latter. Furthermore, the excess cathode used design exhibits better tolerance to high voltage and overcharge than the $LiNi_{0.5}Mn_{1.5}O_4$ -limited cell, owing to taking advantage of the extra capacity of LTO below 1 V. Figure 5.24 compares the cycling performance of LTO-based LIBs with different battery design. When using excess cathode design, the retention rate almost reaches 90 % after 3000 cycles, while the retention rate is only about 60 % when the excess anode is used.

5.5.2 *Gas Generation in LTO Batteries*

Severe gasing during charge–discharge cycles, especially cycling at high temperatures, is a critical but often overlooked problem of LTO batteries. To date, this issue is at present unsettled but clearly bears heavily on electrode materials,

Table 5.2 Gas composition (wt%) of LTO cells stored at different temperature [130, 131]

Temperature (°C)	CO_2	C_2H_4	C_2H_6	C_3H_6	C_3H_8	H_2	CH_4	CO
40	1.42	2.56	0.65	0.00	0.05	86.83	3.50	5.19
55	1.42	2.56	0.65	0.00	0.05	86.63	3.50	5.19
70	1.29	1.49	0.42	0.01	0.04	90.04	2.71	4.00
80	0.77	2.08	0.36	0.01	0.03	91.11	2.29	3.36

electrolyte, and battery system [21, 22]. The gas evolved in LTO batteries mainly consists of H_2, CO, CO_2, and alkanes.

For LMO/LTO battery system, Belharouak et al. [21] confirmed that there was significant gas generation inside of the cells. The species of gas generated in the cell tested at 60°C was much more than those of gas generated in the cell at 45 and 30° C. And concentration of hydrogen was much higher than other gases generated in the cells such as CO, CO_2, and other hydrocarbon species. It is suggested that H_2 was a dominant gas possibly derived from a trace of H_2O and electrolyte decomposition products.

Bernhard et al. [134] tried to determine a direct correlation between the amount of evolved hydrogen and the water content of the electrolyte using online electrochemical mass spectrometry. The analysis indicated that the one-electron reduction reaction of water leads to hydrogen evolution and formation of hydroxide anions. The latter triggers a ring-opening reaction of cyclic carbonates and decomposes solvent into CO_2 and carbonate oligomers. They proposed that the gasing of LTO batteries is caused by water intrusion during cell assembly or by water contained in improperly dried electrodes [134]. CO and CO_2 of the gases result from the decomposition of electrolyte solution initiated at relatively high temperatures by PF_5 that is a strong Lewis acid and one of the decomposition products of the electrolyte, $LiPF_6$ [21].

Wu et al. [130, 131] investigated gas composition of LNMCO/LTO during cycling and it was shown in Table 5.2. Among the detectable gases, the dominant specie was over 80 % hydrogen. With the stored temperature increasing, the amount of H_2 increased and CO_2 decreased.

Wu et al. [130, 131] investigated the possible reason of gas forming and its effect on swelling in LTO cells. They suggested gasing attributed to anode potential, instead of moisture from chemically catalytic reaction. The reaction mechanisms of carbonates were shown in Fig. 5.25. Their studies indicated that linear carbonates brought about H_2 and soluble species, while cyclic carbonates tend to produce alkene gas and insoluble salts [130].

However, in LNMCO/LTO system, the gasing reactions, including decarboxylation, decarbonylation, and dehydrogenation reactions of solvents, are generated not by PF_5 but by LTO [22]. The results showed that the generated gases mainly consist of H_2, CO, and CO_2, which cause swelling of battery pack. The gasing involves the plane transformation of LTO from (111) to (222) and the formation of (101) plane of anatase TiO_2 and the outermost surface Li^+, and O^{2-} ions of the LTO (111) plane are removed from LTO by the interfacial reactions. They suggested that

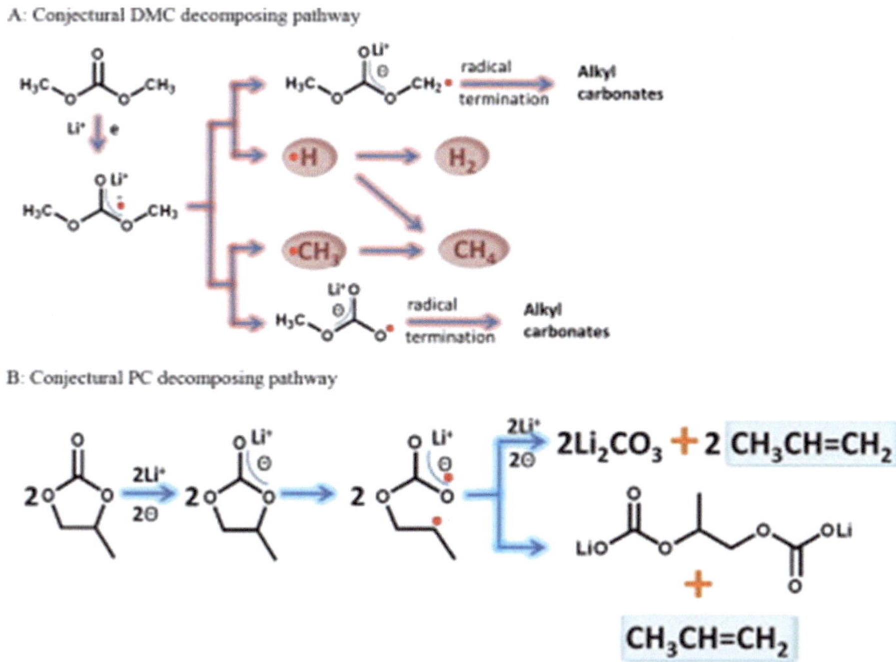

Fig. 5.25 Decomposition mechanism of DMC and PC (Reprinted from [130] with permission from Elsevier)

the modification of the LTO surface is probably a very effective way, which can suppress the gasing generation of LTO battery and improve the high-rate charge–discharge performance of batteries. For instance, carbon layer of carbon coated LTO prevents the interfacial reactions between LTO and the surrounding electrolyte solution [125].

Figure 5.26 shows gasing reaction mechanisms of LTO electrodes. When LTO is soaked in solvents or electrolytes, Ti^{4+} ions of LTO coordinate with the unshared electron pairs of O^{2-} ions of carbonyl groups. The outermost surface O^{2-} of LTO attacks the carbon atom of CH_3CH_2O group and flexible Li^+ ions coordinate with O^{2-} of CH_3CH_2O group. As shown in Fig. 5.26a, these reactions result in the formation of CO_2, CH_3COOLi, anatase TiO_2 (101 plane), and $C_2H_5OC_2H_5$. LTO may also initiate the ring-opening polymerization of EC, which results in the formation of PEO-like oligomers $(-CH_2-CH_2-O-)_n$ and CO_2 (Fig. 5.26b). Dehydrogenation of the alkoxy groups may be promoted by LTO and they are responsible for the generation of H_2 (Fig. 5.26c). CO maybe released by the following reactions: (1) intermediate of solvent dehydrogenation can further accept electrons and Li ions leading to the generation of CO (Fig. 5.26c), and (2) CO_2 can be reduced to form CO (Fig. 5.26d).

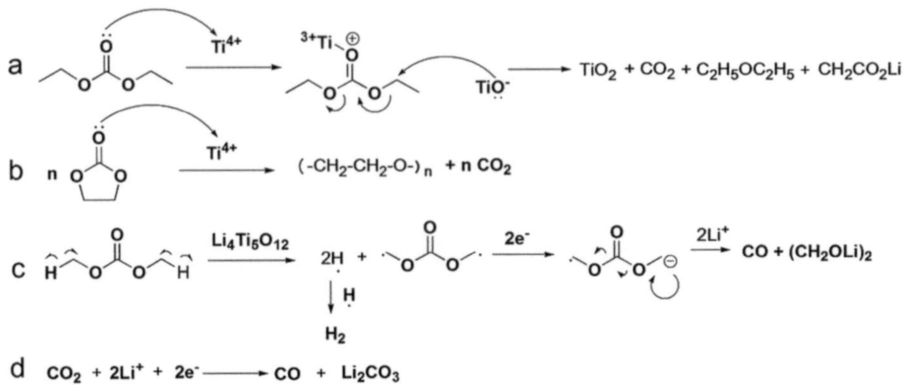

Fig. 5.26 Gasing reactions of LTO electrodes (Reprinted by permission from Macmillan Publishers Ltd:[125] Copyright 2012)

Stable cycle performance is more likely to occur when the cathode has high stability in electrolyte, such as $LiCoO_2$ and $LiNi_{0.8}Co_{0.2}O_2$. This is because gasing and electrolyte salt decomposition product PF_5 can react with water to form phosphorous oxyfluoride species and HF. The most serious gasing occurs only in LMO cathode battery system. The attack of HF on LMO electrodes generates Mn^{2+} ions into the electrolyte, which in turn results in more severe gasing [135]. However, to further understand how gasing occurs and find an effective way to remedy it, more investigation is needed for synthesis of LTO materials and reaction in LTO-based battery system.

5.5.3 Suppressing of Swelling in LTO Batteries

Due to high lithium intercalation voltage, it is generally believed that there is no SEI forming on the surface of LTO anode. It is well known that SEI is an essential prerequisite for proper electrode performance in LIBs. For the conventional graphite anode, reduction of electrolyte results in the formation of SEI, which prevents further side reactions during cycling. According to Shu et al. [136], LTO is an SEI film-free material when it is cycled in the range of 1.0 $\sim$ 3.0 V. When cycled in the range below 1.0 V, unlike graphite anode materials, dense and uniform SEI film does not form on the surface of LTO materials, which does not prevent continuous electrolyte decomposition on the catalytic active sites. On the other hand, when charged at high rates, the potential of LTO may also down to 1.0 V. Under these circumstances, electrolyte may be reduced. According to X-ray photoelectron spectroscopy (XPS) and Fourier transform infrared spectroscopy (FTIR) analysis, organic lithium alkyl carbonates are the main components of SEI film formed during the reduction process, which is similar to that of carbon graphite materials [136]. Dedryvere et al. [137] investigated the $LiMn_{1.6}Ni_{0.4}O_4$/LTO interface by

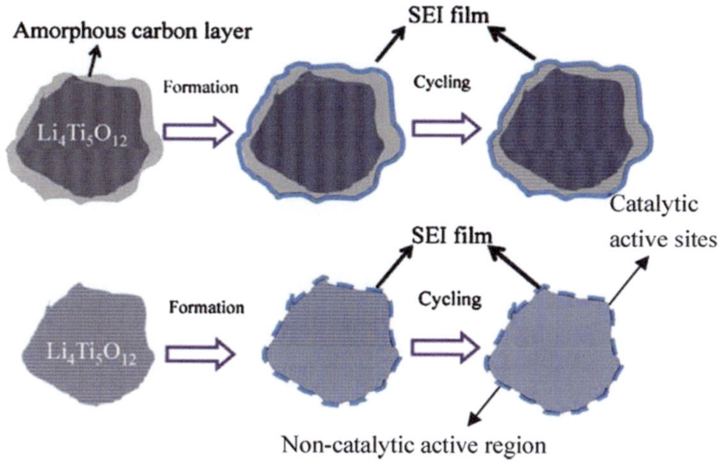

Fig. 5.27 SEI formation mechanism on LTO and LTO/C composite electrodes (Reprinted from [22] with permission from Elsevier)

XPS. It was found that a great amount of inorganic species (LiF and phosphates) and organic species formed during cycling. They discussed that these species are first formed at the positive electrode and then adsorbed at the LTO electrode surface [137]. In order to suppress swelling of LTO batteries, formation of SEI is important to inhibit side reactions.

Carbon coating is an effective way for suppressing side reactions of LTO batteries. He et al. [22] found that carbon layer on the surface of LTO particles can cover the catalytic active sites, and a successive SEI film can be formed on the carbon coating layer of the LTO particles. Successive SEI layer separates the catalytic active sites from the surrounding electrolyte and prevents the electrolyte from continuous decomposition. Whereas, the SEI film formed on the pure LTO particles may be thinner and porous, which are open ways for a new reduction of electrolyte in contact with LTO surface. The function of carbon is illustrated in Fig. 5.27 [22].

Besides traditional carbon coating, Li et al. [128] developed a novel AlF_3 coating method to suppress gas generation. It is found that for AlF_3-modified LTO, only part of Al^{3+} and F^- have co-doped into the LTO particles, the rest of Al^{3+} and F^- form an AlF_3 coating layer on the surface of LTO particles. AlF_3 coating layer is very effective in suppressing the gasing of LTO/LMO battery, as shown in Fig. 5.28. The improved electrochemical performance was tentatively attributed to the "buffer" layer provided by the AlF_3 coating, which reduced the activity of the extracted oxygen and suppressed the electrolyte decomposition.

Although some progress has been made to improve the stability of LTO materials and batteries, there are still only a few reports that specifically refer to the gasing behavior of LTO electrodes. In the future, more investigations on gasing mechanisms and remedies should be studied and developed.

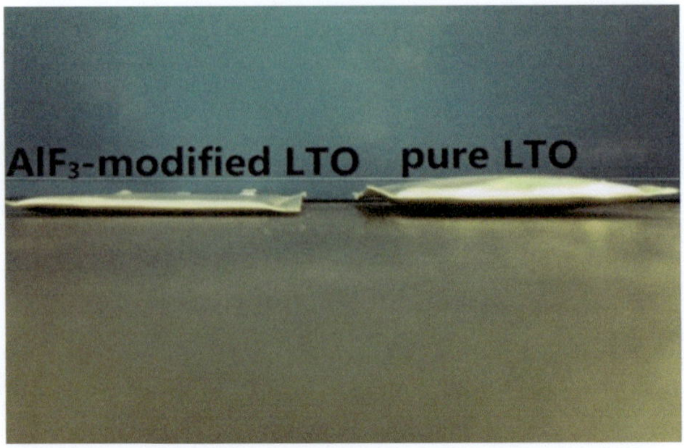

Fig. 5.28 Photograph of the cycled soft-packed LTO/LMO batteries (Reprinted from [128] with permission from Elsevier)

5.6 Promising Applications of LTO Materials

5.6.1 Energy Storage Systems

As a high-potential anode material, LTO has outstanding properties, such as excellent safety characteristics and cyclic life. All these performances make LTO a promising candidate for large-scale energy system. One main disadvantage of LTO batteries is its low energy density. Table 5.3 compares major battery systems, which have potential applications for energy storage systems.

The majority of commercial LTO batteries are LTO/LMO and LTO/$LiNi_{1/3}Co_{1/3}Mn_{1/3}O_2$ systems. On the other hand, due to severe gasing and lower energy density, LTO/$LiNi_{1/3}Co_{1/3}Mn_{1/3}O_2$ batteries have found a wide variety of practical applications.

LTO paired with high-voltage and high-capacity cathode battery systems should be developed, including $LiNi_{0.5}Mn_{1.5}O_4$ and $LiMnPO_4$ cathode systems in the future. Because the operating voltage of spinel $LiNi_{0.5}Mn_{1.5}O_4$ can reach 4.7 V, the full batteries can output a voltage of over 3 V if using $LiNi_{0.5}Mn_{1.5}O_4$ as cathode and LTO as anode, respectively. Figure 5.29 shows the charge–discharge curves and CV of LNMTO/LTO batteries. LNMTO represents Ti-doped compound $LiNi_{0.3}Mn_{1.2}Ti_{0.3}O_4$. It can be seen that this battery displays voltage profile at around 3.2 V [139].

Another important promising field is LTO paired with high-capacity manganese-rich composite materials. The electrodes can be represented in two-component notation as $xLi_2MnO_3(1-x)LiMO_2$ and $xLi_2MnO_3(1-x)LiM_2O_4$. It can be activated by Li_2O and/or lithium removal form of the Li_2MnO_3, $LiMO_2$, and LMO components. The electrodes provide an initial capacity $> 250\,mAhg^{-1}$ when discharged between 5.0 and 2.0 V. Initial charge and discharge profiles (5.0 ~ 2.0 V) and cycling stability of a lithium cell containing an acid-treated $xLi_2MnO_3 \cdot 0.7LiMn_{0.5}Ni_{0.5}O_2$ are shown in Fig. 5.30.

Table 5.3 Typical LIBs systems used for energy storage system [138]

Battery systems	Theoretical energy density (Wh/Kg)	Practical energy density (Wh/Kg)	Cycle life	Safety
C/$LiMn_2O_4$	412.78	80 ∼ 100	∼ 1000	High
C/$LiFePO_4$	390.69	90 ∼ 110	∼ 2000	Good
C/$LiNi_{0.5}Mn_{1.5}O_4$	479.75	130	∼ 500	Low
C/$LiNi_{1/3}Co_{1/3}Mn_{1/3}O_2$	556.65	200 ∼ 220	∼ 1500	Average
LTO/$LiMn_2O_4$	195.89	∼ 70	∼ 6000	Excellent
LTO/$LiFePO_4$	163.85	50 ∼ 70	∼ 20000	Excellent
LTO/$LiNi_{0.5}Mn_{1.5}O_4$	248.28	∼ 120	∼ 3000	Excellent
LTO/$LiNi_{1/3}Co_{1/3}Mn_{1/3}O_2$	220.22	∼ 140	∼ 10000	Excellent

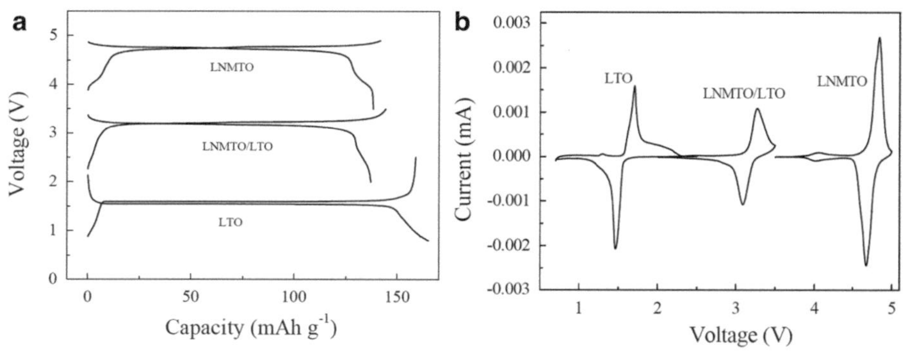

Fig. 5.29 (**a**) Charge–discharge curves; (**b**) CVs [139] With kind permission from Springer Science and Business Media

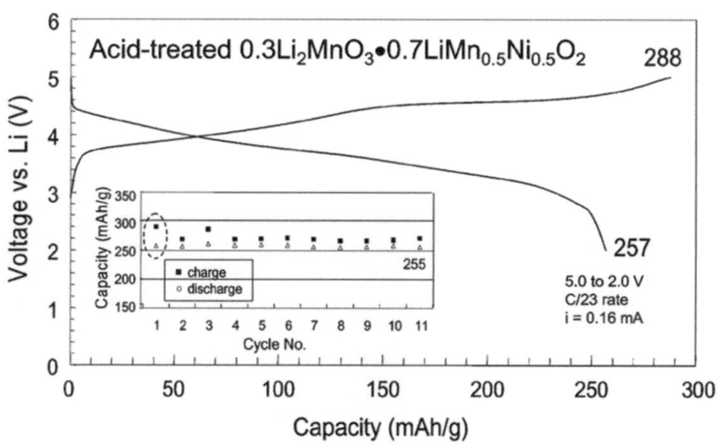

Fig. 5.30 Initial charge and discharge profiles (5.0 ∼ 2.0 V) and cycling performance of a lithium cell containing an acid-treated $xLi_2MnO_3 \cdot 0.7\ LiMn_{0.5}Ni_{0.5}O_2$ electrode (Reproduced [140] with permission of The Royal Society of Chemistry)

Although manganese-rich materials have high voltage and capacity, it still has some problems to be solved. These problems include (1) capacity fading, (2) voltage decline during cycling, and (3) low columbic efficiency. In order to pair with LTO anode materials, these should be investigated in future.

5.6.2 Application of LTO in Lithium-Ion Capacitors (LIC)

A lithium-ion capacitor is a hybrid electrochemical energy storage device which combines the intercalation mechanism of a lithium-ion battery with the cathode of an electric double-layer capacitor (EDLC). The packaged energy density of an LIC is approximately 20 Wh kg^{-1} generally four times higher than an EDLC and five times lower than a lithium-ion battery. The power density however has been shown to match that of EDLC able to completely discharge in seconds. The cathode often employs activated carbon material at which charges are stored in an electric double layer that is developed at the interface between the carbon and the electrolyte. Figure 5.31 shows principles of a LIC. The use of an intercalation compound as the negative electrode offers the opportunity to effectively pin the electrode voltage at potentials very negative with respect to standard hydrogen electrode and maintains acceptable high gravimetric and volumetric energy densities as opposed to a pseudocapacitive reaction.

Nowadays, the anode electrode of LIC was made with LTO or graphite. LIC with LTO as anode has higher safety and long cycle life [141]. In 2001, Plitz et al. [141] reported a nonaqueous hybrid LIC utilizing LTO as a negative electrode

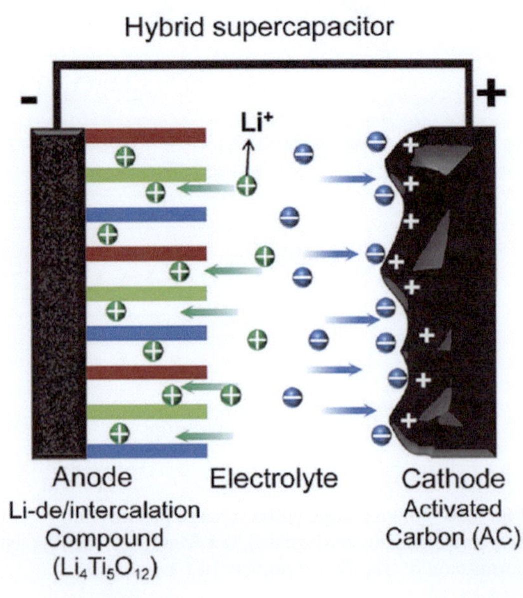

Fig. 5.31 Principles of LIC, cathode is activated carbon and anode is LTO (Reprinted from [141, 142] with permission from Wiley)

and AC as a positive electrode. The fabricated AC/2 M $LiBF_4$(AN)/LTO showed an energy density of 25 Wh kg^{-1}.

Naoi et al. developed new generation LICs based on nano-LTO materials and relevant works can be found in several excellent reviews [143–145]. They used "ultra-centrifuging treatment" to synthesize nano-LTO materials. As shown in Fig. 5.31, the composite comprises highly dispersed nanocrystalline LTO particles hyper-dispersed on the carbon nano-fibers (CNFs). It can discharge in 15 s ($\sim$ 300 C rate). On the other hand, compared with graphite/AC lithium-ion capacitor, an LIC comprised of LTO does not require pre-doping with lithium ions. So, there is a large selection of electrolytes for LTO-based LIC because of its narrower voltage window ($2.7 - 3.0$ V). Acetonitrile (AN), ionic liquids, and linear carbonates can be used in LTO-based LICs. Based upon the above technologies, a novel hybrid capacitor realizes a triply enhanced energy density compared to the conventional EDLC without sacrificing the power performance, cyclability, and float durability. In the future, LTO-based LIC with higher energy density should be investigated and developed.

References

1. Shi Y, Wen L, Li F et al (2011) Nanosized $Li_4Ti_5O_{12}$/graphene hybrid materials with low polarization for high rate lithium ion batteries. J Power Sources 196(20):8610–8617
2. Ferg E, Gummow RJ, Dekock A et al (1994) Spinel anodes for lithium-ion batteries. J Electrochem Soc 141(11):L147–L150
3. Park KS, Benayad A, Kang DJ et al (2008) Nitridation-driven conductive $Li_4Ti_5O_{12}$ for lithium ion batteries. J Am Chem Soc 130(45):14930–14932
4. Li BH, Han CP, He YB et al (2012) Facile synthesis of $Li_4Ti_5O_{12}$/C composite with super rate performance. Energy Environ Sci 5(11):9595–9602
5. Prakash AS, Manikandan P, Ramesha K et al (2010) Solution-combustion synthesized nanocrystalline $Li_4Ti_5O_{12}$ as high-rate performance Li-ion battery anode. Chem Mater 22 (9):2857–2863
6. Rahman MM, Wang JZ, Hassan MF et al (2010) Basic molten salt process a new route for synthesis of nanocrystalline $Li_4Ti_5O_{12} - TiO_2$ anode material for Li-ion batteries using eutectic mixture of $LiNO_3 - LiOH - Li_2O_2$. J Power Sources 195(13):4297–4303
7. Raja MW, Mahanty S, Kundu M et al (2009) Synthesis of nanocrystalline $Li_4Ti_5O_{12}$ by a novel aqueous combustion technique. J AlloyS Compd 468(1-2):258–262
8. Lu HW, Zeng W, Li YS et al (2007) Fabrication and electrochemical properties of three-dimensional net architectures of anatase TiO_2 and spinel $Li_4Ti_5O_{12}$ nanofibers. J Power Sources 164(2):874–879
9. Sorensen EM, Barry SJ, Jung HK et al (2006) Three-dimensionally ordered macroporous $Li_4Ti_5O_{12}$: effect of wall structure on electrochemical properties. Chem Mater 18(2):482–489
10. Tang YF, Yang L, Qiu Z et al (2008) Preparation and electrochemical lithium storage of flower-like spinel $Li_4Ti_5O_{12}$ consisting of nanosheets. Electrochem Commun 10 (10):1513–1516
11. Tang YF, Yang L, Qiu Z et al (2009) Template-free synthesis of mesoporous spinel lithium titanate microspheres and their application in high-rate lithium ion batteries. J Mater Chem 19 (33):5980–5984

12. Cheng L, Li XL, Liu HJ et al (2007) Carbon-coated $Li_4Ti_5O_{12}$ as a high rate electrode material for Li-ion intercalation. J Electrochem Soc 154(7):A692–A697
13. Cheng L, Yan J, Zhu GN et al (2010) General synthesis of carbon-coated nanostructure $Li_4Ti_5O_{12}$ as a high rate electrode material for Li-ion intercalation. J Mater Chem 20 (3):595–602
14. Dominko R, Gaberscek M, Bele A et al (2007) Carbon nanocoatings on active materials for Li-ion batteries. J Eur Ceram Soc 27(2-3):909–913
15. Capsoni D, Bini M, Massarotti V et al (2009) Cr and Ni doping of $Li_4Ti_5O_{12}$: cation distribution and functional properties. J Phys Chem C 113(45):19664–19671
16. Capsoni D, Bini M, Massarotti V et al (2008) Cations distribution and valence states in Mn-substituted $Li_4Ti_5O_{12}$ structure. Chem Mater 20(13):4291–4298
17. Nakayama M, Ishida Y, Ikuta H et al (1999) Mixed conduction for the spinel type $(1 - x)$ $Li_{4/3}Ti_{5/3}O_4 - xLiCrTiO_4$ system. Solid State Ion 117(3-4):265–271
18. Gao J, Jiang CY, Wan CR (2010) Synthesis and characterization of spherical La-Doped nanocrystalline $Li_4Ti_5O_{12}$/C compound for lithium-ion batteries. J Electrochem Soc 157(2): K39–K42
19. Huang SH, Wen ZY, Zhang JC et al (2006) $Li_4Ti_5O_{12}$/Ag composite as electrode materials for lithium-ion battery. Solid State Ion 177(9-10):851–855
20. Huang SH, Wen ZY, Zhu XJ et al (2005) Research on $Li_4Ti_5O_{12}/Cu_xO$ composite anode materials for lithium-ion batteries. J Electrochem Soc 152(7):A1301–A1305
21. Belharouak I, Koenig GM, Yumoto H et al (2012) Performance degradation and gassing of $Li_4Ti_5O_{12}/LiMn_2O_4$ lithium-ion cells. J Electrochem Soc 159(8):A1165–A1170
22. He YB, Ning F, Li BH et al (2012) Carbon coating to suppress the reduction decomposition of electrolyte on the $Li_4Ti_5O_{12}$ electrode. J Power Sources 202:253–261
23. Thackeray MM (1995) Structural considerations of layered and spinel lithiated oxides for lithium ion batteries. J Electrochem Soc 142(8):2558–2563
24. Ariyoshi K, Yamamoto S, Ohzuku T (2003) Three-volt lithium-ion battery with Li $Ni_{1/2}Mn_{3/2}O_4$ and the zero-strain insertion material of Li $Li_{1/3}Ti_{5/3}O_4$. J Power Sources 119:959–963
25. Amatucci GG, Badway F, Du Pasquier A et al (2001) An asymmetric hybrid nonaqueous energy storage cell. J Electrochem Soc 148(8):A930–A939
26. Peramunage D, Abraham KM (1998) Preparation of micron-sized $Li_4Ti_5O_{12}$ and its electrochemistry in polyacrylonitrile electrolyte-based lithium cells. J Electrochem Soc 145 (8):2609–2615
27. Liu GQ, Wen L, Liu GY et al (2011) Synthesis and electrochemical properties of $Li_4Ti_5O_{12}$. J Alloys Compd 509(22):6427–6432
28. Yuan T, Cai R, Shao ZP (2011) Different effect of the atmospheres on the phase formation and performance of $Li_4Ti_5O_{12}$ prepared from ball-milling-assisted solid-phase reaction with pristine and carbon-precoated TiO2 as starting materials. J Phys Chem C 115(11):4943–4952
29. Huang Z, Wang D, Lin Y et al (2014) Enhancing the high-rate performance of $Li_4Ti_5O_{12}$ anode material for lithium-ion battery by a wet ball milling assisted solid-state reaction and ultra-high speed nano-pulverization. J Power Sources 266:60–65
30. Lai C, Wu ZZ, Zhu YX et al (2013) Ball-milling assisted solid-state reaction synthesis of mesoporous $Li_4Ti_5O_{12}$ for lithium-ion batteries anode. J Power Sources 226:71–74
31. Zaghib K, Simoneau M, Armand M et al (1999) Electrochemical study of $Li_4Ti_5O_{12}$ as negative electrode for Li-ion polymer rechargeable batteries. J Power Sources 81:300–305
32. Wen L, Liu G, Liu GY et al (2012) Oxygen deficient $Li_4Ti_5O_{12}$ for high-rate lithium storage. J Chin Chem Soc-Taip 59(10):1201–1205
33. Shin JW, Hong CH, Yoon DH (2012) Effects of TiO_2 starting materials on the solid-state formation of $Li_4Ti_5O_{12}$. J Am Ceram Soc 95(6):1894–1900
34. Nugroho A, Kim SJ, Chung KY et al (2012) Synthesis of $Li_4Ti_5O_{12}$ in supercritical water for Li-ion batteries: reaction mechanism and high-rate performance. Electrochim Acta 78:623–632

35. Kavan L, Gratzel M (2002) Facile synthesis of nanocrystalline $Li_4Ti_5O_{12}$ (spinel) exhibiting fast Li insertion. Electrochem Solid State Lett 5(2):A39–A42
36. Yuan T, Cai R, Ran R et al (2010) A mechanism study of synthesis of $Li_4Ti_5O_{12}$ from TiO_2 anatase. J Alloys Compd 505(1):367–373
37. Ning F, He YB, Li BH et al (2012) Effects of TiO_2 crystal structure on the performance of $Li_4Ti_5O_{12}$ anode material. J Alloys Compd 513:524–529
38. Li J, Jin Y-L, Zhang X-G et al (2007) Microwave solid-state synthesis of spinel $Li_4Ti_5O_{12}$ nanocrystallites as anode material for lithium-ion batteries. Solid State Ion 178 (29–30):1590–1594
39. Matsui E, Abe Y, Senna M et al (2008) Solid-state synthesis of 70 nm $Li_4Ti_5O_{12}$ particles by mechanically activating intermediates with amino acids. J Am Ceram Soc 91(5):1522–1527
40. Shen CM, Zhang XG, Zhou YK et al (2003) Preparation and characterization of nanocrystalline $Li_4Ti_5O_{12}$ by sol-gel method. Mater Chem Phys 78(2):437–441
41. Liu DQ, Lai QY, Hao YJ et al (2004) Study on synthesis and mechanism of $Li_4Ti_5O_{12}$ by sol-gel method. Chin J Inorg Chem 20(7):829–832
42. Hao YJ, Lai QY, Xu ZH et al (2005) Synthesis by TEA sol-gel method and electrochemical properties of $Li_4Ti_5O_{12}$ anode material for lithium-ion battery. Solid State Ion 176 (13-14):1201–1206
43. Xiang HF, Tian BB, Lian PC et al (2011) Sol-gel synthesis and electrochemical performance of $Li_4Ti_5O_{12}$/graphene composite anode for lithium-ion batteries. J Alloys Compd 509 (26):7205–7209
44. Zhou XL, Huang RA, Wu ZC, et al (2010) Sol-gel synthesis and electrochemical properties of spinel $Li_4Ti_5O_{12}$ for lithium-ion batteries. In: Proceedings of the 7th national conference on Chinese functional materials and applications 2010, vols 1–3, pp 570–574
45. Bilecka I, Niederberger M (2010) New developments in the nonaqueous and/or non-hydrolytic sol-gel synthesis of inorganic nanoparticles. Electrochim Acta 55 (26):7717–7725
46. Zhang C, Zhang Y, Wang J et al (2013) $Li_4Ti_5O_{12}$ prepared by a modified citric acid sol-gel method for lithium-ion battery. J Power Sources 236:118–125
47. Hao YJ, Lai QY, Lu JZ et al (2007) Influence of various complex agents on electrochemical property of $Li_4Ti_5O_{12}$ anode material. J Alloys Compd 439(1–2):330–336
48. Hao YJ, Lai QY, Liu DQ et al (2005) Synthesis by citric acid sol–gel method and electrochemical properties of $Li_4Ti_5O_{12}$ anode material for lithium-ion battery. Mater Chem Phys 94 (2–3):382–387
49. Hao YJ, Lai QY, Lu JZ et al (2006) Synthesis and characterization of spinel $Li_4Ti_5O_{12}$ anode material by oxalic acid-assisted sol–gel method. J Power Sources 158(2):1358–1364
50. Kalbac M, Zukalova M, Kavan L (2003) Phase-pure nanocrystalline $Li_4Ti_5O_{12}$ for a lithium-ion battery. J Solid State Electrochem 8(1):2–6
51. Rho YH, Kanamura K (2004) Preparation of $Li_{4/3}Ti_{5/3}O_4$ thin film electrodes by a PVP sol-gel coating method and their electrochemical properties. J Electrochem Soc 151(1): A106–A110
52. Yi TF, Jiang LJ, Shu J et al (2010) Recent development and application of $Li_4Ti_5O_{12}$ as anode material of lithium ion battery. J Phys Chem Solids 71(9):1236–1242
53. Roy R (1994) Accelerating the kinetics of low-temperature inorganic syntheses. J Solid State Chem 111(1):11–17
54. Li N, Mei T, Zhu YC et al (2012) Hydrothermal synthesis of layered $Li_{1.81}H_{0.19}Ti_2O_5 \cdot xH_2O$ nanosheets and their transformation to single-crystalline $Li_4Ti_5O_{12}$ nanosheets as the anode materials for Li-ion batteries. Crystengcomm 14(20):6435–6440
55. Shen LF, Uchaker E, Zhang XG et al (2012) Hydrogenated $Li_4Ti_5O_{12}$ nanowire arrays for high rate lithium ion batteries. Adv Mater 24(48):6502–6506
56. Li JR, Tang ZL, Zhang ZT (2005) Controllable formation and electrochemical properties of one-dimensional nanostructured spinel $Li_4Ti_5O_{12}$. Electrochem Commun 7(9):894–899

57. Fattakhova D, Krtil P (2002) Electrochemical activity of hydrothermally synthesized Li-Ti-O cubic oxides toward Li insertion. J Electrochem Soc 149(9):A1224–A1229
58. Lee SC, Lee SM, Lee JW et al (2009) Spinel $Li_4Ti_5O_{12}$ nanotubes for energy storage materials. JPhys Chem C 113(42):18420–18423
59. Fang W, Ma YL, Zuo PJ et al (2013) Nano-$Li_4Ti_5O_{12}$ pore microspheres: a high power electrode material for lithium ion batteries. Int J Electrochem Sci 8(2):1949–1956
60. Singhal A, Skandan G, Amatucci G et al (2004) Nanostructured electrodes for next generation rechargeable electrochemical devices. J Power Sources 129(1):38–44
61. Sides CR, Li NC, Patrissi CJ et al (2002) Nanoscale materials for lithium-ion batteries. MRS Bull 27(8):604–607
62. Bai Y, Wang F, Wu F et al (2008) Influence of composite LiCl-KCl molten salt on microstructure and electrochemical performance of spinel $Li_4Ti_5O_{12}$. Electrochim Acta 54 (2):322–327
63. Cheng L, Liu HJ, Zhang JJ et al (2006) Nanosized $Li_4Ti_5O_{12}$ prepared by molten salt method as an electrode material for hybrid electrochemical supercapacitors. J Electrochem Soc 153 (8):A1472–A1477
64. Guo Q, Li S, Wang H et al (2014) Molten salt synthesis of nano-sized $Li_4Ti_5O_{12}$ doped with Fe_2O_3 for use as anode material in the lithium-ion battery. RSC Adv 4(104):60327–60333
65. Prakash AS, Khadar AMA, Patil KC et al (2002) Hexamethylenetetramine: a new fuel for solution combustion synthesis of complex metal oxides. J Mater Synth Process 10 (3):135–141
66. Bellakki MB, Shivakumara C, Baidya T et al (2008) Synthesis, structure and oxygen-storage capacity of $Pr_{1-x}Zr_xO_{2-\delta}$ and $Pr_{1-x-y}Pd_yZr_xO_{2-\delta}$. Mater Res Bull 43(10):2658–2667
67. Patil K, Hegde M, Rattan T et al (2008) Chemistry of nanocrystalline oxide materials-combustion synthesis, properties and applications. World Scientifc, Singapore/London
68. Patil KC, Aruna ST, Ekambaram S (1997) Combustion synthesis. Curr Opinion Solid State Mater Sci 2(2):158–165
69. Aruna ST, Kini NS, Rajam KS (2009) Solution combustion synthesis of $CeO_2 - CeAlO_3$ nano-composites by mixture-of-fuels approach. Mater Res Bull 44(4):728–733
70. Yuan T, Cai R, Wang K et al (2009) Combustion synthesis of high-performance $Li_4Ti_5O_{12}$ for secondary Li-ion battery. Ceram Int 35(5):1757–1768
71. Li X, Lin HC, Cui WJ et al (2014) Fast solution-combustion synthesis of nitrogen-modified $Li_4Ti_5O_{12}$ nanomaterials with improved electrochemical performance. Acs Appl Mater Interfaces 6(10):7895–7901
72. Eslamian M, Ashgriz N (2011) Spray drying, spray pyrolysis and spray freeze drying. In: Ashgriz N (ed) Handbook of atomization and sprays. Springer, New York, pp 849–860
73. Yoshikawa D, Kadoma Y, Kim J-M et al (2010) Spray-drying synthesized lithium-excess $Li_{4+x}Ti_{5-x}O_{12-\delta}$ and its electrochemical property as negative electrode material for Li-ion batteries. Electrochim Acta 55(6):1872–1879
74. Wu F, Wang Z, Li X et al (2012) Characterization of spherical-shaped $Li_4Ti_5O_{12}$ prepared by spray drying. Electrochim Acta 78:331–339
75. Zhang XY, Xu HR, Zhao YY et al (2014) A facile one-step spray pyrolysis method to synthesize spherical $Li_4Ti_5O_{12}$ for lithium-ion battery. Mater Lett 129:101–103
76. Yang KM, Ko YN, Yun JY et al (2014) Preparation of $Li_4Ti_5O_{12}$ yolk-shell powders by spray pyrolysis and their electrochemical properties. Chem-Asian J 9(2):443–446
77. Du GD, Winton BR, Hashim IM et al (2014) Mass production of $Li_4Ti_5O_{12}$ with a conductive network via in situ spray pyrolysis as a long cycle life, high rate anode material for lithium ion batteries. Rsc Advances 4(73):38568–38574
78. Doan TNL, Yoo K, Hoang TKA, et al (2014) Recent developments in synthesis of $xLi_2MnO_3 \cdot (1-x)LiMO_2$ ($M = Ni, Co, Mn$) cathode powders for high-energy lithium rechargeable batteries. Front Energy Res 2(36):1–7
79. Balaji S, Mutharasu D, Sankara Subramanian N et al (2009) A review on microwave synthesis of electrode materials for lithium-ion batteries. Ionics 15(6):765–777

80. Dong HY, Yin YH, Zhang ZJ et al (2012) Synthesis and properties of $Li_4Ti_5O_{12}$/C composite by a microwave-assisted method using PAM as both the template and the carbon source. Phys Scr 86(5)
81. Liu J, Li X, Yang J et al (2012) Microwave-assisted hydrothermal synthesis of nanostructured spinel Li4Ti5O12 as anode materials for lithium ion batteries. Electrochim Acta 63:100–104
82. Shi Y, Gao J, Abruna HD et al (2014) Rapid synthesis of $Li_4Ti_5O_{12}$ /graphene composite with superior rate capability by a microwave-assisted hydrothermal method. Nano Energy 8:297–304
83. Wu Y, Reddy MV, Chowdari BVR et al (2012) Electrochemical studies on electrospun Li $(Li_{1/3}Ti_{5/3})O_4$ grains as an anode for Li-ion batteries. Electrochim Acta 67:33–40
84. Yin SY, Song L, Wang XY et al (2009) Synthesis of spinel $Li_4Ti_5O_{12}$ anode material by a modified rheological phase reaction. Electrochim Acta 54(24):5629–5633
85. Liu GY, Wang HY, Liu GQ et al (2012) Facile synthesis of nanocrystalline $Li_4Ti_5O_{12}$ by microemulsion and its application as anode material for Li-ion batteries. J Power Sources 220:84–88
86. Nugroho A, Chang W, Kim SJ et al (2012) Superior high rate performance of core-shell $Li_4Ti_5O_{12}$ /carbon nanocomposite synthesized by a supercritical alcohol approach. Rsc Adv 2 (29):10805–10808
87. Nugroho A, Kim SJ, Chung KY et al (2011) Facile synthesis of nanosized $Li_4Ti_5O_{12}$ in supercritical water. Electrochem Commun 13(6):650–653
88. Nugroho A, Kim SJ, Chang W et al (2013) Facile synthesis of hierarchical mesoporous $Li_4Ti_5O_{12}$ microspheres in supercritical methanol. J Power Sources 244:164–169
89. Nugroho A, Chung KY, Kim J (2014) A facile supercritical alcohol route for synthesizing carbon coated hierarchically mesoporous $Li_4Ti_5O_{12}$ microspheres. J Phys Chem C 118 (1):183–193
90. Borghols WJH, Wagemaker M, Lafont U et al (2009) Size effects in the $Li_{4+x}Ti_5O_{12}$ spinel. J Am Chem Soc 131(49):17786–17792
91. Kim JH, Kang YC (2013) Electrochemical properties of nano-sized $Li_4Ti_5O_{12}$ powders prepared by flame spray pyrolysis. Int J Electrochem Sci 8(3):3379–3389
92. Venkateswarlu M, Chen CH, Do JS et al (2005) Electrochemical properties of nano-sized $Li_4Ti_5O_{12}$ powders synthesized by a sol-gel process and characterized by X-ray absorption spectroscopy. J Power Sources 146(1-2):204–208
93. Zhang N, Liu Z, Yang T et al (2011) Facile preparation of nanocrystalline $Li_4Ti_5O_{12}$ and its high electrochemical performance as anode material for lithium-ion batteries. Electrochem Commun 13(6):654–656
94. Kim DK, Muralidharan P, Lee HW et al (2008) Spinel $LiMn_2O_4$ nanorods as lithium ion battery cathodes. Nano Lett 8(11):3948–3952
95. Chen MM, Sun X, Qiao ZJ et al (2014) Anatase-TiO_2 nanocoating of $Li_4Ti_5O_{12}$ nanorod anode for lithium-ion batteries. J Alloys Compd 601:38–42
96. Luo HJ, Shen LF, Rui K et al (2013) Carbon coated $Li_4Ti_5O_{12}$ nanorods as superior anode material for high rate lithium ion batteries. J Alloys Compd 572:37–42
97. Wang XY, Shen LF, Li HS et al (2014) PEDOT coated $Li_4Ti_5O_{12}$ nanorods: soft chemistry approach synthesis and their lithium storage properties. Electrochim Acta 129:283–289
98. Wu SC, Guo YX, Zhou JH et al (2011) Effect of heat-treatment temperature on the structure and properties of $Li_4Ti_5O_{12}$ nanorods prepared by the hydrothermal ion exchange method. J Inorg Mater 26(2):123–128
99. Song K, Seo DH, Jo MR et al (2014) Tailored oxygen framework of $Li_4Ti_5O_{12}$ nanorods for high-power li ion battery. J Phys Chem Lett 5(8):1368–1373
100. Kim J, Cho J (2007) Spinel $Li_4Ti_5O_{12}$ nanowires for high-rate Li-ion intercalation electrode. Electrochem Solid State Lett 10(3):A81–A84
101. Jo MR, Jung YS, Kang YM (2012) Tailored $Li_4Ti_5O_{12}$ nanofibers with outstanding kinetics for lithium rechargeable batteries. Nanoscale 4(21):6870–6875

102. Xiao L, Chen G, Sun J et al (2013) Facile synthesis of $Li_4Ti_5O_{12}$ nanosheets stacked by ultrathin nanoflakes for high performance lithium ion batteries. J Mater Chem A 1 (46):14618–14626
103. Chen S, Xin Y, Zhou Y et al (2014) Self-supported $Li_4Ti_5O_{12}$ nanosheet arrays for lithium ion batteries with excellent rate capability and ultralong cycle life. Energy Environ Sci 7 (6):1924–1930
104. Zhao YM, Liu GQ, Liu L et al (2009) High-performance thin-film $Li_4Ti_5O_{12}$ electrodes fabricated by using ink-jet printing technique and their electrochemical properties. J Solid State Electrochem 13(5):705–711
105. Yu L, Wu HB, Lou XW (2013) Mesoporous $Li_4Ti_5O_{12}$ hollow spheres with enhanced lithium storage capability. Adv Mater 25(16):2296–2300
106. Chen JZ, Yang L, Fang SH et al (2010) Synthesis of sawtooth-like $Li_4Ti_5O_{12}$ nanosheets as anode materials for Li-ion batteries. Electrochim Acta 55(22):6596–6600
107. Zhao L, Hu YS, Li H et al (2011) Porous $Li_4Ti_5O_{12}$ coated with n-doped carbon from ionic liquids for li-ion batteries. Adv Mater 23(11):1385–1388
108. Lin CF, Fan XY, Xin YL et al (2014) Monodispersed mesoporous $Li_4Ti_5O_{12}$ submicrospheres as anode materials for lithium-ion batteries: morphology and electrochemical performances. Nanoscale 6(12):6651–6660
109. Lv Y, Zhang H, Cao GP et al (2011) Phenol-formaldehyde resin-assisted synthesis of pure porous $Li_4Ti_5O_{12}$ for rate capability improvement. Mater Res Bull 46(12):2312–2316
110. Sun L, Wang J, Jiang K et al (2014) Mesoporous $Li_4Ti_5O_{12}$ nanoclusters as high performance negative electrodes for lithium ion batteries. J Power Sources 248:265–272
111. Shen LF, Yuan CZ, Luo HJ et al (2010) Facile synthesis of hierarchically porous $Li_4Ti_5O_{12}$ microspheres for high rate lithium ion batteries. J Mater Chem 20(33):6998–7004
112. Haetge J, Hartmann P, Brezesinski K et al (2011) Ordered large-pore mesoporous $Li_4Ti_5O_{12}$ spinel thin film electrodes with nanocrystalline framework for high rate rechargeable lithium batteries: relationships among charge storage, electrical conductivity, and nanoscale structure. Chem Mater 23(19):4384–4393
113. Rahman MM, Wang JZ, Hassan MF et al (2011) Amorphous carbon coated high grain boundary density dual phase $Li_4Ti_5O_{12} - TiO_2$: a nanocomposite anode material for li-ion batteries. Adv Energy Mater 1(2):212–220
114. Guo X, Xiang HF, Zhou TP et al (2014) Morphologies and structures of carbon coated on $Li_4Ti_5O_{12}$ and their effects on lithium storage performance. Electrochim Acta 130:470–476
115. Sun XC, Hegde M, Wang J et al (2014) Structural analysis and electrochemical studies of carbon coated $Li_4Ti_5O_{12}$ particles used as anode for lithium ion battery. Ecs Transactions 58 (14):79–88
116. Li TH, Shao LY, Lin XT et al (2014) High rate $Li_4Ti_5O_{12}$@ C anode material fabricated by a facile carbon coating method. J Electroanal Chem 722:54–59
117. Zhu ZQ, Cheng FY, Chen J (2013) Investigation of effects of carbon coating on the electrochemical performance of $Li_4Ti_5O_{12}$/C nanocomposites. J Mate Chem A 1 (33):9484–9490
118. Guo XF, Wang CY, Chen MM et al (2012) Carbon coating of $Li_4Ti_5O_{12}$ using amphiphilic carbonaceous material for improvement of lithium-ion battery performance. J Power Sources 214:107–112
119. Zhu GN, Wang CX, Xia YY (2011) A comprehensive study of effects of carbon coating on $Li_4Ti_5O_{12}$ anode material for lithium-ion batteries. J Electrochem Soc 158(2):A102–A109
120. Li N, Zhou GM, Li F et al (2013) A self-standing and flexible electrode of $Li_4Ti_5O_{12}$ nanosheets with a n-doped carbon coating for high rate lithium ion batteries. Adv Funct Mater 23(43):5429–5435
121. Shen LF, Yuan CZ, Luo HJ et al (2011) In situ growth of $Li_4Ti_5O_{12}$ on multi-walled carbon nanotubes: novel coaxial nanocables for high rate lithium ion batteries. J Mater Chem 21 (3):761–767

122. Poizot P, Laruelle S, Grugeon S et al (2000) Nano-sized transition-metaloxides as negative-electrode materials for lithium-ion batteries. Nature 407(6803):496–499
123. Jansen AN, Kahaian AJ, Kepler KD et al (1999) Development of a high-power lithium-ion battery. J Power Sources 81:902–905
124. Belharouak I, Sun YK, Lu W et al (2007) On the safety of the $Li_4Ti_5O_{12}/LiMn_2O_4$ lithium-ion battery system. J Electrochem Soc 154(12):A1083–A1087
125. He YB Li BH Liu M et al (2012) Gassing in $Li_4Ti_5O_{12}$-based batteries and its remedy. Sci Rep 2, article number 913
126. Gao J, Ying JR, Jiang CY et al (2009) Preparation of spherical nanocrystal $LiFePO_4$ and $Li_4Ti_5O_{12}$ and investigation of the $LiFePO_4/Li_4Ti_5O_{12}$ cell. J Inorg Mater 24(1):139–142
127. Wu HM, Belharouak I, Deng H et al (2009) Development of $LiNi_{0.5}Mn_{1.5}O_4/Li_4Ti_5O_{12}$ system with long cycle life. J Electrochem Soc 156(12):A1047–A1050
128. Li W, Li X, Chen M et al (2014) AlF_3 modification to suppress the gas generation of $Li_4Ti_5O_{12}$ anode battery. Electrochim Acta 139:104–110
129. Lu W, Liu J, Sun YK et al (2007) Electrochemical performance of $Li_{4/3}Ti_{5/3}O_4/Li_{1+x}\left(Ni_{1/3}Co_{1/3}Mn_{1/3}\right)_{(1-x)}O_2$ cell for high power applications. J Power Sources 167(1):212–216
130. Wu K, Yang J, Liu Y et al (2013) Investigation on gas generation of $Li_4Ti_5O_{12}/LiNi_{1/3}Co_{1/3}Mn_{1/3}O_2$ cells at elevated temperature. J Power Sources 237:285–290
131. Wu K, Yang J, Zhang Y et al (2012) Investigation on $Li_4Ti_5O_{12}$ batteries developed for hybrid electric vehicle. J Appl Electrochem 42(12):989–995
132. Panero S, Satolli D, Salomon M et al (2000) A new type of lithium-ion cell based on the $Li_4Ti_5O_{12}/Li_2Co_{0.4}Fe_{0.4}Mn_{3.2}O_8$ high-voltage, electrode combination. Electrochem Commun 2(11):810–813
133. Xiang HF, Zhang X, Jin QY et al (2008) Effect of capacity matchup in the $LiNi_{0.5}Mn_{1.5}O_4/Li_4Ti_5O_{12}$ cells. J Power Sources 183(1):355–360
134. Bernhard R, Meini S, Gasteiger HA (2014) On-line electrochemical mass spectrometry investigations on the gassing behavior of $Li_4Ti_5O_{12}$ electrodes and its origins. J Electrochem Soc 161(4):A497–A505
135. Wen L, Wu ZY, Luo HZ et al (2015) Dual functions of carbon in $Li_4Ti_5O_{12}$/C microspheres. J Electrochem Soc 162(2):A3038–A3044
136. Shu J (2008) Study of the interface between $Li_4Ti_5O_{12}$ electrodes and standard electrolyte solutions in 0.0-0.5 v. Electrochem Solid State Lett 11(12):A238–A240
137. Dedryvere R, Foix D, Franger S et al (2010) Electrode/electrolyte interface reactivity in high-voltage spinel $LiMn_{1.6}Ni_{0.4}O_4/Li_4Ti_5O_{12}$ lithium-ion battery. J Phys Chem C 114 (24):10999–11008
138. Zu C-X, Li H (2011) Thermodynamic analysis on energy densities of batteries. Energ Environ Sci 4(8):2614–2624
139. Liu GQ, Wen L, Liu GY et al (2011) Synthesis and electrochemical properties of $LiNi_{0.4}Mn_{1.5}Cr_{0.1}O_4$ and $Li_4Ti_5O_{12}$. Met Mater Int 17(4):661–664
140. Thackeray MM, Johnson CS, Vaughey JT et al (2005) Advances in manganese-oxide 'composite' electrodes for lithium-ion batteries. J Mater Chem 15(23):2257–2267
141. Plitz I, DuPasquier A, Badway F et al (2006) The design of alternative nonaqueous high power chemistries. Appl Phys A-Mater Sci Process 82(4):615–626
142. Kim H, Park K-Y, Cho M-Y et al (2014) High-performance hybrid supercapacitor based on graphene-wrapped $Li_4Ti_5O_{12}$ and activated carbon. Chemelectrochem 1(1):125–130
143. Naoi K, Naoi W, Aoyagi S et al (2013) New generation "Nanohybrid supercapacitor". Acc Chem Res 46(5):1075–1083
144. Naoi K, Ishimoto S, Miyamoto J-i et al (2012) Second generation 'nanohybrid supercapacitor': evolution of capacitive energy storage devices. Energ Environ Sci 5 (11):9363–9373
145. Naoi K (2010) 'Nanohybrid capacitor': the next generation electrochemical capacitors. Fuel Cells 10(5):825–833

Chapter 6
Anodes and Anode/Electrolyte Interfaces for Rechargeable Magnesium Batteries

Timothy S. Arthur and Nikhilendra Singh

6.1 Introduction

Rechargeable batteries, coupled with other alternative energy sources (hydrogen, solar, wind, etc.), remain a key component in realizing a truly sustainable society. Electrical energy generated by such alternative (primary) energy sources could be stored within tangible rechargeable battery (secondary) sources, enabling us to transport energy on demand to any location while also enhancing the energy efficiency of the system. Lithium-ion (Li-ion) batteries are the most notable example of recently discovered and applied rechargeable batteries and have revolutionized portable consumer electronic devices. Over the last few decades, the automotive industry has also begun focusing on alternative technologies to traditional combustion engines, which currently utilize carbon-based fossil fuels. The introduction of electric vehicles (EVs) and plug-in hybrid vehicles (PHVs) via the use of such alternative technologies (e.g., rechargeable batteries) as stand-alone or tandem energy sources is also slowly revolutionizing the face of the automotive industry [1, 2].

However, while batteries are a viable alternative to fossil fuels, concerns over their range performance in automobiles have hampered their widespread use within the industry. Hence, high-performance battery systems which meet automobile energy use, and especially space requirements, remain paramount in establishing the next generation of EVs and PHVs. In response to such concerns, researchers around the world have recently focused their attention beyond existing Li-ion battery technology, which powers most current-day EVs and PHVs. Post Li-ion

T.S. Arthur (✉) • N. Singh (✉)
Toyota Research Institute of North America, 1555 Woodridge Avenue,
Ann Arbor, MI 48105, USA
e-mail: tim.arthur@toyota.com; nikhilendra.singh@toyota.com

K.I. Ozoemena, S. Chen (eds.), *Nanomaterials in Advanced Batteries and Supercapacitors*, Nanostructure Science and Technology,
DOI 10.1007/978-3-319-26082-2_6

batteries, defined as batteries with a higher theoretical energy density than existing Li-ion batteries containing a carbon anode and a metal oxide cathode (~ 2000 Wh /L based on active materials), require the replacement of the lithiated graphite anode with metal alloy insertion anodes or pure metallic anodes. Via exploration of new chemistries (especially electrochemistries) and materials, one side of such investigations had led researchers to contemplate the feasibility of Li-air and Li-sulfur systems [2–4].

Both these systems offer advantages in terms of energy density over current Li-ion systems and have been reviewed in great detail. However, commercialization of such technologies would still have to overcome the inherent instability of Li metal, which is known to result in short circuits due to dendritic growth. Such properties can lead to internal heating and thermal runaway from exothermic chemical reactions involving multiple battery components, creating an unsafe environment for battery applications. While numerous approaches to stabilizing Li metal anodes and preventing dendrite formation are currently being explored (both from perspectives of anode protection and changes to the battery electrolytes), such concerns remain relevant to the eventual applications of Li-based batteries [2–4].

The other side of such investigations has led researchers to contemplate the feasibility of utilizing the transfer of multiple electrons per atom as an inherent route to improve the energy density. Such systems have come to be known as multivalent systems, of which Mg^{2+}, Ca^{2+}, Zn^{2+}, and Al^{3+} have garnered the most attention (Fig. 6.1). Each of these systems requires multiple electrons to deposit and dissolve the metallic anode during battery operation: $A^{x+} + xe^- \leftrightarrow A_{(m)}$, although

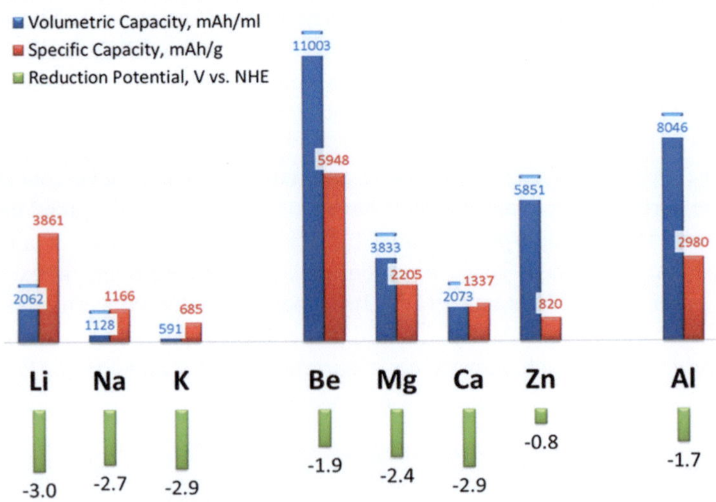

Fig. 6.1 Capacities and reductive potentials for various metal anodes (Reprinted (adapted) with permission from (Muldoon J, Bucur CB, Gregory T (2014) Quest for nonaqueous multivalent secondary batteries: magnesium and beyond. Chemical Reviews 114: 11683-11720). Copyright (2014) American Chemical Society [3])

the equilibrium potential for this reaction will vary from system to system. The scope of this chapter is the development of nonaqueous magnesium (Mg) electrochemistry which is capable of delivering a higher volumetric energy density (3833 mAh cm^{-3}) than Li (2061 mAh cm^{-3}). A high volumetric energy density is more desirable for mobile devices than for stationary energy storage. The more portable the device, the greater the degree of space management necessary to accommodate the battery, and volumetric energy density can play a crucial role in providing the necessary energy within the given space confinements [1–3, 5, 6].

The development of nonaqueous Mg batteries is hindered by two major challenges. First, new cathode materials capable of intercalating Mg^{2+} ions need to be developed. The prototypical intercalation cathode for magnesium batteries is a Chevrel-phase molybdenum sulfide (Mo_6S_8) cathode. However, the low reduction potential ($\sim$ 1.1 V vs. Mg) and reversible capacity (120 mAh g^{-1}) inhibit high-energy density. Recently, incremental increases to energy density have been shown by two-dimensional cathodes, polyanion cathodes, and thin-film sputtered cathodes. However, typical insertion cathodes are still plagued by slow diffusion of the Mg^{2+} ion and reaction mechanisms which differ from Li^+ [1, 3, 7]. The second issue involves the interactions between the anode and the electrolyte and is the main focus of this chapter.

6.2 Magnesium (Mg) Metal Anodes

Apart from system advantages in terms of volumetric energy density, the use of Mg metal as the anode in itself provides certain advantages. Mg metal anodes do not appear to be plagued by dendrite formation to the same degree as Li metal anodes. While Mg batteries had been investigated as far back as the 1980s, recent reports on the advantages of high-energy density Mg batteries and greater interest in the field of new cathodes and electrolytes for Mg batteries have steadily increased [3]. However, despite the potential of Mg batteries, several key challenges need to be overcome for this technology to become competitive with existing Li-ion technologies. This section of the chapter will focus on summarizing the various studies which have benchmarked the Mg anode along with its capabilities and restrictions.

In terms of energy density, high-voltage and high-capacity Mg systems must be developed. To date, various organohaloaluminates $[(MgR_2)_x - (AlCl_{3-n}R_n)_y]$ have been utilized as alternative electrolytes for Mg systems, since conventional battery electrolytes ($TFSI^-$, ClO_4^-, PF_6^-) form a Mg^{2+} blocking layer on the Mg metal anode surface. However, recent reports have shown that these organohaloaluminate electrolytes provide a limited operating voltage window when tested against typical battery current collectors and that it may be possible to utilize conventional battery electrolytes with Mg metal anodes under certain conditions [2]. The initial breakthroughs for rechargeable Mg batteries were realized by Gregory et al. (1990) and Aurbach et al. (2000). Gregory et al. referred to the possibility of nonaqueous Mg

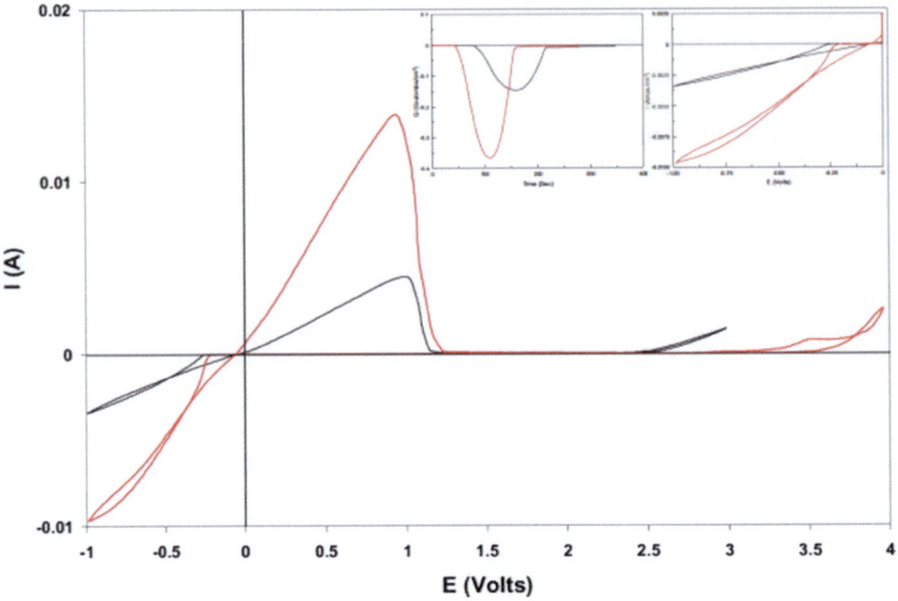

Fig. 6.2 Comparison between the voltammetric behaviors of THF solutions containing 0.25 M of the reaction product between 1:2 $MgBu_2$ and $AlCl_2Et$ (DCC) and 0.4 M of the reaction product between 1:2 $AlCl_3$ and PhMgCl (designated APC) as indicated; 25 mV s^{-1}, Pt wire working electrode, 25 °C. (*Right insert*) Enlargement of the cyclic voltammograms near 0 V vs. Mg reference electrode, comparing the overpotential for Mg deposition in the two solutions. (*Left insert*) Charge balance upon typical Mg deposition/dissolution cycles in these solutions (100 % cycling efficiency of Mg deposition) (Reprinted (adapted) with permission from (Muldoon J, Bucur CB, Gregory T (2014) Quest for nonaqueous multivalent secondary batteries: magnesium and beyond. Chemical Reviews 114: 11683-11720). Copyright (2014) American Chemical Society [3])

batteries using Grignard electrolytes and reported some candidates for electrolytes as well as cathodes. However, these Grignard electrolytes highlighted very low operating voltage stabilities, which would have been inadequate for battery utilization [3]. Aurbach et al. built off of this work and demonstrated the first prototype Mg battery comprising of the Chevrel-phase (Mo_6S_8) cathode, a Grignard-reagent-based electrolyte, and a Mg metal anode. They were able to exhibit excellent cycling performances of over 2000 cycles. Aurbach et al. further improved upon this Grignard-reagent-based electrolyte by developing a 0.25 M 1:2 M ratio of $MgBu_2$ and $AlCl_2Et$ in THF followed by 0.4 M 1:2 M ratio between $AlCl_3$ and PhMgCl in THF, showcasing 100 % coulombic efficiency for Mg plating and dissolution (Fig. 6.2) [3, 6].

Inspired by these pioneering successes, several research efforts around the world began focusing on the development of electrolytes and cathodes. Since the electrolyte is usually the key component for a battery and determines the voltage window for the battery, it was most extensively studied over the past several years. In the

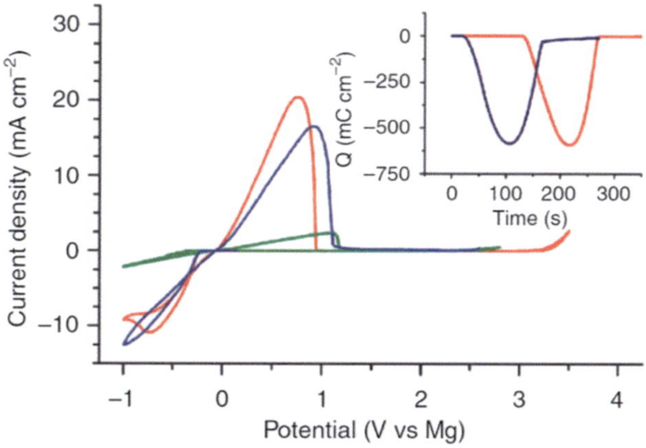

Fig. 6.3 Cyclic voltammograms of HMDSMgCl (*green*), the reaction product generated in situ from a 3:1 mixture of HMDSMgCl to $AlCl_3$ (*blue*), and the crystal obtained from a 3:1 mixture of HMDSMgCl to $AlCl_3$ (designated GEN1) (*red*). (Inset) Charge balance during deposition and subsequent dissolution of Mg (Reprinted (adapted) with permission from (Muldoon J, Bucur CB, Gregory T (2014) Quest for nonaqueous multivalent secondary batteries: magnesium and beyond. Chemical Reviews 114: 11683-11720). Copyright (2014) American Chemical Society [3])

context of Mg batteries, these electrolytes would have to show good Mg deposition and dissolution characteristics along with higher oxidative stabilities to account for high-voltage cathodes. Muldoon et al. were the next group to find success in expanding the operating voltage window of Grignard-reagent-based electrolytes, which are also known to the community as organohaloaluminates. They used a HMDS derivative to produce a 3:1 mixture between HMDSMgCl and $AlCl_3$ in THF (Fig. 6.3), also showing near 100 % coulombic efficiencies for Mg deposition and dissolution. However, as has been the case with previous version of organohaloaluminates, the HMDS version was also subject to being corrosive toward commonly used current collectors for batteries (e.g., stainless steel and Ni), as well as materials far more resistant toward corrosion (e.g., Pt, Au, and glassy carbon). Muldoon et al. captured this information via running linear sweep voltammograms with various working electrode material (current collector materials) and observing the oxidative stability (Fig. 6.4). The observation of an increase in current densities at various voltages suggested corrosion of the working electrode (current collector) materials themselves, rendering such organohaloaluminates difficult to apply in commercial battery applications [3].

An alternate to the corrosive organohaloaluminate electrolytes are the commonly used conventional or commercially available electrolytes, i.e., magnesium analogue salts of Li-ion battery electrolytes in organic solvents (e.g., $Mg(TFSI)_2$, $Mg(ClO_4)_2$, and $Mg(PF_6)_2$). However, until very recently, such electrolytes could not be used directly with Mg metal anodes due to the formation of a passivation layer on the Mg metal surface during the use of such electrolytes. In essence, no Mg deposition/dissolution was found to be possible from such electrolytes. Obrovak et al. recently published a report on utilizing $Mg(TFSI)_2$ in acetonitrile and reported

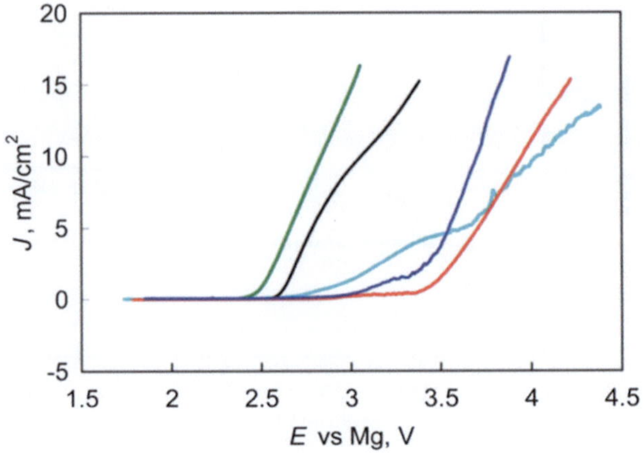

Fig. 6.4 Linear scan voltammograms of $(Mg_2(\mu - Cl)_3 6THF) - (HMDS_nAlCl_{4-n})$ (n = 1, 2) (GEN1) with a scan rate of 25 mV s^{-1} depicting characteristic voltage stabilities on various working electrodes (*green*, stainless steel; *black*, Ni; *blue*, Pt; *turquoise*, Au; *red*, glassy carbon) with areas of 0.02 cm^2 and a magnesium reference and counter-electrode at a temperature of 21 °C (Reprinted (adapted) with permission from (Muldoon J, Bucur CB, Gregory T (2014) Quest for nonaqueous multivalent secondary batteries: magnesium and beyond. Chemical Reviews 114: 11683-11720). Copyright (2014) American Chemical Society [3])

that while the voltage window was found to be stable between 0 – 2.8 V vs. Mg, no Mg deposition was actually observed. Further, exceeding 2.8 V resulted in corrosion of the stainless steel current collector, while exceeding 0 V in the negative direction resulted in electrolyte decomposition (Fig. 6.5) [8]. Uchimoto et al. recently reported the use of $Mg(TFSI)_2$-triglyme (1:5 M ratio) against a Pt working electrode and displayed successful Mg deposition and dissolution, but only after raising the temperature of their electrolyte bath to 100°C (Fig. 6.6) [9]. Similarly, Choi et al. also reported the use of $Mg(TFSI)_2$ from a glyme/diglyme solvent mixture against an Al current collector and displayed successful Mg deposition/dissolution at temperatures as low as 30 °C (Fig. 6.7) [10]. Further investigations into the role played by glymes in assisting Mg deposition and dissolution are warranted to support such new information.

While researchers were discovering that organohaloaluminates and conventional electrolytes both possess certain advantages and disadvantages toward Mg deposition and dissolution, a new family of plausible Mg battery electrolytes was discovered and reported by Mohtadi et al. in 2012. These Mg battery electrolytes were composed of borohydride complexes in DME as a solvent, more specifically, a 1:3 mixture of $Mg(BH_4)_2$ and $LiBH_4$ in both THF and DME. Mohtadi et al. were able to show successful Mg deposition and dissolution from this electrolyte for the very first time, with coulombic efficiencies as high as 94 % (Fig. 6.8). While the oxidative stability of such an electrolyte (1.7 V vs. Mg) currently limits its applicability in a high-voltage Mg battery, this work certainly opens up a new avenue of

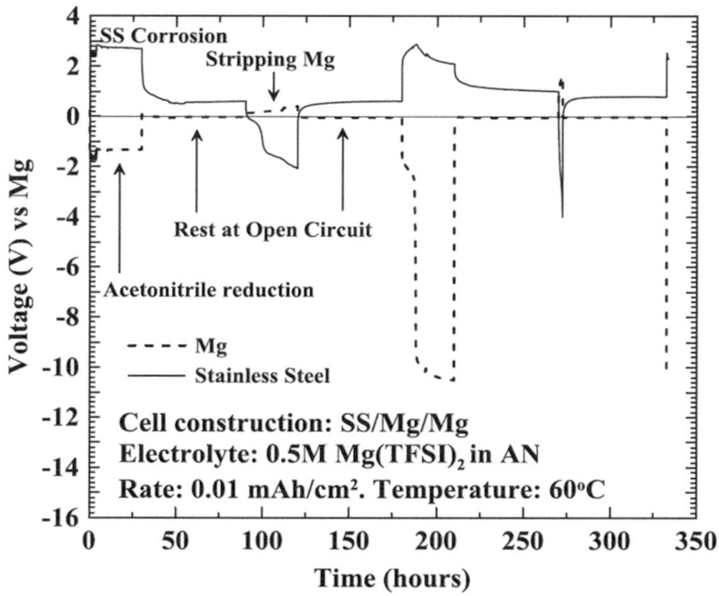

Fig. 6.5 Voltage profile of a SS/Mg/Mg cell with 0.5 M $Mg(TFSI)_2$/AN electrolyte operating at 60°C. The cell was started in the direction corresponding to oxidation at the stainless steel working electrode (Reproduced by permission of The Electrochemical Society (Tran TT, Lamanna WM, Obrovac MN (2012) Evaluation of $Mg\{N(SO_2CF_3)_2\}_2$/acetonitrile electrolyte for use in Mg-ion cells. Journal of The Electrochemical Society 159: A2005-A2009) [8])

exploration into a whole new class of Mg electrolytes which could circumvent the disadvantages posed by both organohaloaluminate and conventional Mg battery electrolytes today [11].

In terms of Mg metal anodes and their corresponding electrolytes, recent studies by Mizuno et al. have shown that the Mg metal anode itself does not seem to restrain Mg dissolution and deposition. However, the interfacial resistance between a Mg metal anode and its corresponding electrolyte was found to drastically change over time and with operating conditions (Fig. 6.9). This phenomenon was observed regardless of the electrolyte compatibility to Mg metal. Furthermore, the processes of Mg deposition and dissolution are known to be mechanistically complicated, so a root cause for the interface characteristics observed by Mizuno et al. still remains unclear. Nonetheless, the Mg metal anode/electrolyte interface properties would affect the proper operation of even the cathodes for rechargeable Mg batteries. Compared to individual studies on the electrolyte and cathode in Mg battery systems, only a few investigations of the Mg metal anode and its reaction process exist. Hence, mechanistic understanding of the interface between Mg metal and the electrolytes is remarkably important for future material design and enhancement of battery performance [1].

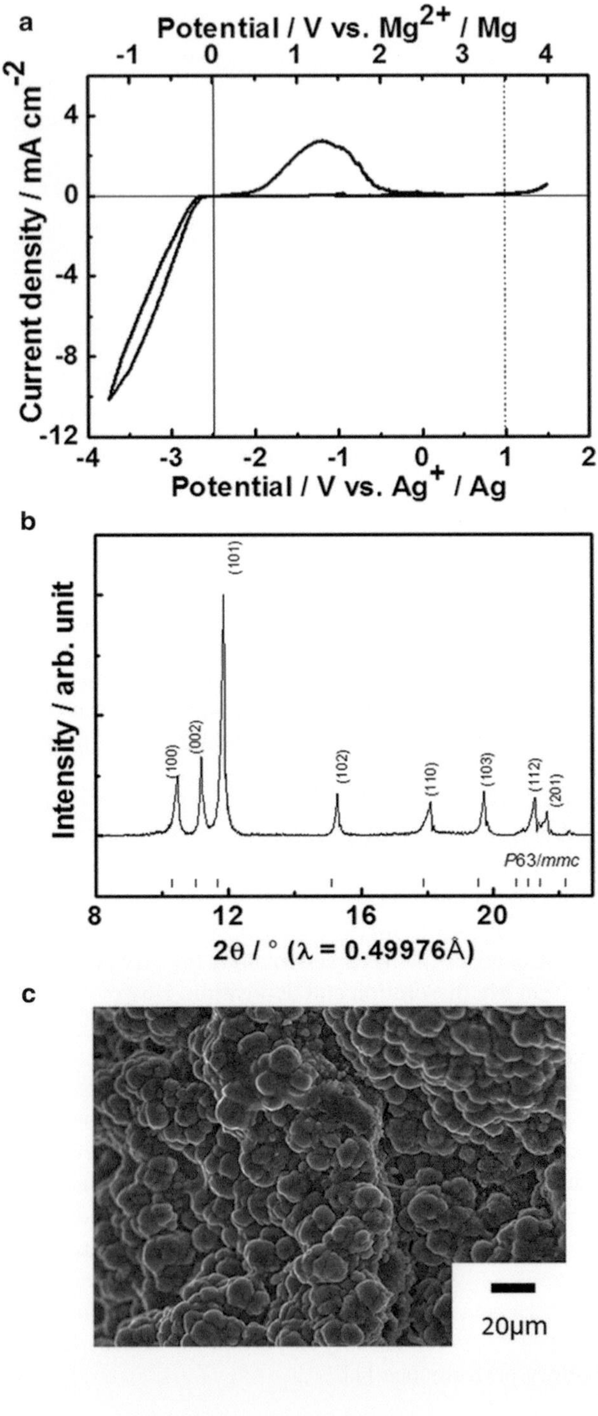

Fig. 6.6 Mg deposition and dissolution in the $Mg(TFSI)_2$-triglyme electrolyte. (**a**) Cyclic voltammograms of platinum electrode in $Mg(TFSI)_2$/triglyme (1:5 M ratio). Three-electrode cells using Mg metal counter-electrode and silver reference electrode were used. Potential sweep rate was set at 1.0 mV s^{-1}, and measurements were conducted at 100 °C. (**b**) XRD pattern of the deposited products. (**c**) SEM image of the deposited products (Reproduced by permission of Creative Commons CC-BY Attribution 4.0 International License (Orikasa Y, Masese T, Koyama Y, Mori T, Hattori M, Yamamoto K, Okado T, Huang Z-D, Minato T, Tassel C, Kim J, Kobayashi Y, Abe T, Kageyama H, Uchimoto Y (2014) High energy density rechargeable magnesium battery using earth-abundant and non-toxic elements. Scientific Reports 4: 5622 DOI:10.1038/srep05622) [9])

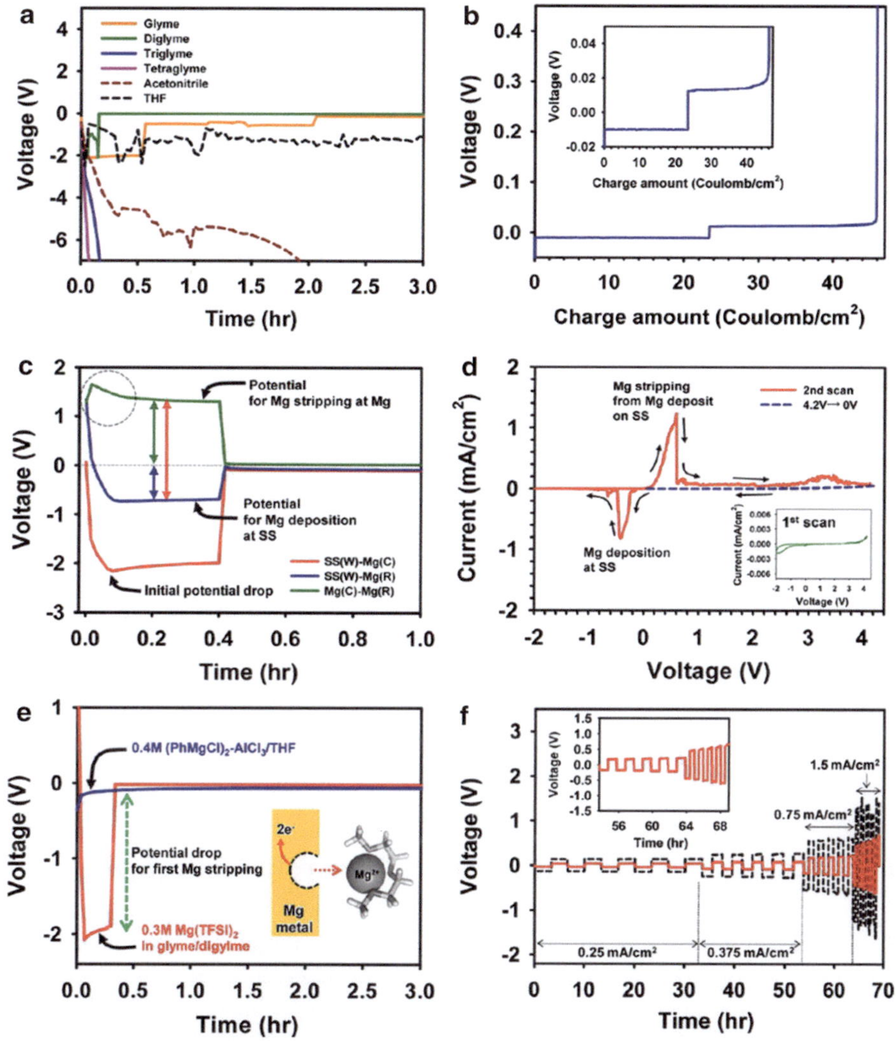

Fig. 6.7 (**a**) Relation between the initial overpotential of Mg/Cu cells and electrolytes with 0.1 M $Mg(TFSI)_2$ for the current density of 0.0188 mA/cm^2. (**b**) Voltage profiles of a Mg/SS cell in glyme/diglyme (1/1, v/v) with 0.3 M $Mg(TFSI)_2$ for a fixed charge amount of 23.4 coulomb/cm^2(0.325 mA/cm^2). (**c**) Voltage profiles of a SS/Mg/Mg cell (three-electrode cell) with 0.3 M Mg $(TFSI)_2$ in glyme/diglyme (1/1, v/v) at 30 °C. *W* working electrode, *C* counter-electrode, *R* reference electrode. (**d**) Cyclic voltammetry showing Mg deposition/stripping on a SS electrode in 0.3 M $Mg(TFSI)_2$ in glyme/diglyme (1/1, v/v) at a scan rate of 0.2 mV/s. (**e**) Initial potential drop of Mg/Cu cells in 0.4 M $(PhMgCl)_2 - AlCl_3$/THF (*blue line*) or glyme/diglyme (1/1, v/v) with 0.3 M $Mg(TFSI)_2$ (*red line*) for the current density of 0.1625 mA/cm^2. The onset shows a schematic illustration of Mg stripping by diglyme forming a stable solvation sheath. (**f**) Rate capability of Mg/Mg cells with 0.3 $Mg(TFSI)_2$ in glyme/diglyme (1/1, v/v) (*red line*) or diglyme (*dashed black line*) for current densities of 0.25, 0.375, 0.75, and 1.5 A/cm^2. The inset displays

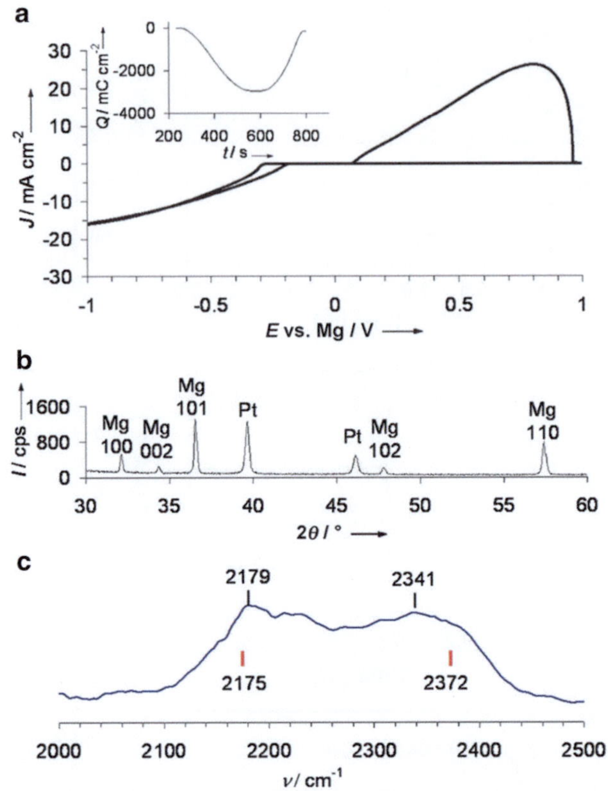

Fig. 6.8 For $LiBH_4$(0.6 M)/$Mg(BH_4)_2$ (0.18 M) in DME: (**a**) cyclic voltammogram (inset shows deposition/stripping charge balance). (**b**) XRD results following galvanostatic deposition of Mg on a Pt disk. (**c**) IR spectra (*red*) indicates band maxima for $Mg(BH_4)_2$/DME (Reprinted (adapted) with permission from (Mohtadi R, Matsui M, Arthur TS, Hwang S-J (2012) Magnesium borohydride: from hydrogen storage to magnesium battery. Angewandte Chemie International Edition 51: 1-5). Copyright (2012) John Wiley and Sons [11])

6.3 Magnesium (Mg) Metal Anode/Electrolyte Interface

Utilization of the Mg metal anode is the ultimate target for nonaqueous Mg batteries. As opposed to Li systems, Mg metal deposition and dissolution are hindered by the electrolyte decomposition layers found on the anode. The high reactivity of the anode compounds the issue by dissuading ex situ analysis of the surface, even with exhaustive attempts to keep the environment inert. Therefore, analysis of the interface and reaction products is best suited for in situ/operando electrochemical and spectroelectrochemical experiments.

Fig. 6.7 (continued) potential profiles for the current densities of 0.75 and 1.5 mA/cm^2 (Reprinted (adapted) with permission from (Ha S-Y, Lee Y-W, Woo SW, Koo B, Kim J-S, Cho J, Lee KT, Choi N-S (2014) Magnesium (II) Bis(trifluoromethane sulfonyl) Imide-based electrolytes with wide electrochemical windows for rechargeable magnesium batteries. ACS Applied Materials & Interfaces 6: 4063-4073). Copyright (2014) American Chemical Society [10])

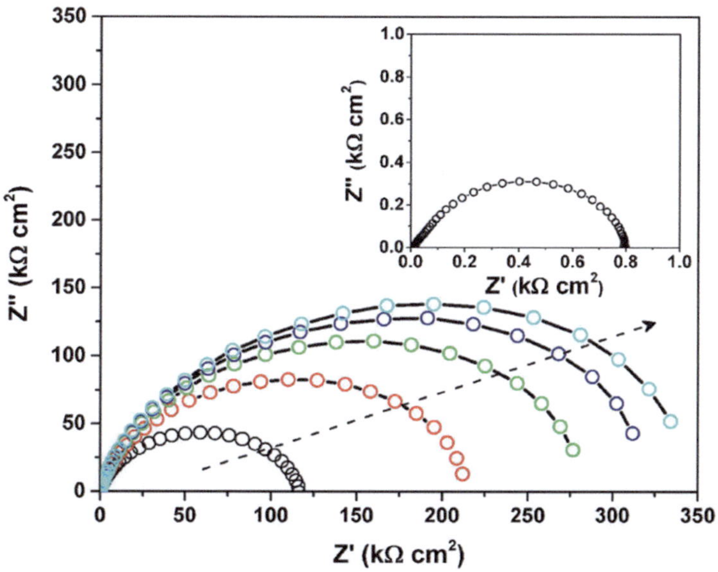

Fig. 6.9 Nyquist plot of the symmetrical Mg/Mg cell (as a function of holding time) under OCV. The inset shows the Nyquist plot of the Mg/Mg symmetrical cell measured under a bias voltage of + 150 mV (Reproduced by permission of Creative Commons CC-BY Attribution 4.0 International License (Mizuno F, Singh N, Arthur TS, Fanson PT, Ramanathan M, Benmayza A, Prakash J, Liu Y-S, Glans P-A, Guo J (2014) Understanding and overcoming the challenges posed by electrode/electrolyte interfaces in rechargeable magnesium batteries. Frontiers in Energy Research 2: 1-11) [1])

Aurbach et al. first realized the importance of using in situ techniques to understand the nature of the Mg metal/electrolyte interface and pioneered the use of in situ scanning tunneling microscopy (STM) to understand the morphology changes during Mg deposition/dissolution [12]. Figure 6.10 shows STM images of a pristine gold electrode at OCV and the same electrode after magnesium deposition $\sim 0.88\ C/cm^2$ in a BuMgCl 1 M/THF solution. The image of the magnesium deposits clearly shows micrometric-sized, pyramid-shaped magnesium crystals. Figure 6.10c presents an image of the magnesium deposits at a higher, sub-micrometric resolution. This image clearly shows some features of nanometric size on the surface.

Figure 6.11a shows an image of a pristine gold electrode at open-circuit voltage (OCV) in a THF solution of 0.25 M $Mg(AlCl_2BuEt)_2$, and Fig. 6.11b and c shows STM images of two resolutions of the same electrode in the same solution after 0.88 C/cm^2 of magnesium was deposited. Crystals of magnesium deposits are clearly seen in the images in Fig. 6.11b and c; however, their shape is less regular and the crystals are smaller, compared with the Mg deposits in the Grignard salt solutions. Figure 6.11d shows an image of the same gold electrode of Fig. 6.11a–c after the Mg dissolution process $\sim 0.88\ C/cm^2$, at 1.2 V vs. Mg reference electrode. It appears that the electrode lost all the typical features of Mg deposition. Its

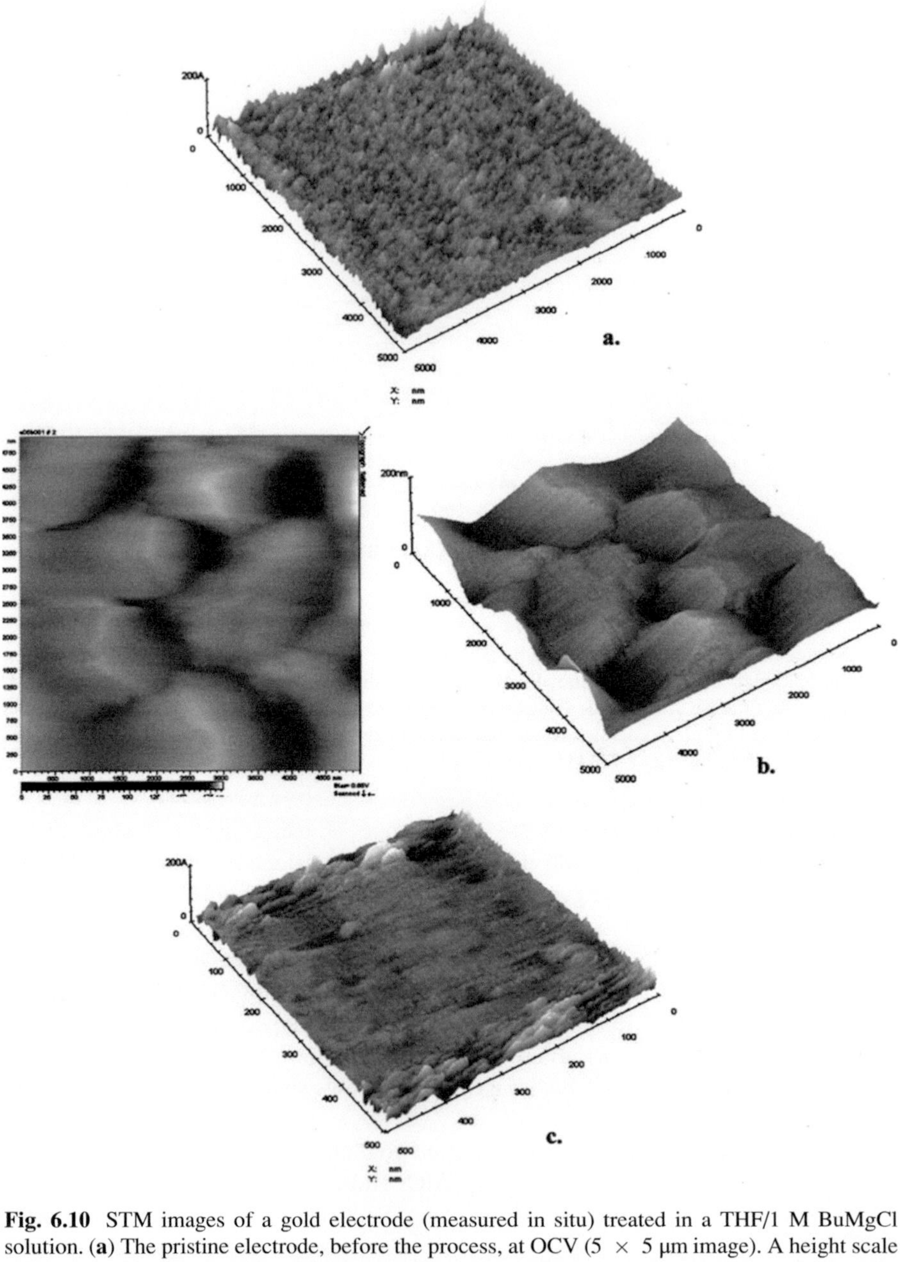

Fig. 6.10 STM images of a gold electrode (measured in situ) treated in a THF/1 M BuMgCl solution. (**a**) The pristine electrode, before the process, at OCV (5 × 5 μm image). A height scale (200 nm) is attached. (**b**) 3D and 2D images (5 × 5 μm) of the same electrode after galvanostatic Mg deposition (0.9 C/cm^2). (**c**) 3D image, same as **b**, 0.5 × 0.5 μm (Reproduced by permission of The Electrochemical Society (Aurbach D, Cohen Y, Moshkovich M (2001) The study of reversible magnesium deposition by in situ scanning tunneling microscopy. Electrochemical and Solid-State Letters 4: A113-A116) [12])

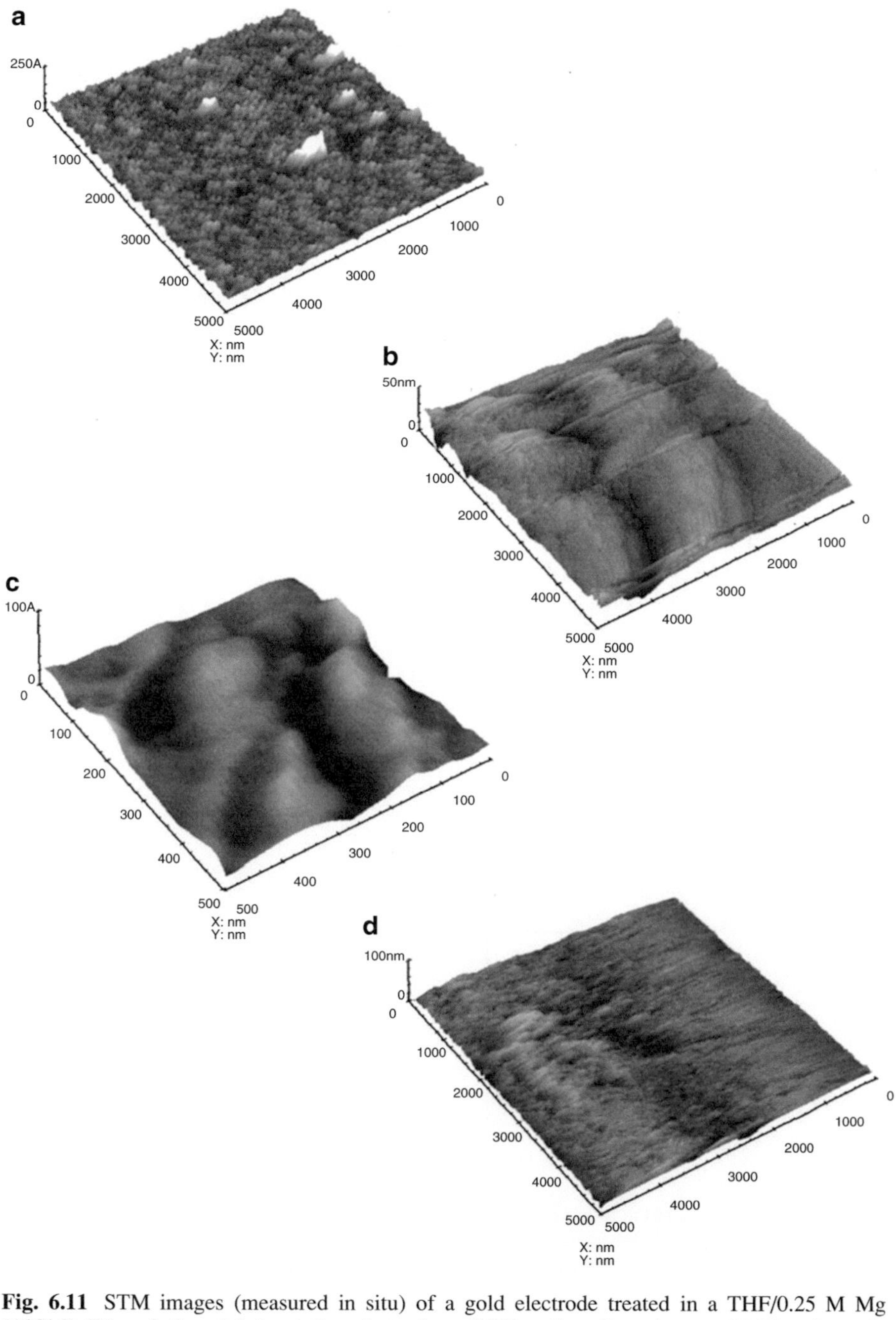

Fig. 6.11 STM images (measured in situ) of a gold electrode treated in a THF/0.25 M Mg $(AlCl_2BuEt)_2$ solution. (**a**) A pristine electrode at OCV, a 5 × 5 μm image. (**b**) The electrode after galvanostatic Mg deposition (0.88C/cm^2), a 5 × 5 μm image. (**c**) Same as b, a 0.5 × 0.5 μm image. (**d**) The electrode after Mg dissolution (0.88 C/cm^2), a 5 × 5 μm image (Reproduced by permission of The Electrochemical Society (Aurbach D, Cohen Y, Moshkovich M (2001) The study of reversible magnesium deposition by in situ scanning tunneling microscopy. Electrochemical and Solid-State Letters 4: A113-A116) [12])

morphology resembles that of the pristine electrode, but is smoother. Hence, the successful STM imaging of Mg electrodes reported herein seems to indicate that in contrast to Li electrodes, whose passivation is due to stable surface film formation, Mg electrodes in ethereal solutions do not develop stable, electronically insulating surface films.

The elegant in situ STM experiments highlighted that Mg metal deposition lacks the formation of Solid-Electrolyte Interface (SEI) layers normally observed on Li-ion anodes. However, further in situ work by Aurbach et al. showed that the electrochemical reduction and oxidation followed a complex mechanism [13]. Figure 6.12 is an example of a cyclic voltammogram-electrochemical quartz-crystal microbalance (CV-EQCM) experiment with a gold-on-quartz electrode using a

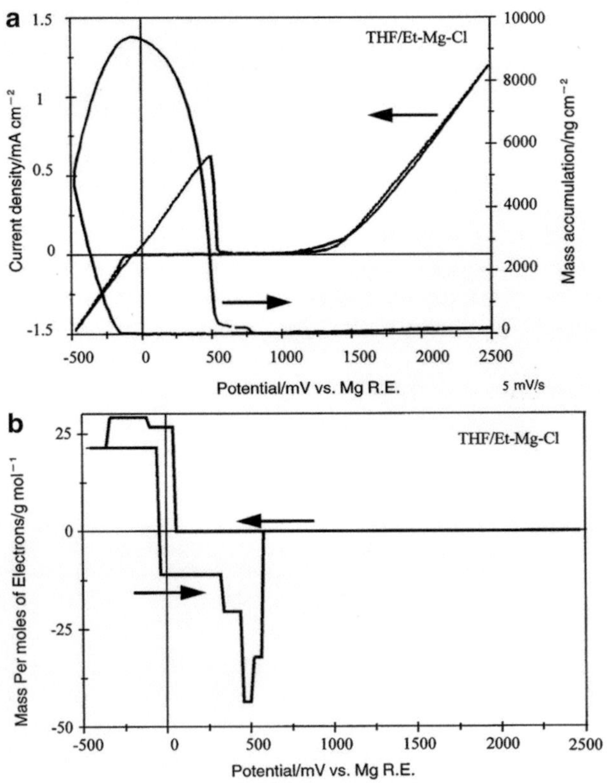

Fig. 6.12 Results of an EQCM experiment of an Au quartz electrode with THF + 2 M EtMgCl solution. The reference electrode was Mg wire in solution. The potential scan rate was 5 mV s^{-1}. 0 V versus Mg RE (corresponds to 1.15 V vs. Li/Li^{+}). (**a**) A presentation of the CV and the mass accumulation, as indicated. (**b**) Calculation of the mass accumulated on the electrodes per moles of electrons (m.p.e./ g mol^{-1}) during the experiments (m.p.e. vs. E). The negative sign indicates mass depletion (Reprinted with permission from (Lu Z, Schechter A, Moshkovich M, Aurbach D (1999) On the electrochemical behavior of magnesium electrodes in polar aprotic electrolyte solutions. Journal of Electroanalytical Chemistry 466: 203-217). Copyright (1999) Elsevier [13])

BuMgCl electrolyte. The CV reflects reversible Mg deposition and dissolution, as well as the oxidation of the alkyl group at potentials above 1.3 V versus Mg RE (around 1.2 V vs. Ag/Ag^{+}), which limits the electrochemical window of these solutions. The CV also reflects the nucleation overpotential required to initiate the Mg deposition (note the hysteresis in the CV at the low potentials, which is typical of deposition processes with nucleation overpotential). This figure also shows the variations in the mass accumulation during the experiment. It is significant that after a completed Mg deposition/dissolution cycle in these systems, the total mass balance is close to zero. This means that the deposited Mg does not react with the solution species and remains electrochemically active. Figure 6.12b shows the calculation of the mass accumulated per mole of electrons transferred (m.p.e) during the experiment described in Fig. 6.12b The m.p.e. values calculated for Mg deposition/dissolution should be close to 12 g/mol (which is the molar mass of ½ Mg). However, as seen in Fig. 6.12b, the m.p.e. values calculated as Mg deposition starts are higher than 25 g/mol. However, it should be noted that in prolonged Mg deposition, m.p.e. values close to 12 g/mol are calculated. As seen in Fig. 6.12b, upon Mg dissolution, the m.p.e. values calculated are indeed close to 12 g/mol. At the end of the process, toward complete dissolution of the magnesium deposited, the m.p.e. values calculated jump to high values (between 40 and 50 g/mol). The EQCM experiments, which clearly showed that Mg deposition/dissolution cycles in Grignard-based electrolytes have both a mass and charge balance close to zero, seem to prove that Mg deposition in these solutions is not accompanied by irreversible surface film formation, as is the case for Li deposition in ethereal solutions. However, variations in the m.p.e. values during the EQCM experiments seem to show that the Mg deposition/dissolution processes accompany adsorption and desorption processes. Recent experiments by Gewirth et al. imply that high efficiency of the Mg deposition/dissolution process is dependent on the cycle number during CV [14].

As shown in Fig. 6.13a and b, the overpotentials associated with Mg deposition and dissolution decrease with cycle number. Additionally, across early cycles, the amount of charge passed during dissolution is less than the charge passed during deposition (Fig. 6.13c and d). It was found that the charge efficiencies for Mg deposition and dissolution from $Mg(AlCl_2EtBu)_2$ range from 78 % to 98 % and 55 % to 100 % on Au and Pt electrodes, respectively, depending upon the cycle number. In general, the charge efficiency increases and approaches unity as the electrode is cycled. Nonetheless, the lack of charge balance implies that some charge is lost to irreversible decomposition processes during the voltammetry, especially during early cycles.

The in situ STM and CV-EQCM results highlight that Mg deposition and dissolution proceeds through adsorbed reaction intermediates; however, little is known about the chemical composition of these intermediates. In situ spectroelectrochemical experiments have since afforded insights into the compounds which may be responsible for the reversible Mg deposition and dissolution. Aurbach et al. have reported on in situ Fourier-transform infrared (FTIR) and Raman spectroscopies [15]. Although highly insightful, an element-specific technique would complement these previous reports by directly observing Mg species at the

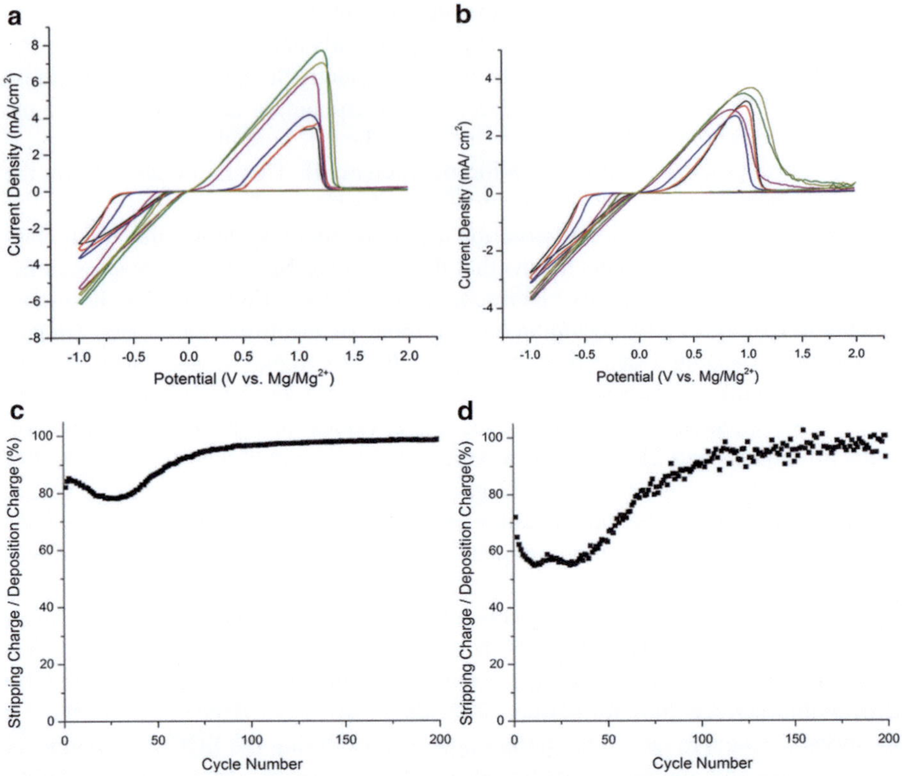

Fig. 6.13 Cyclic voltammograms of the 1st cycle (*black*), 2nd cycle (*red*), 10th cycle (*blue*), 50th cycle (*purple*), 100th cycle (*green*), and 200th cycle (*yellow*) of Mg deposition and stripping from $Mg(AlCl_2EtBu)_2$ in THF at 10 mV/s on Au (**a**) and Pt (**b**) electrodes and the ratios of the stripping charge to the deposition charge as a function of cycle number on Au (**c**) and Pt (**d**) (Reprinted (adapted) with permission from (Barile C, Spatney R, Zavadil K, Gerwith A (2014) Investigating the reversibility of in situ generated magnesium organohaloaluminates for magnesium deposition and dissolution. Journal of Physical Chemistry C 118: 10694-10699). Copyright (2014) American Chemical Society [14])

interface. X-ray absorption near edge structure, XANES, is an element-specific process that empirically determines the oxidation state of the absorbing species, and is ideal for looking at the deposition of Mg at the working electrode. Previously, Nakayama et al. used ex situ XAS to study the Mg and Al *K*-edge of magnesium organohaloaluminate and Grignard electrolytes [16]. As shown in Fig. 6.14, an in situ electrochemical XAS cell has been developed to directly study the interface during Mg deposition [17].

The electrolyte consisted of a 0.45 M solution of the electrolyte salt dissolved in THF. As shown in Fig. 6.15a, deposition and dissolution of magnesium was monitored through two- and three-electrode cyclic voltammograms between the sputtered Pt layer (Fig. 6.14avi, working) and the Mg foil (Fig. 6.14aii, counter). To generate the three-electrode cell, a Mg reference was galvanostatically deposited

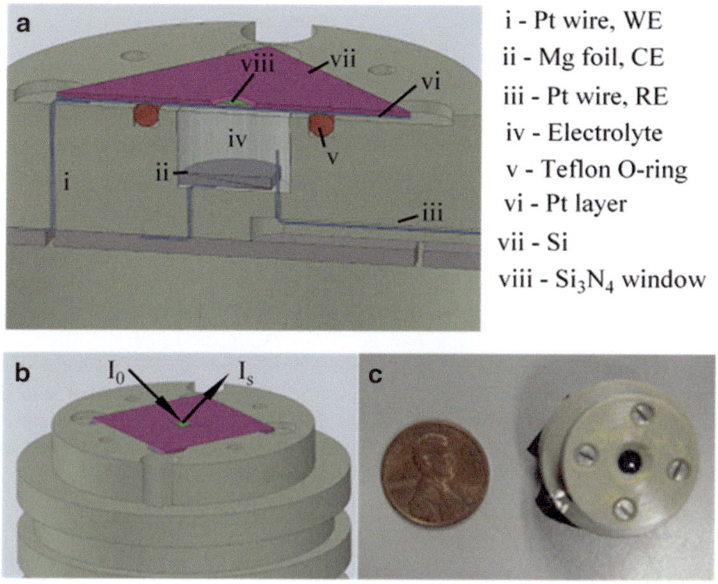

Fig. 6.14 (**a**) Cross section of the in situ electrochemical/XAS cell with annotations. (**b**) Drawing and (**c**) photograph of the assembled cell (Reprinted with permission from (Arthur T, Glans P-A, Matsui M, Zhang R, Ma B, Guo J (2012) Mg deposition observed by in situ electrochemical Mg K-edge X-ray absorption spectroscopy. Electrochemistry Communications 24: 43-46). Copyright (2012) Elsevier [17])

onto the Pt wire electrode (Fig. 6.14aiii) using the Mg foil as a counter electrode. For in situ XAS analysis, the deposition of the Mg reference was performed without the two-electrode cyclic voltammograms to avoid a possible parasitic byproduct. The start of Mg deposition ($j = -1$ mA cm^{-2}) was determined to be −0.2 V vs. Mg RE in the three-electrode configuration. Figure 6.15b shows XANES evidence, normalized to the edge maximum, for the formation of Mg. After a galvanostatic reduction at $J = -15$ mA cm^{-2} for 3600 s, the edge energy shifted from 1311 eV to 1301 eV, the same value as Mg foil. An oxidation current of $j = +15$ mA cm^{-2} applied for 3600 s caused the edge to shift back to the original, as-prepared energy. Furthermore, the high efficiency of the process effectively removes Mg through an equivalent anodic charge with minimal decomposition of the electrolyte. These results are consistent with Aurbach et al.'s results showing the complete removal of Mg metal from the surface of the electrode to reform the interface [12, 13].

In situ potentiostatic X-ray absorption (PSXA) experiments were designed to gain deeper insights on the interface between the electrode and the electrolyte during deposition. Figure 6.16a shows three potentials, −100 mV, −200 mV, and −400 mV vs. Mg RE, chosen from the cathodic wave for the PSXA experiments. The potential $V = -200$ mV was the start of Mg deposition, and −100 mV and −400 mV represent potentials before and after Mg deposition, respectively. Figure 6.16b represents the XANES region for the PSXA spectra. The growth of

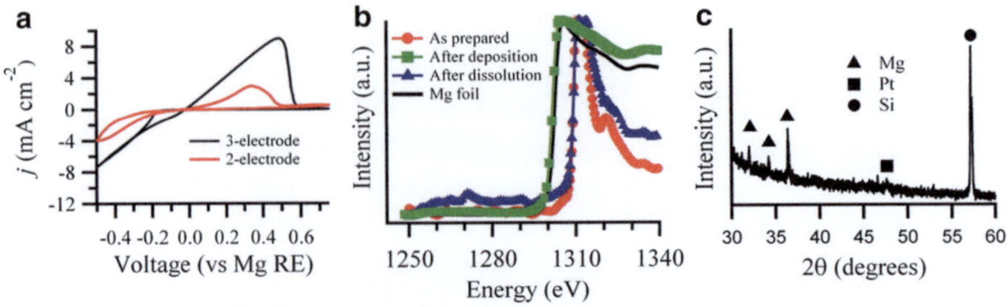

Fig. 6.15 (**a**) CVs of the two- and three-electrode in situ electrochemical/XAS cell ($\nu = 0.025$ V s^{-1}). (**b**) XANES region for the Mg *K*-edge of the as-prepared cell and after galvanic Mg deposition/dissolution. Mg metal foil is shown as a reference. (**c**) XRD pattern of the Si_3N_4 window after galvanic reduction at $j = -15$ mA cm^{-2} for 3600 s (Reprinted with permission from (Arthur T, Glans P-A, Matsui M, Zhang R, Ma B, Guo J (2012) Mg deposition observed by in situ electrochemical Mg K-edge X-ray absorption spectroscopy. Electrochemistry Communications 24: 43-46). Copyright (2012) Elsevier [17])

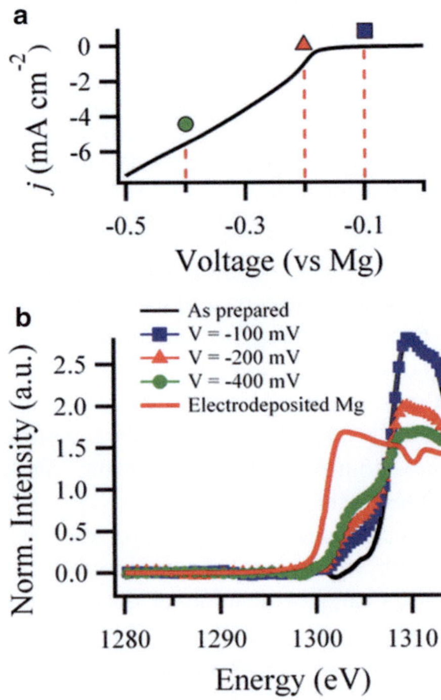

Fig. 6.16 (**a**) Cathodic wave for the deposition of Mg from a 0.45 M electrolyte solution with marked potentials for the PSXA experiments. (**b**) XANES region of the Mg *K*-edge marked in (**a**) (Reprinted with permission from (Arthur T, Glans P-A, Matsui M, Zhang R, Ma B, Guo J (2012) Mg deposition observed by in situ electrochemical Mg K-edge X-ray absorption spectroscopy. Electrochemistry Communications 24: 43–46). Copyright (2012) Elsevier [17])

a low-energy edge was evidence of an additional magnesium absorber at the interface of the platinum working electrode and the electrolyte. By applying lower potentials, the size of the wave increased and the edge energy decreased toward the position for Mg. A potentiostatic deposition at $V = -100$ mV vs. Mg RE for $t = 10800$ s did not show a diffraction pattern for crystalline Mg or an XAS

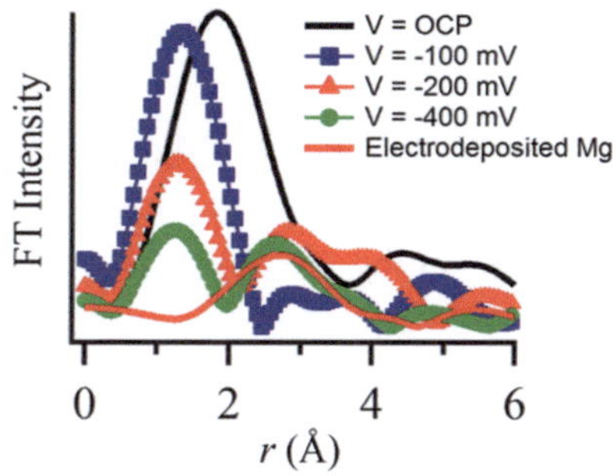

Fig. 6.17 Fourier-transformed EXAFS spectra of the Mg *K*-edge at the Mg metal and electrolyte interface (Reprinted (adapted) with permission from (Benmayza A, Ramanathan M, Arthur T, Matsui M, Mizuno F, Guo J, Glans P-A, Prakash J (2013) Effect of electrolytic properties of a magnesium organohaloaluminate electrolyte on magnesium deposition. Journal of Physical Chemistry C 117: 26881–26888). Copyright (2013) American Chemical Society [18])

signature for Mg metal. The presence of a XANES signal $V = -100$ mV vs. Mg RE was an interesting result, because an additional edge was observed before $V = -200$ mV vs. Mg RE, the start of Mg deposition. These XANES results indicated that the presence of a Mg species besides the bulk electrolyte was required before the deposition of Mg.

The in situ analysis was then applied to the extended X-ray absorption fine-structure (EXAFS) region [18]. Figure 6.17 is the Fourier-transformed EXAFS spectra of the Mg *K*-edge at the interface. At first, spectra of the electrolyte, 0.4 M $\left[Mg_2(\mu - Cl)_3 \cdot 6(OC_4H_8)\right](AlCl_4)$ / THF, at open-circuit potential (OCP) showed a broad peak of $r = 1.9\,\text{Å}'$. When a potential of -100 mV was applied for the interface, a shift of the first-shell peak, $\Delta r = -0.6\,\text{Å}'$, was clearly observed and was interpreted as a significant decrease in the first-shell scattering distances around Mg at the interface. The same peak was present in the spectra measured at $V = -200$ mV and -400 mV vs. Mg/Mg^{2+}, but with a reduced signal intensity. The EXAFS spectrum for electrodeposited $Mg_{(m)}$ is shown as a comparison to the in situ experiments. There are two possible explanations for the loss of signal intensity: a loss of ligands and thus scattering of coordinating atoms around the Mg, or signal attenuation from the deposition of Mg. However, holding the potential at -100 mV vs. Mg/Mg^{2+} did not show a XANES signal or diffraction pattern representative of $Mg_{(m)}$. Hence, presence of the intermediate species at the interface was found to be a vital step before Mg deposition.

From a combination of the electrochemical and XAS measurements, possible Mg deposition pathways were determined as follows:

$$\begin{aligned}&\left[Mg_2(\mu - Cl)_3 \cdot 6(OC_4H_8)\right]^+ + 3(OC_4H_8)\\&\quad\leftrightarrow \left[MgCl \cdot 5(OC_4H_8)\right]^+ + \left[MgCl_2 \cdot 4(OC_4H_8)\right]\end{aligned} \tag{6.1}$$

$$\mathrm{M\ (substrate)} + [\mathrm{MgCl} \cdot 5(\mathrm{OC_4H_8})]^+ \rightarrow [\mathrm{M{-}{-}{-}MgCl} \cdot 5(\mathrm{OC_4H_8})]^+{}_{\mathrm{ads}} \quad (6.2)$$

$$[\mathrm{M{-}{-}{-}MgCl} \cdot 5(\mathrm{OC_4H_8})]^+{}_{\mathrm{ads}} + 2e^- \rightarrow [\mathrm{M{-}{-}{-}Mg}]_{\mathrm{ads}} + \mathrm{Cl}^- + 5(\mathrm{OC_4H_8}) \quad (6.3)$$

$$[\mathrm{M{-}{-}{-}Mg}]_{\mathrm{ads}} \rightarrow \mathrm{M{-}{-}{-}Mg_{(m)}} \quad (6.4)$$

Equation (6.1) is the proposed dimer formation equilibrium, including the coordinated THF molecules [18]. Equation (6.2) is the adsorption of $[\mathrm{MgCl} \cdot 5(\mathrm{OC_4H_8})]^+$ at $V = -100$ mV. Then, under more negative cathodic potentials, $[\mathrm{MgCl} \cdot 5(\mathrm{OC_4H_8})]^+$ was ascertained to be reduced and dissociated on the surface of the electrode to form $[\mathrm{M{-}{-}{-}Mg}]_{\mathrm{ads}}$ (6.3). Finally, the adsorbed Mg^0 diffuses along the surface of the electrode to form $Mg_{(m)}$, the electro-crystallized magnesium layer (6.4). In situ/operando XAS data clearly indicated the presence of an intermediate state from the original species in the bulk electrolyte, which would be $[\mathrm{M{-}{-}{-}MgCl} \cdot 5(\mathrm{OC_4H_8})]^+{}_{\mathrm{ads}}$. Hence, a key step is to keep the Mg^{2+} ion adsorbed on the substrate. The combination of electrochemical and in situ XAS results forms a similar mechanism proposed by Aurbach et al. and clearly distinguishes an active magnesium species at the interface [6].

While applying the positive potentials, the in situ absorption spectra for the Mg dissolution process were also collected as in Fig. 6.18 [1]. When a potential of + 0.5 V vs. Mg was applied, a sudden increase in edge energy due to the complete removal of Mg from the surface of the Pt electrode was observed. The obtained spectrum looked very similar in intensity and edge position of the as-made electrolyte, indicating that all of the Mg metal had been successfully removed. During this experiment, it was difficult to capture any electrolyte intermediates during Mg dissolution due to the very short lifetimes of such species at + 0.5 V vs. Mg. To observe the dissolution process of Mg metal from the surface in more detail, the applied potential was reduced to + 0.1 V vs. Mg. The edge shifted to higher energies indicating the removal of Mg metal from the surface. More importantly, the edge position did not completely shift back to that of the as-made electrolyte. According to careful comparisons, at very positive applied potentials, the Mg *K*-edge returned to the original, as-made position of the electrolyte, before Mg metal deposition. However, at less positive potentials, an additional edge at approximately 1303 eV remained present. The presence of this additional edge could be explained by (1) the presence of Mg metal that was not removed at + 0.1 V vs. Mg or (2) an adsorbed intermediate species necessary for dissolution of Mg metal.

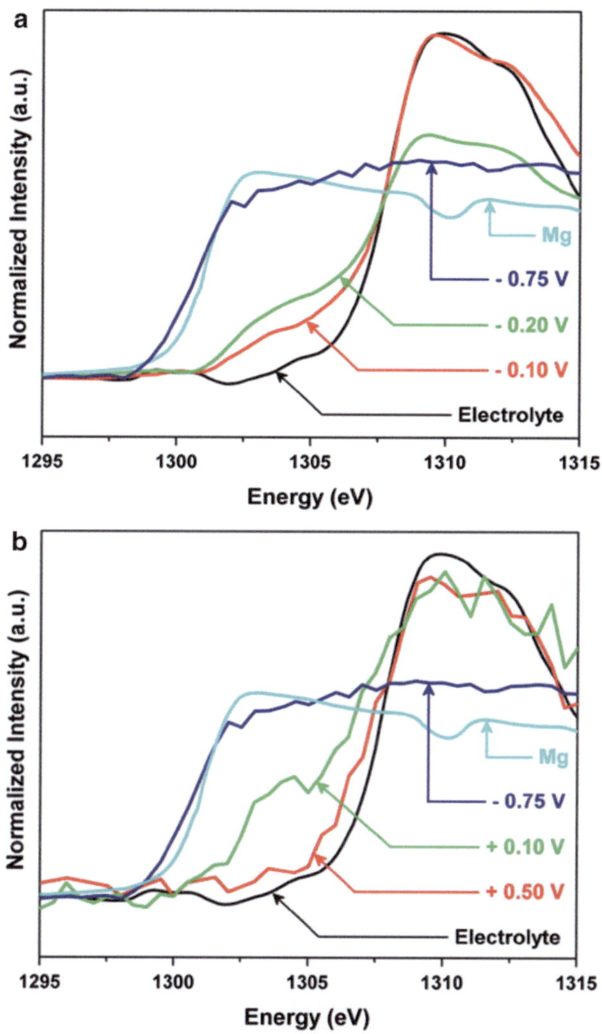

Fig. 6.18 In situ X-ray absorption near edge structure (XANES) spectra of the Mg *K*-edge for the interface between Mg metal and the organohaloaluminate electrolyte (**a**) during Mg deposition on a Pt electrode and (**b**) during Mg dissolution from the electrochemically deposited Mg metal onto the Pt electrode. The applied potentials were –0.1, –0.2, and –0.75V vs. Mg for deposition and +0.1 and +0.5V vs. Mg for dissolution (Reproduced by permission of Creative Commons CC-BY Attribution 4.0 International License (Mizuno F, Singh N, Arthur TS, Fanson PT, Ramanathan M, Benmayza A, Prakash J, Liu Y-S, Glans P-A, Guo J (2014) Understanding and overcoming the challenges posed by electrode/electrolyte interfaces in rechargeable magnesium batteries. Frontiers in Energy Research 2: 1-11) [1])

6.4 Alternate Anodes for Magnesium (Mg)-Ion Batteries

While investigations into the anode/electrolyte interface for Mg batteries continue along with new electrolyte development, there remains one other possible way to develop high-voltage Mg systems via the use of existing conventional battery electrolytes. This can be made possible if the anode system is changed from a Mg metal anode to a Mg-ion insertion-type anode. This change would negate the challenge posed by the Mg^{2+} blocking layer on the Mg metal anode surface. The use of such insertion-type anodes has been abundant in Li-ion battery technology, where the use of a Li metal anode is deemed impractical, primarily due to dendrite formation.

6.4.1 Bismuth (Bi), Antimony (Sb), and Bi-Sb Alloy Anodes

In 2012, T. S. Arthur et al. first studied Bi, Sb, and $Bi_{1-x}Sb_x$ alloys as candidate host materials for the alternate anode active materials because (6.1) Bi and Sb were known to form intermetallics with Mg having high volumetric capacities, and (6.2) both Bi and Sb presented rhombohedral structures which would result in the formation of alloys with a wide composition range. T. S. Arthur et al. employed electrodeposition to prepare the Bi, Sb, and $Bi_{1-x}Sb_x$ alloy anodes and were able to show the electrochemical compatibility of a Bi anode in a $Mg(N(SO_2CF_3)_2)_2$ conventional electrolyte [5].

In their study, Bi, Sb, and $Bi_{1-x}Sb_x$ alloys were electrodeposited by following a previously reported procedure: Here, a 0.025 M Sb_2O_3, 0.4 M citric acid/H_2O solution was brought to a pH of 6.0 with 5 M KOH. Nano-Sb was deposited in a three-electrode cell at -1.2 V vs Ag/AgCl using a high-surface area Pt-mesh counter-electrode. Working electrodes (15 mm diameter, 1.77 cm^2) were made from copper foil with 5 nm of Ti and 200 nm of Pt sputtered onto one surface. All electrochemical half-cells were constructed in an Ar-filled glove box since the electrolyte was acutely sensitive to oxygen and water. The electrolyte was a mixture of 2 mL of ethylmagnesium chloride, 1 mL of diethylaluminum chloride, and 13 mL of anhydrous THF. Magnesium trifluoromethanesulfonimide and acetonitrile were kept in an Ar-filled glove box. Mg (m) disks were used as the anodes in all half-cell experiments. Cyclic voltammetry of the Bi electrode was done in a three-electrode cell with a platinum foil counter-electrode and a Ag/Ag^+ reference (0.01 M $AgNO_3$/acetonitrile) [5]. T. S. Arthur et al. first showed the galvanostatic half-cell reduction of the electrodeposited anodes at a 0.01 C rate, highlighted in Fig. 6.19a. The obtained capacities of Bi and Sb suggested the composition of the magnesiated Bi as $0.44(Mg_3Bi_2) + 0.13Bi$ and the magnesiated Sb as $0.49(Mg_3Sb_2) + 0.02Sb$, respectively. For the electrodeposited $Bi_{0.88}Sb_{0.12}$ and $Bi_{0.55}Sb_{0.45}$ alloys, the Mg^{2+} insertion occurred at multiple potential plateaus, one between 0.27 and 0.29 V and one at 0.23 V vs. Mg. Since Sb magnesiated at

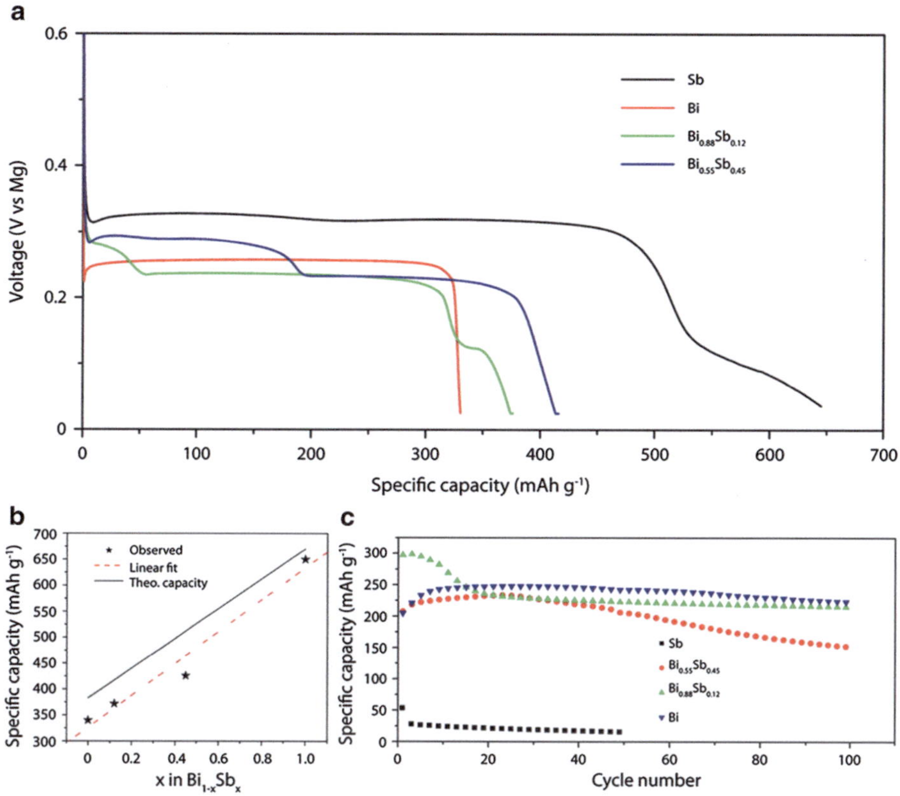

Fig. 6.19 (**a**) Galvanostatic discharge of Bi, Sb, $Bi_{0.88}Sb_{0.12}$, and $Bi_{0.55}Sb_{0.45}$ anodes at a 0.01 C rate. (**b**) Observed capacity of the alloy anodes with a linear fit of these capacities; *dashed line* compared with the theoretical capacity; *solid line*. (**c**) Cycling capabilities of Bi, Sb, $Bi_{0.88}Sb_{0.12}$, and $Bi_{0.55}Sb_{0.45}$ anodes at a 1 C rate (Reprinted with permission from (Arthur TS, Singh N, Matsui M (2012) Electrodeposited Bi, Sb and $Bi_{1-x}Sb_x$ alloys as anodes for Mg-ion batteries. Electrochemistry Communications 16: 103-106). Copyright (2012) Elsevier [5])

a more positive voltage than Bi, the 0.27–0.29 V plateau for the $Bi_{1-x}Sb_x$ alloys was assigned as magnesiation of the Sb. The capacity of the individual plateaus reaffirmed their claim that if the composition of the alloy was calculated from the discharge capacity, the alloy formulas would be $Bi_{0.86}Sb_{0.14}$ and $Bi_{0.53}Sb_{0.47}$.

As expected, Fig. 6.19a and b shows that the capacity of the anode increased as x in $Bi_{1-x}Sb_x$ increased. The capacity deviation from the theoretical value indicated that the electrodeposited alloy anodes were partially affected by the substrate [5]. T. S. Arthur et al. also showed the discharge capacity vs. cycle number for the electrodeposited Bi, Sb, $Bi_{0.88}Sb_{0.12}$, and $Bi_{0.55}Sb_{0.45}$ alloys at a 1 C rate (1.5 Mg^{2+} inserted/metal, $\sim$ 0.2 mA/cm^2), as shown in Fig. 6.19c. $Bi_{0.88}Sb_{0.12}$, having the largest initial specific capacity, reached a maximum capacity of 298 mAh g^{-1}, before dropping to 215 mAh g^{-1} by the 100th cycle. The highest specific capacity of the $Bi_{0.55}Sb_{0.45}$ alloy was 235 mAh g^{-1} at the 22nd cycle, fading to 151 mAh g^{-1} at

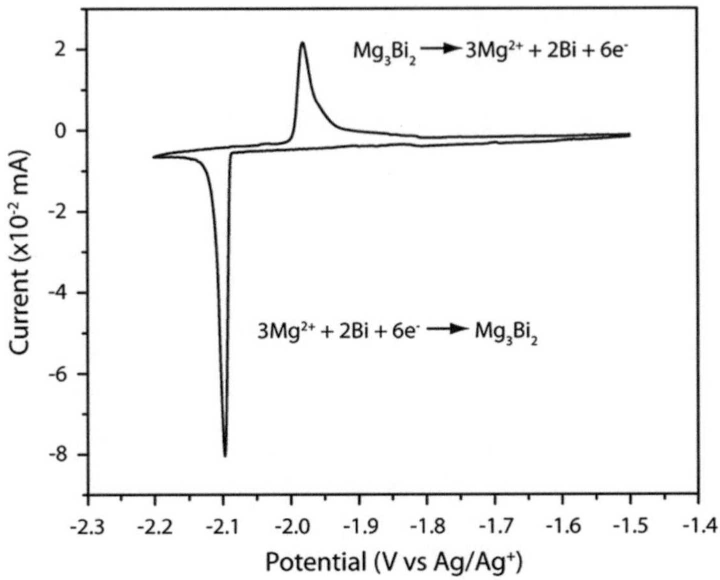

Fig. 6.20 Cyclic voltammogram of a Bi anode cycled in a $Mg(N(SO_2CF_3)_2)_2$/ CH_3CN electrolyte at 10 μV s^{-1}. The electrochemical reactions associated with the peaks are shown in the graph (Reprinted with permission from (Arthur TS, Singh N, Matsui M (2012) Electrodeposited Bi, Sb and $Bi_{1-x}Sb_x$ alloys as anodes for Mg-ion batteries. Electrochemistry Communications 16: 103-106). Copyright (2012) Elsevier [5])

the 100th cycle. The Bi anode showed very good cycling properties by retaining a specific capacity of 222 mAh g^{-1} at the 100th cycle from a maximum specific capacity of 247 mAh g^{-1} at the 20th cycle. The Sb did not cycle well at a 1 C rate, displaying a very small specific capacity of 16 mAh g^{-1} at the 50th cycle [5].

In order to highlight the basis of using insertion-type anodes like Bi, T. S. Arthur et al. showed a proof-of-concept CV of a Bi electrode cycled in a $Mg(N(SO_2CF_3)_2)_2$/acetonitrile solution. The combination of a reduction peak at -2.09 V (vs. Ag/Ag^{+}) and an oxidation peak at -1.98 V (vs. Ag/Ag^{+}) in Fig. 6.20 was found as evidence that Bi could be used as an anode in conventional Mg-ion electrolytes.

In their study, T. S. Arthur et al. employed electrodeposition as a simple, low-cost technique to control and study the effect of electrode morphology on the magnesiation properties of Bi, Sb, and $Bi_{1-x}Sb_x$ alloy anodes for the very first time in Mg battery systems. By electrochemically alloying Bi with Sb, T. S. Arthur et al. were able to improve the energy density of the anode by coupling the low reduction/oxidation potential of Bi with the high theoretical capacity of Sb. However, as per their highlighted results, Sb failed to cycle well at fast rates, regardless of the as-deposited morphology. Most importantly, T. S. Arthur et al. were able to showcase the compatibility of the Bi anode with conventional electrolytes. Based on their results, other researchers have been able to better design Bi-based anodes for Mg-ion batteries.

6.4.2 Advanced Bismuth (Bi) Anodes

Recently, Shao et al. noted that insertion-type anode materials to date displayed poor rate performance and fast capacity fading upon cycling. Further, they noted that the coulombic efficiency for Mg^{2+} insertion/extraction was also very low. Shao et al. attributed these problems largely with the slow solid-state diffusion of Mg^{2+} ions in the host and the large volume expansion/shrinkage during Mg^{2+} insertion/extraction processes. They suggested that substantial improvements and structural optimizations were still required. In order to highlight this, Shao et al. turned to nanostructured materials which have shown great promise in improving the ion diffusion and structural stability of electrode materials in Li-ion batteries. They demonstrated that nanostructured materials could significantly improve the properties of anodes for Mg-ion batteries [19].

In their study, Shao et al. synthesized Bi nanotubes (Bi-NT) by a hydrothermal reaction method to obtain materials with desirable properties. These nanotubes formed a bundled structure because of the van der Waals interactions between the small tubes (like in the case of carbon nanotubes). They assembled the testing cells using Mg metal as the anode and a Bi-NT electrode as the cathode. For comparison, control electrodes using Bi microparticles (Bi-Micro, $\sim 100\ \mu m$) or bismuth nanoparticles (Bi-NP, $30-50$ nm) were also assembled and tested using identical protocols. For the half-cell testing, they used a Mg electrolyte with highly efficient Mg plating/dissolution (with $Mg(BH_4)_2$, $LiBH_4$, and diglyme) [19]. In Fig. 6.21a, Shao et al. compare the cyclic voltammograms (CVs) of Mg^{2+} insertion $(0-0.25\ V)$ and extraction $(0.3-0.5\ V)$ in Bi-NT and Bi-Micro. The Bi-NT electrode exhibited highly symmetric cathodic/anodic sharp peaks with narrower peak separation than Bi-Micro (0.264 V vs. 0.313 V), indicating that the insertion/extraction of Mg^{2+} in Bi-NT was highly reversible and fast. Further, Fig. 6.21b showed the galvanostatic discharge/charge profile of a $Mg-Bi$ half-cell. The plateaus at ~ 0.2 V were marked as the Mg^{2+} insertion process in Bi, corresponding to the $0-0.25$ V peaks in CVs, and the plateaus at ~ 0.35 V were marked as the Mg^{2+} extraction process in Bi, corresponding to the $0.3-0.5$ V peaks in CVs. At the same discharge/charge rate (2C, discharge/charge in 30 min), the overvoltage was found to be lower and the specific capacity higher for Bi-NT [19].

Shao et al. studied the rate performance of electrodes made with Bi-NTs and compared them with Bi-Micro. Their capacities at varied rates from 0.02C to 5C were summarized in Fig. 6.21c. They found that the high rate performance of Bi-NTs was substantially better than Bi-Micro. The capacities of Bi-NTs and Bi-Micro were comparable at low rates (e.g., both were ~ 350 mAh/g Bi at 0.05C, corresponding to 3430 mAh/cm^3 Bi). As the charging/discharging rates were increased, the capacity of Bi-Micro anode decreased dramatically, retaining only 51 mAh/g at 5C ($< 15\ \%$ retention), indicating that the inner-most part of the Bi was not magnesiated due to the large diffusion length of Mg^{2+}. In contrast, the Bi-NTs anode was able to maintain a capacity as high as 216 mAh/g at 5C ($> 60\ \%$ retention) [19].

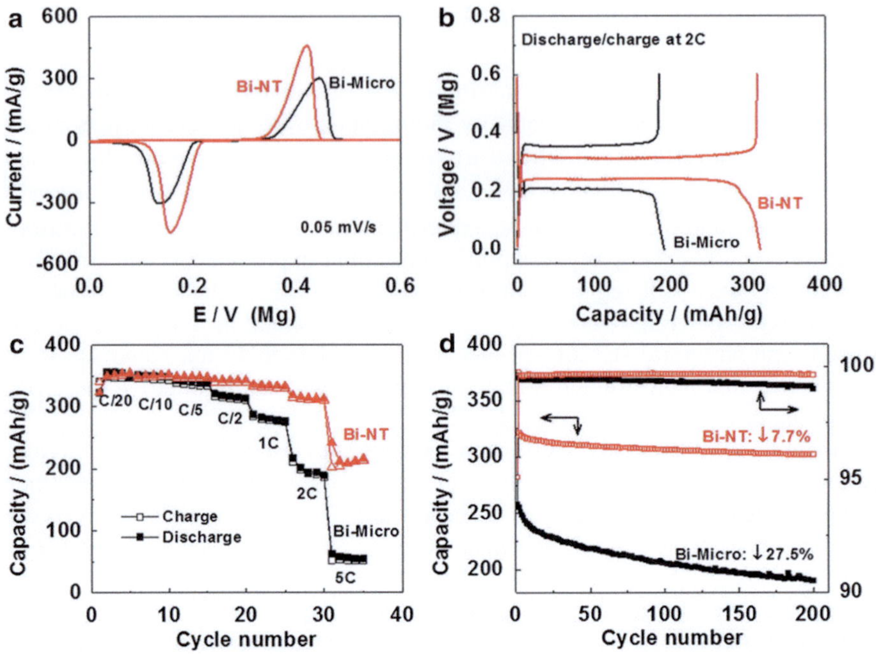

Fig. 6.21 (**a**) Cyclic voltammograms of Mg insertion/deinsertion in bismuth; (**b**) discharge/charge profile of a Mg-Bi cell; (**c**) rate performance of a Mg-Bi cell; (**d**) cycling stability and coulombic efficiency (CE) of bismuth electrode for reversible Mg insertion/deinsertion. Cell configuration: $Mg/0.1\ M\ Mg(BH_4)_2 - 1.5\ M\ LiBH_4 - diglyme/Bi$ (Reprinted (adapted) with permission from (Shao Y, Gu M, Li X, Nie Z, Zuo P, Li G, Liu T, Xiao J, Cheng Y, Wang C, Zhang J-G, Liu J (2014) Highly reversible Mg insertion in nanostructured Bi for Mg ion batteries. Nano Letters 14: 255-260). Copyright (2014) American Chemical Society [19])

Cycling stability is another important factor for battery electrode materials. In their work, Shao et al. studied the cycling stability of Bi-NTs and compared it with Bi-Micro, as shown in Fig. 6.21d. The electrode with Bi-NTs exhibited excellent cycling stability, with only 7.7 % of capacity fading after 200 cycles (maintaining 303 mAh/g at the 200th cycle). In contrast, the Bi-Micro electrodes experienced substantial capacity decay of 27.5 % under the same testing conditions (only 188 mAh/g at the 200th cycle). The coulombic efficiency of Bi-NTs was found to be stable and close to 100 % during the cycling. In contrast, the coulombic efficiency for Bi-Micro was lower and kept decreasing during cycling. Interestingly, Shao et al. found that most of the capacity fading of both electrodes happened during the initial 20 cycles. Considering the fact that the insertion of Mg^{2+} ions into Bi would lead to volume expansion and introduce substantial stress to the crystals, their results indicated that the Bi-Micros were unable to withstand the volume change, eventually leading to material pulverization and capacity fading. Such an observation agrees well with typical insertion-type electrode materials with large volume expansion, such as silicon for Li-ion batteries. Compared with Bi-Micros, the substantially higher cycling stability and higher/stable coulombic efficiency of Bi-NT indicate that the tubular structure was effective in withstanding the volume expansion/shrinkage of Bi when hosting Mg^{2+} [19].

To further demonstrate the unique advantages of Bi-NTs for hosting Mg^{2+}, Shao et al. also studied the behavior of Bi-NP for Mg^{2+} insertion/extraction. Again, the Bi-NP electrodes showed better performance than Bi-Micros in that their capacity only faded 16.2 % after 200 cycles. These results indicate that the morphology and size of active Bi had a significant influence on Mg^{2+} insertion/extraction behavior [19].

Shao et al. also highlighted that the Bi anode material could be used with a conventional electrolyte (previously also shown by T. S. Arthur et al.) as shown in Fig. 6.22. A proof-of-concept Mg-ion battery with a Mg intercalation material Mo_6S_8 cathode and a premagnesiated Bi electrode as the anode was assembled with a conventional electrolyte 0.4 M $Mg(TFSI)_2$ -diglyme for testing (cell A in Fig. 6.22). Mo_6S_8 was chosen by Shao et al. as the cathode to demonstrate the compatibility of the anode material with the conventional electrolyte because it was the most widely studied cathode material, despite its low specific capacity. For comparison, they also assembled a cell with the Mg electrolyte $Mg(BH_4)_2 - LiBH_4$ -diglyme (cell B in Fig. 6.22). Figure 6.22 shows the discharge/charge profiles and cycling stability of these two cells (the capacity was based on the weight of cathode Mo_6S_8). The two cells showed comparable specific capacities and good cycling stability. The discharge/charge plateaus were also similar with the one with conventional electrolyte showing a slightly larger overvoltage.

In summary, Shao et al. developed a high-performance anode material based on Bi nanotubes for a rechargeable Mg-ion battery. The Mg cell demonstrated by coupling a premagnesiated Bi (Mg_3Bi_2) anode with a Mo_6S_8 cathode in a

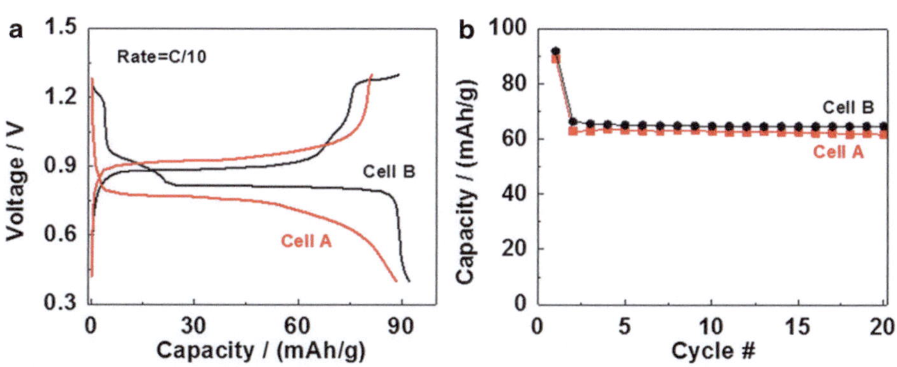

Fig. 6.22 (**a**) Discharge/charge profile of a $Mg_3Bi_2 - Mo_6S_8$ cell and (**b**) cycling stability of a $Mg_3Bi_2 - Mo_6S_8$ cell. Cell configuration: (A) Mg_3Bi_2/0.4 M $Mg(TFSI)_2$-diglyme/ Mo_6S_8, (B) Mg_3Bi_2/0.1 M $Mg(BH_4)_2$ – 1.5 M $LiBH_4$-diglyme/ Mo_6S_8 (Reprinted (adapted) with permission from (Shao Y, Gu M, Li X, Nie Z, Zuo P, Li G, Liu T, Xiao J, Cheng Y, Wang C, Zhang J-G, Liu J (2014) Highly reversible Mg insertion in nanostructured Bi for Mg ion batteries. Nano Letters 14: 255-260). Copyright (2014) American Chemical Society [19])

conventional electrolyte (0.4 M $Mg(TFSI)_2$ –diglyme) indicated the anodes' compatibility with conventional electrolytes. This work suggested that other high-energy density alloy compounds may also be considered for Mg-ion chemistry, which, coupled with high-capacity high-voltage cathodes using conventional electrolytes, could lead to a high-energy density Mg-ion battery.

6.4.3 *Tin (Sn) Anodes*

Based on the previously reported use of bismuth (Bi) as a Mg^{2+} insertion-type anode material and its compatibility with conventional battery electrolytes, Singh et al. studied a higher-energy density electrode material in order to improve the energy density of Mg-ion batteries utilizing insertion-type anodes. Since the ability to deliver high-energy density depends on cell voltage and capacity, Singh et al. suggested that it was essential that the scientific community develop next-generation Mg-ion battery anode materials which remain compatible with high-voltage conventional battery electrolytes while displaying lower Mg^{2+} insertion/extraction voltages and higher capacities than Bi [2]. In their study, Singh et al. presented a high-energy density tin (Sn) anode which was capable of displaying the abovementioned necessary characteristics. Figure 6.23 shows the 1 st cycle galvanostatic half-cell (anode/Mg) magnesiation/demagnesiation curves for a Sn and a Bi anode at a 0.002 C rate. This experiment was executed in an organohaloaluminate electrolyte to test the feasibility of using an insertion-type Sn anode in Mg-ion batteries. The magnesiation (defined as insertion of Mg^{2+}) plateau observed at + 0.15 V corresponded to the formation of Mg_2Sn according to binary-phase diagrams and reaction (6.5):

$$Sn + 2Mg^{2+} + 4e^- \rightarrow Mg_2Sn \tag{6.5}$$

The experimental result of the electrochemical voltage of formation of Mg_2Sn was found to concur with theoretical DFT calculations for the work presented by Singh et al. The demagnesiation (defined as extraction of Mg^{2+}) plateau was observed at +0.20 V and corresponded to the extraction of Mg^{2+} from Mg_2Sn [2].

According to the authors, an interesting feature to note in Fig. 6.23 was the low hysteresis between the magnesiation and demagnesiation curves for the Sn anode. The voltage required to extract Mg^{2+} from Mg_2Sn was only 50 mV more positive than the voltage needed to insert Mg^{2+} into Sn. Further, the experimental capacity for the formation of Mg_2Sn corresponded closely to the theoretical capacity that could be obtained using a Sn anode (903 mA h g^{-1}). In comparison, Singh et al. also compared the performance of the Sn anode to that of previously reported Bi anodes. In Fig. 6.23, the magnesiation plateau observed at + 0.23 V corresponded to the formation of Mg_3Bi_2, and consequently, the demagnesiation plateau is observed at + 0.32 V and corresponded to the extraction of Mg^{2+} from Mg_3Bi_2. The hysteresis

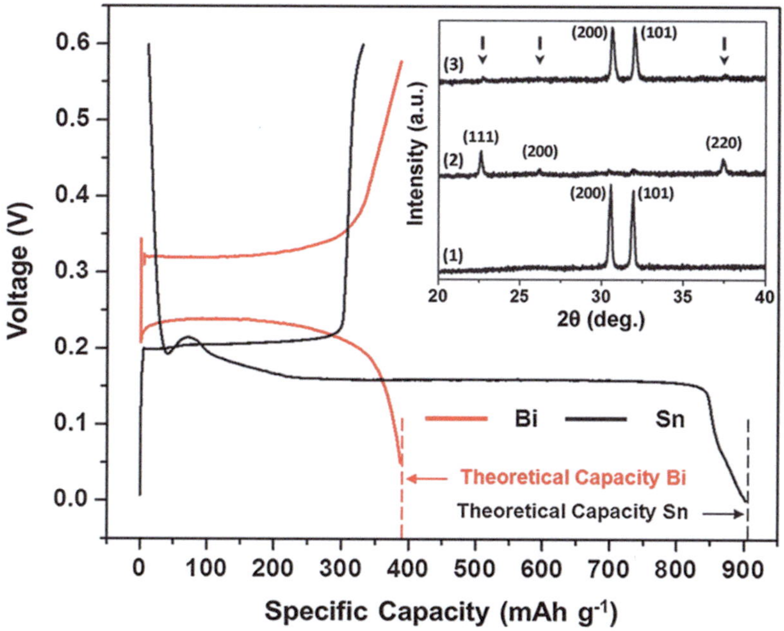

Fig. 6.23 First cycle galvanostatic magnesiation/demagnesiation curves for Sn/Mg (*black*) and Bi/Mg (*red*) half-cells (using organohaloaluminate electrolyte) highlighting their achievable theoretical capacities. Inset – XRD spectra for (*1*) as-fabricated Sn, (*2*) magnesiated Sn (or Mg_2Sn – peak positions marked with arrows), and (*3*) demagnesiated Mg_2Sn (Singh N, Arthur TS, Ling C, Matsui M, Mizuno F (2013) A high energy-density tin anode for rechargeable magnesium-ion batteries. Chemical Communications 49: 149-151) (Reproduced by permission of The Royal Society of Chemistry [2])

between the magnesiation and demagnesiation curves for the Bi anode in Fig. 6.23 was much larger (90 mV) than that for the Sn anode [2].

Singh et al. utilized XRD spectra to confirm magnesiation and demagnesiation of the Sn anode. The inset in Fig. 6.23 shows the XRD spectra of the (6.1) as-fabricated Sn, (6.2) magnesiated Sn, and (6.3) demagnesiated Mg_2Sn anodes discussed above. The as-fabricated Sn film was crystalline and indexed to the tetragonal phase. Upon magnesiation, crystalline peaks associated with the formation of Mg_2Sn were observed along with remnant Sn peaks with negligible intensities. No metallic Mg was observed in the Mg_2Sn spectra. The demagnesiated Mg_2Sn film displayed the reformation of the crystalline Sn phase along with remnant Mg_2Sn peaks with negligible intensities [2].

The rate capability for both Sn and Bi anodes was also tested in an organohaloaluminate electrolyte by varying the galvanostatic cycling rates between 0.002 C and 0.05 C, by Singh et al. Figure 6.24 shows the comparison of the 1st cycle galvanostatic half-cell magnesiation and demagnesiation capacities obtained for Sn and Bi anodes at the various C rates. Fresh cells were used to investigate each C rate. The Sn anodes displayed dramatic decreases in both magnesiation and

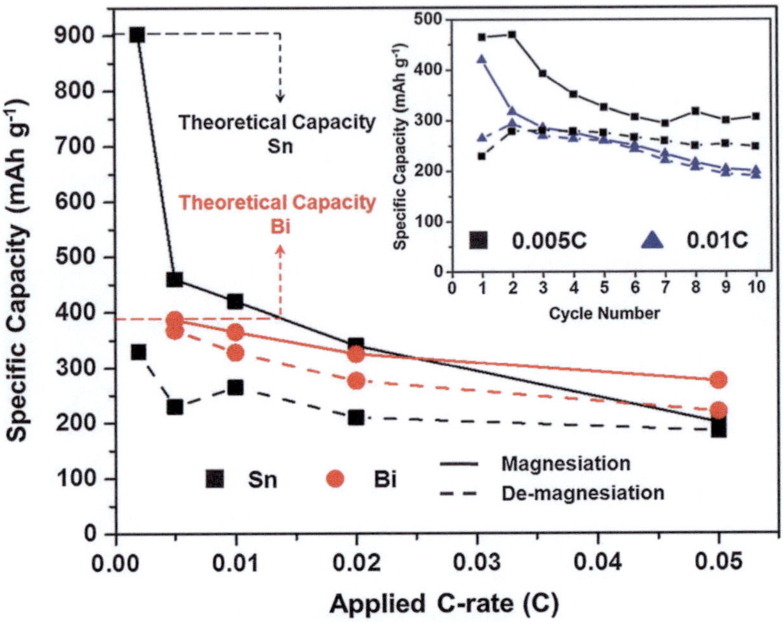

Fig. 6.24 Magnesiation/demagnesiation capacities for Sn/Mg (*black*) and Bi/Mg (*red*) half-cells at various C rates. All half-cells were tested using an organohaloaluminate electrolyte. Inset – 10 cycles of a Sn/Mg half-cell at 0.005 C (*black*) and 0.01 C (*blue*) rates in the same electrolyte [Singh N, Arthur TS, Ling C, Matsui M, Mizuno F (2013) A high energy-density tin anode for rechargeable magnesium-ion batteries. Chemical Communications 49: 149-151] (Reproduced by permission of The Royal Society of Chemistry [2])

demagnesiation capacities at higher C rates. While it was evident that the Sn anodes suffered from poor kinetics (limiting their capacities at faster C rates), it was also noted that at faster C rates, the hysteresis between the magnesiation and demagnesiation capacities diminished. Further, the Sn anodes displayed decent cycling characteristics past the first cycle at various C rates (Fig. 6.24, inset). Unlike Sn, Bi displayed acceptable magnesiation/demagnesiation capacities, and hysteresis at the same C rates, as also reported by T. S. Arthur et al. and Shao et al [2, 5, 19].

In order to highlight the compatibility and performance of a Sn anode in conventional battery electrolytes, Singh et al. utilized a premagnesiated Sn (Mg_2Sn) anode and a Chevrel-phase (Mo_6S_8) cathode, which were coupled with 0.5 M $Mg(N(SO_2CF_3)_2)_2$ in DME, as a full cell. The experiment was repeated with an organohaloaluminate electrolyte, so as to compare the performance of the two cells. Figure 6.25 shows the first 10 cycles of both cells with 1st cycle capacities of 82 mA h g^{-1} (using a conventional electrolyte) and 83 mA h g^{-1} (using an organohaloaluminate electrolyte), followed by a drop in capacities from the second cycle. The 1st cycle magnesiation/demagnesiation (of Mo_6S_8) curves are shown in Fig. 6.25 inset, for both cells. The performance of both cells during these 10 cycles was found to be comparable [2].

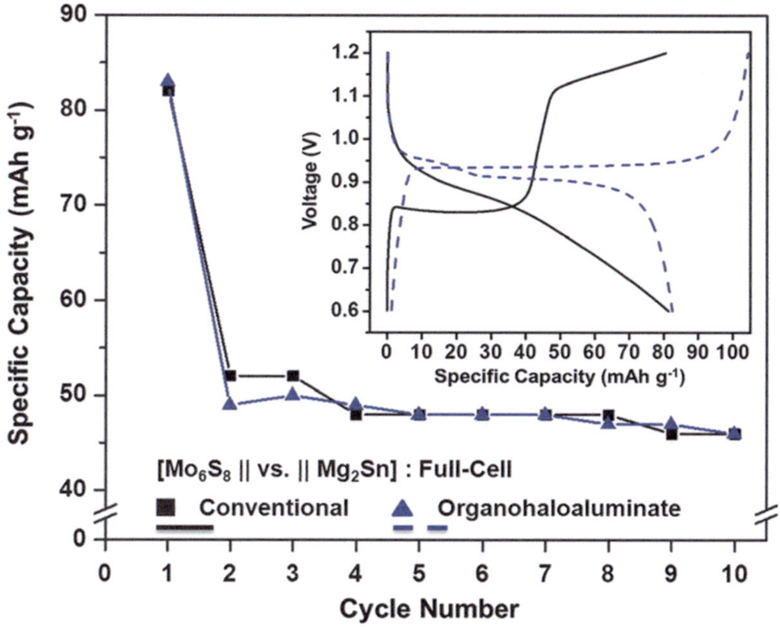

Fig. 6.25 The first 10 cycles for a [Mo_6S_8-conventional electrolyte-Mg_2Sn] full cell (*black*) and a [Mo_6S_8-organohaloaluminate electrolyte-Mg_2Sn] full cell (blue); inset – 1st cycle voltage profiles for each full cell [Singh N, Arthur TS, Ling C, Matsui M, Mizuno F (2013) A high energy-density tin anode for rechargeable magnesium-ion batteries. Chemical Communications 49: 149-151] (Reproduced by permission of The Royal Society of Chemistry [2])

Since the capacity observed in Fig. 6.25 was limited by the weight of the cathode active material (Mo_6S_8) and not by the weight of the anode active material (Sn), Singh et al. fabricated two additional Mo_6S_8/Mg_2Sn full cells using both previously used electrolytes (organohaloaluminate and conventional). Here, the loading content of Mo_6S_8 was increased so that its gravimetric capacity matched that of Sn. Mg^{2+} was extracted from Mg_2Sn and inserted into Mo_6S_8 via the application of + 0.1 V for 72 h. The authors once again utilized XRD spectra (shown in Fig. 6.26) to provide evidence for the complete magnesiation of Mo_6S_8 (inset) and complete demagnesiation of Mg_2Sn, for both cells. The data suggested that the entire capacity of Sn may be utilized for Mg-ion batteries under the correct circumstances [2].

In this report, Singh et al. used Sn powder films as simple, low-cost electrodes to study the possibility of using high-energy density insertion-type anode materials for a Mg-ion battery system. They were able to show that a Sn anode could attain higher capacities (903 mA h g^{-1}) and lower Mg^{2+} insertion/extraction voltages (+ 0.15/0.20 V) than those of previously reported Bi (384 mA h g^{-1}, + 0.23/0.32 V). More importantly, they were able to highlight the necessary compatibility and cyclability of the higher-energy density Sn anode with conventional battery electrolytes. As also suggested by Shao et al., the use of Sn as an insertion-type anode would allow for the evaluation of future, high-voltage and high-capacity Mg-ion battery cathodes using conventional battery electrolytes.

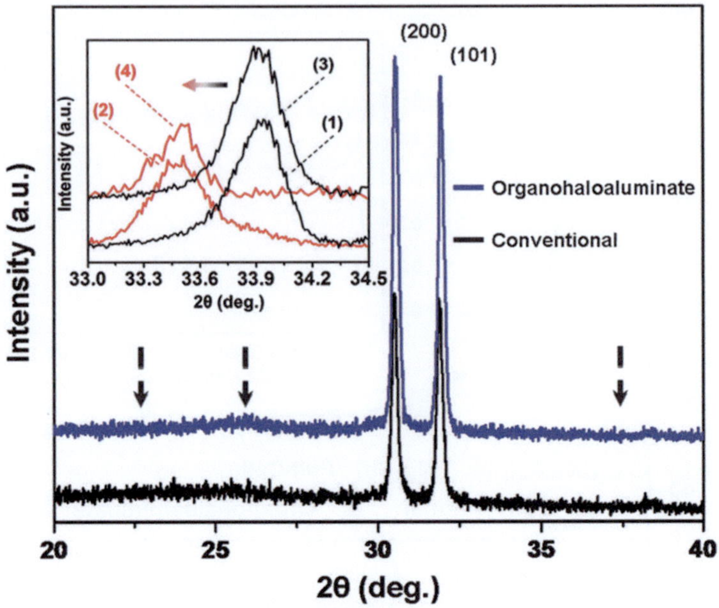

Fig. 6.26 XRD spectra of demagnesiated Mg_2Sn from organohaloaluminate (*blue*) and conventional (*black*) electrolytes (Mg_2Sn peak positions marked with *arrows*). Inset – XRD spectra showing a peak shift due to magnesiation (*red*) of Mo_6S_8 (*black*) in organohaloaluminate (peak shift 1 → 2) and conventional (peak shift 3 → 4) electrolytes [Singh N, Arthur TS, Ling C, Matsui M, Mizuno F (2013) A high energy-density tin anode for rechargeable magnesium-ion batteries. Chemical Communications 49: 149-151] (Reproduced by permission of The Royal Society of Chemistry [2])

6.5 Concluding Remarks

As considerable research efforts in the field of Mg batteries continue, it remains essential to find the correct combination of cathode, electrolyte, and anode materials in order to attain a high-energy density Mg battery system. Achieving such a Mg battery system requires the use of high-voltage cathodes and electrolytes, as well as high-capacity cathodes and anodes, as energy density is dependent upon both voltage and capacity. While the use of Mg metal as an anode is preferable due to its optimum energy density, the studies summarized in this chapter provide significant insight into the multitude of interactions occurring at the Mg metal/electrolyte interface, which tend to hinder the performance of Mg metal. Likewise, the use of alternate anodes like Bi and Sn is hindered due to the significantly lower performance characteristics (in terms of rate capability and cyclability) and lower energy densities than Mg metal. Nonetheless, these anodes offer a viable alternative Mg-ion battery system.

More specifically, achieving high-voltage Mg battery systems which result in high-energy density Mg battery systems seems possible, only if we solve the current issues at the Mg anode/electrolyte interface. In this chapter, we outlined two possible strategies by which this can be achieved: (1) directly observing, analyzing, and understanding the Mg anode/electrolyte interface in order to isolate the active species involved in Mg deposition and dissolution and (2) changing the anode system to an alternative (insertion-type) anode system and enhancing the alternate anode performance characteristics via engineering solutions (e.g., morphology control, etc.). Addressing such strategies would allow Mg-based batteries to achieve their maximum potential in terms of energy density, in turn allowing them to be competitive with current-day Li battery technologies.

References

1. Mizuno F, Singh N, Arthur TS, Fanson PT, Ramanathan M, Benmayza A, Prakash J, Liu Y-S, Glans P-A, Guo J (2014) Understanding and overcoming the challenges posed by electrode/electrolyte interfaces in rechargeable magnesium batteries. Front Energy Res 2:1–11
2. Singh N, Arthur TS, Ling C, Matsui M, Mizuno F (2013) A high energy-density tin anode for rechargeable magnesium-ion batteries. Chem Commun 49:149–151
3. Muldoon J, Bucur CB, Gregory T (2014) Quest for nonaqueous multivalent secondary batteries: magnesium and beyond. Chem Rev 114:11683–11720
4. Bruce PG, Freunberger SA, Hardwich LJ, Tarascon J-M (2012) Li–O_2 and Li-S batteries with high energy storage. Nat Mater 11:19–29
5. Arthur TS, Singh N, Matsui M (2012) Electrodeposited Bi, Sb and $Bi_{1-X}Sb_X$ alloys as anodes for Mg-ion batteries. Electrochem Commun 16:103–106
6. Yoo HD, Shterenberg I, Gofer Y, Gershinsky G, Pour N, Aurbach D (2013) Mg rechargeable batteries: an on-going challenge. Energy Environ Sci 6:2265–2279
7. Saha P, Datta MK, Velikokhatnyi OI, Manivannan A, Almon D, Kumta PN (2014) Rechargeable magnesium battery: current status and key challenges for the future. Prog Mater Sci 66:1–86
8. Tran TT, Lamanna WM, Obrovac MN (2012) Evaluation of Mg $\{N(SO_2CF_3)_2\}_2$/acetonitrile electrolyte for use in Mg-ion cells. J Electrochem Soc 159:A2005–A2009
9. Orikasa Y, Masese T, Koyama Y, Mori T, Hattori M, Yamamoto K, Okado T, Huang Z-D, Minato T, Tassel C, Kim J, Kobayashi Y, Abe T, Kageyama H, Uchimoto Y (2014) High energy density rechargeable magnesium battery using earth-abundant and non-toxic elements. Sci Rep 4:5622. doi:10.1038/srep05622
10. Ha S-Y, Lee Y-W, Woo SW, Koo B, Kim J-S, Cho J, Lee KT, Choi N-S (2014) Magnesium (II) bis(trifluoromethane sulfonyl) imide-based electrolytes with wide electrochemical windows for rechargeable magnesium batteries. ACS Appl Mater Interfaces 6:4063–4073
11. Mohtadi R, Matsui M, Arthur TS, Hwang S-J (2012) Magnesium borohydride: from hydrogen storage to magnesium battery. Angew Chem Int Ed 51:1–5
12. Aurbach D, Cohen Y, Moshkovich M (2001) The study of reversible magnesium deposition by in situ scanning tunneling microscopy. Electrochem Solid-State Lett 4:A113–A116
13. Lu Z, Schechter A, Moshkovich M, Aurbach D (1999) On the electrochemical behavior of magnesium electrodes in polar aprotic electrolyte solutions. J Electroanal Chem 466:203–217
14. Barile C, Spatney R, Zavadil K, Gerwith A (2014) Investigating the reversibility of in situ generated magnesium organohaloaluminates for magnesium deposition and dissolution. J Phys Chem C 118:10694–10699
15. Aurbach D, Turgeman R, Chusid O, Gofer Y (2001) Spectroelectrochemical studies of magnesium deposition by in situ FTIR spectroscopy. Electrochem Commun 3:252–261

16. Nakayama Y, Kudo Y, Oki H, Yamamoto K, Kitajima Y, Noda K (2008) Complex structures and electrochemical properties of magnesium electrolytes. J Electrochem Soc 155:A754–A759
17. Arthur T, Glans P-A, Matsui M, Zhang R, Ma B, Guo J (2012) Mg deposition observed by in situ electrochemical Mg K-edge X-ray absorption spectroscopy. Electrochem Commun 24:43–46
18. Benmayza A, Ramanathan M, Arthur T, Matsui M, Mizuno F, Guo J, Glans P-A, Prakash J (2013) Effect of electrolytic properties of a magnesium organohaloaluminate electrolyte on magnesium deposition. J Phys Chem C 117:26881–26888
19. Shao Y, Gu M, Li X, Nie Z, Zuo P, Li G, Liu T, Xiao J, Cheng Y, Wang C, Zhang J-G, Liu J (2014) Highly reversible Mg insertion in nanostructured Bi for Mg ion batteries. Nano Lett 14:255–260

Chapter 7
Nanostructured Oxides as Cathode Materials for Supercapacitors

Y. Liu, F. Yu, X.W. Wang, Z.B. Wen, Y.S. Zhu, and Y.P. Wu

7.1 Introduction

With the development of the global economy, the depletion of fossil fuels, and the increasing environmental pollution, there is an urgent need for clean, efficient, and sustainable sources of energy, as well as new technologies associated with energy conversion and storage [1].

Currently, the fast-growing interest in portable electronic devices and electric vehicles has stimulated extensive research in high-performance energy storage devices, such as supercapacitors, which are also named ultracapacitors (UCs) or

Y. Liu
College of Energy and Institute for Electrochemical Energy Storage, Nanjing Tech University, Nanjing 211816, China

College of Chemistry and Chemical Engineer, Jiangxi Normal University, Nanchang 330022, China

F. Yu • Z.B. Wen (✉)
College of Chemistry and Chemical Engineer, Jiangxi Normal University, Nanchang 330022, China
e-mail: zbwen@jxnu.edu.cn

X.W. Wang • Y.S. Zhu
College of Energy and Institute for Electrochemical Energy Storage, Nanjing Tech University, Nanjing 211816, China

Y.P. Wu (✉)
College of Energy and Institute for Electrochemical Energy Storage, Nanjing Tech University, Nanjing 211816, China

New Energy and Materials Laboratory (NEML), Department of Chemistry & Shanghai Key Laboratory of Molecular Catalysis and Innovative Materials, Fudan University, Shanghai 200433, China
e-mail: wuyp@fudan.edu.cn

K.I. Ozoemena, S. Chen (eds.), *Nanomaterials in Advanced Batteries and Supercapacitors*, Nanostructure Science and Technology,
DOI 10.1007/978-3-319-26082-2_7

electrochemical capacitors (ECs), are a new class of energy storage devices that can store a large amount of charge, and deliver it at high power densities [2, 3]. However, we need to improve their performance substantially to meet the higher requirements of future systems, ranging from portable electronics to hybrid electric vehicles and large industrial equipments, by developing new materials and advancing our understanding of the electrochemical interfaces at the nanoscale [4]. The earliest EC patent was filed in 1957. However, not until the 1990s did EC technology begin to draw some attention in the field of hybrid electric vehicles [5]. It was found that the main function of ECs could promote the electrochemical performance of a battery or fuel cell in a hybrid electric vehicle by providing necessary power for acceleration and recuperating brake energy [6]. Thus, the capacitor technology is regarded as a promising means for storing electricity [7].

Nanomaterials and nanostructures play a pivotal role in the recent advancement of some key technologies. Nanomaterials differ from microsized and bulk materials not only in the scale of their characteristic dimensions but also in the fact that they may possess new physical properties and offer new possibilities for various technical applications [8]. Moreover, nanomaterials have higher surface energy compared with micro- or submillimeter materials, which means higher surface activity, and thus more electroactive sites in the nanostructured electrodes, leading to high capacity utilization of the electrode materials [9, 10]. In addition, their intrinsic inner porous structures allow for effective electrolyte infiltration, and the reduced dimensions of particle size can shorten the transport and diffusion path lengths of electrolyte ions, facilitating fast kinetics and high charge–discharge rates [2, 11]. As mentioned above, nanomaterials have been shown to possess a lot of favorable characteristics and therefore are being extensively investigated as electrode materials in energy systems [12].

Metal oxides, especially in the form of nanomaterials, have long been studied as potential electrode materials for pseudocapacitors because of the ease of large-scale fabrication and their rich redox reactions involving different ions, which contributes to higher capacitances than double-layer carbonaceous materials [13]. Typically transition metal oxides may be classified as noble transition metal oxides which exhibit excellent capacitive properties but have the drawback of generally being relatively expensive. Alternatively, base metal oxides may also be utilized which are considerably cheaper and more environment friendly than noble transition metals as well as exhibiting good capacitive properties [14]. In addition, intercalation compounds and conversion compounds have attracted a lot of interest thanks to their large specific charge. They have been typically used in organic electrolytes. However, aqueous electrolytes are more interesting [15].

Given that nanostructure materials can help ameliorate the electrochemical performances of metal oxides, the aim of this chapter is to summarize the recent results of metal oxides as cathode materials for supercapacitors as well as their electrochemical performance, which are advantageous to achieve high energy density [16]. Moreover, future developments of electrode materials are also discussed.

7.2 Fundamentals of Supercapacitors

Supercapacitors may be divided by several criteria such as energy storage mechanism, electrode material, electrolyte, or cell design [17]. Based on the energy storage mechanisms, there are two main types: electric double-layer capacitors (EDLCs) and faradic pseudocapacitors.

7.2.1 Two Types of Supercapacitors

In the case of EDLC, its capacitance is from the pure electrostatic charge accumulated at the electrode/electrolyte interface. It is strongly dependent on the surface area of the electrode materials that is accessible to the electrolyte ions [18]. The thickness of the double layer depends on the concentration of the electrolyte and the size of the ions and is in the order of 5–10 Å for concentrated electrolytes. The double-layer capacitance is about 10–20 μF/cm^2 for a smooth electrode in concentrated electrolyte solution, which is estimated according to equation Eq. (7.1):

$$\frac{C}{A} = \frac{\varepsilon_o{}^* \varepsilon_r}{d} \tag{7.1}$$

assuming a relative dielectric constant ε_r of ten for water in the double layer and d being the thickness of the double layer with surface area A. The corresponding electric field in the electrochemical double layer is very high, which can be easily up to 106 V/cm [5].

EDLCs are constructed much like a battery in that there are two electrodes immersed in an electrolyte, with ion permeable separator located between the electrodes (Fig. 7.1) [19]. In such a device, each electrode/electrolyte interface represents a capacitor so that the complete cell can be considered as two capacitors in series. For a symmetrical capacitor (similar electrodes), the cell capacitance (C_{cell}) will therefore be:

$$\frac{1}{C_{cell}} = \frac{1}{C_1} + \frac{1}{C_2} \tag{7.2}$$

where C_1 and C_2 represent the capacitance of the first and second electrodes, respectively [20]. In the case of EDLCs, carbon materials are the most used because of their low electrical resistance, easy processing ability, chemical inertness, stability, and low cost. Typical materials include activated carbon, carbon aerogels, and carbide-derived carbon/ordered mesoporous carbon, carbon nanotubes, and graphene [8].

The other one is pseudocapacitance, which is originated from the redox reaction of the electrode material with the electrolyte [21]. The accumulation of electrons at

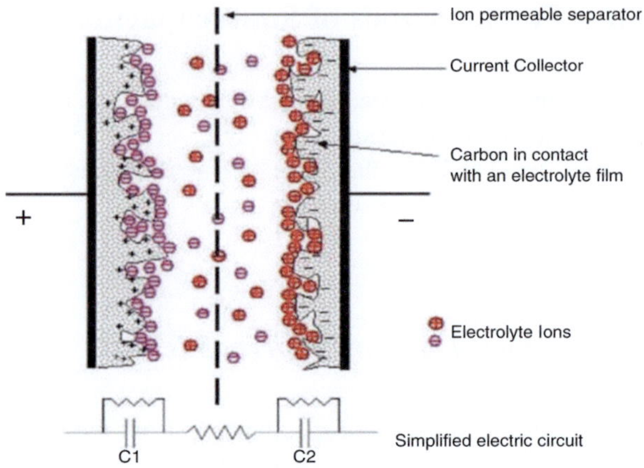

Fig. 7.1 Schematic representation of an electrochemical double-layer capacitor in its charged state (Modified from Ref. [19] Copyright 2006, with permission from Elsevier)

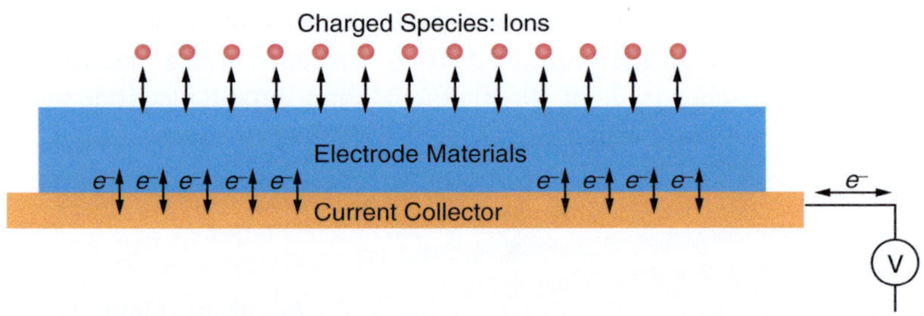

Fig. 7.2 Charge storage mechanism of pseudocapacitive materials showing the surface charge-transfer reaction (Modified from Ref. [11] Copyright 2011, with permission from RSC)

the electrode is a faradic process where the electrons produced by redox reactions are transferred across the electrolyte/electrode interface (Fig. 7.2) [11]. The theoretical pseudocapacitance of metal oxide can be calculated as:

$$C = \frac{(n \times F)}{(M \times V)} \tag{7.3}$$

where n is the mean number of the electrons transferred in the redox reaction, F is the Faraday constant, M is the molar mass of the metal oxide, and V is the operating voltage window [22]. Common materials for pseudocapacitors include conducting polymers and metal oxides or nitrides. Redox reactions occur in the electroactive material because of several oxidation states. Often, reactions do not propagate into the bulk material and occur only at the electrode/electrolyte interface [23]. Typical pseudocapacitive materials include conducting polymers such as polypyrrole, polythiophene, and polyaniline [24] and metal oxides such as RuO_2, MnO_2, and Co_3O_4 [25]. Subsequently, metal oxides will be discussed predominantly in this chapter.

7.2.2 Main Parameters of Supercapacitors

The key performance parameters of supercapacitors include specific capacitance (normalized by electrode mass, volume, or area), energy density, power density, rate capability (retained capacitance at a high current loading) and cycling stability [26]. The energy density (E) of a supercapacitor is expressed as:

$$E = \frac{1}{2CV^2} \tag{7.4}$$

The maximum power density (P_{max}) of a supercapacitor is determined by:

$$P_{max} = \frac{V^2}{(4R)} \tag{7.5}$$

where C is the direct current capacitance in Farad, V is the nominal voltage, and R is the equivalent series resistance (ESR) in ohms [27]. In order to increase the energy and power densities of a supercapacitor, it is desirable to increase the specific capacitance (C) and the operating voltage window (V) as well as reduce the equivalent series resistance (R).

The capacitance of a supercapacitor is largely dependent on the characteristics of the electrode material, specifically the surface area and the pore-size distribution. Due to the high porosity and correspondingly low density of electrode materials, it is generally accepted that the volumetric capacitance of each electrode determines the energy density [19].

The maximum operating voltage window (V_m) is mainly dependent on the electrolyte, which is limited by the electrochemical window of the electrolyte. Aqueous electrolytes have the advantage of high ionic conductivity (up to $\sim$1 S cm $^{-1}$), low cost, and wide acceptance. On the other hand, they have the inherent disadvantage of a restricted voltage range with a relatively low decomposition voltage of $\sim$1.23 V [28]. Nonaqueous electrolytes of various types have been developed and a cell voltage can be up to 3.5 V [29]. The electrical resistivity of nonaqueous electrolytes is, however, at least an order of magnitude higher than that of aqueous electrolytes, and therefore the resulting capacitors generally have a higher internal resistance.

A high internal resistance limits the power capability of a supercapacitor and its application. In supercapacitors, a number of sources contribute to the internal resistance and are collectively measured and referred to as the equivalent series resistance (ESR) [20]. There are several contributors to the ESR of supercapacitors as shown in the following:

1. Electrolyte resistance
2. Ionic resistance of ions moving through the separator
3. Ionic diffusion resistance of ions moving in small pores

4. Electronic resistance of the electrode materials
5. Interfacial resistance between the electrode and the current collector

7.3 Preparation Methods of Nanomaterials for Supercapacitors

With the development of nanoscience and technology, various metal oxide nanomaterials including nanoporous [30], nanorod [31], nanobelt [32], nanowire [33], nanotube [34], and nanoplate [35] have been synthesized. Different preparation routes to nanomaterials have been employed. Such approaches can be classified into physical routes (e.g., sonication, microwaves) and chemical ones (e.g., hydrothermal, sol–gel).

7.3.1 Physical Routes

7.3.1.1 Ball Milling Method

Ball milling, also named as mechanochemistry, has become an important way to prepare ultrafine materials. Traditionally, the generation of new material, crystal-type transformation, or lattice deformation is achieved by high temperature or chemical changes. However, mechanical energy is directly used to induce a chemical reaction, which may provide a new way to synthesize novel materials. As a new technology, it has significantly reduced the reaction activation energy and the grain size, greatly improved the activity of powder and the uniformity of particle size distribution, enhanced the combination of the interface between the body and matrix, and promoted solid-state ion diffusion as well as chemical reaction induced by low temperature; thereby, the compactness and electrical and thermal performance of the material are improved. This method is energy saving and high efficient.

7.3.1.2 Sonochemical Method

Sonochemistry deals with the understanding of the effect of sonic wave properties on chemical systems. It has several interesting features. Ultrasound remarkably enhances mass transport, reduces the diffusion layer thickness, and may also affect the surface morphology of the treated materials. It typically enhances the surface contact area [36]. Deposition and reduction of the particles (induced by ultrasonic waves) almost consecutively take place so that the heating step normally employed in other protocols can be avoided [37]. The chemical effects of ultrasound are not from a direct interaction with precursors but from the acoustic cavitation. When a liquid is exposed to strong ultrasound stimulation, the bubbles will experience the process of formation, growth, and implosive collapse. The bubble collapse brings intense local heating (~5000 K), high pressure (over 1800 atm), and enormous cooling rates (~10^{10} K s^{-1}), which enable many chemical reactions to occur [38]. The major advantages of sonochemical method include rapid reaction rate, controllable reaction parameters, and the ability to generate highly pure nanocrystals with uniform shapes and narrow size distributions [39–41].

7.3.1.3 Microwave Method

Microwave chemistry has been well expanded for the liquid-phase preparation of various metal oxide nanomaterials due to its specific advantages such as high reaction rate (reaction time can be reduced by orders of magnitude), low processing cost, high yield, and depression of side reactions [42]. The efficient heating for materials is processed with energy by electromagnetic radiation in the frequency range of 0.3–2.45 GHz. The microwave dielectric heating mechanism contains two main processes, dipolar polarization and ionic conduction [42], which are different from the direct absorption of high-energy electromagnetic radiation to induce chemical reaction. Considering the efficiency of heating materials, a microwave frequency of 2.45 GHz with energy of only about 1 J mol^{-1} is optimized to convert the microwave energy into thermal energy, which, however, is too low to cleave chemical bonds [43].

7.3.2 *Chemical Routes*

7.3.2.1 Hydrothermal Method

The hydrothermal method has been widely used for the synthesis of a variety of functional nanomaterials with specific sizes and shapes since the 1960s [44]. It uses water as the reaction medium in a sealed steel pressure vessel with Teflon liner, which is then heated to a designed temperature to promote the reaction. The temperature adopted is often higher than 100 °C in order to reach the pressure of vapor saturation, so that autogenously pressure will be developed in a closed system [45]. The generated pressure within the reactor not only strongly relies on the reaction temperature but also depends on other experimental factors, such as the added amounts of solvents and dissolved solutes. Moreover, the main advantages of hydrothermal processes contain fast reaction kinetics, short processing time, high-phase purity, good crystallinity, low cost, and so on [46]. Furthermore, the process is environmentally benign and versatile since it does not involve any organic solvents or posttreatments such as calcinations [47]. In a hydrothermal reaction system, water is used as the main reaction medium, and many inorganic salts containing the source of the metal ions can be well dissolved. In addition, water is also beneficial to introduce small coordinating molecules to adjust the growth of the final nanocrystals [48].

7.3.2.2 Sol–Gel Method

Sol–gel method generally refers to the hydrolysis and condensation of metal alkoxides or alkoxide precursors, leading to dispersions of oxide particles in a sol. The sol is then dried or gelled by solvent removal or by chemical reaction. The used solvent is generally water, but the precursors can also be hydrolyzed by an acid or base. Basic catalysis induces the formation of a colloidal gel, whereas acid catalysis yields a polymeric form of the gel [49]. The rates of hydrolysis and condensation are important parameters that affect the properties of the final products. Smaller particle sizes are obtained at slower and more controlled hydrolysis rates. The particle size also depends on the solution composition, pH, and temperature. Magnetic ordering in the sol–gel system can determine the formed phases and the particle volume fraction including the size distribution and the dispersion of the particles [50]. In the case of nanocomposites derived from the gels, structural parameters and material porosity are determined by the rates of hydrolysis and condensation of the gel precursors and other oxidation–reduction reactions during the gelling and subsequent heat treatment stages [50, 51].

7.3.2.3 Template Method

Template synthesis has become very popular for the fabrication of functional materials with various nanostructures [52, 53]. It is a straightforward way to fabricate nanostructures by inducing the target materials to grow according to the patterns of the templates. This strategy provides an easy way for the synthesis of nanomaterials with desired shape and size and has been widely applied in the construction of 1D nanostructure [54]. Generally, templates employed in the synthesis can be classified into two categories: (1) soft templates, which contain ligands, surfactants, polymers, and organogelators, and (2) hard templates, which are either used as physical scaffolds for the next deposition of desired coating materials or employed not only as shape-defining templates but also as chemical reagents that react with other chemicals to create desired nanostructures [55]. Possin is the first to use this technique to prepare nanowires in the 1970s [56], and Martin extended this technique and first put forward the term "template synthesis" in the 1990s [57]. Now the commonly used templates include but are not limited to ordered porous membranes prepared with anodized aluminum oxide (AAO) [58], silica [59], nanochannel glass [60], and ion-track-etched polymers [61].

7.3.2.4 Electrodeposition Method

The electrodeposition method is attractive to produce a metallic coating on a surface by the cathodic reduction reaction in aqueous or organic solvents. The substrate material is used as a cathode and is immersed into a solution containing a salt of the metal to be deposited. The dissolved metallic ions are attracted to the cathode and then reduced to the metallic form [62]. The distinctive advantages of the electrodeposition method are that it contains the capacity to enable the conformal deposition of structures and it can easily control the thickness of deposited films [63]. Experiment data show that the pore diameters of the substrate have little effect on the crystallite sizes of nanomaterials at higher overpotential. The effect, however, is very large at lower overpotential [64].

7.3.2.5 Coprecipitation Method

The coprecipitation method involves the simultaneous precipitation of the metal and the support. This approach has been expanded to include other inorganic oxide supports [65]. However, the approach has its associated drawbacks. The presence of the metallic precursors in solution may interfere with the polymerization chemistry of the material, which often results in samples with undesirable properties including less well-defined pore size and shape. Furthermore, this approach has limited applicability to polymeric supports [66].

7.3.3 Other Methods

In addition to the synthetic approaches discussed above, some other emerging methods for fast and facile preparation of nanomaterials have been reported in the past few years such as hot-injection method [67], microemulsions method [68], and chemical vapor deposition [69]. As to these methods, however, they are rarely used to prepare nanostructure metal oxides and will not be discussed here.

7.4 Nanostructure Oxides as Cathode Materials for Supercapacitors

Generally, the electrode materials of ECs can be categorized into three types [70, 71]: (1) carbon materials with high specific surface area, (2) conducting polymers, and (3) metal oxides. However, metal oxides can provide higher energy density for supercapacitors than conventional carbon materials and better electrochemical stability than polymer materials. They not only store energy like electrostatic carbon materials but also exhibit electrochemical faradic reactions between electrode materials and ions within appropriate potential windows [72]. Hence, we focus on the most recent work regarding metal oxides as cathode for supercapacitors. To date, these nanostructure oxides include transition metal oxides, intercalation compounds, and conversion compounds.

7.4.1 Transition Metal Oxides

7.4.1.1 RuO_2

Among the transition metal oxides, ruthenium oxide (RuO_2) is the most promising cathode material for supercapacitors owing to its high specific capacitance, good thermal stability, highly reversible redox reactions in a wide potential range, high rate capability, and long cycling life [73]. There are two phases of RuO_2: a crystal phase (RuO_2, rutile phase) and an amorphous hydrous phase ($RuO_2{\cdot}xH_2O$). The annealing temperature is a key factor affecting the morphology, which exists usually in the form of nanostructure such as nanoparticle, nanorod, and nanofiber. High annealing temperature, for example, from 300 °C to 800 °C, leads to the crystalline phase of RuO_2 and less water content, resulting in the change of the number of active reaction sites, as well as the electron and proton conductivity. Electron conductivity of RuO_2 nanocrystals is balanced with the proton conductivity of the ambient structural water when one mole of RuO_2 is hydrated by 0.58 mol of water ($RuO_2{\cdot}0.58H_2O$). Primary particulates will aggregate into secondary particles to form 3D (three dimension) framework porosity at above 200 °C. High specific surface area with nanoporous structure is also important to achieve high capacitance [15].

As for nanostructure RuO_2, it can be synthesized via several methods including hydrothermal synthesis [74], template [75], microwave d [76], electrodeposition [77], solid-state method [78], and oxidative precipitation [79]. For instance, RuO_2 nanoparticles are synthesized by instant method using Li_2CO_3 as stabilizing agent, under microwave irradiation at 60 °C [80], and hollow fusiform $RuO_2{\cdot}xH_2O$ nanostructures were prepared by using α-Fe_2O_3 nanospindles as a sacrificial template via a hydrothermal approach [81].

It is well known that RuO_2 is a good cathode material for electrochemical capacitors; however, the lack of abundance and the expensive cost of the noble metallic element (Ru) are the drawbacks for its commercial application in supercapacitors. Consequently, as an alternative material for pure RuO_2, many nanocomposites containing RuO_2 are synthesized by various above-mentioned methods. For example, the composites of amorphous $Ru_{1-y}Cr_yO_2/TiO_2$ nanotubes are synthesized by loading various amount of $Ru_{1-y}Cr_yO_2$ on TiO_2 nanotubes through a redox reaction of aqueous $K_2Cr_2O_7$ with $RuCl_3$. The 3D nanotube framework of TiO_2 offers a solid support for $Ru_{1-y}Cr_yO_2$ active materials, allowing the active material to be readily available for electrochemical reactions, and an increase in the efficient use of active materials [15]. Different morphologies of RuO_2 can be prepared to get more efficient use of RuO_2. For instance, core–shell structured RuO_2@MnO_2 nanorods and RuO_2@TiO_2 nanotube can be prepared [82, 83]. The nanocomposites of RuO_2@MnO_2 show excellent capacitive behavior which is attributed to the conductive RuO_2 template as well as its 1D nanowire morphology.

Moreover, it can also form composites with carbon materials such as AC (activated carbon), CNTs (carbon nanotubes), NPC (nanoporous carbon), and graphene. For example, the RuO_2@AC composite can be synthesized by a sol–gel method on the surface of AC [84]. CNTs acts as a support due to its good electron conductivity and accessible surface textures, and RuO_2 can be dispersed on CNTs with different arrays [85, 86]. The composites of nanocrystalline RuO_2 on vertically aligned few-layered graphene (FLG) nanoflakes are synthesized on bare n-type heavily doped silicon substrates by microwave plasma chemical vapor deposition. The RuO_2 nanoparticles were deposited using a combination of low base pressure radio frequency magnetron sputtering and subsequent electrochemical cycling in acidic media. In this case, the porous structure of FLGs not only provides a large surface area for the deposition of RuO_2 nanoparticles but also facilitates electrolyte infiltration, thereby helping ion transport to support redox processes (Fig. 7.3a). The direct sputter deposition of Ru onto FLGs results in uniform and well-separated RuO_2 nanoparticles along the FLG surface. The strong interactions between individual RuO_2 nanoparticles with the FLG surface can help in the efficient collection of capacitance charge from the individual RuO_2 nanoparticles along the conductive graphene backbone of the FLGs, thereby enhancing the overall performance of RuO_2–FLG nanocomposites. The nanoscale integration of pseudocapacitive RuO_2 nanoparticles with FLGs leads to a high specific capacitance of nearly 650 F g^{-1} which is nearly twice the value measured for pristine RuO_2 films (~320 F g^{-1}) for similar mass loadings. What's more, the capacitance retention is nearly 70 % after 4000 cycles at a high charge–discharge rate (Fig. 7.3b). The superior capacitive nature of the RuO_2–FLGs is attributed to the uniform distribution of RuO_2 nanoparticles on FLGs and synergistic effects between them [87]. In addition, a simple and scalable way of preparing a 3D sub-5 nm hydrous ruthenium oxide (RuO_2) anchored on graphene and CNT hybrid foam (RGM) architecture for high-performance supercapacitor electrodes is reported. Supercapacitors based on RGM show superior gravimetric and per-area capacitive performance (specific capacitance, 503 F g^{-1}; areal capacitance, 1.11 F cm^{-2}) which leads to an exceptionally high energy density of 39.3 Wh kg^{-1} and power density of 128 kW kg^{-1} [88].

The mechanism of charge storage in RuO_2 is dominated by a coupled electron–proton transfer process. The electrochemical process involves a reversible change in the valence state of RuO_2 via an exchange of protons with the solution, as shown in Eq. (7.6) [89, 90]:

$$RuO_2 + \delta H^+ + \delta e^- \leftrightarrow RuO_{2-\delta}(OH)_\delta \ (0 \le \delta \le 2) \tag{7.6}$$

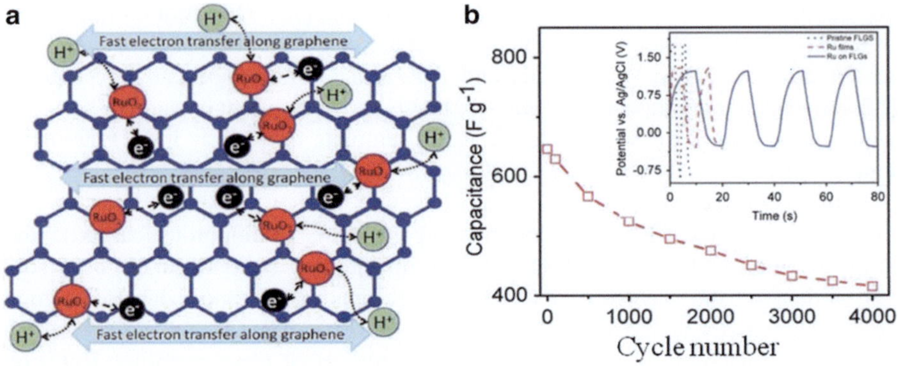

Fig. 7.3 (**a**) Schematic of charge storage in RuO_2 nanoparticles and fast electron transfer on the FLG (few-layered graphene) backbone and (**b**) cycling lifetime of RuO_2-FLG nanocomposites tested for 4000 cycles at a scan rate of 500 mV s^{-1} in 1 M H_2SO_4. The inset shows the charging–discharging characteristics of pristine FLGs, bare RuO_2 films, and RuO_2-FLG nanocomposites (Modified from Ref. [87] Copyright 2012, with permission from RSC)

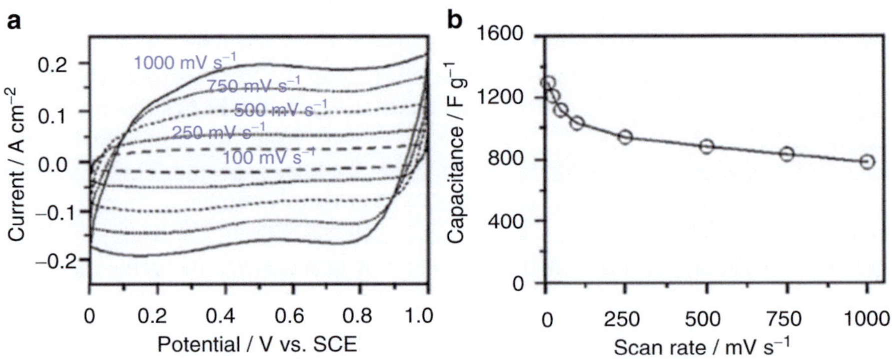

Fig. 7.4 (**a**) CV curves at different scan rates and (**b**) dependence of the capacitance on the scan rate for an annealed $RuO_2 \cdot xH_2O$ nanotube arrayed electrode (0.19 mg cm^{-2}) in 1.0 mol l^{-1} H_2SO_4 solution (Modified from Ref. [73] Copyright 2006, with permission from ACS)

RuO_2 can present a very good rectangular shape in the CV curves. For example, the 3D mesoporous $RuO_2 \cdot xH_2O$ nanotubular arrayed electrode can retain this shape up to 1000 mV s^{-1} (Fig. 7.4a). Its capacitance does not fade much with the scan rate (Fig. 7.4b). It shows a specific power of 4320 kW kg^{-1} and specific energy of 7.5 Wh kg^{-1} in 1.0 mol l^{-1} H_2SO_4 solution at the potential range of 0–1.0 V (vs. Ag/AgCl), respectively [73]. The specific capacitance and charge–discharge performance of the thin-film RuO_2 are dependent on the electrode thickness. The thinner the electrode thickness is, the higher the specific capacitance will be. For instance, a maximum specific capacitance of 788 F g^{-1} is achieved with an electrode thickness of 0.0014 g cm^{-2} in 0.5 mol l^{-1} H_2SO_4 solution at the potential range of 0–0.6 V (vs. Ag/AgCl) [91]. The supercapacitor with anthraquinone-modified carbon fabric as the anode and RuO_2 as the cathode shows a maximum energy density of 26.7 Wh kg^{-1} in 1 mol l^{-1} H_2SO_4 solution in the voltage range of 0–1.3 V. It can arrive at an energy density of 12.7 Wh kg^{-1} at 0.8 A cm^{-2} discharge rate with average power density of 17.3 kW kg^{-1}.

In order to increase the electrochemical charge stored per RuO_2, a common method is to add a second metal oxide such as VO_2. In 0.5 M H_2SO_4 acidic solution, its specific capacitance based on RuO_2 is >1200 F g^{-1} for $Ru_{0.35}V_{0.65}O_2$ in the potential range of 0.1–1.3 V (vs. NHE) [92]. The RuO_2@TiO_2 nanotube composite has good electrochemical performance in 1 mol l^{-1} KOH solution in the voltage range of 0–1.4 V. The energy density reaches 12.5 Wh kg^{-1} at a power density of 150 W kg^{-1}. When the power density increases up to 1207 W kg^{-1}, its energy density decreases to 5.7 Wh kg^{-1}. Comparing with $RuO_2 \cdot xH_2O$, $RuO_2 \cdot xH_2O$@TiO_2 nanocomposites exhibit the weakly mass-dependent specific capacitance and high-power capacitive characteristics [83].

7.4.1.2 MnO_2

MnO_2 has been widely investigated as an electrode material for applications in supercapacitors due to its environmental friendly, low cost, and abundant availability on earth, which functions as cathode of electrochemical capacitors as Eq. (7.7):

$$MnO_2 + \delta M^+ + \delta e^- \leftrightarrow MnO_2/M_\delta \qquad (7.7)$$

To date, MnO_2 possesses several different crystallographic forms, such as α−, β−, γ−, δ−, λ−, and ramsdellite type, which made of MnO_6 octahedral with different connectivities, and their different crystalline phases are described by the size of the tunnel formed with the number of octahedral subunits (n × m), and the tunnel types as well as sizes are presented in Table 7.1 [15].

Although MnO_2 has a high theoretical specific capacitance value of 1370 F g^{-1} for storing one electron per Mn center [93], its actual specific capacitance often lies in the range from 100 to 400 F g^{-1} [94, 95]. Therefore, the capacitance enhancement of MnO_2 cathode material is very important for MnO_2-based supercapacitors. Results demonstrate that MnO_2 nanoscale cathode material can enhance the

Table 7.1 Tunnel sizes of different crystallographic forms of MnO_2

Crystallographic forms	Tunnel type	Tunnel size (Å)
α	(1 × 1), (2 × 2)	1.89, 4.6
β	(1 × 1)	1.89
γ	(1 × 1), (1 × 2)	1.89, 2.3
δ	Interlayer distance	7.0
Ramsdellite	(1 × 2)	2.3

electrochemical utilization and promote high-rate operation for both charge and discharge because the nanostructures with a large specific surface area and a high surface-to-bulk ratio can offer short diffusion distance, favorable kinetics, and high specific capacitance [96, 97]. Up to now, many efforts have been made to synthesize manganese oxide nanostructures with different morphologies including hierarchical nanoflower [98], nanosheets [99], nanorods [100], and hollow urchin-like nanostructures [101]. For example, MnO_2 nanostructures, composed of interleaving nanoplates, exhibit mesoporous characteristics and the morphologies are clew-like. The MnO_2 with a mean pore size of 10.6 nm has a high specific capacitance of 404 F g^{-1} in 1 mol L^{-1} Na_2SO_4 solution at a 2 mV s^{-1} scan rate and 2 % capacitance loss after 800 cycles at a 10 mV s^{-1} scan rate. In addition, cyclic voltammogram in a mild aqueous electrolyte shows that the charge storage process is poorly related to double-layer charging processes and is mainly faradic. The linear relationship between the response current density and scan rate indicates that the adsorption of cations on the MnO_2 surface is the controlling process in the faradic reaction [102].

In the case of the composites of MnO_2 with other materials, most of the latest reports are related to carbonaceous materials such as carbon aerogels [103], AC [104], CNTs, and graphene. However, there are a few reports on their composites with conductive polymers [105, 106], which will not be expounded upon here. As we know, CNTs present higher electron conductivity than AC, and there are some composites of CNTs with MnO_2. For instance, a hybrid of MnO_2 nanowires well entangled with MWCNTs is prepared by a versatile hydrothermal method. The capacitance shows only a small fading after 13,000 cycles even when oxygen is not removed. The major cause is that the uniform MWCNTs network in the hybrid provides good electron conductivity and the small diameter of MnO_2 nanowires ensures fast gain/loss of Li^+ ions. In addition, the MnO_2 nanowire structure of good resilient property can keep structural stability during the charge and discharge processes even at very high current density [107]. In the case of ultralight and flexible MnO_2/carbon foam (MnO_2/CF) hybrids, their synthesis and capacitance performance are systematically studied. Flexible carbon foam with a low mass density of 6.2 mg cm^{-3} and high porosity of 99.66 % is simply obtained by carbonization of commercially available and low-cost melamine resin foam. Attributed to the good flexibility and ultralight weight, the MnO_2/CF nanomaterials can be directly fabricated into supercapacitor electrodes without any binder and conductive agents. Supercapacitor based on the MnO_2/CF composite with 3.4 % weight percent of MnO_2 shows a high specific capacitance of 1270.5 F g^{-1} (92.7 % of the theoretical specific capacitance of MnO_2) and high energy density of 86.2 Wh kg^{-1} [108].

Amorphous or poorly crystalline MnO_2 nanoparticles can be deposited onto ultraporous activated graphene from $KMnO_4$ solution [109]. A solution-phase assembly can be used to prepare a composite of MnO_2 nanowires@graphene sheets [110]. Graphene can be decorated with flower-like MnO_2 nanostructures by electrodeposition to achieve a composite [111]. The MnO_2@CNT-RGO composite can be prepared by the process shown in Fig. 7.5a. Firstly, a hierarchical architecture

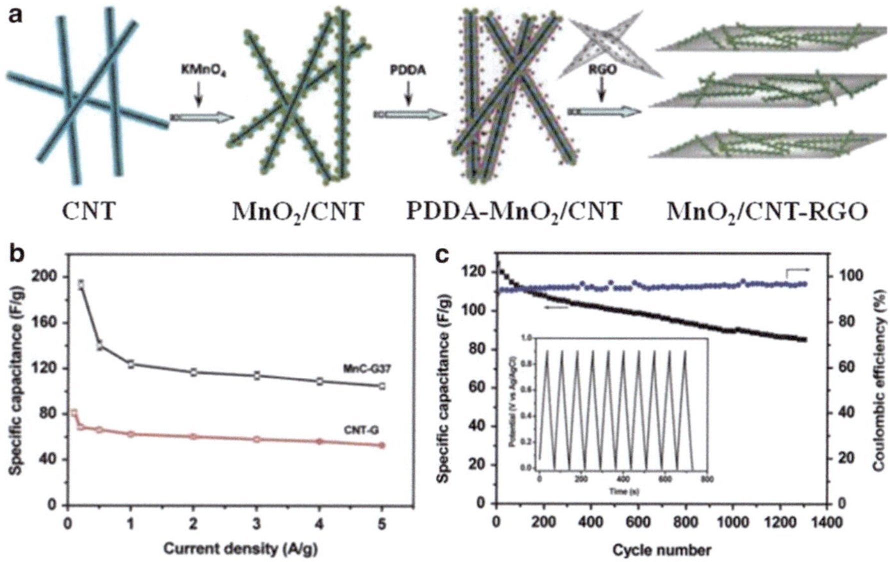

Fig. 7.5 (**a**) Schematic illustration of the synthetic procedures for MnO_2@CNTs-RGO composite, (**b**) variation of specific capacitance against current density for MnC-37 and CNT-G electrode, and (**c**) cycling performance and Coulombic efficiency of MnC-G37 electrode measured at current density of 2.0 A/g. The inset shows the last ten cycles of galvanostatic charge–discharge (Modified from Ref. [112] Copyright 2011, with permission from Elsevier)

consisting of MnO_2@CNTs and RGO sheets is prepared based on the electrostatic self-assembly. The positively charged MnO_2@CNTs functionalized with poly (diallyldimethylammonium chloride) (PDDA) and negatively charged RGO creates a strong electrostatic interaction, forming a graphene-based nanostructure with MnO_2@CNTs uniformly incorporated between RGO sheets. The second is a maximum specific capacitance of 193 F g^{-1} was achieved for the MnO_2@CNTs-RGO composite with 37 % RGO, which was almost threefold higher than 69 F g^{-1} of carbon nanotubes/RGO and twofold higher than 89 F g^{-1} of MnO_2/RGO composite.

Moreover, an excellent rate performance, good capacitance retention (~70%), and a superior Coulombic efficiency (94–96 %) were also observed during the continuous 1300 galvanostatic charge–discharge cycles [112]. Conductive polymers can be further coated on MnO_2@MWCNT composites [113]. Other kinds of conductive matrix can also be a support for MnO_2 such as Zn_2SnO_4 [114], TiN [115], and nanoporous gold [16].

In the case of the MnO_2 nanowires@graphene composite, the nanostructured MnO_2 can prevent the aggregation of graphene sheets caused by van der Waals interactions, consequently leading to an increase in the available electrochemical active surface area and a suitable porous structure for energy storage [110]. The graphene@MnO_2 composite is a flexible film which achieves a high specific capacitance (372 F g^{-1}) with excellent rate capability without the need of current collectors and binders [116]. The unique characteristics of the MnO_2@graphene nanostructured textiles make them promising candidates for high-performance supercapacitor electrode materials. For example, 3D porous microstructure of the polyester textiles allows conformal coating of graphene nanosheets (GNSs) and subsequent loading of MnO_2 and facilitates easy access of electrolyte ions to electrode surfaces. Moreover, the GNS coating serves as high-surface-area, conductive paths for the deposition of MnO_2, providing excellent interfacial contact between MnO_2 and graphene for fast electron transport. In addition, nanoflower architecture of the electrodeposited MnO_2 offers large electrochemically active surface area for charge transfer and reduced ion diffusion length during the charge–discharge process [117].

As to the composites with conductive carbon materials, the movement of electrons is greatly improved, leading to an enhancement of the capacitance and rate capability. For instance, hierarchical MnO_2 nanospheres@CNTs@conducting polymer ternary composite achieves a specific capacitance of 427 F g^{-1} in 1 mol L^{-1} Na_2SO_4 solution [118]. The binder-free MnO_2@CNT array composite shows excellent rate capability (50.8 % capacity retention at 77 A g^{-1}) and high capacitance (199 F g^{-1} and 305 F cm^{-3}) [119]. The PANI@MnO_2-MWCNT composite delivers remarkably enhanced specific capacitance and cycling stability compared to MnO_2-MWCNTs, where a specific capacitance of 350 F g^{-1} is obtained at a current density of 0.2 A g^{-1}, higher than 92 F g^{-1} for pristine MWCNTs and 306 F g^{-1} for MnO_2-MWCNTs [113].

In the case of composites of MnO_2 with other conductive materials, their electrochemical performance is also greatly improved. For instance, the mass loading of MnO_2 can be up to 60 % in the MnO_2@Zn_2SnO_4 composites, which show a capacitance as high as 200 F g^{-1} including excellent charge–discharge rate and cycling stability [114]. The coaxial array of MnO_2@TiN nanotube composite shows a capacitance of 681 F g^{-1} at a current density of 2 A g^{-1} and excellent rate capability (267 F g^{-1} at an impressing scan rate of 2000 mV s^{-1}) in 1 mol L^{-1} Na_2SO_4 solution due to the electron conducting TiN framework [115]. In addition, Fig. 7.6a shows a simple and efficient method for fabricating ultrathin and flexible supercapacitor electrodes containing a manganese dioxide (MnO_2) nanostructure deposited onto 3D nickel nanocone arrays (NCAs). The electrode exhibits outstanding mechanical robustness, high specific capacitance (632 F g^{-1}), enhanced energy density (52.2 W h kg^{-1}), and excellent cycling performance (95.3 % retention after 20,000 cycles), which are shown in Fig. 7.6b and c, respectively. Moreover, the MnO_2-NCAs supercapacitor as an ultrathin power source can drive a LED indicator (Fig. 7.6a). This technology may find vast applications in future wearable electronics [120]. The MnO_2@nanoporous gold composite can show a surprising high specific capacitance (about 1145 F g^{-1}) that is close to the theoretical value [93].

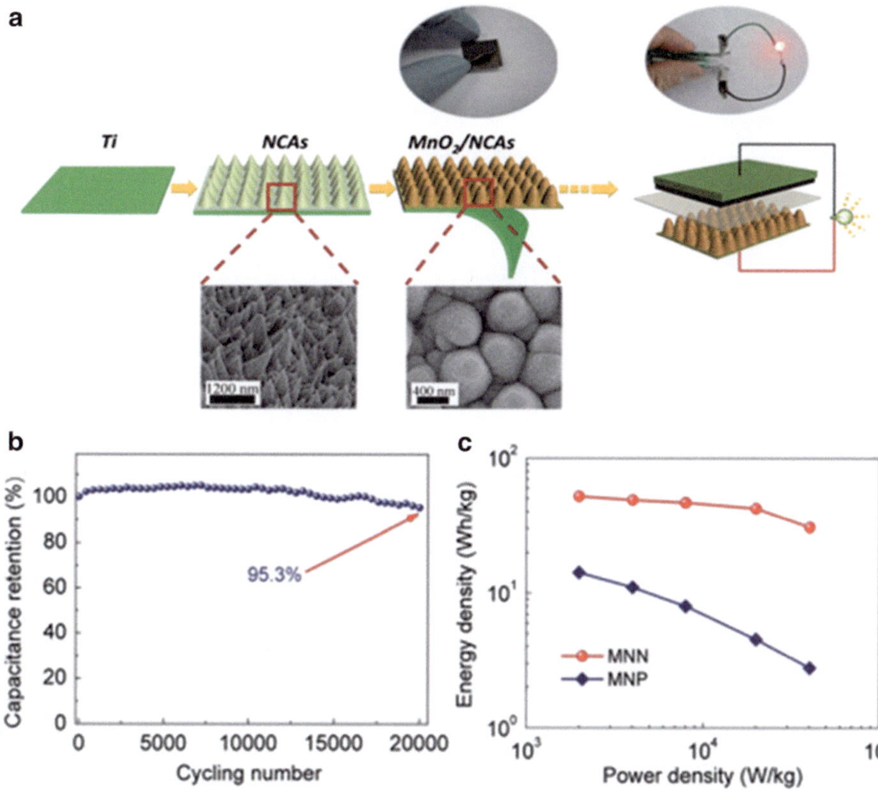

Fig. 7.6 (**a**) Schematic illustration of the fabrication process for an ultrathin supercapacitor, (**b**) capacitance retention of MNN as a function of the cycling number, and (**c**) Ragone plots of nickel nanocone arrays (*NCAs*) electrodeposited with MnO_2 nanostructure (*MNN*) and MnO_2 deposited on a Ni plate (*MNP*) (Modified from Ref. [120] Copyright 2008, with permission from ACS)

7.4.1.3 MoO_3

Of the various transition metal oxides, MoO_3 is a layered n-type semiconductor [121, 122]. There are three main polymorphs of MoO_3: orthorhombic α-MoO_3 (thermodynamically stable phase), monoclinic β-MoO_3, and hexagonal h-MoO_3 (low temperature metastable phases). The α-MoO_3 phase possesses a unique layered structure: each layer is composed of two sub-layers which are formed by corner-sharing MoO_6 octahedra along [001] and [100] directions, and two sub-layers are stacked together by sharing the edges of octahedra along the [001] direction. An alternate stacking of these layers along the [010] direction with van der Waals interaction leading to the formation of α-MoO_3 with a two-dimensional structure allows guest atoms, ions, or molecules to be introduced between the layers through intercalation [123].

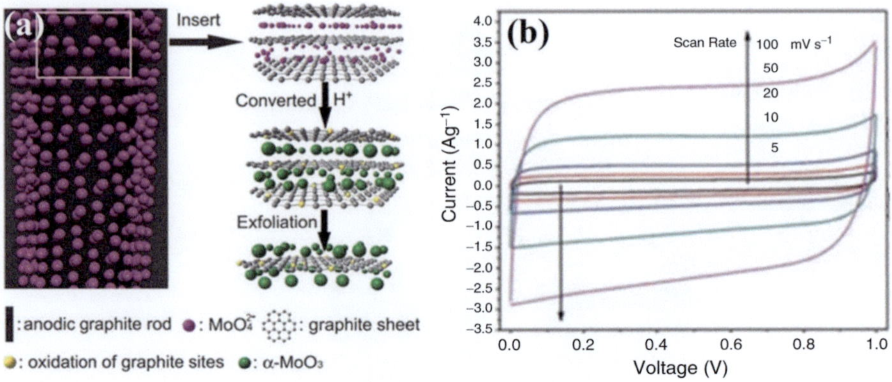

Fig. 7.7 (**a**) Illustration of the formation of the composite of α-MoO_3-decorated graphene sheets and (**b**) CV curves of the composite at different scan rates using 6 M KOH aqueous electrolyte (Modified from Ref. [126] Copyright 2011, with permission from RSC)

To achieve good capacitive behavior, nanostructured materials such as nanoparticles, nanorods, and nanoplates are prepared to favor the rapid charge and discharge process. Nanostructured molybdenum oxide can be prepared by potentiodynamic electrodeposition onto stainless steel. The deposit consists of particulates in the range of 30–80 nm [124]. Nanorod α-MoO_3 can be synthesized on a large scale by a hydrothermal method [125], and MoO_3 nanoplates can be conveniently prepared by a sol–gel method [35].

Since MoO_3 has a wide redox potential, it can be cathode materials for supercapacitors. However, its redox behavior at high potential is not good. Thus, it is modified to form composites with conductive materials such as graphite [122] and graphene [126]. Figure 7.7a shows that the composite of MoO_3@graphene sheets is prepared by electrolytic exfoliation from graphite with the assistance of molybdate. During the electrochemical reaction process, the anions of MoO_4^{2-} move to the graphite electrode under the influence of an applied electric field and insert into the graphite layers, which subsequently are converted to MoO_3 nanoparticles under strong acid conditions, leading to the electrolytic exfoliation of the graphite rod. At the same time, anodic oxidation of water produces hydroxyl and oxygen radicals. These radicals can result in hydroxylation or oxidation of graphite and the dissolution of carbon nanocrystals from the anode. These processes occur initially at edge sites, grain boundaries, or defect sites [126].

As we can see from Fig. 7.7b, the redox peaks are not clear in neutral or alkaline solutions. However, the redox peaks can be clearly identified in acid solutions such as 1 M H_2SO_4 [122]. After the heat treatment at 200 °C, the composite of MoO_3@graphene sheets can exhibit a high specific capacitance (up to 86.3 F g^{-1}) in the voltage range of 0–1 V in 6 M KOH solution due to very high electron conductivity (up to 5367 S m^{-1}) [126].

In the case of composites of MoO_3 with other conductive polymer materials, they can also show enhanced electrochemical performance. For example, the unique hybrid of polyaniline (PANI)-intercalated molybdenum oxide nanocomposites was prepared through the simultaneous reaction process, which includes the generation of metal oxide layers, the oxidation polymerization of monomers, and the in situ formation of polymer–metal oxides sandwich structure. Moreover, the prepared hybrids of PANI with MoO_3 nanosheet and nanotube have a flexible lamellar structure, swelled interlayer, typical nanosize, and high conducting electron content and also exhibit superior electron conductivity [127]. In addition, a large scale of MoO_3/PANI coaxial heterostructure nanobelt has been fabricated for high-performance supercapacitors via a simple and green approach without any surfactant. The as-synthesized MoO_3/PANI coaxial heterostructure nanobelts exhibit excellent supercapacitor performance with high specific capacitances of 714 F g^{-1} at a scan rate of 1 mV s^{-1} and 632 F g^{-1} at a current density of 1 A g^{-1} in 1 M H_2SO_4 electrolyte [128]. As for the SnO_2/MoO_3 core–shell composite nanowires which are synthesized by a combination of hydrothermal and wet chemical routes, its specific capacitance is 295 F g^{-1}, which is much higher than those of the pure individual MoO_3 (69 F g^{-1}) and SnO_2 (96.6 F g^{-1}). Moreover, the core–shell composite also exhibits excellent long-term cycling stability (only 3 % loss of its initial capacitance after 1000 cycles) which may lead to its multifarious usage in high-performance energy storage devices for future potential applications [129].

7.4.1.4 Co_3O_4

Co_3O_4 seems to exhibit excellent reversible redox behavior, large surface area, high conductivity, long-term performance, and good corrosion stability [130]. Therefore, it has been considered as an alternative electrode material for supercapacitors [131]. The pseudocapacitance of Co_3O_4 originates from the following redox reaction [132]:

$$Co_3O_4 + H_2O + OH^- \leftrightarrow 3CoOOH + e^- \tag{7.8}$$

Various methods including solid-state synthesis, wet chemical processes, electrodeposition, radio frequency magnetron sputtering, and microwave methods can be used to prepare tailored Co_3O_4. To achieve good electrochemical performance, the specific surface area is an important factor. That is to say, nanostructured Co_3O_4 are preferred [133]. For example, a Co_3O_4 nanoparticle with 14 nm in the mean grain size was synthesized through simple microwave method. Maximum specific capacitance of 519 F g^{-1} is obtained from charge–discharge studies. After 1000 continuous charge–discharge cycles, only about 1.3 % degradation in specific capacitance could be noticed [134]. The uniform Co_3O_4 nanocubes with a size of 50 nm were prepared by a simple and one-step shape-controlled process. In this process, meso carbon nanorods not only act as a heat receiver to directly obtain Co_3O_4 eliminating the high-temperature post-calcination but also control the morphology of the resulting Co_3O_4 to form nanocubes with uniform distribution

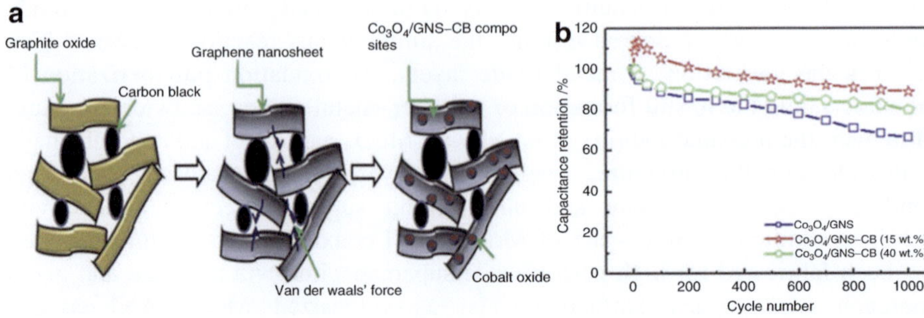

Fig. 7.8 (**a**) Schematic representation of the microstructure diagrams of Co_3O_4/GNS–CB composites and (**b**) cycling performances of Co_3O_4/GNS, Co_3O_4/GNS–CB (15 wt%), and Co_3O_4/GNS–CB (40 wt%) (Modified from Ref. [138] Copyright 2013, with permission from Elsevier)

[135]. Utilizing sawdust as a bio-template, Co_3O_4 nanoparticles with the average diameter of about 40 and 60 nm are prepared [133].

Co_3O_4 can form composites with conductive materials such as CNTs and graphene. For example, composites of cobalt oxides/CNTs are prepared by thermally decomposing cobalt nitrates directly onto the surface of CNTs, which exhibit a specific capacitance of 569 F g^{-1} at 10 mA cm^{-2} [136]. The graphene/Co_3O_4 composites are prepared by using Na_2CO_3 simultaneously as a reducing agent and a depositing agent. The electrochemical properties of Co_3O_4 are significantly improved after deposition on graphene sheets and it exhibits a high specific capacitance of 434 F g^{-1} at 10 A g^{-1} and good cycling behavior [137]. In the case of composites of Co_3O_4/GNSs by adding carbon black (CB) fillers, they are prepared by ultrasonication and microwave-assisted methods. Figure 7.8a represents microstructure diagram of the three types of composites: (1) formation of spacers and carbon black links between layers of graphite oxide, (2) simultaneous reduction of graphite oxide and precursors of cobalt, and (3) crystallization of cobalt oxide on surfaces of GNSs through an oxidation process. Microstructure measurements show that CB could be conductive links among layers of graphene and cobalt oxide particles (5–7 nm in size), which is deposited on the edge surfaces of nanosheets. The optimized Co_3O_4/GNS–CB (15 wt%) electrode exhibits good rate capability, good cycling performance (~89% of initial value for 1000 cycles), and high specific capacitance of 341 F g^{-1} (Fig. 7.8b). Especially, the capacitance retention (89 %) for Co_3O_4/GNS–CB (15 wt%) is greatly enhanced in comparison with that (67 %) for Co_3O_4/GNS. This means that CB fillers improve electrode structural stability and result in good reversibility and cycling stability. The electrochemical properties of the composite are enhanced significantly, because the conductively linked CB could prevent graphene-based composites from aggregation, and improve electron conductivity [138]. GNSs can effectively restrain the volume expansion of Co_3O_4 nanoplates during cycling and improve the electron conductivity of the electrode, whereas the Co_3O_4 nanoplates prevent the restacking of the GNSs. As a result, the composite of Co_3O_4 nanoplates with 7.0 % GNSs shows a very high specific capacitance of 668 F g^{-1} at 1.25 A g^{-1} in 2 M KOH solution in the potential range of −0.1–0.5 V (vs. SCE) [139].

As for composites of Co_3O_4 with other conductive materials, their electrochemical performance is also greatly improved. In terms of Co_3O_4/Ni$(OH)_2$ composite, its mesoporous nanosheet networks (NNs) grown on conductive substrates are synthesized by heat treatment of Co$(OH)_2$/Ni$(OH)_2$ NNs that were prepared on Ti substrates by a facile electrochemical deposition route. The Co_3O_4/Ni$(OH)_2$ mesoporous NNs achieve a high capacitance of 1144 F g^{-1} at 5 mV s^{-1} and long-term cyclability. Its excellent capacitive behaviors is attributed to the following five factors: (i) the porous network structures that can accelerate the diffusion of active species, (ii) the mesopores within nanosheets that can favor proton transfer and provide a large OH^- and cation accessible area, (iii) the interconnections among nanosheets that can facilitate electron transportation, (iv) the combination of cobalt and nickel cations that afford binary redox couples of Co^{2+}/Co^{3+} and Ni^{2+}/Ni^{3+}, and (v) the direct growth on the current collector in a good solid contact that can greatly enhance the conductivity [140]. In addition, composites of Co_3O_4 nanoflower (NF) on Ni foam are prepared via a general two-step synthesis. Through a controllable solvothermal process and hereafter a post-calcination process in air, the NFs are grown firmly on the Ni foam, which is convenient for the construction of supercapacitors without any extra process. The NFs have a high specific capacitance of 1937 F g^{-1} at a current density of 0.2 A g^{-1} and a capacity retention of 78.2 % after 1000 cycles at a high current density of 3 A g^{-1} (1309 F g^{-1}) in 6 M KOH solution [141]. Moreover, porous Co_3O_4 nanowires grown on a nickel fiber were evaluated as the positive electrodes of fiber-based all-solid-state asymmetric supercapacitors with a high capacitance of 2.1 F cm^{-3} at a current density of 20 mA cm^{-3}. The potential window is improved from 0.6 to 1.5 V, and the corresponding stored energy and delivered power are also enhanced at least by 1860 % compared with those of the supercapacitor with a potential window of 0–0.6 V [142].

7.4.1.5 VO_2

Among all the cathode materials of supercapacitors, vanadium oxides have attracted great attention. Vanadium exists with a wide range of oxidation states from +2 to +5. As an important functional metal oxide, vanadium dioxide with a layered structure possesses excellent physical and chemical properties. VO_2 exhibits four different polymorphic structures, including the most stable VO_2(R) with rutile structure, the monoclinic VO_2(R) with a slightly distorted rutile structure, a tetragonal structure of VO_2(A), and the metastable VO_2(B) with a monoclinic structure. Especially, VO_2(B) with metastable monoclinic structure is a promising cathode material for both organic and aqueous supercapacitors owing to its proper electrode potential and tunnel structure [143].

So far, there are a few methods such as hydrothermal method which could be used to prepare VO_2, but many morphologies can be achieved, such as nanorods [144], nanoflowers [145], nanobelts [146], etc. For instance, VO_2 (B) bundles of nanorods were synthesized by a facile one-step quinol-assisted hydrothermal method [147]. Monodispersed vanadium dioxide VO_2 hollow microspheres aggregated by elliptical nanorods were synthesized via a one-step template-free method. The inside-out Ostwald ripening mechanism is responsible for the formation of the hollow microspheres [148]. A hexangular starfruit-like VO_2 was prepared by a one-step hydrothermal treatment of NH_4VO_3, HCOOH, and P123 solutions. The reversible capacitance of VO_2 electrode at the current density of 1 A g^{-1} is 216 F g^{-1}, and the specific capacitance of VO_2 electrode during extended electrochemical cycles maintains about 95 % of its initial value after 500 cycles, signifying a good stability of VO_2 material [149].

In the case of the composites of VO_2 with CNTs, their electrochemical performance is enhanced. For instance, a coaxial-structured hybrid material of vanadium dioxide (VO_2(B)) and multiwalled carbon nanotubes (MWCNTs) is synthesized by a facile sol–gel method assisted with freeze-drying process. A few layers of VO_2(B) sheath are firmly coated on the CNTs surfaces, resulting in the formation of a network morphology with abundant pores and good electron conductivity. A specific capacitance of 250 F g^{-1} is obtained in 1 M Na_2SO_4 solution at a current density of 0.5 A g^{-1}, with a capacitance retention up to 71 % when the current density is increased to 10 A g^{-1}. The excellent electrochemical performance can be ascribed to its unique coaxial structure with the VO_2(B) as the shell and the CNTs as the core. The CNT core offers fast electron transport due to its high electron conductivity, while the VO_2(B) shell provides large faradic pseudocapacitance for the composite [150]. As shown in Fig. 7.9a, a facile one-step strategy, which can be readily scaled up for mass production by using commercial V_2O_5 and graphene oxide as precursors, has been developed to prepare 3D graphene/VO_2 nanobelt composite hydrogels. Figure 7.9b shows typical CV curves of graphene hydrogels, VO_2 nanobelts, and graphene/VO_2 nanobelt hydrogels within the potential range from −0.6 to 0.6 V at a scan rate of 5 mV s^{-1}. For the VO_2 nanobelts and graphene/VO_2 composite hydrogels, CV curves show two pairs of redox peaks, which are associated with the electrochemical K^+ intercalation/deintercalation process. The area under the CV curve is clearly much larger for the graphene/VO_2 composite hydrogels than that for the graphene hydrogels or VO_2 nanobelts at the same scan rate, indicating that the graphene/VO_2 composite hydrogels have higher capacitance than that of the graphene hydrogels or VO_2 nanobelts. Figure 7.9c shows the cycling behavior of graphene hydrogels, VO_2 nanobelts, and graphene/VO_2 composite hydrogels for 5000 cycles. Interestingly, the specific capacitance of graphene/VO_2 composite hydrogels increases with the cycle number, indicating that the additional 1000 cycles are needed to fully activate the sample. After a 5000-cycle test, the graphene/VO_2 composite hydrogels electrode can still keep 92 % of the initial value. This strongly indicates that the integration of vanadium oxides into 3D graphene hydrogels can reduce the chemical dissolution of amphoteric vanadium oxides by providing an electrolyte buffer and serve as a matrix to maintain the vanadium oxide structure [151].

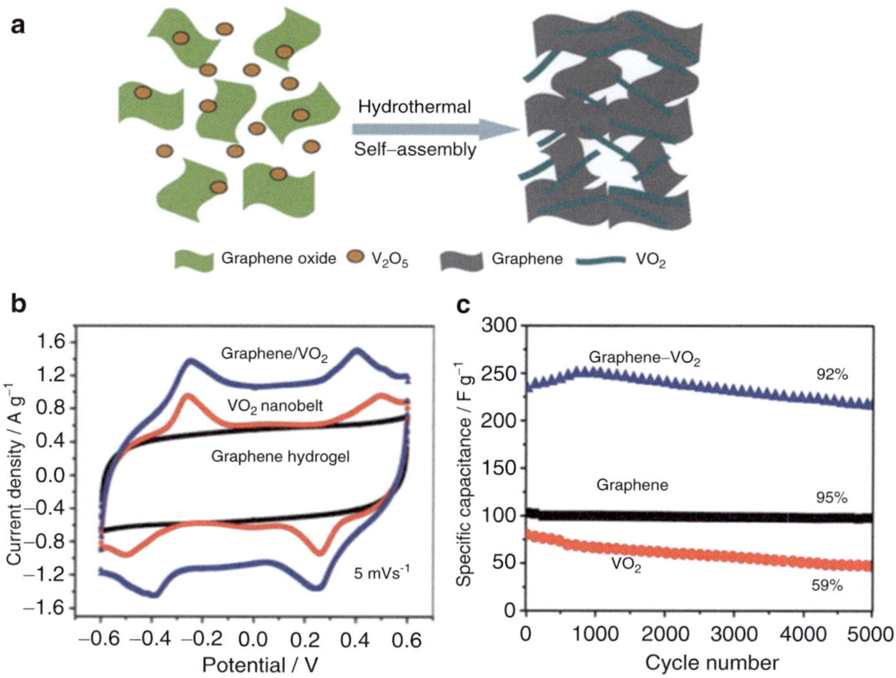

Fig. 7.9 (**a**) The formation process of 3D graphene/VO_2 nanobelt composite hydrogels, (**b**) CV curves of the samples at 5 mV s^{-1}, and (**c**) cycling performance of graphene hydrogels, VO_2 nanobelts, and graphene/VO_2 composite hydrogels at 10 A g^{-1} (Modified from Ref. [151] Copyright 2013, with permission from Elsevier)

7.4.1.6 V_2O_5

Vanadium pentaoxide (V_2O_5) has been intensively studied for energy storage due to its high theoretical capacity and structural reversibility [152, 153]. Its redox potential is also very wide. When it is higher than 0 V (vs. normal hydrogen electrode: NHE), it can act as a cathode material for supercapacitors in aqueous electrolytes [154]. V_2O_5 crystallizes with an orthorhombic layered structure belonging to the P_{mnm} space group with lattice parameters $a = 11.510$ Å, $b = 3.563$ Å, and $c = 4.369$ Å. It is composed of distorted trigonal bipyramidal coordination polyhedra of O atoms around V atoms. In addition, V_2O_5 also exists in an amorphous phase.

The common methods to prepare nano-V_2O_5 include a sol–gel process, which is the predominant and oldest one, and quenching and hydrothermal method [155]. The morphology and crystallinity of V_2O_5 are directly dependent on the as-used methods and conditions. For example, nanoparticles can be prepared by a precipitation process, nanoroll V_2O_5 by a ligand-assisted templating method, nanoporous V_2O_5 by a sol–gel method, and nanobelt V_2O_5 via a hydrothermal

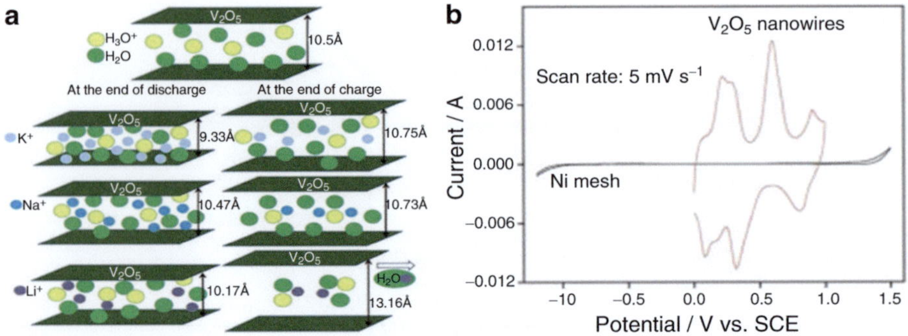

Fig. 7.10 (**a**) Schematic of the different electrochemical behavior and structural changes of $V_2O_5{\cdot}0.6H_2O$ electrodes during charge–discharge in K_2SO_4, Na_2SO_4, and Li_2SO_4 aqueous electrolyte solutions and (**b**) CV curve of V_2O_5 nanowires and Ni mesh (current collector) (Modified from Refs. [154, 156] Copyright 2013 and 2009, with permission from Elsevier)

process [156]. The structure, surface morphology, and diameter of the fibers can be easily controlled by adjusting the variables of the electrospinning method, such as the applied potential, precursor concentration, and flow rate of the solution. It is used to prepare V_2O_5 nanofibers (VNF) [157].

In terms of the virginal V_2O_5 nanostructure, they can also exhibit a high electrochemical performance. To be more specific, V_2O_5 nanoporous network was synthesized via simple capping-agent-assisted precipitation technique, and it is further annealed at different temperatures. The introduction of nanoporous network enhances the ion diffusion to the electrode material and leads to superior specific capacitance of 316 F g^{-1} and an energy density of 43.8 W h kg^{-1} [158]. For the electrospun V_2O_5 nanofibers, they yield a maximum capacitance of 190 F g^{-1} in neutral KCl solution [157]. As shown in Fig. 7.10a, the original interlayer spacing of (001) planes of $V_2O_5{\cdot}0.6H_2O$ is calculated to be 10.5 Å according to Bragg's law with H_2O molecules and some H_3O^+ ions located between these layered structures. Among the three alkaline metal cations (Li^+, Na^+, and K^+), intercalation/deintercalation of K^+ into/from the interlayered space of $V_2O_5{\cdot}0.6H_2O$ is the most facile, which leads to the highest capacity in the K_2SO_4 solution. Intercalation/deintercalation of Li^+ is difficult due to its strong interaction with the interlayer H_2O of $V_2O_5{\cdot}0.6H_2O$ [154]. The $V_2O_5{\cdot}0.6H_2O$ nanobelts show four distinct couples of redox peaks in the potential range of 0–1.0 V (vs. SCE) in 0.5 M K_2SO_4 (Fig. 7.10b). The redox peaks of $V_2O_5{\cdot}0.6H_2O$ electrode in K_2SO_4 can be ascribed to the intercalation/deintercalation of K^+ ions into/from $V_2O_5{\cdot}0.6H_2O$ lattice accompanied with the electrochemical conversion of V^{5+} to different valence states. Its specific capacitance can be up to 181 F g^{-1} at 2 C [156].

Conductive materials are also introduced to modify V_2O_5. For instance, the specific capacitance of the composite of V_2O_5@mesoporous carbon is increased by 61 % in 1 M KNO_3 solution comparing with that of the virginal mesoporous carbon [159]. A composite of V_2O_5@CNTs is achieved by using arc-ion plating under Ar gas flow. The electrochemical performance of V_2O_5@CNTs is significantly improved as compared with those of the bare oxide films and virginal CNTs. The CNTs covered with uniformly dispersed oxides lead to a significantly improved capacitive performance in comparison with the bare oxide films [160]. In addition, flexible all-solid-state thin-film supercapacitors with high electrochemical performance based on two-dimensional $V_2O_5{\cdot}H_2O$/graphene nanocomposites were successfully fabricated. They exhibit high areal capacitance (11,700 μF cm^{-2} at a current density of 0.25 A m^{-2}), long cycle life (>2000 cycles), superior rate capability, small charge-transfer resistance and ultra-flexibility, as well as the high energy density of 1.13 μW h cm^{-2} at a power density of 10.0 μW cm^{-2} [161]. The excellent electrochemical performance can be attributed to the unique composite structure, which results in maximum integration of the merits of each component.

As for the V_2O_5@PPy composite, a new 3D network built from V_2O_5@PPy nanoribbons with thin nature was successfully fabricated via a combined hydrothermal and freeze-drying approach and a subsequent nanocasting process. Remarkably, a high capacitance of 502 F g^{-1} is achieved at a current density of 0.25 A g^{-1} in the case of the 3D V_2O_5@PPy network. This value is much higher than those of the pure 3D V_2O_5 network (335 F g^{-1}) and V_2O_5 particle@PPy (84 F g^{-1}). Moreover, the 3D V_2O_5@PPy network exhibits a good capacitance retention (~74%) compared with the pure 3D V_2O_5 network sample (only 52 %) after 10,000 cycles at a current density of 2.5 A g^{-1}. These excellent electrochemical performances for supercapacitor can be attributed to a unique 3D V_2O_5@PPy network, which not only can provide the high surface area for enhancement of the electrolyte/electrode interaction but also can reduce the diffusion length of ions. Moreover, the PPy coating can efficiently maintain the high electrical conductivity of the overall electrode [162].

The capacitance of the V_2O_5@TiO_2 nanotubes can reach to 220 F g^{-1} which is stable during cycling. The supercapacitive behavior can be ascribed to the V^{4+}/V^{5+} redox switching of the V_2O_5 phase embedded in the TiO_2 matrix. The good cycling stability of the mixed V_2O_5–TiO_2 nanotube arrays is due to the highly ordered 3D TiO_2-stabilized V_2O_5 nanotube structure which restrains the strain during ion intercalation [163]. However, the cycling performance of V_2O_5 still needs improvement due to the dissolution of V into the aqueous electrolytes leading to capacity fading [164], which is an urgent problem to be solved.

7.4.1.7 ZnO

Zinc oxide (ZnO) possesses several excellent physical properties, such as wide direct band gap (3.37 eV), large free exciton binding energy (60 mV), and wide-range resistivity (10^{-4}–10^{12} Ω cm), and has been applied in electronic and optoelectronic devices. However, available information on the suitability of the ZnO for supercapacitors application is very little [165]. Recently, various methods like ultrasonic spray pyrolysis (USP), microwave-assisted synthesis, and mechanical mixing were employed for the preparation of ZnO electrode for achieving high-performance supercapacitor [166]. Moreover, 1D, 2D, or 3D self-assembled ZnO such as nanobelts [167], nanotubes [168], aligned nanonails [169], and oriented ZnO nanorods [170] have been prepared.

In order to improve electrochemical performance, carbonaceous materials are also introduced to decorate ZnO. More specifically, the maximum capacitance of the ZnO/carbon aerogel supercapacitor was 500 F g^{-1} at 100 mA cm^{-2} [171]. The ZnO/C composite is synthesized by self-propagating solution combustion method using dextrose as fuel and carbon source. In dilute 0.1 M KNO_3 solution, ZnO/C shows a specific capacitance of 21.7 F g^{-1} at a scan rate of 50 mV s^{-1}, and a further increase in concentration of electrolyte solution leads to a decrease in the electrochemical capacitance [172]. The ZnO nanodots are deposited onto CNT films by ultrasonic spray pyrolysis with different times. The results of electrochemical measurements show a very high capacitance of 324 F g^{-1} [173]. A ZnO/CNT nanocomposite is prepared by using reactive magnetron sputtering in Ar/O_2 environment. This approach leads to more uniform coating with tuneable thickness, which alters the electrochemical performance of the nanocomposite electrodes [166].

A carboxylated graphene–ZnO (G-COOZn) composite is prepared by growing ZnO nanoparticles onto graphene sheets by a one-step hydrothermal method using carboxylated graphene (G-COOH) and $Zn(NO_3)_2$ as precursors. The composite achieves a specific capacitance as high as 238 F g^{-1} at 50 mA g^{-1} in the potential range of 0–1.0 V [174]. A hybrid structure of zinc oxide (ZnO) on 3D graphene foam has been synthesized by chemical vapor deposition (CVD) growth of graphene followed by a facile in situ precipitation of ZnO nanorods under hydrothermal condition. The graphene–ZnO hybrids display superior capacitive performance with high specific capacitance (~400 F g^{-1}) as well as excellent cycle life [175].

In the case of ZnO compound with other metal oxides, Al-doped ZnO (AZO)@MnO_2 hybrid electrodes in core–shell geometries have been synthesized on stainless steel substrates by a scalable low-cost solution route. The specific capacitance based on total electrode mass is estimated to be 275 F g^{-1} at a current density of 2 A g^{-1}. The AZO@MnO_2 electrode also displays an excellent long-term cycling stability at a current density of 10 A g^{-1} with less than 3 % loss after 2000 charge–discharge cycles. A ~35 % capacity retention is obtained at 60 A g^{-1}, which is over 400 % specific capacitance increment compared with the ZnO@MnO_2 electrode [176]. MoO_3@ZnO core–shell nanocables have been synthesized in large quantities by a simple electrochemical method at room temperature, and the MoO_3 shell thickness can be controlled by changing the deposition time. The core–shell nanocables give a specific capacitance of 236 F g^{-1} at the scan rate of 5 mV s^{-1}, much larger than that of MoO_3 nanoparticles [177].

As shown in Fig. 7.11a, aligned and ordered ZnO nanowires (NWs) are grown on carbon cloth using a wet chemical method. Hydrogenation of the ZnO NWs in a commercial tube furnace improves the electrical properties of ZnO. The uniform hydrogenated ZnO NWs are then soaked in a glucose aqueous solution and subsequently annealed in Ar gas. An amorphous carbon is formed and serves as a

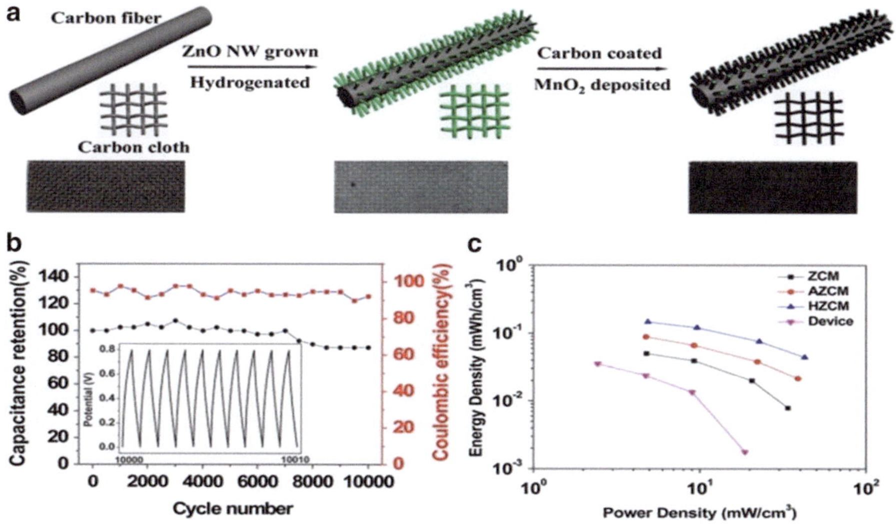

Fig. 7.11 (**a**) Schematic illustrating the synthesis procedure of HZM core–shell nanostructure on carbon cloth and the corresponding sample photos, (**b**) cycling performance and Coulombic efficiency of the HZM supercapacitor device measured at a current density of 0.5 mA cm^{-2}, the inset in (**b**) is the galvanostatic charge–discharge curve for 10,000 cycles, and (**c**) Ragone plots of the ZM, AZM, HZM electrodes, and HZM-based supercapacitor device (Modified from Ref. [178] Copyright 2013, with permission from ACS)

reducing agent for the deposition of MnO_2. Hydrogenated single-crystal ZnO@amorphous ZnO-doped MnO_2 core–shell nanocables (HZM) on carbon cloth show excellent performance such as areal capacitance of 139 mF cm^{-2} and specific capacitance of 1261 F g^{-1}. The working devices achieve very high areal capacitance of 26 mF cm^{-2} and retain 87.5 % of the original capacitance even after 10,000 charge–discharge cycles (Fig. 7.11b). A power density of 2.44 mW cm^{-3} and an energy density of 0.04 mWh cm^{-3} are achieved. In particular, the energy density of this supercapacitor is higher than that of the other solid-state supercapacitors (shown in Fig. 7.11c) such as CNT-based supercapacitors (0.008 mWh cm^{-3}) and MnO_2-coated ZnO NWs (0.005 mWh cm^{-3}) [178].

In addition to the above-mentioned conditions, an efficient method for the fabrication of novel, flexible electrodes based on ZnO nanoflakes and nickel–cobalt layered double hydroxides (denoted as ZnONF/NiCoLDH) as a core–shell nanostructure on textile substrates for wearable energy storage devices, is reported. As an electrode for supercapacitors, ZnONF/NiCoLDH exhibits a high specific capacitance of 1624 F g^{-1}, which is nearly 1.6 times greater than NiCoLDH-coated ZnO nanowire (denoted as ZnO NW/NiCoLDH) counterparts. It also shows a maximum energy density of 48.3 Wh kg^{-1} at a power density of 27.5 kW kg^{-1} and an excellent cycling stability with capacitance retention of 94 % and a Coulombic efficiency of 93 % over 2000 cycles. The superior performance of the ZnONF/NiCoLDH hybrids is primarily due to the large surface area of the nanoflake structure and the open spaces between nanoflakes. Both of them provide a large space for the deposition of NiCoLDH, resulting in reduced internal resistance and improved capacitance performance [179].

7.4.1.8 NiO

Nickel oxide (NiO), known to be cheap and environment friendly, is a promising cathode material for supercapacitors. The theoretical capacitances of NiO in a potential window of 0.5 V is up to 2583 F g^{-1} [180]. To date, there have been a variety of reports on the synthesis of different NiO nanostructures including nanospheres, nanoflowers, nanosheets, nanofibers, nanowires, and nanofilms with open pores. Different morphologies and nanostructures show substantial differences in electrochemical performance due to differences in the electrode/electrolyte interface properties and ion transfer rates during the charge storage processes. In addition, the supercapacitive performance of NiO is dependent on heating, intercalating anions, and the ionic nature of the surfactant during the synthesis process [181]. From this point of view, a variety of porous nanostructures with large surface areas have been synthesized via surfactant-template, sol–gel, anodization, and hard-template methods. For instance, porous NiO samples are prepared by conventional refluxing with microwave-assisted heating method under homogeneous precipitation conditions [182]. Porous NiO nanoslices, nanoplates, and nanocolumns with different specific surface areas have been prepared by thermal decomposition of β-$Ni(OH)_2$ nanostructures [183].

In terms of pure NiO nanostructures, they also show an excellent electrochemical performance. For example, NiO nanoflake can be prepared at different calcination temperatures by using cetyltrimethylammonium bromide (CTAB) as surfactant via microwave method. The NiO nanoflakes exhibit maximum specific capacitance of 401 F g^{-1} at a current density of 0.5 mA cm^{-2} [184]. - One-dimensional hierarchical hollow nanostructures composed of NiO nanosheets are successfully synthesized through a facile carbon nanofiber directed solution method followed by a simple thermal annealing treatment. A high capacitance of 642 F g^{-1} is achieved when the charge–discharge current density is 3 A g^{-1} and the total capacitance loss is only 5.6 % after 1000 cycles [185]. Short NiO nanotubes with mesopores in sidewalls deliver a specific capacitance of 903 F g^{-1} at 1 A g^{-1}, much larger than those of long NiO nanotubes and NiO nanoparticles. A capacitance degradation of ca. 6 % after 2000 charge–discharge cycles at 4 A g^{-1} demonstrates their desirable electrochemical stability [186]. Mesoporous NiO with nickel oxide nanoparticles of 3–6 nm crystalline walls and 4–5 nm pore diameter is synthesized by a simple and scalable strategy. The product has a moderately high surface area of 294 m^2 g^{-1} and a high specific capacitance of 2418 F g^{-1} at a scan rate of 5 mV s^{-1} [187].

In the case of the composite of NiO with carbon and graphene, their electrochemical performance is improved. To be more precise, nanowhisker-like NiO/carbon aerogel microbead (NiO/CAMB) composites with chestnut-like core–shell structure are prepared by in situ encapsulating method. The specific capacitance of the NiO/CAMB composite is up to 356 F g^{-1}, which is the combination of electrical double-layer capacitance of CAMB and pseudocapacitance based on the redox reaction of NiO [188]. The electrochemical performance of NiO/C nanocomposites is dependent on the calcination temperature by using starch as a stabilizing agent. The sample calcified at 500 °C exhibits a highly porous nature and a specific capacitance of 644 F g^{-1} at a scan rate of 2 mV s^{-1} [189]. As for the NiO/graphene nanocomposites, the nickel foam-graphene/NiO (NF-G/NiO) composite shows excellent properties as a pseudocapacitive device with a high specific capacitance value of 783 F g^{-1} at a scan rate of 2 mV s^{-1}. This is attributed mainly to the 3D nature of the NF which allows for 3D graphene growth, the incorporated NiO, and the conductivity of the graphene [190]. The graphene sheet/porous NiO hybrid film exhibits excellent pseudocapacitive behavior with pseudocapacitances of 400 and 324 F g^{-1} at 2 and 40 A g^{-1}, respectively, which are higher than those of the porous NiO film (279 and 188 Fg^{-1} at 2 and 40 A g^{-1}). The enhancement of the pseudocapacitive properties is due to reinforcement of the electrochemical activity of the graphene sheet film [191]. In addition, a pulsed laser deposition process using ozone as an oxidant is developed to grow NiO nanoparticles on highly conductive 3D graphene foam (GF). The excellent electron conductivity and the interconnected pore structure of the NiO/GF hybrid facilitate fast electron and ion transportation. The NiO/GF electrode displays a high specific capacitance (1225 F g^{-1} at 2 A g^{-1}) and a superb rate capability (68 % capacity retention at 100 A g^{-1}) [192].

To further enhance its electrochemical performance, nickel foam can be used. For example, the resulting NiO/Ni composite could achieve a high specific capacitance of 760 F g^{-1} at 20 A g^{-1}, significantly higher than the 480 F g^{-1} for NiO. Additionally, the specific capacitance of NiO/Ni composite remains as high as 816 F g^{-1} after 1000 cycles at 4 A g^{-1}, revealing superb electrochemical characteristics [193]. NiO nanorod arrays on Ni foam have a combination of ultrahigh specific capacitance (2018 F g^{-1} at 2.27 A g^{-1}), high power density (1536 F g^{-1} at 22.7 A g^{-1}), and good cycling stability (only 8 % of capacitance loss in the first 100 cycles and no further change in the subsequent 400 cycles). This leads to an improvement in the reversible capacitance record for NiO by 50 % or more, reaching 80 % of the theoretical value, and demonstrates that a 3D regular porous array structure can afford all of these virtues in a supercapacitor. The excellent performance can be attributed to the slim (<20 nm) rod morphology, high crystallinity, regularly aligned array structure, and strong bonding of the nanorods to the metallic Ni substrate. More specifically, the slim rod morphology reduces the diffusion distance, through which the electrolyte has to penetrate, to less than 10 nm, and thus maximizes the active surface area for insertion and extrusion of OH^-, as schematically shown in Fig. 7.12a, and consequently maximizes the pseudocapacitance (Fig. 7.12b—Zone 1). At the same time, it also maximizes the surface area exposed to the bulk solution, which enlarges the electric double-layer capacitance, as shown in the discharge curves (Fig. 7.12b—Zone 2). In addition, the 3D porous space between the slim NiO nanorods facilitates the diffusion of electrolytes, and the good crystallinity of the nanorods ensures a good cycling stability and

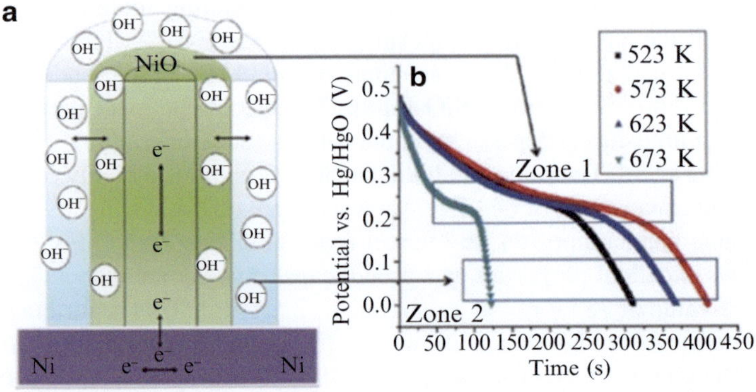

Fig. 7.12 (**a**) A schematic image showing how the NiO nanorod generates electric double-layer capacitance and pseudocapacitance and (**b**) the discharge curves with a current density of 5 mA cm $^{-2}$ of the NiO/Ni films calcified at various temperatures (Modified from Ref. [194] Copyright 2011, with permission from Springer)

electron conductivity. The strong chemical bonding of the rods to the substrate not only ensures good mechanical stability but also facilitates the transport of electrons from the nanorods to the metallic substrate, the current collector [194]. In addition to the above mentioned, a facile mixed solvothermal method has been developed to prepare porous NiO nanoflake films on an ITO (indium tin oxide) glass substrate. The porous $Ni(OH)_2$ intermediate, which is first obtained from the solvothermal reaction, is converted into highly crystalline NiO without noticeable changes in morphology via annealing in air. Owing to a well-defined porous network structure resulting from intercrossing ultrathin porous nanoflakes, such NiO-based cathode manifests exceptional performance with a high pseudocapacitance of 960 F g^{-1} and Coulombic efficiency of 100 % at very high current density of 20 A g^{-1} [195].

7.4.1.9 PbO_2

Lead dioxide (PbO_2) has been widely used as a positive electrode material of lead–acid batteries. Supercapacitor of AC//PbO_2 in sulfuric acid solution is considered more economically viable to replace the negative electrode of Pb in a lead–acid battery. It presents an improved performance in terms of power and lifespan over the lead–acid battery [196].

As a surface reaction, the redox activity of the PbO_2 relates to the morphology and structure (α-PbO_2 or β-PbO_2), and its electrochemically active surface area should be increased as the cathode of a supercapacitor. It has been recognized that porous PbO_2 presents higher electrochemical activities. As a result, thin-film, porous, and nanostructured PbO_2 have been prepared, and different substrates have been investigated [197, 198]. For example, highly ordered macroporous films of both α-PbO_2 and β-PbO_2 are prepared by electrodeposition through

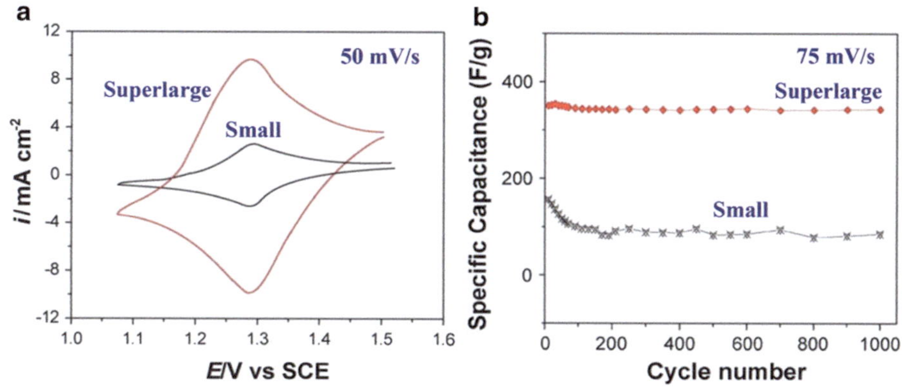

Fig. 7.13 (**a**) Typical CV curves and (**b**) cycling performance of super-large PbO_2 dendrites composed of trigonal nanoplates and the small dendrites in 5.0 M H_2SO_4 solution recorded a scan rate of 75 mV s^{-1} (Modified from Ref. [200] Copyright 2012, with permission from RSC)

templates assembled from submicron diameter PS (polystyrene) spheres assembled on either gold or ITO substrates [199]. Super-large dendrites (about 500 mm) composed of trigonal PbO_2 nanoplates are synthesized by electrodeposition [200]. These prepared super-large PbO_2 dendrites composed of orderly arranged trigonal nanoplates.

The CV curve of PbO_2 exhibits one couple of well-defined redox peaks. The potential difference between the oxidation and reduction peaks is usually very large due to the poor kinetics of the PbO_2/Pb^{2+} couple. The thin-film PbO_2 on Ti/SnO_2 foil can deliver a discharge specific capacitance of 79.9 F g^{-1} at a current of 0.75 mA cm^{-2} and 74.1 F g^{-1} at 10 mA cm^{-2}, based on the total weight of both active electrode materials, and the capacitance retains 83 % of its initial value after 3000 deep cycles at the 4 C rate [197]. The electrochemical activity of the resulting macroporous β-PbO_2 is greater than that of the corresponding plain film [199]. The super-large PbO_2 dendrites composed of orderly arranged trigonal nanoplates can remarkably promote the electrochemical performances of the prepared supercapacitors. As shown in Fig. 7.13, the current response and capacitance of the super-large PbO_2 dendrites are much larger than that of the small dendrites in 5.0 M H_2SO_4 solution [200].

In the case of the C//PbO_2 nanowire supercapacitor, its specific capacitance is 34 F g^{-1}, and the maximum energy density is 29 Wh kg^{-1} at a current density of 10 mA cm^{-2}. These values are seven times larger than those measured on a C// PbO_2 thin-film supercapacitor. There is no sign of degradation over 5000 cycles [201]. When PbO_2 thin films are electrodeposited on a graphite substrate from a Pb $(NO_3)_2$ solution, and the assembled AC//PbO_2 supercapacitor can operate from 1.88 to 0.65 V in an H_2SO_4 solution and delivers a specific capacitance of 63.2 F g^{-1} based on the mass of two active materials at a discharge current of 150 mA g^{-1} (5 C rate) when the mass ratio of AC to PbO_2 is set at 2:1. The energy density reaches 27 Wh kg^{-1} at 152 W kg^{-1} and 18 Wh kg^{-1} at 691 W kg^{-1}. The use of a graphite current collector also ensures stability of the electrode in the H_2SO_4 environment. After 3000 deep cycles under 10 C, the capacity only fades 20 % from its initial value [202].

In terms of composites of PbO_2 with nano-Mn_3O_4, they are prepared by anode composite deposition method from the Pb^{2+} plating solution containing suspended nano-Mn_3O_4 particles from one-step homogeneous precipitation at low temperature. The optimal deposition potential for the nano-Mn_3O_4/PbO_2 composite is 1.4 V. It is composed of tetragonal γ-Mn_3O_4 and β-PbO_2. The composite shows high specific capacitance up to ~340 F g^{-1}. This could be attributed to the intimate electronic interaction between PbO_2 and nano-Mn_3O_4 and the porous quasi-3D structure of the composite [203].

7.4.1.10 SnO_2

Tin oxide (SnO_2) is a typical wide band gap *n*-type semiconductor with $E_g = 3.6$ eV at room temperature and has applications in solar cells, catalysis, and gas sensing and supercapacitors [204–206]. In recent years, research on SnO_2 is mostly focused on its use in supercapacitor with particular attention on SnO_2/graphene or SnO_2/carbon nanotube (CNT) composites [207].

Evidently, the nanocrystalline SnO_2 thin films as cathodes for supercapacitors also exhibit satisfactory electrochemical performance. They can be deposited by various chemical deposition methods such as chemical vapor deposition [208], sol–gel [209], spray pyrolysis [210], and electrodeposition [211]. More precisely, self-organized nanoporous SnO_2 films are fabricated by anodizing a tin substrate in an aqueous electrolyte containing oxalic or phosphoric acid. Compared to the irregular array of nanoporous structure, the film with more open and interconnected nanoporous architecture demonstrates better capacitive performance with the maximum specific capacitance of 274 F g^{-1} and nearly 85 % capacity retention even after 10,000 charge–discharge cycles [212]. Nanocrystalline SnO_2 thin films with the average size of 5–10 nm are deposited by simple and inexpensive chemical route. The SnO_2 shows a maximum specific capacitance of 66 F g^{-1} in 0.5 M Na_2SO_4 electrolyte at the scan rate of 10 mV s^{-1} [213].

In the case of the composites of SnO_2 with carbonaceous and conductive polymers, for example, carbon aerogels are doped with pseudocapacitive tin oxide either by impregnating tin oxide sol into resorcinol–formaldehyde (RF) wet gels (Method I) or by impregnating tin tetrachloride solution into carbon aerogel electrodes (Method II). The tin oxide/carbon aerogel composites synthesized by both methods have similar specific capacitances (66–70 Fg^{-1}). The composites synthesized via Method II show better cycling stability compared with those via Method I. The highest capacitances of the composite are 69.8 and 119.2 F g^{-1}, respectively, which are measured according to the galvanostatic charge–discharge method and cyclic voltammetry. These high specific capacitances are attributed to the surface faradic redox reactions of SnO_2 additives, in addition to the original EDLC of carbon aerogels [214]. A facile solvent-based synthesis route via the oxidation–reduction reaction between graphene oxide (GO) and $SnCl_2{\cdot}2H_2O$ has been developed to prepare SnO_2/graphene (SnO_2/G) nanocomposites. The electrochemical performance of the nanocomposite shows an excellent specific capacitance of 363 F g^{-1}, which is fivefold higher than that of the as-synthesized graphene (68.4 F g^{-1}). The improvement is from the synergistic effects of the excellent conductivity of graphene and the SnO_2 nanoparticles with an average particle size of 10 nm [215]. A pulse microwave-assisted deposition method is adopted to grow SnO_2 crystals, which are homogeneously inserted into the rGO sheets, forming SnO_2/rGO composites. The specific capacitance of the supercapacitors fabricated with the SnO_2/rGO composites can reach 348 F g^{-1} at a current density of 50 mA g^{-1}, showing an increase of 98 % in comparison with that of a fresh rGO electrode. This enhanced capacitance is attributed to the SnO_2 crystals that not only serve as spacers to create more active sites on the rGO surface but also maximize the fraction of hydrophilic surface, which is more accessible for the formation of the electric double layer [216].

Polyaniline (PANI)/tin oxide (SnO_2) fibrous nanocomposites are successfully prepared by an in situ chemical polymerization method. The composite exhibits a specific capacitance of 173 F g^{-1} at a scan rate 25 mV s^{-1} in an H_2SO_4 solution, which is a threefold increase compared to the pure SnO_2 [206]. A nanocomposite of SnO_2@polyaniline can be prepared as the follows: SnO_2 is produced through ultrasonication in the presence of ethanolamine (ETA), while polyaniline (PANI) is in situ polymerized. The nanocomposite of SnO_2@PANI (SP-2) has a high specific surface area (91.63 m^2 g^{-1}) and exhibits remarkable pseudocapacitive activity including a high capacitance (335.5 F g^{-1} at 0.1 A g^{-1}), good rate capability (108.8 F g^{-1} at 40 A g^{-1}), and excellent cycling stability (no capacitance loss after 10,000 cycles) [217].

In terms of ternary nanocomposites (GSPs) based on graphene, tin oxide (SnO_2), and polypyrrole (PPy), they are prepared via one-pot synthesis. The CV curves of the GSP composite in 1 M H_2SO_4 at various scan rates are shown in Fig. 7.14a.

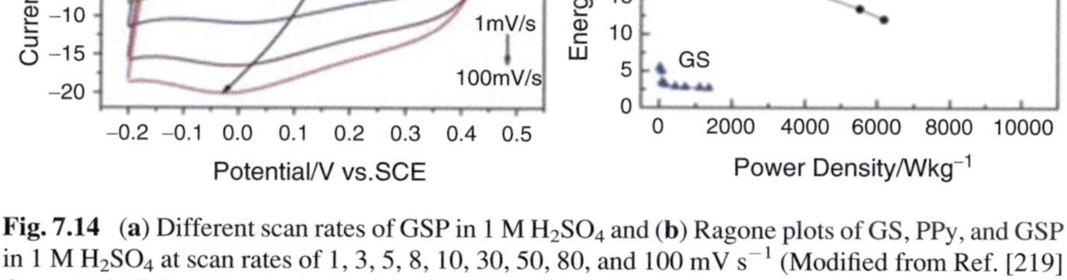

Fig. 7.14 (**a**) Different scan rates of GSP in 1 M H_2SO_4 and (**b**) Ragone plots of GS, PPy, and GSP in 1 M H_2SO_4 at scan rates of 1, 3, 5, 8, 10, 30, 50, 80, and 100 mV s^{-1} (Modified from Ref. [219] Copyright 2012, with permission from RSC)

Notably, all curves show a similar rectangular shape with redox peaks. The redox peaks shift with the increment of scan rates, indicative of EDL capacitive behavior and good ion response [218]. The high currents of the CV curves mean that the electrode materials show high conductivity and low internal resistance. Moreover, the current increases with the scan rate, indicating that there is a good rate capability for GSP in 1 M H_2SO_4, which may be due to efficient combination of the three kinds of electrode materials and the synergistic effect between components in the nanostructured GSP composite [219], which consists of thin conducting film of PPy on the surface of graphene/SnO_2 (GS). An enhanced capacitance of 616 F g^{-1} is obtained at 1 mV s^{-1} in 1 M H_2SO_4 solution compared with 80.2 F g^{-1} for GS and 523 F g^{-1} for PPy. The GSP electrode shows better cycling stability and no obvious decay after 1000 cycles at 1 A g^{-1}. Ragone plots of GS, PPy, and GSP are presented in Fig. 7.14b. Compared with GS and PPy, the ternary composite GSP is more suitable as supercapacitor since it has higher energy and power densities. Its energy density is estimated to be 19.4 Wh kg^{-1} at 9970 W kg^{-1} and about 31.2 Wh kg^{-1} at 818 W kg^{-1}. In contrast, the energy density of GS is only 2.7 Wh kg^{-1} at 1366 W kg^{-1} [219].

7.4.2 *Intercalation Compounds*

Intercalation compounds have attracted a lot of interest thanks to their large specific capacitance. They have been typically used in organic electrolytes but aqueous examples are more interesting.

7.4.2.1 $LiCoO_2$

$LiCoO_2$ is a hexagonal layered structure belonging to *P*nm space group with $a = 0.2816$ nm and $c = 1.4056$ nm. It is widely applied in lithium-ion batteries due to its stable structure during the charging and discharging process. Due to the good reversibility of intercalation/deintercalation of Li^+ ions, which is shown in Eq. (7.9), it could be used as cathode for aqueous supercapacitors [220].

$$LiCoO_2 \leftrightarrow Li_{1-x}CoO_2 + xLi^+ + xe^- \quad (x \leq 0.5) \tag{7.9}$$

Its traditional preparation method is via solid-state reaction. In order to get better electrochemical performance, nanostructured $LiCoO_2$ has been prepared by sol–gel methods [155, 221]. CV curves (Fig. 7.15) of $LiCoO_2$ in a saturated aqueous Li_2SO_4 solution show that the intercalation and deintercalation of lithium ions are similar to those in organic electrolytes [222]. There are also three couples of redox peaks for $LiCoO_2$ in the aqueous solution, located at 0.87/0.71, 0.95/0.90, and 1.06/1.01 V (vs. SCE), which agree well with those in the organic electrolyte solution at 4.08/3.83, 4.13/4.03, and 4.21/4.14 V (vs. Li^+/Li). The evolution of oxygen (about 1.7 V vs. SCE) in the aqueous electrolyte occurs at much higher potential, indicating the

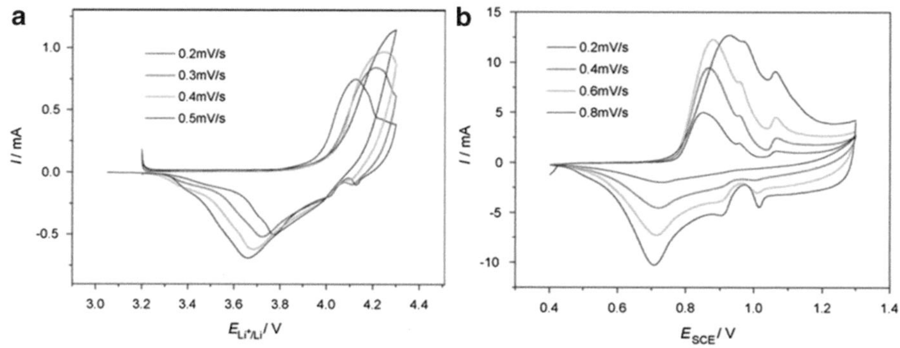

Fig. 7.15 Cyclic voltammograms of the $LiCoO_2$ electrode at different scan rates (**a**) in the organic electrolyte and (**b**) in the saturated Li_2SO_4 solution (Modified from Ref. [222] Copyright 2009, with permission from Elsevier)

good electrochemical stability of $LiCoO_2$ as the cathode for the supercapacitors. The diffusion coefficient of the lithium ions is of the same magnitude as that in the organic electrolyte. However, the current response and the reversibility of redox behavior in the aqueous solutions are better than those in the organic electrolytes due to their higher ionic conductivity [223, 224].

The supercapacitor of AC//nano-$LiCoO_2$ in 0.5 M Li_2SO_4 solution at 1 A g^{-1} between 0 and 1.8 V shows high capacitance and efficiency. The efficiency of this system increases to nearly 100 % after the initial cycle, and the capacity does not change much after 40 cycles [225]. Due to the high cost of cobalt, there is not much work on its redox behavior in aqueous solutions.

7.4.2.2 $LiMn_2O_4$

$LiMn_2O_4$, whose reaction mechanism is shown in Eq. (7.10), is found to be a good cathode for aqueous supercapacitors due to its low cost compared to $LiCoO_2$ [226]. It mainly exists in a spinel structure. Manganese cations occupy half of the octahedral interstitial sites and Li^+ ions occupy one eighth of tetrahedral sites. The Mn_2O_4 framework provides 3D interstitial space for Li^+ ion transport, maintaining its structure over the compositional range Li_xMnO_4 ($0 < x < 1$) by changing the average Mn oxidation state between 3.5 and 4.0 [227]. As early as 1994, it was reported for the first time that $LiMn_2O_4$ can deintercalate and intercalate Li^+ ions in aqueous electrolytes [228].

$$LiMn_2O_4 \leftrightarrow Li_{1-x}Mn_2O_4 + xLi^+ + xe^- \ (x \leq 1) \tag{7.10}$$

The $LiMn_2O_4$ spinel can be prepared by solid-state reaction, sol–gel method, and hydrothermal methods [155]. Among them, 1D nanostructures including nanowires, nanotubes, and nanorods have attracted special attention [229, 230]. To improve the

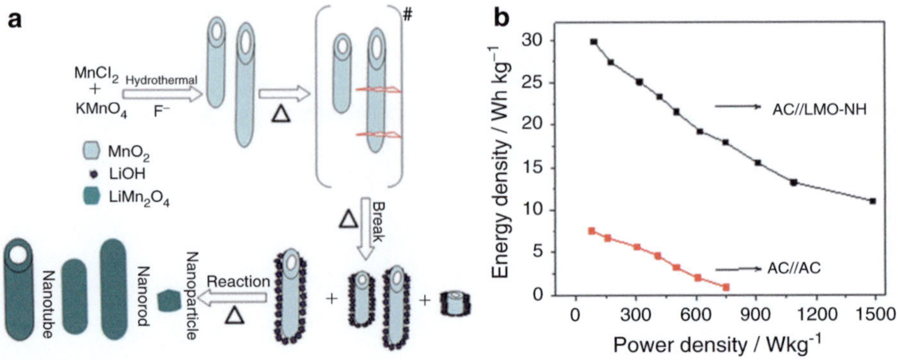

Fig. 7.16 (**a**) The formation process of $LiMn_2O_4$ nanohybrid and (**b**) Ragone plots of the supercapacitor AC//LMO-NH and AC//AC (Modified from Ref. [234] Copyright 2014, with permission from Elsevier)

structural stability of $LiMn_2O_4$ from solid-state reaction, it can also be doped by heteroatoms such as Al, Cr, Cr–Fe, and Ni like in organic electrolytes [231–233]. As shown in Fig. 7.16a, $LiMn_2O_4$ nanohybrid (LMO-NH) consisting of nanotubes, nanorods, and nanoparticles has been synthesized using α-MnO_2 nanotubes from hydrothermal reaction as a precursor. This $LiMn_2O_4$ nanohybrid exhibits a high specific discharge capacitance of 415 F g^{-1} at 0.5 A g^{-1} in 0.5 M Li_2SO_4 aqueous solution. Even at 10 A g^{-1}, it still has a specific discharge capacitance of 208 F g^{-1}. The Ragone plots of the asymmetric supercapacitor based on AC/0.5 M Li_2SO_4/LMO-NH and the symmetric supercapacitor based on AC/0.5 M Li_2SO_4/AC are shown in Fig. 7.16b. The asymmetric supercapacitor presents an energy density of 29.8 Wh kg^{-1} at power density of 90 W kg^{-1}, much higher than that of the symmetric AC//AC capacitor. Moreover, it keeps an energy density of 11.2 Wh kg $^{-1}$ at 1500 W kg^{-1}. Moreover, the capacity retention is 91 % after 1000 cycles [234].

In the case of porous or nanostructured materials as electrode materials for supercapacitors, they have shown some advantages such as higher capacities and better rate capability over traditional micrometer-sized materials [31, 235]. For example, $LiMn_2O_4$ nanochains exhibit a high reversible capacity of 110 mAh g^{-1} at 4.5 C and 95 mAh g^{-1} even at 91 C in 0.5 M Li_2SO_4 aqueous electrolyte. When charged at 136 C, 84 % capacity could be obtained [235]. As for the nanostructured $LiMn_2O_4$, when they are assembled into supercapacitors, power density and cycling behavior are greatly improved. For example, the supercapacitor of AC//$LiMn_2O_4$ nanorod presents very high power density in 0.5 M Li_2SO_4, up to 14.5 kW kg^{-1}, and there is still 94 % capacity retention after 1200 full cycles [229]. The supercapacitor of AC//porous $LiMn_2O_4$ in 0.5 M Li_2SO_4 aqueous solution shows excellent cycling performance at the rate of 9 C (1000 mA g^{-1}), with no more than 7 % capacity loss after 10,000 full cycles [230]. The cycling behavior of AC//porous $LiMn_2O_4$ is the best among the reported $LiMn_2O_4$. The nanograins in the porous $LiMn_2O_4$ could restrain the Mn^{3+} on the surface from dissolution, which can also accommodate effectively the strain caused by Jahn–Teller distortion during the charge and discharge process and favor the morphological and structural stability [230]. Nanoporous $LiMn_2O_4$ spinel with a pore size of about 40–50 nm exhibits a high specific capacitance of 189 F g^{-1} at 0.3 A g^{-1}. Even at 12 A g^{-1}, it still has a capacitance of 166 F g^{-1}. After 1500 cycles, there is no evident capacitance fading [236].

Doping the $LiMn_2O_4$ with several cations is also a good way to effectively reduce the Jahn–Teller distortion and further improve the cycling stability [237]. In the CV curve of $LiCr_{0.15}Mn_{1.85}O_4$ in the saturated aqueous $LiNO_3$ (9 M) solution, faster "CV response" of $LiCr_{0.15}Mn_{1.85}O_4$ is in correlation with higher capacitance retention in comparison to the undoped $LiMn_2O_4$ [231]. The Ni-doped $LiMn_2O_4$ presents much better rate capability and cycling behavior [232].

7.4.2.3 $Li[Ni_xCo_yMn_{1-x-y}]O_2$

$Li[Ni_xCo_yMn_{1-x-y}]O_2$ (NCM: $0 < x, y < 1$) has a rhombohedral structure belonging to the *R*3m space group of a hexagonal α-$NaFeO_2$ structure. The lattice is formed by oxygen atoms in ABC stacking with alternating layers containing mixtures of nickel (+2), cobalt (+3), and manganese (+4) atoms. Its reaction mechanism as a cathode for supercapacitors is as shown in Eq. (7.11). During the deintercalation of Li^+ ions, the valence of Ni is changed from +2 to +3 and that of Co from +3 to +4. In the meanwhile, Mn stays at +4. NCM is traditionally prepared by heat-treating the coprecipitated spherical metal hydroxides with lithium salt or hydroxide [238].

$$\begin{aligned}&Li[Ni(II)]_x[Co(III)]_y[Mn(IV)]_{1-x-y}O_2 \leftrightarrow Li_{1-a}Li[Ni(III)]_x[Co(IV)]_y\\&[Mn(IV)]_{1-x-y}O_2 + aLi^+ + ae^-\end{aligned} \tag{7.11}$$

The electrochemical stability of NCM in a Li^+-containing aqueous solution is critically dependent on the pH value [239]. As shown in Fig. 7.17a, one couple of its redox peaks in 2 M Li_2SO_4 solution is situated at 0.48 and 0.68 V (vs. SCE), corresponding to the intercalation and deintercalation of lithium ions in NCM, respectively. The oxygen evolution potential shifts to 1.29 V (vs. SCE), which is much lower than those for $LiCoO_2$ and $LiMn_2O_4$ systems. It shows that it is possible to extract lithium ions from the host before the evolution of oxygen [240]. The cycling performance of NCM is good as illustrated in Fig. 7.17b. In the case of NCM, $NH_2NH_2{\cdot}H_2O$ which is used as controlling agent is used to prepare the precursor $Ni_xCo_yMn_{1-x-y}(OH)_2$ in the hydroxide coprecipitation method. The results indicate that species of aqueous electrolyte, current density, scan rate, and potential limit have influence on the capacitive property of NCM cathode. The initial discharge capacitance of 298 F g^{-1} is obtained in 1 M Li_2SO_4 solution within potential range 0–1.4 V at the current density of 100 mA g^{-1}, and its capacitance fading is less than 0.058 F g^{-1} per cycle in 1000 cycles [241]. In addition, NCM

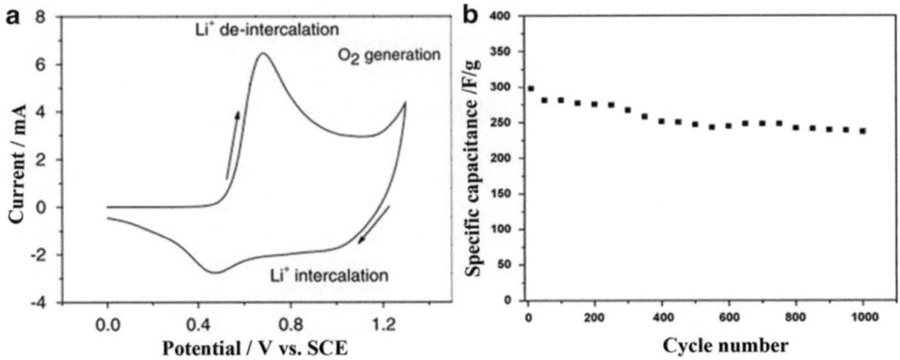

Fig. 7.17 (**a**) CV curve of NCM in 2 M Li_2SO_4 aqueous solution and (**b**) cycling ability of AC/NMO supercapacitor at 100 mA g^{-1} in 1 M Li_2SO_4 (Modified from Refs. [240, 241] Copyright 2008 and 2009, with permission from Springer and Elsevier, respectively)

prepared by a sol–gel method can also exhibit a good cycling performance in a 2 M $LiNO_3$ aqueous solution at different charge–discharge rates from 80 to 200 C in spite of low capacitance [242].

The electrochemical performance of NCM in aqueous solution can be modified by mixing with PPy. The redox reversibility and capacity retention are much improved compared to those of the pristine NCM. The main reason is that the added PPy could enhance the electron conductivity of NCM cathode [243].

7.4.2.4 Na_xMnO_2

The known phases of Na_xMnO_2 (x = 0.2, 0.40, 0.44, 0.70, 1) have been summarized [244], and there are two phases for $NaMnO_2$. Low temperature α-$NaMnO_2$ has an O_3 layered structure with a monoclinic structural distortion due to the Jahn–Teller distortion of the Mn^{3+} ion. At high temperature, the orthorhombic β-$NaMnO_2$ forms in a different layered structure containing MnO_2 sheets consisting of a double stack of edge-sharing MnO_6 octahedra. Na occupies the octahedral sites between two neighboring sheets [245]. First-principle computations indicate that the monoclinic $NaMnO_2$ is energetically more stable than other competing phases. This is in contrast to $LiMnO_2$, which prefers an orthorhombic structure [246, 247]. Both α- and β-$NaMnO_2$ have been electrochemically tested as a cathode in 1985. Their results show that only 0.22 and 0.15 Na could be reversibly extracted and reintercalated in α- and β- $NaMnO_2$, respectively. Besides $NaMnO_2$, $Na_{0.4}MnO_2$ and $Na_{0.6}MnO_2$ have also been studied as cathode [247].

Among the small number of oxides of potential interest, $Na_{0.44}MnO_2$ is particularly attractive because of its adequate crystal structure forming suitable large-size tunnels for sodium cations [248]. Moreover, in previous reports, the materials prepared by solid-state method contain Mn_2O_3 impurities. To avoid the presence of such parasitic phases, acidic treatment using HCl is used for their dissolution.

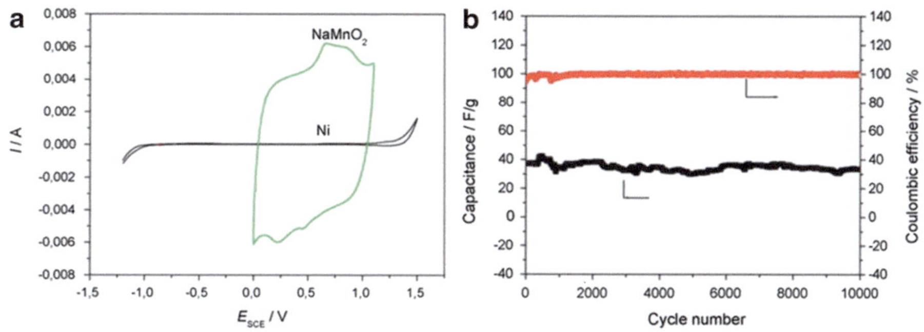

Fig. 7.18 (**a**) CV curves of $NaMnO_2$ in 0.5 M Na_2SO_4 aqueous solution and (**b**) the cycling behavior of the asymmetric AC//$NaMnO_2$ supercapacitor (Modified from Rref. [250] Copyright 2009, with permission from Elsevier)

However, this procedure concomitantly induces sodium leaching, leading to an isostructurally deficient sodium phase of approximate composition $Na_{0.2}MnO_2$ [248]. Recently, some researchers pointed out the possibility of preparing pure $Na_{0.44}MnO_2$ powders by accurately adjusting the solid-state synthesis parameters [249]. In addition, $NaMnO_2$ can be synthesized by ball milling mixtures of Na_2CO_3 and MnO_2 at a molar ratio of 1:2 for 12 h followed by heating at 870°C for 10 h [250]. $Na_{0.35}MnO_2$ nanowire is achieved by a simple and low energy consumption hydrothermal method [251].

Cyclic voltammograms (CV) of $NaMnO_2$ electrode in 0.5 M Na_2SO_4 aqueous solution at a scan rate of 5 mV s^{-1} are shown in Fig. 7.18a. The current collector (Ni mesh) is very stable in the range of $-1.0 - 1.2$ V (vs. SCE) in Na_2SO_4 aqueous solution. Hydrogen and oxygen evolution occur at < -1.0 V and > 1.2 V (vs. SCE), respectively, because of significant overpotentials consistent with results previously reported for aqueous Li_2SO_4 solution [223]. In the potential range of 0–1.1 V, the behavior of the $NaMnO_2$ electrode slightly deviates from the ideal rectangular shape with two small redox couples, indicative of the pseudocapacitance properties of $NaMnO_2$ cathode. These redox peaks are similar to that of $Na_xMnO_2{\cdot}H_2O$, which can be definitely ascribed to the intercalation/deintercalation of Na^+ into and from the solid lattice [252]. The supercapacitor exhibits a sloping voltage–time curve in the entire voltage region of 0–1.9 V and delivers an energy density of 19.5 Wh kg^{-1} at a power density of 130 W kg^{-1} based on the total mass of the two electrode materials. It can be seen from Fig. 7.18b that it also shows excellent cycling behavior with not more than 3 % capacitance loss after 10,000 cycles at a current rate of 10 C [250].

In the case of the nanowire $Na_{0.35}MnO_2$ (157 F g^{-1}), its capacitance is much higher than that of the rodlike $Na_{0.95}MnO_2$ (92 F g^{-1}). It presents excellent cycling performance even the oxygen in the aqueous electrolyte is not removed, no evident capacitance fading after 5000 cycles. The $Na_{0.35}MnO_2$ nanowire delivers an energy density of 42.6 Wh kg^{-1} at a power density of 129.8 W kg^{-1} when testing using activated carbon as the anode, higher than that of the rodlike $Na_{0.95}MnO_2$, and 27.3 Wh kg^{-1} at a power density of 74.8 W kg^{-1} [251].

7.4.2.5 K_xMnO_2

Lithium and sodium intercalation compounds can be a promising cathode for supercapacitors as mentioned above; potassium intercalation compounds can also be used as cathodes. The lamellar structure of $LiMnO_2$ has been demonstrated to be unstable during cycling and easily transforms into the spinel form since the lamellar compound has a cubic-close-packed arrangement of oxide atoms, which is identical to that of a spinel [253]. Therefore, introduction of large alkali ions and H_2O molecules into the interlayer space of manganese oxide could stabilize the lamellar structure [254]. For example, lamellar K_xMnO_2, with the large K^+ ions as pillars, shows better cycling performance than Li_xMnO_2 during the Li^+ intercalation/deintercalation in organic electrolytes. This can be ascribed to the expansion of the interlayer space. At the same time, large K^+ ions make manganese cations diffusion into the interlayer region to form spinel structure less favorable [255]. When $K_xMnO_2{\cdot}yH_2O$ is used as the cathode material in an aqueous supercapacitor, the existence of these interlayer H_2O molecules is not supposed to influence the performance of the capacitor, unlike commercial lithium-ion batteries using organic electrolytes, which require anhydrous environment for the assembly of battery. In addition, lamellar $K_xMnO_2{\cdot}yH_2O$ lattice possesses a large interlayer distance of about 0.7 nm that could be intercalated by large quaternary ammonium cations and other alkaline cations [256–258].

There are many scientific reports on routes to prepare K_xMnO_2. For instance, sol–gel method usually involves various sugars and other organic polyalcohols as well as further heating to develop good crystallinity and remove the organic additives [259]. Precipitation route usually involves not only the oxidation of aqueous Mn^{2+} cations but also the oxidation of amorphous solid precursors via the redox reaction between Mn^{2+} and MnO_4^- [260]. A soft-template method for the layered manganate (K_xMnO_2, birnessite) with a relatively good control at the nanoscale regime on the number of octahedral layers is described [261]. Its precursor is prepared by ball milling the mixture of K_2CO_3 and MnO_2 in a molar ratio of 1:2 for 12 h, followed by calcination at 550 °C for 8 h. The precursor is further washed several times with water to remove residual K_2CO_3 [262]. Nanowire $K_{0.19}MnO_2{\cdot}0.2H_2O$ can be prepared by a hydrothermal method. It presents better electrochemical behavior than $K_{0.45}MnO_2{\cdot}0.3H_2O$ from the solid-state method as the cathode for an asymmetric supercapacitor in 0.5 mol L^{-1} K_2SO_4 aqueous solution using activated carbon as the anode [263].

A scheme is shown in Fig. 7.19a to illustrate the different movement behaviors of $K_xMnO_2{\cdot}nH_2O$ electrodes in Li_2SO_4, Na_2SO_4, and K_2SO_4 aqueous electrolytes. The molar ionic conductivity of the three alkaline-cation aqueous solutions (K^+, 73.5 S cm^2 mol^{-1}; Na^+, 50.1 S cm^2 mol^{-1}; Li^+, 38.6 S cm^2 mol^{-1}) [264], namely, their migration rate, is supposed to account for the different electrolyte resistance. Since the hydrated ionic radius of the three alkaline metal cations is similar (K^+, 3.31 Å; Na^+, 3.58 Å; and Li^+, 3.82 Å) [264], the different capacitive features should be mainly affected by their different solvation interactions. K^+ possesses the weakest solvation interaction with H_2O due to its smallest charge density, making

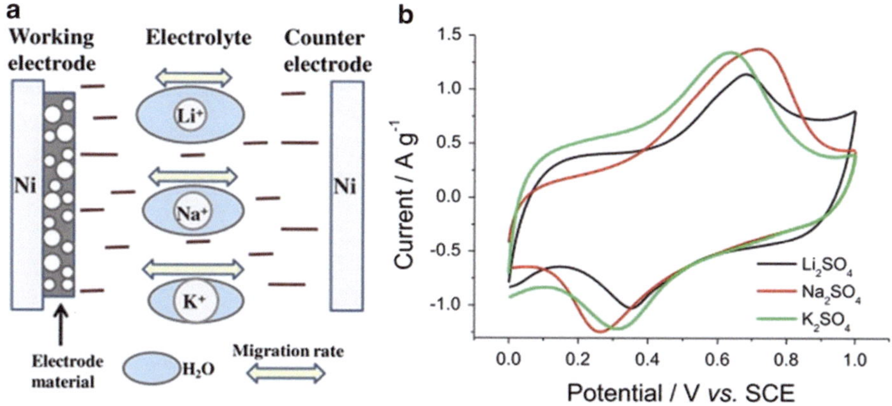

Fig. 7.19 (**a**) Schematic illustration of the hydrated ionic radius and migration rate of three types of alkaline metal cations and (**b**) CV curves of $K_xMnO_2{\cdot}nH_2O$ electrodes in the three aqueous electrolytes at the scan rate of 5 mV s^{-1} (Modified from Ref. [265] Copyright 2013, with permission from Elsevier)

it dehydrate readily in the interior of the porous electrode. As a result, the charge-transfer process occurs quickly, and the double-layer capacitance is formed rapidly on the electrode/electrolyte interface. All the above results explain the best power ability of $K_xMnO_2{\cdot}nH_2O$ in the K_2SO_4 electrolyte [265].

CV curves of $K_xMnO_2{\cdot}nH_2O$ electrodes in the three aqueous electrolytes at the scan rate of 5 mV s^{-1} are presented in Fig. 7.19b. A couple of reversible redox peaks can be observed distinctly for all the three electrolytes, suggesting the faradic pseudocapacitive nature of $K_xMnO_2{\cdot}nH_2O$ material. Previous work on the crystalline structure and chemical composition of $K_xMnO_2{\cdot}nH_2O$ during cycling in K_2SO_4 electrolyte indicates that this couple of redox peaks is concerned with the reversible intercalation/deintercalation of K^+ into/from crystalline $K_xMnO_2{\cdot}nH_2O$ lattice [266]. In this case, the three kinds of electrolytes lead to slightly different potentials of these redox peaks. In addition, the redox peaks of $K_xMnO_2{\cdot}nH_2O$ electrodes are considerably more distinct than those of MnO_2 reported in literatures, which can be ascribed to the high crystallinity of $K_xMnO_2{\cdot}nH_2O$ material [265]. Studies of the electrochemical behavior of $K_{0.27}MnO_2{\cdot}0.6H_2O$ in K_2SO_4 show an energy density of 25.3 Wh kg^{-1} at power density of 140 W kg^{-1} based on the total mass of the active electrode materials. It also shows excellent cycling behavior with no more than 2 % capacitance loss after 10,000 cycles at a current rate of 25C [262].

The capacitance of the nanowire $K_{0.19}MnO_2{\cdot}0.2H_2O$ (148 F g^{-1}) is much higher than that of $K_{0.45}MnO_2{\cdot}0.3H_2O$ (132 F g^{-1}) though its potassium amount is less. Both of them present excellent cycling performance and there is no evident capacitance fading after 2500 cycles. The $K_{0.19}MnO_2{\cdot}0.2H_2O$ delivers an energy density of 41.3 Wh kg^{-1} (based on the total mass of the two electrode materials) at a power density of 156.8 W kg^{-1} with activated carbon as the anode, higher than that of $K_{0.45}MnO_2{\cdot}0.3H_2O$, and 28.4 Wh kg^{-1} at a power density of 115.1 W kg^{-1} [263].

7.4.3 *Conversion Compounds*

So far, various inexpensive transition metal oxides as electrode materials have been investigated for supercapacitors. Few research work concerns on the preparation of conversion compounds toward supercapacitors in the literatures. Therefore, it is significant and challenging to exploit multi-metal oxide materials for supercapacitors.

7.4.3.1 $NiCo_2O_4$

$NiCo_2O_4$ is generally regarded as a mixed valence oxide that adopts a pure spinel structure in which nickel occupies the octahedral sites and cobalt is distributed over both octahedral and tetrahedral sites [267]. Spinel nickel cobaltite ($NiCo_2O_4$) has received tremendous interest recently for supercapacitor applications due to better electron conductivity (two orders of magnitude higher than conventional transitional metal oxides), higher electrochemical activity, lower cost, higher availability, and environmental friendliness [268].

Several methods have been developed to synthesize $NiCo_2O_4$, including coprecipitation of oxalates, hydroxide carbonates, and metal hydroxides [269, 270], nanocasting [271], sol–gel [268], combustion [272], and hydrothermal [273]. For example, spinel $NiCo_2O_4$ nanocrystals have been synthesized through a thermal decomposition method. The highly porous $NiCo_2O_4$ exhibits high specific capacitance (764 F g^{-1} at 2 mV s^{-1}). In addition, its specific capacitance shows no significant decay after the 10,000 cycles in 1 M NaOH [274]. The $NiCo_2O_4$ controlled by CTAB exhibits highly specific capacitance of 1440 F g^{-1} at a current density of 5 mA cm^{-2}. Remarkably, it still displays desirable cycling retention of 94.1 % over 1000 cycles at a current density of 20 mA cm^{-2} [275]. The superior electrochemical performance could be attributed to the combination of the following factors: (1) The effect of CTAB in modifying the microstructure and surface morphology of $NiCo_2O_4$ is superior to that of PVA. (2) The $NiCo_2O_4$ modified by CTAB generates more pores, which can provide passages for the transport of electrolyte, especially for OH^- to have a fast diffusion rate within the redox phase. This may facilitate the improvement of power performance. (3) The smaller particles and better dispersion can generate higher specific surface area and more active sites, which are in favor of electrode material to contact the electrolyte. In addition, the electrolyte can easily permeate the smaller particles for redox reaction, which can increase the utilization ratio of electrode materials and increase the specific capacitance [275]. A facile hydrothermal and thermal decomposition process which is successfully developed to grow 3D $NiCo_2O_4$ microspheres constructed with radial chain-like $NiCo_2O_4$ nanowires with different exposed crystal planes is reported. Chain-like $NiCo_2O_4$ nanowires exhibit excellent electrochemical performance in supercapacitors with high specific capacitance (1284 F g^{-1} at 2 A g^{-1}), good rate capability, and excellent cycling stability (only 2.5 % loss after 3000 cycles) [276].

In order to further improve its electrochemical performance, conductive materials should be introduced to form hybrid with $NiCo_2O_4$, such as carbon fibers and graphene. To be more exact, two 1D hierarchical hybrid nanostructures composed of $NiCo_2O_4$ nanorods and ultrathin nanosheets on carbon nanofibers (CNFs) are controllably synthesized through facile solution methods in combination with a simple thermal treatment. A high capacitance of 902 F g^{-1} at 2 A g^{-1} is obtained for the CNF@$NiCo_2O_4$ nanosheet composite with a low capacitance loss of only 3.6 % after 2400 cycles. This enhanced performance is attributed to several important structural factors of the CNF@$NiCo_2O_4$ nanosheet composite: ultrathin and porous nanosheets, CNF support, and porous hierarchical structure [277]. In addition, a series of flexible nanocomposites can be fabricated by facile electrodeposition of cobalt and nickel double hydroxide (DH) nanosheets on porous $NiCo_2O_4$ nanowires grown radially on carbon fiber paper (CFP). This nano-architecture composite shows the best cycling stability while maintaining high capacitance of about 1.64 F cm^{-2} at 2 mA cm^{-2}. Of course, it also exhibits excellent rate capability. The areal capacitance decrease is less than 33 % as the current density is increased from 2 to 90 mA cm^{-2}, offering excellent specific energy density (about 33 Wh kg^{-1}) and power density (41.3 kW kg^{-1}) at high rates (up to 150 mA cm^{-2}) [278]. $NiCo_2O_4$ nanosheet arrays deposited on flexible carbon fabric (CF) can act as a high-performance cathode for supercapacitors. The $NiCo_2O_4$ arrays are constructed by interconnected ultrathin nanosheets (10 nm) with many interparticle pores. The porous feature of $NiCo_2O_4$ nanosheets increases the amount of electroactive sites and facilitates the electrolyte penetration. The capacitances of the $NiCo_2O_4$/CF composite are 2658, 2200, 2170, 2000, and 1866 F g^{-1} at 2, 5, 10, 15, and 20 A g^{-1}, respectively (Fig. 7.20a). About 70 % of the initial capacitances is remained even at a high current density of 20 A g^{-1}, indicating that the prepared $NiCo_2O_4$/CF composites have excellent rate performance. This might be attributed to high surface area of the ultrathin porous nanosheet structure, which can provide massive electroactive sites and facilitate transport of the electrolyte. More importantly, each interconnected nanosheet can

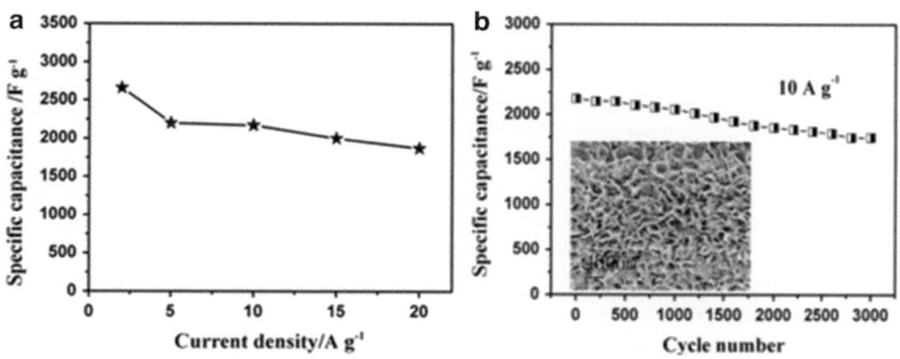

Fig. 7.20 (**a**) Changes of capacitance at different current densities and (**b**) cycling performance at a current density of 10 A g^{-1} (*inset*: SEM micrograph) of a $NiCo_2O_4$/CF composite (Modified from Ref. [279] Copyright 2013, with permission from ACS)

directly contact with CF substrate, which is of good electron conductivity and shortens the diffusion paths of the electrons. The cycling performance of the $NiCo_2O_4$ nanosheet arrays supported on CF is shown in Fig. 7.20b. Its capacitance decreases approximately 20 % after 3000 cycles at 10 A g^{-1} [279]. In the case of $NiCo_2O_4$@RGO composite, it is prepared by a facile hydrothermal method without any surfactant. It has a high capacitance of 737 F g^{-1} at a current density of 1 A g^{-1} and only 6 % loss of the initial specific capacitance after 3000 cycles at 4 A g^{-1} in 2 M KOH electrolyte [280].

$NiCo_2O_4$ can also form composites with nickel foam or other metal oxides to enhance its electrochemical performance. For instance, $NiCo_2O_4$ nanowire arrays supported on Ni foams are prepared by a facile and highly efficient hydrothermal method and can be used to fabricate highly flexible solid-state supercapacitors. They demonstrate excellent electrochemical performance with a high areal capacitance of 161 mF cm^{-2} at 1 mA cm^{-2} and also superior flexibility and cycling stability that can sustain over 3000 cycles under harsh mechanical conditions including twisted and bent tests [281]. Porous $NiCo_2O_4$ heterostructure arrays on nickel foam prepared by a facile hydrothermal method exhibit excellent pseudocapacitive properties in 2 M KOH solution, with a high capacitance of 891 F g^{-1} at 1 A g^{-1} and 619 F g^{-1} at 40 A g^{-1} before activation as well as excellent cycling stability. Its capacitance can achieve a maximum of 1089 F g^{-1} at a current density of 2 A g^{-1}, which can still retain 1058 F g^{-1} (97.2 % retention) after 8000 cycles. The enhanced pseudocapacitive performance is mainly attributed to its unique heterostructure which provides fast ion and electron transfer, large reaction surface area, and good strain accommodation [282]. $NiCo_2O_4$ multiple hierarchical structures (MHSs) composed of 1D nanowire and 2D nanosheets are simply grown on Ni foam using an ordered two-step hydrothermal synthesis followed by annealing processing. The $NiCo_2O_4$ MHS shows superior capacitances of 2620 and 1786 F g^{-1} under current densities of 1 and 40 A g^{-1}, respectively, with an excellent cycling stability (only 6.6 % loss after 3000 cycles). The remarkable electrochemical performance is likely due to the vast electroactive surface area and favorable mesoporous architecture, which are formed during the integrated assemblies [283]. A facile and scalable strategy has been developed to construct hierarchical $NiCo_2O_4$@MnO_2 core–shell nano-architectures with high electrochemical performance for supercapacitors. The overall capacitance loss for the $NiCo_2O_4$ nanowire arrays is about 15 % after 1000 cycles, while it is only 12 % even after 2000 cycles for the core–shell nanowire arrays [284].

In terms of the mesoporous $NiO/NiCo_2O_4/Co_3O_4$ nanocomposite, it is prepared by a sol–gel process followed by calcifying at 250 °C. A high capacitance of 1717 F g^{-1} is achieved at 5 mA cm^{-2} and 94.9 % of the initial specific capacitance is remained after 1000 cycles [285]. Two types of homogeneous $NiCo_2O_4$ nanorod@$NiCo_2O_4$ nanosheet and heterogeneous NiO nanoflake@$NiCo_2O_4$ nanosheet hierarchical core–shell arrays are synthesized via facile solution methods in combination with a simple thermal treatment. In both cases, the $NiCo_2O_4$ nanosheets serve as the core backbone for anchoring the shell materials. The two as-prepared hierarchical nanoarrays demonstrate excellent electrochemical performance with high specific capacitance (1925 and 2210 F g^{-1} for $NiCo_2O_4$@$NiCo_2O_4$ and NiO@$NiCo_2O_4$ at 0.5 A g^{-1}, respectively), good rate capability, and superior cycling stability. The superior capacitive performance is mainly due to the unique hierarchical core–shell architecture with faster ion/electron transfer, improved reactivity, and enhanced structural stability [286].

7.4.3.2 $ZnCo_2O_4$

Spinel zinc cobaltite ($ZnCo_2O_4$) is a low-cost, binary metal oxide where Zn occupies tetrahedral sites in the cubic spinel structure and Co occupies the octahedral sites [287]. In recent years, $ZnCo_2O_4$ nanostructures with different structures, such as 3D mesostructures of ordered crystalline spinel-type $ZnCo_2O_4$ nanoparticles [288, 289], porous nanotubes [290], porous nanoflakes [291], porous nanowires [292], $ZnCo_2O_4$ nanowire arrays/carbon cloth [293] have been successfully fabricated by a variety of methods. For example, $ZnCo_2O_4$ nanorods are prepared by a hydrothermal method [294]. Porous $ZnCo_2O_4$ nanotubes can be prepared by an easy single-nozzle electrospinning strategy combined with subsequent heat treatment [290]. Nano-$ZnCo_2O_4$ is prepared by using molten salt method at low temperature of 280 °C for 30 min or 3 h in air [295]. Porous $ZnCo_2O_4$ nanowires are prepared by annealing $ZnCo_2(C_2O_4)_3$ nanowires as sacrificial templates. The morphology and surface area of the as-prepared porous $ZnCo_2O_4$ nanowires can be tuned by the annealing temperature [292].

As for electrochemical performance of $ZnCo_2O_4$ nanomaterials, there are only few studies on $ZnCo_2O_4$ as cathode for supercapacitors. For example, when $ZnCo_2O_4$ nanomaterial from the coprecipitation method is mixed with equal proportion of carbon nanofoam, its capacitance is up to 77 F g^{-1} at a scan rate of 5 mV s^{-1} [287]. In addition, highly crystalline $ZnCo_2O_4$ nanocrystals of about 4 nm in diameter show a capacitance approaching 700 F g^{-1} at 5 mV s^{-1}. Nearly 95 % of the capacitance is retained after 3000 cycles, indicating outstanding durability [296]. One-dimensional (1D) $ZnCo_2O_4$ porous nanotubes (PNTs), which are prepared by using a simple electrospinning followed by calcination in air, exhibit a high capacitance of 770 F g^{-1} at 10 A g^{-1}, excellent rate property (84 % of the capacitance retention at 60 A g^{-1}) and good cycling stability of 10.5 % loss after 3000 cycles [297].

In order to further improve its electrochemical performance, nickel foam can also be used. For example, mesoporous $ZnCo_2O_4$ nanosheet arrays are uniformly grown on Ni foam via a facile two-step method and are directly used as an integrated electrode for supercapacitors. As shown in Fig. 7.21a, they exhibit excellent pseudocapacitive behavior with capacitance of 2468, 2382, 2217, 2128, 1904, 1740, 1616 and 1482 F g^{-1} at current densities of 5, 8, 15, 20, 40, 60, 80 and 100 A g^{-1}, respectively. Additionally, 60.0 % of initial capacitance could be retained when current density is increased from 5 to 100 A g^{-1}, indicating that

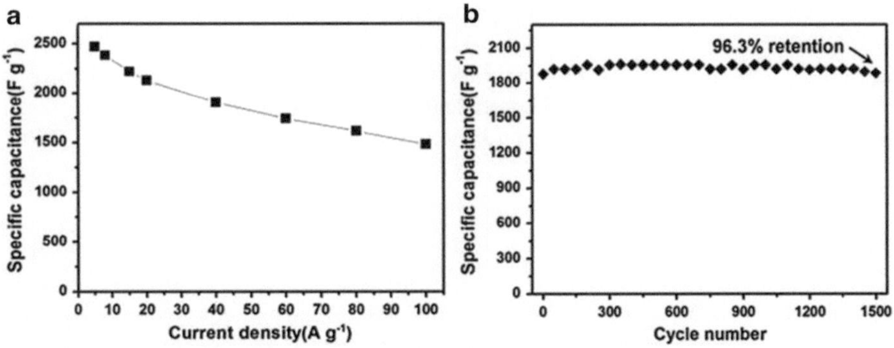

Fig. 7.21 (**a**) Change of capacitance with current density and (**b**) cycling performance at a constant current density of 30 A g^{-1} for the mesoporous $ZnCo_2O_4$ nanosheet arrays on Ni foam (Modified from Ref. [298] Copyright 2014, with permission from RSC)

the $ZnCo_2O_4$ nanosheets on Ni foam prepared via hydrothermal method have great potential applications in supercapacitors. As shown in Fig. 7.21b, even at the high current density of 30 A g^{-1}, the $ZnCo_2O_4$ nanosheet arrays/Ni foam integrated electrode only fades 3.7 % of its original capacitance (~1900 F g^{-1}) after 1500 cycles numbers, indicating its excellent electrochemical stability [298]. The excellent electrochemical performance of the $ZnCo_2O_4$ nanosheet arrays/Ni foam electrode is ascribed to its unique nanostructure. First, the nanosheets are staggered with their neighbors and directly attached to the surface of conductive Ni substrate, thus an open 3D nanoarray network with several merits is formed: (1) staggered and separated growth of nanosheets not only reduces the free space on the surface of Ni substrate but also increases the effective contact area between electrode and electrolyte, which significantly improves the utilization of active materials; (2) nanosheets are connected with each other to form a nanoarray network that is directly adhered to the surface of Ni foam, thus the nanostructure could be maintained during long-term cycling process and the contact resistance of the whole electrode could be reduced. Secondly, mesoporous nanosheets not only provide more electroactive sites for the faradic reactions but also shorten the distance of OH^- ion/electron transformation, which leads to faster kinetics and enhanced electrochemical performance [298]. In addition, uniform $ZnCo_2O_4$ nanowire arrays can be directly grown on nickel foam through a facile hydrothermal method and subsequent thermal treatment process. The $ZnCo_2O_4$ nanowires have a diameter of about 100 nm and length of up to 5 μm. The $ZnCo_2O_4$ nanowire/nickel foam electrode exhibits a high capacitance of 1625 F g^{-1} at 5 A g^{-1}. Furthermore, the retention of capacitance is 94 % after 5000 cycles at a high current density of 20 A g^{-1} [299].

7.4.3.3 $CoMoO_4$

$CoMoO_4$ has received much research interest due to its photocatalytic, magnetic, and electrochemical properties [300]. It has three phases, which are formed under atmospheric pressure: the low temperature α-phase (α-$CoMoO_4$), the high temperature β-isomorph (β-$CoMoO_4$), and the hydrate ($CoMoO_4{\cdot}nH_2O$) [301]. Co^{2+} is at octahedral sites in all these isomorphs whereas Mo^{6+} is at octahedral coordination in α-$CoMoO_4$ and at tetrahedral coordination in β and hydrate phases. Since the Mo atoms are in an octahedral environment in α-$CoMoO_4$ and MoO_3 oxides, they have similar electronic properties [302].

Several preparation methods for nanoscale $CoMoO_4$ with different shapes have been employed. For example, $CoMoO_4$ nanoplates are synthesized via citrate complex route assisted by microwave irradiation [303]. One-dimensional $CoMoO_4$ nanorods with monoclinic crystal structure are successfully fabricated by a facile chemical coprecipitation method. It presents high capacitance of 286 F g^{-1} at 5 mA cm^{-2} and 97.5 % of its initial capacitance is remained after 2000 cycles. The excellent stability is mainly due to the stable crystal structure and good structure stability of 1D $CoMoO_4$ nanorods [304]. $CoMoO_4{\cdot}0.9H_2O$ nanorods is successfully fabricated by a facile hydrothermal process, and exhibits a high capacitance of 326 F g^{-1} at 5 mA cm^{-2} and 96 % of its initial capacitance is remained after 1000 cycles. The results also show that too high crystallinity is unfavorable to supercapacitive performance of some metal oxides [305].

In the case of the composites of $CoMoO_4$ with CNTs and graphene, their electrochemical performance is improved. For instance, $CoMoO_4$ nanoplatelets with a crystalline-amorphous core–shell structure anchored via MWCNTs are prepared by a solvent-free microwave synthesis method. The nanocomposite shows a capacitance of 170 F g^{-1} with a potential window of 0.8 V, and its fading is only 6.8 % after 1000 cycles [306]. As for the composite of $CoMoO_4$/graphene, it is prepared by growing $CoMoO_4$ nanoparticles on the graphene sheet by a hydrothermal reaction. The composite possesses high surface/body ratios and large electroactive regions, which could facilitate easy accession of OH^- ions and fast charge transport. As a result, the composite achieves a capacitance of about 395 F g^{-1} and an energy density of around 54.8 Wh kg^{-1} at 1 mV s^{-1}, which are higher than those of the virginal $CoMoO_4$ (72.0 F g^{-1}, and 10.0 Wh Kg^{-1}) [307]. In addition, $CoMoO_4{\cdot}0.9H_2O$ nanorods can be grown on reduced graphene oxide to achieve another composite ($CoMoO_4{\cdot}0.9H_2O$–rGO), which has good electrochemical properties with exceptionally high capacitance of 802 F g^{-1} at 1 A g^{-1} and good cycling stability with capacitance retention of ~86.3 % after 5000 cycles [308].

As shown in Fig. 7.22a, $MnMoO_4$ nanowires can be used as the 'substrate', which can guide the $CoMoO_4$ self-assembling growth in aqueous solution without surfactant and stabilizer. Then, the 'oriented attachment' can guide the oriented growth of nanoparticles. That is to say, supersaturated solution with plenty of $CoMoO_4$ small crystals are formed by adding Co resource. Because of the high

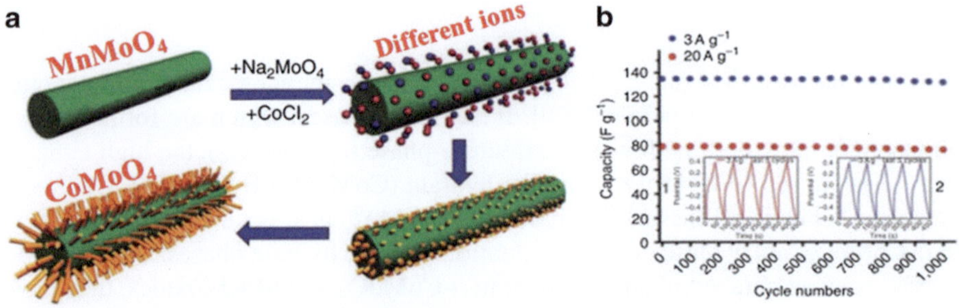

Fig. 7.22 (**a**) Construction of hierarchical $MnMoO_4/CoMoO_4$ nanowires and (**b**) its charge–discharge cycling behavior in 2 M NaOH aqueous solution (Modified from Ref. [309] Copyright 2011, with permission from Nature)

surface energy and thermodynamics instability, nanoparticles can attach to the surface of $MnMoO_4$ to decrease surface energy. The crystallographic orientation of the particles with respect to each other is determined by the minimization of the highest surface energy. Therefore, with a matching lattice, the lattice fringes' orientation and crystal growth direction are uniform to some extent [309]. The hierarchical $MnMoO_4/CoMoO_4$ electrodes can reach to 204, 187, 163, 135 F g^{-1} and 28.4, 26.0, 22.7, 18.7 Wh kg^{-1} at 0.5, 1, 2, 3 A g^{-1}, respectively. The capacitance for hierarchical $MnMoO_4/CoMoO_4$ heterostructured nanowires is significantly higher than that for the virginal nanorods ($MnMoO_4$ (9.7 F g^{-1}, 8.8 Wh kg^{-1}), $CoMoO_4$ (62.8 F g^{-1}, 1.4 Wh kg^{-1}), $MnMoO_4/CoMoO_4$ nanocomposite (69.2 F g^{-1}, 9.6 Wh kg^{-1}) at a charge–discharge current density of 1 A g^{-1}). The hierarchical $MnMoO_4/CoMoO_4$ heterostructured nanowires exhibits a good reversibility with cycling efficiency of 98 % after 1000 cycles (Fig. 7.22b) [309].

When $CoMoO_4$ nanoplate arrays (NPAs) are grown directly on Ni foam via a template-free hydrothermal route, such unique array architecture exhibits remarkable electrochemical performance with a high capacitance of 1.26 F cm^{-2} at a charge and discharge current density of 4 mA cm^{-2} and 0.78 F cm^{-2} at 32 mA cm^{-2} with an excellent cycling ability (79.5 % of the initial capacitance after 4000 cycles). The superior electrochemical performances could be attributed to the open network structure consisting of the interconnected $CoMoO_4$ NPAs directly grown on current collectors that could improve electron transport and electrolyte diffusion [310].

7.4.3.4 Other Conversion Compounds

Beside conversion compounds as mentioned above, other conversion compounds can also be used as cathode materials for supercapacitors such as Zn_2SnO_4 [311], $NiMoO_4$ [312] and $MnMoO_4$ [313]. To date, only few articles reported these conversion compounds as cathodes for supercapacitors. For example, a nickel–cobalt layered double hydroxides (LDHs) on conducting Zn_2SnO_4 (ZTO) heterostructure is synthesized through a combined CVD and electrochemical deposition method. This material demonstrates outstanding electrochemical performance with a high capacitance of 1805 F g^{-1} at 0.5 A g^{-1}, and an excellent rate performance of 1275 F g^{-1} can be achieved at 100 A g^{-1}. Moreover, an asymmetric supercapacitor based on the Ni_xCo_{1-x} LDH-ZTO heterostructure exhibits a high energy density of 23.7 Wh kg^{-1} at a power density of 284 W kg^{-1}. Meanwhile, a high power density of 5817 W kg^{-1} can be achieved at an energy density of 9.7 Wh kg^{-1}. More importantly, this device exhibits long-term cycling stability, with 92.7 % capacity retention after 5000 cycles [311].

As for $NiMoO_4$ nanomaterials, they have attracted great research interest because of its chemical stability, low cost, and enhanced electrochemical performance [314]. For example, $NiMoO_4{\cdot}nH_2O$ nanorods can be synthesized by a facile chemical coprecipitation method. A maximum capacitance of 1136 F g^{-1} is achieved at 5 mA cm^{-2}. To enhance the energy density and enlarge the potential window, an asymmetric supercapacitor is assembled using $NiMoO_4{\cdot}nH_2O$ as the cathode that exhibits a maximum capacitance of 96.7 F g^{-1} and an energy of 34.4 W h kg^{-1} [315]. $NiMoO_4$ nanowires (NWs) supported on Ni foam are fabricated via a template-free hydrothermal route and could be directly used as integrated electrodes for supercapacitors. Impressively, the as-prepared $NiMoO_4$ NWs deliver a high capacitance of 1.96 F cm^{-2} (1308 F g^{-1}) even at a very high current density of 112 mA cm^{-2} (74.7 A g^{-1}), desirable rate performance, and electrochemical stability. Such intriguing capacitive behavior is attributed to the one-dimensional array structure and synergistic effect between nickel and molybdenum ions in the binary oxide [312].

In the case of $MnMoO_4$, we have discussed $MnMoO_4/CoMoO_4$ nanocomposites in the above section. However, effect of electrolyte on the capacitive behavior of α-$MnMoO_4$ nanorods has been studied including in para-toluene sulfonic acid (p-TSA), sulfuric acid (H_2SO_4), and hydrochloric acid (HCl) solutions. Here, α-$MnMoO_4$ nanorods are prepared on conducting glass substrate via sol–gel spin coating method. At the optimum doping level, the as-prepared α-$MnMoO_4$ nanorods exhibit a capacitance of 998 F g^{-1} at a scan rate of 5 mV s^{-1} in H_2SO_4 electrolyte, while capacitances of 784 F g^{-1} and 530 F g^{-1} have been obtained in p-TSA and HCl electrolytes, respectively. At higher scan rates, p-TSA electrolyte exhibits superior capacitive behavior than H_2SO_4 [313].

7.5 Conclusions and Outlook

As the above illustrated, many kinds of typical and mostly used methods for various metal oxides with different nanostructures are discussed. In general, nanostructures provide higher surface area, easier access of electrolyte to the active material, and shorter diffusion distances, leading to improved energy storage performance in supercapacitors. Moreover, the performance in terms of stability, capacitance, and Coulombic efficiency of the metal oxide materials can be further ameliorated by modifying with other functional nanostructure materials. The newly emerging nanostructure conversion compounds as well as metal oxide-based nanocomposites are now offering significant opportunities for the development of cost-effective, efficient, and environmentally benign supercapacitor devices.

It is worth noting that by using nanostructure materials, the inner stress can be alleviated and the ion diffusion length is reduced. Therefore, better rate capability and cycling performance can be expected. Electrode materials with porous nanostructures can accommodate large volume changes during the charge–discharge process, and the hollow structures may provide extra space for ion storage [164, 316]. Hierarchical nanostructures, combining nanoscale and microscale materials, may capitalize on the advantages and restrain the shortcomings of the two components [317, 318]. Besides, nanostructure current collectors and flexible paper/textile electrodes will gain attention in the future [319, 320]. Nanostructure current collectors, which have much larger surface areas and good mechanical robustness, can provide both efficient pathways for ion and electron transport through the entire electrode architecture.

Although the last decade has witnessed great progress in the preparation of nanostructure metal oxide/metal oxide-based nanocomposites and their applications for supercapacitor devices, challenges still exist at the current stage of technology. For example, one of the key challenges for supercapacitors is their limited energy density, which has hindered their wider application in the field of energy storage. To overcome this challenge, a major focus on research and development for supercapacitors should be directed to discover new electrode materials with high capacitance and a wide potential window. The porosity of their electrode materials should be particularly emphasized. Nano-micropores are necessary to achieve higher specific surface area, and these micropores must be ensured to be electrochemically accessible for ions. Hence, pore network and the availability and wettability of pores, with dimensions matching the size of solvated anions and cations, are crucial for supercapacitor electrode materials [1]. While metal oxide nanomaterials can achieve a much higher capacitance than EDLCs, they suffer from decreased cycling stability. One emerging research direction is combining metal oxides with carbon-based materials in order to utilize capacitance contributions from both double-layer capacitance and pseudocapacitance. Moreover, two or more metal elements with mixed valence can further boost energy density and cycling stability, which may provide a new path to explore excellent supercapacitors.

In the future, research directions on metal oxide nanomaterials for supercapacitors applications would be proposed as the following. At first, not only the existing preparation methods should be utilized but also new and advanced synthetic strategies should be developed to prepare high-quality metal oxide nanomaterials. It is necessary to make more efforts to understand the relationship between the reaction parameters and the final structure and properties of the metal oxides as well as their underlying nucleation and growth mechanism in solution, which could guide us to tailor metal oxide nanomaterials. Secondly, high attention on the material engineering of metal oxides should be paid to generate combined composite/hybrid nanomaterials. The modification techniques could help us access a huge number of new metal oxide-based nanocomposites, which not only integrate the properties of the virginal metal oxide but also bring new collective functions. Although recent progress has shown successful cases such as metal oxide/CNTs, metal oxide/graphene, intercalation compounds, and conversion compounds, a wide range of other conversion compounds is highly pursued for broadening and enhancing the applications of metal oxide-based nanomaterials.

Lastly, it is important to understand the fundamental of the principles of supercapacitors. For example, an in-depth understanding of the pseudocapacitance mechanisms and their relationship with the structure and composition of the active sites is beneficial to prepare new nanomaterials and develop new theories. Furthermore, more further investigations into the electrochemical properties and surface structure as well as synergetic effects of metal oxide/modified-metal oxide nanomaterials should be done, and the combination with smart experimental strategies will greatly shorten the development process of highly efficient metal oxide-based nanocomposites for supercapacitor applications.

Acknowledgment Financial supports from China National Funds for Distinguished Young Scientists (NSFC No. 51425301), NSFC (21073046 and 21463013) and STCSM (14520721800) are greatly appreciated.

References

1. Wang GP, Zhang L, Zhang JJ (2012) A review of electrode materials for electrochemical supercapacitors. Chem Soc Rev 41:797–828
2. Liu C, Li F, Ma LP et al (2010) Advanced materials for energy storage. Adv Mater 22: E28–62
3. Chen S, Xing W, Duan J et al (2013) Nanostructured morphology control for efficient supercapacitor electrodes. J Mater Chem A 1:2941–2954
4. Simon P, Gogotsi Y (2008) Materials for electrochemical capacitors. Nature Mater 7:845–854
5. Kötz R, Carlen M (2000) Principles and applications of electrochemical capacitors. Electrochim Acta 45:2483–2498
6. Kötz R, Mu ̈ller S, Ba ̈rtschi M et al (2001) Super capacitors for peak-power demand in fuel-cell-driven cars. Electrochem Soc Proc 21:564–575
7. Naoi K, Naoi W, Aoyagi S et al (2013) New generation "nanohybrid supercapacitor". Accounts Chem Res 46:1075–1083
8. Zhang QF, Uchaker E, Candelaria SL et al (2013) Nanomaterials for energy conversion and storage. Chem Soc Rev 42:3127–3171
9. Zhang Q, Cao G (2011) Nanostructured photoelectrodes for dye-sensitized solar cells. Nano Today 6:91–109
10. Cheng C, Fan HJ (2012) Branched nanowires: synthesis and energy applications. Nano Today 7:327–343
11. Liu R, Duay J, Lee SB (2011) Heterogeneous nanostructured electrode materials for electrochemical energy storage. Chem Commun 47:1384–1404
12. Chandra A, Roberts AJ, Lam How Yee E et al (2009) Nanostructured oxides for energy storage applications in batteries and supercapacitors. Pure Appl Chem 81:1489–1498

13. Jiang J, Li YY, Liu JP et al (2012) Recent advances in metal oxide-based electrode architecture design for electrochemical energy storage. Adv Mater 24:5166–5180
14. Deng W, Ji X, Chen Q et al (2011) Electrochemical capacitors utilising transition metal oxides: an update of recent developments. RSC Adv 1:1171–1178
15. Wang FX, Xiao SY, Hou YY et al (2013) Electrode materials for aqueous asymmetric supercapacitors. RSC Adv 3:13059–13084
16. Lang XY, Hirata A, Fujita T et al (2011) Nanoporous metal/oxide hybrid electrodes for electrochemical supercapacitors. Nat Nanotechnol 6:232–236
17. Lokhande CD, Dubal DP, Joo OS (2011) Metal oxide thin film based supercapacitors. Curr Appl Phys 11:255–270
18. Zhang LL, Zhao XS (2009) Carbon-based materials as supercapacitor electrodes. Chem Soc Rev 38:2520–2531
19. Pandolfo AG, Hollenkamp AF (2006) Carbon properties and their role in supercapacitors. J Power Sources 157:11–27
20. Andrieu X (2000) New trends electrochemical technology. In: Osaka T, Datta M (eds) Energy storage systems for electronics, vol 1. Gordon and Breach, London, p 521
21. Jiang H, Ma J, Li C (2012) Mesoporous carbon incorporated metal oxide nanomaterials as supercapacitor electrodes. Adv Mater 24:4197–4202
22. Zhi M, Xiang C, Li J et al (2013) Nanostructured carbon-metal oxide composite electrodes for supercapacitors: a review. Nanoscale 5:72–88
23. Hall PJ, Mirzaeian M, Fletcher SI et al (2010) Energy storage in electrochemical capacitors: designing functional materials to improve performance. Energy Environ Sci 3:1238–1251
24. Ghosh S, Inganas O (1999) Conducting polymer hydrogels as 3D electrodes: applications for supercapacitors. Adv Mater 11:1214–1218
25. Cottineau T, Toupin M, Delahaye T et al (2006) Nanostructured transition metal oxides for aqueous hybrid electrochemical supercapacitors. Appl Phys A 82:599–606
26. Stoller MD, Ruoff RS (2010) Best practice methods for determining an electrode material's performance for ultracapacitors. Energy Environ Sci 3:1294–1301
27. Nishihara H, Kyotani T (2012) Templated nanocarbons for energy storage. Adv Mater 24:4473–4498
28. Haas O, Cairns EJ (1999) Electrochemical energy storage. Annu Rep Sect C 95:163–198
29. Balducci A, Dugas R, Taberna PL et al (2007) High temperature carbon–carbon supercapacitor using ionic liquid as electrolyte. J Power Sources 165:922–927
30. Hu Y, Wang J, Jiang X et al (2013) Facile chemical synthesis of nanoporous layered $\delta-MnO_2$ thin film for high-performance flexible electrochemical capacitors. Appl Surf Sci 271:193–201
31. Tang W, Liu LL, Tian S et al (2011) $LiMn_2O_4$.. nanorods as a super-fast cathode material for aqueous rechargeable lithium batteries. Electrochem Commun 13:1159–1162
32. Rui X, Sim D, Xu C et al (2012) One-pot synthesis of carbon-coated VO_2(B) nanobelts for high-rate lithium storage. RSC Adv 2:1174–1180
33. Tang W, Gao XW, Zhu YS et al (2012) A hybrid of V_2O_5 nanowires and MWCNTs coated with polypyrrole as an anode material for aqueous rechargeable lithium batteries with excellent cycling performance. J Mater Chem 22:20143–20145
34. Tang W, Hou YY, Wang FX et al (2013) $LiMn_2O_4$ nanotube as cathode material of second-level charge capability for aqueous rechargeable batteries. Nano Lett 13:2036–2040
35. Tang W, Liu LL, Tian S et al (2011) Aqueous supercapacitors of high energy density based on MoO_3 nanoplates as anode material. Chem Commun 47:10058–10060
36. Gedanken A (2004) Using sonochemistry for the fabrication of nanomaterials. Ultrason Sonochem 11:47–55
37. Perkas N, Zhong Z, Grinblat J et al (2008) Deposition of gold particles on mesoporous catalyst supports by sonochemical method, and their catalytic performance for CO oxidation. Catal Lett 120:19–24

38. Wang H, Zhu JJ, Zhu JM et al (2002) Sonochemical method for the preparation of bismuth sulfide nanorods. J Phys Chem B 106:3848–3854
39. Li B, Xie Y, Liu Y et al (2001) Sonochemical synthesis of nanocrystalline silver tellurides Ag_2Te and Ag_7Te_4. J Solid State Chem 158:260–263
40. Li B, Xie Y, Huang J et al (2000) Sonochemical synthesis of nanocrystalline copper tellurides Cu_7Te_4 and Cu_4Te_3 at room temperature. Chem Mater 12:2614–2616
41. Salinas EP, Sánchez EM (2010) Preparation of Sb_2S_3 Nanostructures by the ionic liquid-assisted sonochemical method. Cryst Growth Des 10:3917–3924
42. Bilecka I, Niederberger M (2010) Microwave chemistry for inorganic nanomaterials synthesis. Nanoscale 2:1358–1374
43. Gao MR, Xu YF, Jiang J et al (2013) Nanostructured metal chalcogenides: synthesis, modification, and applications in energy conversion and storage devices. Chem Soc Rev 42:2986–3017
44. Zhuang Z, Peng Q, Li Y (2011) Controlled synthesis of semiconductor nanostructures in the liquid phase. Chem Soc Rev 40:5492–5513
45. Gao MR, Jiang J, Yu SH (2012) Solution-based synthesis and design of late transition metal chalcogenide materials for oxygen reduction reaction (ORR). Small 8:13–27
46. Devaraju MK, Honma I (2012) Hydrothermal and solvothermal process towards development of $LiMPO_4$ (M=Fe, Mn) nanomaterials for lithium-ion batteries. Adv Energy Mater 2:284–297
47. Sue K, Kimura K, Arai K (2004) Rapid hydrothermal synthesis of ZnO nanorods without organics. Mater Lett 58:3229–3231
48. Lu AH, Salabas EL, Schuth F (2007) Magnetic nanoparticles: synthesis, protection, functionalization, and application. Angew Chem Int Ed 46:1222–1244
49. Lam UT, Mammucari R, Suzuki K et al (2008) Processing of iron oxide nanoparticles by supercritical fluids. Ind Eng Chem Res 47:599–614
50. Tavakoli A, Sohrabi M, Kargari A (2007) A review of methods for synthesis of nanostructured metals with emphasis on iron compounds. Chem Pap 61:151–170
51. Teja AS, Koh PY (2009) Synthesis, properties, and applications of magnetic iron oxide nanoparticles. Prog Cryst Growth Charact Mater 55:22–45
52. Liang HW, Liu S, Yu SH (2010) Controlled synthesis of one-dimensional inorganic nanostructures using pre-existing one-dimensional nanostructures as templates. Adv Mater 22:3925–3937
53. Lou XW, Archer LA, Yang Z (2008) Hollow micro-/nanostructures: synthesis and applications. Adv Mater 20:3987–4019
54. Zhao YS, Fu H, Peng A et al (2008) Low-dimensional nanomaterials based on small organic molecules: preparation and optoelectronic properties. Adv Mater 20:2859–2876
55. Fan H, Zhang YG, Zhang MF et al (2008) Glucose-assisted synthesis of CoTe nanotubes in situ templated by Te nanorods. Cryst Growth Des 8:2838–2841
56. Possin GE (1970) A method for forming very small diameter wires. Rev Sci Instrum 41:772–774
57. Martin CR (1994) Nanomaterials-A membrane-based synthetic approach. Science 266:1961–1966
58. Murakami H, Kobayashi M, Takeuchi H et al (1999) Preparation of poly(dl-lactide-co-glycolide) nanoparticles by modified spontaneous emulsification solvent diffusion method. Int J Pharm 187:143–152
59. Kim TW, Park I, Ryoo R (2003) A synthetic route to ordered mesoporous carbon materials with graphitic pore walls. Angew Chem Int Ed 115:4511–4515
60. Berry AD, Tonucci RJ, Fatemi M (1996) Fabrication of GaAs and InAs wires in nanochannel glass. Appl Phys Lett 69:2846–2848
61. Chakarvarti SK, Vetter J (1998) Template synthesis-a membrane based technology for generation of nano-/micro materials: a review. Radiat Meas 29:149–159

62. Chen XB, Mao SS (2007) Titanium dioxide nanomaterials: synthesis, properties, modifications, and applications. Chem Rev 107:2891–2959
63. She GW, Zhang XH, Shi WS et al (2008) Template-free electrochemical synthesis of single-crystal CuTe nanoribbons. Cryst Growth Des 8:1789–1791
64. Pinisetty D, Davis D, Podlaha-Murphy EJ et al (2011) Characterization of electrodeposited bismuth–tellurium nanowires and nanotubes. Acta Mater 59:2455–2461
65. Barau A, Budarin V, Caragheorgheopol A et al (2008) A simple and efficient route to active and dispersed silica supported palladium nanoparticles. Catal Lett 124:204–214
66. White RJ, Luque R, Budarin VL et al (2009) Supported metal nanoparticles on porous materials. Methods and applications. Chem Soc Rev 38:481–494
67. Donega´ CM, Liljeroth P, Vanmaekelbergh D (2005) Physicochemical evaluation of the hot-injection method, a synthesis route for monodisperse nanocrystals. Small 1:1152–1162
68. Martínez A, Prieto G (2007) The key role of support surface tuning during the preparation of catalysts from reverse micellar-synthesized metal nanoparticles. Catal Commun 8:1479–1486
69. Serp P, Kalck P, Feurer R (2002) Chemical vapor deposition methods for the controlled preparation of supported catalytic materials. Chem Rev 102:3085–3128
70. Lee H, Cho MS, Kim IH et al (2010) RuO_x/polypyrrole nanocomposite electrode for electrochemical capacitors. Synth Met 160:1055–1059
71. Choi D, Kumta PN (2006) Nanocrystalline TiN derived by a two-step halide approach for electrochemical capacitors. J Electrochem Soc 153:A2298–2303
72. Zhao DD, Bao SJ, Zhou WJ et al (2007) Preparation of hexagonal nanoporous nickel hydroxide film and its application for electrochemical capacitor. Electrochem Commun 9:869–874
73. Hu CC, Chang KH, Lin MC et al (2006) Design and tailoring of the nanotubular arrayed architecture of hydrous RuO_2 for next generation supercapacitors. Nano Lett 6:2690–2695
74. Gao B, Hao L, Fu Q et al (2010) Hydrothermal synthesis and electrochemical capacitance of $RuO_2{\cdot}xH_2O$ loaded on benzenesulfonic functionalized MWCNTs. Electrochim Acta 55:3681–3686
75. Zhang J, Ma J, Zhang LL et al (2010) Template synthesis of tubular ruthenium oxides for supercapacitor applications. J Phys Chem C 114:13608–13613
76. Kim JY, Kim KH, Park SH et al (2010) Microwave-polyol synthesis of nanocrystalline ruthenium oxide nanoparticles on carbon nanotubes for electrochemical capacitors. Electrochim Acta 55:8056–8061
77. Sieben JM, Morallón E, Cazorla-Amorós D (2013) Flexible ruthenium oxide-activated carbon cloth composites prepared by simple electrodeposition methods. Energy 58:519–526
78. Liang YY, Li HL, Zhang XG (2007) Solid state synthesis of hydrous ruthenium oxide for supercapacitors. J Power Sources 173:599–605
79. Chen IL, Wei YC, Chen TY et al (2014) Oxidative precipitation of ruthenium oxide for supercapacitors: enhanced capacitive performances by adding cetyltrimethylammonium bromide. J Power Sources 268:430–438
80. Devadas A, Baranton S, Napporn TW et al (2011) Tailoring of RuO_2 nanoparticles by microwave assisted "Instant method" for energy storage applications. J Power Sources 196:4044–4053
81. Wu X, Zeng Y, Gao H et al (2013) Template synthesis of hollow fusiform $RuO_2{\cdot}xH_2O$ nanostructure and its supercapacitor performance. J Mater Chem A 1:469–472
82. Chou JC, Chen YL, Yang MH et al (2013) RuO_2/MnO_2 core–shell nanorods for supercapacitors. J Mater Chem A 1:8753–8758
83. Hu CC, Guo HY, Chang KH et al (2009) Anodic composite deposition of $RuO_2{\cdot}xH_2O - TiO_2$ for electrochemical supercapacitors. Electrochem Commun 11:1631–1634
84. Su YF, Wu F, Bao LY et al (2007) RuO_2/activated carbon composites as a positive electrode in an alkaline electrochemical capacitor. New Carbon Mater 22:53–57

85. Lee Y, Kim B, Jung HJ et al (2012) Hierarchically grown single crystalline RuO_2 nanorods on vertically aligned few-walled carbon nanotubes. Mater Lett 89:115–117
86. Chen PC, Shen G, Shi Y et al (2010) Preparation and characterization of flexible asymmetric supercapacitors based on transition-metal-oxide nanowire/single-walled carbon nanotube hybrid thin-film electrodes. ACS Nano 4:4403–4411
87. Soin N, Roy SS, Mitra SK et al (2012) Nanocrystalline ruthenium oxide dispersed few layered graphene (FLG) nanoflakes as supercapacitor electrodes. J Mater Chem 22:14944–14950
88. Wang W, Guo S, Lee I et al (2014) Hydrous ruthenium oxide nanoparticles anchored to graphene and carbon nanotube hybrid foam for supercapacitors. Sci Rep 4:1–9
89. Zhao X, Sanchez BM, Dobson PJ et al (2011) The role of nanomaterials in redox-based supercapacitors for next generation energy storage devices. Nanoscale 3:839–855
90. Conway BE, Birss V, Wojtowicz J (1997) The role and utilization of pseudocapacitance for energy storage by supercapacitors. J Power Sources 66:1–14
91. Park BO, Lokhande CD, Park HS et al (2004) Performance of supercapacitor with electrodeposited ruthenium oxide film electrodes-effect of film thickness. J Power Sources 134:148–152
92. Sugimoto W, Shibutani T, Murakami Y et al (2002) Charge storage capabilities of rutile-type $RuO_2 - VO_2$ solid solution for electrochemical supercapacitors. Electrochem Solid State Lett 5:A170–172
93. Toupin M, Brousse T, Be´langer D (2004) Charge storage mechanism of MnO_2 electrode used in aqueous electrochemical capacitor. Chem Mater 16:3184–3190
94. Nam KW, Kim MG, Kim KB (2007) In situ Mn K-edge X-ray absorption spectroscopy studies of electrodeposited manganese oxide films for electrochemical capacitors. J Phys Chem C 111:749–758
95. Nakayama M, Tanaka A, Sato Y et al (2005) Electrodeposition of manganese and molybdenum mixed oxide thin films and their charge storage properties. Langmuir 21:5907–5913
96. Yuan CZ, Zhang XG, Hou LR et al (2010) Lysine-assisted hydrothermal synthesis of urchin-like ordered arrays of mesoporous $Co(OH)_2$ nanowires and their application in electrochemical capacitors. J Mater Chem 20:10809–10816
97. Guo YG, Hu JS, Wan LJ (2008) Nanostructured materials for electrochemical energy conversion and storage devices. Adv Mater 20:2878–2887
98. Zhu G, Deng L, Wang J et al (2013) Hydrothermal preparation and the capacitance of hierarchical MnO_2 nanoflower. Colloid Surf A 434:42–48
99. Zhang X, Yu P, Zhang H et al (2013) Rapid hydrothermal synthesis of hierarchical nanostructures assembled from ultrathin birnessite-type MnO_2 nanosheets for supercapacitor applications. Electrochim Acta 89:523–529
100. Song Z, Liu W, Zhao M et al (2013) A facile template-free synthesis of $\alpha - MnO_2$ nanorods for supercapacitor. J Alloy Compd 560:151–155
101. Zhou M, Zhang X, Wei J et al (2011) Morphology-controlled synthesis and novel microwave absorption properties of hollow urchinlike $\alpha - MnO_2$ nanostructures. J Phys Chem C 115:1398–1402
102. Wan C, Cheng M, Zhang Q et al (2013) Preparation of MnO_2 nanostructures by controlled crystal growth and its pseudocapacitive properties. Powder Technol 235:706–711
103. Li GR, Feng ZP, Ou YN et al (2010) Mesoporous MnO_2/carbon aerogel composites as promising electrode materials for high-performance supercapacitors. Langmuir 26:2209–2213
104. Wang YT, Lu AH, Li WC (2012) Mesoporous manganese dioxide prepared under acidic conditions as high performance electrode material for hybrid supercapacitors. Microporous Mesoporous Mater 153:247–253
105. Snook GA, Kao P, Best AS (2011) Conducting-polymer-based supercapacitor devices and electrodes. J Power Sources 196:1–12

106. Iranagh SA, Eskandarian L, Mohammadi R (2013) Synthesis of MnO_2-polyaniline nanofiber composites to produce high conductive polymer. Synthetic Met 172:49–53
107. Tang W, Hou YY, Wang XJ et al (2012) A hybrid of MnO_2 nanowires and MWCNTs as cathode of excellent rate capability for supercapacitors. J Power Sources 197:330–333
108. He S, Chen W (2014) High performance supercapacitors based on three-dimensional ultra-light flexible manganese oxide nanosheets/carbon foam composites. J Power Sources 262:391–400
109. Lei Z, Shi F, Lu L (2012) Incorporation of MnO_2-coated carbon nanotubes between graphene sheets as supercapacitor electrode. ACS Appl Mater Interfaces 4:1058–1064
110. Zhao X, Zhang LL, Murali S et al (2012) Incorporation of manganese dioxide within ultraporous activated graphene for high-performance electrochemical capacitors. ACS Nano 6:5404–5412
111. Wu ZS, Ren WC, Wang DW et al (2010) High-energy MnO_2 nanowire/graphene and graphene asymmetric electrochemical capacitors. ACS Nano 4:5835–5842
112. Cheng Q, Tang J, Ma J et al (2011) Graphene and nanostructured MnO_2 composite electrodes for supercapacitors. Carbon 49:2917–2925
113. Kim KS, Park SJ (2012) Synthesis and high electrochemical performance of polyaniline/ MnO_2-coated multi-walled carbon nanotube-based hybrid electrodes. J Solid State Electrochem 16:2751–2758
114. Bao L, Zang J, Li X (2011) Flexible Zn_2SnO_4/MnO_2 core/shell nanocable-carbon microfiber hybrid composites for high-performance supercapacitor electrodes. Nano Lett 11:1215–1220
115. Dong S, Chen X, Gu L et al (2011) One dimensional MnO_2/titanium nitride nanotube coaxial arrays for high performance electrochemical capacitive energy storage. Energy Environ Sci 4:3502–3508
116. Cheng Y, Lu S, Zhang H et al (2012) Synergistic effects from graphene and carbon nanotubes enable flexible and robust electrodes for high-performance supercapacitors. Nano Lett 12:4206–4211
117. Yu G, Hu L, Vosgueritchian M et al (2011) Solution-processed graphene/MnO_2 nanostructured textiles for high-performance electrochemical capacitors. Nano Lett 11:2905–2911
118. Su Z, Yang C, Xie B et al (2014) Scalable fabrication of MnO_2 nanostructure deposited on free-standing Ni nanocone arrays for ultrathin, flexible, high-performance micro-supercapacitor. Energy Environ Sci 7:2652–2659
119. Hou Y, Cheng Y, Hobson T et al (2010) Design and synthesis of hierarchical MnO_2 nanospheres/carbon nanotubes/conducting polymer ternary composite for high performance electrochemical electrodes. Nano Lett 10:2727–2733
120. Zhang H, Cao GP, Wang ZY et al (2008) Growth of manganese oxide nanoflowers on vertically-aligned carbon nanotube arrays for high-rate electrochemical capacitive energy storage. Nano Lett 8:2664–2668
121. Liu Y, Zhang BH, Yang YQ et al (2013) Polypyrrole-coated $\alpha-MoO_3$ nanobelts with good electrochemical performance as anode materials for aqueous supercapacitors. J Mater Chem A 1:13582–13587
122. Tao T, Chen Q, Hu H et al (2012) MoO_3 nanoparticles distributed uniformly in carbon matrix for supercapacitor applications. Mater Lett 66:102–105
123. Brezesinski T, Wang J, Tolbert SH et al (2010) Ordered mesoporous $\alpha-MoO_3$ with iso-oriented nanocrystalline walls for thin-film pseudocapacitors. Nat Mater 9:146–151
124. Farsi H, Gobal F, Raissi H et al (2010) On the pseudocapacitive behavior of nanostructured molybdenum oxide. J Solid State Electrochem 14:643–650
125. Shakir I, Shahid M, Yang HW et al (2010) Structural and electrochemical characterization of $\alpha-MoO_3$ nanorod-based electrochemical energy storage devices. Electrochim Acta 56:376–380
126. Hu JB, Ramadan A, Luo F et al (2011) One-step molybdate ion assisted electrochemical synthesis of $\alpha-MoO_3$-decorated graphene sheets and its potential applications. J Mater Chem 21:15009–15014

127. Zheng L, Xu Y, Jin D et al (2011) Polyaniline-intercalated molybdenum oxide nanocomposites: simultaneous synthesis and their enhanced application for supercapacitor. Chem Asian J 6:1505–1514
128. Jiang F, Li W, Zou R et al (2014) MoO_3/PANI coaxial heterostructure nanobelts by in situ polymerization for high performance supercapacitors. Nano Energy 7:72–79
129. Shakir I, Shahid M, Nadeem M et al (2012) Tin oxide coating on molybdenum oxide nanowires for high performance supercapacitor devices. Electrochim Acta 72:134–137
130. Srinivasan V, Weidner JW (2002) Capacitance studies of cobalt oxide films formed via electrochemical precipitation. J Power Sources 108:15–20
131. Liu TC, Pell WG, Conway BE (1999) Stages in the development of thick cobalt oxide films exhibiting reversible redox behavior and pseudocapacitance. Electrochim Acta 44:2829–2842
132. Shan Y, Gao L (2007) Formation and characterization of multi-walled carbon nanotubes/ Co_3O_4 nanocomposites for supercapacitors. Mater Chem Phys 103:206–210
133. Gong LY, Liu XH, Su LH et al (2012) Synthesis and electrochemical capacitive behaviors of Co_3O_4 nanostructures from a novel biotemplating technique. J Solid State Electrochem 16:297–304
134. Vijayakumar S, Ponnalagi AK, Nagamuthu S et al (2013) Microwave assisted synthesis of # nanoparticles for high-performance supercapacitors. Electrochim Acta 106:500–505
135. Liu X, Long Q, Jiang C et al (2013) Facile and green synthesis of mesoporous Co_3O_4 nanocubes and their applications for supercapacitors. Nanoscale 5:6525–6529
136. Fan Z, Chen J, Cui K et al (2007) Preparation and capacitive properties of cobalt–nickel oxides/carbon nanotube composites. Electrochim Acta 52:2959–2965
137. Huang S, Jin YH, Jia MQ (2013) Preparation of graphene/Co_3O_4 composites by hydrothermal method and their electrochemical properties. Electrochim Acta 95:139–145
138. Park S, Kim S (2013) Effect of carbon blacks filler addition on electrochemical behaviors of Co_3O_4/graphene nanosheets as a supercapacitor electrodes. Electrochim Acta 89:516–522
139. Wang XW, Liu SQ, Wang HY et al (2012) Facile and green synthesis of Co_3O_4 nanoplates/ graphene nanosheets composite for supercapacitor. J Solid State Electrochem 16:3593–3602
140. Zhong JH, Wang AL, Li GR et al (2012) Co_3O_4/$Ni(OH)_2$ composite mesoporous nanosheet networks as a promising electrode for supercapacitor applications. J Mater Chem 22:5656–5665
141. Qing XX, Liu SQ, Huang KL et al (2011) Facile synthesis of Co_3O_4 nanoflowers grown on Ni foam with superior electrochemical performance. Electrochim Acta 56:4985–4991
142. Wang X, Liu B, Liu R et al (2014) Fiber-based flexible all-solid-state asymmetric supercapacitors for integrated photodetecting system. Angew Chem Int Ed 53:1849–1853
143. Zhao QQ, Jiao LF, Peng WX et al (2012) Facile synthesis of VO_2(B)/carbon nanobelts with high capacity and good cyclability. J Power Sources 199:350–354
144. Wu CZ, Zhang XD, Dai J et al (2011) Direct hydrothermal synthesis of monoclinic VO_2(M) - single-domain nanorods on large scale displaying magnetocaloric effect. J Mater Chem 21:4509–4517
145. Ni J, Jiang WT, Yu K et al (2011) Hydrothermal synthesis of VO_2(B) nanostructures and application in aqueous Li-ion battery. Electrochim Acta 56:2122–2126
146. Sediri F, Gharbi N (2009) Controlled hydrothermal synthesis of VO_2(B) nanobelts. Mater Lett 63:15–18
147. Ganganagappa N, Siddaramanna A (2012) One step synthesis of monoclinic VO_2(B) bundles of nanorods: cathode for Li ion battery. Mater Charact 68:58–62
148. Kong FY, Li M, Yao XY et al (2012) Template-free hydrothermal synthesis of VO_2 hollow microspheres. Cryst Eng Comm 14:3858–3861
149. Shao J, Li XY, Qu QT et al (2012) One-step hydrothermal synthesis of hexangular starfruit-like vanadium oxide for high power aqueous supercapacitors. J Power Sources 219:253–257

150. Wang H, Yi H, Chen X et al (2014) One-step strategy to three-dimensional graphene/VO_2 nanobelt composite hydrogels for high performance supercapacitors. J Mater Chem A 2:1165–1173
151. Liang LY, Liu HM, Yang WS (2013) Fabrication of VO_2(B) hybrid with multiwalled carbon nanotubes to form a coaxial structure and its electrochemical capacitance performance. J Alloys Compd 559:167–173
152. Sathiya M, Prakash AS, Ramesha K et al (2011) V_2O_5-anchored carbon nanotubes for enhanced electrochemical energy storage. J Am Chem Soc 133:16291–16299
153. Rui XH, Zhu JX, Liu WL et al (2011) Facile preparation of hydrated vanadium pentoxide nanobelts based bulky paper as flexible binder-free cathodes for high-performance lithium ion batteries. RSC Adv 1:117–122
154. Qu QT, Liu LL, Wu YP et al (2013) Electrochemical behavior of $V_2O_5 \cdot 0.6H_2O$ nanoribbons in neutral aqueous electrolyte solution. Electrochim Acta 96:8–12
155. Fu L, Liu H, Li C et al (2005) Electrode materials for lithium secondary batteries prepared by sol–gel methods. Prog Mater Sci 50:881–928
156. Qu QT, Shi Y, Li LL et al (2009) $V_2O_5 \cdot 0.6H_2O$ nanoribbons as cathode material for asymmetric supercapacitor in K_2SO_4 solution. Electrochem Commun 11:1325–1328
157. Wee G, Soh HZ, Cheah YL et al (2010) Synthesis and electrochemical properties of electrospun V_2O_5 nanofibers as supercapacitor electrodes. J Mater Chem 20:6720–6725
158. Saravanakumar B, Purushothaman KK, Muralidharan G (2012) Interconnected V_2O_5 nanoporous network for high-performance supercapacitors. ACS Appl Mater Interfaces 4:4484–4490
159. Yu L, Zhao CX, Long X et al (2009) Ultrasonic synthesis and electrochemical characterization of V_2O_5/mesoporous carbon composites. Microporous Mesoporous Mater 126:58–64
160. Fang WC (2008) Synthesis and electrochemical characterization of vanadium oxide/carbon nanotube composites for supercapacitors. J Phys Chem C 112:11552–11555
161. Bao J, Zhang X, Bai L et al (2014) All-solid-state flexible thin-film supercapacitors with high electrochemical performance based on a two-dimensional $V_2O_5 \cdot H_2O$/graphene composite. J Mater Chem A 2:10876–10881
162. Cao L, Zhu J, Li Y et al (2014) Ultrathin single-crystalline vanadium pentoxide nanoribbon constructed 3D networks for superior energy storage. J Mater Chem A 2:13136–13142
163. Yang Y, Kim D, Yang M et al (2011) Vertically aligned mixed $V_2O_5 - TiO_2$ nanotube arrays for supercapacitor applications. Chem Commun 47:7746–7748
164. Qu QT, Zhu YS, Gao XW et al (2012) Core-shell structure of polypyrrole grown on V_2O_5 nanoribbon as high performance anode material for supercapacitors. Adv Energy Mater 2:950–955
165. Chen CY, Chiang CY, Shih SJ et al (2013) High supercapacitive performance of sol–gel ZnO-doped manganese oxide coatings. Thin Solid Films 528:61–66
166. Aravinda LS, Nagaraja KK, Nagaraja HS et al (2013) ZnO/carbon nanotube nanocomposite for high energy density supercapacitors. Electrochim Acta 95:119–124
167. Chen YX, Zhao XQ, Sha B et al (2008) Stacking fault directed growth of thin ZnO nanobelt. Mater Lett 62:2369–2371
168. Jeong JS, Lee JY, Cho JH et al (2005) Single-crystalline ZnO microtubes formed by coalescence of ZnO nanowires using a simple metal-vapor deposition method. Chem Mater 17:2752–2756
169. Lao JY, Huang JY, Wang DZ et al (2003) ZnO nanobridges and nanonails. Nano Lett 3:235–238
170. Li C, Fang GJ, Liu NS et al (2007) Structural, photoluminescence, and field emission properties of vertically well-aligned ZnO nanorod arrays. J Phys Chem C 111:12566–12571
171. Kalpana D, Omkumar KS, Kumar SS et al (2006) A novel high power symmetric ZnO/carbon aerogel composite electrode for electrochemical supercapacitor. Electrochim Acta 52:1309–1315

172. Jayalakshmi M, Palaniappa M, Balasubramanian K (2008) Single step solution combustion synthesis of ZnO/carbon composite and its electrochemical characterization for supercapacitor application. Int J Electrochem Sci 3:96–103
173. Zhang YP, Sun XW, Pan LK et al (2009) Carbon nanotube–ZnO nanocomposite electrodes for supercapacitors. Solid State Ionics 180:1525–1528
174. Park KW, Jung JH (2012) Spectroscopic and electrochemical characteristics of a carboxylated graphene–ZnO composites. J Power Sources 199:379–385
175. Dong XC, Cao YF, Wang J et al (2012) Hybrid structure of zinc oxide nanorods and three dimensional graphene foam for supercapacitor and electrochemical sensor applications. RSC Adv 2:4364–4369
176. Yu MP, Sun HT, Sun X et al (2013) Hierarchical Al-doped and hydrogenated ZnO nanowire@MnO_2 ultra-thin nanosheet core/shell arrays for high-performance supercapacitor electrode. Int J Electrochem Sci 8:2313–2329
177. Li GR, Wang ZL, Zheng FL et al (2011) ZnO@MoO_3 core/shell nanocables: facile electrochemical synthesis and enhanced supercapacitor performances. J Mater Chem 21:4217–4221
178. Yang PH, Xiao X, Li YZ et al (2013) Hydrogenated ZnO core shell nanocables for flexible supercapacitors and self powered systems. ACS Nano 7:2617–2626
179. Trang NT, Ngoc HV, Lingappan N et al (2014) A comparative study of supercapacitive performances of nickel cobalt layered double hydroxides coated on ZnO nanostructured arrays on textile fibre as electrodes for wearable energy storage devices. Nanoscale 6:2434–2439
180. Pang H, Zhang B, Du JM et al (2012) Porous nickel oxide nanospindles with huge specific capacitance and long-life cycle. RSC Adv 2:2257–2261
181. Wang HW, Yi H, Chen X et al (2013) Facile synthesis of a nano-structured nickel oxide electrode with outstanding pseudocapacitive properties. Electrochim Acta 105:353–361
182. Meher SK, Justin P, Rao GR (2011) Microwave-mediated synthesis for improved morphology and pseudocapacitance performance of nickel oxide. ACS Appl Mater Interfaces 3:2063–2073
183. Zhang XJ, Shi WH, Zhu JX et al (2010) Synthesis of porous NiO nanocrystals with controllable surface area and their application as supercapacitor electrodes. Nano Res 3:643–652
184. Vijayakumar S, Nagamuthu S, Muralidharan G (2013) Supercapacitor studies on NiO nanoflakes synthesized through a microwave route. ACS Appl Mater Interfaces 5:2188–2196
185. Zhang G, Yu L, Hoster HE et al (2013) Synthesis of one-dimensional hierarchical NiO hollow nanostructures with enhanced supercapacitive performance. Nanoscale 5:877–881
186. Yuan CZ, Hou LR, Feng YL et al (2013) Sacrificial template synthesis of short mesoporous NiO nanotubes and their application in electrochemical capacitors. Electrochim Acta 88:507–512
187. Chen H, Xu JL, Xu XW et al (2013) Preparation of mesoporous NiO with excellent pseudocapacitive behavior. Eur J Inorg Chem 2013:1105–1108
188. Wang XY, Yi LH, Liu L et al (2013) Preparation and capacitive properties of the core–shell structure carbon aerogel microbeads- nanowhisker-like NiO composites. J Power Sources 224:317–323
189. Vijayakumar S, Nagamuthu S, Muralidharan G (2013) Porous NiO/C nanocomposites as electrode material for electrochemical supercapacitors. ACS Sustainable Chem Eng 1:1110–1118
190. Bello A, Makgopa K, Fabiane M et al (2013) Chemical adsorption of NiO nanostructures on nickel foam-graphene for supercapacitor applications. J Mater Sci 48:6707–6712
191. Xia X, Tu J, Mai Y et al (2011) Graphene sheet/porous NiO hybrid film for supercapacitor applications. Chemistry 17:10898–10905
192. Wang H, Yi H, Chen X et al (2014) Asymmetric supercapacitors based on nano-architectured nickel oxide/graphene foam and hierarchical porous nitrogen-doped carbon nanotubes with ultrahigh-rate performance. J Mater Chem A 2:3223–3230

193. Liu MM, Chang J, Sun J et al (2013) A facile preparation of NiO/Ni composites as high-performance pseudocapacitor materials. RSC Adv 3:8003–8008
194. Lu ZY, Chang Z, Liu JF et al (2011) Stable ultrahigh specific capacitance of NiO nanorod arrays. Nano Res 4:658–665
195. Zhang Y, Guo Z (2014) Green synthesis of open porous NiO films with an excellent capacitance performance. Chem Commun 50:3443–3446
196. Lam LT, Louey R (2006) Development of ultra-battery for hybrid-electric vehicle applications. J Power Sources 158:1140–1148
197. Yu N, Gao L, Zhao S et al (2009) Electrodeposited PbO_2 thin film as positive electrode in PbO2/AC hybrid capacitor. Electrochim Acta 54:3835–3841
198. Shen PK, Wei XL (2003) Morphologic study of electrochemically formed lead dioxide. Electrochim Acta 48:1743–1747
199. Bartlett PN, Dunford T, Ghanem MA (2002) Templated electrochemical deposition of nanostructured macroporous PbO_2. J Mater Chem 12:3130–3135
200. Ding LX, Zheng FL, Wang JW et al (2012) Super-large dendrites composed of trigonal PbO_2 nanoplates with enhanced performances for electrochemical devices. Chem Commun 48:1275–1277
201. Perret P, Khani Z, Brousse T et al (2011) Carbon/PbO_2 asymmetric electrochemical capacitor based on methanesulfonic acid electrolyte. Electrochim Acta 56:8122–8128
202. Ni JF, Wang HB, Qu YH et al (2013) PbO_2 electrodeposited on graphite for hybrid supercapacitor applications. Phys Scr 87:1–4
203. Dan YY, Lin HB, Liu XL et al (2012) Porous quasi three-dimensional nano-$Mn_3O_4 + PbO_2$ composite as supercapacitor electrode material. Electrochim Acta 83:175–182
204. Su D, Ahn HJ, Wang G (2013) SnO_2@graphene nanocomposites as anode materials for Na-ion batteries with superior electrochemical performance. Chem Commun 49:3131–3133
205. Ding H, Zhu JH, Jiang J et al (2012) Preparation and gas-sensing property of ultra-fine NiO/SnO_2 nano-particles. RSC Adv 2:10324–10329
206. Manivel P, Ramakrishnan S, Kothurkar NK et al (2013) Optical and electrochemical studies of polyaniline/SnO_2 fibrous nanocomposites. Mater Res Bull 48:640–645
207. Rao CRK, Vijayan M, Anwar S et al (2011) Hydrolysis of $SnCl_2$ on polyaniline: Formation of conducting PAni-SnO_2 composite with enhanced electrochemical properties. J Appl Polym Sci 124:4819–4826
208. Davazoglou D (1997) Optical properties of SnO_2 thin films grown by atmospheric pressure chemical vapour deposition oxidizing $SnCl_4$. Thin Solid Films 302:204–213
209. Mishra S, Ghanshyam C, Ram N et al (2002) Alcohol sensing of tin oxide thin film prepared by sol-gel process. Bull Mater Sci 25:231–234
210. Jiao Z, Wan XJ, Zhao B et al (2008) Effects of electron beam irradiation on tin dioxide gas sensors. Bull Mater Sci 31:83–86
211. Patil UM, Salunkhe RR, Gurav KV et al (2008) Chemically deposited nanocrystalline NiO thin films for supercapacitor application. Appl Surf Sci 255:2603–2607
212. Shinde DV, Lee DY, Patil SA et al (2013) Anodically fabricated self-organized nanoporous tin oxide film as a supercapacitor electrode material. RSC Adv 3:9431–9435
213. Pusawale SN, Deshmukh PR, Lokhande CD (2011) Chemical synthesis of nanocrystalline SnO_2 thin films for supercapacitor application. Appl Surf Sci 257:9498–9502
214. Hwang SW, Hyun SH (2007) Synthesis and characterization of tin oxide/carbon aerogel composite electrodes for electrochemical supercapacitors. J Power Sources 172:451–459
215. Lim SP, Huang NM, Lim HN (2013) Solvothermal synthesis of SnO_2/graphene nanocomposites for supercapacitor application. Ceram Int 39:6647–6655
216. Hsieh CT, Lee WY, Lee CE et al (2014) Electrochemical capacitors fabricated with tin oxide/graphene oxide nanocomposites. J Phys Chem C 118:15146–15153
217. Wang L, Chen L, Yan B et al (2014) In situ preparation of SnO_2@polyaniline nanocomposites and their synergetic structure for high-performance supercapacitors. J Mater Chem A 2:8334–8341

218. Xu CH, Sun J, Gao L (2011) Synthesis of novel hierarchical graphene/polypyrrole nanosheet composites and their superior electrochemical performance. J Mater Chem 21:11253–11258
219. Wang WJ, Hao QL, Lei W et al (2012) Graphene/SnO_2/polypyrrole ternary nanocomposites as supercapacitor electrode materials. RSC Adv 2:10268–10274
220. Wang YG, Luo JY, Wang CX et al (2006) Hybrid aqueous energy storage cells using activated carbon and lithium-ion intercalated compounds. J Electrochem Soc 153: A1425–1431
221. Tang W, Liu LL, Tian S et al (2010) Nano-$LiCoO_2$ as cathode material of large capacity and high rate capability for aqueous rechargeable lithium batteries. Electrochem Commun 12:1524–1526
222. Wang GJ, Qu QT, Wang B et al (2009) Electrochemical behavior of $LiCoO_2$ in a saturated aqueous $LiSO_4$ solution. Electrochim Acta 54:1199–1203
223. Wang GJ, Fu LJ, Zhao NH et al (2007) An aqueous rechargeable lithium battery with good cycling performance. Angew Chem Int Ed 119:299–301
224. Wang GJ, Yang LC, Qu QT et al (2010) An aqueous rechargeable lithium battery based on doping and intercalation mechanisms. J Solid State Electrochem 14:865–869
225. Winter M, Besenhard JO, Spahr ME et al (1998) Insertion electrode materials for rechargeable lithium batteries. Adv Mater 10:725–763
226. Wu YP, Yuan XY, Dong C et al (2012) Lithium ion batteries: practice and applications, 2nd edn. Chemical Industry Press, Beijing
227. Shaju KM, Bruce PG (2008) A stoichiometric nano-$LiMn_2O_4$ spinel electrode exhibiting high power and stable cycling. Chem Mater 20:5557–5562
228. Wu L, Dahn JR, Wainwright DS (1994) Rechargeable lithium batteries with aqueous electrolytes. Science 264:1115–1118
229. Zhao MS, Song XP, Wang F et al (2011) Electrochemical performance of single crystalline spinel $LiMn_2O_4$ nanowires in an aqueous $LiNO_3$ solution. Electrochim Acta 56:5673–5678
230. Qu QT, Fu LJ, Zhan XY et al (2011) Porous $LiMn_2O_4$ as cathode material with high power and excellent cycling for aqueous rechargeable lithium batteries. Energy Environ Sci 4:3985–3990
231. Stojković IB, Cvjetićanin ND, Mentus SV (2010) The improvement of the Li-ion insertion behaviour of $Li_{1.05}Cr_{1.05}Mn_{1.85}O_4$ in an aqueous medium upon addition of vinylene carbonate. Electrochem Commun 12:371–373
232. Wang FX, Xiao SY, Shi Y et al (2013) Spinel $LiNi_xMn_{2-x}O_4$ as cathode material for aqueous rechargeable lithium batteries. Electrochim Acta 93:301–306
233. Yuan A, Tian L, Xu W et al (2010) Al-doped spinel $LiAl_{0.1}Mn_{1.9}O_4$ with improved high-rate cyclability in aqueous electrolyte. J Power Sources 195:5032–5038
234. Wang FX, Xiao SY, Zhu YS et al (2014) Spinel $LiMn_2O_4$ nanohybrid as high capacitance positive electrode material for supercapacitors. J Power Sources 246:19–23
235. Tang W, Tian S, Liu LL et al (2011) Nanochain $LiMn_2O_4$ as ultra-fast cathode material for aqueous rechargeable lithium batteries. Electrochem Commun 13:205–208
236. Wang FX, Xiao SY, Gao XW et al (2013) Nanoporous $LiMn_2O_4$ spinel prepared at low temperature as cathode material for aqueous supercapacitors. J Power Sources 242:560–565
237. Wang GJ, Zhang HP, Fu LJ et al (2007) Aqueous rechargeable lithium battery (ARLB) based on LiV_3O_8 and $LiMn_2O_4$ with good cycling performance. Electrochem Commun 9:1873–1876
238. Liu L, Tian F, Wang X et al (2012) Electrochemical behavior of spherical $LiNi_{1/3}Co_{1/3}Mn_{1/3}O_2$ as cathode material for aqueous rechargeable lithium batteries. J Solid State Electrochem 16:491–497
239. Wang YG, Lou JY, Wu W et al (2007) Hybrid aqueous energy storage cells using activated carbon and lithium-ion intercalated compounds. J Electrochem Soc 154:A228–234
240. Wang GJ, Fu LJ, Wang B et al (2007) An aqueous rechargeable lithium battery based on LiV_3O_8 and $Li[Ni_{1/3}Co_{1/3}Mn_{1/3}]O_2$. J Appl Electrochem 38:579–581

241. Zhao Y, Wang YY, Lai QY et al (2009) Pseudocapacitance properties of AC/$LiNi_{1/3}Co_{1/3}Mn_{1/3}O_2$ asymmetric supercapacitor in aqueous electrolyte. Synthetic Met 159:331–337
242. Zheng J, Chen JJ, Jia X et al (2010) Electrochemical performance of the $LiNi_{1/3}Co_{1/3}Mn_{1/3}O_2$ in aqueous electrolyte. J Electrochem Soc 157:A702–706
243. Shivashankaraiah RB, Manjunatha H, Mahesh KC et al (2012) Electrochemical characterization of polypyrrole–$LiNi_{1/3}Co_{1/3}Mn_{1/3}O_2$ composite cathode material for aqueous rechargeable lithium batteries. J Solid State Electrochem 16:1279–1290
244. Parant JP, Olazcuaga R, Devalette M et al (1971) Sur quelques nouvelles phases de formule Na_xMnO_2. J Solid State Chem 3:1–11
245. Mendiboure A, Delmas C, Hagenmuller P (1985) Electrochemical intercalation and deintercalation of Na_xMnO_2 bronzes. J Solid State Chem 57:323–331
246. Hoppe R, Brachtel G, Jansen M (1975) Zur kenntnis der oxomanganate(III): Über $LiMno_2$ und $\beta-NaMnO_2$. Z Anorg Allg Chem 417:1–10
247. Ma XH, Chen HL, Ceder GB (2011) Electrochemical properties of monoclinic $NaMnO_2$. J Electrochem Soc 158:A1307–1312
248. Doeff MM, Richardson TJ, Kepley L (1996) Lithium insertion processes of orthorhombic Na_xMnO_2 -based electrode materials. J Electrochem Soc 143:2507–2516
249. Sauvage F, Laffont L, Tarascon JM et al (2007) Study of the insertion/deinsertion mechanism of sodium into $Na_{0.44}MnO_2$. Inorg Chem 46:3289–3294
250. Qu QT, Shi Y, Tian S et al (2009) A new cheap asymmetric aqueous supercapacitor: activated carbon//$NaMnO_2$. J Power Sources 194:1222–1225
251. Zhang BH, Liu Y, Chang Z et al (2014) Nanowire $Na_{0.35}MnO_2$ from a hydrothermal method as a cathode material for aqueous asymmetric supercapacitors. J Power Sources 253:98–103
252. Athouël L, Moser F, Dugas R et al (2008) Variation of the MnO_2 birnessite structure upon charge/discharge in an electrochemical supercapacitor electrode in aqueous Na_2SO_4 electrolyte. J Phys Chem C 112:7270–7277
253. Lu YL, Wei M, Wang ZQ et al (2004) Characterization of structure and electrochemical properties of lithium manganese oxides for lithium secondary batteries hydrothermally synthesized from $\delta-K_xMnO_2$. Electrochim Acta 49:2361–2367
254. Lu Y, Yang L, Wei M et al (2007) Studies on structure and electrochemical properties of pillared $M-MnO_2$ ($M=Ba^{2+}$, Sr^{2+}, ZrO^{2+}). J Solid State Electrochem 11:1157–1162
255. Chen RJ, Whittingham MS (1997) Cathodic behavior of alkali manganese oxides from permanganate. J Electrochem Soc 144:L64–67
256. Yoshitomo O, Takayoshi S, Wang LZ et al (2003) Redoxable nanosheet crystallites of MnO_2 derived via delamination of a layered manganese oxide. J Am Chem Soc 125:3568–3575
257. Wang LZ, Omomo Y, Sakai N et al (2003) Fabrication and characterization of multilayer ultrathin films of exfoliated MnO_2 nanosheets and polycations. Chem Mater 15:2873–2878
258. Nakayama M, Konishi S, Tagashira H et al (2005) Electrochemical synthesis of layered manganese oxides intercalated with tetraalkylammonium ions. Langmuir 21:354–359
259. Ching S, Landrigan JA, Jorgensen ML (1995) Sol-Gel synthesis of birnessite from $KMnO_2$ and simple sugars. Chem Mater 7:1604–1606
260. Shchukin DG, Sukhorukov GB (2004) Nanoparticle synthesis in engineered organic nanoscale reactors. Adv Mater 16:671–682
261. Tartaj P (2012) Layered manganates from soft-templates: preparation, characterization and enhanced dye demethylation capabilities. J Mater Chem 22:17718–17723
262. Qu QT, Li L, Tian S et al (2010) A cheap asymmetric supercapacitor with high energy at high power: activated carbon//$K_{0.27}MnO_2{\cdot}0.6H_2O$. J Power Sources 195:2789–2794
263. Zhang BH, Liu Y, Chang Z et al (2014) Nanowire $K_{0.19}MnO_2$ from hydrothermal method as cathode material for aqueous supercapacitors of high energy density. Electrochim Acta 130:693–698
264. Reddy RN, Reddy RG (2003) Sol–gel MnO_2 as an electrode material for electrochemical capacitors. J Power Sources 124:330–337

265. Shao J, Li XY, Qu QT et al (2013) Study on different power and cycling performance of crystalline $K_xMnO_2 \cdot nH_2O$ as cathode material for supercapacitors in Li_2SO_4, Na_2SO_4, and K_2SO_4 aqueous electrolytes. J Power Sources 223:56–61
266. Kim SH, Kim SJ, Oh SM (1999) Preparation of layered MnO_2 via thermal decomposition of $KMnO_2$ and its electrochemical characterizations. Chem Mater 11:557–563
267. Wu YQ, Chen XY, Ji PT et al (2011) Sol–gel approach for controllable synthesis and electrochemical properties of $NiCo_2O_4$ crystals as electrode materials for application in supercapacitors. Electrochim Acta 56:7517–7522
268. Wei TY, Chen CH, Chien HC et al (2010) A cost-effective supercapacitor material of ultrahigh specific capacitances: spinel nickel cobaltite aerogels from an epoxide-driven sol–gel process. Adv Mater 22:347–351
269. Alca´ ntara R, Jaraba M, Lavela P et al (2002) $NiCo_2O_4$ spinel: first report on a transition metal oxide for the negative electrode of sodium-ion batteries. Chem Mater 14:2847–2848
270. Kobayashi Y, Ke XL, Hata H et al (2008) Soft chemical conversion of layered double hydroxides to superparamagnetic spinel platelets. Chem Mater 20:2374–2381
271. Cabo M, Pellicer E, Rossinyol E et al (2009) Mesoporous $NiCo_2O_4$ spinel: influence of calcination temperature over phase purity and thermal stability. Cryst Growth Des 9:4814–4821
272. Verma S, Joshi HM, Jagadale T et al (2008) Nearly monodispersed multifunctional $NiCo_2O_4$ spinel nanoparticles: magnetism, infrared transparency, and radiofrequency absorption. J Phys Chem C 112:15106–15112
273. Li YN, Zhang P, Guo ZP et al (2008) $NiCo_2O_4$/C nanocomposite as a highly reversible anode material for lithium-ion batteries. Electrochem Solid-State Lett 11:A64–67
274. Hsu CT, Hu CC (2013) Synthesis and characterization of mesoporous spinel $NiCo_2O_4$ using surfactant-assembled dispersion for asymmetric supercapacitors. J Power Sources 242:662–671
275. Kong LB, Lu C, Liu MC et al (2013) Effect of surfactant on the morphology and capacitive performance of porous $NiCo_2O_4$. J Solid State Electrochem 17:1463–1471
276. Zou RJ, Xu KB, Wang T et al (2013) Chain-like $NiCo_2O_4$ nanowires with different exposed reactive planes for high-performance supercapacitors. J Mater Chem A 1:8560–8566
277. Zhang G, Lou XW (2013) Controlled growth of $NiCo_{(2)}O_{(4)}$ nanorods and ultrathin nanosheets on carbon nanofibers for high-performance supercapacitors. Sci Rep 3:1–6
278. Huang L, Chen D, Ding Y et al (2013) Nickel-cobalt hydroxide nanosheets coated on $NiCo_2O_4$ nanowires grown on carbon fiber paper for high-performance pseudocapacitors. Nano Lett 13:3135–3139
279. Du J, Zhou G, Zhang H et al (2013) Ultrathin porous $NiCo_2O_4$ nanosheet arrays on flexible carbon fabric for high-performance supercapacitors. ACS Appl Mater Interfaces 5:7405–7409
280. He GY, Wang L, Chen HQ et al (2013) Preparation and performance of $NiCo_2O_4$ nanowires-loaded graphene as supercapacitor material. Mater Lett 98:164–167
281. Wang QF, Wang XF, Liu B et al (2013) $NiCo_2O_4$ nanowire arrays supported on Ni foam for high-performance flexible all-solid-state supercapacitors. J Mater Chem A 1:2468–2473
282. Liu XY, Zhang YQ, Xia XH et al (2013) Self-assembled porous $NiCo_2O_4$ hetero-structure array for electrochemical capacitor. J Power Sources 239:157–163
283. Zhou Q, Xing J, Gao Y et al (2014) Ordered assembly of $NiCo_2O_4$ multiple hierarchical structures for high-performance pseudocapacitors. ACS Appl Mater Interfaces 6:11394–11402
284. Yu L, Zhang G, Yuan C et al (2013) Hierarchical $NiCo_2O_4$@MnO_2 core-shell heterostructured nanowire arrays on Ni foam as high-performance supercapacitor electrodes. Chem Commun 49:137–139
285. Liu MC, Kong LB, Lu C et al (2012) A sol-gel process for fabrication of NiO/$NiCo_2O_4$/Co_3O_4 composite with improved electrochemical behavior for electrochemical capacitors. ACS Appl Mater Interfaces 4:4631–4636

286. Zhou W, Kong D, Jia X et al (2014) $NiCo_2O_4$ nanosheet supported hierarchical core–shell arrays for high-performance supercapacitors. J Mater Chem A 2:6310–6315
287. Karthikeyan K, Kalpana D, Renganathan NG (2008) Synthesis and characterization of $ZnCo_2O_4$ nanomaterial for symmetric supercapacitor applications. Ionics 15:107–110
288. Sharma Y, Sharma N, Subba Rao GV et al (2007) Nanophase $ZnCo_2O_4$ as a high performance anode material for Li-ion batteries. Adv Funct Mater 17:2855–2861
289. Deng D, Lee JY (2011) Linker-free 3D assembly of nanocrystals with tunable unit size for reversible lithium ion storage. Nanotechnology 22:1–9
290. Luo W, Hu XL, Sun YM et al (2012) Electrospun porous $ZnCo_2O_4$ nanotubes as a high-performance anode material for lithium-ion batteries. J Mater Chem 22:8916–8921
291. Qiu YC, Yang SH, Deng H et al (2010) A novel nanostructured spinel $ZnCo_2O_4$ electrode material: morphology conserved transformation from a hexagonal shaped nanodisk precursor and application in lithium ion batteries. J Mater Chem 20:4439–4444
292. Du N, Xu Y, Zhang H et al (2011) Porous $ZnCo_2O_4$ nanowires synthesis via sacrificial templates: high-performance anode materials of Li-ion batteries. Inorg Chem 50:3320–3324
293. Liu B, Zhang J, Wang X et al (2012) Hierarchical three-dimensional $ZnCo_{(2)}O_{(4)}$ nanowire arrays/carbon cloth anodes for a novel class of high-performance flexible lithium-ion batteries. Nano Lett 12:3005–3011
294. Liu HW, Wang J (2013) One-pot synthesis of $ZnCo_2O_4$ nanorod anodes for high power lithium ions batteries. Electrochim Acta 92:371–375
295. Reddy MV, Kenrick KYH, Wei TY et al (2011) Nano-$ZnCo_2O_4$ material preparation by molten salt method and its electrochemical properties for lithium batteries. J Electrochem Soc 158:A1423–1430
296. Davis M, Gümeci C, Black B et al (2012) Tailoring cobalt doped zinc oxide nanocrystals with high capacitance activity: factors affecting structure and surface morphology. RSC Adv 2:2061–2066
297. Zhou G, Zhu J, Chen Y et al (2014) Simple method for the preparation of highly porous $ZnCo_2O_4$ nanotubes with enhanced electrochemical property for supercapacitor. Electrochim Acta 123:450–455
298. Bao F, Wang X, Zhao X et al (2014) Controlled growth of mesoporous $ZnCo_2O_4$ nanosheet arrays on Ni foam as high-rate electrodes for supercapacitors. RSC Adv 4:2393–2397
299. Wang S, Pu J, Tong Y et al (2014) $ZnCo_2O_4$ nanowire arrays grown on nickel foam for high-performance pseudocapacitors. J Mater Chem A 2:5434–5440
300. Ding Y, Wan Y, Min YL et al (2008) General synthesis and phase control of metal molybdate hydrates $MMoO_4 \cdot nH_2O$ (M = Co, Ni, Mn, n = 0, 3/4, 1) nano/microcrystals by a hydrothermal approach: magnetic, photocatalytic, and electrochemical properties. Inorg Chem 47:7813–7823
301. Livage C, Hynaux A, Marrot J et al (2002) Solution process for the synthesis of the "high-pressure" phase $CoMoO_4$ and X-ray single crystal resolution. J Mater Chem 12:1423–1425
302. Robertson L, Duttine M, Gaudon M et al (2011) Cobalt–zinc molybdates as new blue pigments involving Co^{2+} in distorted trigonal bipyramids and octahedra. Chem Mater 23:2419–2427
303. Cherian CT, Reddy MV, Haur SC et al (2013) Interconnected network of $CoMoO_4$ submicrometer particles as high capacity anode material for lithium ion batteries. ACS Appl Mater Interfaces 5:918–923
304. Liu MC, Kong LB, Lu C et al (2013) Facile fabrication of $CoMoO_4$ nanorods as electrode material for electrochemical capacitors. Mater Lett 94:197–200
305. Liu MC, Kong LB, Ma XJ et al (2012) Hydrothermal process for the fabrication of $CoMoO_4 \cdot 0.9H_2O$ nanorods with excellent electrochemical behavior. New J Chem 36:1713–1716
306. Xu ZW, Li Z, Tan XH et al (2012) Supercapacitive carbon nanotube-cobalt molybdate nanocomposites prepared via solvent-free microwave synthesis. RSC Adv 2:2753–2755

307. Xia XF, Lei W, Hao QL et al (2013) One-step synthesis of $CoMoO_4$/graphene composites with enhanced electrochemical properties for supercapacitors. Electrochim Acta 99:253–261
308. Xu K, Chao J, Li W et al (2014) $CoMoO_4 \cdot 0.9H_2O$ nanorods grown on reduced graphene oxide as advanced electrochemical pseudocapacitor materials. RSC Adv 4:34307–34314
309. Mai LQ, Yang F, Zhao YL et al (2011) Hierarchical $MnMoO_{(4)}$/$CoMoO_4$ heterostructured nanowires with enhanced supercapacitor performance. Nat Commun 2:1–5
310. Guo D, Zhang HM, Yu XZ et al (2013) Facile synthesis and excellent electrochemical properties of $CoMoO_4$ nanoplate arrays as supercapacitors. J Mater Chem A 1:7247–7254
311. Wang X, Sumboja A, Lin M et al (2012) Enhancing electrochemical reaction sites in nickel-cobalt layered double hydroxides on zinc tin oxide nanowires: a hybrid material for an asymmetric supercapacitor device. Nanoscale 4:7266–7272
312. Guo D, Zhang P, Zhang HM et al (2013) $NiMoO_4$ nanowires supported on Ni foam as novel advanced electrodes for supercapacitors. J Mater Chem A 1:9024–9027
313. Purushothaman KK, Cuba M, Muralidharan G (2012) Supercapacitor behavior of $\alpha -$ $MnMoO_4$ nanorods on different electrolytes. Mater Res Bull 47:3348–3351
314. Park KS, Seo SD, Shim HW et al (2012) Electrochemical performance of $Ni_xCo_{1-x}MoO \cdot$ $(0 \leq x \leq 1)$ nanowire anodes for lithium-ion batteries. Nanoscale Res Lett 7:1–7
315. Liu MC, Kang L, Kong LB et al (2013) Facile synthesis of $NiMoO_4$ center dot xH_2O nanorods as a positive electrode material for supercapacitors. RSC Adv 3:6472–6478
316. Li W, Zhang F, Dou YQ et al (2011) A self-template strategy for the synthesis of mesoporous carbon nanofibers as advanced supercapacitor electrodes. Adv Energy Mater 1:382–386
317. Zhou W, Lin LJ, Wang WJ et al (2011) Hierarchical mesoporous hematite with "electron-transport channels" and its improved performances in photocatalysis and lithium ion batteries. J Phys Chem C 115:7126–7133
318. Luo Y, Luo J, Jiang J et al (2012) Seed-assisted synthesis of highly ordered $TiO_2@Fe_2O_3$ core/shell arrays on carbon textiles for lithium-ion battery applications. Energy Environ Sci 5:6559–6566
319. Taberna PL, Mitra S, Poizot P et al (2006) High rate capabilities Fe_3O_4-based Cu nano-architectured electrodes for lithium-ion battery applications. Nature Mater 5:567–573
320. Jiang J, Liu JP, Zhou WW et al (2011) CNT/Ni hybrid nanostructured arrays: synthesis and application as high-performance electrode materials for pseudocapacitors. Energy Environ Sci 4:5000–5007

Chapter 8
Carbon Materials for Supercapacitors

Weijia Zhou, Xiaojun Liu, Kai Zhou, and Jin Jia

8.1 Porous Carbon

8.1.1 Random Porous Structure

Random porous carbons are the mostly widely used as electrode materials due to their large surface area, relatively good electrical properties, and moderate cost. The majority of random porous carbons are derived from carbon-rich organic precursors by heat treatment in inert atmospheres. The ultimate properties of these carbons are mainly dependent on carbon precursor and heat treatment. Some carbon precursors pass through a fluid stage during carbonization that allows large aromatic molecules to align with each other and form a more extensive pre-graphitic structure. Other carbon precursors can retain a solid phase during carbonization, and the limited mobility of the crystallites leads to the formation of a rigid amorphous structure that consists of randomly oriented graphene layers. Random porous carbons are generally produced from physical (thermal) and/or chemical activation of various carbon precursors.

8.1.1.1 Physical Activation

Carbon precursor is first exposed to pyrolysis in an inert atmosphere at 400–900 °C to eliminate the bulk of volatile matter, followed by partial gasification using an oxidizing gas at 350–1000 °C. Also, CO_2 and steam are used as activating agents. Generally, activation with CO_2 makes it possible to achieve a greater uniformity of

W. Zhou (✉) • X. Liu • K. Zhou • J. Jia
New Energy Research Institute, School of Environment and Energy, Guangzhou Higher Education Mega Center, South China University of Technology, Guangzhou 510006, China
e-mail: eszhouwj@scut.edu.cn

K.I. Ozoemena, S. Chen (eds.), *Nanomaterials in Advanced Batteries and Supercapacitors*, Nanostructure Science and Technology,
DOI 10.1007/978-3-319-26082-2_8

pores compared with activation using steam. The reaction mechanism is actually complex, involving the fixation and then the dissociation of CO_2 on active sites to form unstable intermediates. In physical activation process, the physical nature of physically activated carbons is dependent on the precursor, the oxidizing agent employed, the temperature of activation, and the degree of activation. Depending on these factors, activated carbons with high porosity and varying surface chemistry can be achieved. As a general trend, the higher activation temperature/activation time, the larger porosity was developed. However, higher porosity developments are usually accompanied by a broadening of the pore size distribution. It is well known that the porous structure of activated carbons produced by activation processes have a broad pore size distribution consisting of micropores (<2 nm), mesopores (2–50 nm), and macropores (>50 nm). Several researchers have pointed out the discrepancy between the capacitance of the activated carbons and their specific surface area. With a high surface area up to 3000 $m^2\ g^{-1}$, only a relatively small specific capacitance of 10 mF cm^{-2} was obtained, much smaller than the theoretical electrical double-layer capacitance (15–25 mF cm^{-2}), indicating that not all pores are effective in charge accumulation [1]. Therefore, although the specific surface area is an important parameter for the performance of electrochemical double-layer capacitors (EDLCs), some other aspects of the carbon materials such as pore size distribution, pore shape and structure, electrical conductivity, and surface functionality can also influence their electrochemical performance to a great extent. Efforts have been made to search for the relationship between the nanoporous structure of activated carbon and their capacitance performance in different electrolytes. Xia et al. [2] reported that the prepared activated carbons by CO_2 activation were used as supercapacitor electrode materials. The effects of CO_2 treatment on their pore structures were discussed. Compared to the pristine mesoporous carbons, the samples subjected to CO_2 treatment exhibited remarkable improvement in textural properties. The electrochemical measurement in 6 M KOH electrolyte showed that CO_2 activation leads to better capacitive performances. The activated carbons, obtained after CO_2 treatment for 6 h at 950 °C using CMK-3 as a precursor, showed the best electrochemical behavior with a specific gravimetric capacitance of 223 F g^{-1} and volumetric capacitance of 54 F cm^{-3} at a scan rate of 2 mV s^{-1} and 73 % retained ratio at 50 mV s^{-1}. The good capacitive behavior of activated carbons may be attributed to the hierarchical pore structure (abundant micropores and interconnected mesopores with the size of 3–4 nm), high surface area (2749 $m^2\ g^{-1}$), large pore volume (2.09 $cm^3\ g^{-1}$), as well as well-balanced microporosity and mesoporosity. Fang et al. [3] reported activated carbon aerogels obtained from resorcinol–formaldehyde under a CO_2 flow, which resulted in a doubling of surface area and increase in the micropore volume from 0.02 to 0.61 $cm^3\ g^{-1}$. As a result, the specific capacitance was also doubled up to 100 F g^{-1} in Et_4NBF_4/PC. Zhu et al. [4] have also applied a CO_2 activation step to carbon aerogels synthesized from cresol–formaldehyde, increasing the surface area from 245 $m^2\ g^{-1}$ to between 401 and 1418 $m^2\ g^{-1}$ depending on the activation conditions (temperature and time). This enhancement in porosity was reflected in an increase of capacitance from 78 F g^{-1} up to 146 F g^{-1} in KOH electrolyte. Additionally, the activated carbon aerogel showed a good rate performance, only a 10 % capacitance decrease with the increase of the current density from 1 to 20 mA cm^{-2}. Recently, Kumagai et al. [5] prepared biomass-derived carbons by CO_2 activation. The biomass-derived activated carbon, exhibiting BET (Brunauer–Emmett–Teller) specific surface area of 1357 $m^2\ g^{-1}$, total pore volume of 0.99 mL g^{-1}, and mesopore volume fraction of 44.4 %, showed superior capacitive performance in both aqueous (1 M H_2SO_4) and organic (1 M TEMA · BF_4/PC) electrolytes.

8.1.1.2 Chemical Activation

The required temperatures of chemical activation of the porous structure are generally lower than those prevailing in physical activation with CO_2 or H_2O. It is usually carried out at lower temperatures (from 400 °C to 900 °C) with activating agents. Of the many reagents proposed for chemical activation ($ZnCl_2$, H_3PO_4, $AlCl_3$, $MgCl_2$, KOH, NaOH, etc.), the most common chemical agents are $ZnCl_2$, H_3PO_4, and KOH. $ZnCl_2$ and H_3PO_4 act as dehydrating agents, whereas KOH acts as an oxidant. Chemical activation offers several advantages over physical activation: (a) it is usually carried out only one step, (b) lower activation temperatures, (c) much higher carbon yield, (d) high surface area (up to 3600 $m^2\ g^{-1}$), and (e) the microporosity can be very well developed with a well-controlled distribution.

H_3PO_4 is widely used as an impregnation agent. Because of the highly polar character of H_3PO_4 and hence the control of the physical and chemical interactions occurring in the bulk of the solution and with the substratum during the impregnation treatment, the solution concentration is likely to be the primary factor of the activation process. During the heat treatment, the activating agent present in the interior of the particles produces a dehydrating effect on the carbon precursor. It is presumed that, during this dehydrating step, cross-linking reactions are predominant. The dehydration of the carbon precursor causes reduction in the dimensions of the particle, although such reduction is partially inhibited because the activating agent remains within the particle during the thermal treatment, thus acting as a template for the creation of microporosity. In addition, H_3PO_4 combines with organic species forming phosphate and polyphosphate bridges that connect biopolymer fragments, thereby partially hindering the contraction in the material as the temperature increases. Heterogeneity in the microporosity is observed, which is mostly independent of the amount of H_3PO_4 used, and it is due to the fact that during activation there are no phosphoric acid molecules, but a mixture of molecules (i.e., H_3PO_4, $H_4P_2O_5$, and $H_{13}P_{11}O_{34}$) [6]. Using H_3PO_4 as an activating agent, most of the numerous studies carried out so far on the preparation of porous carbon have focused on the influence of concentration of the impregnation solution and soaking temperature on the porous structure. Recently, Xu et al. [7] observed that activation of reedy grass leaves with H_3PO_4 in N_2 atmosphere and their characteristics were investigated. The surface area of the activated carbons produced at 500 °C for 2 h was 1474 $m^2\ g^{-1}$. Ding et al. [8] reported hydrochar-based porous carbons synthesized by H_3PO_4 activation. The porous carbon showed a hierarchical porous structure; the corresponding BET surface areas and pore volumes were 1498 m^2 g^{-1} and 1.27 cm^3 g^{-1}. The highest specific capacitance was 106.6 F g^{-1} at a current density of 6.25 A g^{-1}. Wang et al. [9] prepared P/N co-doped porous carbon by a simple method through direct carbonization of phosphoric acid-doped polyaniline. P/N co-doped carbons exhibited typical microporous characters with a surface area of 353.0 m^2 g^{-1} and a narrow pore size distribution, which had high specific capacitance (154.4 F g^{-1}) and excellent stability. On the other hand, activated carbon prepared by KOH activation had well-defined pore size distribution and high specific surface area [10, 11]. Although the KOH activation is a well-known method to generate the pore network in carbons, the activation mechanism has not been well understood because of the complexity due to the large number of variables in both the experimental parameters and the reactivity of different precursors used.

Among the various activating agents, KOH generates activated carbons with an extremely large surface area up to 4000 m^2 g^{-1}, a high pore volume up to 2.7 cm^3 g^{-1}, and tunable and narrow pore size distribution through the control of the activating conditions (i.e., temperature and amount of KOH). For other activating agents, the surface areas are normally between 1500 and 2000 m^2 g^{-1} with pore volumes <1.5 cm^3 g^{-1} and broad pore size distribution. Kang et al. [12] investigated the effect of pre-carbonization of mesophase pitch-based porous carbons on the textural properties and electrochemical performance of the supercapacitor. Pre-carbonization was beneficial for growing graphitic microcrystallites and hindered the crystallites from the intersection during KOH activation, and thus the specific surface area and the pore size range of 1.5–2.4 nm of the porous carbon decreased with increasing pre-carbonization temperature. The maximum specific capacitance of 145 F g^{-1} in organic electrolytes was obtained for the activated carbon with 2583 m^2 g^{-1} which was applied in the supercapacitors having high energy density (31 W h kg^{-1}) and power density (12 kW kg^{-1}). Raymundo-Piñero et al. [13] prepared a series of activated carbons with progressively changing nanotextural characteristics by heat treatment of a bituminous coal at temperatures in the range 520–1000 °C and subsequent activation by KOH at 700 °C or 800 °C. The surface area of the activated carbons varied from 800 to 3000 m^2 g^{-1} and the average micropore between 0.65 and 1.51 nm depending on the carbonization temperature. The specific capacitance of the carbons was analyzed in H_2SO_4, KOH, and $TEABF_4$/acetonitrile solutions, measuring values of 160–310, 124–286, and 120–180 F g^{-1}, respectively. Furthermore, KOH-activated carbons inevitably contain the surface oxygen groups, making a positive contribution to the capacitance [14, 15]. Ling et al. [15] adopted a quick KOH activation of coal to prepare oxygen-rich activated carbons with the specific surface area of 1950 m^2 g^{-1}, good conductivity, and high oxygen content of up to 12 wt%. Compared to high specific surface area activated carbons prepared by the conventional KOH activation, oxygen-rich activated carbons presented a higher specific capacitance of 370 F g^{-1} in KOH electrolytes. Nitrogen-doped porous carbons prepared through KOH activation of a nonporous nitrogen-enriched carbon which is synthesized by pyrolysis of the polymerized ethylenediamine and carbon tetrachloride have well-developed specific surface area reaching from 521 to 1913 $m^2\ g^{-1}$ and 10.8 to 1.1 wt% of nitrogen. Increased capacity and charge exchange were observed due to the nitrogen heteroatoms introduced. The porous carbon materials exhibited the largest specific capacitance of 363 F g^{-1} at a current density of 0.1 A g^{-1} in 1 M H_2SO_4 aqueous electrolyte [16].

8.1.2 Ordered Pore Structure

As above discussion, the synthesis of porous carbons with high surface areas could be simply achieved by direct carbonization of organic precursors with physical or chemical activations [10]. However, the carbon materials obtained from these methods often contain disordered structures with wide size distributions, which greatly limit thier broad applications, especially in dealing with selective molecular diffusions or discriminations. To control the ordered pore structures with relatively narrow pore size distribution, the use of the template carbonization route is an effective approach. A templating method offers an effective way to produce nanostructured carbons with well-controlled narrow pore size distribution, ordered pore structures, large specific surface area, and an interconnected pore network, making them promising candidates for supercapacitor electrode materials. There has been remarkable progress in the synthesis of ordered nanostructured carbons through template techniques [17]. In general, the preparation procedure of templated carbons is infiltration of a carbon precursor into the pores of the template, followed by a carbonization treatment and finally the removal of the template to leave behind a porous carbon structure. The schematic representation in Fig. 8.1 shows the general concept of the template technique and the porous carbons obtained from different templates.

Template-assisted synthesis, which typically involves various inorganic or organic nanostructures as templates, is one of the most effective methods to prepare porous carbon materials. We mainly divide these approaches into two categories: hard-template synthesis and soft-template synthesis.

8.1.2.1 Hard Templates

The hard templates, mostly nanoscaled silica materials, are infiltrated with carbon precursors (e.g., sucrose, furfuryl alcohol, phenolic resin, pitches, acetonitrile), which are then polymerized and finally carbonized. The preparation of ordered porous carbon by hard templates generally involves four major steps: (1) preparation of hard templates, (2) functionalization/modification of template surface to achieve favorable surface properties, (3) coating the templates by various approaches, and (4) selective removal of the templates. The silica particles, MgO platelet, nickel foam and powder, ice, and block copolymer films have been used as templates for the fabrication of ordered porous carbons [18–22]. Since the synthesis

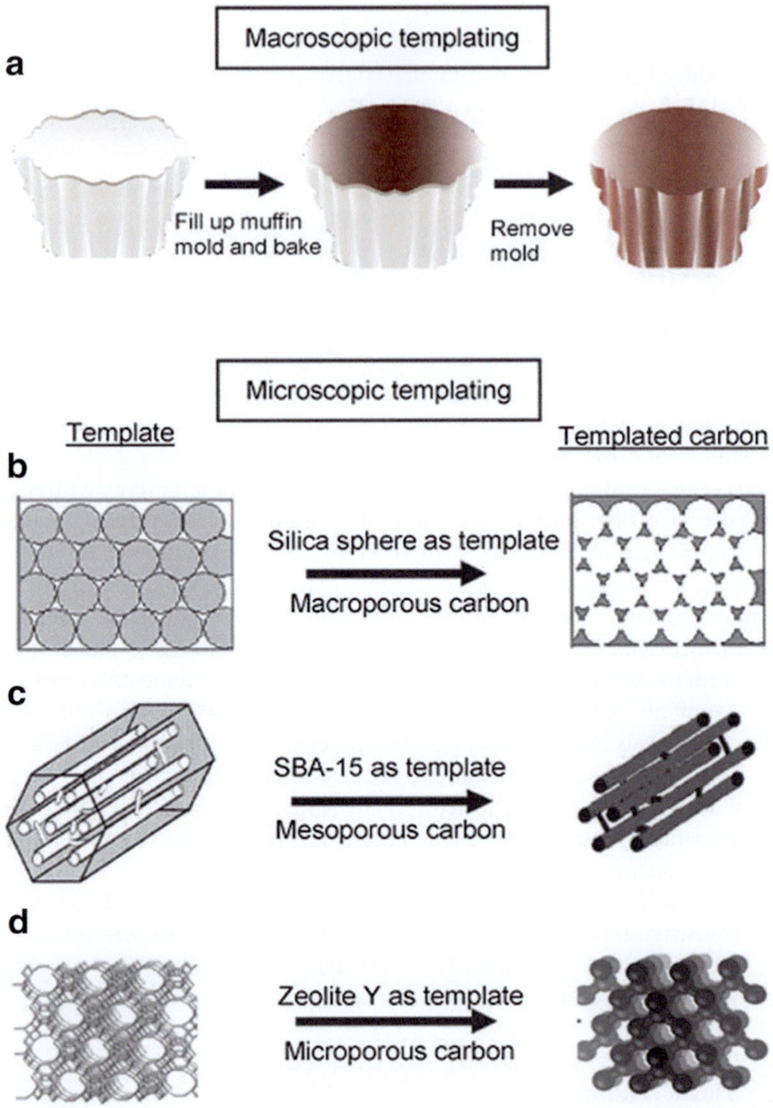

Fig. 8.1 (**a**) Macroscopic representation showed the general concept of the templating technique, microscopic synthesis of (**b**) macroporous carbons using silica spheres as template, (**c**) mesoporous carbons using SBA-15 as template, and (**d**) microporous carbons using zeolite Y as template [17] (Reprinted with permission from Ref. [17]. Copyright (2009) The Royal Society of Chemistry)

of ordered mesoporous carbons using the hard-templating strategy [23], ordered porous carbons with various pore symmetries, pore diameters/lengths, and chemical composition have been intensively investigated as electrode materials for supercapacitors. For example, Zhao et al. [24] prepared ordered porous carbon derived from ferrocene as the precursor and mesoporous silica nanospheres used as the template. The as-obtained materials with specific surface area of 2396 $m^2\ g^{-1}$, as the symmetric supercapacitor electrode, demonstrated a specific capacitance of 343 F g^{-1} and delivered energy densities of 62.8 and 9.1 Wh kg^{-1} at power densities of 0.16 and 32 kW kg^{-1}, respectively, in a 1.5 mol L^{-1} tetraethylammonium tetrafluoroborate in acetonitrile electrolyte. Zhang et al. [25] synthesized ordered porous carbon with morphology fiber-like through a hard-template method. The conventional fiber-like SBA-15 was employed as hard template and sucrose as carbon precursor. The as-prepared carbon has a high specific surface area up to 1358 $m^2\ g^{-1}$ and exhibited high capacitance (151 F g^{-1} at a scan rate of 2 mV s^{-1}). Built by this method, nitrogen doping reduced graphite oxide (rGO) was also assembled and showed improved electrical properties. The specific capacitances of the as-pyrolyzed and N-doped macroporous rGO assemblies were measured to be 86.7 and 103.2 F g^{-1} in 1 M aqueous H_2SO_4 aqueous electrolyte, respectively. In contrast, the planar rGO electrode only showed 62.9 F g^{-1} in capacitance [26]. Besides chemical vapor deposition (CVD) methods, direct carbonization of carbon source precursor was also realizable. For example, 3D pillared-porous carbon nanosheets with supporting carbon pillars in the adjacent carbon layers had been prepared using MgO hard template. The unique structure endows the high-rate transportation of electrolyte ions and electrons throughout the electrode matrix, resulting in an excellent electrochemical performance [27]. The Fuertes group also fabricated an ordered mesoporous graphitic carbon material using iron-impregnated polypyrrole as a carbon source and SBA-15 as a template [28]. $FeCl_3$ was used not only as an oxidant for the polymerization but also as a catalyst which promotes the formation of a graphitic structure during the carbonization step. When used as electrode materials for EDLCs, graphitic carbon showed a superior performance to other non-graphitic mesoporous carbons at high current densities. This superior electrode performance seemed to be derived from highly accessible pores and the high conductivity of the graphitic framework.

8.1.2.2 Soft Templates

Hard (solid)-template method is inarguably the most effective and common employed approach for synthesizing ordered porous carbon. However, hard templates have several intrinsic disadvantages, such as the inherent difficulty of achieving high product yields, the multistep synthetic process, and the lack of structural robustness of the shells upon template removal. The issues have prompted interests in developing simpler synthetic approaches to prepare ordered porous carbon that permit easy interaction and release of guest species. Among these approaches, soft (liquid or gaseous) templates have attracted more attentions and achieved some positive progress. The soft-templating approach uses surfactants as structure-directing agents and carbon precursors that interact with the surfactant and assemble around the formed micelles. Varying the surfactant/precursor ratio gives access to different mesostructures. In this approach, the pore size can also be changed by changing the length of the hydrophobic chains of the surfactants. For example, the pore size can be increased from 2.8 to 4.3 nm, if the hydrophobic chain length is enlarged from ($EO_{132}PO_{50}EO_{132}$) to ($EO_{106}PO_{70}EO_{106}$) [29].

Reducing the synthesis temperature and applying swelling agents can further increase the micelle size and the carbon pore size [30]. Furthermore, the size of the carbon precursor and the pre-polymerization degree determine the final pore size. The precursor assembles with the hydrophilic chains giving small mesopores (2–5 nm) or if it is larger, it assembles around the chains giving larger pores (<6 nm). Very large mesopores (>20 nm) are accessible using special high-molecular surfactants such as polystyrene-block-poly(4-vinylpyridine) or poly(ethylene oxide)-polystyrene diblock copolymers. In addition, the pore symmetry (cubic, hexagonal, wormlike) [31] as well as the pore connectivity [32, 33] can be controlled to lead to advanced ion transport properties. In 2004, Dai's group reported direct soft-template synthesis of ordered porous carbon films by using resorcinol–formaldehyde as a carbon source and lab-prepared diblock copolymer polystyrene-*block*-poly(4-vinylpyridine) (PS-P4VP) as a soft template [34]. Resorcinol monomers can interact with the P4VP segment of the copolymer via hydrogen bonds. They were pre-organized into well-ordered mesostructured films with the assistance of amphiphilic PS-P4VP self-assembly through spin coating and solvent annealing and then in situ polymerized with formaldehyde vapor to form a carbon precursor. Upon carbonization under N_2 atmosphere, ordered porous carbons films with a hexagonal structure and large pore size of *ca.* 34 nm were obtained.

Xia's group used a natural crab shell, which contains a number of relatively uniform ellipse-type macropores with dimensions of *ca.* 1×0.5 μm, as a confined space for Pluronic P123-templated self-assembly of phenolic resol (Fig. 8.2) [35]. The obtained mesoporous carbon nanofibers retain an interpenetrated ordered array that is replicated from the crab shell hard template, and all the carbon nanofibers are composed of ordered mesopores with closed to 2D hexagonal structure. The hierarchical pore structures ensure the retention and immersion of electrolyte, as well as greatly reducing the resistance of ion transportation. Their BET surface area and total pore volume are 1270 $m^2\ g^{-1}$ and 4.3 $cm^3\ g^{-1}$. The specific capacitance is estimated to be ~12 $\mu F\ cm^{-2}$ (~152 $F\ g^{-1}$) in 1.0 M [$(C_2H_5)_4N$] [BF_4] electrolyte, and 95 % of initial capacitance can be maintained even after 1000 cycles.

Recently, a novel soft-template method is developed to synthesize a mesoporous carbon/graphene composite. The specific capacitance of composite is up to 242 $F\ g^{-1}$ at the current density of 0.5 $A\ g^{-1}$, which is much higher than mesoporous carbon, graphene, and a sample made by mechanical mixing of mesoporous carbon with graphene. When the current densities were increased to 1, 2, and 4 $A\ g^{-1}$, the specific capacitances of the sample decreased to 203, 168, and 154 $F\ g^{-1}$, respectively [36]. Hence, the soft-template method is one of the most suitable techniques to mass-scale preparation of ordered porous carbons with various pore diameters and structures. Furthermore, soft-templated carbons are also important reference materials providing information about the effect of pore size, morphology, channel structures, and other parameters on the ion diffusion and charge storage in the nano-confined system.

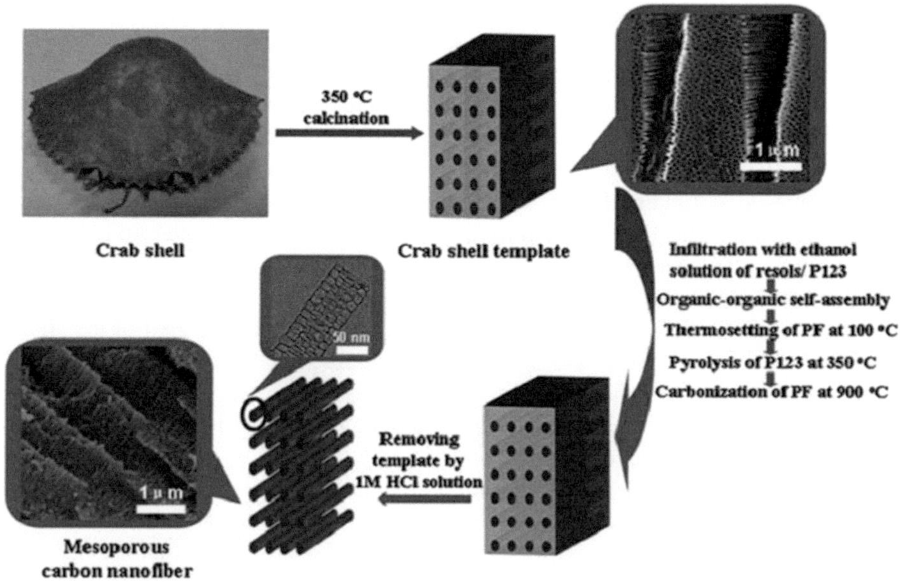

Fig. 8.2 Schematic representation of the highly ordered porous carbon nanofiber arrays prepared by combining the soft-template synthesis of phenolic resin with the hard-template synthesis of carbon in a crab shell [35] (Reproduced with permission from Ref. [35]. Copyright (2010), The Royal Society of Chemistry)

8.1.3 Shell Structure

Hierarchical nanostructured materials that contain interconnected macroporous/mesoporous/microporous structures have demonstrated enhanced properties compared with single-pore-sized porous materials due to the improved mass transport, the enhanced selectivity and maintenance of specific surface area on the level of fine porous systems [37–41]. Therefore, from the viewpoint of practical applications, hierarchical nanostructured carbons (HNCs) with combined macro-/mesoporous porosity are of particular significance and interest because particular applications can be tailored through controllable synthesis of HNCs with designed nanostructures. The most common technique for creating HNCs is "nanocasting," that involves first the creation of a sacrificial hierarchical nanostructured material, i.e., liquid crystal, polymer lattices, or silica template with a hierarchical porous nanostructure, followed by the impregnation of the template with an appropriate carbon source, carbonization, and subsequent removal of the template. Shell structure carbon is one of the most interesting HNCs, which has been demonstrated very promising application in supercapacitor. However, traditional fabrication of shell carbon spheres involves prior preparation of solid core–mesoporous shell (SCMS) silica [42–44], which makes the whole synthesis process complicated and time-consuming. Fu et al. reported simpler synthesis of hollow core–porous shell carbon spheres by the pyrolysis of core–shell polystyrene (PS)-cross-linked polyphosphazene composites [45]. The core size of the carbon spheres was controlled by the PS template size, which is generally so large (i.e., usually larger than 300 nm for commercially available PS spheres) that it may result in high electrical resistance as an electrode material. Alternatively, large quantity of micropores (i.e., less than 0.7 nm in diameter) could be produced through the pyrolysis of polystyrene-cross-linked polyphosphazene composites, but this would limit their application where fast mass transport is required. Yu et al. prepared hollow core–mesoporous shell carbon (HCMSC) capsules by nanocasting method which involves surfactant-assisted synthesis of silica spheres with mesoporous shell as template and impregnation of carbon precursor into mesoporous shell of the silica, followed by carbonization and silica template removal. The resulting HCMSC capsules exhibited excellent capacitive performance (162 F g^{-1} at 0.3 A g^{-1}). Furthermore, by an in situ constructing strategy, shell structure macroporous graphene can be fabricated using graphene oxide (GO) as a carbon precursor and polymer as template such as polymethyl methacrylate (PMMA) spheres and polystyrene (PS) colloidal particles [46–48]. For example, Wang and coworkers prepared doped hierarchically porous graphene (DHPG) on nickel foam using GO, sulfonated polystyrene (S-PS) spheres, and poly(vinylpyrrolidone) (PVP) as precursors [48]. GO sheets could be deposited on the surface of S-PS and assemble into 3D continuous structure with the assistance of PVP. Moreover, nitrogen atoms from PVP and sulfur atoms from S-PS were doped into the graphene during pyrolysis of the precursor (Fig. 8.3). Due to its macroporous structure, graphene electrodes with hierarchically porous architectures were beneficial for ion transport and also provided sufficient contact area between the electrode and electrolyte. Additionally, N, S-co-doping simultaneously increased electrical conductivity and electrochemical activity of graphene in the high-rate electrochemical process.

8.2 Graphene

Graphene, a two-dimensional (2D) form of carbon atoms with the hexagonal lattice structure, has been considered as one of the most important materials in the current century. Since the direct investigation of mechanically exfoliated graphene by Novoselov et al. [49] in 2004, tremendous research efforts have been devoted to the graphene as well as graphene-based materials. Several effective and facile approaches have been developed to synthesize graphene-based materials. Ruoff and coworkers [50] first explored graphene-based EDLCs utilizing this kind of chemically modified graphene (CMG) as electrode materials. Although the individual graphene sheets partially agglomerated into particles approximately 15–25 μm in diameter during the reduction process, the relatively high specific surface area of the graphene-based material (GBM) aggregation (705 m^2 g^{-1}) still allows these CMG electrodes to have high electrochemical performance. Large specific capacitance values of 135 and 99 F g^{-1} for aqueous and organic

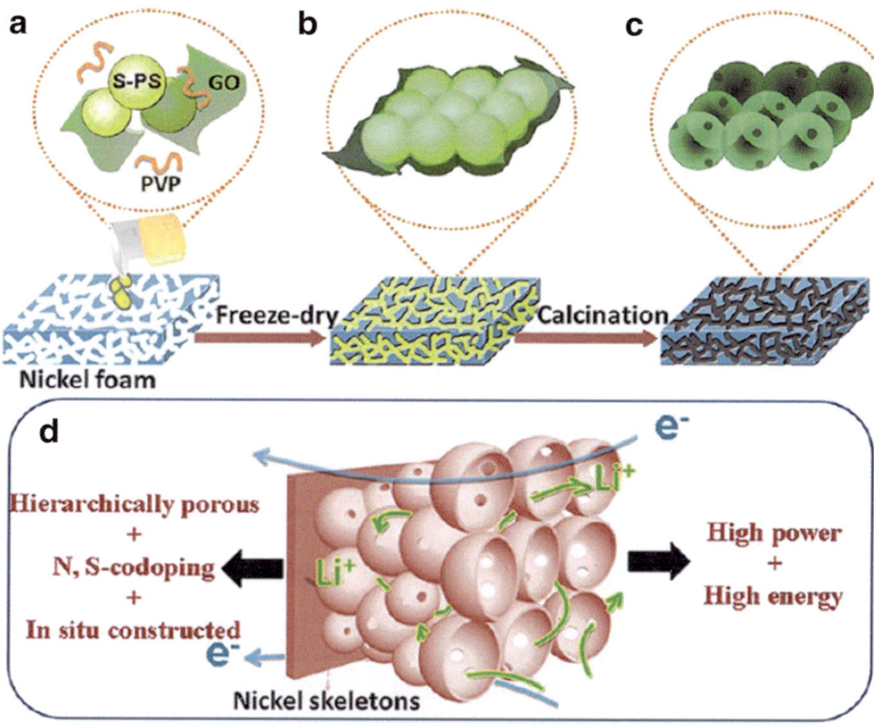

Fig. 8.3 Schematic illustration of the synthesis procedures of the DHPG electrodes: (**a**) impregnating the precursor solution containing GO, PVP, and S-PS into the porous collector of nickel foam; (**b**) freeze-drying the precursor solution; (**c**) in situ calcining the precursor gel; (**d**) illustration of the features in the interior structure of DHPG electrodes [48] (Reproduced with permission from Ref. [48]. Copyright 2013, American Chemical Society)

electrolytes, respectively, were achieved by these CMG materials. In order to avoid the agglomeration of graphene nanosheets, the graphene sols were prepared by CVD, hydrothermal treatment, etc.

8.2.1 Graphene Nanosheets

Graphene nanosheets (GNSs) have emerged as a new class of promising materials for potential applications in supercapacitors. Graphene nanosheet is an ideal carbon electrode material because it is capable of storing an EDLC capacitance up to 550 F g^{-1}, provided that the surface area 2675 m^2 g^{-1} surface area is fully utilized. Many methods have been reported to synthesize graphene sheets over the past few years, including micromechanical exfoliation of graphite [49]; unzipping carbon nanotubes [51]; reduction from graphene oxide (GO), which is synthesized by

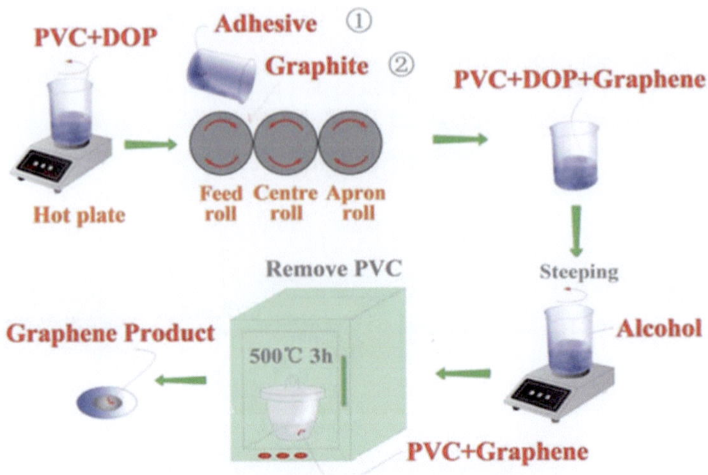

Fig. 8.4 Schematic of continuous mechanical exfoliation of graphene [59] (Reproduced with permission from Ref. [59]. Copyright 2010, The Royal Society of Chemistry)

Hummers methods or modified ones [52]; solvothermal synthesis [53]; epitaxial growth on SiC surfaces [54] and epitaxial growth and CVD growth of graphene on matched substrate [55] [56], bottom-up organic synthesis [57]; etc. Taghioskoui [58] recently summarized various graphene preparation methods. One of the greatest challenges is the stabilization of single graphene sheets with controllable size and morphology. Fundamentals of chemistry and physics are especially important and preparation methods with or without polymeric and surfactant dispersants have been developed to prepare exfoliated graphene. Herein, we mainly focus on two methods, namely, mechanical exfoliation that produces few-layer graphene of high quality and a chemical method that has been demonstrated to give high throughput.

Mechanically exfoliated single-layer graphene was developed by Geim and coworkers [49], using a technique called micromechanical cleavage. Typically, a cellophane tape is used to peel off graphene layers from a graphite flake, followed by pressing the tape against a substrate. Upon removing the tape, a single-sheet graphene is obtained. Advantages of this method include the ability to produce high-quality graphene typically used for laboratory research due to little graphene processing steps required. However, limitations of this method include its unsuitability for large-scale production as it is slow and labor intensive. Chen and coworkers attempted to resolve this limitation by investigating a novel continuous mechanical exfoliation process (Fig. 8.4) [59]. This process uses a three-roll mill machine, common in the rubber industry, to produce single-/few-layer graphene from natural graphite. This approach could also be used to prepare graphene–polymer composites and could be further enhanced by using the right types of polymer adhesives as well as rolling time, temperature, and speed of rolling.

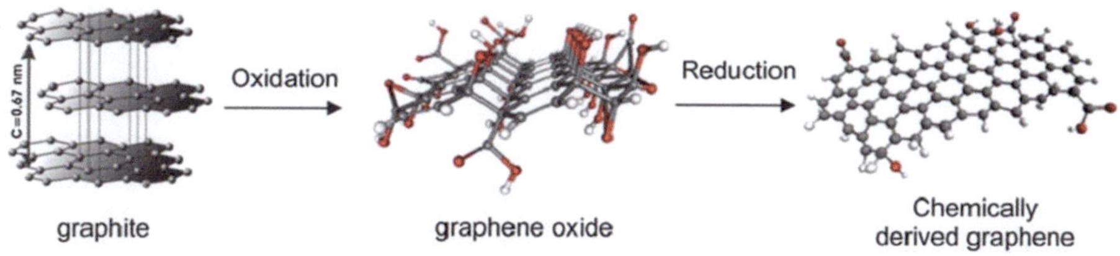

Fig. 8.5 An illustration of the chemical route to the synthesis of chemically derived graphene [52] (Reproduced with permission from Ref. [52]. (2009) Nature Publishing Group)

However, the low throughput of the mechanical exfoliation method largely limits its applications for mass production in supercapacitors. Thereafter, alternative approaches affording a high yield are highly desirable. On the other hand, the chemical method that is considered a scalable approach to synthesize graphene has been used widely to synthesize chemically derived graphene. As is schematically illustrated in Fig. 8.5, graphite is first oxidized to graphene oxide (GO) using either the Hummers method or the modified Hummers method [52]. Chemically derived graphene can then be obtained after reduction of the GO using hydrazine solution or other reducing agents. GO is an excellent precursor to synthesize graphene nanosheets [60]. The wet chemical synthesis by modified Hummers method is also the simplest method to synthesize large-size graphene. The graphite oxide was firstly obtained by oxidizing graphite with strong oxidant ($KMnO_4$, P_2O_5, etc.). Suspending graphene oxide sheets in water were reduced with reducing agent, which is a simple yet versatile method to prepare reduced graphene-based materials. All kinds of reducing methods were used, such as hydrazine hydrate [61], $NaBH_4$ [62], urea [63], UV light [64], and laser [65].

One research study has expanded the idea of single-/few-layer graphene (FLG) [66]. It is believed that with fewer layers, there would be a lower likelihood of aggregation. In this study, FLG was synthesized via intercalation of graphite and pre-graphite oxide followed by the reduction of graphite oxide. Such FLG produced a specific capacitance of 180 F g^{-1} in 1 M Na_2SO_4 with a BET measured area of 1400 m^2 g^{-1}. A key limitation in using graphite oxide produced from oxidized graphite includes significant defects which are irreversible, and this reduces the electrical conductivity of the electrode produced. Looking ahead, one of the key research areas is the discovery of novel synthesis methods that can produce single-/few-layer graphene in bulk quantities and without the presence of defects that would reduce its superior electrochemical properties. In addition, John R. Miller et al. [67] reported the vertically oriented graphene nanosheets grown directly on metal current collectors (Fig. 8.6) using radio frequency (RF) plasma-enhanced CVD, which is used in EDLCs. The vertically oriented structure minimized electronic and ionic resistances and produced capacitors with resistor-capacitor (RC) time constants of less than 200 microseconds, in contrast with ~1 s for typical EDLCs. However, the capacitance values are small (175–375 μF) due to the low loading of graphene.

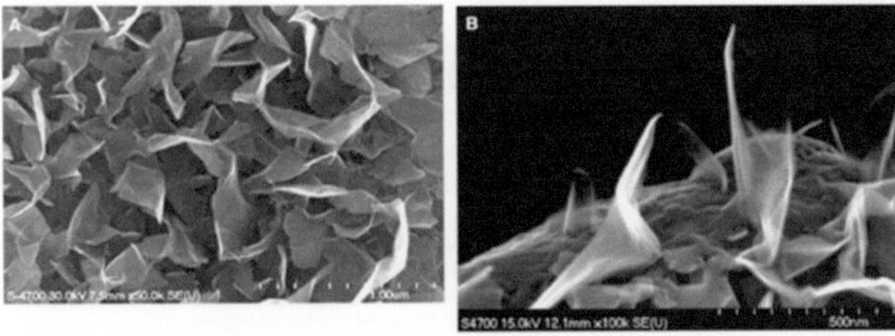

Fig. 8.6 (**a**) SEM of coated electrode and (**b**) a coated fiber showed plan and shallow angle views [67] (Reproduced with permission from Ref. [67]. (2010) AAAS)

8.2.2 Graphene-Like Carbon Nanosheets

The carbonization from organic matters or biomasses with the assist of the metal catalysts is the important methods to obtain graphene-like carbon nanosheets. The sheetlike graphitic carbon with porous structure can provide low-resistant pathways and short ion diffusion channels for energy storage and thus is expected to be an excellent material for high-power supercapacitor. Biomass waste is widely distributed in earth. Importantly, it is cheap, abundant, environmental safe, commercially available, and sustainable biological resource. All kinds of organic matters or biomasses were used to synthesize the porous carbon or carbon nanosheets, such as coconut shell [68], tree leaves [69], and cellulose [70] and surplus sludge [71] used as supercapacitor electrodes. The catalysis of metal is fatal to form the graphene-like carbon nanosheets. Without the metal as catalysts, the amorphous carbons were produced. For example, Biswal et al. [69] reported the synthesis of microporous conducting carbon by single-step pyrolysis of a variety of dead plant leaves, which exhibited a very high specific capacitance of 400 F g^{-1} and an energy density of 55 W h kg^{-1} at a current density of 0.5 A g^{-1} in aqueous 1 M H_2SO_4. Fu et al. [68] reported the porous graphene-like nanosheets with large surface area (1874 m^2 g^{-1}) were synthesized via an easy and cost-effective simultaneous activation–graphitization route from renewable biomass waste coconut shell, as shown in Fig. 8.7. In the synthesis, the graphitic catalyst precursor ($FeCl_3$) and activating agent ($ZnCl_2$) were introduced simultaneously into the skeleton of the coconut shell through coordination of metal precursor with the functional groups in coconut shell, thus making simultaneous realization of activation and graphitization of carbon resource under the heat treatment. Notably, the iron catalyst in the framework of the coconut shell can generate carburized phase which plays a key role in the formation of graphene-like structure during pyrolytic process. When using as supercapacitor electrode, the PGNS exhibited a high specific capacitance of 268 F g^{-1}, much higher than the activated carbon (210 F g^{-1}) fabricated by only activation and graphitic carbon (117 F g^{-1}) by only graphitization at 1 A g^{-1}, with superior cycle durability and columbic efficiency over 99.5 % after 5000 cycle.

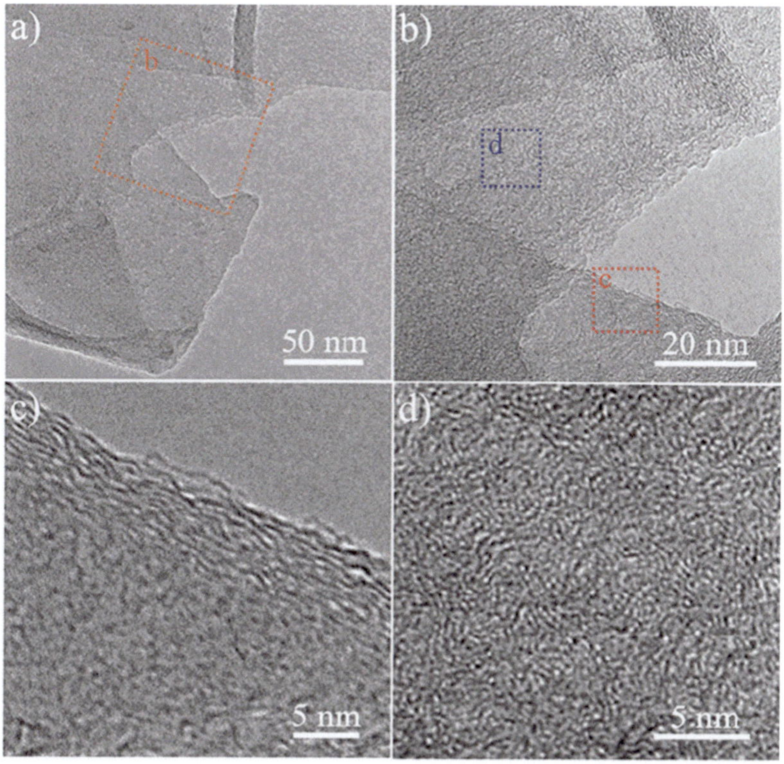

Fig. 8.7 (**a**, **b**) are the TEM images of porous graphene-like nanosheets under different magnifications; (**c**) and (**d**) are the HRTEM images of the selected areas of (**b**) [68] (Reproduced with permission from Ref. [68]. (2013) Royal Society of Chemistry)

8.2.3 Carbide-Derived Carbons

Carbide-derived carbons (CDCs) are a new large type of carbon materials derived from carbide precursors that are converted into carbon by means of physical processes or chemical reactions [72]. There are three major ways used to prepare carbide-derived carbon materials, namely, selective etching of carbides by halogens, selective etching of carbides by melts, and supercritical water or thermal decomposition of carbides. Besides, some new routes for the preparation of carbide-derived carbons via electrochemical process have also drawn wide attention of academic world recently [73, 74]. In general, carbide-derived carbons are produced from carbides by etching non-carbon elements in the process of selective etching by chlorination. The derived carbon structures synthesized by removing the metal and/or metalloid atom(s) from the carbides depend on the synthesis technology, applied temperature, and the carbide precursor. The carbide-derived carbons range from amorphous carbon to graphite [75], graphene [76], onion-like structure [77],

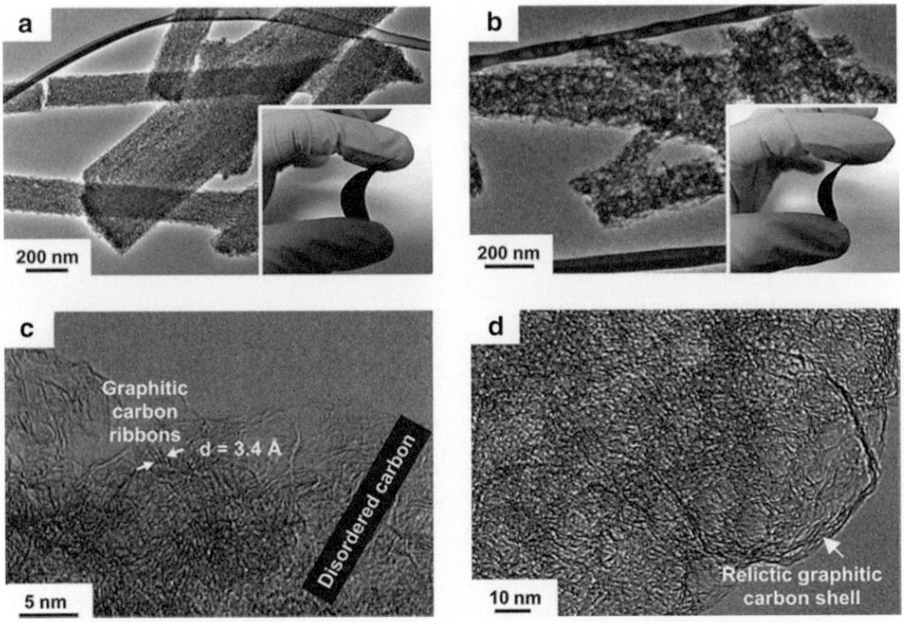

Fig. 8.8 TEM images of the TiC–CDC nanofelts after chlorination at 400 °C (**a**, **c**) and 600 °C (**b**, **d**). In addition to newly formed CDC, graphitic carbon could be identified in the interface regions of previous TiC nanocrystallites (**d**). The insets in (**a**) and (**b**) demonstrate the flexibility of the developed TiC–CDC nanofelts [89] (Reproduced with permission from Ref. [89]. (2011) John Wiley and Sons)

open fullerene-like structure [78], or nanodiamonds [79]. Carbide-derived carbons as electrode material in EDLCs have attained much attention over recent years.

Adjusting the synthesis parameters provide an accurate method for the control of pore size distribution with sub-Ångstrom. Carbide-derived carbons with high specific surface area and tunable pore size distribution produced from various carbide materials are well known as the new breakthrough for supercapacitors [80, 81], such as Al_4C_3 [82], TiC [83], SiC [84], VC [85], Mo_2C [86], SiCN [87], $WTiC_2$ [88], etc. Meanwhile, the kind of carbide precursor and electrolyte, chlorination temperature, particle size, activation, catalyst, and post-synthesis treatment and other parameters have significant influence on performance of carbide-derived carbon supercapacitor. For example, electrochemical measurements of the flexible nanobelts of carbide-derived carbons from electrospun TiC nanobelts showed a high gravimetric capacitance of 110 F g^{-1} in aqueous electrolyte (1 M H_2SO_4) and 65 F g^{-1} in organic electrolyte (1.5 M TEA-BF_4 in acetonitrile) (Fig. 8.8) [89]. Presser et al. reported Al_4C_3-CDC showed a high gravimetric capacitance (260 F g^{-1}) in KOH electrolyte [72]. Thomberg et al. showed that micro-/mesoporous Mo_2C-CDC yielded a higher gravimetric capacitance of 143 F g^{-1} as electric double-layer capacitor than vanadium carbide (VC)-CDC in 1 M $(C_2H_5)_3CH_3NBF_4$ and acetonitrile solution, where VC-CDC obtained by chlorination at 900 °C, and with pores of an average diameter of 1.18 nm, gave the highest capacitance value of 133 F g^{-1} [85, 86]. The carbide-derived carbons derived from TiC/TiO_2 composite employed in the excellent nonaqueous electric double-layer capacitor possess the highest volumetric capacitance of over 90 F cm^{-3} at that time [90]. Portet et al. [91] testified that as high as 180 F g^{-1} has been measured for KOH-activated TiC–CDC in an organic electrolyte, which is caused by a markedly increased surface area and micropores. Ordered mesoporous SiC-CDC was used in high-rate supercapacitors and exhibited a specific surface area up to 2430 m^2 g^{-1}, a specific capacitance up to 170 F g^{-1}, outstanding capacitance retention of 85 % at current densities up to ~20 A g^{-1}, and energy density (~140 Wh kg^{-1}) in 1 M tetraethylammonium tetrafluoroborate when tested in acetonitrile solution [92].

8.3 Doped Carbons

Heteroatom-doped carbon materials represent one of the most outstanding families of materials that are used in supercapacitors [93]. Due to heteroatom (N, S, B, P, etc.) with the lower size and electronegativity than those of carbon, structural distortions and changes of the charge densities are induced in the carbon materials. While doping carbons with nitrogen atoms have experienced great progress throughout the past decades and yielded promising material concepts, also other doping candidates, such as boron and sulfur, have gained the researchers' interest in the last few years.

8.3.1 N-Doped Carbons

The obtained N-doped carbons are multitudinous, including synthetic pathways and nitrogen sources [94]. Classical process is the pyrolysis of crude carbons with reactive nitrogen source, which can be from the foreign or self-contained [95], such as ammonia [18], heterocycles [96], melamine [97], or nitrogen-containing carbohydrate biomass [33, 98–100]. Chen et al. [97] reported that N-doped graphene with a pore volume as high as 3.42 cm^3 g^{-1} was synthesized by decomposing cyanamide on graphene oxide nanosheets. The specific capacity of N-doped graphene (248.4 F g^{-1} at 5 mV s^{-1}) is four times and twice higher than those of thermally reduced graphene sheets (51.7 F g^{-1}) and chemically reduced graphene (106.3 F g^{-1}), respectively. Ruoff et al. [101] reported the increase in bulk capacitance with increasing N concentration and the increase of the quantum capacitance in the N-doped monolayer graphene versus pristine monolayer graphene, which suggest the increase in the electrical double-layer type of capacitance. It was further found that the quantum capacitance is closely related to the N-dopant concentration, and N-doping provides an effective way to increase the

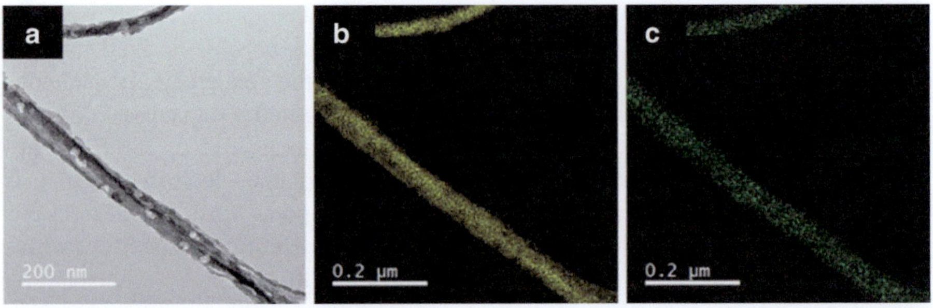

Fig. 8.9 TEM image and elemental mapping of A-p-BC–N-25. (**a**) Typical TEM image (**b**) and (**c**) corresponding elemental mapping images of (**b**) C and (**c**) N [33] (Reproduced with permission from Ref. [33]. (2013) Royal Society of Chemistry)

density of the states of monolayer graphene. Mitlin et al. [102] reported that biomass-derived proteins (egg white) serve as an ideal precursor for synthesizing nitrogen-doped carbon materials for energy applications. The carbons with a high specific surface area (805.7 m^2 g^{-1}), partial graphitization, and very high bulk nitrogen content (10.1 wt%) exhibited a capacitance of 390 F g^{-1}, with an excellent cycle life (7 % loss after 10,000 cycles).

Meanwhile, nitrogen-rich organic networks and frameworks have been thermally treated to give nitrogen-containing carbon materials with promising properties for supercapacitor. Yu et al. [33, 103] firstly present a high-capacity supercapacitor material based on the nitrogen-doped porous carbon nanofibers synthesized by carbonization of macroscopic-scale carbonaceous nanofibers coated with polypyrrole at an appropriate temperature, as shown in Fig. 8.9. The composite nanofibers exhibit a reversible specific capacitance of 202.0 F g^{-1} at the current density of 1.0 A g^{-1} in 6.0 M aqueous KOH electrolyte. Then, they fabricated a flexible all-solid-state supercapacitor device with nitrogen-doped pyrolyzed bacterial cellulose as the electrode material via a low-cost, eco-friendly, low-temperature, and scalable fabrication hydrothermal synthesis [33]. The pliable device can reversibly deliver a maximum power density of 390.53 kW kg^{-1} and exhibit a good cycling durability with 95.9 % specific capacitance retained after 5000 cycles.

8.3.2 B-Doped Carbons

Boron is already widely studied, and as its electronic situation is contrary to the one of nitrogen, co-doping carbons with both heteroatoms can probably create synergistic effects. Application of the derived hierarchically porous materials as supercapacitor electrodes has been reported, indicating that the favorable pseudocapacitance effects are achievable by nitrogen/boron doping. Gao et al. [104] have followed a relatively unconventional synthetic pathway using citric acid as the

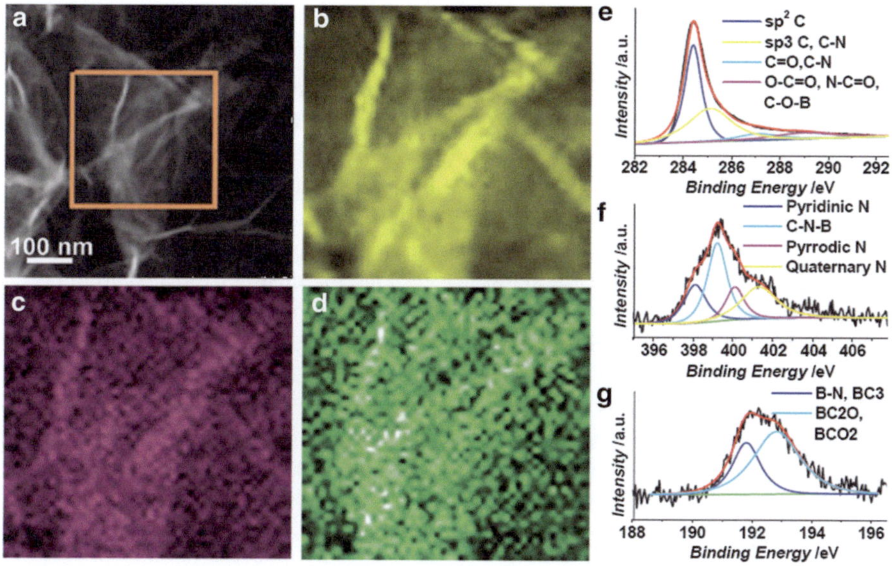

Fig. 8.10 (**a**) Scanning transmission electron microscopy (STEM) image and (**b**) carbon, (**c**) boron, and (**d**) nitrogen element mapping images of BN-Gas taken in the square region in (**a**). (**e**) C1s, (**f**) N1s, and (**g**) B1s XPS spectra of BN-GAs [106] (Reproduced with permission from Ref. [106]. (2012) John Wiley and Sons)

carbon source and boric acid as the boron source, and claim to have succeeded in the incorporation of boron into their structures. Park et al. [105] reported that novel B-doped nanoplatelets (borane-reduced graphene oxide, B-rGO) were produced on a large scale via the reduction of graphene oxide by a borane–tetrahydrofuran adduct under reflux, which as supercapacitor electrodes had a high specific surface area of 466 $m^2\ g^{-1}$ and showed excellent supercapacitor performance including a high specific capacitance of 200 $F\ g^{-1}$ in aqueous electrolyte as well as superior surface area-normalized capacitance to typical carbon-based supercapacitor materials and good stability after 4500 cycles. Wu et al. [106] reported a novel prototype device of all-solid-state supercapacitors with an electrode–separator–electrolyte integrated structure based on 3D nitrogen and boron co-doped monolithic graphene aerogels (BN-GAs), as shown in Fig. 8.10. As a consequence, the resulting BN-GAs-based all-solid-state supercapacitors exhibit not only minimized device thickness but also high specific capacitance ($\approx$62 $F\ g^{-1}$), good rate capability, and enhanced energy density ($\approx$8.65 $W\ h\ kg^{-1}$) or power density ($\approx$1600 $W\ kg^{-1}$) with respect to un-doped, nitrogen-doped, and boron-doped GAs or layer-structured graphene paper.

8.3.3 S-Doped Carbons

In comparison to boron, sulfur doping in carbon materials is hitherto still quite rare and represents an emerging field within carbon material research. Gu et al. [107] reported that synthesis of S-doped activated carbons by carbonization and simultaneous activation of S-based polymers (2-thiophenemethanol) was found to be an efficient route to produce porous carbons for EDLCs with high specific energy (200 F g^{-1}) combined with low self-discharge. More importantly, these capacitance values remained virtually identical for a sweep rate increasing from 1 to 50 mV s^{-1}. Kanamori et al. [108] have used divinylbenzene as a precursor and polydimethylsiloxane as a phase separation inducing agent, enabling a spinodal decomposition and thus macroporous monolith formation upon polymerization of divinylbenzene, yielding a complex porous structure of a sulfur-functionalized carbon monolith (with 0.83 atom% S) with a surface area >2400 m^2 g^{-1}. This monolithic carbon showed significantly higher specific capacitances (103 F g^{-1} at a scan rate of 200 mV s^{-1}) than crude activated carbon electrodes without sulfur functionalities, further having negligible degradation over 2000 cycles. Yan et al. [99] reported that N- and S-doped porous carbon flakes were prepared via carbonization of Chinese human hair fibers and employed for high-performance supercapacitor electrode materials with a specific capacitance of 340 F g^{-1} in 6 M KOH at a current density of 1 A g^{-1} and good stability over 20,000 cycles.

8.4 3D Graphene Networks

Graphene network is a 3D graphene architecture by a self-assembly of the 2D flexible graphene nanosheet building blocks or by chemical vapor deposition (CVD) methodology. Furthermore, the pore sizes of the graphene network are in the range of submicrometer to several micrometers, which bring about ultralight, high mechanical strength and compressibility, excellent conductivity, and adsorption characteristics. The macroporous morphology not only prevents individual graphene sheets from aggregating and restacking during the process of assembling but also ensures a sufficient contact area between the electrolyte and electrode with the large adsorption space. Moreover, 3D porous frameworks endowed high-rate transportation of the electrolyte ions and provided multidimensional electron transport pathways, resulting in excellent electrochemical performance.

8.4.1 Hydrothermal Method

Recently, Shi and coworkers reported a self-assembled 3D porous network graphene using a hydrothermal method at 180 °C for 12 h, as shown in Fig 8.11

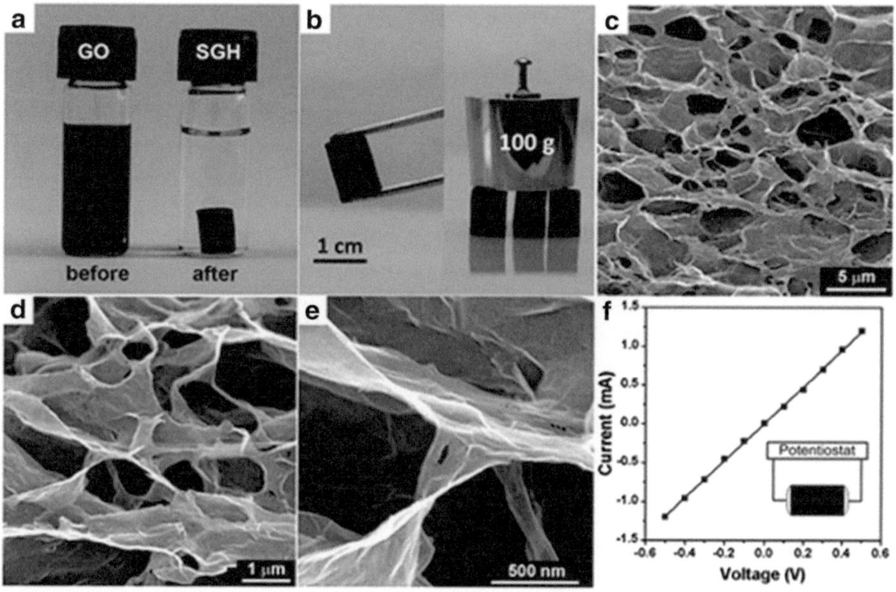

Fig. 8.11 (**a**) Photographs of a 2 mg mL^{-1} homogeneous GO aqueous dispersion before and after a hydrothermal reduction at 180 °C for 12 h; (**b**) photographs of a strong GF allowing easy handling and supporting weight; (**c–e**) SEM images with different magnifications of the SGH interior microstructures; (**f**) room temperature–V curve of the SGH exhibiting Ohmic characteristics, inset shows a two-probe method for conductivity measurements [109] (Reproduced with permission from Ref. [109]. (2010) American Chemical Society)

[109]. The framework was formed by partial overlapping or coalescing of flexible graphene sheets with physical cross-linking sites. The formation of foam was driven by π–π stacking interactions of graphene sheets. Using GO as a precursor, the morphology and structure of the foam greatly depended on the GO concentration and hydrothermal reduction time. When the GO concentration was lower than 0.5 mg mL^{-1}, there were few cross-linking sites that cannot form the monolithic cylinder. As an electrode material for supercapacitors, the obtained 3D network graphene had electrical conductivity of 5×10^3 S cm^{-1} and exhibited a specific capacitance of 175 F g^{-1} in an aqueous electrolyte.

Most recently, 3D graphene structures as well as the doping of N into graphene have been the focal point of much research as it was discovered that the addition of nitrogen atoms can positively improve the performance of graphene electrodes. In an investigation on such novel structures, a facile hydrothermal synthesis process was devised to produce nitrogen-doped graphene hydrogel (NGH) for supercapacitor applications. Yu et al. used GO and organic amine as precursors to synthesize N-doped 3D network graphene by a hydrothermal method at 180 °C for 12 h [110]. The introduced amine acted not only as nitrogen sources to achieve N-doped 3D network graphene but also as a modifier for the reaction with oxygen-containing functional groups to reduce GO and increase the activity of GO sheets for assembling 3D interconnected foam. The tail chain of the organic amine and forming of hydrogen bond on the graphene sheets could effectively prevent graphene sheets from π–π stacking interactions and simultaneously expand the space between the graphene sheets. Therefore, a 3D porous graphene network with submicrometer and micrometer pores was advantageous for the electrolyte transport and storage, resulting in ultrahigh supercapacitor performances (specific capacitance of 113.8 F g^{-1} and power density of 205.0 kW kg^{-1}). Three-dimensional N-doped graphene/carbon nanotubes (CNT) networks can be obtained by hydrothermal treatment, freeze-drying, and subsequent carbonization of graphene oxide-dispersed pristine CNTs in the presence of pyrrole. The resulting sample used as supercapacitor electrode shows specific capacitance of 180 F g^{-1} at a current density of 0.5 A g^{-1} and retains 96 % of the initial capacitance after 3000 cycles [111]. Moreover, Wu et al. similarly used GO and ammonia boron trifluoride (NH_3BF_3) as precursors to fabricate nitrogen and boron co-doped monolithic graphene aerogels (BN-GAs) with 3D interconnected macroporous architecture through a hydrothermal reaction and freeze-drying processes. The BN-GAs acted as an additive-/binder-free electrode material, and a polyvinyl alcohol (PVA)/ H_2SO_4 gel was used as solid-state electrolyte and separator. Therefore, the as-assembled all-solid-state supercapacitors (ASSSs) showed high specific capacitance (62 F g^{-1}), good rate capability, and enhanced energy density (8.65 W h kg^{-1}) or power density (1600 W kg^{-1}). To some extent, the ASSSs are ideal candidates for miniaturized electronics [106].

8.4.2 Chemical Reduction Method

Because there are some oxygen-containing functional groups (hydroxyl, carboxyl, and epoxy groups) on GO sheets, the multifunctional reduction reagent such as hydrazine, $NaHSO_3$, Na_2S, L-ascorbic acid (vitamin C), HI, hydroquinone, sodium ascorbate, hypophosphorous acid–iodine, and ethylene diamine (EDA) can combine with these oxygen-containing functional groups, which not only reduce the GO into graphene but also can impel graphene layer cross-link together by chemical bonding. Zhang et al. fabricated graphene aerogels with hierarchically porous structure by only heating GO with vitamin C followed by supercritical CO_2 drying or freeze-drying [112]. Vitamin C served as a reducing agent for the reduction of GO, and the as-prepared foam showed lightweight (12–96 mg cm^{-3}), large specific surface area (512 m^2 g^{-1}), and large pore volume (2.48 cm^3 g^{-1}). The graphene hydrogel (GH) was mechanically strong and electrically conductive (10^2 S m^{-1}), could support at least 14,000 times its own weight and showed a specific capacitance of 128 F g^{-1} at a current density of 50 mA g^{-1}. Subsequently, in order to obtain higher conductivity of graphene hydrogel, Shi and coworkers further used hydrazine or hydroiodic acid (HI) for reduction of the graphene hydrogels [113]. The chemical reduction of graphene hydrogels was carried out by immersing it into an aqueous solution of HI (55 %) or hydrazine monohydrate (50 %) in a sealed cuvette. The reduction was conducted for 3 or 8 h at 100 °C for HI and 95 °C for hydrazine, respectively. The chemically reduced graphene hydrogels possess high conductivities of 1.3–3.2 S m^{-1} and specific surface areas in the range from 780 to 950 m^2 g^{-1}. Moreover, the further treatment of graphene hydrogel with reducing agent has less impact on the microscopic pore structure of graphene hydrogel. The supercapacitor based on the HZ-reduced graphene hydrogel exhibited a high specific capacitance of 220 F g^{-1} at 1 A g^{-1} in 5 M KOH aqueous electrolyte, and this capacitance can be maintained for 74 % as the discharging current density was increased up to 100 A g^{-1}. Furthermore, the capacitor exhibited a high power density of 30 kW kg^{-1} and energy density of 5.7 W h kg^{-1} at 100 A g^{-1}. This is possibly attributed to the improved conductivity of reduced hydrogels accelerating its charge transfer during the discharge processes at high current densities. Moreover, Hur and coworkers demonstrated that a graphene foam was fabricated by the cross-linking reaction between GO and ethylenediamine (EDA) and then submerged into hydrazine at 90 °C for 2 days followed by freeze-drying [114]. The as-prepared graphene foam showed high electrical conductivity of 1351 S m^{-1}, high SSA of 745 m^2 g^{-1}, high break strength of 10.3 MPa, and high specific capacitance of 232 F g^{-1} in the supercapacitor application.

8.4.3 Chemical Vapor Deposition

3D graphene networks can also be synthesized via the chemical vapor deposition (CVD) process. An investigation into this novel process using Ni foam as a template and ethanol as the carbon source. The prepared graphene has a unique 3D configuration that allows rapid access to electrolytes due to its large surface area. In addition, graphene is directly synthesized onto the Ni current collector that would allow rapid transfer of charges from the active materials to the current collectors through the graphene 3D configuration [115]. Notably, CVD synthesis of graphene should not be conducted on flat metal surfaces or thin films to avoid problems of low yield. Furthermore, depending on the metal on which graphene is grown, optimal conditions for CVD vary. Other metals on which graphene can be grown via CVD include Co, Rh, Pd, Pt, Cu, Au, and a number of other alloys such as Co–Ni, Au–Ni, Ni–Mo, and stainless steel. Recently, Cheng and coworkers first obtained 3D graphene foam (GF) using porous nickel foam as a template for the deposition of graphene by CVD using CH_4 at 1000 °C under ambient pressure [20]. The as-prepared graphene foam inherited the interconnected 3D scaffold of the nickel foam template with a macroporous structure, which can be used as a gas-sensing device due to its high sensitivity detection of NH_3 and NO_2 at room temperature and atmospheric pressure [116]. Since the groundbreaking work for synthesis of 3D structure done by Cheng and coworkers, some other scientists have also used this foam as the basis for a more far-reaching research. For example, Zhang and coworkers synthesized 3D NiO–graphene hybrid foam using NiO electrochemically deposited on ethanol CVD-grown graphene foam [115]. The obtained NiO–graphene hybrid foam as a supercapacitor electrode exhibited a high specific capacitance of 816 F g^{-1} at 5 mV s^{-1} and a stable cycling performance without any decrease of its specific capacitance after 2000 cycles. Similar to the previous experiment, Matsushima et al. used a pulsed plasma CVD technique to create nanoporous carbon films containing graphene to use as electrodes for EDLC. For this synthesis process, a metal Ni as catalyst is required. When the porous graphene films were grown on Ni films that have 50 nm thickness, the capacitance was discovered to be approximately 13 mF cm^{-2}. However, when the thickness of the Ni surface was increased from 50 to 100 nm, the capacitance increased to 152 mF cm^{-2}. This significant increase was attributed to the breaking into larger Ni particles, and this facilitated the absorption of carbon atoms resulting in the formation of more desirable graphene structures. Their experiment showed how graphene–nanoporous carbon composites are suitable materials for electrodes in EDLCs [117]. Chen and coworkers [118] also demonstrated that 3D graphene–polyaniline (PANI) hybrid foam was synthesized by in situ polymerization of aniline monomers using ammonium persulfate (APS) as catalyst on ethanol CVD-grown 3D graphene network. The graphene network hybrid material was highly hydrophobic, while hydrophilic-conducting polymer PANI was favorable for ion diffusion between the electrode and electrolyte. Additionally, as the electrode for supercapacitor application, it also exhibited high specific capacitance (346 F g^{-1} at 4 A g^{-1}) and good cycling stability.

8.5 Flexible Supercapacitors with Planar Structures

Recently, flexible energy storage devices have attracted a rapid growing interest for their potential applications in various portable electronic devices such as mobile phones, wearable electronics, and flexible displays. Consequently, compared to the conventional supercapacitors, the flexible solid-state supercapacitors hold great promises as favorite energy storage devices, which shows significant advantages such as high flexibility, lightweight, and reduced interfacial resistance. Typically, flexible solid-state supercapacitor devices generally include flexible electrodes, solid-state electrolyte, separator, and bendable packaging material. To this point, various efforts have been devoted to prepare flexible electrodes with outstanding electrochemical, mechanical, and interfacial properties for flexible solid-state supercapacitors. Using flexible conductive substrates as support structures for active materials are expected to an inexpensive and promising strategy to develop the novel flexible solid-state supercapacitors.

To date, metal substrates and carbon-based electrodes have received increasing attention as conductive substrates. However, metal substrates are easily corroded in electrolytes and less flexible and increase the total weight of supercapacitor devices, which limits the use of the devices. Carbon-based materials, which include fabric, cellulose networks, textile substrates and conventional paper, etc., have been widely investigated for flexible electrodes due to their good electrical conductivity,

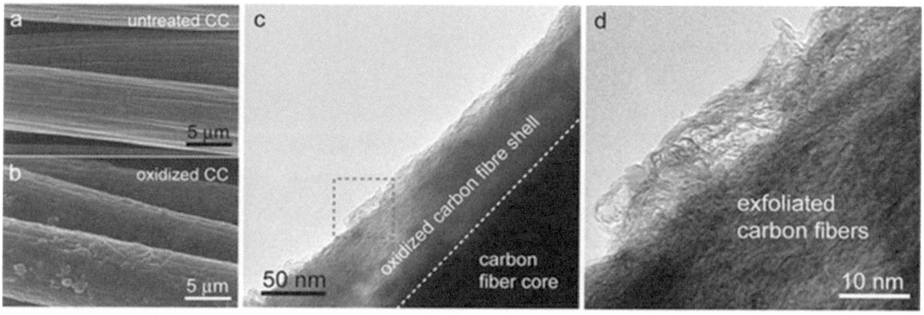

Fig. 8.12 SEM images of (**a**) untreated and (**b**) oxidized carbon fiber cloth substrates. (**c**) Low-magnification TEM image collected at the edge of the oxidized carbon fiber. (**d**) High-resolution TEM image showed the exfoliated carbon fibers; it shows the edge of the oxidized carbon fiber, as indicated by the dashed box in (**c**) [121] (Reprinted with kind permission from Ref. [121]. Copyright 2014, John Wiley and Sons)

outstanding chemical stability, and excellent mechanical properties, which can serve as both the active materials and the current collectors.

8.5.1 Carbon Fiber-Based Fabric

Carbon fiber-based fabric, including carbon fiber cloths and carbon fiber paper, etc., consists of carbon microfibers manufactured into a highly conductive network with excellent mechanical flexibility and strength. As a cost-effective commercially product, carbon fiber cloth is preferable for an electrode material for flexible supercapacitors. However, because of the smaller surface area than graphene and carbon nanotubes, carbon fiber cloths present a relatively low specific capacitance (1–2 F g^{-1}). Consequently, several studies have been carried out on the functionalization of carbon fiber cloths to improve the performance of carbon fiber cloth for electrode materials or current collectors [119–121]. For example, Wang et al. [121] reported an effective three-step method to activate carbon fiber cloth electrode, using the mixed solutions of H_2SO_4/HNO_3 and $KMnO_4$ to chemically functionalize the carbon fiber cloth, followed by reduction in hydrazine gas and thermal annealing in ammonia atmosphere (Fig. 8.12). The activated electrode exhibited a remarkable areal capacitance of 88 mF cm^{-2} at a scan rate of 10 mV s^{-1} in a three-electrode system. The symmetric supercapacitor based on the activated carbon fiber cloth electrode showed an excellent rate capability and still yielded an area capacitance of 15.3 mF cm^{-2} at the ultrafast charging rate of 10,000 mV s^{-1}.

Furthermore, carbon materials such as porous carbon, CNTs, and graphene have also been fabricated on carbon fiber fabric as electrochemical active materials for flexible ECs. These types of composite electrodes are expected to facilitate the rapid electron transport ascribed to the large active surface area and pore sizes. Qian et al. [122] proposed a modification route that structural carbon fiber fabrics are

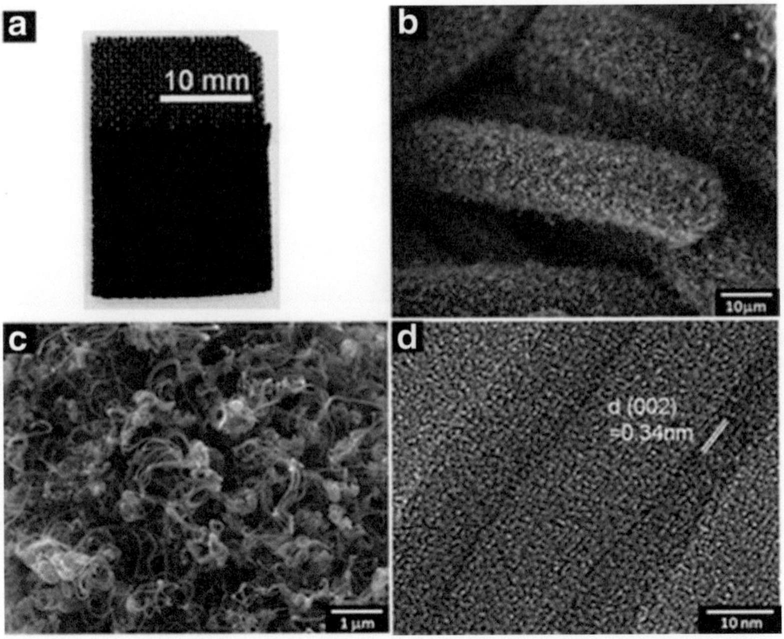

Fig. 8.13 (**a**) Photograph of CNTs on carbon cloth; (**b**) and (**c**) FESEM images of CNTs/carbon cloth electrodes taken under different magnifications; (**d**) HRTEM image of CNT [123] (Reprinted with kind permission from Ref. [123]. Copyright 2012, Royal Society of Chemistry)

embedded in a continuous network of carbon aerogel (CAG) to form a coherent but porous monolith. The incorporation of CAG significantly increased the surface area of carbon fiber fabrics, and hence the electrochemical performance, resulting in a specific capacitance of around 62 F g^{-1} in an aqueous electrolyte, around hundred-fold than pure carbon fiber fabrics. As shown in Fig. 8.13, Chen et al. [123] employed CNTs directly grown on the carbon fabrics prepared by nickel-catalyst assisted microwave plasma-enhanced CVD technique. The direct growth approach for fabricating a composite CNTs/carbon cloth electrode not only results in strong adhesion between the CNTs and flexible carbon cloth, forming a good electron pathway with no boundary loss, but also provides a hierarchical hybrid structure with the larger surface area that allows rapid penetration of the electrolyte. Moreover, the symmetric supercapacitor of CNTs/carbon cloth exhibits an excellent cyclability over 10,000 cycles and high energy density of 27.8 Wh kg^{-1}, with a wide cell potential of 2 V and a large specific capacitance of 210 F g^{-1}. Wang et al. [124] prepared a highly porous graphene on carbon cloth (PG/CC) via an electrophoretic deposition process, and it was employed as flexible electrode materials for flexible all-solid-state supercapacitors. The porous structure of PG/CC electrode significantly increased the surface area of graphene and thus the specific capacitance of 79.19 F g^{-1} for PG/CC. Wang et al. [125] developed a GO-assisted electrophoretic deposition method to co-deposit GO and CNTs onto carbon fabrics (G-CNT/CC). The as-fabricated flexible supercapacitor based on the G-CNT/CC electrodes shows significantly enhanced supercapacitor performance.

8.5.2 CNT Networks

CNTs have been widely explored for using as electrode materials in supercapacitors, because of their controllable regular porous network, high specific surface area, and good electrical conductivity. Moreover, due to their characteristic of high aspect ratio and mechanical flexibility, CNTs have been directly assembled into flexible films by various methods, including vacuum filtration, self-assembly, dry drawing, blade coating, and so on. The obtained freestanding membrane/paperlike CNT-based materials with excellent flexibility, high conductivity, and lightweight are proposed as a promising electrode material for flexible supercapacitors. Kaempgen et al. [126] reported a printable thin-film supercapacitor using single-walled CNTs sprayed on polyethylene terephthalate (PET) films as both electrodes and charge collectors. Two species of CNT electrodes and a PVA/H_3PO_4 gel electrolyte were sandwiched together to fabricate the flexible solid-state supercapacitor device, which showed the specific capacitance of ~36 F g^{-1}. The solid-state SWCNTs-supercapacitor device achieved a high energy density of 6 W h kg^{-1} and power density of 23 kW kg^{-1}. Recently, Niu et al. [127] developed an all-solid-state stretchable supercapacitor based on a highly stretchable buckled SWCNT film in combination with a H_2SO_4/PVA gel electrolyte (Fig. 8.14a). The highly stretchable buckled SWCNT films were fabricated by directly combining SWCNT films having continuous reticulate architecture with an enhanced prestrained elastomeric substrate (PDMS) (Fig. 8.14b–d); the electrical resistance of the resultant electrode had no significant change even under a high strain of 140 % (Fig. 8.14e and f). The specific capacitances of the resulting supercapacitors calculated from the discharge slopes were 48 F g^{-1} for the supercapacitor without applied strain and 53 F g^{-1} for the supercapacitor subject to 120 % strain. Apart from the polymeric film materials [128], CNTs deposited onto office papers via a drop-dry method have also been used as electrodes in flexible supercapacitors [129]. Activated carbon powders [130] and granulated CNTs [131] were incorporated into CNT papers to increase the specific surface area which further improve the performance of the CNT-based flexible supercapacitors.

8.5.3 Graphene-Based Aerosols

Besides CNTs, the filtrated graphene papers have also been intensively studied as freestanding and flexible electrochemical capacitance electrodes. Moreover, graphene can be produced by various methods on a large scale and relatively low

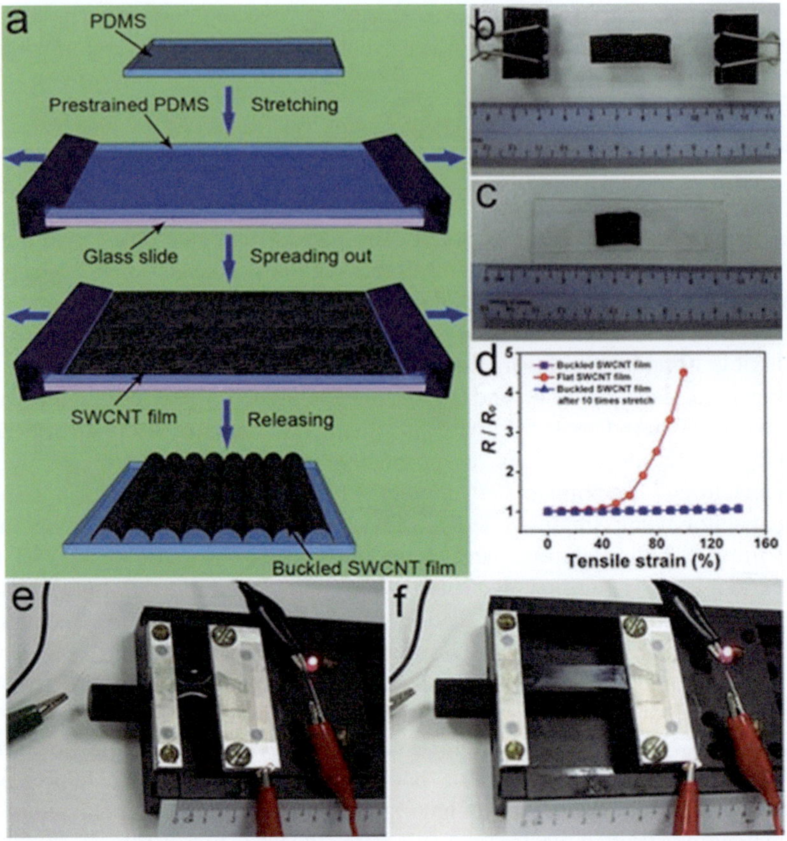

Fig. 8.14 (**a**) Schematic of preparing buckled single-walled carbon nanotube (SWCNT) film on PDMS. Optical images of SWCNT film on (**b**) prestrained PDMS and (**c**) buckled SWCNT film obtained by releasing the SWCNT film on prestrained PDMS. (**d**) The normalized sheet resistance of SWCNT films on PDMS with and without buckled structure at different strain levels and buckled SWCNT film after 10 times stretching at different strain levels, where R_0 is the resistance of the unstretched SWCNT films and R is the resistance of the SWCNT films at different strain levels. Optical images of an LED illuminated by using a buckled SWCNT film as the connecting wire under (**e**) no strain and (**f**) about 140 % strain [127] (Reprinted with kind permission from Ref. [127]. Copyright 2013, WILEY-VCH Verlag GmbH & Co. KGaA, Weinheim)

cost. Graphene papers are composed of graphene nanosheets, which are seriously restacked due to the interplanar π–π interaction and the van der Waals attraction between them during the filtration process. However, the restacking of graphene films will lead to reduced effective surface area and limited diffusion of the electrolyte ions which greatly affect the capacitive properties of graphene paper-based flexible EC. Choi et al. [132] reported the Nafion-functionalized reduced graphene oxide-based supercapacitors showed a higher specific capacitance (118.5 F g^{-1} at 1 A g^{-1}), the pristine rGO-based solid-state supercapacitors (62.3 F g^{-1} at 1 A g^{-1}).

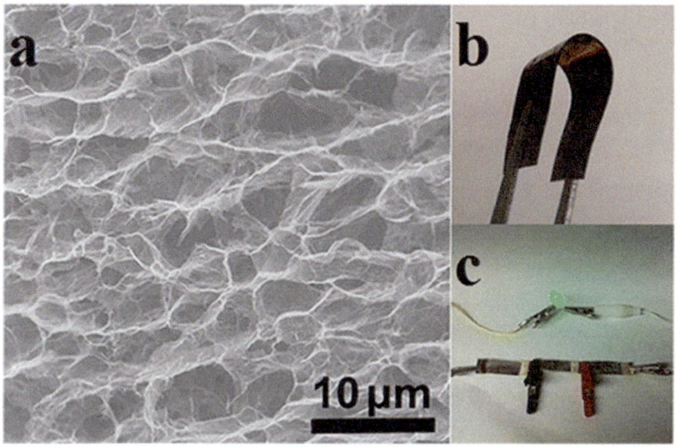

Fig. 8.15 (**a**) Low-magnification SEM images of the interior microstructure of the graphene hydrogel before pressing. (**b**) Flexible solid-state supercapacitors based on graphene hydrogel films. (**c**) Photograph of a green LED powered by the three supercapacitors in series [135] (Reprinted with kind permission from Ref. [135]. Copyright 2013, American Chemical Society)

Therefore, significant efforts have been dedicated to overcome the limitation of the restack of graphene. Porous graphene structure with a remarkably high specific surface area would facilitate electrolyte diffusion which improved the performance of flexible supercapacitors. Considerable strategies, including no-template-directed assembly method [106, 125, 133–135], template-directed assembly method [136], and so on, have been used to fabricate porous graphene structure. Yang et al. [137] developed an effective bioinspired approach to prevent the restacking of graphene sheets in multilayered films. These face-to-face-stacked multilayered graphene sheet films obtained by vacuum filtration can provide a highly open pore structure, allowing the electrolyte solution to easily access to the surface of individual sheets. The flexible supercapacitors based on graphene paper electrode achieved a high specific capacitance of 273.1 F g^{-1} and a substantially high energy density up to 150.9 W h kg^{-1}. As shown in Fig. 8.15, Duan et al. [135] prepared 3D reduced graphene hydrogel for the fabrication of high-performance solid-state flexible supercapacitors. The flexible supercapacitors constructed of a graphene hydro-gel thin film with thickness of 120 μm can exhibit excellent capacitive characteristics, including a high gravimetric specific capacitance of 186 F g^{-1} and an unprecedented area-specific capacitance of 372 mF cm^{-2}, excellent cycling stability (8.4 % capacitance decay after 10,000 charge–discharge cycles), and extraordinary mechanical flexibility (CV curves almost unchanged with different bending angles) (Fig. 8.15b). El-Kady et al. [65] reported a standard LightScribe DVD optical drive for laser-induced reduction of graphite oxide films to greatly lower restacking of graphene film electrodes as flexible electrodes in flexible supercapacitors (Fig. 8.16). Consequently, the electrodes have an extremely large specific surface area of 1520 m^2 g^{-1} and high conductivity of 1738 S m^{-1}. Flexible SCs based on these electrodes and an ionic liquid electrolyte exhibited a maximum energy density of 1.36 mW h cm^{-3} and power density of up to 20 W cm^{-3}. Recently, the 3D

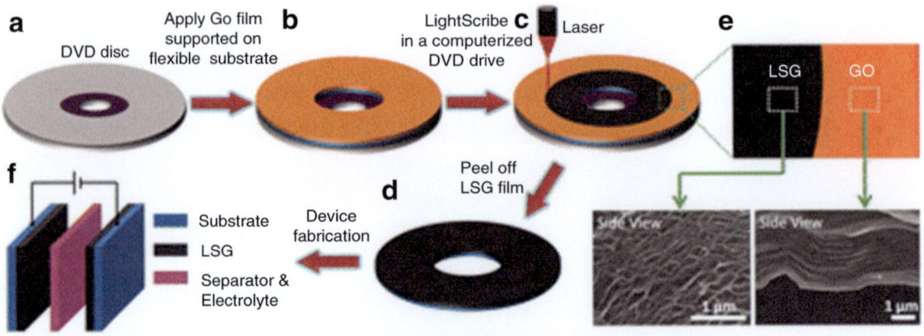

Fig. 8.16 Schematic illustration of the fabrication of laser-scribed graphene-based electrochemical capacitors. (**a–d**) A GO film supported on a flexible substrate is placed on top of a LightScribe-enabled DVD media disc, and a computer image is then laser irradiated on the GO film in a computerized LightScribe DVD drive. (**e**) As shown in the photograph, the GO film changes from golden brown color to black as it reduced to laser-scribed graphene. The low-power infrared laser changes the stacked GO sheets immediately into well-exfoliated few-layered LSG film, as shown in the cross-sectional SEM images. (**f**) A symmetric EC is constructed from two identical LSG electrodes, ion-porous separator, and electrolyte [65] (Reprinted with kind permission from Ref. [65]. Copyright 2012, AAAS)

graphene networks can be synthesized by using Ni foams with template-directed CVD. Su et al. [138] have fabricated a three-dimensional bubble graphene film, with controllable and uniform macropores and tailorable microstructure, using monodisperse polymethyl methacrylate (PMMA) spheres as hard templating by vacuum filtration. The as-prepared macroporous bubble graphene film exhibits a SSA of 128.2 $m^2\ g^{-1}$. Furthermore, the CV profiles of MGF continue to approach a rectangular shape with scan rates increasing from 3 to 1000 mV s^{-1}.

Besides, nanostructured carbon materials such as carbon black [139] and CNTs [140–142] have been introduced into graphene sheets to serve as spacers to create pores which facilitate electrolyte ion diffusion and accordingly improve electrochemical performances of the flexible supercapacitors. Lian et al. [139] processed flexible graphene paper (GP) pillared by carbon black nanoparticles using a simple vacuum filtration method. The pillared GP-based supercapacitors exhibit excellent electrochemical performances (the specific capacitance of 138 and 83.2 F g^{-1} in aqueous and organic electrolytes at a scan rate of 10 mV s^{-1}) and cyclic stabilities. The pillared GP has great potential in the development of promising flexible and ultralightweight supercapacitors for electrochemical energy storage.

8.5.4 Other Conductive Carbon-Based Networks

Apart from the highly conductive carbon-based materials discussed above, paper/textiles can also be used as a flexible support for electroactive materials with high flexibility, mechanical strength, and low cost. To realize this goal, a key step is to

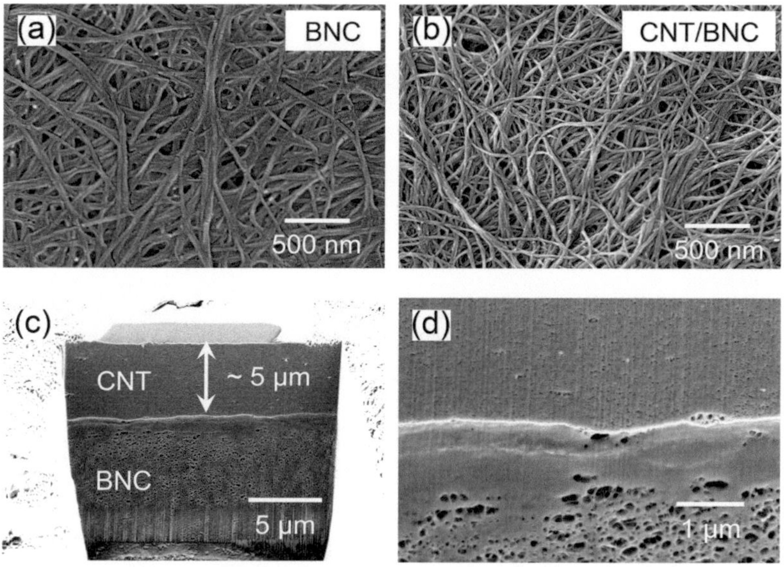

Fig. 8.17 Characteristics of bacterial nanocellulose (BNC) papers and CNT-coated BNC papers: scanning electron microscopy (SEM) images of (**a**) a BNC paper and (**b**) CNTs coated on a BNC paper. (**c**) Cross-sectional view of a CNT/BNC paper and (**d**) magnified SEM image of the interface between the CNT and the BNC layer [145] (Reprinted with kind permission from Ref. [145]. Copyright 2012, American Chemical Society)

make them conductive to form an electrically conductive composite paper with high electrochemical performance.

The rough surface of cellulose papers has large surface area, and the large number of surface functional groups on cellulose paper is beneficial for binding foreign matter, which is good candidate for substrates of flexible energy storage applications [33, 143, 144]. Kang et al. [145] reported a CNT layer was deposited onto the bacterial nanocellulose (BNC) (Fig. 8.17) via a vacuum filtering process to form conductive papers. All-solid-state flexible supercapacitors fabricated with BNC, CNTs, and ionic liquid-based polymer gel electrolytes showed a specific capacitance, energy, and power of 50.5 F g^{-1} (~20.2 mF cm^{-2}), 15.5 mWh g^{-1}, and 1.5 W g^{-1}, respectively. Moreover, they showed excellent cyclic stability over 5000 charge–discharge cycles. Chen et al. [146] fabricated graphene–cellulose paper (GCP) as electrode for flexible supercapacitors. The GCP electrode consisted of a unique three-dimensional interwoven structure of graphene nanosheets and cellulose fibers and has excellent mechanical flexibility, good specific capacitance and power performance, and excellent cyclic stability. This flexible GCP electrode has a high capacitance of 81 mF cm^{-2} (120 F g^{-1} of graphene) and retains >99 % capacitance over 5000 cycles. The supercapacitors showed a high capacitance of 46 mF cm^{-2} for the complete devices under highly flexible conditions.

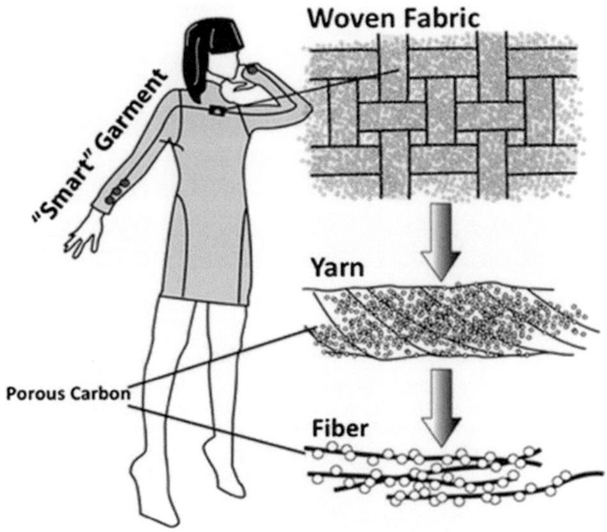

Fig. 8.18 Design concept of a porous textile supercapacitor integrated into a smart garment, demonstrating porous carbon impregnation from the weave, to the yarn, to the fibers [147] (Reprinted with kind permission from Ref. [147]. Copyright 2011, Royal Society of Chemistry)

Because of cotton textiles with high porosity, large surface area and hydrophilic functional groups on the surface, it is an ideal electrode material for flexible supercapacitors after processed by loading conductive carbon. Gogotsi's group [147] reported a carbon-coated flexible fabric supercapacitors by uniformly screen printing porous carbon into woven cotton (Fig. 8.18). The porous structure of cotton could improve the ion diffusion between electrodes and ions. The electrodes achieved specific capacitances as high as 85 F g^{-1} (0.43 F cm^{-2}) at 0.25 A g^{-1} and a good cyclic stability over 10,000 cycles.

8.6 Fiber Supercapacitors

Recently, wire-shaped supercapacitors have gained increasing attention due to the rapid development of wearable electronics. Various carbon-based fiber materials have been explored as the substrates, including carbon fiber, carbon nanotube (CNT) fibers, and graphene fiber.

8.6.1 Carbon Fiber

Due to their high mechanical flexibility and electrical conductivity, carbon fibers are thus attractive as promising electrode materials for flexible substrates. However, a relatively low specific surface area of carbon fibers often leads to reduce the performance of flexible supercapacitors due to the limited reaction kinetics for the

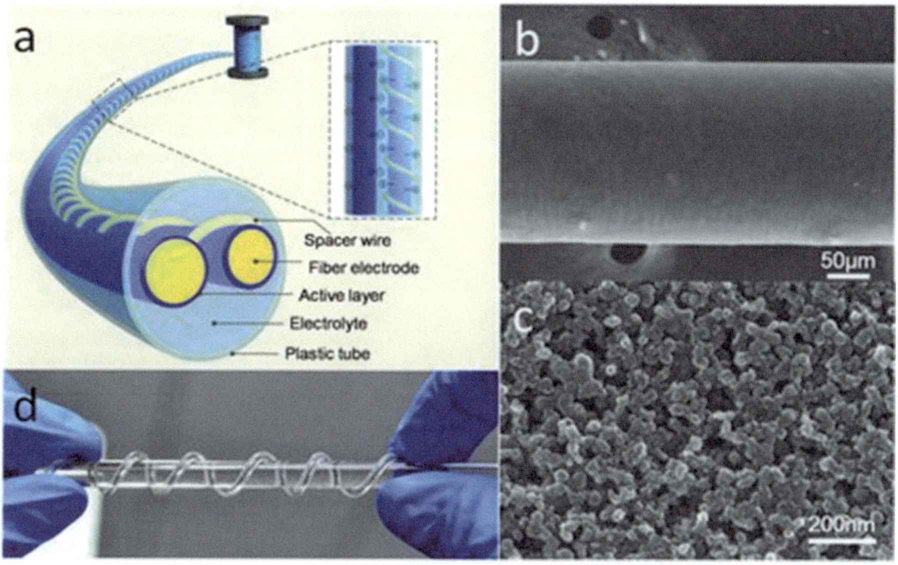

Fig. 8.19 Architecture of the fiber supercapacitor (FSC) and morphology of the electrode. (**a**) Schematic diagram of the FSC. (**b**) SEM image of the plastic fiber electrode coated with pen ink film. (**c**) SEM image of ink nanoparticles at high magnification, with a particle size of around 20 nm. (**d**) Photograph of a flexible FSC packaged using plastic tube (Φ inner = 0.9 mm, Φ outer = 1.8 mm). The excellent flexibility of the FSC was demonstrated by wrapping the FSC around a 7.5 mm-diameter glass rod [150] (Reprinted with kind permission from Ref. [150]. Copyright 2012, WILEY-VCH Verlag GmbH & Co. KGaA, Weinheim)

charge transfer and ion diffusion. This drawback could be alleviated by building electroactive material layer on carbon fibers, simultaneously enlarging the surface area and providing both nanoporous channels for low-resistant ion diffusion and nano-sized skeletons for fast electron transfer.

Zhou et al. [148] have reported porous core–shell carbon fibers synthesized by an acid oxidation treatment. The resulting porous core–shell carbon fibers showed an outstanding electrochemical performance and excellent mechanical properties, with a specific capacitance of 20.4 F g^{-1} at a current density of 1 A g^{-1} based on the entire device (0.41 F cm^{-1}) and good cyclic stability (about 96 % retention of capacitance after 3000 charge–discharge cycles at 1 A g^{-1}). Commercial pen ink was introduced as an active electrode material [149]. As shown in Fig. 8.19, a uniform film with fine dispersion and strong adhesion on substrates can be achieved by a simple dip-coating method. A highly flexible solid-state micro-supercapacitor based on a pen ink–carbon–fiber (Ink-CF) thread structure was fabricated by Xi's group [150]. This fabricated structure shows excellent electrochemical performance such as a high capacitance of 4.31 mF cm^{-2} and an energy density of 3.8×10^{-7} W h cm^{-2} at a power density of 5.6×10^{-6} W cm^{-2}. Le et al. [151] reported a coaxial fiber supercapacitor, which consisted of carbon microfiber

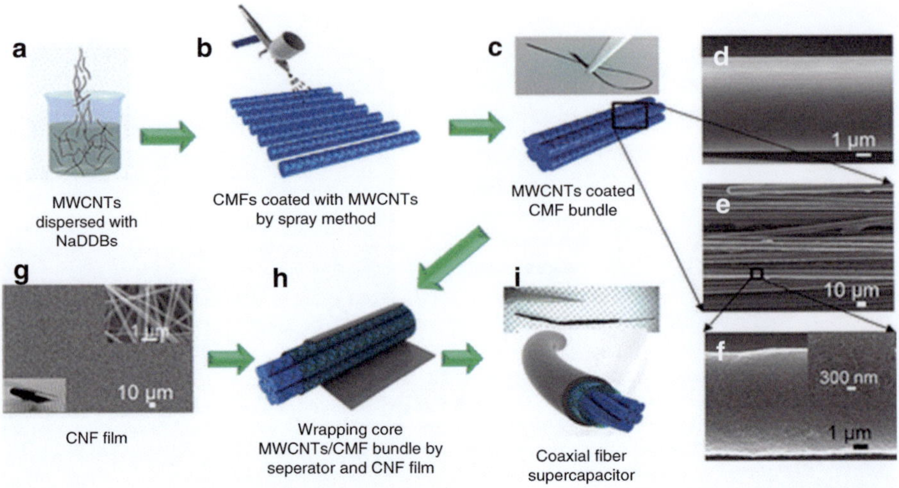

Fig. 8.20 Schematic of the coaxial fiber supercapacitor fabrication process. (**a**) Multiwalled carbon nanotubes (MWCNTs) were dispersed in a sodium dodecylbenzenesulfonate solution. (**b**) MWCNTs are deposited onto planar CMFs by spray coating. (**c**) The MWCNTs/CMFs are assembled into bundles after removing surfactant. (**d** and **f**) SEM images of single uncoated CMF and single CMF coated with MWCNTs. (**e**) SEM image of a MWCNTs/CMF bundle. (**g**) SEM image of a CNF film and its enlargement on upper right (the inset is a digital photo of the bendable CNF film). (**h**) After soaking with polymer electrolyte, the core MWCNTs/CMF bundle was wrapped with separator and CNF film. (**i**) Schematic and digital photo of a coaxial fiber supercapacitor [151] (Reprinted with kind permission from Ref. [151]. Copyright 2013, American Chemical Society)

bundles coated with multiwalled carbon nanotubes by spray coating as a core electrode and carbon nanofiber paper as an outer electrode (Fig. 8.20) and exhibited a high capacitance of 6.3 mF cm^{-1} (86.8 mF cm^{-2}) with an excellent cycling performance and an energy density of 0.7 μW h cm^{-1} (9.8 μW h cm^{-2}) at a power density of 13.7 μW cm^{-1} (189.4 μW cm^{-2}). Zhu et al. [152] processed electrochemically reduced graphene oxide (ERGO) which was electrochemically deposited on carbon fibers (CFs). Upon addition of only a small amount (~1 wt%) of ERGO, the flexible solid hybrid fibers showed superior electrochemical capacitances, which was measured to be up to 13.5 mF cm^{-1} (307 mF cm^{-2}) at a current density of 0.05 mA cm^{-1}. The maximum power density of such a solid supercapacitor could reach 0.74 mW cm^{-1} (8.5 mW cm^{-2}), and the maximum energy density was measured to be 1.9 μW h cm^{-1} (21.4 μW h cm^{-2}). The superior capacitive performance could be attributed to the synergistic effect between wrinkled and porous ERGO sheets and highly conductive CFs.

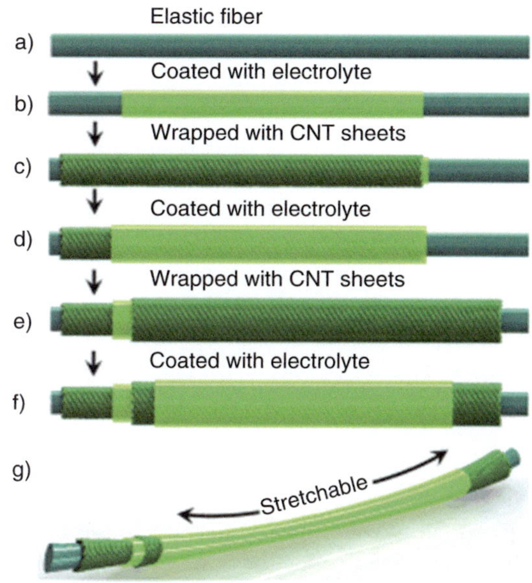

Fig. 8.21 Illustration to the fabrication of a highly stretchable, fiber-shaped supercapacitor with a coaxial structure [155] (Reprinted with kind permission from Ref. [155]. Copyright 2013, WILEY-VCH Verlag GmbH & Co. KGaA, Weinheim)

8.6.2 Carbon Nanotube (CNT) Fibers

As one of the carbon-based fibers, carbon nanotube (CNT) fibers/yarns have been manufactured by dry- and wet-spinning methods, which have attracted tremendous attention in the field of energy storage as a result of their excellent electrical conductivity, good mechanical properties, and outstanding flexibility. Dalton et al. [153] used two twisted CNT fibers to fabricate a wire-shaped supercapacitor (WSS) and reported that it can be woven into textiles. However, it is a big challenge for the wire and fiber-shaped supercapacitors not only with high electrochemical performance but also with remarkable stretchability, which is critically important to avoid the resulting electronic textiles to easily break. A great breakthrough has been made recently by Peng's group [154, 155]. The springlike fiber electrodes comprised of twisted aligned multiwalled carbon nanotubes (CNTs) with uniform coiled loops, which provided the electrode with a high stretchability and elongation of over 300 % (Fig. 8.21). An elastic supercapacitor was fabricated by placing two springlike fibers in parallel; the specific capacitances 18.12 F cm^{-3} in an optimal diameter occurred at 65 μm for the fiber-shaped supercapacitor. A highly stretchable, fiber-shaped supercapacitor has been developed by winding aligned CNT sheets on elastic fibers. The resultant fiber-shaped supercapacitor exhibited a high stretchability and stability; its CV curve remained unchanged even under 75 % strain, while its specific capacitance was maintained by more than 95 % over 100 stretching cycles with a strain of up to 75 % without any obvious structural damage.

Very recently, Chou and coworkers [156] reported an all-solid-state flexible and stretchable WSS consisting of two CNT fiber electrodes and H_2SO_4–PVA gel electrolyte. As demonstrated, WSS was closely attached to a prestrained spandex fiber using PDMS. After drying for 24 h, the resulting WSS device exhibited excellent stretchability which is more applicable in practical situations. Notably, the performances of the device at the stretched state were better than that at the relaxed state. This may be due to the better wettability of CNT fiber electrodes with easier diffusion of ions at the kink sites of the WSS. Area-specific capacitance of 4.63–4.99 mF cm^{-2} and excellent electrochemical stability over 10,000 charge–discharge cycles were observed.

Mesoporous carbon materials are more favorable for the transport and storage of charges. The active carbon on the surface of SWCNT bundles not only provides a relatively large specific surface area of electrode material but also acts as the binder between the crossed SWCNT bundles to improve the yarn electrical conductivity. Peng et al. [157] fabricated a flexible and weavable electric double-layer capacitor wire developed by twisting two aligned carbon nanotube/ordered mesoporous carbon composite fibers with remarkable mechanical and electronic properties as electrodes. This capacitor wire exhibited high specific capacitance and long life stability. The highest specific capacitance of 39.67 mF cm^{-2} (or 1.907 mF cm^{-1}) was found at the weight percentage of 87 %. It was appropriately 20 times of the supercapacitor wire based on bare MWCNT fibers (1.97 mF cm^{-2} or 0.017 mF cm^{-1}) under the same condition. Wei et al. [158] used a wet-spinning method to prepare SWNT and chitosan composite yarn. The capacitance of the yarn micro-supercapacitor with single-walled carbon nanotubes and activated carbon electrodes is 48.5 F cm^{-3} (74.6 F g^{-1}) at a scan rate of 2 mV s^{-1} and 42.4 F cm^{-3} (65.2 F g^{-1}) at a current density of 0.05 mA cm^{-2}. Besides, the energy density reaches 3.7 mWh cm^{-3} at a power density of 45.7 mW cm^{-3}. The device shows high stability retaining 98.5 % of the capacitance after 10,000 charge–discharge cycles.

8.6.3 Graphene Fibers

Besides carbon-fiber backbones, graphene-based cable-type supercapacitors have also been successfully prepared. Qu et al. [159] fabricated an all-graphene core–sheath fiber combining the high electrical conductivity and mechanical flexibility of graphene fiber core with the highly exposed surface area of 3D graphene network sheath. The core–sheath graphene fiber @3D-G offers a great advantage as flexible, lightweight electrodes for efficient fiber-based electrochemical supercapacitor. The all-solid-state fiber supercapacitor built on the basis of H_2SO_4–PVA gel electrolyte can be managed to spring-shaped supercapacitor with highly compressible and stretchable properties; the measured area-specific capacitance is 1.2–1.7 mF cm^{-2} and a mass-specific capacitance of 25–40 F g^{-1}. Gao et al. [160] used the aqueous solution of sodium carboxymethyl cellulose (CMC) as the outer spinning dope to

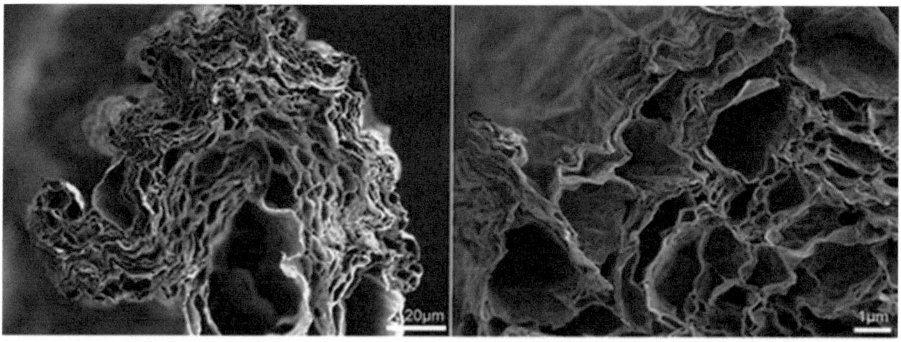

Fig. 8.22 The slight acidic condition of liquid crystal GO dopants (pH ~3) further changes the difference in imbibition rate, resulting in much higher water extraction rate and consequently more porous geometry [161] (Reprinted with kind permission from Ref. [161]. Copyright 2014, American Chemical Society)

spin polyelectrolyte-wrapped graphene/carbon nanotube core–sheath fibers. The fibers are used directly as safe electrodes to assembly two-ply yarn supercapacitors with liquid and solid electrolytes that show ultrahigh capacitances of 269 and 177 mF cm^{-2} and energy densities of 5.91 and 3.84 μWh cm^{-2}, respectively. CMC is an ionically conductive while electrically insulative polyelectrolyte. Such a polymer sheath ensures fiber electrodes free of short circuit when intertwined together, while allowing ions to smoothly penetrate from the electrolyte matrix to electrodes simultaneously.

As shown in Fig. 8.22, a highly porous graphene oxide (GO) and rGO fibers and yarns were fabricated by simple wet-spinning route using acetone as a coagulation bath, followed by a heat treatment [161]. The as-prepared final architectures were found to be mechanically robust (Young's modulus in excess of 29 GPa) and exhibited high native electrical conductivity (2508 ± 632 S m^{-1}) and exceptionally high specific surface area (2605 m^2 g^{-1} before reduction and 2210 m^2 g^{-1} after reduction), which were highly desirable for a range of wearable energy storage applications. The flexible fabrication exhibited extraordinary charge storage capacity values up to 409 F g^{-1} in a practical two-electrode configuration.

CNT or graphene fibers show many combined advantages, such as high electrochemical activity, conductivity, flexibility, strength, and lightweight. Graphene/CNT composite fibers will be a promising solution for wire-shaped electrochemical supercapacitors. Peng et al. [162] reported graphene/CNT composite fibers were synthesized with graphene sheets being incorporated neighboring CNTs to serve as effective bridges to improve the charge transport because of strong π–π interactions between CNT and graphene sheets. The supercapacitors possess specific capacitances up to ~31.50 F g^{-1} (4.97 mF cm^{-2} or 27.1 μF cm^{-1}), much higher than ~5.83 F g^{-1} (0.90 mF cm^{-2} or 5.1 μF cm^{-1}) based on the bare CNT fibers under the same condition. Recently, Chen et al. [163] developed a scalable method to continuously produce the hierarchically structured carbon microfibers using a silica capillary column functioning as a hydrothermal microreactor. The microfiber made of an interconnected network of aligned single-walled carbon nanotubes with interposed nitrogen-doped reduced graphene oxide sheets. The nanomaterials form mesoporous structures of large specific surface area (396 $m^2\ g^{-1}$) and high electrical conductivity (102 $S\ cm^{-1}$). The resultant fibers show a specific volumetric capacity as high as 305 $F\ cm^{-3}$ in sulfuric acid (measured at 73.5 $mA\ cm^{-3}$ in a three-electrode cell) or 300 $F\ cm^{-3}$ in polyvinyl alcohol (PVA)/H_3PO_4 electrolyte (measured at 26.7 $mA\ cm^{-3}$ in a two-electrode cell).

References

1. Kierzek K, Frackowiak E, Lota G et al (2004) Electrochemical capacitors based on highly porous carbons prepared by KOH activation. Electrochim Acta 49:515–523
2. Xia K, Gao Q, Jiang J et al (2008) Hierarchical porous carbons with controlled micropores and mesopores for supercapacitor electrode materials. Carbon 46:1718–1726
3. Fang B, Wei YZ, Maruyama K et al (2005) High capacity supercapacitors based on modified activated carbon aerogel. J Appl Electrochem 35:229–233
4. Zhu Y, Hu H, Li W-C et al (2006) Cresol–formaldehyde based carbon aerogel as electrode material for electrochemical capacitor. J Power Sources 162:738–742
5. Kumagai S, Sato M, Tashima D (2013) Electrical double-layer capacitance of micro- and mesoporous activated carbon prepared from rice husk and beet sugar. Electrochim Acta 114:617–626
6. Molina-Sabio M, Rodrıguez-Reinoso F (2004) Role of chemical activation in the development of carbon porosity. Colloids Surf A Physicochem Eng Asp 241:15–25
7. Xu J, Chen L, Qu H et al (2014) Preparation and characterization of activated carbon from reedy grass leaves by chemical activation with H_3PO_4. Appl Surf Sci 320:674–680
8. Ding L, Zou B, Li Y et al (2013) The production of hydrochar-based hierarchical porous carbons for use as electrochemical supercapacitor electrode materials. Colloids Surf A Physicochem Eng Asp 423:104–111
9. Wang C, Sun L, Zhou Y et al (2013) P/N co-doped microporous carbons from H_3PO_4-doped polyaniline by in situ activation for supercapacitors. Carbon 59:537–546
10. Wang J, Kaskel S (2012) KOH activation of carbon-based materials for energy storage. J Mater Chem 22:23710–23725
11. Zhang LL, Zhao X, Stoller MD et al (2012) Highly conductive and porous activated reduced graphene oxide films for high-power supercapacitors. Nano Lett 12:1806–1812
12. Zhai D, Li B, Du H et al (2011) The effect of pre-carbonization of mesophase pitch-based activated carbons on their electrochemical performance for electric double-layer capacitors. J Solid State Electrochem 15:787–794
13. Raymundo-Piñero E, Kierzek K, Machnikowski J et al (2006) Relationship between the nanoporous texture of activated carbons and their capacitance properties in different electrolytes. Carbon 44:2498–2507
14. Zhang C, Long D, Xing B et al (2008) The superior electrochemical performance of oxygen-rich activated carbons prepared from bituminous coal. Electrochem Commun 10:1809–1811
15. Ruiz V, Blanco C, Granda M et al (2008) Enhanced life-cycle supercapacitors by thermal treatment of mesophase-derived activated carbons. Electrochim Acta 54:305–310
16. Zhou M, Pu F, Wang Z et al (2014) Nitrogen-doped porous carbons through KOH activation with superior performance in supercapacitors. Carbon 68:185–194
17. Zhang LL, Zhao X (2009) Carbon-based materials as supercapacitor electrodes. Chem Soc Rev 38:2520–2531

18. Huang X, Qian K, Yang J et al (2012) Functional nanoporous graphene foams with controlled pore sizes. Adv Mater 24:4419–4423
19. Ning G, Fan Z, Wang G et al (2011) Gram-scale synthesis of nanomesh graphene with high surface area and its application in supercapacitor electrodes. Chem Commun 47:5976–5978
20. Chen Z, Ren W, Gao L et al (2011) Three-dimensional flexible and conductive interconnected graphene networks grown by chemical vapour deposition. Nat Mater 10:424–428
21. Shan C, Tang H, Wong T et al (2012) Facile synthesis of a large quantity of graphene by chemical vapor deposition: an advanced catalyst carrier. Adv Mater 24:2491–2495
22. Vickery JL, Patil AJ, Mann S (2009) Fabrication of graphene–polymer nanocomposites with higher-order three-dimensional architectures. Adv Mater 21:2180–2184
23. Lee J, Yoon S, Hyeon T et al (1999) Synthesis of a new mesoporous carbon and its application to electrochemical double-layer capacitors. Chem Commun 21:2177–2178
24. Lei Z, Christov N, Zhang LL et al (2011) Mesoporous carbon nanospheres with an excellent electrocapacitive performance. J Mater Chem 21:2274–2281
25. Xiao Y, Dong H, Lei B et al (2015) Ordered mesoporous carbons with fiber- and rod-like morphologies for supercapacitor electrode materials. Mater Lett 138:37–40
26. Lee SH, Kim HW, Hwang JO et al (2010) Three-dimensional self-assembly of graphene oxide platelets into mechanically flexible macroporous carbon films. Angew Chem Int Ed 49:10084–10088
27. Fan Z, Liu Y, Yan J et al (2012) Template-directed synthesis of pillared-porous carbon nanosheet architectures: high-performance electrode materials for supercapacitors. Adv Energy Mater 2:419–424
28. Fuertes AB, Centeno TA (2005) Mesoporous carbons with graphitic structures fabricated by using porous silica materials as templates and iron-impregnated polypyrrole as precursor. J Mater Chem 15:1079–1083
29. Zhai Y, Dou Y, Zhao D et al (2011) Carbon materials for chemical capacitive energy storage. Adv Mater 23:4828–4850
30. Feng G, Cummings PT (2011) Supercapacitor capacitance exhibits oscillatory behavior as a function of nanopore size. J Phys Chem Lett 2:2859–2864
31. Sun G, Wang J, Liu X et al (2010) Ion transport behavior in triblock copolymer-templated ordered mesoporous carbons with different pore symmetries. J Phys Chem C 114:18745–18751
32. Liang Y, Wu D, Fu R (2009) Preparation and electrochemical performance of novel ordered mesoporous carbon with an interconnected channel structure. Langmuir 25:7783–7785
33. Chen L-F, Huang Z-H, Liang H-W et al (2013) Flexible all-solid-state high-power supercapacitor fabricated with nitrogen-doped carbon nanofiber electrode material derived from bacterial cellulose. Energy Environ Sci 6:3331–3338
34. Liang C, Hong K, Guiochon GA et al (2004) Synthesis of a large-scale highly ordered porous carbon film by self-assembly of block copolymers. Angew Chem Int Ed 43:5785–5789
35. Liu H-J, Wang X-M, Cui W-J et al (2010) Highly ordered mesoporous carbon nanofiber arrays from a crab shell biological template and its application in supercapacitors and fuel cells. J Mater Chem 20:4223–4230
36. Liu Y, Deng R, Wang Z et al (2012) Carboxyl-functionalized graphene oxide-polyaniline composite as a promising supercapacitor material. J Mater Chem 22:13619–13624
37. Fang B, Kim M, Kim JH et al (2008) Controllable synthesis of hierarchical nanostructured hollow core/mesopore shell carbon for electrochemical hydrogen storage. Langmuir 24:12068–12072
38. Fang B, Kim JH, Lee C et al (2007) Hollow macroporous core/mesoporous shell carbon with a tailored structure as a cathode electrocatalyst support for proton exchange membrane fuel cells. J Phys Chem C 112:639–645
39. Kim JH, Fang B, Song MY et al (2012) Topological transformation of thioether-bridged organosilicas into nanostructured functional materials. Chem Mater 24:2256–2264

40. Chai GS, Shin IS, Yu JS (2004) Synthesis of ordered, uniform, macroporous carbons with mesoporous walls templated by aggregates of polystyrene spheres and silica particles for use as catalyst supports in direct methanol fuel cells. Adv Mater 16:2057–2061
41. Manthiram A, Vadivel Murugan A, Sarkar A et al (2008) Nanostructured electrode materials for electrochemical energy storage and conversion. Energy Environ Sci 1:621–638
42. Fang B, Kim M, Yu J-S (2008) Hollow core/mesoporous shell carbon as a highly efficient catalyst support in direct formic acid fuel cell. Appl Catal Environ 84:100–105
43. Kim JH, Fang B, Yoon SB et al (2009) Hollow core/mesoporous shell carbon capsule as an unique cathode catalyst support in direct methanol fuel cell. Appl Catal Environ 88:368–375
44. Kim JH, Yu J-S (2010) Erythrocyte-like hollow carbon capsules and their application in proton exchange membrane fuel cells. Phys Chem Chem Phys 12:15301–15308
45. Fu J, Xu Q, Chen J et al (2010) Controlled fabrication of uniform hollow core porous shell carbon spheres by the pyrolysis of core/shell polystyrene/cross-linked polyphosphazene composites. Chem Commun 46:6563–6565
46. Chen K, Chen L, Chen Y et al (2012) Three-dimensional porous graphene-based composite materials: electrochemical synthesis and application. J Mater Chem 22:20968–20976
47. Choi BG, Yang M, Hong WH et al (2012) 3D macroporous graphene frameworks for supercapacitors with high energy and power densities. ACS Nano 6:4020–4028
48. Wang Z-L, Xu D, Wang H-G et al (2013) In situ fabrication of porous graphene electrodes for high-performance energy storage. ACS Nano 7:2422–2430
49. Novoselov KS, Geim AK, Morozov SV et al (2004) Electric field effect in atomically thin carbon films. Science 306:666–669
50. Stoller MD, Park S, Zhu Y et al (2008) Graphene-based ultracapacitors. Nano Lett 8:3498–3502
51. Cai J, Ruffieux P, Jaafar R et al (2010) Atomically precise bottom-up fabrication of graphene nanoribbons. Nature 466:470–473
52. Tung VC, Allen MJ, Yang Y et al (2009) High-throughput solution processing of large-scale graphene. Nat Nanotechnol 4:25–29
53. Choucair M, Thordarson P, Stride JA (2009) Gram-scale production of graphene based on solvothermal synthesis and sonication. Nat Nanotechnol 4:30–33
54. Emtsev KV, Bostwick A, Horn K et al (2009) Towards wafer-size graphene layers by atmospheric pressure graphitization of silicon carbide. Nat Mater 8:203–207
55. Sutter PW, Flege J-I, Sutter EA (2008) Epitaxial graphene on ruthenium. Nat Mater 7:406–411
56. Kim KS, Zhao Y, Jang H et al (2009) Large-scale pattern growth of graphene films for stretchable transparent electrodes. Nature 457:706–710
57. Yang X, Dou X, Rouhanipour A et al (2008) Two-dimensional graphene nanoribbons. J Am Chem Soc 130:4216–4217
58. Taghioskoui M (2009) Trends in graphene research. Mater Today 12:34–37
59. Chen J, Duan M, Chen G (2012) Continuous mechanical exfoliation of graphene sheets via three-roll mill. J Mater Chem 22:19625–19628
60. Li D, Muller MB, Gilje S et al (2008) Processable aqueous dispersions of graphene nanosheets. Nat Nanotechnol 3:101–105
61. Park S, Ruoff RS (2009) Chemical methods for the production of graphenes. Nat Nanotechnol 4:217–224
62. Shen J, Hu Y, Shi M et al (2009) Fast and facile preparation of graphene oxide and reduced graphene oxide nanoplatelets. Chem Mater 21:3514–3520
63. Lei Z, Lu L, Zhao XS (2012) The electrocapacitive properties of graphene oxide reduced by urea. Energy Environ Sci 5:6391–6399
64. Huang H-C, Huang C-W, Hsieh C-T et al (2011) Photocatalytically reduced graphite oxide electrode for electrochemical capacitors. J Phys Chem C 115:20689–20695
65. El-Kady MF, Strong V, Dubin S et al (2012) Laser scribing of high-performance and flexible graphene-based electrochemical capacitors. Science 335:1326–1330

66. Li ZJ, Yang BC, Zhang SR et al (2012) Graphene oxide with improved electrical conductivity for supercapacitor electrodes. Appl Surf Sci 258:3726–3731
67. Miller JR, Outlaw RA, Holloway BC (2010) Graphene double-layer capacitor with ac line-filtering performance. Science 329:1637–1639
68. Sun L, Tian C, Li M et al (2013) From coconut shell to porous graphene-like nanosheets for high-power supercapacitors. J Mater Chem A 1:6462–6470
69. Biswal M, Banerjee A, Deo M et al (2013) From dead leaves to high energy density supercapacitors. Energy Environ Sci 6:1249–1259
70. Long C, Qi D, Wei T et al (2014) Nitrogen-doped carbon networks for high energy density supercapacitors derived from polyaniline coated bacterial cellulose. Adv Funct Mater 24:3953–3961
71. Chen Z, Wen J, Yan C et al (2011) High-performance supercapacitors based on hierarchically porous graphite particles. Adv Energy Mater 1:551–556
72. Presser V, Heon M, Gogotsi Y (2011) Carbide-derived carbons – from porous networks to nanotubes and graphene. Adv Funct Mater 21:810–833
73. Lukatskaya MR, Halim J, Dyatkin B et al (2014) Room-temperature carbide-derived carbon synthesis by electrochemical etching of MAX phases. Angew Chem Int Ed 53:4877–4880
74. Zhang L, Qin X, Shao G et al (2014) A new route for preparation of titanium carbide derived carbon and its performance for supercapacitors. Mater Lett 122:78–81
75. Dimovski S, Nikitin A, Ye H et al (2004) Synthesis of graphite by chlorination of iron carbide at moderate temperatures. J Mater Chem 14:238–243
76. Dash R, Chmiola J, Yushin G et al (2006) Titanium carbide derived nanoporous carbon for energy-related applications. Carbon 44:2489–2497
77. Zheng J, Ekstrom TC, Gordeev SK et al (2000) Carbon with an onion-like structure obtained by chlorinating titanium carbide. J Mater Chem 10:1039–1041
78. Cambaz ZG, Yushin GN, Gogotsi Y et al (2006) Formation of carbide-derived carbon on β-silicon carbide whiskers. J Am Ceram Soc 89:509–514
79. Welz S, Gogotsi Y, McNallan MJ (2003) Nucleation, growth, and graphitization of diamond nanocrystals during chlorination of carbides. J Appl Phys 93:4207–4214
80. Gao P C, Tsai W Y, Daffos B et al (2015) Graphene-like carbide derived carbon for high-power supercapacitors. Nano Energy 12:197–206
81. Gogotsi Y, Nikitin A, Ye H et al (2003) Nanoporous carbide-derived carbon with tunable pore size. Nat Mater 2:591–594
82. Lätt M, Käärik M, Permann L et al (2010) A structural influence on the electrical double-layer characteristics of Al4C3-derived carbon. J Solid State Electrochem 14:543–548
83. Chmiola J, Largeot C, Taberna P-L et al (2010) Monolithic carbide-derived carbon films for micro-supercapacitors. Science 328:480–483
84. Portet C, Yushin G, Gogotsi Y (2008) Effect of carbon particle size on electrochemical performance of EDLC. J Electrochem Soc 155:A531–A536
85. Thomberg T, Jänes A, Lust E (2009) Energy and power performance of vanadium carbide derived carbon electrode materials for supercapacitors. J Electroanal Chem 630:55–62
86. Thomberg T, Jänes A, Lust E (2010) Energy and power performance of electrochemical double-layer capacitors based on molybdenum carbide derived carbon. Electrochim Acta 55:3138–3143
87. Yeon S-H, Reddington P, Gogotsi Y et al (2010) Carbide-derived-carbons with hierarchical porosity from a preceramic polymer. Carbon 48:201–210
88. Tallo I, Thomberg T, Kurig H et al (2014) Novel micromesoporous carbon materials synthesized from tantalum hafnium carbide and tungsten titanium carbide. Carbon 67:607–616
89. Presser V, Zhang L, Niu JJ et al (2011) Flexible nano-felts of carbide-derived carbon with ultra-high power handling capability. Adv Energy Mater 1:423–430
90. Arulepp M, Leis J, Lätt M et al (2006) The advanced carbide-derived carbon based supercapacitor. J Power Sources 162:1460–1466

91. Portet C, Lillo-Rodenas MA, Linares-Solano A et al (2009) Capacitance of KOH activated carbide-derived carbons. Phys Chem Chem Phys 11:4943–4945
92. Korenblit Y, Rose M, Kockrick E et al (2010) High-rate electrochemical capacitors based on ordered mesoporous silicon carbide-derived carbon. ACS Nano 4:1337–1344
93. Paraknowitsch JP, Thomas A (2013) Doping carbons beyond nitrogen: an overview of advanced heteroatom doped carbons with boron, sulphur and phosphorus for energy applications. Energy Environ Sci 6:2839–2855
94. Wood KN, O'Hayre R, Pylypenko S (2014) Recent progress on nitrogen/carbon structures designed for use in energy and sustainability applications. Energy Environ Sci 7:1212–1249
95. Wu D, Li Z, Zhong M et al (2014) Templated synthesis of nitrogen-enriched nanoporous carbon materials from porogenic organic precursors prepared by ATRP. Angew Chem 126:4038–4041
96. Zhao Y, Hu C, Hu Y et al (2012) A versatile, ultralight, nitrogen-doped graphene framework. Angew Chem Int Ed 51:11371–11375
97. Wen Z, Wang X, Mao S et al (2012) Crumpled nitrogen-doped graphene nanosheets with ultrahigh pore volume for high-performance supercapacitor. Adv Mater 24:5610–5616
98. Zhao L, Fan L-Z, Zhou M-Q et al (2010) Nitrogen-containing hydrothermal carbons with superior performance in supercapacitors. Adv Mater 22:5202–5206
99. Qian W, Sun F, Xu Y et al (2014) Human hair-derived carbon flakes for electrochemical supercapacitors. Energy Environ Sci 7:379–386
100. Dutta S, Bhaumik A, Wu KCW (2014) Hierarchically porous carbon derived from polymers and biomass: effect of interconnected pores on energy applications. Energy Environ Sci 7:3574–3592
101. Zhang LL, Zhao X, Ji H et al (2012) Nitrogen doping of graphene and its effect on quantum capacitance, and a new insight on the enhanced capacitance of N-doped carbon. Energy Environ Sci 5:9618–9625
102. Deng D, Yu L, Chen X et al (2013) Iron encapsulated within pod-like carbon nanotubes for oxygen reduction reaction. Angew Chem Int Ed 52:371–375
103. Chen L-F, Zhang X-D, Liang H-W et al (2012) Synthesis of nitrogen-doped porous carbon nanofibers as an efficient electrode material for supercapacitors. ACS Nano 6:7092–7102
104. Guo H, Gao Q (2009) Boron and nitrogen co-doped porous carbon and its enhanced properties as supercapacitor. J Power Sources 186:551–556
105. Han J, Zhang LL, Lee S et al (2012) Generation of B-doped graphene nanoplatelets using a solution process and their supercapacitor applications. ACS Nano 7:19–26
106. Wu Z-S, Winter A, Chen L et al (2012) Three-dimensional nitrogen and boron Co-doped graphene for high-performance all-solid-state supercapacitors. Adv Mater 24:5130–5135
107. Gu W, Sevilla M, Magasinski A et al (2013) Sulfur-containing activated carbons with greatly reduced content of bottle neck pores for double-layer capacitors: a case study for pseudocapacitance detection. Energy Environ Sci 6:2465–2476
108. Hasegawa G, Aoki M, Kanamori K et al (2011) Monolithic electrode for electric double-layer capacitors based on macro/meso/microporous S-containing activated carbon with high surface area. J Mater Chem 21:2060–2063
109. Xu Y, Sheng K, Li C et al (2010) Self-assembled graphene hydrogel via a one-step hydrothermal process. ACS Nano 4:4324–4330
110. Chen P, Yang J-J, Li S-S et al (2013) Hydrothermal synthesis of macroscopic nitrogen-doped graphene hydrogels for ultrafast supercapacitor. Nano Energy 2:249–256
111. You B, Wang L, Yao L et al (2013) Three dimensional N-doped graphene-CNT networks for supercapacitor. Chem Commun 49:5016–5018
112. Zhang X, Sui Z, Xu B et al (2011) Mechanically strong and highly conductive graphene aerogel and its use as electrodes for electrochemical power sources. J Mater Chem 21:6494–6497
113. Worsley MA, Olson TY, Lee JRI et al (2011) High surface area, sp2-cross-linked three-dimensional graphene monoliths. J Phys Chem Lett 2:921–925

114. Luan VH, Tien HN, Hoa LT et al (2013) Synthesis of a highly conductive and large surface area graphene oxide hydrogel and its use in a supercapacitor. J Mater Chem A 1:208–211
115. Cao X, Shi Y, Shi W et al (2011) Preparation of novel 3D graphene networks for supercapacitor applications. Small 7:3163–3168
116. Yavari F, Chen Z, Thomas AV et al (2011) High sensitivity gas detection using a macroscopic three-dimensional graphene foam network. Sci Rep 1:166
117. Masahiro M, Mikio N, Golap K et al (2012) Formation of graphene-containing porous carbon film for electric double-layer capacitor by pulsed plasma chemical vapor deposition. Jpn J Appl Phys 51:045103
118. Yong Y-C, Dong X-C, Chan-Park MB et al (2012) Macroporous and monolithic anode based on polyaniline hybridized three-dimensional graphene for high-performance microbial fuel cells. ACS Nano 6:2394–2400
119. Jiang S, Shi T, Zhan X et al (2014) High-performance all-solid-state flexible supercapacitors based on two-step activated carbon cloth. J Power Sources 272:16–23
120. Jin HY, Peng ZH, Tang WM et al (2014) Controllable functionalized carbon fabric for high-performance all-carbon-based supercapacitors. RSC Adv 4:33022–33028
121. Wang G, Wang H, Lu X et al (2014) Solid-state supercapacitor based on activated carbon cloths exhibits excellent rate capability. Adv Mater 26:2676–2682
122. Qian H, Kucernak AR, Greenhalgh ES et al (2013) Multifunctional structural supercapacitor composites based on carbon aerogel modified high performance carbon fiber fabric. ACS Appl Mater Interfaces 5:6113–6122
123. Hsu Y-K, Chen Y-C, Lin Y-G et al (2012) High-cell-voltage supercapacitor of carbon nanotube/carbon cloth operating in neutral aqueous solution. J Mater Chem 22:3383–3387
124. Wang S, Pei B, Zhao X et al (2013) Highly porous graphene on carbon cloth as advanced electrodes for flexible all-solid-state supercapacitors. Nano Energy 2:530–536
125. Xu Y, Lin Z, Huang X et al (2013) Functionalized graphene hydrogel-based high-performance supercapacitors. Adv Mater 25:5779–5784
126. Kaempgen M, Chan CK, Ma J et al (2009) Printable thin film supercapacitors using single-walled carbon nanotubes. Nano Lett 9:1872–1876
127. Niu Z, Dong H, Zhu B et al (2013) Highly stretchable, integrated supercapacitors based on single-walled carbon nanotube films with continuous reticulate architecture. Adv Mater 25:1058–1064
128. Niu Z, Zhou W, Chen J et al (2013) A repeated halving approach to fabricate ultrathin single-walled carbon nanotube films for transparent supercapacitors. Small 9:518–524
129. Yu Jin K, Haegeun C, Chi-Hwan H et al (2012) All-solid-state flexible supercapacitors based on papers coated with carbon nanotubes and ionic-liquid-based gel electrolytes. Nanotechnology 23:065401
130. Zheng C, Qian W, Cui C et al (2012) Hierarchical carbon nanotube membrane with high packing density and tunable porous structure for high voltage supercapacitors. Carbon 50:5167–5175
131. Xu G, Zheng C, Zhang Q et al (2011) Binder-free activated carbon/carbon nanotube paper electrodes for use in supercapacitors. Nano Res 4:870–881
132. Choi BG, Hong J, Hong WH et al (2011) Facilitated ion transport in all-solid-state flexible supercapacitors. ACS Nano 5:7205–7213
133. Maiti UN, Lim J, Lee KE et al (2014) Three-dimensional shape engineered, interfacial gelation of reduced graphene oxide for high rate, large capacity supercapacitors. Adv Mater 26:615–619
134. Niu Z, Chen J, Hng HH et al (2012) A leavening strategy to prepare reduced graphene oxide foams. Adv Mater 24:4144–4150
135. Xu Y, Lin Z, Huang X et al (2013) Flexible solid-state supercapacitors based on three-dimensional graphene hydrogel films. ACS Nano 7:4042–4049

136. Huang H, Tang Y, Xu L et al (2014) Direct formation of reduced graphene oxide and 3D lightweight nickel network composite foam by hydrohalic acids and its application for high-performance supercapacitors. ACS Appl Mater Interfaces 6:10248–10257
137. Yang X, Zhu J, Qiu L et al (2011) Bioinspired effective prevention of restacking in multilayered graphene films: towards the next generation of high-performance supercapacitors. Adv Mater 23:2833–2838
138. Chen C-M, Zhang Q, Huang C-H et al (2012) Macroporous 'bubble' graphene film via template-directed ordered-assembly for high rate supercapacitors. Chem Commun 48:7149–7151
139. Wang G, Sun X, Lu F et al (2012) Flexible pillared graphene-paper electrodes for high-performance electrochemical supercapacitors. Small 8:452–459
140. Seo DH, Yick S, Han ZJ et al (2014) Synergistic fusion of vertical graphene nanosheets and carbon nanotubes for high-performance supercapacitor electrodes. ChemSusChem 7:2317–2324
141. Fan Z, Yan J, Zhi L et al (2010) A three-dimensional carbon nanotube/graphene sandwich and its application as electrode in supercapacitors. Adv Mater 22:3723–3728
142. Wang W, Guo S, Penchev M et al (2013) Three dimensional few layer graphene and carbon nanotube foam architectures for high fidelity supercapacitors. Nano Energy 2:294–303
143. Gao K, Shao Z, Li J et al (2013) Cellulose nanofiber-graphene all solid-state flexible supercapacitors. J Mater Chem A 1:63–67
144. Hu L, Pasta M, Mantia FL et al (2010) Stretchable, porous, and conductive energy textiles. Nano Lett 10:708–714
145. Kang YJ, Chun S-J, Lee S-S et al (2012) All-solid-state flexible supercapacitors fabricated with bacterial nanocellulose papers, carbon nanotubes, and triblock-copolymer ion gels. ACS Nano 6:6400–6406
146. Weng Z, Su Y, Wang D-W et al (2011) Graphene–cellulose paper flexible supercapacitors. Adv Energy Mater 1:917–922
147. Jost K, Perez CR, McDonough JK et al (2011) Carbon coated textiles for flexible energy storage. Energy Environ Sci 4:5060–5067
148. Zhou W, Zhou K, Liu X et al (2014) Flexible wire-like all-carbon supercapacitors based on porous core-shell carbon fibers. J Mater Chem A 2:7250–7255
149. Fu Y, Cai X, Wu H et al (2012) Fiber supercapacitors utilizing pen ink for flexible/wearable energy storage. Adv Mater 24:5713–5718
150. Dai S, Guo H, Wang M et al (2014) A flexible micro-supercapacitor based on a pen ink-carbon fiber thread. J Mater Chem A 2:19665–19669
151. Le VT, Kim H, Ghosh A et al (2013) Coaxial fiber supercapacitor using all-carbon material electrodes. ACS Nano 7:5940–5947
152. Cao Y, Zhu M, Li P et al (2013) Boosting supercapacitor performance of carbon fibres using electrochemically reduced graphene oxide additives. Phys Chem Chem Phys 15:19550–19556
153. Dalton AB, Collins S, Munoz E et al (2003) Super-tough carbon-nanotube fibres. Nature 423:703–703
154. Zhang Y, Bai W, Cheng X et al (2014) Flexible and stretchable lithium-Ion batteries and supercapacitors based on electrically conducting carbon nanotube fiber springs. Angew Chem Int Ed 53:14564–14568
155. Yang Z, Deng J, Chen X et al (2013) A highly stretchable, fiber-shaped supercapacitor. Angew Chem Int Ed 52:13453–13457
156. Xu P, Gu T, Cao Z et al (2014) Carbon nanotube fiber based stretchable wire-shaped supercapacitors. Adv Energy Mater 4:n/a–n/a
157. Ren J, Bai W, Guan G et al (2013) Flexible and weaveable capacitor wire based on a carbon nanocomposite fiber. Adv Mater 25:5965–5970
158. Meng Q, Wu H, Meng Y et al (2014) High-performance all-carbon yarn micro-supercapacitor for an integrated energy system. Adv Mater 26:4100–4106

159. Meng Y, Zhao Y, Hu C et al (2013) All-graphene core-sheath microfibers for all-solid-state, stretchable fibriform supercapacitors and wearable electronic textiles. Adv Mater 25:2326–2331
160. Lee JA, Shin MK, Kim SH et al (2013) Ultrafast charge and discharge biscrolled yarn supercapacitors for textiles and microdevices. Nat Commun 4:1970
161. Aboutalebi SH, Jalili R, Esrafilzadeh D et al (2014) High-performance multifunctional graphene yarns: toward wearable all-carbon energy storage textiles. ACS Nano 8:2456–2466
162. Sun H, You X, Deng J et al (2014) Novel graphene/carbon nanotube composite fibers for efficient wire-shaped miniature energy devices. Adv Mater 26:2868–2873
163. Yu D, Goh K, Wang H et al (2014) Scalable synthesis of hierarchically structured carbon nanotube-graphene fibres for capacitive energy storage. Nat Nano 9:555–562

Chapter 9
Transition Metal Oxides as Supercapacitor Materials

Zhibin Wu, Yirong Zhu, Xiaobo Ji, and Craig E. Banks

9.1 Introduction

The rapidly growing commercial markets in electric vehicles and portable electronic products have created a seemingly insatiable demand for high-performance energy storage devices [1]. However, it seems hard to meet the increasingly higher standards by using the current energy conversation or storage devices alone, such as lithium-ion batteries or fuel cells. Supercapacitors (SCs), also called electrochemical capacitors, store charges through either ion adsorption/desorption (electrochemical double-layer capacitors, EDLCs) or fast surface redox reactions (pseudocapacitors), which can complement or replace batteries in electrical energy storage when high-power delivery or uptake is needed [2]. According to charge storage mechanisms, SCs may be placed into two distinctive classes: EDLCs and pseudocapacitors [3]. As is well known, SCs have the advantages of high power density, ultralong cycle life, and rapid recharge capability, devoting to meeting the ever-growing consumption demands and alleviating the energy crisis [4, 5]. The Ragone plot [2] in Fig. 9.1 (power density against energy density) shows that SCs shorten the gap between conventional capacitors and batteries.

Z. Wu • Y. Zhu • X. Ji (✉)
College of Chemistry and Chemical Engineering, Central South University, Changsha 410083, China
e-mail: xji@csu.edu.cn

C.E. Banks
College of Chemistry and Chemical Engineering, Central South University, Changsha 410083, China

Faculty of Science and Engineering, School of Chemistry and the Environment, Division of Chemistry and Environmental Science, Manchester Metropolitan University, Chester Street, M1 5GD Lancs, UK

K.I. Ozoemena, S. Chen (eds.), *Nanomaterials in Advanced Batteries and Supercapacitors*, Nanostructure Science and Technology,
DOI 10.1007/978-3-319-26082-2_9

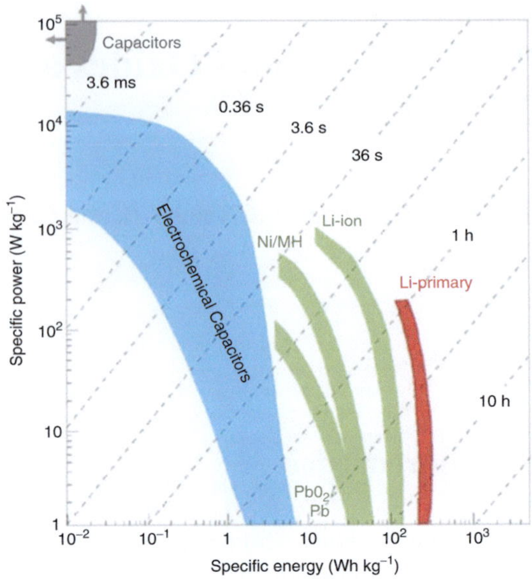

Fig. 9.1 Ragone plot (specific power against specific energy) for various electrical energy storage devices (Reprinted from Ref. [2] with permission. Copyright 2008, Macmillan Publishers Limited)

SCs were first reported in 1957 Becker's patent [6] and then emerged in the commercial market in 1978 (gold capacitors from Panasonic/Matsushita) and 1980 (supercap from NEC Tokin) [7]. Generally, they can serve as load-leveling and uninterruptible power supplies, such as backup sources for memories and microcomputers, and have great potential applications in electric vehicles or hybrid electric vehicles. Since the 1990s, electrochemical capacitors ushered in its development with materials used as EDLCs mainly based on carbonaceous materials such as activated carbon [8, 9], carbon nanotube [10, 11], and graphene [12–14] which provide large surface areas for electrical charges to reside. Alternatively, pseudocapacitors utilizing transition metal oxides and conductive polymers combine electrosorption with fast surface redox reactions to store energy, where the amount of charge held is proportional to the applied voltage [7]. Researchers had no clear concept regarding pseudocapacitance until Trasatti and Buzzanca discovered the capacitive properties of RuO_2 [15].

Note that electrode materials have become the core competence for supercapacitors and could be typically divided into three types including carbon materials, conducting polymers, and transition metal oxides. Carbon materials with low cost, extremely developed surface area (up to 3000 m^2g^{-1}), and good electrical conductivity and chemical stability seem to be ideal materials for electrochemical double-layer capacitors, but it usually suffers from low specific capacitance (commonly lower than 400 F g^{-1}) [16–18]. While conducting polymers have higher specific capacitance (usually lower than 530 F g^{-1}) and excellent intrinsic conductivity, however, their cycle life is extremely poor because of substantial expansion and contraction during the charging–discharging process [19, 20]. Compared with these two types mentioned above, transition metal oxides usually possess multiple oxidation states that are in favor of fast redox reaction resulting in a much higher specific capacitance [21]. Impressively, specific capacitance of pseudocapacitors obtained from conducting polymers and transition metal oxides can be 10–100 times higher than EDLCs obtained from carbon materials [22].

While supercapacitors benefit from the large power capability (as high as 10 kW kg^{-1}), high current capability, ease of maintenance, and long-term cycling stability ($> 10^6$ cycles) [23], their moderate energy density for the widespread deployment in everyday technology using clean and renewable energy media is still far away to meet the performances and cost requirements, in particular providing power boosts to start the engine or to assist acceleration. In order to make capacitors commercially viable and extensively used, the energy that they can store needs to increase significantly and to endow supercapacitors with sufficient energy for higher requirements in the future is still a challenge. Extensive efforts have been devoted to improving the specific capacitance of supercapacitors by introducing pseudocapacitive metal oxides. In this way, cheap transition metal oxides with high energy density are very attractive electrode materials for supercapacitors to realize commercialization quickly.

9.2 Energy Storage Mechanism

Transition metal oxides store its charges both on the surface and in the bulk near the surface of the solid electrode where the electrochemical processes occur. Moreover, it should be noted that the metal oxides possess not only double-layer capacitance but also high pseudocapacitance (typically 10~100 times higher than double-layer capacitance), whereas the double-layer capacitances shown by CV curves for the electrodes are relatively too small to be observed [24]. In general, two kinds of electrolytes including alkaline solution and acid solution are used to analyze their corresponding electrochemical performances. Typically, RuO_2 and MnO_2 electrodes are measured in H_2SO_4 solution, while NiO, Co_3O_4, and $NiCo_2O_4$ use KOH as electrolyte, thus obtaining remarkable pseudocapacitances; their pseudocapacitive behaviors in the electrolyte can be described by the following equations [24, 25]:

$$RuO_2 + \delta H^+ + \delta e^- \leftrightarrow RuO_{(2-\delta)}(OH)_\delta \tag{9.1}$$

$$MnO_2 + H^+ + e^- \leftrightarrow MnOOH \tag{9.2}$$

$$NiO + OH^- + e^- \leftrightarrow NiOOH \tag{9.3}$$

$$\begin{cases} Co_3O_4 + OH^- + H_2O \leftrightarrow 3CoOOH + e^- \\ CoOOH + OH^- \leftrightarrow CoO_2 + H_2O + e^- \end{cases} \tag{9.4}$$

$$\begin{cases} NiCo_2O_4 + OH^- + H_2O \rightleftharpoons NiOOH + 2CoOOH + e^- \\ CoOOH + OH^- \rightleftharpoons CoO_2 + H_2O + e^- \end{cases} \tag{9.5}$$

As typical pseudocapacitor electrode materials, ruthenium oxides have a completely different mechanism with that of electrochemical double-layer

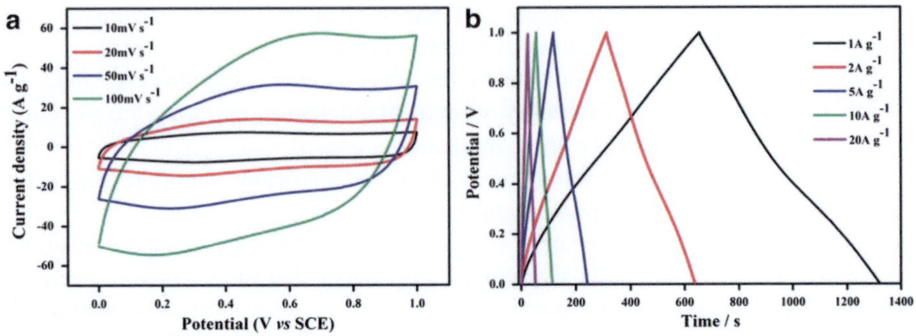

Fig. 9.2 (**a**) The cyclic voltammetry (CV) curve and (**b**) galvanostatic charge–discharge curve of RuO_2 (Reprinted from Ref. [29] with permission. Copyright 2014, The Royal Society of Chemistry)

capacitors and possess multiple redox states and good electrical conductivity. RuO_2 has three distinct oxidation states accessible within 1.2 V, and its pseudocapacitive behavior in acidic solutions can be described by Eq. (9.1). Protons can participate in the redox reaction, where RuO_2 acts as a proton condenser and the reaction is electrochemically reversible. Amorphous structures are the key parameter in determining the value of specific capacitances, as the proton can be easily intercalated into the bulk of $RuO_2 \cdot xH_2O$ but not into the crystalline phase [26, 27]. It is also proposed that hydrous ruthenium dioxide has an improved capacitive properties, better than those observed from anhydrous structures, as hydrous regions are more permeable to protons and proton conduction inside ruthenium dioxide is dominant when compared with the electron conduction [28]. The cyclic voltammetry (CV) curve and galvanostatic charge–discharge curve [29] are displayed in Fig. 9.2a and b. It shows a nearly rectangular shape and good coulombic efficiency as high as ~100%, which is closely related to the superlong cycle life and excellent rate capacity.

Very recently, spinel $NiCo_2O_4$ has sparked worldwide concern as an attractive pseudocapacitive electrode material for supercapacitor application due to its many intrinsic advantages. It possesses a ferrimagnetic cubic spinel structure with multiple redox states and good electrical conductivity as a P-type semiconductor with the band gap of 2.1 eV [30]. Note that $NiCo_2O_4$ adopts a spinel structure in which all the nickel cations occupy the octahedral interstices, whereas the cobalt cations are distributing among the tetrahedral and octahedral interstices [31]. The redox reactions offered by nickel cobaltite, including contributions from both nickel and cobalt ions, are richer than those of the monometallic nickel oxides and cobalt oxides as illustrated in Eq. (9.5). In addition, the electrical conductivity of nickel cobaltite is at least two orders of magnitude higher than that of pure Co_3O_4 and NiO, which facilities to acquire better cycling performance and fast charge–discharge capability [32, 33]. It is noteworthy that, in the charging–discharging process (Fig. 9.3), there exist valence state changes of Co^{3+}/Co^{4+} as well as M^{2+}/M^{3+} ($M = Co$ or Ni) on the surface of the electrode materials, where

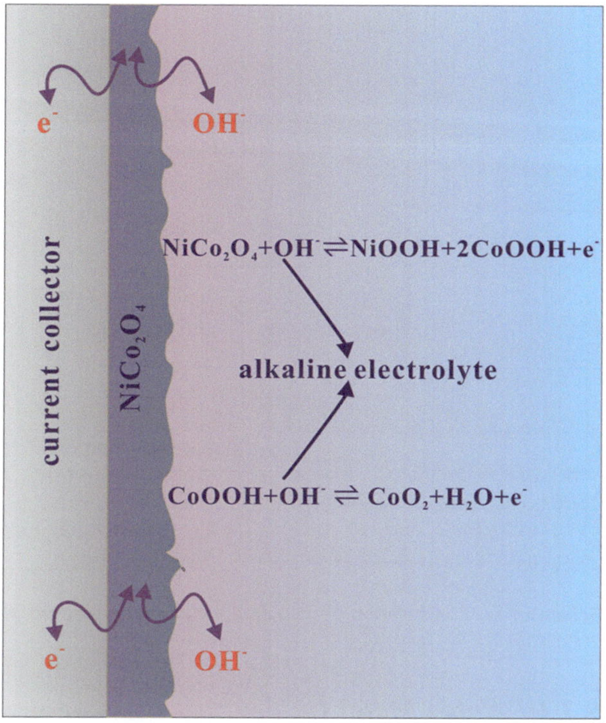

Fig. 9.3 Schematic of charge storage mechanism for supercapacitors based on $NiCo_2O_4$ (Reprinted from Ref. [24] with permission. Copyright 2014, The Royal Society of Chemistry)

fast and reversible faradaic reactions occur [33, 34]. The electrochemical redox potentials of M^{2+}/M^{3+} and Co^{3+}/Co^{4+} transitions are so close that the redox peaks observed overlap as one. Intriguingly, the specific capacitance of $NiCo_2O_4$ is observed to rise after several hundred cycles in some work which attributed to their unique morphologies and process of activation (potential cycling). Other transition metal oxides have the same storage mechanism and are not covered here.

9.3 Transition Metal Oxides for Supercapacitors

Transition metal oxides as the basis for supercapacitors can be typically divided into two types: noble metal oxides and base transition metal oxides. Noble metal oxides include RuO_2, IrO_2, etc., while base transition metal oxides for supercapacitors are developing from monometallic oxides (such as NiO, Co_3O_4, Fe_3O_4, and Mn_3O_4) to single-phase bimetallic oxides (such as $NiCo_2O_4$, $MnCo_2O_4$, $NiMoO_4$, $NiFe_2O_4$, and $CoMn_2O_4$). The supercapacitor materials based on transition metal oxides are summarized and classified in Scheme 9.1.

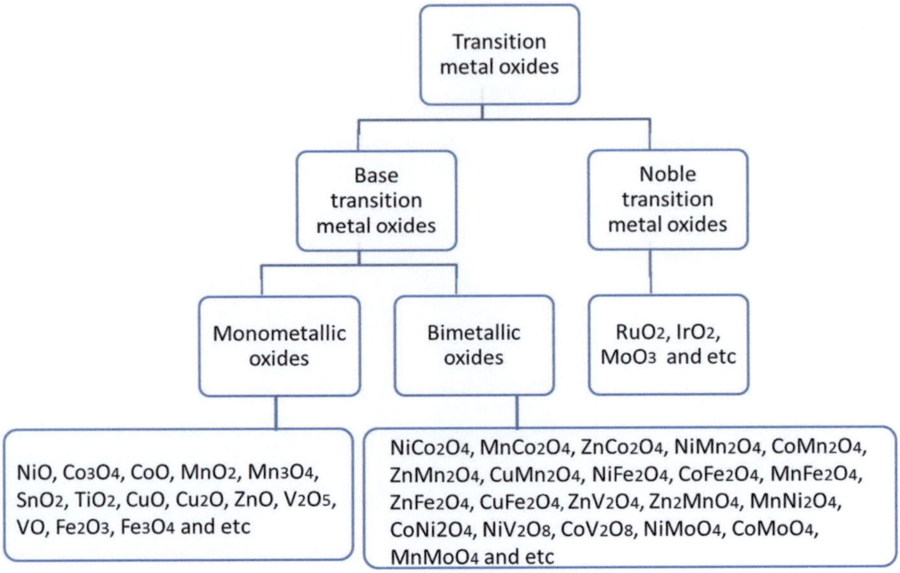

Scheme 9.1 Classification of transition metal oxides utalised as supercapacitor materials

9.3.1 Noble Metal Oxides

Nanostructured materials facilitate to help ameliorate the electrochemical performances of transition metal oxides, taking advantage of revealing the essentials during the charge–discharge process and grasping the key element in fabricating nanomaterials by transforming bulk transition metal oxides into nano-porous structures. Noble metal oxides such as RuO_2, IrO_2 and so forth were first found exhibiting good capacitive properties with remarkable electrochemical performances. In 1971, Trasatti and Buzzanca first reported that RuO_2 films could form a rectangular shaped cyclic voltammogram which closely resembled the carbon-based electrochemical double-layer capacitor [15]. Since this pioneering report, researchers have explored the capacitive properties of RuO_2 extensively since its theoretical value of specific capacitance could be as high as $1300 - 2200\ \mathrm{F\ g^{-1}}$ [35]. RuO_2 is a rare stoichiometric oxide which possesses wide potential window, remarkably high specific capacitance, high rate capability, long cycle life, thermal and chemical stability, and metallic-type conductivity [36]. The key methodology to boost the specific capacitance is their morphological and chemical composition design, as high surface-to-volume ratio with suitable pore sizes is desirable for the penetration of electrolytes and reactants into the whole electrode matrix and can promote the electrochemical double-layer capacitances and provide a large amount of superficial electroactive species for faradaic redox reactions. More specifically, the factors playing key roles in the electrochemical behavior of Ru oxides mainly

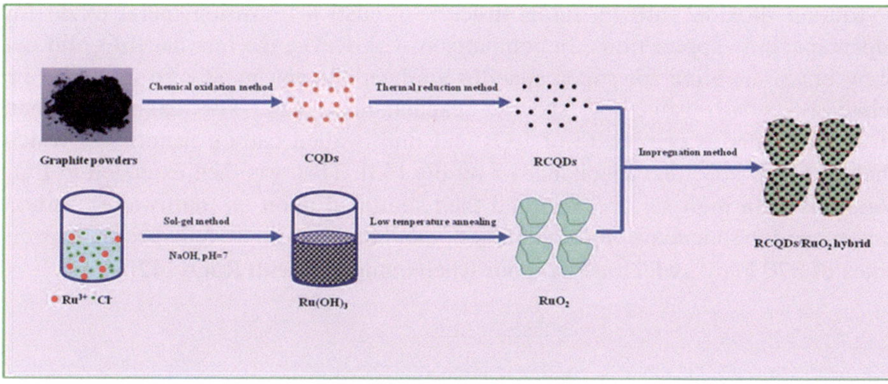

Fig. 9.4 Preparation procedure for the RCQD/RuO_2 hybrid (Reprinted from Ref. [40] with permission. Copyright 2013, The Royal Society of Chemistry)

include specific surface area, the combined water in RuO_x, the crystallinity of RuO_2, the size of RuO_2. xH_2O, and electrolytes used [37].

Generally, there are two common methods to prepare ruthenium oxides for supercapacitor applications. One is the chemical vapor deposition that synthesizes nanomaterials by thermal decomposition and then oxidization of the precursor at high temperatures [38, 39]. The other is the sol–gel method which firstly produces a sol–gel of precursor followed by annealing at low temperatures. It was reported that amorphous ruthenium oxides obtained at low temperatures of 150 °C show specific capacitance as high as $720\ F\ g^{-1}$, while in the crystalline phase annealing with temperature beyond 175°C exhibited rapidly dropped specific capacitance [26].

Recently, it was reported that the group of Ji et al. developed a sol–gel method fabricating carbon spheres at RuO_2 [29] and carbon quantum dots at RuO_2 [40] hybrid for supercapacitors, demonstrating ultrahigh rate capability and excellent cycling stability. The excellent electrochemical performance was mainly attributed to the formation of the hybrid structure, which facilitates the fast charge transportation and ionic motion during the charge–discharge process and greatly improves the utilization of RuO_2, leading to rapid redox reactions. The process of fabricating carbon quantum dots at RuO_2 is elaborated in Fig. 9.4. In addition, Zhang et al. synthesized nanotubular ruthenium oxides by using manganite nanorods as a morphology sacrificial template and obtained a high specific capacitance of 860 F g^{-1} with a good capacitive retention [41]. Extensive research has been undertaken by researchers in this field with the specific capacitance reported to reach more than $1300\ F\ g^{-1}$. For example, Hu et al. [35] used the membrane-templated synthesis route to obtain hydrous RuO_2 ($RuO_2 \cdot xH_2O$) nanotubular structures by means of an anodic deposition technique. The obtained ruthenium oxide exhibited an ultrahigh specific capacitance of $1300\ F\ g^{-1}$, excellent charge–discharge behavior at $1000\ mV\ s^{-1}$, and high-frequency (4.0–7.8 kHz) capacitive responses, which were much higher than any value ever reported before.

Iridium dioxide with its rutile structure is also a transition metal oxide for supercapacitors applications. In comparison with RuO_2, the fact that IrO_2 did not draw much attention for supercapacitor applications can mostly be owing to its relatively small value of specific capacitance [42]. For example, Chen et al. fabricated a composite of IrO_2 and multiwalled carbon nanotubes, which showed a low specific capacitance of 69 F/g [43]. This was then extended to IrO_x nanofoils with high surface area and then sputtered upon on multiwalled carbon nanotubes for supercapacitor application, exhibiting an improved specific capacitance of 370 F g^{-1}, which is still poor when compared with RuO_2 [42].

9.3.2 Base Transition Metal Oxides

9.3.2.1 Monometallic Oxides

NiO

NiO is a very versatile material and has shown to exhibit good electrochemical activity based on the transformation process of NiO/NiOOH. Various morphologies such as flowers [44], hollow spheres [45, 46], and nanosheets [47] have been successfully prepared by different methodologies.

A number of chemical methods have been used to prepare nanostructured nickel oxides such as chemical precipitation, chemical bath deposition, microwave route, sol–gel process, and hydrothermal method [25]. In traditional chemical precipitation methods, $Ni(OH)_2$ precursor is first obtained by adding precipitants into a soluble nickel salt such as $Ni(CH_3COO)_2$ and $Ni(NO_3)_2$, and then $Ni(OH)_2$ is calcinated into NiO at high temperatures. The preparation method is facile, low cost, and suitable for large-scale production of NiO. For example,Zhang et al. [48] reported a facile hydrothermal process to grow various porous NiO nanostructures including nanoslices, nanoplates, and nanocolumns by calcinating $\beta - Ni(OH)_2$. Consequently, the samples exhibited specific capacitances of 176, 285, and 390 F g^{-1} at a discharge current of 5 A g^{-1}, respectively.

Sol–gel process is another efficient way to obtain desired nanostructures. Typically, the first step is to prepare a sol–gel and the second step is to apply heat treatment. The sol–gel process seems very simple; however, in fact it is difficult to control. As reported, Yang et al. [49] obtained hexagonal and single crystalline NiO nanowires by a sol–gel process followed by calcination. In addition, NiO_x xerogels were formed by a sol–gel method followed by annealing at 250 °C, obtaining a maximum specific capacitance of 696 F g^{-1} [50].

There are some other methods such as chemical bath deposition [51], hydrothermal methods [52], and microwave routes [53] that are widely used in preparing NiO for supercapacitor applications. Nickel oxide is considered as an alternative electrode material for SC in alkaline electrolytes due to the advantages of high specific capacitance (theoretical specific capacitance of 3750 F g^{-1}), environment friendliness, and low cost; however, it faces the challenges involving poor cycle performance and high resistivity [37]. The cycle performance and conductivity of nickel oxide could be improved by introducing cobalt ions into the nickel oxide matrix or combining with carbon materials [54, 55].

MnO_2

Manganese oxide (MnO_2) is advantageous to its low cost, no toxicity, easy accessibility, and large energy density and consequently has been considered as one of the most promising materials in application of SCs. MnO_2 has not only a larger energy density but also is more economical than noble transition metal oxides. Over the past years, significant efforts have been directed toward the synthesis of nanomaterials with controlled sizes and ordered morphologies owing to the close relationship between the morphology and the capacitive performances.

The most common method to prepare MnO_2 is a hydrothermal synthesis in which Mn^{2+} is often used as raw material, and the as-prepared mixture solution is placed in a Teflon-lined pressure vessel. It has been demonstrated that the morphology of MnO_2 depends much on the dwell time of hydrothermal process, showing an evolution of crystallinity of the nanostructures as the hydrothermal dwell time increased from 1 to 18 h. Scanning electron microscopy (SEM) images of nanostructure MnO_2 exhibiting different morphologies are depicted in Fig. 9.5, and the MnO_2 nanostructures prepared at 140 °C for 6 h showed the best performance [56]. Electrochemical deposition is another effective methodology to prepare nanostructured MnO_2. For example, Tao et al. [57] have successfully prepared a solid-state supercapacitor based on a PPy – MnO_2 nanoflake-carbon fiber hybrid structure by electrochemical deposition process. The device exhibited a high-volume capacitance of 69.3 F cm^{-3} at 0.1 A cm^{-3} and a large energy density of 6.16×10^{-3} Wh cm^{-3} at 0.04 W m^{-3}. Furthermore Li et al. [58] developed an electrospinning approach to fulfill scale and thickness requirements of nanostructured MnO_2 electrodes. The MnO_2 electrode for supercapacitors could also be fabricated by a sol–gel method. For instance, Pang et al. [59] fabricated an MnO_2 film via dip coating or drop coating with manganese dioxide suspensions (sols) onto nickel foils, showing high specific capacitance of 698 F g^{-1}.

Regarding MnO_2 as SC materials, several challenges should be addressed, including partial dissolution problem of MnO_2 in the acidic electrolyte and its poor electronic conductivity and ionic conductivity. Impressively, Kang et al. [60] improved the electronic conductivity of MnO_2 by nonequilibrium Au doping changing the electronic structure of MnO_2, demonstrating a high specific capacitance (626 F g^{-1} at 5mV s^{-1}) and excellent cycling stability (7 % increment after 15,000 cycles). Furthermore, Zhang et al. [61] investigated the electrochemical properties of MnO_2 in 1-butyl-3-methylimidazolium hexafluorophosphate ([Bmim] PF_6)/N,N-dimethyformamide (DMF) electrolyte, fabricating a

Fig. 9.5 SEM images of MnO_2 prepared at different hydrothermal dwell times (Reprinted from Ref. [56] with permission. Copyright 2005, American Chemical Society)

high-voltage (3 V) asymmetric supercapacitor with a maximum specific energy of 67.5 Wh kg^{-1} and a maximum specific power of 20.4 kW kg^{-1}. The impressive results revealed that [Bmim] PF_6/DMF can be a promising electrolyte for MnO_2 -based supercapacitors.

Extensive work is ongoing to prepare a better modality of MnO_2 achieving higher specific capacitances, and the whole electrochemical performances of asymmetric supercapacitors have also been improved by enhancing the electrochemical properties of the negative electrode and the voltage of the whole cell.

Co_3O_4

Co_3O_4 is an important transition metal oxide for supercapacitors with excellent reversible redox behavior, high conductivity, and good stability. Different methods have been performed to synthesize Co_3O_4 nanostructures with morphologies from one dimension to three dimensions. In general, these methodologies consist of two steps: formation of precursor and calcination. Great efforts have been made to improve the electrochemical performance of Co_3O_4 by constructing different morphologies and combining with other novel materials.

For example, Co_3O_4 nanocrystals with well-controlled shapes, including nanosheets and microspheres assembled from nanosheets, have been successfully prepared by a facile ethanolamine-directed solvothermal method and sequential thermal decomposition at atmospheric pressure [62]. Moreover, Xia et al. [63] developed a facile hydrothermal synthesis method to prepare self-supported hollow Co_3O_4 nanowire arrays on various conductive substrates, revealing high specific capacitances with 599 F g^{-1} at 2 A g^{-1} and 439 F g^{-1} at 40 A g^{-1}, thus making it suitable for high-rate supercapacitor applications. The enhanced capacitive performance is due to its unique porous structure and hollow structure which could provide fast ion and electron transfer, good strain accommodation, and large reaction surface area. Chen et al. [64] reported a simple hydrothermal method through in situ growing Co_3O_4 nanowires on three-dimensional graphene foam grown by chemical vapor deposition, demonstrating high specific capacitance of $\sim$ 1100 F g^{-1} at a current density of 10 A g^{-1} with remarkable cycling stability. These recent studies indicate that appropriate morphology, microstructure, and chemical composition are important in achieving high-performance SC based on Co_3O_4.

9.3.2.2 Bimetallic Oxides

$NiCo_2O_4$

Nickel cobaltite ($NiCo_2O_4$) has become a new class of energy storage material for supercapacitors with outstanding electrochemical performance, delivering not only large power density but also high energy densities [65]. This material possesses richer electroactive sites and at least two magnitudes higher electrical conductivity than that of Co_3O_4 and NiO. Moreover, it shows comparable capacitive performances with noble metal oxides of RuO_2, but with much lower cost and more abundant resources. It is believed that the redox reactions offered by nickel cobaltite, including contributions from both nickel and cobalt ions, are richer than those of the monometallic nickel oxides and cobalt oxides. More importantly, multiple oxidation states and various nanostructures can also enable rich redox reactions for spinel nickel cobaltite to store more charges [32, 33, 66]. Last but not least, there are abundant nickel and cobalt elements on the earth, and it seems to be a promising

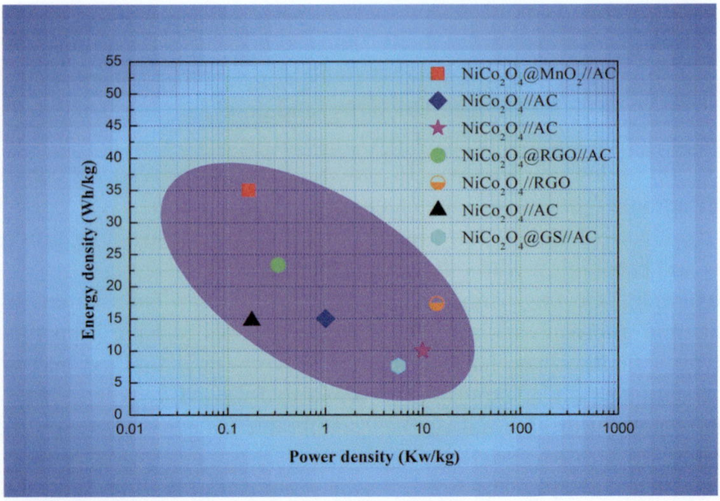

Fig. 9.6 Ragone plot of the current electrochemical performance for $NiCo_2O_4$-based asymmetric supercapacitors (Reprinted from Ref. [24] with permission. Copyright 2014, The Royal Society of Chemistry)

opportunity for nickel cobaltite to be widely applied and quickly industrialized as electrode material for supercapacitors due to its environmental friendliness and low cost [67]. To some extent, $NiCo_2O_4$ is playing an important role in complementing or replacing electrode materials based on Co_3O_4, NiO, and RuO_2 in energy storage field. A Ragone plot, as shown in Fig. 9.6, is derived from $NiCo_2O_4$-based asymmetrical supercapacitors' work which illustrates the performances in terms of energy density versus power density indicating this material is beneficial.

The methods for preparing $NiCo_2O_4$ for supercapacitors can primarily be divided into three categories: (i) sol–gel method, (ii) hydrothermal synthesis, and (iii) electrodeposition process. In brief, the synthetic route of $NiCo_2O_4$ nanomaterial can be thought as the formation of precursor followed by an annealing process. These three methods are summarized in Fig. 9.7 and are discussed in detail below; other pertinent methods are also briefly mentioned.

In general, the sol–gel method can be briefly defined as the conversion of a precursor solution into an inorganic solid by a chemical means [68]. Herein, the precursor solution consists of either an inorganic metal salt or a metal organic species, while nickel and cobalt acetates, halides, and nitrates are typically employed here for preparing $NiCo_2O_4$ [69, 70]. It is well known that the sol–gel method is simple and cheap with products of high purity, homogeneity, and porosity, mainly including four steps: (a) preparation of the precursor solutions, (b) formation of the intermediates as sol, (c) transformation from sol to gel, and (d) calcination at high temperature. Note that solvents, surfactants, reaction time, and temperature are crucial regulatory factors for obtaining appropriate structures with outstanding electrochemical performances [68].

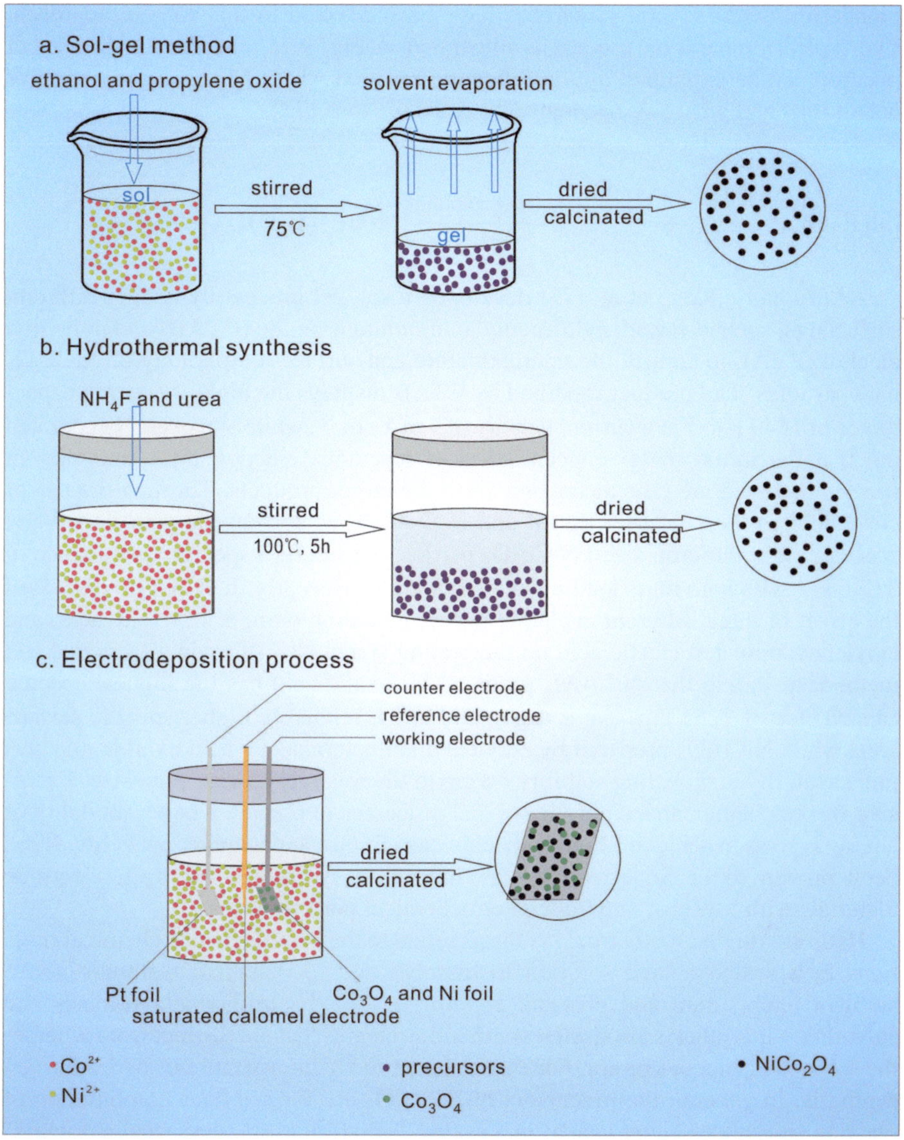

Fig. 9.7 Typical synthetic approaches of $NiCo_2O_4$ for supercapacitor application (Reprinted from Ref. [24] with permission. Copyright 2014, The Royal Society of Chemistry)

Hu and coworkers [71] first reported the preparation of $NiCo_2O_4$ aerogels via an epoxide-driven sol–gel process in 2010, which showed an ultrahigh specific capacitance of 1400 F g^{-1} with a mass loading of 0.4 mg cm^{-2}. The majority of the pores were in the optimal sizes of 2–5 nm for supercapacitor applications, and the as-obtained aerogels were of excellent reversibility and cycle stability. Since this pioneering work, several researches have been devoted to this sol–gel approach. The typical preparation process is illustrated in Fig. 9.7a, and the mechanism of gelation can be explained by the following reaction where M represents nickel or cobalt ions and where A^- denotes Cl^- or NO_3^- ions [72]:

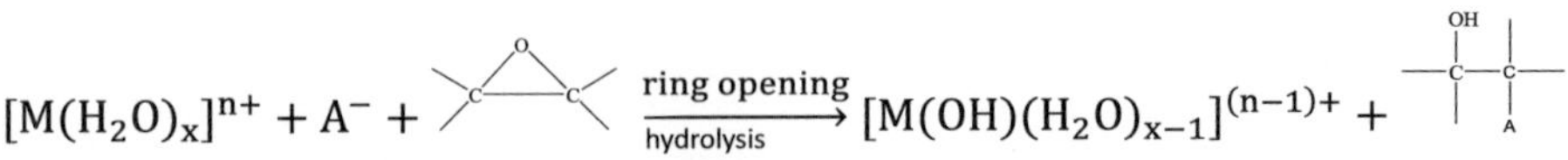

Additionally, Kong et al. [73] developed a sol–gel process by adding different surfactants, such as hexadecyltrimethylammonium bromide (CTAB) and polyvinyl alcohol (PVA), to control the microstructure and surface morphology of $NiCo_2O_4$ nanoparticles. The product modified by CTAB displays the highest specific capacitance of 1440 F g^{-1} at a current density of 5 mA cm^{-2}, while Wu et al. [74] pointed out that the initial molar concentration of reactants, reaction time, and solvent species involved are crucial for preparing the target products. Through the use of citric acid as the chelating ligand and H_2O–DMF as solvent via a facile sol–gel process, the submicron-sized $NiCo_2O_4$ particles exhibited a specific capacitance of 217 F g^{-1} with high mass loading of 5.6 mg cm^{-2}. Recently, Ji et al. [75] discussed the effect of three different chelating agents including citric acid, oxalic acid, and ethylenediaminetetraacetic acid on fabricating spinel $NiCo_2O_4$ via a facile sol–gel method, revealing that $NiCo_2O_4$ prepared by oxalic acid has the highest specific capacitance of 1254 F g^{-1} at 2 A g^{-1} due to the relatively higher specific surface area, while $NiCo_2O_4$ prepared by ethylenediaminetetraacetic acid exhibits the best rate capability and cycling stability owing to the relatively larger pore size. Extensive work is being carried out via sol–gel processes to prepare a better modality of $NiCo_2O_4$ so as to achieve a higher specific capacitance and a longer cycle life. It has been proven to be an effective way indeed to synthesize $NiCo_2O_4$ electrode materials with low cost and high electrochemical performances.

Hydrothermal/solvothermal synthesis refers to the heterogeneous chemical reactions in a sealed heated solution to dissolve and recrystallize materials above ambient temperature and pressure conditions [76]. Hydrothermal synthesis and solvothermal synthesis are similar synthetic strategies that are distinct from whether the solution is aqueous or not. For convenience and habit, we call this hydrothermal synthesis. In general, the precursors of $NiCo_2O_4$ are formed by a combination of reaction substances being heated in a sealed Teflon-lined stainless steel autoclave. As the most common method of synthesizing nickel cobaltite, hydrothermal method is advantageous to form highly monodispersed nanoparticles with a control over size and morphology.

A large number of groups have made attempts to optimize the reaction conditions and form special morphologies so as to enhance the electrochemical performances of $NiCo_2O_4$ electrodes [77–84]. For example, Chen et al. [77] reported that a 3D hierarchical $NiCo_2O_4$ nanosheet–nanowire cluster arrays were manufactured via a facile hydrothermal method, exhibiting an ultrahigh specific capacitance of 2000 F g^{-1} at 10 A g^{-1} with 93.8 % retention (more than 10,000 cycles) and a high power density of 26.1 kW kg^{-1} at a current density of 80 A g^{-1}. As illustrated in Fig. 9.7b, the process is based on Ni–Co nitrates induced by NH_4F and urea being heated in a sealed Teflon-lined stainless steel autoclave at 100 °C for 5 h. Zou et al. [79] also reported a facile hydrothermal method that successfully grew 3D $NiCo_2O_4$ microspheres constructed by radial chain-like $NiCo_2O_4$ nanowires with different exposed crystal planes, demonstrating high specific capacitance (1284 F g^{-1} at 2 A g^{-1}), good rate capability, and excellent cycling stability (only 2.5 % loss after 3000 cycles). It was proposed that the chain-like nanowires with different exposed crystal planes exhibited higher electronic conductivity than other mesoporous $NiCo_2O_4$ nanostructures. In comparison, solvothermal method just occurs in nonaqueous solvents such as ethanol and ethylene glycol. Very recently, the 3D flowerlike $NiCo_2O_4$ hierarchical architectures have been fabricated by An et al. [78] via a solvothermal method using polyvinylpyrrolidone (PVP) as the structure-directing agent in a polytetrafluoroethylene (PTFE) reaction environment followed by a simple thermal annealing treatment, displaying a large surface specific area of 212.6 m^{2}g^{-1} and a high specific capacitance of 1191.2 F g^{-1} at current densities of 1 A g^{-1}; however, it suffered from a relatively poor cycling performance due to the electric isolation caused by collapse of the nanosheets. Meanwhile, Ji et al. [84] successfully fabricated 3D network-like mesoporous $NiCo_2O_4$ nanostructures through a solvothermal route coupled with a post-annealing treatment. Benefiting from the large specific surface area (170.6 m^{2}g^{-1}) and abundant mesoporosity (5–10 nm), the as-obtained $NiCo_2O_4$ manifests high specific capacitance of 931 F g^{-1} at 3 A g^{-1}.

The electrodeposition process occurs in the electrolysis of plating solutions where micron- or submicron-sized particles are suspended, and variable amounts of these particles are embedded in the electrochemically produced solid phase with special properties [85]. As illustrated in Fig. 9.7c, the process can be described in three steps: (1) formation of precursor solution, (2) co-electrodeposition of nanoparticles, and (3) thermal decomposition. Due to cooperative formation mechanism of the composite electrode, as a general rule, ultrahigh specific capacitances and good cycling performances will be achieved through this novel method.

For instance, Du et al. [86] electrodeposited $NiCo_2O_4$ nanosheet arrays onto flexible carbon fabric successfully, which displayed high specific capacitance of 2658 F g^{-1} (2 A g^{-1}), good rate property, and superior cycling life. Note that carbon fabric (CF) is an ideal conductive substrate for directly growing active materials owing to its superior electrical conductivity, good flexibility, and high mechanical strength, through which we can fabricate some distinctive flexible supercapacitors or lithium-ion batteries. Moreover, Lou and his coworkers [87] reported the synthesis involving the co-electrodeposition of a bimetallic (Ni, Co) hydroxide precursor on an Ni foam and subsequently thermal transformation to spinel mesoporous $NiCo_2O_4$, and the electrode exhibited an ultrahigh specific capacitance of 1450 F g^{-1} even at a very high current density of 20 A g^{-1}. The sizes of interparticle mesopores ranged from 2 to 5 nm, which were important indeed to facilitate the mass transport of electrolytes within the electrodes for fast redox

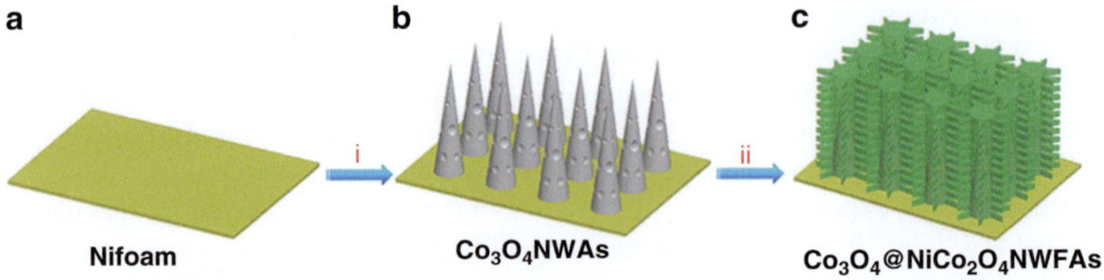

Fig. 9.8 Illustration of the formation process of Co_3O_4@$NiCo_2O_4$ NWFA hierarchical heterostructures. (**a**) Ni foam substrate; (**b**) hydrothermal synthesis of aligned Co_3O_4 nanowire arrays; (**c**) formation of hierarchical Co_3O_4@$NiCo_2O_4$ nanoflake arrays via a co-electrodeposition process (Reprinted from Ref. [89] with permission. Copyright 2013, Elsevier Ltd)

reactions and double-layer charging–discharging. Composite electrodes such as Ni $(OH)_2$@ $NiCo_2O_4$ [88] and Co_3O_4@ $NiCo_2O_4$ [89] via the electrodeposition method were also reported in the literature. And the formation process of the nanoforest of hierarchical Co_3O_4@ $NiCo_2O_4$ [89] nanowire arrays is elaborated in Fig. 9.8 as a typical example.

The electrodeposition process is widely used to fabricate composite electrodes with different nanomaterials. It is very important to choose the befitting materials that should possess good electrical conductivity or unique nanostructure to assist the target material in pursuing for better electrochemical performances. However, the ambiguous relationship between the composite structures and electrochemical performances has not been clarified yet.

As elaborated above, all of the three major methods are distinctive. The sol–gel method tends to acquire nanoparticles of high purity, homogeneity, and porosity. And the hydrothermal synthesis is controllable of size and nanostructure by adjusting temperature parameters or types and concentration of auxiliaries. By contrast, the electrodeposition process has some distinct advantages in depositing $NiCo_2O_4$ onto various conductive substrates such as nickel foams and carbon textiles and combining $NiCo_2O_4$ with other excellent materials that show intriguing synergistic effect, which can be manipulated through adjusting solution concentration, solution pH value, and applied overpotential or current density in the electrodeposition procedure. In addition, other methods, such as coprecipitation method, template approach, and microwave technique, are also simply mentioned below, which are not widely used or just assistant measures for fabricating $NiCo_2O_4$ material, and the following examples are only constructed for the purpose of comprehensively illustrating the diversity and novelty of processes as reported.

Recently, coprecipitation method is getting popular owing to its simple and facile process [90–92]. For example, Ji et al. [92] reported a facile coprecipitation way using $NaHCO_3$ as precipitant, and the as-obtained uniform porous $NiCo_2O_4$ nanoparticles showed high specific capacitance of $726.8\ F\ g^{-1}$ ($1\ A\ g^{-1}$) and good cycle stability of 72.7 % retention at a current density of $5\ A\ g^{-1}$ after 2000 cycles. In addition, it has been reported that microwave technique has been used to help prepare $NiCo_2O_4$ electrodes for supercapacitors. What is more, the microwave-assisted process has been demonstrated to be advantageous to extremely fast kinetics of crystallization, very rapid heating to the required temperature, and possible formation of new metastable phases, which facilitate to tune the crystal size and phase transformation degree of binary Ni–Co precursors [93, 94]. As a flexible and efficient approach, the template approach, using silica spheres [95] and sodium dodecyl sulfate [96] as templates, has also been used to synthesize hierarchical mesoporous hollow $NiCo_2O_4$ sub-microspheres and hexagonal mesoporous $NiCo_2O_4$ for supercapacitor electrodes, respectively. Both of the electrodes show remarkable electrochemical performances. Besides, sonochemical precipitation technique [97], chemical bath deposition [98], reflux synthesis [99], room-temperature solid-state reaction [100], and single-spinneret electrospinning technique [101] have also been utilized to synthesize $NiCo_2O_4$ nanostructures for supercapacitors. These approaches deserve a mention and readers are directed to these for further details.

$MnCo_2O_4$

As a potential candidate for supercapacitors, spinel $MnCo_2O_4$ has been restricted by the poor ionic conductivity and partial dissolution in alkaline electrolytes [102]. Despite some disappointing physicochemical properties of $MnCo_2O_4$, it is believed that $MnCo_2O_4$ -based supercapacitors can be modified by combining with certain carbonaceous materials which possess both light weight and good electronic conductivity [103]. It is fascinating that the faradaic redox reactions of $MnCo_2O_4$ occurring in the charge–discharge process are speculated to be both $Co^{2+}/Co^{3+}/Co^{4+}$ and $Mn^{2+}/Mn^{3+}/Mn^{4+}$, which facilitate to achieve much higher energy density with multiple valence changes [104].

As reported by Kong et al. [103], the spinel $MnCo_2O_4$ fabricated by a facile sol–gel method displayed high specific capacitance of $405\ F\ g^{-1}$ at $5\ mA\ cm^{-2}$, remarkable rate capability of 67.9 % capacity retention at $40\ mA\ cm^{-2}$, and good cycle stability of 4.9 % loss after 1000 cycles, owing to the porous structure and fairly stable spinel structure. In addition, Li et al. [105] developed a new strategy to prepare a flowerlike C@ $MnCo_2O_4$ hybrid composite using a facile hydrothermal method combined with a post-calcination process. Due to the unique hollow structure, high specific surface area, and suitable pore channels, the C@ $MnCo_2O_4$ composite exhibits high specific capacitance, low charge transfer resistance, and superior capacitance retention. A high specific capacitance of $728.4\ F\ g^{-1}$ has been achieved at a current density of $1\ A\ g^{-1}$, and 95.9 % of the initial capacitance is still retained after 1000 cycles. Recently, Xu et al. [106] successfully prepared porous $MnCo_2O_4$ nanowires by hydrothermal method followed by a calcinating process, showing excellent capacitance and desirable rate performance ($1342\ F\ g^{-1}$ at $1\ A\ g^{-1}$ and $988\ F\ g^{-1}$ at $20\ A\ g^{-1}$).

$NiFe_2O_4$

Spinel $NiFe_2O_4$ can be a promising candidate for use as supercapacitor materials due to its low cost and environmental friendliness. However, it exhibits low electrical conductivity, small energy density, and poor cycling performance.

For example, Wang et al. [107] reported a facile synthesis of RGO– $NiFe_2O_4$ composites by a hydrothermal process. The synthesized electrodes were measured in a 1 M Na_2SO_4 aqueous solution, showing the best capacitive properties of 459.6 F g^{-1} at 1 A g^{-1} due to the suitable particle size and good dispersion property of the active material. Senthilkumar et al. [108] reported the PANI–$NiFe_2O_4$ composites prepared by an in situ chemical oxidative polymerization method, which provided high specific capacitance of 448 F g^{-1} and good capacitance retention of 80 % after 1000 cycles. In addition, Sen et al. [109] developed a sol–gel method to prepare nanocrystalline nickel ferrites (5–20 nm), exhibiting good electrochemical performance.

$NiMoO_4$

$NiMoO_4$ has drawn significant research attention because of its excellent electrical conductivity and outstanding electrochemical performance resulting from the high electrochemical activity of the Ni element and the good electrical conductivity of the Mo element [110]. The crystalline structure of $NiMoO_4$ is depicted in Fig. 9.9, and it displays that $NiMoO_4$ occupies the monoclinic structure with molybdenum atoms and nickel atoms adopting the tetrahedral and octahedral sites, respectively. Note that the edge-sharing NiO_6 octahedra interconnected by MoO_4 tetrahedra facilitate to constitute a 3D network structure with open channels, which are beneficial for ion diffusion [111]. Different morphologies of $NiMoO_4$, such as nanosheets [112, 113], nanorods [114], nanowires [115–117], and nanotubes [118], have been developed for pseudocapacitive application, and typical

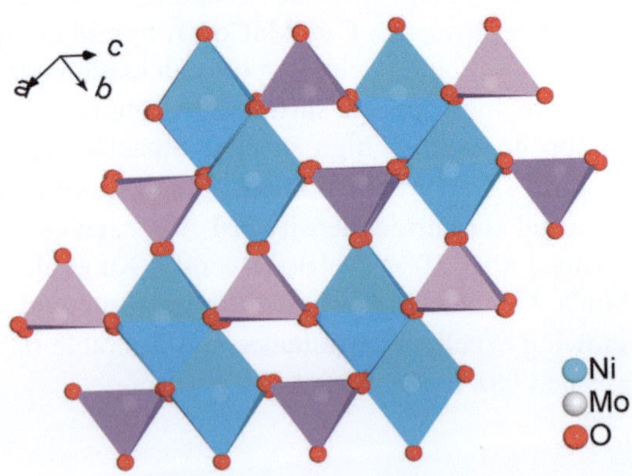

Fig. 9.9 Crystal structure of $NiMoO_4$ (Reprinted from Ref. [111] with permission. Copyright 2014, WILEY-VCH Verlag GmbH & Co. KGaA, Weinheim)

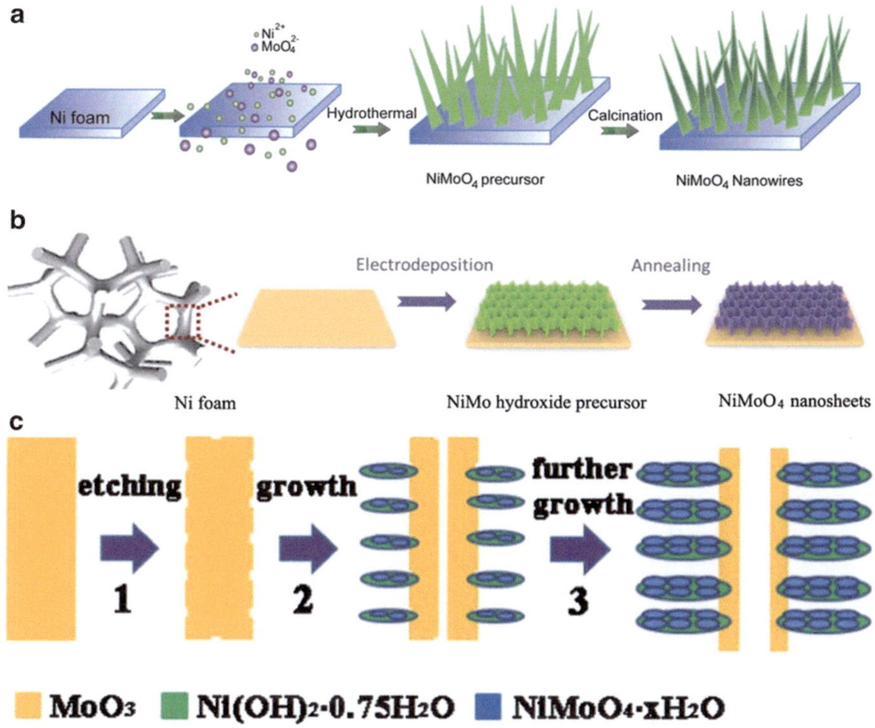

Fig. 9.10 Schematic illustration of the preparation process of $NiMoO_4$. (**a**) Hydrothermal method; (**b**) electrodeposition process; (**c**) hydrothermal method (Reprinted from Ref. [110] with permission. Copyright 2015, The Royal Society of Chemistry. Reprinted from Ref. [115] with permission. Copyright 2013, The Royal Society of Chemistry. Reprinted from Ref. [118] with permission. Copyright 2015, The Royal Society of Chemistry)

approaches [110, 115, 118] for synthesizing $NiMoO_4$ electrode materials are described in Fig. 9.10.

For example, Wang and his coworkers [110] have reported that amorphous honeycomb-like $NiMoO_4$ nanosheets can be prepared by electrodeposition process followed by thermal treatment. The as-prepared $NiMoO_4$ nanosheets with interconnecting pore channels and an ultrathin structure showed a high specific capacitance (1694 F g^{-1} at 1 A g^{-1} and 1220 F g^{-1} at 50 A g^{-1}) and excellent cycle performance (92.7 % after 9000 cycles). Peng et al. [111] developed a facile hydrothermal method to grow hierarchical $NiMoO_4$ architectures assembled from well-aligned nanosheets or nanorods on various conductive substrates for supercapacitor applications. It was found that $NiMoO_4$ nanosheets manifest higher specific capacitance and better cycling stability than $NiMoO_4$ nanorods and the assembled asymmetric supercapacitors based on $NiMoO_4$ nanosheets and activated carbon delivered a high energy density of 60.9 Wh kg^{-1} at a power density of 850 W kg^{-1}. Wang and his coworkers [114] synthesized $NiMoO_4$ nanospheres by a facile hydrothermal method, demonstrating high specific capacitance of 974.4 F g^{-1} at 1 A g^{-1}.

By virtue of the synergistic effect, the composite electrode based on $NiMoO_4$ has been extensively investigated. Composite electrodes for supercapacitors, including Co_3O_4@$NiMoO_4$ [119–121], MnO_2@$NiMoO_4$ [122], $NiCo_2O_4$@$NiMoO_4$ [123], $Co(OH)_2$@$NiMoO_4$ [124], $CoMoO_4$@$NiMoO_4$ [125, 126], and graphene@$NiMoO_4$ [117], have been reported recently. For example, Liu et al. [125] designed $CoMoO_4$–$NiMoO_4$ bundle composite material for supercapacitors by a chemical coprecipitation method, and the composite electrode showed a maximum specific capacitance of 1039 F g^{-1} at 2.5 mA cm^{-2} and an excellent rate capability ascribing to its unique microstructure, large specific surface area, multiple oxidation states, and synergistic effects between $NiMoO_4$ and $CoMoO_4$. In addition, Zhang et al. [115] developed a 3D sea urchin-like $NiCo_2O_4$@$NiMoO_4$ core–shell structure constructed by interconnected ultrathin nanosheets via a facile two-step hydrothermal method using a post-annealing procedure. The $NiCo_2O_4$@$NiMoO_4$ material revealed a high specific capacitance of 2474 F g^{-1} at 1 A g^{-1} and 2080 F g^{-1} at 20 A g^{-1} and maintained 95.0 % of the initial capacitance after 1000 cycles at 10 A g^{-1}, which can be attributed to its unique 3D core–shell and hierarchical mesoporous structures.

Considering the promising electrochemical performance of $NiMoO_4$ and easy fabrication process, it can be one of the most potential electrode materials applied for supercapacitor in the future energy storage/conversion fields.

9.4 Summary and Outlook

Significant progress has clearly been made in electrochemical supercapacitors which has sparked worldwide attention due to the rapid development of supercapacitive electrode materials, which is of great significance to fast industrialize this important energy storage/conversion device. Since the first report of supercapacitors in 1957, electrode materials have developed from carbon materials to pseudocapacitive materials, while transition metal oxides for supercapacitors have developed from noble transition metal oxides to base transition metal oxides, from monometallic oxides to single-phase bimetallic oxides, and from monocomponent materials to multicomponent composites. Note that bimetallic transition metal oxides usually have great advantage over monometallic transition oxides, including higher electrical conductivity, multiple valence changes, and potential synthetic effects. In addition, multicomponent composites such as transition metal oxides at carbon materials utilizing as supercapacitive electrode material have the advantages in obtaining more remarkable rate capability and much better cycle performance. It tends to combine the different virtues together, thus enhancing the electrochemical performances of supercapacitors.

Inspiringly, transition metal oxides have already been demonstrated to display striking performance as electrode materials for electrochemical supercapacitors. However, the majority of the reports regarding the applications of transition metal oxides for supercapacitors are still retained in the laboratory, with few solutions yet for a mass market. Most researches only focus on preparing various nanostructures for transition metal oxides, while little studies are devoted to improving the whole production process of the whole device of supercapacitor. Optimizing the manufacturing process and reducing the production cost are also vital to realize large-scale production for the industrial applications, and designing products based on market demand can more effectively and healthily promote the development of supercapacitor. Meanwhile, a comprehensive understanding of the relationship between the property/performance and the structure/composition of these transition metal oxides for supercapacitors should be systematically established. Developing reliable and efficient methods and criteria to solve this problem is urgently needed. The establishment of theoretical simulation and mathematical modeling is also highly anticipated to investigate the detailed energy storage mechanism and the rational design of electrode materials, directing the purposeful design of facile, low-cost, and large-scale preparation of transition metal oxides with excellent electrochemical performance.

At present, electrochemical supercapacitors based on transition metal oxides should provide improved energy density on the premise of keeping high power density. However, there is still considerable distance to meet the demand of commercial application. Narrow stable potential window of aqueous electrolytes greatly restricts the enhancement of energy density, and limited mechanism study cannot effectively guide the direction of investigation. To maximize the operating voltage output, some nonaqueous electrolytes such as ionic liquids and organic electrolytes that provide a two or three times wider working voltage window than the aqueous ones have been studied as desirable candidates. It is of high importance to explore nonaqueous electrolytes to obtain higher energy/power density, which will greatly pave the way for the practical production of electrochemical supercapacitors. In order to fully exploit the potential of transition metal oxide-based electrode materials for supercapacitors, it is imperative to optimize both synthesis parameters and material properties. The engineering factors of electrodes, including current collectors, electrolytes, membrane separators, counter electrodes, and packaging, still need a deeper investigation from fundamental investigation in the future.

Despite the challenges ahead, transition metal oxides utilizing as supercapacitive electrode material will be promising candidates for energy storage/conversation in the future. On one hand, systematical work should be performed on the current remarkable transition metal oxides to improve the electrochemical performance to the utmost. On the other hand, novel ideas such as discovering new metal oxide that is of excellent electrochemical performances are also urgently needed. In-depth understanding of the mechanisms of transition metal oxides for supercapacitors will greatly promote the whole industrial development of the supercapacitors. All in all, with sustained and dedicated research efforts, these fascinating transition metal oxides nanomaterials will offer a new avenue to make the attractive electrochemical supercapacitors commercially viable.

References

1. Aricò AS, Bruce P, Scrosati B, Tarascon J-M, Van Schalkwijk W (2005) Nanostructured materials for advanced energy conversion and storage devices. Nat Mater 4(5):366–377
2. Simon P, Gogotsi Y (2008) Materials for electrochemical capacitors. Nat Mater 7 (11):845–854
3. Conway B, Pell W (2003) Double-layer and pseudocapacitance types of electrochemical capacitors and their applications to the development of hybrid devices. J Solid State Electrochem 7(9):637–644
4. Lang X, Hirata A, Fujita T, Chen M (2011) Nanoporous metal/oxide hybrid electrodes for electrochemical supercapacitors. Nat Nanotechnol 6(4):232–236
5. Simon P, Gogotsi Y, Dunn B (2014) Where do batteries End and supercapacitors begin? Sci Mag 343:1210–1211
6. Becker HI (1957) Low voltage electrolytic capacitor. Google Patents
7. Kötz R, Carlen M (2000) Principles and applications of electrochemical capacitors. Electrochim Acta 45(15):2483–2498
8. Largeot C, Portet C, Chmiola J, Taberna P-L, Gogotsi Y, Simon P (2008) Relation between the ion size and pore size for an electric double-layer capacitor. J Am Chem Soc 130 (9):2730–2731
9. Xia K, Gao Q, Jiang J, Hu J (2008) Hierarchical porous carbons with controlled micropores and mesopores for supercapacitor electrode materials. Carbon 46(13):1718–1726
10. Yu C, Masarapu C, Rong J, Wei B, Jiang H (2009) Stretchable supercapacitors based on buckled single-walled carbon-nanotube macrofilms. Adv Mater 21(47):4793–4797
11. Futaba DN, Hata K, Yamada T, Hiraoka T, Hayamizu Y, Kakudate Y, Tanaike O, Hatori H, Yumura M, Iijima S (2006) Shape-engineerable and highly densely packed single-walled carbon nanotubes and their application as super-capacitor electrodes. Nat Mater 5 (12):987–994
12. Liu C, Yu Z, Neff D, Zhamu A, Jang BZ (2010) Graphene-based supercapacitor with an ultrahigh energy density. Nano Lett 10(12):4863–4868
13. Deng W, Ji X, Gómez-Mingot M, Lu F, Chen Q, Banks CE (2012) Graphene electrochemical supercapacitors: the influence of oxygen functional groups. Chem Commun 48 (22):2770–2772
14. Song W, Ji X, Deng W, Chen Q, Shen C, Banks CE (2013) Graphene ultracapacitors: structural impacts. Phys Chem Chem Phys 15(13):4799–4803
15. Trasatti S, Buzzanca G (1971) Ruthenium dioxide: a new interesting electrode material. Solid state structure and electrochemical behaviour. J Electroanal Chem Interfacial Electrochem 29 (2):A1–A5
16. Alonso A, Ruiz V, Blanco C, Santamaria R, Granda M, Menendez R, De Jager S (2006) Activated carbon produced from Sasol-Lurgi gasifier pitch and its application as electrodes in supercapacitors. Carbon 44(3):441–446
17. Frackowiak E, Beguin F (2001) Carbon materials for the electrochemical storage of energy in capacitors. Carbon 39(6):937–950
18. Huang J, Sumpter BG, Meunier V (2008) A universal model for nanoporous carbon supercapacitors applicable to diverse pore regimes, carbon materials, and electrolytes. Chem Eur J 14(22):6614–6626

19. Snook GA, Kao P, Best AS (2011) Conducting-polymer-based supercapacitor devices and electrodes. J Power Sources 196(1):1–12
20. Snook GA, Chen GZ (2008) The measurement of specific capacitances of conducting polymers using the quartz crystal microbalance. J Electroanal Chem 612(1):140–146
21. Yuan C, Wu HB, Xie Y, Lou XWD (2014) Mixed transition-metal oxides: design, synthesis, and energy-related applications. Angew Chem Int Ed 53(6):1488–1504
22. Conway BE (1991) Transition from "supercapacitor" to "battery" behavior in electrochemical energy storage. J Electrochem Soc 138(6):1539–1548
23. Zhang LL, Zhao X (2009) Carbon-based materials as supercapacitor electrodes. Chem Soc Rev 38(9):2520–2531
24. Wu Z, Zhu Y, Ji X (2014) $NiCo_2O_4$-based materials for electrochemical supercapacitors. J Mater Chem A 2(36):14759–14772
25. Deng W, Ji X, Chen Q, Banks CE (2011) Electrochemical capacitors utilising transition metal oxides: an update of recent developments. Rsc Adv 1(7):1171–1178
26. Zheng J, Cygan P, Jow T (1995) Hydrous ruthenium oxide as an electrode material for electrochemical capacitors. J Electrochem Soc 142(8):2699–2703
27. Zheng J, Jow T (1995) A new charge storage mechanism for electrochemical capacitors. J Electrochem Soc 142(1):L6–L8
28. Sugimoto W, Iwata H, Yokoshima K, Murakami Y, Takasu Y (2005) Proton and electron conductivity in hydrous ruthenium oxides evaluated by electrochemical impedance spectroscopy: the origin of large capacitance. J Phys Chem B 109(15):7330–7338
29. Zhou Z, Zhu Y, Wu Z, Lu F, Jing M, Ji X (2014) Amorphous RuO_2 coated on carbon spheres as excellent electrode materials for supercapacitors. Rsc Adv 4(14):6927–6932
30. Cui B, Lin H, Liu Y-Z, Li J-B, Sun P, Zhao X-C, Liu C-J (2009) Photophysical and photocatalytic properties of core-ring structured $NiCo_2O_4$ nanoplatelets. J Phys Chem C 113(32):14083–14087
31. Marco J, Gancedo J, Gracia M, Gautier J, Rıos E, Berry F (2000) Characterization of the nickel cobaltite, $NiCo_2O_4$, prepared by several methods: an XRD, XANES, EXAFS, and XPS study. J Solid State Chem 153(1):74–81
32. Wang X, Liu WS, Lu X, Lee PS (2012) Dodecyl sulfate-induced fast faradic process in nickel cobalt oxide–reduced graphite oxide composite material and its application for asymmetric supercapacitor device. J Mater Chem 22(43):23114–23119
33. Li Y, Hasin P, Wu Y (2010) $NixCo_{3\_}xO_4$ nanowire arrays for electrocatalytic oxygen evolution. Adv Mater 22(17):1926–1929
34. Rasiyah P, Tseung A, Hibbert D (1982) A mechanistic study of oxygen evolution on $NiCo_2O_4$ I. Formation of higher oxides. J Electrochem Soc 129(8):1724–1727
35. Hu C-C, Chang K-H, Lin M-C, Wu Y-T (2006) Design and tailoring of the nanotubular arrayed architecture of hydrous RuO_2 for next generation supercapacitors. Nano Lett 6 (12):2690–2695
36. Ozolins V, Zhou F, Asta M (2013) Ruthenia-based electrochemical supercapacitors: insights from first-principles calculations. Acc Chem Res 46(5):1084–1093
37. Wang G, Zhang L, Zhang J (2012) A review of electrode materials for electrochemical supercapacitors. Chem Soc Rev 41(2):797–828
38. Han JH, Lee SW, Kim SK, Han S, Hwang CS, Dussarrat C, Gatineau J (2010) Growth of RuO_2 thin films by pulsed-chemical vapor deposition using RuO_4 precursor and 5% H_2 reduction gas. Chem Mater 22(20):5700–5706
39. Vetrone J, Foster C, Bai G, Wang A, Patel J, Wu X (1998) Growth, microstructure, and resistivity of RuO_2 thin films grown by metal-organic chemical vapor deposition. J Mater Res 13(08):2281–2290
40. Zhu Y, Ji X, Pan C, Sun Q, Song W, Fang L, Chen Q, Banks CE (2013) A carbon quantum dot decorated RuO_2 network: outstanding supercapacitances under ultrafast charge and discharge. Energy Environ Sci 6(12):3665–3675

41. Zhang J, Ma J, Zhang LL, Guo P, Jiang J, Zhao X (2010) Template synthesis of tubular ruthenium oxides for supercapacitor applications. J Phys Chem C 114(32):13608–13613
42. Chen Y-M, Cai J-H, Huang Y-S, Lee K-Y, Tsai D-S, Tiong K-K (2011) A nanostructured electrode of IrOx foil on the carbon nanotubes for supercapacitors. Nanotechnology 22 (35):355708
43. Chen Y, Cai J, Huang Y, Lee K, Tsai D (2011) Preparation and characterization of iridium dioxide–carbon nanotube nanocomposites for supercapacitors. Nanotechnology 22 (11):115706
44. Du D, Hu Z, Liu Y, Deng Y, Liu J (2014) Preparation and characterization of flower-like microspheres of nano-NiO as electrode material for supercapacitor. J Alloys Compd 589:82–87
45. Yu W, Jiang X, Ding S, Li BQ (2014) Preparation and electrochemical characteristics of porous hollow spheres of NiO nanosheets as electrodes of supercapacitors. J Power Sources 256:440–448
46. Yang Z, Xu F, Zhang W, Mei Z, Pei B, Zhu X (2014) Controllable preparation of multishelled NiO hollow nanospheres via layer-by-layer self-assembly for supercapacitor application. J Power Sources 246:24–31
47. Huang M, Li F, Ji JY, Zhang YX, Zhao XL, Gao X (2014) Facile synthesis of single-crystalline NiO nanosheet arrays on Ni foam for high-performance supercapacitors. CrystEngComm 16(14):2878–2884
48. Zhang X, Shi W, Zhu J, Zhao W, Ma J, Mhaisalkar S, Maria TL, Yang Y, Zhang H, Hng HH (2010) Synthesis of porous NiO nanocrystals with controllable surface area and their application as supercapacitor electrodes. Nano Res 3(9):643–652
49. Yang Q, Sha J, Ma X, Yang D (2005) Synthesis of NiO nanowires by a sol–gel process. Mater Lett 59(14):1967–1970
50. Cheng J, Cao G-P, Yang Y-S (2006) Characterization of sol–gel-derived NiO x xerogels as supercapacitors. J Power Sources 159(1):734–741
51. Xia X-H, Tu J-P, Wang X-l, Gu C-D, Zhao X-B (2011) Hierarchically porous NiO film grown by chemical bath deposition via a colloidal crystal template as an electrochemical pseudocapacitor material. J Mater Chem 21(3):671–679
52. Justin P, Meher SK, Rao GR (2010) Tuning of capacitance behavior of NiO using anionic, cationic, and nonionic surfactants by hydrothermal synthesis. J Phys Chem C 114 (11):5203–5210
53. Vijayakumar S, Nagamuthu S, Muralidharan G (2013) Supercapacitor studies on NiO nanoflakes synthesized through a microwave route. ACS Appl Mater Interfaces 5 (6):2188–2196
54. Fan Z, Chen J, Cui K, Sun F, Xu Y, Kuang Y (2007) Preparation and capacitive properties of cobalt–nickel oxides/carbon nanotube composites. Electrochim Acta 52(9):2959–2965
55. Zhao B, Song J, Liu P, Xu W, Fang T, Jiao Z, Zhang H, Jiang Y (2011) Monolayer graphene/NiO nanosheets with two-dimension structure for supercapacitors. J Mater Chem 21 (46):18792–18798
56. Subramanian V, Zhu H, Vajtai R, Ajayan P, Wei B (2005) Hydrothermal synthesis and pseudocapacitance properties of MnO_2 nanostructures. J Phys Chem B 109(43):20207–20214
57. Tao J, Liu N, Ma W, Ding L, Li L, Su J, Gao Y (2013) Solid-state high performance flexible supercapacitors based on polypyrrole-MnO_2-carbon fiber hybrid structure. Sci Rep 3
58. Li X, Wang G, Wang X, Li X, Ji J (2013) Flexible supercapacitor based on MnO_2 nanoparticles via electrospinning. J Mater Chem A 1(35):10103–10106
59. Pang SC, Anderson MA, Chapman TW (2000) Novel electrode materials for thin-film ultracapacitors: comparison of electrochemical properties of sol-gel-derived and electrodeposited manganese dioxide. J Electrochem Soc 147(2):444–450
60. Kang J, Hirata A, Kang L, Zhang X, Hou Y, Chen L, Li C, Fujita T, Akagi K, Chen M (2013) Enhanced supercapacitor performance of MnO_2 by atomic doping. Angew Chem 125 (6):1708–1711

61. Zhang X, Zhao D, Zhao Y, Tang P, Shen Y, Xu C, Li H, Xiao Y (2013) High performance asymmetric supercapacitor based on MnO_2 electrode in ionic liquid electrolyte. J Mater Chem A 1(11):3706–3712
62. Xiong S, Yuan C, Zhang X, Xi B, Qian Y (2009) Controllable synthesis of mesoporous Co_3O_4 nanostructures with tunable morphology for application in supercapacitors. Chem-Eur J 15(21):5320–5326
63. Xia X-H, Tu J-P, Mai Y-J, Wang X-l, Gu C-D, Zhao X-B (2011) Self-supported hydrothermal synthesized hollow Co_3O_4 nanowire arrays with high supercapacitor capacitance. J Mater Chem 21(25):9319–9325
64. Dong X-C, Xu H, Wang X-W, Huang Y-X, Chan-Park MB, Zhang H, Wang L-H, Huang W, Chen P (2012) 3D graphene–cobalt oxide electrode for high-performance supercapacitor and enzymeless glucose detection. ACS Nano 6(4):3206–3213
65. Xu K, Li W, Liu Q, Li B, Liu X, An L, Chen Z, Zou R, Hu J (2014) Hierarchical mesoporous $NiCo_2O_4$@ MnO_2 core–shell nanowire arrays on nickel foam for aqueous asymmetric supercapacitors. J Mater Chem A 2(13):4795–4802
66. Chen H, Jiang J, Zhang L, Qi T, Xia D, Wan H (2014) Facilely synthesized porous $NiCo_2O_4$ flowerlike nanostructure for high-rate supercapacitors. J Power Sources 248:28–36
67. Dalvi AD, Bacon WG, Osborne RC (2004) The past and the future of nickel laterites. In: PDAC 2004 international convention, trade show & investors exchange. The Prospectors and Developers Association of Canada, Toronto, pp 1–27
68. Hench LL, West JK (1990) The sol–gel process. Chem Rev 90(1):33–72
69. Niederberger M (2007) Nonaqueous sol–gel routes to metal oxide nanoparticles. Acc Chem Res 40(9):793–800
70. Livage J, Henry M, Sanchez C (1988) Sol–gel chemistry of transition metal oxides. Prog Solid State Chem 18(4):259–341
71. Wei TY, Chen CH, Chien HC, Lu SY, Hu CC (2010) A cost-effective supercapacitor material of ultrahigh specific capacitances: spinel nickel cobaltite aerogels from an epoxide-driven sol–gel process. Adv Mater 22(3):347–351
72. Liu M-C, Kong L-B, Lu C, Li X-M, Luo Y-C, Kang L (2012) A sol–gel process for fabrication of $NiO/NiCo_2O_4/Co_3O_4$ composite with improved electrochemical behavior for electrochemical capacitors. ACS Appl Mater Interfaces 4(9):4631–4636
73. Kong L-B, Lu C, Liu M-C, Luo Y-C, Kang L (2013) Effect of surfactant on the morphology and capacitive performance of porous $NiCo_2O_4$. J Solid State Electrochem 17(5):1463–1471
74. Wu YQ, Chen XY, Ji PT, Zhou QQ (2011) Sol–gel approach for controllable synthesis and electrochemical properties of $NiCo_2O_4$ crystals as electrode materials for application in supercapacitors. Electrochim Acta 56(22):7517–7522
75. Zhu Y, Ji X, Wu Z, Song W, Hou H, Wu Z, He X, Chen Q, Banks CE (2014) Spinel $NiCo_2O_4$ for use as a high-performance supercapacitor electrode material: Understanding of its electrochemical properties. J Power Sources 267:888–900
76. Feng S, Xu R (2001) New materials in hydrothermal synthesis. Acc Chem Res 34(3):239–247
77. Chen Y, Qu B, Hu L, Xu Z, Li Q, Wang T (2013) High-performance supercapacitor and lithium-ion battery based on 3D hierarchical NH4F-induced nickel cobaltate nanosheet–nanowire cluster arrays as self-supported electrodes. Nanoscale 5(20):9812–9820
78. An C, Wang Y, Huang Y, Xu Y, Xu C, Jiao L, Yuan H (2014) Novel three-dimensional $NiCo_2O_4$ hierarchitectures: solvothermal synthesis and electrochemical properties. CrystEngComm 16(3):385–392
79. Zou R, Xu K, Wang T, He G, Liu Q, Liu X, Zhang Z, Hu J (2013) Chain-like $NiCo_2O_4$ nanowires with different exposed reactive planes for high-performance supercapacitors. J Mater Chem A 1(30):8560–8566
80. Liu X, Zhang Y, Xia X, Shi S, Lu Y, Wang XL, Gu C, Tu J (2013) Self-assembled porous $NiCo_2O_4$ hetero-structure array for electrochemical capacitor. J Power Sources 239:157–163

81. Wang X, Han X, Lim M, Singh N, Gan CL, Jan M, Lee PS (2012) Nickel cobalt oxide-single wall carbon nanotube composite material for superior cycling stability and high-performance supercapacitor application. J Phys Chem C 116(23):12448–12454
82. Shen L, Che Q, Li H, Zhang X (2014) Mesoporous $NiCo_2O_4$ nanowire arrays grown on carbon textiles as binder-free flexible electrodes for energy storage. Adv Funct Mater 24 (18):2630–2637
83. Zhang G, Lou XWD (2013) Controlled growth of $NiCo_2O_4$ nanorods and ultrathin nanosheets on carbon nanofibers for high-performance supercapacitors. Sci Rep 3:1470
84. Zhu Y, Wu Z, Jing M, Song W, Hou H, Yang X, Chen Q, Ji X (2014) 3D network-like mesoporous $NiCo_2O_4$ nanostructures as advanced electrode material for supercapacitors. Electrochim Acta 149:144–151
85. Musiani M (2000) Electrodeposition of composites: an expanding subject in electrochemical materials science. Electrochim Acta 45(20):3397–3402
86. Du J, Zhou G, Zhang H, Cheng C, Ma J, Wei W, Chen L, Wang T (2013) Ultrathin porous $NiCo_2O_4$ nanosheet arrays on flexible carbon fabric for high-performance supercapacitors. ACS Appl Mater Interfaces 5(15):7405–7409
87. Yuan C, Li J, Hou L, Zhang X, Shen L, Lou XWD (2012) Ultrathin mesoporous $NiCo_2O_4$ nanosheets supported on Ni foam as advanced electrodes for supercapacitors. Adv Funct Mater 22(21):4592–4597
88. Huang L, Chen D, Ding Y, Wang ZL, Zeng Z, Liu M (2013) Hybrid composite Ni (OH) 2@ $NiCo_2O_4$ grown on carbon fiber paper for high-performance supercapacitors. ACS Appl Mater Interfaces 5(21):11159–11162
89. Zhang G, Wang T, Yu X, Zhang H, Duan H, Lu B (2013) Nanoforest of hierarchical Co_3O_4@ $NiCo_2O_4$ nanowire arrays for high-performance supercapacitors. Nano Energy 2(5):586–594
90. Chang S-K, Lee K-T, Zainal Z, Tan K-B, Yusof NA, Yusoff WMDW, Lee J-F, Wu N-L (2012) Structural and electrochemical properties of manganese substituted nickel cobaltite for supercapacitor application. Electrochim Acta 67:67–72
91. Shakir I, Sarfraz M, Rana UA, Nadeem M, Al-Shaikh MA (2013) Synthesis of hierarchical porous spinel nickel cobaltite nanoflakes for high performance electrochemical energy storage supercapacitors. Rsc Adv 3(44):21386–21389
92. Wu Z, Pu X, Zhu Y, Jing M, Chen Q, Jia X, Ji X (2015) Uniform porous spinel $NiCo_2O_4$ with enhanced electrochemical performances. J Alloys Compd 632:208–217
93. Tseng C-C, Lee J-L, Liu Y-M, Ger M-D, Shu Y-Y (2013) Microwave-assisted hydrothermal synthesis of spinel nickel cobaltite and application for supercapacitors. J Taiwan Inst Chem Eng 44(3):415–419
94. Hu C-C, Hsu C-T, Chang K-H, Hsu H-Y (2013) Microwave-assisted hydrothermal annealing of binary Ni–Co oxy-hydroxides for asymmetric supercapacitors. J Power Sources 238:180–189
95. Yuan C, Li J, Hou L, Lin J, Pang G, Zhang L, Lian L, Zhang X (2013) Template-engaged synthesis of uniform mesoporous hollow $NiCo_2O_4$ sub-microspheres towards high-performance electrochemical capacitors. Rsc Adv 3(40):18573–18578
96. Ding R, Qi L, Jia M, Wang H (2013) Hydrothermal and soft-templating synthesis of mesoporous $NiCo_2O_4$ nanomaterials for high-performance electrochemical capacitors. J Appl Electrochem 43(9):903–910
97. Padmanathan N, Selladurai S (2013) Sonochemically precipitated spinel Co_3O_4 and $NiCo_2O_4$ nanostructures as an electrode materials for supercapacitor. In: American Institute of Physics conference series, pp 1216–1217
98. Salunkhe RR, Jang K, Yu H, Yu S, Ganesh T, Han S-H, Ahn H (2011) Chemical synthesis and electrochemical analysis of nickel cobaltite nanostructures for supercapacitor applications. J Alloys Compd 509(23):6677–6682
99. Zhu Y, Wu Z, Jing M, Hou H, Yang Y, Zhang Y, Yang X, Song W, Jia X, Ji X (2015) Porous $NiCo_2O_4$ spheres tuned through carbon quantum dots utilised as advanced materials for an asymmetric supercapacitor. J Mater Chem A 3(2):866–877

100. Zhu Y, Pu X, Song W, Wu Z, Zhou Z, He X, Lu F, Jing M, Tang B, Ji X (2014) High capacity $NiCo_2O_4$ nanorods as electrode materials for supercapacitor. J Alloys Compd 617:988–993
101. Li L, Peng S, Cheah Y, Teh P, Wang J, Wee G, Ko Y, Wong C, Srinivasan M (2013) Electrospun porous $NiCo_2O_4$ nanotubes as advanced electrodes for electrochemical capacitors. Chem Eur J 19(19):5892–5898
102. Hsieh Y-C, Lee K-T, Lin Y-P, Wu N-L, Donne SW (2008) Investigation on capacity fading of aqueous $MnO_2 \cdot nH_2O$ electrochemical capacitor. J Power Sources 177(2):660–664
103. Kong L-B, Lu C, Liu M-C, Luo Y-C, Kang L, Li X, Walsh FC (2014) The specific capacitance of sol–gel synthesised spinel $MnCo_2O_4$ in an alkaline electrolyte. Electrochim Acta 115:22–27
104. Chuang P-Y, Hu C-C (2005) The electrochemical characteristics of binary manganese–cobalt oxides prepared by anodic deposition. Mater Chem Phys 92(1):138–145
105. Li L, He F, Gai S, Zhang S, Gao P, Zhang M, Chen Y, Yang P (2014) Hollow structured and flower-like C@ $MnCo_2O_4$ composite for high electrochemical performance in a supercapacitor. CrystEngComm 16(42):9873–9881
106. Xu Y, Wang X, An C, Wang Y, Jiao L, Yuan H (2014) Facile synthesis route of porous $MnCo_2O_4$ and $CoMn_2O_4$ nanowires and their excellent electrochemical properties in supercapacitors. J Mater Chem A 2(39):16480–16488
107. Wang Z, Zhang X, Li Y, Liu Z, Hao Z (2013) Synthesis of graphene–$NiFe_2O_4$ nanocomposites and their electrochemical capacitive behavior. J Mater Chem A 1 (21):6393–6399
108. Senthilkumar B, Vijaya Sankar K, Sanjeeviraja C, Kalai Selvan R (2013) Synthesis and physico-chemical property evaluation of PANI–$NiFe_2O_4$ nanocomposite as electrodes for supercapacitors. J Alloys Compd 553:350–357
109. Sen P, De A (2010) Electrochemical performances of poly (3, 4-ethylenedioxythiophene)–$NiFe_2O_4$ nanocomposite as electrode for supercapacitor. Electrochim Acta 55 (16):4677–4684
110. Xiao K, Xia L, Liu G, Wang S, Ding L-X, Wang H (2015) Honeycomb-like $NiMoO_4$ ultrathin nanosheet arrays for high-performance electrochemical energy storage. J Mater Chem A 3 (11):6128–6135
111. Peng S, Li L, Wu HB, Madhavi S, Lou XWD (2014) Controlled growth of $NiMoO_4$ nanosheet and nanorod arrays on various conductive substrates as advanced electrodes for asymmetric supercapacitors. Adv Energy Mater 5(2):1401172
112. Cai D, Liu B, Wang D, Liu Y, Wang L, Li H, Wang Y, Wang C, Li Q, Wang T (2014) Enhanced performance of supercapacitors with ultrathin mesoporous $NiMoO_4$ nanosheets. Electrochim Acta 125:294–301
113. Cai D, Liu B, Wang D, Liu Y, Wang L, Li H, Wang Y, Wang C, Li Q, Wang T (2014) Facile hydrothermal synthesis of hierarchical ultrathin mesoporous $NiMoO_4$ nanosheets for high performance supercapacitors. Electrochim Acta 115:358–363
114. Cai D, Wang D, Liu B, Wang Y, Liu Y, Wang L, Li H, Huang H, Li Q, Wang T (2013) Comparison of the electrochemical performance of $NiMoO_4$ nanorods and hierarchical nanospheres for supercapacitor applications. ACS Appl Mater Interfaces 5(24):12905–12910
115. Guo D, Zhang P, Zhang H, Yu X, Zhu J, Li Q, Wang T (2013) $NiMoO_4$ nanowires supported on Ni foam as novel advanced electrodes for supercapacitors. J Mater Chem A 1 (32):9024–9027
116. Guo D, Luo Y, Yu X, Li Q, Wang T (2014) High performance $NiMoO_4$ nanowires supported on carbon cloth as advanced electrodes for symmetric supercapacitors. Nano Energy 8:174–182
117. Liu X, Zhang K, Yang B, Song W, Liu Q, Jia F, Qin S, Chen W, Lib J, Zhang Z (2015) Three-dimensional graphene supported nickel molybdate nanowires as novel ultralight and flexible electrode for supercapacitors. arXiv preprint arXiv:150201059

118. Yin Z, Zhang S, Chen Y, Gao P, Zhu C, Yang P, Qi L (2015) Hierarchical nanosheet-based $NiMoO_4$ nanotubes: synthesis and high supercapacitor performance. J Mater Chem A 3 (2):739–745
119. Cai D, Wang D, Liu B, Wang L, Liu Y, Li H, Wang Y, Li Q, Wang T (2014) Three-dimensional Co_3O_4@ $NiMoO_4$ core/shell nanowire arrays on Ni foam for electrochemical energy storage. ACS Appl Mater Interfaces 6(7):5050–5055
120. Ma X-J, Kong L-B, Zhang W-B, Liu M-C, Luo Y-C, Kang L (2014) Design and synthesis of 3D Co_3O_4@ $MMoO_4$ (M = Ni, Co) nanocomposites as high-performance supercapacitor electrodes. Electrochim Acta 130:660–669
121. Hong W, Wang J, Gong P, Sun J, Niu L, Yang Z, Wang Z, Yang S (2014) Rational construction of three dimensional hybrid Co_3O_4@ $NiMoO_4$ nanosheets array for energy storage application. J Power Sources 270:516–525
122. Guo D, Ren W, Chen Z, Mao M, Li Q, Wang T (2015) $NiMoO_4$ nanowire@ MnO_2 nanoflake core/shell hybrid structure aligned on carbon cloth for high-performance supercapacitors. Rsc Adv 5(14):10681–10687
123. Zhang Q, Deng Y, Hu Z, Liu Y, Yao M, Liu P (2014) Seaurchin-like hierarchical $NiCo_2O_4$@ $NiMoO_4$ core–shell nanomaterials for high performance supercapacitors. Phys Chem Chem Phys 16(42):23451–23460
124. Ren W, Guo D, Zhuo M, Guan B, Zhang D, Li Q (2015) $NiMoO_4$@ $Co(OH)_2$ core/shell structure nanowire arrays supported on Ni foam for high-performance supercapacitors. Rsc Adv 5(28):21881–21887
125. Liu M-C, Kong L-B, Lu C, Ma X-J, Li X-M, Luo Y-C, Kang L (2013) Design and synthesis of $CoMoO_4$–$NiMoO_4 \cdot x$ H2O bundles with improved electrochemical properties for supercapacitors. J Mater Chem A 1(4):1380–1387
126. Senthilkumar B, Meyrick D, Lee Y-S, Selvan RK (2013) Synthesis and improved electrochemical performances of nano β-$NiMoO_4$–$CoMoO_4 \cdot xH_2O$ composites for asymmetric supercapacitors. Rsc Adv 3(37):16542–16548

Chapter 10
Nanostructured Manganese Oxides in Supercapacitors

Katlego Makgopa, Paul M. Ejikeme, and Kenneth I. Ozoemena

10.1 Introduction

The development of energy storage systems (ESSs) comes as a crucial factor when it comes to addressing the problem of energy in the world due to the rapid development of the global economy, the depletion of fossil fuels and the increase of environmental pollution. Hence there is an urgent need for clean, cheap, efficient and sustainable sources of energy, as well as new technologies associated with energy storage and conversion [1–3]. The electrochemical capacitor (EC), also known as ultracapacitor or supercapacitor, is an energy storage device that utilises two closely spaced layers composed of two opposing charges to store energy which is used to power hybrid electric vehicles, portable electronic equipment and other devices [4].

Conventional capacitors are capacitors characterised by high power density with very low energy density, whereas the opposite applies to batteries [2]. A high energy density obtained in a battery technology and high power of conventional

K. Makgopa (✉)
Department of Chemistry, University of Pretoria, Pretoria 0002, South Africa
e-mail: katlego.makgopa@up.ac.za

P.M. Ejikeme
Department of Chemistry, University of Pretoria, Pretoria 0002, South Africa

Department of Pure and Industrial Chemistry, University of Nigeria, Nsukka 410001, Nigeria

K.I. Ozoemena (✉)
Department of Chemistry, University of Pretoria, Pretoria 0002, South Africa

Energy Materials Unit, Materials Science and Manufacturing, Council for Scientific & Industrial Research (CSIR), Pretoria 0001, South Africa

School of Chemistry, University of the Witwatersrand, Johannesburg 2050, South Africa
e-mail: kozoemena@csir.co.za

K.I. Ozoemena, S. Chen (eds.), *Nanomaterials in Advanced Batteries and Supercapacitors*, Nanostructure Science and Technology,
DOI 10.1007/978-3-319-26082-2_10

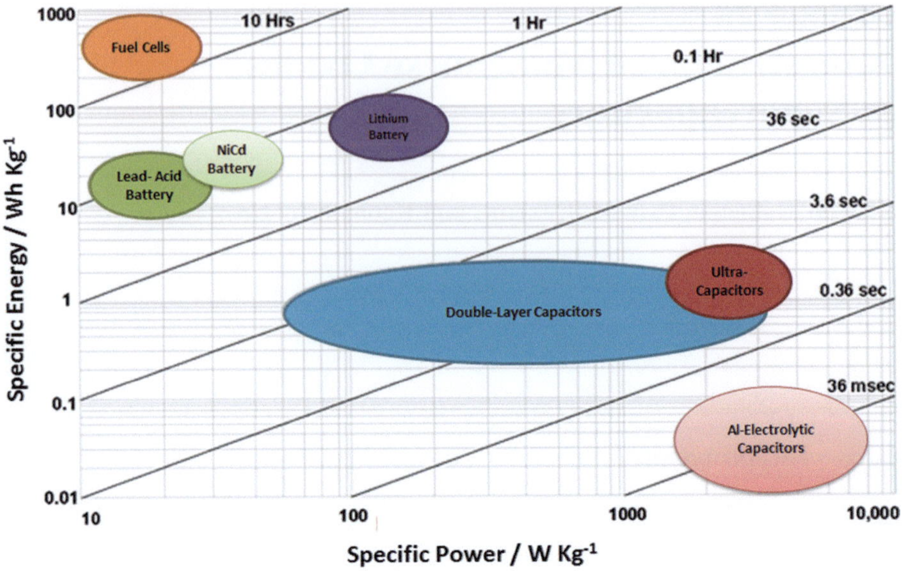

Fig. 10.1 Ragone plot indicating specific power versus specific energy of various electrochemical energy storage devices (Source: http://www.mpoweruk.com/performance.htm)

capacitors resulted in the development of supercapacitors. EC technology manage to obtain relatively high energy density as compared to the conventional capacitors even though its energy density is still lower than that supplied by a battery while still maintaining the high power output found in conventional capacitors; hence, ECs serve as a gap between the conventional capacitor and battery technologies. ECs, with a combined high power density and much reasonable high energy density, are a versatile solution to a variety of emerging energy applications. The variation of these energy storage and conversion devices is best illustrated using Ragone plot shown in Fig. 10.1.

Due to different mechanisms of energy storage at the electrode–electrolyte interface, ECs are divided into three categories, based on design of the electrodes: (1) electrical double-layer capacitors (EDLCs), (2) pseudocapacitors and (3) hybrid capacitors. The electrical double-layer capacitor (EDLC), in which capacitance is due to a non-Faradaic process, is based on charge separation at the electrode–electrolyte interface made of a high surface porous carbon electrode. Whereas, in pseudocapacitor, the capacitance is due to a Faradaic process that relies on electrosorption processes and redox reactions (transfer of electrons from one species to another) that occur within the electrode made of metal oxide or conducting polymer [2, 5, 6]. Hybrid capacitor is made from electrodes that exhibit both major double-layer capacitance and pseudocapacitance, such as lithium-ion capacitors and

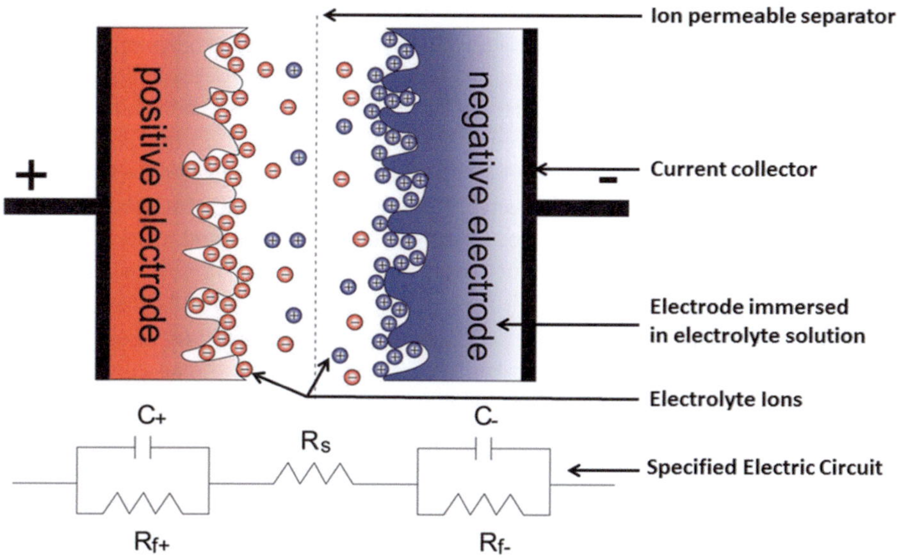

Fig. 10.2 Representation of the charged state of a symmetric electrical double-layer capacitor using porous electrodes and of the corresponding equivalent circuit (with modification). *R* denotes resistors, *C* denotes capacitors (Reprinted with permission from Ref. [9], copyright 2014, WILEY-VCH Verlag GmbH & Co. KGaA, Weinheim)

asymmetric capacitors [7, 8]. Figure 10.2 illustrates a typical design of supercapacitors showing the charged state and the corresponding equivalent circuit [9].

Nanomaterials of about the length scale of less than 100 nm have received increasing interest, owing to their fundamental scientific significance as well as their potential applications that derive from their fascinating electrical, magnetic and catalytic properties [10, 11]. As compared to the bulk active electrode materials, the corresponding nanomaterials possess more excellent electrochemical activity, such as higher capacitance, larger surface areas and relatively good conductivity; thereby, nanomaterials have vast potential application in the electrochemistry field [12]. The most widely used active electrode materials are carbon, conducting polymers and transition metal oxides [9, 13, 14]. Transition metal oxides gained much interest from various researchers as electrode materials for ECs. Among transition metal oxides, ruthenium oxide, RuO_2, seems to be a promising electrode material due to its high capacitance, good conductivity, excellent electrochemical reversibility, high rate capability and long cyclability. However, the cost, lack of abundance and toxic character of RuO_2 necessitate seeking alternative electrode materials for EC devices. Therefore, great attention has been devoted to other types of transition metal oxides such as manganese dioxide (MnO_2)-based materials and its derivatives, due to their opposing properties as compared to RuO_2. MnO_2-based materials and its derivatives have demonstrated to be environmentally friendly in character while being relatively cheap and abundant in nature. Apart from the socio-economic benefits of the MnO_2-based materials and its derivatives, these electrode materials are also characterised by their relatively high theoretical specific surface area (1370m^2g^{-1}), multiple crystal phases, high theoretical capacitance and good electrochemical response such as cyclability [12, 15–21]. Several studies on electrochemical behaviour of MnO_2-based materials and its derivatives in an aqueous electrolyte have been conducted in order to better understand their mechanism of charge storage on the surface of the electrode. Hydrated manganese oxides exhibit specific capacitances within the range of 100 -200 F g^{-1} in alkali salt solutions, which are much lower than those for RuO_2ECs. Even though amorphous, hydrated ruthenium oxide, RuO_2, is capable of giving high specific capacitance of 720 F g^{-1} with excellent cyclability [22], unfortunately RuO_2 is considered too expensive to be commercially attractive; hence, much interest in the development of ECs that use cheap transition metal oxides such as manganese dioxides continues to grow [23]. In addition, these RuO_2 capacitor-type materials are only suitable for aqueous electrolytes, thus limiting the nominal cell voltage to 1.0 V. It has been established that pseudocapacitive (Faradaic) reactions occurring on the surface and in the bulk of the electrode are the major charge storage mechanisms for MnO_2-based materials. The surface reaction involves the adsorption of electrolyte cations ($C^+ = H^+, Li^+, Na^+$ and K^+) on the MnO_2-based materials, whereas the bulk Faradaic reaction relies on the intercalation or de-intercalation of electrolyte cations in the bulk of the manganese oxide and are illustrated by Eqs. 10.1 and10.2, respectively [24, 25]:

$$(MnO_2)_{surface} + C^+ + e^- \leftrightarrow (MnO_2{}^- C^+)_{surface} \tag{10.1}$$

$$MnO_2 + C^+ + e^- \leftrightarrow MnOOC \tag{10.2}$$

In the established charge storage mechanisms, a redox reaction between the III and IV oxidation states of Mn ions occurs [24]. As Mn is a transition metal element, there are several types of manganese oxide existing as a variety of stable oxide, and these include manganese (II) oxide (MnO), manganese (II, III) oxide (Mn_3O_4), manganese (III) oxide (Mn_2O_3), manganese (IV) oxide (MnO_2) and manganese (VII) oxide (Mn_2O_7). They can also be identified in their various mineral forms, which are birnessite, hausmannite, pyrolusite, manganite, manganosite and psilomelane [26–28]. Manganese oxide exists in various types of crystal structures, as shown in an idealised representation in Fig. 10.3, having alternate layers of full and empty octahedral sites [29]. Dedicated approach on the synthesis of MnO_2-based materials and the electrolyte-type plays a major role on the electrochemical outcome of these EC electrode materials. Therefore, a careful decision has to be made from the beginning when preparing MnO_2-based materials for EC applications.

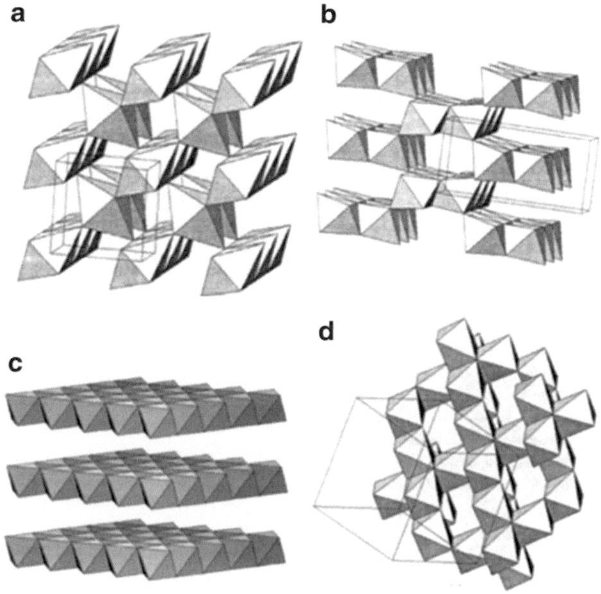

Fig. 10.3 Typical schematic representation of an assembled full-cell EC design based on MnO_2/graphene composite as positive electrode and graphene as negative electrode. Different MnO_2 crystal structures: (**a**) pyrolusite β-MnO_2 (1×1), (**b**) ramsdellite (1×2), (**c**) birnessite δ-MnO_2 (layered structure $1 \times \infty$) and (**d**) spinel λ-MnO_2 (3D tunnels) (Reproduced from Ref. [29] by permission of The Electrochemical Society)

10.1.1 Fundamental Calculations of Supercapacitive Parameters

As shown in Fig. 10.2, a supercapacitor consists of two porous electrode materials which are in direct contact with the current collector and separated by a porous separator impregnated with an electrolyte solution. These two porous electrodes are equivalent to two capacitors connected in series. In the charged state, the electrolyte cations and anions move towards the negative and positive electrode, respectively. The separation of ions leads to a potential difference across the cells. Each electrode–electrolyte interface is a capacitor, and thus the complete cell is considered as two capacitors connected in series. The resulting cell capacitance (C_{cell}) obtained after polarisation of electrodes by applying potential difference (voltage) between them can be expressed according to the following equation:

$$\frac{1}{C_{cell}} = \frac{1}{C_+} + \frac{1}{C_-} \tag{10.3}$$

where C, C_+ and C_- are the capacitance (in Farad = Coulomb/V) of the resulting device or cell, of the positive electrode and of the negative electrode, respectively, with the capacitance (C_{cell}) being determined from the cyclic voltammetry

(CV) and/or the slope of the discharge curve of galvanostatic charge–discharge (GCD) profile by Eq. (10.4):

$$C_{cell}(F) = \frac{i}{\Delta E/\Delta t} \tag{10.4}$$

where i (A) is the applied current and $\Delta E/\Delta t$(V s^{-1}) the slope of the discharge curve after the initial *iR* drop or scan rate in terms of CV. In a symmetrical system where the two electrodes (positive and negative electrode) are similar with similar morphological and electronic properties ($C_+ = C_-$), the capacitance of the complete cell becomes half of the capacitance of each individual electrode, i.e.:

$$C_{cell} = \frac{C_e}{2} \tag{10.5}$$

where $C_e = C_+ = C_-$.

The specific (gravimetric) capacitance of an electrode ($C_{e,sp}$) is given by:

$$C_{e,sp}\left(Fg^{-1}\right) = \frac{2.C_{cell}}{m_e} \tag{10.6}$$

where m_e (g) is the mass of the single electrode. Note that Eqs. 10.5 and 10.6 simply remind the researchers that when reporting or comparing the values of capacitance from different literature, one must clearly specify if the values are those of the electrodes or cells. The value of a single electrode derived from a *three-electrode* (half-cell) measurement will be higher than the actual cell capacitance derived from a *two-electrode* (full-cell) measurement.

For a symmetric system, the specific (gravimetric) capacitance of the two electrodes (C_{sp}) is given by:

$$C_{sp}\left(F\ g^{-1}\right) = \frac{4\ C_{cell}}{M} \tag{10.7}$$

where M is the total mass of the active materials of the two electrodes (i.e. M = 2.m_e since the mass of each electrode is the same). The multiplier of 4 simply adjusts the capacitance of the cell and the combined mass of two electrodes to the capacitance and mass of a single electrode [30, 31].

The other two important parameters of a capacitor apart from the C_{sp} are its energy and power density [32]. The energy (E) stored in a capacitor is related to the charge (Q) at each interface and the potential difference between the two plates and therefore is directly proportional to the capacitance as follows:

$$E\ \left(Wh\ kg^{-1}\right) = \frac{C_{cell}\ V^2}{2M} \tag{10.8}$$

where m (kg) is the mass of the ECs and V (V) is the maximum voltage of electrochemical stability. The maximum power of the device can be obtained from the following equation:

$$P\left(W\,kg^{-1}\right) = \frac{V^2}{4R_sM} \tag{10.9}$$

The internal resistance R_s is calculated from the voltage drop at the beginning of a discharge curve and is shown by the following equation:

$$R_s(\Omega) = \frac{\Delta V_{IR}}{2i} \tag{10.10}$$

where ΔV_{IR} is the voltage drop between the first two points from the start of the discharge curve.

For an asymmetric system, Eqs. 10.7, 10.8 and 10.9 are also applied. However, before they can be applied to the asymmetric system, it is very critical to first perform *mass balancing* from the *three-electrode* experiment for each of the electrodes. In the symmetric supercapacitor, the applied voltage is split equally between the two electrodes due to the use of the same material having the same mass in each electrode. In the asymmetric supercapacitors, however, the voltage split is dependent on the capacitance of the active material in each electrode of its two electrodes. The capacitance is usually related to the mass and the specific capacitance of the active material [33, 34]. Thus, in order to split voltage equally, we must optimise the mass balance between the two electrodes using the relationship $q_+ = q_-$, where q_+ means the charges stored at the positive electrode and q_- means the charges stored at the negative electrode. The stored charge can be expressed by the following equation:

$$q = C_{\mathrm{sp}}.m.\Delta E \tag{10.11}$$

or

$$\frac{m_+}{m_-} = \frac{C_{sp_-}}{C_{sp_+}} \times \frac{\Delta E_-}{\Delta E_+} \tag{10.12}$$

where m, C_{sp} and ΔE represent the mass, specific capacitance and potential range obtained from the charging/discharging process of three-electrode configuration of the individual positive and negative electrode, respectively.

10.1.2 Effects of Electrolytes on the Supercapacitive Properties

The electrolyte that is itself an electronic insulator but also an ionic conductor serves as the medium for transfer of charge as ions between the positive and the negative electrodes. The choice of electrolyte solutions plays a crucial role when it comes to the performance of the ECs especially when material such as metal oxides with their poor conductivity is going to be employed as electrodes [1, 30]. The three types of electrolytes mainly used in ECs include aqueous, organic and liquid salts (also known as ionic liquids). There are two main criteria involved for the selection of an electrolyte: (i) the electrochemical stability window (which is crucial in maximising the specific energy values according to Eq. (10.8)) and (ii) the ionic conductivity (which has a major influence on the values of specific power). Aqueous electrolytes include acid (e.g. H_2SO_4), alkali (e.g. KOH) and neutral (e.g. Na_2SO_4) solutions, all of which have higher conductivity (i.e. up to ~ 1 S/cm). Due to the relatively high conductivity of these electrolyte types, the EC systems using them result in a higher-power performance as compared to other systems that use non-aqueous electrolyte solutions. Nevertheless, this type of EC system suffers from the narrow electrochemical stability window due to water electrolysis (~1.23 V), resulting in a relatively low operating voltage and, consequently, limiting the energy that can be stored in the device [30, 35]. However, acid and alkali-based aqueous electrolytes are considered environmentally unfriendly and harmful to work with due to their corrosive nature [36]. In order to mitigate the corrosive nature of the acid and alkali media, electrolytes with neutral pH such as Na_2SO_4 and Li_2SO_4 have been investigated. Also, these neutral electrolytes have high voltage values of up to 2 V when used in ECs made of manganese oxide/carbon composites. Good charge–discharge cycle life has been observed in carbon materials such as AC for symmetric cells and in asymmetric cell using MnO_2 as positive electrode by employing Na_2SO_4 and Li_2SO_4 as electrolyte solution [15, 37–42]. It is worth mentioning that the overpotential for electrolyte decomposition varies, depending on the used carbon with the temperature playing an important role in the degradation mechanism [9]. Organic electrolytes composed of salt dissolved in an organic solvent (i.e. propylene carbonate (PC) and/or acetonitrile (AN)) provide a wider electrochemical stability window (ranging from 2.7 to 2.8 V) as compared to acid–alkali-based aqueous electrolytes due to their resistance against hydrolysis. Nonetheless, they suffer relatively low ionic conductivity and high viscosity that result in lower specific capacitance (100–150 F/g) as compared to acid–alkali-based aqueous electrolytes [43–46]. However, the wide electrochemical stability window in organic electrolytes is advantageous to the ECs as it helps to deliver a higher specific energy stored as compared to acid–alkali-based aqueous EC systems, and for this reason, the majority of industrial systems are currently produced with organic electrolytes. Salts commonly applied in organic electrolytes are quaternary ammonium salts, out of which, tetraethylammonium tetrafluoroborate ($TEA-BF_4$) is the most widely used salt for commercial supercapacitors. Other salts include $EMIM - BF_4$ (1-ethyl-3-methylimidazolium tetrafluoroborate), $MEPY - BF_4$ (1-ethyl-1-methylpyrrolidinium tetrafluoroborate), $TMPY - BF_4$ (tetramethylene-pyrrolidinium tetrafluoroborate) and $TEMA - BF_4$ (triethylmethylammonium tetrafluoroborate) [36, 47]. Recently, ionic electrolytes have been explored by researchers as an alternative by fine-tuning electrolytes that result in an increase energy density of the ECs. Ionic liquids (ILs) also called room-temperature ionic liquids (RTILs) are organic salts that are likely liquids at or around room temperature [48]. This type of salts is called molten salts. Their desirable properties make them promising candidates for EC electrolytes. These electrolytes have a very low vapour pressure due to the fact that no solvent is required, hence, limiting environmental exposure and preventing the risk of explosion. The advantage of ILs is that they possess interesting properties such as high thermal stability at increased temperatures (beyond the $\sim 80°C$ limit of organic electrolytes), a broad electrochemical stability window (ranging from 2 to 6 V, typically about 4.5 V) which is extremely higher than that of organic and aqueous electrolytes and low flammability [1, 49]. However, the ionic conductivity of ILs, specifically at room temperature, is generally lower than that of organic electrolytes, therefore reducing the power performance of IL-based ECs [50–53]. Today, most supercapacitors based on manganese oxides utilise aqueous electrolytes.

10.2 Selection Criteria and Synthesis of Manganese Oxide Nanomaterials

In pursuit for suitable manganese oxide materials for ECs, several important requirements need to be considered by researchers; out of many, here are some of the factors to consider when studying EC electrode materials: (1) High stability in various types of electrolyte solution, since the prerequisite for the application of manganese oxides in ECs is that they do not dissolve in any electrolyte media during the performance testing. (2) High resistance against electrochemical corrosion: manganese oxides should be electrochemically stable under the working conditions of ECs, or metal nanoparticles supported on them will peel off from them and agglomerate, leading to the degradation of catalytic performance. (3) Good/better electronic conductivity: manganese oxides should be conductive enough to potentially allow flow of electrons and not act as insulators. Thus far, further advancements of manganese oxides as EC electrode are constrained by manganese oxide electrode material limitations such as poor conductivity, limited specific capacitance, lack of structural stability and poor rate capability. The persistence on extensive study of manganese oxide materials such as MnO_2 is upon the fact that these materials possess a good environmental compatibility, are relatively cheap, have abundant reserves and possess excellent performance when combined with other conductive material to form composites; hence, several researchers view these materials as having a great potential for development as electrode materials in ECs. There are various ways in which various forms of manganese oxides have been synthesised and reported in the literature.

10.2.1 Synthesis of MnO_2-Based Electrode Materials

Reduced sizes of bulk materials into the nanoscale greatly affect the physical and chemical properties of the materials. Nanomaterials are estimated to bring significant improvements for energy storage devices as the size reduction of materials increases the contact surface area between the electrode and the electrolyte and decreases the transport path length for both electrons and ions [10, 54]. It has been established that MnO_2 nanomaterials can exist in different types of structural phases, $\alpha-$, $\beta-$, $\gamma-$, $\delta-$, $\varepsilon-$ and λ types, when the basic structural unit ([MnO_6] octahedron) is linked in different ways [55]. The properties of MnO_2 are significantly affected by their phases and morphologies; moreover, the operating properties of ECs also depend on the phase of MnO_2. In this regard, more energy has been focused towards the preparation of MnO_2 with different phases and shapes [56]. Several structural forms of MnO_2 with different nanoarchitectures such as nanowires [57, 58], nanorods [59], single crystal nanotubes [55], nanourchins [60] and amorphous [61] have been synthesised using hydrothermal techniques. Some other techniques used for the synthesis of MnO_2 nanostructures include thermal decomposition, co-precipitation [23, 29, 62], simple reduction [63, 64], sol–gel [65, 66], solid-state process and microwave process [67–69].

10.2.2 Synthesis of Mn_3O_4-Based Electrode Materials

Manganese oxide can be presented in various oxidation states and phases as mentioned above. The most often studied phases are MnO_2 with an oxidation state of +4, Mn_2O_3 with an oxidation state of +3 and Mn_3O_4 with the oxidation state of both +2 and +3. Manganese (II, III) oxide is the chemical compound with formula Mn_3O_4. It exists in two oxidation states, +2 and +3, and its formula can sometime be written as MnO. Mn_2O_3. Manganese (II, III) oxide, Mn_3O_4, is found in nature as the mineral hausmannite. Hausmannite Mn_3O_4 is normally known to be made of a spinel structure with tetragonal distortion elongated along the c-axis due to Jahn–Teller effect on the Mn^{3+} ion. Manganese ions occupy the octahedral B-site (Mn^{3+}) and tetrahedral A-site (Mn^{2+}) corresponding to a normal spinel structure. There are 32 oxygens and 24 cations in the unit cell [70]. Several studies have been conducted in order to improve Mn_3O_4 electrode materials in ECs. Subsequently, to improve the electrocapacitive behaviour of Mn_3O_4, the nanosized particles of Mn_3O_4 have been prepared by various methods such as successive ionic layer adsorption and reaction (SILAR) [71, 72], hydrothermal [73, 74], solution combustion [75], chemical bath deposition [76, 77], sonochemical [78], microwave irradiation and microwave-assisted techniques [79–82]. Different nanoarchitectures of Mn_3O_4 particles such as single crystals, nanocrystals, nanocubics, nanorods, microspheres and interlocked cubes have been successfully synthesised and tested for EC behaviour [72, 74, 77, 79, 83, 84].

10.2.3 Synthesis of Mn_2O_3-Based Electrode Materials

Manganese (III) oxide, Mn_2O_3, and manganese (II) oxide, MnO, are other types of the oxides of manganese derivatives. There are substantially few literature reviews on these materials as EC materials. There are two forms that are generally recognisable in Mn_2O_3, $\alpha - Mn_2O_3$ and $\gamma - Mn_2O_3$. The $\alpha - Mn_2O_3$ has the cubic bixbyite structure, while pure Mn_2O_3 has an orthorhombic structure and $\gamma - Mn_2O_3$ has a structure related to the spinel structure of Mn_3O_4 with the oxide ions closely packed in cubic form [85]. MnO, just like any monoxides, adopts the rock salt structure, with anions and cations being both octahedrally coordinated [86]. Just as much as any manganese oxide material, Mn_2O_3 nanospheres have been synthesised using hydrothermal and sonochemical techniques by Nathan et al. as an EC electrode material [87]. Chen et al. have successfully synthesised $\gamma - Mn_2O_3$ nanowire bundles [88]. Recently, Li et al. synthesised nanocubic Mn_2O_3 using hydrothermal technique for the EC application [89]. Nanosized manganese oxide (Mn_2O_3) was synthesised by Chiang et al. using a solvothermal technique [90]. There is limited literature on the synthesis of Mn_2O_3 for EC application even though several of this material have been widely synthesised using various techniques such as chemical oxidation [91], calcination of MnO_2 and hydrothermal [92].

10.3 Application of Manganese Oxides in Supercapacitors

10.3.1 MnO_2-Based Supercapacitors

Interest on manganese dioxide (MnO_2) as a potential electrode material for ECs surfaced very slowly at around 1999, with the study from Lee and Goodenough [23] when they were investigating pseudocapacitive behaviour of an amorphous manganese dioxide (MnO_2) in aqueous electrolyte (2 M KCl) as a way to compensate the cost disadvantage brought by the use of RuO_2. In 2002, there was an improved outcome in terms of the literature regarding synthesis and testing of manganese oxide-based EC electrode materials [93, 94]. An amorphous hydrous manganese oxide ($\alpha - MnO_2 nH_2O$) was anodically deposited onto a graphite substrate showing an improved maximum specific capacitance of 330 F g^{-1} and 320 F g^{-1} achieved from cyclic voltammetry and galvanostatic charge–discharge measurements, respectively, using Na_2SO_4 (0.1 M) as electrolyte [93]. In the same year, 2002, Toupin et al. also managed to synthesise the pure α-phase MnO_2 from co-precipitation technique, and the maximum specific capacitance of *ca.* 180 F g^{-1} was achieved in Na_2SO_4 (0.1 M) electrolyte [94]. It can be said that before the year 2004, the fundamental understanding of the reaction involved in the electrode–electrolyte interface of the manganese oxide-based electrode was not yet clearly understood. Toupin and co-workers later published their work with the aim of getting a better understanding of the charge storage mechanism in manganese dioxide electrodes when cycled in aqueous electrolyte [25]. At this time, it was clear that the crystal structure of the crystallised Mn_xO_y-based materials plays an important role in the improvement of the electrochemical properties of the electrode material, and also, preparation of composites with other materials does improve the electrocapacitive performance [29, 95, 96]. However, owing to the high resistivity of $\alpha - MnO_2 nH_2O$, a conducting additive, such as carbon materials, CNTs or graphene and OLCs, is required for the realisation of EC electrodes [33, 97]. As a result, $\alpha - MnO_2$ has been prepared on the SWCNT by a simple precipitation technique with good cycle power at the current density of 2 A g^{-1} [98]. In the same year, 2007, Ma et al. [99] synthesised a birnessite-type MnO_2 coated uniformly on multiwalled carbon nanotubes (CNTs) with specific surface area of $200 m^2 g^{-1}$ by using a spontaneous direct redox reaction between the multiwalled CNTs and permanganate ions (MnO_4^-) with a high specific capacitance of the MnO_2/CNT nanocomposite in an organic electrolyte of *ca.* 250 F g^{-1} at a high current density of 1 A g^{-1}. In this process, the starting solution of $KMnO_4$ (0.1 M) was heated to 70°C and followed by addition of 1.0 g CNT. Much later, similar work was done by Xia et al. [100], with MnO_2/CNT nanocomposites having the hierarchy architecture of CNT core/porous MnO_2 sheath (see Figs. 10.4 and 10.5), using a simple hydrothermal treatment. Morphologies of the CNTs, the birnessite-type MnO_2 powder and the MnO_2/CNT nanocomposite shown by Fig. 10.4a, b, c, respectively, revealed different morphologies with CNTs showing long tubes and the MnO_2 showing 3D hierarchy architecture of interconnected MnO_2, while the nanocomposite indicated a full decoration of MnO_2 on the CNT support. The well dispersion of MnO_2 together with its crystallinity may be observed from TEM analysis (see Fig. 10.5). The nanocomposite has a displayed surface area of 237.8 $m^2 g^{-1}$. The nanocomposite resulted with a maximum specific capacitance of 223 F g^{-1} at a scan rate of 10 mV s^{-1}. The improved electrocapacitive performance in the above studies is due to the good adhesion of the coated oxide layer on the CNT template as observed from the SEM and TEM. The synergistic effect of the oxide material and the conductive CNT resulted in improved cycling performance by allowing easy access of ions to the bulk of the active material through an improvement of conductivity and a good resiliency of the MnO_2/CNT composite electrodes.

Yang and co-workers fine-tuned the porosity of manganese oxide-based electrode by preparing a porous hydrous MnO_2 with pore sizes from 5 to 30 nm by the

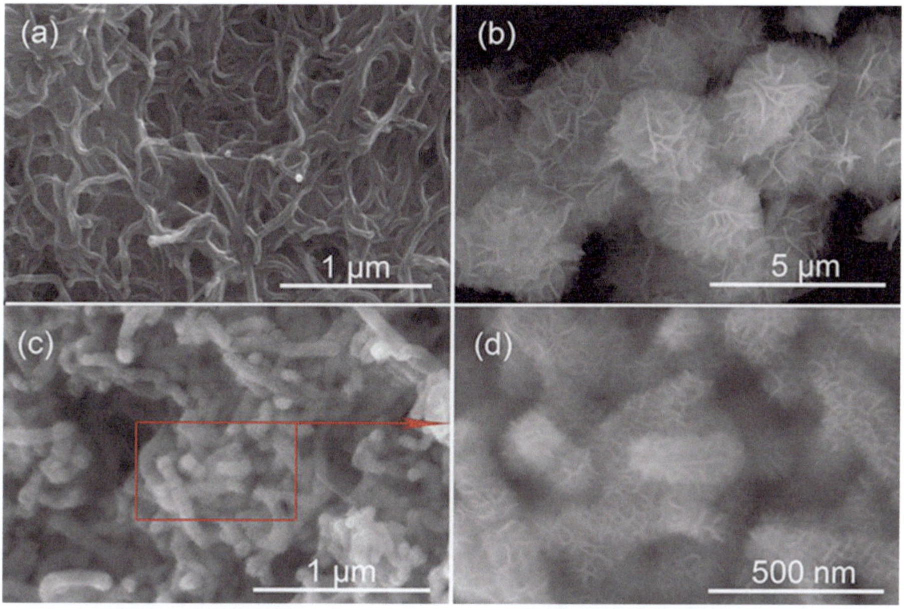

Fig. 10.4 FESEM images of (**a**) the pristine CNTs, (**b**) the flower-like MnO_2 powder and (**c**) the MnO_2/CNT nanocomposite; (**d**) magnified FESEM image of the MnO_2/CNT nanocomposite (Reprinted with permission from Ref. [100], copyright 2012, Springer)

organic–aqueous interfacial method [101]. Interestingly, they showed that the surface area and pore size distribution could be controlled by adjusting the reaction time and the concentration of surfactant in the aqueous phase. The MnO_2 synthesised by this technique exhibited maximum capacitance of 261 F g^{-1} with good cyclic performance but only at a low charge–discharge rate, showing that pure manganese oxide cannot be applied for capacitor application. By the year 2010, much attention and various techniques have been employed in the race of developing better MnO_2-based electrodes, especially the testing and fabrication of different crystal phases of MnO_2. Nearly pure manganese oxide ($\lambda - MnO_2$) has also been successfully obtained by Malak et al. through the removal of lithium from the layered structure of spinel-type $LiMn_2O_4$ with aqueous acid as a means to develop hybrid ECs [102]. In this method, most of the lithium ions were removed from the tetrahedral sites after stirring the mixture for 3 h while preserving the framework of the spinel structure (confirmed by X-ray diffraction measurements). Even though the final product showed a very low specific surface area of about $5m^2g^{-1}$, a maximum value of specific capacitance of about 300 F g^{-1} was obtainable at moderate discharge current density of 0.1 A g^{-1}. It has been shown that MnO_2 embedded on the conducting polymer such as polypyrrole (PPY) results in an improved capacitive performance. Sharma et al. synthesised and tested MnO_2/PPy nanocomposite, thin film electrodes which exhibited a remarkable specific

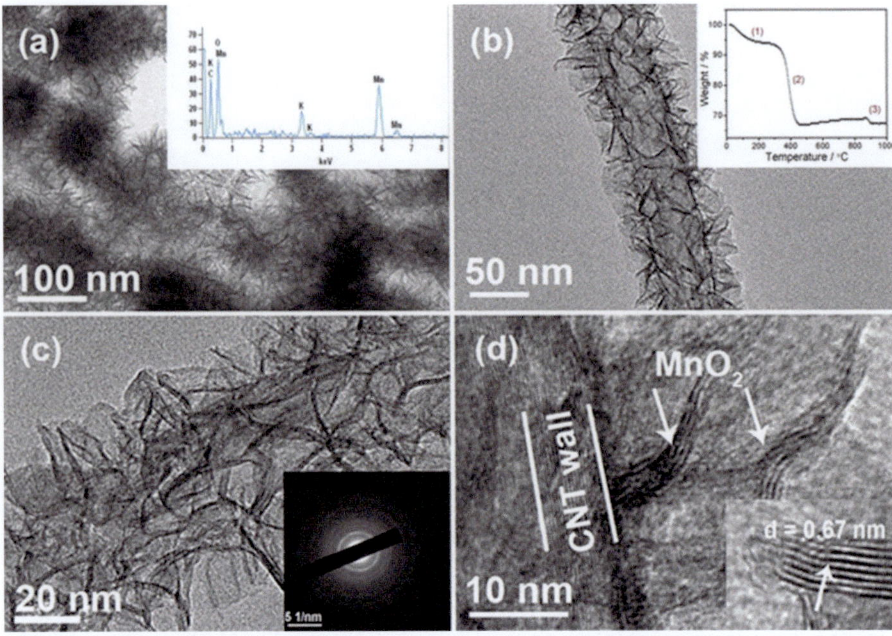

Fig. 10.5 TEM image of the MnO_2/CNT nanocomposite (**a**), TEM (**b**) and magnified TEM (**c**) images of a single CNT coated with a porous MnO_2 layer, HRTEM image of the interface between MnO_2 and CNT (**d**). Inset in **a**, **b**, **c** and **d** shows the EDX, TGA and SAED pattern of the MnO_2/CNT nanocomposite and the interplanar spacing of MnO_2 nanoflake grown on the CNT, respectively (Reprinted with permission from Ref. [100], copyright 2012, Springer)

capacitance as high as 620 F g^{-1} due to the synergistic contribution of MnO_2 and PPY [103]. Other researchers have been developing micro-sized capacitors from the MnO_2 material for the electronic EC application. MnO_2 was electrochemically deposited onto micro-patterned current collectors, and due to the high surface-to-volume ratio of manganese oxide and short diffusion distance between electrodes, the capacitor gave an ultra-high areal capacitance of 56.3 mF cm^{-2} at a current density of 27.2 mA cm^{-2} [104]. In this group we have successfully synthesised different morphologies of MnO_2 as a means of improving its capacitive performance. Jafta et al. [15] synthesised $\alpha - MnO_2$ from the raw electrolytic manganese dioxide (EMD) using hydrothermal technique in the presence of a sodium dodecyl sulphate (SDS) surfactant as outlined in Fig. 10.6 and decorated them on the surface of the graphene oxide (GO).

The electrochemical properties of nanostructured $\alpha - MnO_2/GO$ composite fabricated in an aqueous asymmetric electrochemical capacitor exhibited high energy density of $35 Whkg^{-1}$ and specific capacitance of 280 F g^{-1} at high voltage window of 1.8 V using 1 M Li_2SO_4 as electrolyte. These remarkable electrochemical properties coupled with long-term cycling stability clearly indicated that the nanocomposite may be suitable for future development of low-cost asymmetric

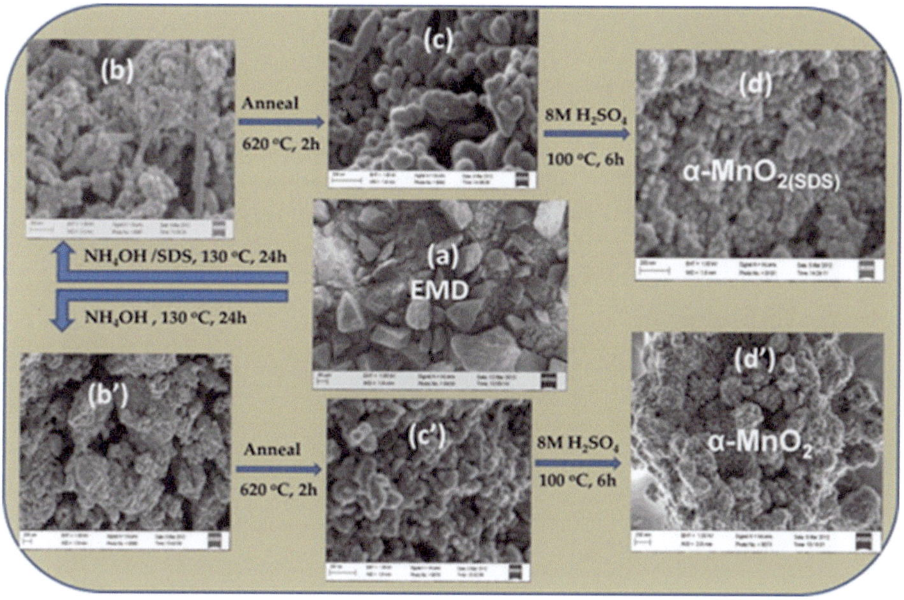

Fig. 10.6 The SDS-based and SDS-free routes for the preparation of α-$MnO_{2(SDS)}$ (**d**) and α-MnO_2 (**d**′) from EMD precursor (**a**) via intermediate nanostructures (**b/b**′) and (**c/c**′). Scale bars are 20 μm for (**a**) and 200 nm for structures (**b/b**′) to (**d/d**′) (Reprinted with permission from Ref. [15], copyright 2013, Elsevier Ltd)

supercapacitors. In 2015, Makgopa et al. [105] showed that the use of birnessite-type MnO_2 decorated on highly graphitised onion-like carbon (OLC) to give a nanocomposite of OLC/MnO_2 fabricated on a symmetrical two-electrode device using the Ni foam as a current collector exhibited a large specific capacitance of 254 F g^{-1} and a remarkably high power density of 74.8 kW kg^{-1}. The surface morphologies of OLC and OLC/MnO_2 nanohybrid are shown in Fig. 10.7a, c (FESEM) and in Fig. 10.7b, d (TEM) which indicate clear interactions of the precursor material with each other for good capacitive performance. The SAED pattern in the inset (Fig. 10.7d) shows lattice fringes for crystalline MnO_2 and circular lattice shells for OLC confirming the synergistic effect from the nanocomposites.

This section elaborated on some reported literature performance of MnO_2 material and its composites with other supporting electrode materials, while other manganese oxide-based electrodes are further explained in the next section.

In a nutshell, MnO_2 nanomaterials have been investigated as supercapacitors in both three-electrode and two-electrode configurations. Tables 10.1 and 10.2 summarise some of the findings as symmetric and asymmetric supercapacitors, respectively.

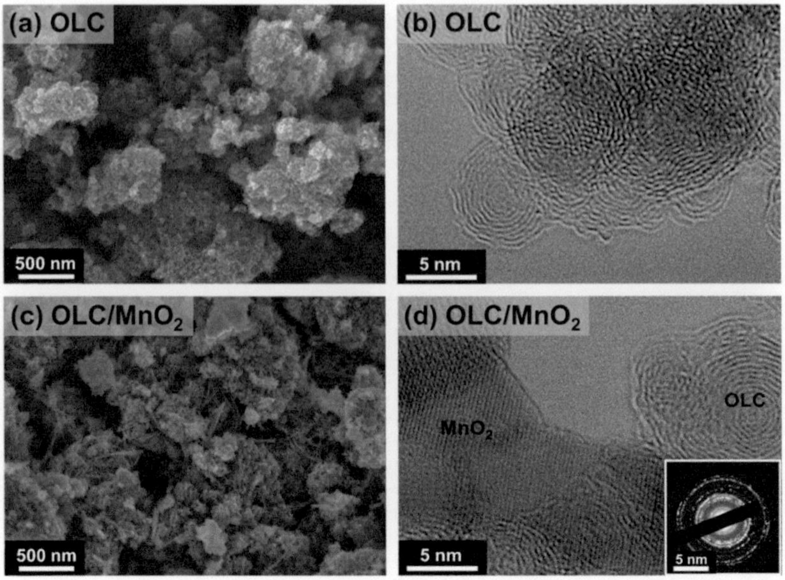

Fig. 10.7 SEM images of (**a**) OLC and (**c**) OLC/MnO_2 nanohybrid, TEM images of (**b**) OLC and (**d**) OLC/MnO_2 nanohybrid. Inset is the corresponding SAED pattern of (**d**)

10.3.2 Mn_xO_y-Based Supercapacitors

The most popular manganese oxide for use in supercapacitor is the MnO_2. However, there are few studies that describe the potential application of other manganese oxides, Mn_xO_y (where $x = 1, 2$ or 3, and $y = 1, 3$ or 4, i.e. MnO, Mn_2O_3 and Mn_3O_4) in supercapacitor. Most of the studies on Mn_xO_y as supercapacitors were based on *three-electrode* configurations.

Manganese (II, III) oxide (Mn_3O_4) as a potential electrode material for ECs showed up just after the researchers discovered MnO_2 as pseudocapacitive electrode material. In 2003, a thin-film Mn_3O_4 electrode material was reported to have displayed specific capacitance of 92 F g^{-1} in aqueous electrolyte and 58 F g^{-1} in organic electrolyte from cyclic voltammetry at the scan rate of 20 mV s^{-1} [145]. In the following years, even though the study of manganese oxide continued, it was still not convincing that this material can find its prime time in the energy storage applications though there was still an ongoing research [146, 147]. In 2005, Wu and Hu [148] studied a mixture, consisting of Mn_3O_4 and MnOOH crystals, and reported specific capacitance ranging between 45 and 71 F g^{-1}. In the same year, Djurfors et al. [149] showed that Mn_3O_4 film prepared by either thermal or electrochemical oxidation of Mn/MnO films has an effect on the capacitive performance of the electrode material. Nagarajan et al. [150] later followed the principle of Djurfors et al. to form a spinel Mn_3O_4 phase at 300 °C with an improved specific capacitance value of 445 F g^{-1} in 0.25 M Na_2SO_4 aqueous solution as electrolyte.

Table 10.1 MnO_2-based nanomaterials in symmetric supercapacitors

Aqueous electrolyte	Electrode material	Cell voltage (V)	Specific capacitance ($F\ g^{-1}$)	Current density/scan rate	Energy ($Wh\ kg^{-1}$)	Power ($kW\ kg^{-1}$)	Ref.
Na_2SO_4	$Ni - MnO_2$	1.0	685	$2\ mA\ cm^{-2}$	–	–	[95]
Na_2SO_4	$Co - MnO_2$	1.0	560	$2\ mA\ cm^{-2}$	–	–	[95]
Na_2SO_4	$a - MnO_2/CNT$	–	140	2 mV/s	–	–	[106]
Na_2SO_4	CNT/PPy/MnO_2		149	$1\ mA\ cm^{-2}$	–	–	[107]
Na_2SO_4	MnO_2	1.0	198	$0.28\ A\ g^{-1}$	–	–	[108]
KOH	MnO_2	1.0	401	$0.28\ A\ g^{-1}$	–	–	[108]
Na_2SO_4	MnO_2/CNT -textile	1.0	410	50 mV/s	20.0	13.0	[109]
H_2SO_4	aMEGO/MnO_2	1.0	256	$0.25A\ g^{-1}$	–	–	[110]
KOH	$GN - (\gamma - MnO_2/CNT)$	1.0	310	20 mV/s	43.0	26.0	[111]
Na_2SO_4	MnO_2/AC	1.2	49	$0.1\ A\ g^{-1}$	9.7	3.0	[112]
Na_2SO_4	GF/MnO_2	1.0	240	$0.1\ A\ g^{-1}$	8.3	20.0	[113]
Na_2SO_4	C/MnO_2 DNTAs	0.8	161	5 mV/s	35.0	16.0	[114]
H_2SO_4	CNOs/MnO_2	1.0	575	$0.5A\ g^{-1}$	19.95[a]	2.25[a]	[115]
KOH	$\alpha - MnO_2$	1.0	775	2 mV/s	76.0	6.5	[116]
Na_2SO_4	MnO_2/PDDA/CNO	0.9	219	–	6.1	–	[17]
Na_2SO_4	OLC/MnO_2	0.8	254	$0.1\ A\ g^{-1}$	5.6	74.8	[105]
Na_2SO_4	$\alpha - MnO_2$/a-MCMB	1.0	357	$0.135\ A\ g^{-1}$	–	–	[117]

Key: *GN* graphene nanosheet, *GF* graphene foam, *CNT* carbon nanotube, *DNTA* double-walled nanotube array, *CNOs* carbon nano-onions, *PDDA* polydiallyldimethylammonium, *aMEGO* activated microwave expanded graphite oxide, *a-MCMB* activated mesocarbon microbeads

[a]Values are based on the published corrections from the authors

Table 10.2 MnO_2-based nanomaterials in asymmetric supercapacitors

Aqueous electrolyte	Positive electrode material	Negative electrode material	Cell voltage (V)	Specific cap. ($F\ g^{-1}$)	Current density/ scan rate	Energy ($Wh\ kg^{-1}$)	Power ($kW\ kg^{-1}$)	Ref.
KCl	MnO_2	AC	2.0	–	–	28.8	8.0	[118]
K_2SO_4	MnO_2	AC	2.2	–	–	19.0	3.6	[119]
KNO_3	MnO_2	AC	2.0	140	–	21.0	123	[96]
K_2SO_4	MnO_2	AC	2.2	31.0	$0.55\ A\ g^{-1}$	17.3	19.0	[33]
KNO_3	MnO_2/CNT	PEDOT	1.8	–	–	13.5	120.1	[120]
K_2SO_4	MnO_2	AC	2.0	21	$1.5\ A\ g^{-1}$	10.0	16.0	[121]
K_2SO_4	MnO_2NR	AC	1.8	53.7	–	17.0	2.0	[122]
Nitrates of Mg, Ba, Ca	MnO_2	AC	2.0	29-37	$2\ mV\ s^{-1}$	21.0	0.3	[123]
Na_2SO_4	$NaMnO_2$	AC	1.9	38.9	–	19.5	1.0	[124]
K_2SO_4	$K_{0.27}MnO_2 \cdot 0.6H_2O$	AC	1.8	57.7	–	26.0	2.0	[125]
Na_2SO_4	MnO_2NW/Gr	Gr	2.0	31.0	$0.5\ A\ g^{-1}$	30.4	5.0	[18]
Na_2SO_4	MnO_2NW/SWNT	In_2O_3 NW/SWNT	2.0	184	$0.001 A\ cm^{-2}$	25.5	50.3	[34]
Na_2SO_4	MnO_2	AC	2.0	100	$0.2\ A\ g^{-1}$	13.9	–	[126]
Na_2SO_4	MnO_2/Gr	AC	1.8	113.5	$1\ mV\ s^{-1}$	51.1	198	[127]
Na_2SO_4	MnO_2/CNT	AC	2.0	40.2	50 mV/s	23.2	3.0	[128]
Na_2SO_4	MnO_2	GHG	2.0	41.7	$1.0\ A\ g^{-1}$	23.2	10.0	[129]
Na_2SO_4	$GHCS - MnO_2$	GHCS	2.0	–	–	22.1	7.0	[130]
Na_2SO_4	MnO_2	CNT	2.0	85.8	$0.2\ A\ g^{-1}$	47.4	2.0	[131]
Na_2SO_4	$MnO_2/e - CMG$	e-CMG	2.0	–	–	44.0	25.0	[132]
Na_2SO_4	aMEGO /MnO_2	aMEGO	2.0	175	$0.25\ A\ g^{-1}$	24.3	32.3	[110]
Li_2SO_4	$\alpha - MnO_2/GO$	GO	1.8	280	$0.5\ A\ g^{-1}$	35.0	7.2	[15]
Li_2SO_4	$\alpha - MnO_2/MWCNT$	MWCNT	1.6	170	$0.5\ A\ g^{-1}$	18.0	4.0	[133]
Na_2SO_4	MnO_2/CNF	AC	2.0	52.1	$5\ mV\ s^{-1}$	30.6	20.8	[134]

Na_2SO_4/PVP	CNT/Gr/MnO_2	CNT/PANI	1.6	–	–	24.8	2.3	[135]
Na_2SO_4	rGO/MnO_2	rGO/MoO_3	2.0	307.0	0.2 A g^{-1}	42.6	> 1	[136]
Na_2SO_4	Gr/MnO_2NR	Gr/Ag	1.8	112.8	1 mV s^{-1}	50.8	90.3	[137]
Na_2SO_4	MnO_2	Gr	2.0	37	5 mA cm^{-2}	25.2	2.5	[138]
Na_2SO_4	MnO_2/3DOM – C	3DOM-C	2.0	159	0.1 A g^{-1}	30.2	14.5	[139]
KCl-gel	Gr–MnO2	Gr	2.0	212	0.5 A g^{-1}	29.4	25.9	[140]
Na_2SO_4	MnO_2/rGO	rGO	1.6	59.6	1 A g^{-1}	21.2	6.61	[141]
Na_2SO_4	MnO_2/GO	HPC	2.0	84	0.1 A g^{-1}	46.7	2.0	[142]
Na_2SO_4	MnO_2/TiO_2	AG	1.8	120	0.1 A g^{-1}	29.8	> 5	[143]
KOH	$ZnCo_2O_4$@MnO_2	3D PFO	1.3	161	2.5 mA cm^{-2}	37.8	> 1	[144]

Key: *GHCS* graphitic hollow carbon spheres, *Gr* graphene, *AG* activated graphene, *NW* nanowires, *3D PFO* 3D porous α-Fe_2O_3, *NR* nanorods, *GHG* graphene hydrogel, *CNF* carbon nanofiber, *HPC* hierarchical porous structure, *aMEGO* activated microwave expanded graphite oxide, *a-MCMB* activated mesocarbon microbeads, *3DOM-C* three dimensional ordered macroporous carbon

As the race in finding a stable and suitable electrode derived from the Mn_3O_4 material continued, researchers began to develop composites of this material from carbon templates. In 2008, Mn_3O_4/MWCNT composite was examined using cyclic voltammetry and obtained a maximum specific capacitance of 330 F g^{-1} [151]. In 2010, Wang et al. [152] used graphene sheet as a template for decoration of Mn_3O_4 by mixing graphene suspension in ethylene glycol with MnO_2 organosol, followed by subsequent ultrasonication processing and heat treatment. The as-prepared Mn_3O_4/graphene nanocomposite gave a specific capacitance of 175 F g^{-1} in 1 M Na_2SO_4 electrolyte and 256 F g^{-1} in 6 M KOH electrolyte. The enhanced capacitive performance of Mn_3O_4/graphene nanocomposites was attributed to the nanoparticulate nature of Mn_3O_4 nanoparticles and functional groups attached to graphene nanosheets. Two other simple techniques such as chemical bath deposition (CBD) and successive ionic layer adsorption and reaction (SILAR) were employed by Dubal et al. to fabricate Mn_3O_4 nanoparticles on a stainless steel. The capacitive performance of Mn_3O_4 thin film synthesised from CBD and SILAR independently exhibited maximum capacitance of 284 and 314 F g^{-1}, respectively, in 1 M Na_2SO_4 electrolyte [70, 77]. Uniform and single-crystalline Mn_3O_4 nano-octahedrons were successfully synthesised by a simple ethylenediaminetetraacetic acid disodium salt (EDTA-2Na)-assisted hydrothermal route (Fig. 10.8). The Mn_3O_4 nano-octahedrons exhibited an enhanced specific capacitance of 322 F g^{-1}, making them a promising electrode material for supercapacitors [153].

In 2013, Dubal and Holze [154] reported a novel kind of all-solid-state flexible electrochemical capacitor configuration consisting of two slightly separated Mn_3O_4 thin films as electrodes and H_2SO_4-PVA gel as solid-state electrolyte. The device showed good electrochemical performances, such as high specific capacitance of about 127 F g^{-1} and well enough energy and power density values of more than 10 Wh Kg^{-1} and 5 kW Kg^{-1}, respectively, with 89 % of capacity retention after 2000 cycles. Recently, in 2015, Qiao et al. [155] synthesised micro-/nanostructured Mn_3O_4 with an open 3D flower-like morphology by a facile solvothermal approach using hexadecyltrimethylammonium bromide as a surfactant and ethanol as a solvent. The Mn_3O_4 microspheres used exhibited electrochemical performance with a specific capacitance of 286 F g^{-1} at a low current density (0.5 A g^{-1}) and still retained 80 % (230 F g^{-1}) and 73 % (210 F g^{-1}) at higher current densities of 5 A g^{-1} and 10 A g^{-1}, respectively. This section elaborated on some reported literature performance of Mn_3O_4 material and its composites with other supporting electrode materials.

The use of Mn_2O_3 and MnO as supercapacitors is still not well reported. The study of Mn_2O_3 as electrode material for supercapacitors surfaced at around 2006, when Chiang et al. [90] reported on the synthesis and the examination of the nanoparticles of Mn_2O_3 as a potential supercapacitor electrode material. The electrode material exhibited a maximum specific capacitance of about 197 F g^{-1} from cyclic voltammetry at the scan rate of 10 mV s^{-1}. In 2008, Yu et al. [156] were able to synthesise Mn_2O_3 particles that were made of micropores. The supercapacitive performance of these microporous particles gave high specific

Fig. 10.8 (**a–c**) different magnification SEM image of the Mn_3O_4 nano-octahedrons, (**d**) the corresponding XRD pattern (Reprinted with permission from Ref. [153], copyright 2010, The Royal Society of Chemistry)

capacitance (i.e. 350 F g^{-1}). The improved performance was attributed to the relatively high specific surface area of 283 m^2g^{-1}. Zhang et al. [157] reported nanosized Mn_2O_3 homogenously incorporated on a mesoporous carbon template. The composite showed an improved electrochemical performance with a maximum specific capacitance of 600 F g^{-1}, a value that is almost twice the one obtained by Yu et al. [156] and also close to the values obtained when using RuO_2 as an electrode material in a *half-cell* three-electrode system. In 2011, Wang and co-workers [158] reported a composite of Mn_2O_3 and carbon aerogel microbead (CAMB), obtained by using an in situ encapsulation technique, that displayed maximum specific capacitance of 368 F g^{-1}, emphasising the fact that indeed this type of manganese oxide material is well suited for supercapacitor applications. Recently, Li et al. [89] developed porous Mn_2O_3 nanocubics through hydrothermal technique that was followed by calcination in air (Fig. 10.9). The as-synthesised Mn_2O_3 nanocubics exhibited a specific capacitance of 191 F g^{-1} at a current density of 0.1 A g^{-1}. This electrode material showed relatively high rate capability at high current density of 5.0 A g^{-1} and excellent long-term cycle stability even after 3000 cycles. The spectroscopic (XRD, Fig. 10.9a) and the morphological profiles (SEM, Fig. 10.9b–d) are shown and compared with the morphological profile (see Fig. 10.9b) of $MnCO_3$ precursor. The enhanced rate capability of this material was associated with the improved porosity as depicted by the SEM images (Fig. 10.9c, d).

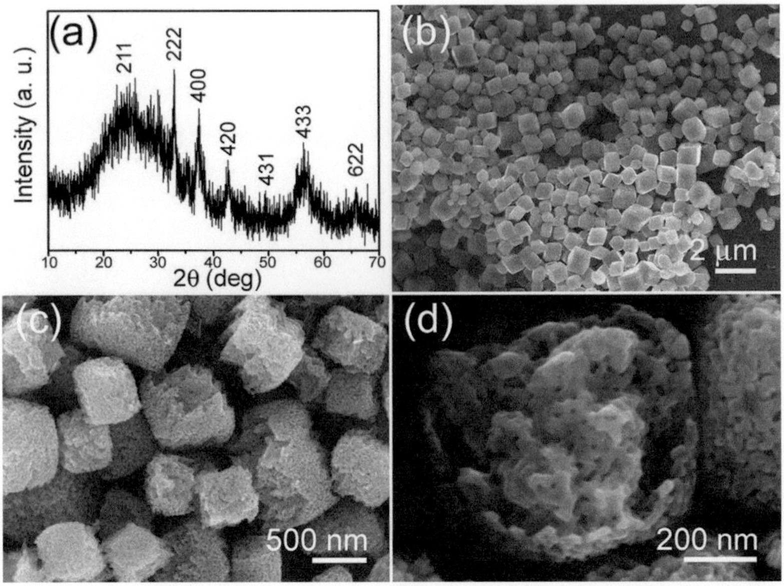

Fig. 10.9 Illustration of (**a**) XRD pattern of the porous Mn_2O_3 nanocubics, (**b**) low magnification SEM image of the $MnCO_3$ precursor, (**c**) SEM image of the porous Mn_2O_3 nanocubics, (**d**) enlarged SEM image of a broken porous Mn_2O_3 nanocubic (Reprinted with permission from Ref. [89], copyright 2015, Elsevier Ltd)

In conclusion, Mn_2O_3-based electrode materials are potentially useful as supercapacitor electrode materials, albeit its growth or attention has been dramatically slow compared to other manganese oxide-based electrodes such as MnO_2 and Mn_3O_4. To the best of our knowledge, there is no literature on the use of manganese mono-oxide (MnO) as an electrode material for supercapacitor application.

10.4 Summary and Future Prospects

The increasing worldwide demands for clean and sustainable energy have made electrochemical capacitors one of the major emerging devices for energy storage and power supply. This follows the fact that fossil fuels have shown potential threat to the environment and thereby deemed unreliable. The advantages of EC systems account for their high power density, high efficiency and long life expectancy. However, the development of these energy storage devices depends upon the innovative ways of fine-tuning their electrode materials. Manganese oxide nanostructured materials with their interesting properties such as low cost, good electrochemical response, various oxidation states and availability present themselves as attractive electrode materials for energy storage. Several manganese oxide nanostructured materials have been shown to be easily synthesised from various synthetic routes even though their synthesis and application are still restricted to the lab-scale production. Nevertheless, an improved energy deliverance and electrochemical performance have been achieved by designing nanostructured crystalline manganese oxide. Note that the hindrance of nanostructured manganese oxide towards the commercial applications is due to its low conductivity as well as agglomeration upon rapid cycling which limits their electrochemical performance to match their theoretical estimation of 1370 F g^{-1}. Recently, much work has been devoted in mitigating the above limitation by incorporation of highly conductive materials such as carbon allotropes to make composite materials with an improved electrochemical performance and stability upon long-time cycling. This approach has drastically improved the rapid use of nanostructured manganese oxide as EC electrode materials due to the added advantages of obtaining more remarkable rate capability and much better cycle performance, thus creating an opportunity for these materials to be upscaled to the commercial applications in industries.

From the literature, a lot of work has been devoted to the MnO_2 nanostructured materials despite the fact that some of the Mn_xO_y-based electrode materials exhibit great potential to be employed as EC materials. There exists little literature reports on the electrochemical performance of Mn_xO_y-based electrode, while the already reported works on these materials are based mainly on the half-cell three-electrode system and not on the full-cell two-electrode systems (i.e. symmetric and/or asymmetric). Therefore, there is still a gap to be filled in terms of interrogating fully the electrochemical capacitive behaviour of other Mn_xO_y-based electrode materials. It is worth mentioning that, for the progression of the manganese oxide in showing high capacitive behaviour with good electrochemical response, the use of aqueous electrolytes should be the choice of interests since the high viscosity of other electrolytes will work against Mn_xO_y-based electrode material properties. Since the main focus of any fundamental research is to reach a point where the lab-scale synthesis or design should be elevated to large-scale industrialization applications, innovative skills such as further improvement of the existing Mn_xO_y-based electrodes and synthesis of novel composites that will result in stable fabrication desirable for bringing a long-lasting solution to meet the market demand of EC technology are required. The developments of one-step synthesis without any additional activation process to yield Mn_xO_y-based electrode or composite materials composing of properties that are relatively close to their theoretical values would be beneficial for the compact design of high power energy sources. This phenomenon will require a broad understanding of the effect played by the electrolytes on the structural composition of these transition metal oxides as well as the nature of their electrochemical behaviour resulting from the original synthesis EC materials. In order to fully succeed on the generation of the reliable, efficient and effective technology that can mitigate the global energy demand, more conductive materials obtained from different sources and various synthetic routes should be explored and be incorporated on the transition metal oxides such as manganese oxides. Another major drawback in the development of supercapacitor technology is the relatively high cost associated with it when compared to other energy devices. Thus, future research should be invested towards the development of fascinating transition metal oxide nanomaterials exhibiting a high charge capacity while demonstrating a minimum electronic and electrolytic resistance in a very cost-effective manner.

References

1. Wang G, Zhang L, Zhang J (2012) A review of electrode materials for electrochemical supercapacitors. Chem Soc Rev 41:797–828. doi:10.1039/c1cs15060j
2. Simon P, Gogotsi Y (2008) Materials for electrochemical capacitors. Nat Mater 7:845–854. doi:10.1038/nmat2297
3. Miller JR (2012) Applied physics. Valuing reversible energy storage. Science 335:1312–1313. doi:10.1126/science.1219134
4. Pech D, Brunet M, Durou H, Huang P, Mochalin V, Gogotsi Y et al (2010) Ultrahigh-power micrometre-sized supercapacitors based on onion-like carbon. Nat Nanotechnol 5:651–654. doi:10.1038/nnano.2010.162
5. Augustyn V, Simon P, Dunn B (2014) Pseudocapacitive oxide materials for high-rate electrochemical energy storage. Energy Environ Sci 7(5):1597–1614. doi:10.1039/c3ee44164d
6. Ike IS, Sigalas I, Iyuke S, Ozoemena KI (2015) An overview of mathematical modeling of electrochemical supercapacitors/ultracapacitors. J Power Sources 273:264–277. doi:10.1016/j.jpowsour.2014.09.071
7. Faraji S, Ani FN (2014) Microwave-assisted synthesis of metal oxide/hydroxide composite electrodes for high power supercapacitors – a review. J Power Sources 263:338–360. doi:10.1016/j.jpowsour.2014.03.144
8. Jiang H, Lee PS, Li C (2013) 3D carbon based nanostructures for advanced supercapacitors. Energy Environ Sci 6:41. doi:10.1039/c2ee23284g
9. Béguin F, Presser V, Balducci A, Frackowiak E (2014) Carbons and electrolytes for advanced supercapacitors. Adv Mater 26(14):2219–51, 2283. doi:10.1002/adma.201304137
10. Burda C, Chen X, Narayanan R, El-sayed MA (2005) Chemistry and properties of nanocrystals of different shapes. Chem Rev 105:1025–1102
11. Tan Y, Meng L, Peng Q, Li Y (2011) One-dimensional single-crystalline Mn_3O_4 nanostructures with tunable length and magnetic properties of Mn_3O_4 nanowires. Chem Commun (Camb) 47:1172–1174. doi:10.1039/c0cc00978d
12. Yan D, Cheng S, Zhuo RF, Chen JT, Feng JJ, Feng HT et al (2009) Nanoparticles and 3D sponge-like porous networks of manganese oxides and their microwave absorption properties. Nanotechnology 20:105706. doi:10.1088/0957-4484/20/10/105706
13. Wang K, Wu H, Meng Y, Wei Z (2014) Conducting polymer nanowire arrays for high performance supercapacitors. Small 10:14–31. doi:10.1002/smll.201301991
14. Chen S, Zhu J, Wu X, Han Q, Wang X (2010) Graphene oxide/MnO_2 nanocomposites for supercapacitors. ACS Nano 4:2822–2830
15. Jafta CJ, Nkosi F, le Roux L, Mathe MK, Kebede M, Makgopa K et al (2013) Manganese oxide/graphene oxide composites for high-energy aqueous asymmetric electrochemical capacitors. Electrochim Acta 110:2–7. doi:10.1016/j.electacta.2013.06.096
16. Chidembo AT, Aboutalebi SH, Konstantinov K, Jafta CJ, Liu HK, Ozoemena KI (2014) In situ engineering of urchin-like reduced graphene oxide–Mn_2O_3–Mn_3O_4 nanostructures for supercapacitors. RSC Adv 4:886. doi:10.1039/c3ra44973d

17. Borgohain R, Selegue JP, Cheng Y-T (2014) Ternary composites of delaminated-MnO_2/PDDA/functionalized-CNOs for high-capacity supercapacitor electrodes. J Mater Chem A 2:20367–20373. doi:10.1039/C4TA04439H
18. Wu Z-S, Ren W, Wang D-W, Li F, Liu B, Cheng H-M (2010) High-energy MnO_2 nanowire/graphene and graphene asymmetric electrochemical capacitors. ACS Nano 4:5835–5842. doi:10.1021/nn101754k
19. Subramanian V, Zhu H, Vajtai R, Ajayan PM, Wei B (2005) Hydrothermal synthesis and pseudocapacitance properties of MnO_2 nanostructures. J Phys Chem B 109:20207–20214
20. Yu Z, Duong B, Abbitt D, Thomas J (2013) Highly ordered MnO_2 nanopillars for enhanced supercapacitor performance. Adv Mater 25:3302–3306. doi:10.1002/adma.201300572
21. Wang Y, Yu SF, Sun CY, Zhu TJ, Yang HY (2012) MnO_2/onion-like carbon nanocomposites for pseudocapacitors. J Mater Chem 22:17584. doi:10.1039/c2jm33558a
22. Zheng JP (1995) Hydrous ruthenium oxide as an electrode material for electrochemical capacitors. J Electrochem Soc 142:2699. doi:10.1149/1.2050077
23. Lee HY, Lee HY, Goodenough JB, Goodenough JB (1999) Brief communication supercapacitor behavior with KCl electrolyte. J Solid State Chem 223:220–223. doi:10.1006/jssc.1998.8128
24. Yang P, Ding Y, Lin Z, Chen Z, Li Y, Qiang P et al (2014) Low-cost high-performance solid-state asymmetric supercapacitors based on MnO_2 nanowires and Fe_2O_3 nanotubes. Nano Lett 14:731–736. doi:10.1021/nl404008e
25. Toupin M, Brousse T, Be D (2004) Charge storage mechanism of MnO_2 electrode used in aqueous electrochemical capacitor. Chem Mater 16:3184–3190
26. Potter R, Rossman G (1979) The tetravalent manganese oxides: identification, hydration, and structural relationships by infrared spectroscopy. Am Mineral 64:1199–1218
27. Fan D, Yang P (1999) Introduction to and classification of manganese deposits of China. Ore Geol Rev 15:1–13
28. Wei W, Cui X, Chen W, Ivey DG (2011) Manganese oxide-based materials as electrochemical supercapacitor electrodes. Chem Soc Rev 40:1697–1721. doi:10.1039/c0cs00127a
29. Brousse T, Toupin M, Dugas R, Athouël L, Crosnier O, Bélanger D (2006) Crystalline MnO_2 as possible alternatives to amorphous compounds in electrochemical supercapacitors. J Electrochem Soc 153:A2171. doi:10.1149/1.2352197
30. Stoller MD, Ruoff RS (2010) Review of best practice methods for determining an electrode material's performance for ultracapacitors. Energy Environ Sci 3:1294–1301. doi:10.1039/c0ee00074d
31. Qu D, Qu D, Shi H, Shi H (1998) Studies of activated carbons used in double-layer capacitors. J Power Sources 74:99–107
32. Zhi M, Xiang C, Li J, Li M, Wu N (2012) Nanostructured carbon-metal oxide composite electrodes for supercapacitors: review. Nanoscale 5(1):72–88. doi:10.1039/c2nr32040a
33. Cottineau T, Toupin M, Delahaye T, Brousse T, Bélanger D (2006) Nanostructured transition metal oxides for aqueous hybrid electrochemical supercapacitors. Appl Phys A Mater Sci Process 82:599–606. doi:10.1007/s00339-005-3401-3
34. Chen P, Shen G, Shi Y, Chen H, Zhou C (2010) Preparation and characterization of flexible asymmetric supercapacitors based on transition-metal-oxide nanowire/single-walled carbon nanotube hybrid thin-film electrodes. ACS Nano 4:4403–4411
35. Ruiz V, Blanco C, Raymundo-Piñero E, Khomenko V, Béguin F, Santamaría R (2007) Effects of thermal treatment of activated carbon on the electrochemical behaviour in supercapacitors. Electrochim Acta 52:4969–4973. doi:10.1016/j.electacta.2007.01.071
36. Béguin F, Frackowiak E (2013) Supercapacitors: materials, systems and applications. Wiley, Weinheim
37. Fic K, Lota G, Meller M, Frackowiak E (2012) Novel insight into neutral medium as electrolyte for high-voltage supercapacitors. Energy Environ Sci 5:5842. doi:10.1039/c1ee02262h

38. Gao Q, Demarconnay L, Raymundo-Piñero E, Béguin F (2012) Exploring the large voltage range of carbon/carbon supercapacitors in aqueous lithium sulfate electrolyte. Energy Environ Sci 5:9611. doi:10.1039/c2ee22284a
39. Staiti P, Arenillas A, Lufrano F, Menéndez JA (2012) High energy ultracapacitor based on carbon xerogel electrodes and sodium sulfate electrolyte. J Power Sources 214:137–141. doi:10.1016/j.jpowsour.2012.04.056
40. Zhao L, Qiu Y, Yu J, Deng X, Dai C, Bai X (2013) Carbon nanofibers with radially grown graphene sheets derived from electrospinning for aqueous supercapacitors with high working voltage and energy density. Nanoscale 5:4902–4909. doi:10.1039/c3nr33927k
41. Fic K, Frackowiak E, Béguin F (2012) Unusual energy enhancement in carbon-based electrochemical capacitors. J Mater Chem 22:24213–24223. doi:10.1039/c2jm35711a
42. Lekitima JN, Ozoemena KI, Jafta CJ, Kobayashi N, Song Y, Tong D et al (2013) High-performance aqueous asymmetric electrochemical capacitors based on graphene oxide/cobalt (ii)-tetrapyrazinoporphyrazine hybrids. J Mater Chem A 1:2821. doi:10.1039/c2ta01325h
43. Arulepp M, Permann L, Leis J, Perkson A, Rumma K, Jänes A et al (2004) Influence of the solvent properties on the characteristics of a double layer capacitor. J Power Sources 133:320–328. doi:10.1016/j.jpowsour.2004.03.026
44. Liu P, Verbrugge M, Soukiazian S (2006) Influence of temperature and electrolyte on the performance of activated-carbon supercapacitors. J Power Sources 156:712–718. doi:10.1016/j.jpowsour.2005.05.055
45. Wang RQ (2011) Studies and characterisations of activated carbons used for carbon/carbon supercapacitors. J Power Sources 5:866–868
46. Jiang D, Wu J (2014) Unusual effects of solvent polarity on capacitance for organic electrolytes in a nanoporous electrode. Nanoscale 6:5545–5550. doi:10.1039/c4nr00046c
47. Kurzweil P, Chwistek M (2008) Electrochemical stability of organic electrolytes in supercapacitors: spectroscopy and gas analysis of decomposition products. J Power Sources 176:555–567. doi:10.1016/j.jpowsour.2007.08.070
48. Wilkes JS (2002) A short history of ionic liquids—from molten salts to neoteric solvents. Green Chem 4:73–80. doi:10.1039/b110838g
49. Balducci A, Dugas R, Taberna PL, Simon P, Plée D, Mastragostino M et al (2007) High temperature carbon-carbon supercapacitor using ionic liquid as electrolyte. J Power Sources 165:922–927. doi:10.1016/j.jpowsour.2006.12.048
50. Oxford Dictionaries (2010) Alternative energy
51. Balducci A, Bardi U, Caporali S, Mastragostino M, Soavi F (2004) Ionic liquids for hybrid supercapacitors. Electrochem Commun 6:566–570. doi:10.1016/j.elecom.2004.04.005
52. Arbizzani C, Biso M, Cericola D, Lazzari M, Soavi F, Mastragostino M (2008) Safe, high-energy supercapacitors based on solvent-free ionic liquid electrolytes. J Power Sources 185:1575–1579. doi:10.1016/j.jpowsour.2008.09.016
53. Weingarth D, Noh H, Foelske-Schmitz A, Wokaun A, Kötz R (2013) A reliable determination method of stability limits for electrochemical double layer capacitors. Electrochim Acta 103:119–124. doi:10.1016/j.electacta.2013.04.057
54. Liu X, Chen C, Zhao Y, Jia B (2013) A review on the synthesis of manganese oxide nanomaterials and their applications on lithium-ion batteries. J Nanomater 2013:1–7
55. Xiao W, Xia H, Fuh JYH, Lu L (2009) Growth of single-crystal ??-MnO_2 nanotubes prepared by a hydrothermal route and their electrochemical properties. J Power Sources 193:935–938. doi:10.1016/j.jpowsour.2009.03.073
56. Zhang Y, Sun C, Lu P, Li K, Song S, Xue D (2012) Crystallization design of MnO_2 towards better supercapacitance. CrystEngComm 14:5892. doi:10.1039/c2ce25610j
57. Wang HY, Xiao FX, Yu L, Liu B, Lou XW (2014) Hierarchical α-MnO_2 nanowires@Ni1-xMn xOy nanoflakes core-shell nanostructures for supercapacitors. Small 10:3181–3186. doi:10.1002/smll.201303836
58. Yin B, Zhang S, Yang J, Yang L, Fengyu Q, Xiang W (2014) Facile synthesis of ultralong MnO_2 nanowires as high performance supercapacitor electrodes and photocatalysts with

enhanced photocatalytic activities. CrystEngComm 16:9999–10005. doi:10.1039/C4CE01302F
59. Wu J, Huang H, Yu L, Hu J (2013) Controllable hydrothermal synthesis of MnO_2 nanostructures. Adv Mater Phys Chem 3:201–205
60. Xu M, Kong L, Zhou W, Li H (2007) Hydrothermal synthesis and pseudocapacitance properties of Î}\^{I}{ ± −MnO_2 hollow spheres and hollow urchins. J Phys Chem C 111:19141–19147
61. Hatzell KB, Fan L, Beidaghi M, Boota M, Pomerantseva E, Kumbur EC et al (2014) Composite manganese oxide percolating networks as a suspension electrode for an asymmetric flow capacitor. ACS Appl Mater Interfaces 6:8886–8893. doi:10.1021/am501650q
62. Nam HS, Kwon JS, Kim KM, Ko JM, Kim JD (2010) Supercapacitive properties of a nanowire-structured MnO_2 electrode in the gel electrolyte containing silica. Electrochim Acta 55:7443–7446. doi:10.1016/j.electacta.2010.02.027
63. Ragupathy P, Park DH, Campet G, Vasan HN, Hwang SJ, Choy JH et al (2009) Remarkable capacity retention of nanostructured manganese oxide upon cycling as an electrode material for supercapacitor. J Phys Chem C 113:6303–6309. doi:10.1021/jp811407q
64. Ni J, Lu W, Zhang L, Yue B, Shang X, Lv Y (2009) Low-temperature synthesis of monodisperse 3D manganese oxide nanoflowers and their pseudocapacitance properties. J Phys Chem 113:54–60. doi:10.1021/jp806454r
65. Wang X, Wang X, Huang W, Sebastian PJ, Gamboa S (2005) Sol–gel template synthesis of highly ordered MnO_2 nanowire arrays. J Power Sources 140:211–215. doi:10.1016/j.jpowsour.2004.07.033
66. Wang X, Yuan A, Wang Y (2007) Supercapacitive behaviors and their temperature dependence of sol–gel synthesized nanostructured manganese dioxide in lithium hydroxide electrolyte. J Power Sources 172:1007–1011. doi:10.1016/j.jpowsour.2007.07.066
67. Bello A, Fashedemi OO, Fabiane M, Lekitima JN, Ozoemena KI, Manyala N (2013) Microwave assisted synthesis of MnO_2 on nickel foam-graphene for electrochemical capacitor. Electrochim Acta 114:48–53. doi:10.1016/j.electacta.2013.09.134
68. Ming B, Li J, Kang F, Pang G, Zhang Y, Chen L et al (2012) Microwave–hydrothermal synthesis of birnessite-type MnO_2 nanospheres as supercapacitor electrode materials. J Power Sources 198:428–431. doi:10.1016/j.jpowsour.2011.10.003
69. Meher SK, Rao GR (2012) Enhanced activity of microwave synthesized hierarchical MnO_2 for high performance supercapacitor applications. J Power Sources 215:317–328. doi:10.1016/j.jpowsour.2012.04.104
70. Dubal DP, Dhawale DS, Salunkhe RR, Pawar SM, Lokhande CD (2010) A novel chemical synthesis and characterization of Mn_3O_4 thin films for supercapacitor application. Appl Surf Sci 256:4411–4416. doi:10.1016/j.apsusc.2009.12.057
71. Dubal DP, Holze R (2012) Successive ionic layer adsorption and reaction (SILAR) method to induce Mn_3O_4 nanospots on CNTs for supercapacitor. New J Chem 37(2):403–408. doi:10.1039/c2nj40862g
72. Dubal DP, Dhawale DS, Salunkhe RR, Pawar SM, Fulari VJ, Lokhande CD (2009) A novel chemical synthesis of interlocked cubes of hausmannite Mn_3O_4 thin films for supercapacitor application. J Alloys Compd 484:218–221. doi:10.1016/j.jallcom.2009.03.135
73. Yang F, Zhao M, Sun Q, Qiao Y (2015) A novel hydrothermal synthesis and characterisation of porous Mn_3O_4 for supercapacitors with high rate capability. RSC Adv 5:9843–9847. doi:10.1039/C4RA10175H
74. Lee JW, Hall AS, Kim JD, Mallouk TE (2012) A facile and template-free hydrothermal synthesis of Mn_3O_4 nanorods on graphene sheets for supercapacitor electrodes with long cycle stability. Chem Mater 24:1158–1164
75. Gao W, Ye S, Shao M (2011) Solution-combusting preparation of mono-dispersed Mn_3O_4 nanoparticles for electrochemical applications. J Phys Chem Solids 72:1027–1031. doi:10.1016/j.jpcs.2011.05.015

76. Xu HY, Le Xu S, Li XD, Wang H, Yan H (2006) Chemical bath deposition of hausmannite Mn_3O_4 thin films. Appl Surf Sci 252:4091–4096. doi:10.1016/j.apsusc.2005.06.011
77. Dubal DP, Dhawale DS, Salunkhe RR, Fulari VJ, Lokhande CD (2010) Chemical synthesis and characterization of Mn_3O_4 thin films for supercapacitor application. J Alloys Compd 497:166–170. doi:10.1016/j.jallcom.2010.02.182
78. Baykal A, Kavas H, Durmuş Z, Demir M, Kazan S, Topkaya R et al (2010) Sonochemical synthesis and chracterization of Mn_3O_4 nanoparticles. Cent Eur J Chem 8:633–638. doi:10.2478/s11532-010-0037-8
79. Sankar KV, Senthilkumar ST, Berchmans LJ, Sanjeeviraja C, Selvan RK (2012) Effect of reaction time on the synthesis and electrochemical properties of Mn_3O_4 nanoparticles by microwave assisted reflux method. Appl Surf Sci 259:624–630. doi:10.1016/j.apsusc.2012.07.087
80. Liu C-L, Chang K-H, Hu C-C, Wen W-C (2012) Microwave-assisted hydrothermal synthesis of Mn_3O_4/reduced graphene oxide composites for high power supercapacitors. J Power Sources 217:184–192. doi:10.1016/j.jpowsour.2012.05.109
81. Zhou T, Mo S, Zhou S, Zou W, Liu Y, Yuan D (2011) Mn_3O_4/worm-like mesoporous carbon synthesized via a microwave method for supercapacitors. J Mater Sci 46:3337–3342. doi:10.1007/s10853-010-5221-x
82. Apte SK, Naik SD, Sonawane RS, Kale BB, Pavaskar N, Mandale AB et al (2006) Nanosize Mn_3O_4 (Hausmannite) by microwave irradiation method. Mater Res Bull 41:647–654. doi:10.1016/j.materresbull.2005.08.028
83. Dong R, Ye Q, Kuang L, Lu X, Zhang Y, Zhang X et al (2013) Enhanced supercapacitor performance of Mn_3O_4 nanocrystals by doping transition-metal ions. ACS Appl Mater Interfaces 5:9508–9516. doi:10.1021/am402257y
84. Wang D, Li Y, Wang Q, Wang T (2012) Facile synthesis of porous Mn_3O_4 nanocrystal-graphene nanocomposites for electrochemical supercapacitors. Eur J Inorg Chem 2012:628–635. doi:10.1002/ejic.201100983
85. Wells AF (1984) Structural inorganic chemistry, 5th edn. Oxford Science Publications, New York
86. Greenwood N (1997) Chemistry of the elements, 2nd edn. Butterworth–Heinemann, Oxford
87. Nathan T, Cloke M, Prabaharan SRS (2008) Electrode properties of Mn_2O_3 nanospheres synthesized by combined sonochemical/solvothermal method for use in electrochemical capacitors. J Nanomater 2008:1–8. doi:10.1155/2008/948183
88. Chen X, Li X, Jiang Y, Shi C, Li X (2005) Rational synthesis of MnO_2 and Mn_2O_3 nanowires with the electrochemical characterization of MnO_2 nanowires for supercapacitor. Solid State Commun 136:94–96. doi:10.1016/j.ssc.2005.06.033
89. Li W, Shao J, Liu Q, Liu X, Zhou X, Hu J (2015) Facile synthesis of porous Mn_2O_3 nanocubics for high-rate supercapacitors. Electrochim Acta 157:108–114. doi:10.1016/j.electacta.2015.01.056
90. Chiang NK, Clokec M, Chena GZ, Engineering E, Capacitor E, Oxide M et al (2006) Nano-sized Mn_2O_3 preparedby a novel solvolysis route as an electrochemical capacitor. Inst Eng Malaysia 69:31–36
91. Chen Z, Zhang S, TAn S, Li F, Wang J, Jin S et al (1997) Preparation and electron spin resonance effect of nanometer-sized Mn_2O_3. J Cryst Growth 180:280–283. doi:10.1016/S0022-0248(97)00215-7
92. He W, Zhang Y, Zhang X, Wang H, Yan H (2003) Low temperature preparation of nanocrystalline Mn_2O_3 via ethanol-thermal reduction of MnO_2. J Cryst Growth 252:285–288. doi:10.1016/S0022-0248(03)00937-0
93. Hu C, Tsou T (2002) Ideal capacitive behavior of hydrous manganese oxide prepared by anodic deposition. Electrochem Commun 4:105–109
94. Toupin M, Toupin M, Brousse T, Brousse T, Bélanger D, Bélanger D (2002) Influence of microstucture on the charge storage properties of chemically synthesized manganese dioxide. Chem Mater 14:3946–3952. doi:10.1021/cm020408q

95. Rajendra Prasad K, Miura N (2004) Electrochemically synthesized MnO_2-based mixed oxides for high performance redox supercapacitors. Electrochem Commun 6:1004–1008. doi:10.1016/j.elecom.2004.07.017
96. Khomenko V, Raymundo-Piñero E, Béguin F (2006) Optimisation of an asymmetric manganese oxide/activated carbon capacitor working at 2 v in aqueous medium. J Power Sources 153:183–190. doi:10.1016/j.jpowsour.2005.03.210
97. Sharma RK, Oh H-S, Shul Y-G, Kim H (2007) Carbon-supported, nano-structured, manganese oxide composite electrode for electrochemical supercapacitor. J Power Sources 173:1024–1028. doi:10.1016/j.jpowsour.2007.08.076
98. Subramanian V, Zhu H, Wei B (2006) Synthesis and electrochemical characterizations of amorphous manganese oxide and single walled carbon nanotube composites as supercapacitor electrode materials. Electrochem Commun 8:827–832. doi:10.1016/j.elecom.2006.02.027
99. Ma S-B, Nam K-W, Yoon W-S, Yang X-Q, Ahn K-Y, Oh K-H et al (2008) Electrochemical properties of manganese oxide coated onto carbon nanotubes for energy-storage applications. J Power Sources 178:483–489. doi:10.1016/j.jpowsour.2007.12.027
100. Xia H, Wang Y, Lin J, Lu L (2012) Hydrothermal synthesis of MnO_2/CNT nanocomposite with a CNT core/porous MnO_2 sheath hierarchy architecture for supercapacitors. Nanoscale Res Lett 7:33. doi:10.1186/1556-276X-7-33
101. Yang XH, Wang YG, Xiong HM, Xia YY (2007) Interfacial synthesis of porous MnO_2 and its application in electrochemical capacitor. Electrochim Acta 53:752–757. doi:10.1016/j.electacta.2007.07.043
102. Malak A, Fic K, Lota G, Vix-Guterl C, Frackowiak E (2010) Hybrid materials for supercapacitor application. J Solid State Electrochem 14:811–816. doi:10.1007/s10008-009-0856-8
103. Sharma RK, Rastogi AC, Desu SB (2008) Manganese oxide embedded polypyrrole nanocomposites for electrochemical supercapacitor. Electrochim Acta 53:7690–7695. doi:10.1016/j.electacta.2008.04.028
104. Wang X, Myers BD, Yan J, Shekhawat G, Dravid V, Lee PS (2013) Manganese oxide micro-supercapacitors with ultra-high areal capacitance. Nanoscale 5:4119–4122. doi:10.1039/c3nr00210a
105. Makgopa K, Ejikeme PM, Jafta CJ, Raju K, Zeiger M, Presser V et al (2015) A high-rate aqueous symmetric pseudocapacitor based on highly graphitized onion-like carbon/birnessite-type manganese oxide nanohybrids. J Mater Chem A 3:3480–3490. doi:10.1039/C4TA06715K
106. Raymundo-Piñero E, Khomenko V, Frackowiak E, Béguin F (2005) Performance of manganese oxide/CNTs composites as electrode materials for electrochemical capacitors. J Electrochem Soc 152:A229. doi:10.1149/1.1834913
107. Sivakkumar SR, Ko JM, Kim DY, Kim BC, Wallace GG (2007) Performance evaluation of CNT/polypyrrole/MnO_2 composite electrodes for electrochemical capacitors. Electrochim Acta 52:7377–7385. doi:10.1016/j.electacta.2007.06.023
108. Wang H-Q, Yang G, Li Q-Y, Zhong X-X, Wang F-P, Li Z-S et al (2011) Porous nano-MnO_2: large scale synthesis via a facile quick-redox procedure and application in a supercapacitor. New J Chem 35:469. doi:10.1039/c0nj00712a
109. Hu L, Chen W, Xie X, Liu N, Yang Y, Wu H et al (2011) Symmetrical MnO_2 carbon nanotube-textile nanostructures for wearable pseudocapacitors with high mass loading. ACS Nano 5:8904–8913. doi:10.1021/nn203085
110. Zhao X, Zhang L, Murali S, Stoller MD, Zhang Q, Zhu Y et al (2012) Incorporation of manganese dioxide within ultraporous activated graphene for high-performance electrochemical capacitors. ACS Nano 6:5404–5412. doi:10.1021/nn3012916
111. Rakhi RB, Chen W, Cha D, Alshareef HN (2012) Nanostructured ternary electrodes for energy-storage applications. Adv Energy Mater 2:381–389. doi:10.1002/aenm.201100609

112. Zhang X, Sun X, Zhang H, Zhang D, Ma Y (2012) Development of redox deposition of birnessite-type MnO_2 on activated carbon as high-performance electrode for hybrid supercapacitors. Mater Chem Phys 137:290–296. doi:10.1016/j.matchemphys.2012.09.023
113. Bello A, Fashedemi OO, Lekitima JN, Fabiane M, Dodoo-Arhin D, Ozoemena KI et al (2013) High-performance symmetric electrochemical capacitor based on graphene foam and nano-structured manganese oxide. AIP Adv 3:0–9. doi:10.1063/1.4819270
114. Li Q, Lu X-F, Xu H, Tong Y-X, Li G-R (2014) Carbon/MnO(2) double-walled nanotube arrays with fast ion and electron transmission for high-performance supercapacitors. ACS Appl Mater Interfaces 6:2726–2733. doi:10.1021/am405271q
115. Azhagan MVK, Vaishampayan MV, Shelke MV (2014) Synthesis and electrochemistry of pseudocapacitive multilayer fullerenes and MnO_2 nanocomposites. J Mater Chem A 2:2152. doi:10.1039/c3ta14076h
116. Maiti S, Pramanik A, Mahanty S (2014) Interconnected network of MnO_2 nanowires with a "cocoonlike" morphology: redox couple-mediated performance enhancement in symmetric aqueous supercapacitor. ACS Appl Mater Interfaces 6:10754–10762. doi:10.1021/am502638d
117. Li Z, Liu Z, Li D, Wang H (2015) Facile synthesis of a -MnO_2 nanowires/spherical activated carbon composite for supercapacitor application in aqueous neutral electrolyte. J Mater Sci 26:353–359. doi:10.1007/s10854-014-2407-z
118. Hong MS, Lee SH, Kim SW (2002) Use of KCl aqueous electrolyte for 2V manganese oxide/activated carbon hybrid capacitor. Electrochem Solid-State Lett 5:A227. doi:10.1149/1.1506463
119. Brousse T, Toupin M, Bélanger D (2004) A hybrid activated carbon-manganese dioxide capacitor using a mild aqueous electrolyte. J Electrochem Soc 151:A614. doi:10.1149/1.1650835
120. Khomenko V, Raymundo-Piñero E, Frackowiak E, Béguin F (2006) High-voltage asymmetric supercapacitors operating in aqueous electrolyte. Appl Phys A Mater Sci Process 82:567–573. doi:10.1007/s00339-005-3397-8
121. Brousse T, Taberna PL, Crosnier O, Dugas R, Guillemet P, Scudeller Y et al (2007) Long-term cycling behavior of asymmetric activated carbon/MnO_2 aqueous electrochemical supercapacitor. J Power Sources 173:633–641. doi:10.1016/j.jpowsour.2007.04.074
122. Qu Q, Zhang P, Wang B, Chen Y, Tian S, Wu Y et al (2009) Electrochemical performance of MnO_2 nanorods in neutral aqueous electrolytes as a cathode for asymmetric supercapacitors. J Phys Chem C 113:14020–14027. doi:10.1021/jp8113094
123. Xu C, Du H, Li B, Kang F, Zeng Y (2009) Asymmetric activated carbon-manganese dioxide capacitors in mild aqueous electrolytes containing alkaline-earth cations. J Electrochem Soc 156:A435. doi:10.1149/1.3106112
124. Qu QT, Shi Y, Tian S, Chen YH, Wu YP, Holze R (2009) A new cheap asymmetric aqueous supercapacitor: activated carbon//$NaMnO_2$. J Power Sources 194:1222–1225. doi:10.1016/j.jpowsour.2009.06.068
125. Qu Q, Li L, Tian S, Guo W, Wu Y, Holze R (2010) A cheap asymmetric supercapacitor with high energy at high power: activated carbon//$K_{0.27}MnO_2 0.6H_2O$. J Power Sources 195:2789–2794. doi:10.1016/j.jpowsour.2009.10.108
126. Demarconnay L, Raymundo-Piñero E, Béguin F (2011) Adjustment of electrodes potential window in an asymmetric carbon/MnO_2 supercapacitor. J Power Sources 196:580–586. doi:10.1016/j.jpowsour.2010.06.013
127. Fan Z, Yan J, Wei T, Zhi L, Ning G, Li T et al (2011) Asymmetric supercapacitors based on graphene/MnO_2 and activated carbon nanofiber electrodes with high power and energy density. Adv Funct Mater 21:2366–2375. doi:10.1002/adfm.201100058
128. Xia H, Huo C (2011) Electrochemical properties of MnO_2/CNT nanocomposite in neutral aqueous electrolyte as cathode material for asymmetric supercapacitors. Int J Smart Nano Mater 2:1–9. doi:10.1080/19475411.2011.623728

129. Gao H, Xiao F, Ching CB, Duan H (2012) High-performance asymmetric supercapacitor based on graphene hydrogel and nanostructured MnO_2. ACS Appl Mater Interfaces 4:2801–2810. doi:10.1021/am300455d
130. Lei Z, Zhang J, Zhao XS (2012) Ultrathin MnO_2 nanofibers grown on graphitic carbon spheres as high-performance asymmetric supercapacitor electrodes. J Mater Chem 22:153. doi:10.1039/c1jm13872c
131. Jiang H, Li C, Sun T, Ma J (2012) A green and high energy density asymmetric supercapacitor based on ultrathin MnO_2 nanostructures and functional mesoporous carbon nanotube electrodes. Nanoscale 4:807. doi:10.1039/c1nr11542a
132. Choi BG, Yang M, Hong WH, Choi JW, Huh YS (2012) 3D macroporous graphene frameworks for supercapacitors with high energy and power densities. ACS Nano 6:4020–4028. doi:10.1021/nn3003345
133. Jafta CJ, Nkosi F, le Roux L, Kebede M, Makgopa K, Mathea MK et al (2013) Tuning electrolytic manganese dioxide for a high-voltage aqueous asymmetric electrochemical capacitor. Electrochem Soc Trans 50:93–101
134. Wang JG, Yang Y, Huang ZH, Kang F (2013) A high-performance asymmetric supercapacitor based on carbon and carbon-MnO_2 nanofiber electrodes. Carbon N Y 61:190–199. doi:10.1016/j.carbon.2013.04.084
135. Jin Y, Chen H, Chen M, Liu N, Li Q (2013) Graphene patched CNT/MnO_2 nanocomposite papers for the electrode of high-performance flexible asymmetric supercapacitors. ACS Appl Mater Interfaces 5:3408–3416. doi:10.1021/am400457x
136. Chang J, Jin M, Yao F, Kim TH, Le VT, Yue H et al (2013) Asymmetric supercapacitors based on graphene/MnO_2 nanospheres and graphene/MoO_3 nanosheets with high energy density. Adv Funct Mater 23:5074–5083. doi:10.1002/adfm201301851
137. Shao Y, Wang H, Zhang Q, Li Y (2013) High-performance flexible asymmetric supercapacitors based on 3D porous graphene/MnO_2 nanorod and graphene/Ag hybrid thin-film electrodes. J Mater Chem C 1:1245. doi:10.1039/c2tc00235c
138. Cao J, Wang Y, Zhou Y, Ouyang JH, Jia D, Guo L (2013) High voltage asymmetric supercapacitor based on MnO_2 and graphene electrodes. J Electroanal Chem 689:201–206. doi:10.1016/j.jelechem.2012.10.024
139. Yang C, Zhou M, Xu Q (2013) Three-dimensional ordered macroporous MnO_2/carbon nanocomposites as high-performance electrodes for asymmetric supercapacitors. Phys Chem Chem Phys 15:19730–19740. doi:10.1039/c3cp53504e
140. Liu M, Tjiu WW, Pan J, Zhang C, Gao W, Liu T (2014) One-step synthesis of graphene nanoribbon-MnO_2 hybrids and their all-solid-state asymmetric supercapacitors. Nanoscale 6:4233–4242. doi:10.1039/c3nr06650a
141. Wu S, Chen W, Yan L (2014) Fabrication of a 3D MnO_2/graphene hydrogel for high-performance asymmetric supercapacitors. J Mater Chem A 2:2765. doi:10.1039/c3ta14387b
142. Zhao Y, Ran W, He J, Huang Y, Liu Z, Liu W et al (2014) High-performance asymmetric supercapacitors based on multilayer MnO_2/graphene oxide nanoflakes and hierarchical porous carbon with enhanced cycling stability. Small 11:1310–1319. doi:10.1002/smll.201401922
143. Zhang YX, Kuang M, Hao XD, Liu Y, Huang M, Guo XL et al (2014) Rational design of hierarchically porous birnessite-type manganese dioxides nanosheets on different one-dimensional titania-based nanowires for high performance supercapacitors. J Power Sources 270:675–683. doi:10.1016/j.jpowsour.2014.07.114
144. Ma W, Nan H, Gu Z, Geng B, Zhang X (2015) Superior performance asymmetric supercapacitors based on $ZnCo_2O_4$ @MnO_2 core–shell electrode. J Mater Chem A 3:5442–5448. doi:10.1039/C5TA00012B
145. Jiang J, Kucernak A (2002) Electrochemical supercapacitor material based on manganese oxide: preparation and characterization. Electrochim Acta 47:2381–2386. doi:10.1016/S0013-4686(02)00031-2

146. Chang J, Tsai W (2004) Effects of temperature and concentration on the structure and effects of temperature and concentration on the structure and specific capacitanceof manganese oxide deposited in manganese acetate solution. J Appl Electrochem 34:953–961
147. Taguchi A, Inoue S, Akamaru S, Hara M, Watanabe K, Abe T (2006) Phase transition and electrochemical capacitance of mechanically treated manganese oxides. J Alloys Compd 414:137–141. doi:10.1016/j.jallcom.2005.02.108
148. Wu Y-T, Hu C-C (2005) Aspect ratio controlled growth of MnOOH in mixtures of Mn_3O_4 and MnOOH single crystals for supercapacitors. Electrochem Solid-State Lett 8:A240–A244
149. Djurfors B, Broughton JN, Brett MJ, Ivey DG (2005) Electrochemical oxidation of Mn/MnO films: formation of an electrochemical capacitor. Acta Mater 53:957–965. doi:10.1016/j.actamat.2004.10.041
150. Nagarajan N, Humadi H, Zhitomirsky I (2006) Cathodic electrodeposition of MnOx films for electrochemical supercapacitors. Electrochim Acta 51:3039–3045. doi:10.1016/j.electacta.2005.08.042
151. An G, Yu P, Xiao M, Liu Z, Miao Z, Ding K et al (2008) Low-temperature synthesis of Mn_3O_4 nanoparticles loaded on multi-walled carbon nanotubes and their application in electrochemical capacitors. Nanotechnology 19:275709. doi:10.1088/0957-4484/19/27/275709
152. Wang B, Park J, Wang C, Ahn H, Wang G (2010) Mn_3O_4 nanoparticles embedded into graphene nanosheets: preparation, characterization, and electrochemical properties for supercapacitors. Electrochim Acta 55:6812–6817. doi:10.1016/j.electacta.2010.05.086
153. Jiang H, Zhao T, Yan C, Ma J, Li C (2010) Hydrothermal synthesis of novel Mn_3O_4 nano-octahedrons with enhanced supercapacitors performances. Nanoscale 2:2195–2198. doi:10.1039/c0nr00257g
154. Dubal DP, Holze R (2013) All-solid-state flexible thin film supercapacitor based on Mn_3O_4 stacked nanosheets with gel electrolyte. Energy 51:407–412. doi:10.1016/j.energy.2012.11.021
155. Qiao Y, Sun Q, Cui H, Wang D, Yanga F, Wang X (2015) Synthesis of micro/nano-structured Mn_3O_4 for supercapacitor electrode with excellent rate performance. RSC Adv 5 (40):31942–31946. doi:10.1039/C4RA04783D
156. Yu C, Zhang L, Shi J, Zhao J, Gao J, Yan D (2008) A simple template-free strategy to synthesize nanoporous manganese and nickel oxides with narrow pore size distribution, and their electrochemical properties. Adv Funct Mater 18:1544–1554. doi:10.1002/adfm.200701052
157. Zhang LL, Wei T, Wang W, Zhao XS (2009) Manganese oxide-carbon composite as supercapacitor electrode materials. Microporous Mesoporous Mater 123:260–267. doi:10.1016/j.micromeso.2009.04.008
158. Wang X, Liu L, Wang X, Yi L, Hu C, Zhang X (2011) Mn_2O_3/carbon aerogel microbead composites synthesized by in situ coating method for supercapacitors. Mater Sci Eng B Solid-State Mater Adv Technol 176:1232–1238. doi:10.1016/j.mseb.2011.07.003

Chapter 11
Suspension Electrodes for Flow-Assisted Electrochemical Systems

Kelsey B. Hatzell and Yury Gogotsi

11.1 Introduction

Energy storage is widely used in small-scale portable electronics and has recently gained increased attention for grid-scale applications [1–3]. Grid energy storage (GES) is pivotal for (1) increasing integration of intermittent renewable energy technologies (wind, solar, etc.) [4], (2) enabling distributed energy generation (decentralized electricity production) and smart grid technologies, and (3) for delivery of electricity in developing economies where electricity transmission services are unreliable. The first application, integrating renewable energy technologies, is especially pertinent because the US alone will require 4–5 terra kilowatt-hours of electricity annually by 2050 [5]. Thus, new (and ideally clean) energy generation systems will be needed to accommodate this demand.

Currently pumped hydroelectric storage is the most widely used and mature grid-scale energy storage technology available, accounting for 95 % (or 24.6 GW) of the stored energy in the United States [6]. This 24.6 GW of stored energy only represents a small percentage (2.3 %) of the total electricity produced. Thus, there is tremendous room for growth. Despite being the most mature GES technology, pumped-hydro requires access to large quantities of water and elevation changes, which makes it geographically limited. Additionally, pumped-hydro systems

K.B. Hatzell
Department of Material Science and Engineering, A.J. Drexel Nanomaterials Institute, Drexel University, 3141 Chestnut Street, Philadelphia, PA 19104, USA

Lawrence Berkeley National Laboratory, Berkeley, CA 94720, USA

Y. Gogotsi (✉)
Department of Material Science and Engineering, A.J. Drexel Nanomaterials Institute, Drexel University, 3141 Chestnut Street, Philadelphia, PA 19104, USA
e-mail: yg36@drexel.edu

K.I. Ozoemena, S. Chen (eds.), *Nanomaterials in Advanced Batteries and Supercapacitors*, Nanostructure Science and Technology,
DOI 10.1007/978-3-319-26082-2_11

usually are large-scale systems (> 10 MW capacity) and have discharge times on the order of tens of hours [6]. At this rate they can be used for bulk power management, but cannot supply energy storage for frequency regulation, load shifting, peak shaving, and other ancillary grid services [7]. The broad range of applications that exist for GES will require a diverse set of energy storage systems. Recently, the US Department of Energy's (DOE) Advanced Research Projects Agency-Energy (ARPA-E) established ambitious research and industrial targets for GES. Specifically, there is a need for systems that can supply 4–20 h of charge/discharge times, costs under $100/kWh/hr of storage, and has a lifetime that exceeds 10 years with minimal maintenance costs [1, 4, 8, 9]. Electrochemical energy storage (EES) has been identified as a promising avenue for achieving these goals and for addressing the growing and diverse grid energy storage needs [1, 2, 4].

Cost and lifetime are the two most important criteria in order for GES to be a viable alternative to fossil fuel-based power generation [8, 10]. The three primary types of electrochemical energy storage devices that are being considered for GES include (1) stationary battery devices, (2) redox flow batteries, and (3) suspension (semi-solid) flow systems (batteries and supercapacitors). Stationary battery devices currently being used for GES include lithium-ion, sodium-sulfur, and sodium-ion battery [9, 11] technologies (Fig. 11.1a). Lithium-ion batteries (LIB) are typically composed of two intercalating materials that serve as the anode and cathode. Battery systems are considered more downstream technologies and offer high energy density, and scalability. Individual batteries can be connected in series and parallel to achieve desired power and energy needs. LIBs represent a GES technology that can be easily tuned for applications ranging from 1 to 10 MW. The scalability and modularity is ideal; however, it is limited by its lifetime and its cost which far exceeds (~$900 – 1700/kWh) ARPA-E's target of $100/kWh [1]. Thus, alternative battery chemistries based on cheap and abundant materials are being explored. Specifically, sodium-ion-based (intercalation) batteries [12] and sodium-sulfur batteries are promising low-cost alternatives to LIBs.

Flow-assisted electrochemical energy storage (FAES) is another route toward scalable energy storage for the grid. FAES utilizes a flowable architecture (Fig. 11.1b) for tunable energy and power densities. The size of the tanks is proportional to the energy stored, while the number of cells (stack-size) is directly related to the power density [13–15]. A vanadium redox flow battery (VRFB) is an example of a GES system that has reached a commercial scale. In a VFRB, the redox-active material (metal ions) are soluble in an electrolyte (aqueous or nonaqueous). While battery systems store energy in the solid material via chemical reactions (and ion intercalation processes) that induce charge transfer, solution-based RFBs store energy in redox species that are reduced and oxidized during subsequent charge/discharge processes. These species (Vanadium, Cerium, or Iron ionic species) are stable and remain in the solution in various oxidation states and undergo reversible electrochemical processes. The typical reactions that occur in a VRFB:

$$VO_2^+ + 2H^+ + e^- \overrightarrow{\text{discharge}}\, VO^{2+} + H_2O \text{ (Catholyte)} \tag{11.1}$$

$$V^{2+} - e^- \overrightarrow{\text{discharge}}\, V^{3+} + H_2O \text{ (Anolyte)} \tag{11.2}$$

with the whole cell reaction

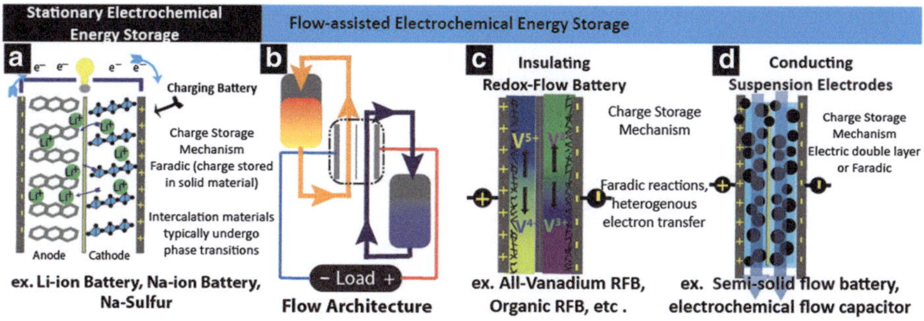

Fig. 11.1 Summary of primary grid energy storage systems. Solid-state batteries (**a**) based on intercalation compounds can be connected in series in parallel to create battery packs suitable for large-scale systems. In contrast to stationary devices, flow-assisted electrochemical energy storage systems that utilize a flow architecture (**b**) offer scalable power and energy densities while minimizing inactive material costs. Two types of flow systems include the redox flow battery (**c**) and flow systems based on suspension (*semi-solid*) electrodes

$$VO_2^+ + 2H^+ + V^{2+} \overrightarrow{\text{discharge}}\, VO^{2+} + H_2O + V^{3+} \tag{11.3}$$

The overall voltage for VRFB systems is dictated by the oxidation and reduction potential of the redox couples and limited by hydrogen and oxygen evolution when in an aqueous environment. In an acidic environment (sulfuric acid or hydrochloric acid), the $VO_2{}^+/VO^{2+}$ redox pair occurs at +0.99V and V^{3+}/V^{2+} redox reaction occurs at −0.26 for an overall system voltage around 1.25 V [14]. Energy density is proportional to the voltage window, so there is a significant amount of research done on identifying redox couples with greater open circuit voltages (e.g., zinc-cerium). Nevertheless, the challenge with metal-based, aqueous redox flow batteries is threefold: (1) energy density is heavily dependent on the solubility of the redox species in the electrolyte, (2) sluggish kinetics limit RFB to applications that do not need high power, and (3) RFBs require costly and expensive temperature control, because solubility of the metal ions decreases at low temperatures.

Recently there has been a push toward nonaqueous flow batteries with higher voltage windows for higher energy density. Moreover, instead of metal-redox species, people have begun investigating (low-cost and abundant) carbonyl-based organic molecules (C-bOMs) as soluble redox-active species in nonaqueous and aqueous flow batteries [16–18]. C-bOMs in an aqueous environment typically undergo a reversible single-step two-electron two-proton reduction process, where protons act to stabilize the di-anion [19, 20]. In a nonaqueous electrolyte, a two-step process occurs, and small-radii cations such as Li^+/Mg^+ (instead of hydrogen) offer the strong stabilization effect [19, 21] which has been shown to shift the formal potential (increasing the voltage window and energy density significantly) [22]. An important and ongoing research challenge is to understand how these ionic interactions occur with C-bOMs with varying reduced carbonyl moieties to enable high gravimetric energy density. Also, there are increased and continuing studies toward tailored-organic materials either by molecular weight or through ligand substitutions for favorable electrochemical properties [23].

The last type of FAESs, which will be the focus of this chapter, is suspension-based flow systems (Fig. 11.1d). Suspension (or semi-solid) electrodes are electronically conducting, in contrast to insulating RFB systems. In a suspension electrode, the active material (charge storing) is not soluble but suspended in an electrolyte, and electron transport occurs through the formation of percolation networks as active materials (particles) become agglomerated and come into contact. There are two types of suspension electrodes, ones that utilize faradic charge storage (battery) and ones that utilize electrostatic charge storage in an electric double layer. The former has been named the semi-solid flow battery (SSFB) [24] and the latter the electrochemical flow capacitor (EFC) [25]. The same capacitive suspension electrodes used in the EFC have been demonstrated to be effective in removing ions from low-concentration solutions via a process called flow-electrode capacitive deionization [26] and for energy generation via flow-electrode capacitive mixing [27].

Currently, there has been array of targeted scientific investigations toward understanding key transport processes in materials with an emphasis on the discovery of new materials to enable new and enhanced energy storage technologies. In parallel with these pursuits, a new area rapidly being explored is materials systems engineering, which takes a holistic approach to understanding how key material properties transition to system level performances. Understanding individual materials and how they interact with other materials is an integral part in enabling new and disruptive technologies. The current state-of-the-art technologies for grid energy storage are too expensive (especially in the United States) to compete with fossil fuel technologies. Much of the future discovery and innovation within grid energy storage relies on designing high-performing, low-cost, and energy-efficient material systems. Here we discuss suspension (semi-solid) flow electrodes as a possible material system for grid energy storage. Moreover, this chapter places emphasis on how capacitive suspension electrodes are broadening the way we think about supercapacitors toward new applications such as water deionization and energy generation.

11.2 Systems That Utilize Suspension-Type Electrodes

Suspension electrodes ("flow electrodes") provide avenues for achieving high capacity and building tunable systems for a range of critical infrastructure-level challenges. The core idea behind flow electrodes is that active material (e.g., carbon) can adsorb ions and transport/store electrons to and from electrodes via conducting percolation networks. Thus, the physical attributes of the percolation network play a large role in the performance of these systems. Furthermore, the dynamic nature of pumping a suspension (exposing the material system to shear dynamics) in order to charge (adsorb) or discharge (desorb) charged particles is the

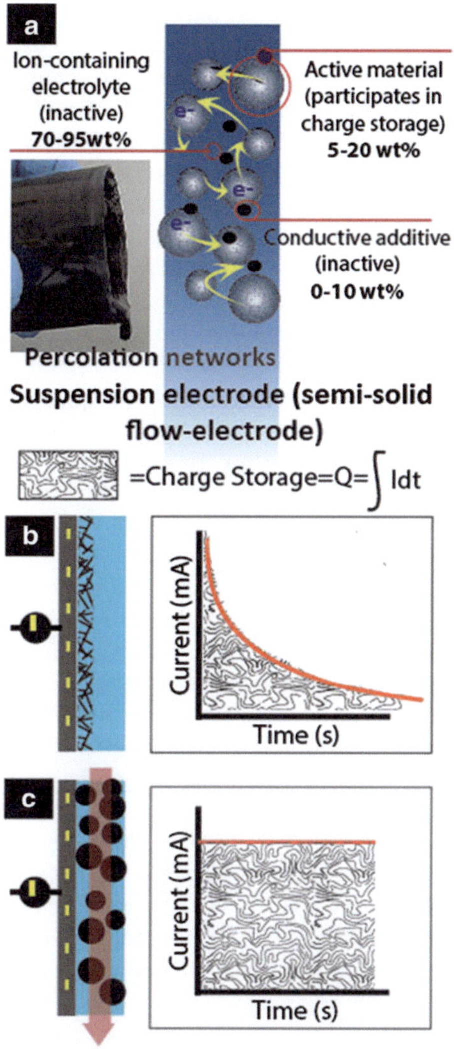

Fig. 11.2 Compositional breakdown for a suspension-type electrode (**a**) and chronoamperometric response characteristic of static electrodes (**b**) and flow electrodes (**c**) where Q is the charge stored (Coulomb), I is the current (Amperes), and t is time (s)

defining difference between suspension electrodes and film electrodes. Suspension electrodes are composed of an active material, an electrolyte, and a conductive additive. Figure 11.2a provides the compositional breakdown for a typical flow electrode: 5–20 wt% active material (charge storing material), 0–10 wt% conductive additive, and 70–95 % ion-containing electrolyte. The majority component is the electrolyte, which does not participate in charge storage but aids in the physical transport (flow) of the active material to and from the electroactive region. The electroactive region is defined as the area where the material can be charged or discharged by an applied current or voltage (the region adjacent to the current collectors). In traditional supercapacitors, the material is constantly in contact with the current collector, but in suspension systems the material flows into and out of the electroactive area to enable more versatile applications.

In contrast to suspension-type electrodes, film electrodes, used in batteries and supercapacitors, are composed of about 90 wt% active material. The remaining 10 wt% is usually a binder and conductive additive. Gravimetrically suspension-type electrodes suffer in terms of active material, but benefit in terms of their functionality. Figure 11.2b demonstrates the chronoamperometric (constant applied potential) response standard to systems based on static film electrodes. In electrochemical capacitors (supercapacitors), as the active material's surface area becomes saturated with adsorbed ions, the current decays with time (signaling full utilization of the surface area with ions). Once the current decays, film electrodes need to be regenerated through an ion desorption process (applied 0 V or negative applied current) also known as a discharge process. In a suspension electrode, since the electroactive material is constantly being replenished with new carbon slurry, the chronomperometric current response is independent of time as demonstrated in Fig. 11.2c. The flowable nature allows suspensions to be charged and discharged in different cells which has advantages for versatile operation and control. Moreover, this provides a means toward scalable battery and electrochemical capacitors, which do not require costly and sometimes inefficient parallel and series electrical connections. The principle of suspension-type electrodes expands the inventory of possible energy storage materials that can be used for large-scale applications beyond soluble redox couples. Examples of electrochemical technologies that have been examined in a suspension-type form include: all-carbon-based supercapacitors [28], asymmetric manganese/activated carbon supercapacitors [29], LIBs [24, 30], sodium-ion-based batteries [31], conducting polymer-based supercapacitors [32, 33], and lithium-polysulfide chemistries [34, 35], to name just a few of the growing material system inventory. Below we focus on describing applications for capacitive/pseudocapacitive suspension electrodes and their applications for energy storage, as well as for deionization and energy generation technologies. This chapter focuses on the key aspects toward designing high-performing material systems (suspension electrodes), and how they enable us to reimagine supercapacitors in array of large-scale applications.

11.2.1 Electrochemical Flow Capacitor (Energy Storage)

The electrochemical flow capacitor (EFC) is a system used for achieving scalable electrostatic energy storage for high-power grid applications [25, 36]. This system has the advantages of both supercapacitors and flow batteries: (1) rapid charge and discharge (i.e., fast response rates and high power) and (2) decoupled energy storage and power output. The EFC operationally is similar to a flow battery. Figure 11.3a demonstrates the basic working principle proposed for semi-solid systems. Four vessels house suspensions, two house uncharged suspensions, and two are filled with charged suspensions. The total charge is close to zero for each container as the charge on the solid carbon is balanced by ions of the opposite charge. The uncharged suspensions are pumped between two current collectors (cell) and are charged at a set potential (or set current). Similarly to traditional supercapacitors, the two electrodes are electrically insulated from each other with an ion-permeable separator (e.g., polyvinylidene fluoride). During this charging process, ions are adsorbed at the electrolyte|active material interface (e.g., carbon) in an electric double layer. Positive ions adsorb to the negatively charged particles and negative ions adsorb onto the positively charged particles in an electric double layer (Fig. 11.3c). The fact that oppositely charge ions are attracted to the carbon particles makes the particles electrically neutral. Once the charging process is complete, the suspension can be flowed into two storage reservoirs. When power is needed, the suspension's flow can be reversed (returned to the cell) and discharged. Again, this process enables decoupling of energy and power densities in the system. The volumes of the reservoirs are proportional to energy density, whereas the number of cells (housing current collectors) is proportional to the power density.

The suspension (or flow electrode) can take on various mechanical forms when composed of varying wt% of active material. At low carbon wt%, the suspension is more liquid-like (Fig. 11.3b) and can easily be flowed or pipetted. As the compositional loading of the dry material (e.g., carbon) is increased in a suspension electrode, it becomes more like a viscous solid (gel). This has implications not only for ion and electron percolation kinetics but also on the viscosity (flowability) of the flow electrode. The electrolyte also has a bearing on the structural ordering of the particles in a suspension. In highly concentrated electrolytes (high salt concentrations), the suspension can become highly viscous due to increased active material aggregation. The aggregation properties of a suspension electrode can be explained by the Derjaguin, Landau, Verwey, and Overbeek theory [37, 38]. In high-concentration solutions, van der Waals forces of attraction dominate over double layer repulsion dynamics, which leads to strong aggregation effects. Diffusion-limited aggregation of active particles across the cross section of the electrode enables efficient electron percolation and is a driving force for the design of these flow-electrode systems (Fig. 11.3d). This theory will be discussed more in detail in our description of the effect of oxidation of carbon material on the electrochemical and rheological properties of a flow electrode.

The EFC's main advantage in comparison to other competing technologies is that it offers a long lifetime ($\tilde{}100,000$ charge/discharge cycles) and high-power densities $\left(> 10 \text{ W l}^{-1}\right)$. Thus, it is targeted for high-power applications on the grid, such as frequency regulation which requires power to be delivered at MW/min rates. Moreover, the EFC eliminates the costly aspect of connecting devices in series and parallel by decreasing the quantity of current collectors and separators and packaging (inactive materials). For large-scale infrastructure applications, cost and lifetime are the guiding parameters for design (see Fig. 11.4), and in these parameters, the EFC hits the mark for high-rate applications.

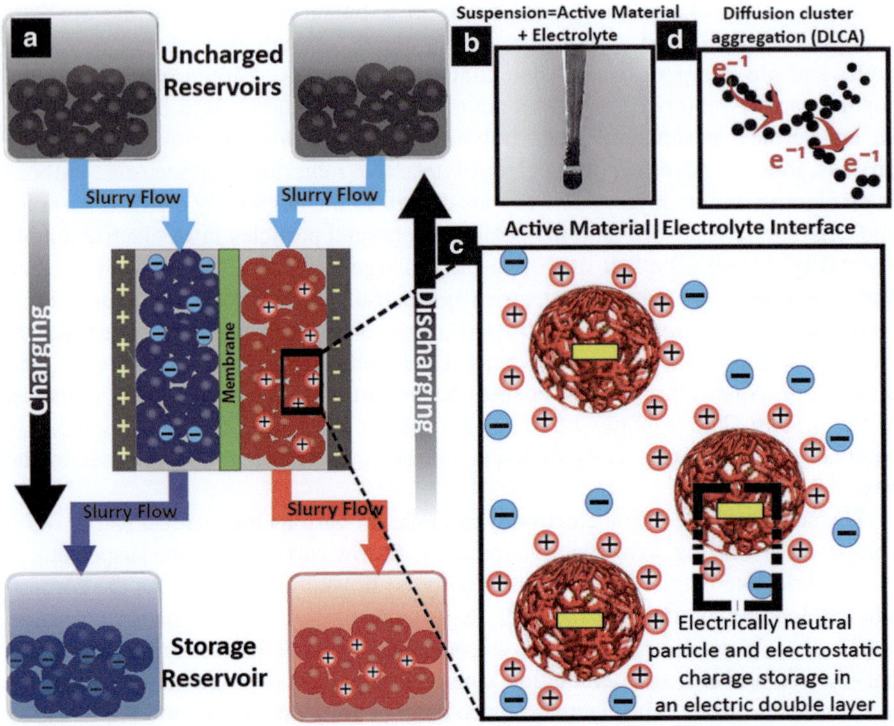

Fig. 11.3 Schematic of an electrochemical flow capacitor (**a**), a pipette tip drop of a suspension electrode (**b**), electrostatic energy storage utilized by an all-carbon EFC, and aggregation effects between active material (**d**)

11.2.2 Flow Capacitive Deionization (Ion Removal/ Desalination)

Access to water, energy, and food are areas of growing concern across the globe [39, 40]. To address water scarcity issues, people are looking to develop new, low-cost, and low-energy methods for desalting sea water, or for ion removal from non-traditional water sources (such as brackish water). As a result of parallel developments in the manufacturing and synthesis processes of nanostructured carbon materials, capacitive deionization has reemerged as a possible water desalination/purification technology [41, 42]. The process of capacitive deionization is similar to a supercapacitor, with one main difference. Instead of an ion-permeable (electron-insulating) separator between the two electrodes, a feed channel exists. This channel "feeds" in brackish water or ionic solutions between the two parallel film electrodes [43, 44]. As the solution passes between the two carbon (film) electrodes, a potential can be applied and ions are adsorbed into an electric double layer at the surface of the active material. The effluent is thus a deionized solution. The primary limitation of CDI is in its operation. As ions saturate the active

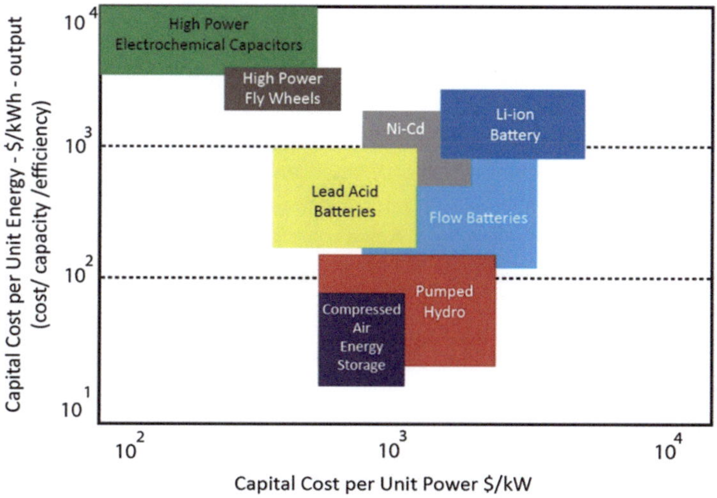

Fig. 11.4 Comparison of electrochemical energy storage systems for the electric grid comparing capital cost per unit energy to capital cost per unit power (Reprinted with permission from [2]. Copyright 2011 American Chemical Society)

material, film electrodes cannot deionize the influent feed solution. Thus, in a single-pass mode, where a set volume is passed between film electrodes, the effluent concentration will experience a transient profile. Figure 11.5b demonstrates the effluent concentration profile of a film electrode, which is characterized by a decline in concentration up to the point of full electrode saturation and then an increase back to the original concentration.

Figure 11.5a demonstrates the working principles of flow capacitive deionization based on membranes (which utilizes flow electrodes instead of static film electrodes) [26, 42, 45–47]. In FCDI there are three flow channels instead of one. The outer channels house the capacitive suspension electrodes, while the inner channel is the feed channel (similar to traditional CDI). The feed channel and flow-electrode channels are separated with ion-exchange membranes. An anion exchange membrane shares an interface with the cathodic flow electrode, whereas a cation exchange membrane shares an interface with the anodic flow electrode. The main advantage of using ion-exchange membranes is to eliminate co-ion repulsion for greater salt removal efficiencies and to enable negative potentials and more effective flushing of the counterions from the surface during regeneration [48]. The use of flow electrodes allows for controlled effluent concentration profiles [49]. The concentration profile at the effluent is constant (Fig. 11.5b) in FCDI, because the electroactive region is constantly being replenished with new material. As a result, this allows for better control, operational flexibility, and scalable technologies.

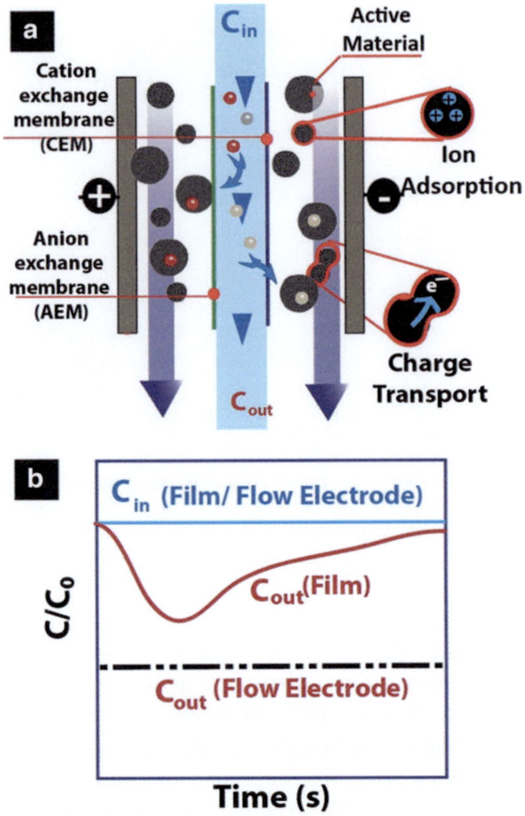

Fig. 11.5 Schematic for flow capacitive deionization with ion-exchange membranes (**a**) and concentration profiles characteristic of FCDI and CDI (**b**) (Reprinted with permission from [45]. Copyright 2015 American Chemical Society)

11.2.3 Flow-Electrode Capacitive Mixing (Energy Generation)

Finally, capacitive suspension electrodes have also been shown to have utility in an energy generation technology known as capacitive mixing (CapMix). Capacitive mixing is a technology currently being explored to capture salinity gradient energy that is released by sea and river water mixing processes. It has been estimated that there is ~2.6 TW of energy that is released during this process annually. Energy extraction from CapMix is completed through a four-step process, involving compression and expansion of the electric double layer at the surface of a carbon-based electrode. Figure 11.6a shows the four-step process and Fig. 11.6b shows the voltage response. Briefly, in step 1, the electrodes are charged through a closed circuit in a high-concentration (sea water) solution. After charging is stabilized (voltage completely decays), the high-concentration solution is replaced by a low-concentration solution (river water), and the cell is changed to an open circuit configuration, leading to EDL expansion. Next in step 3, a resistor is used to discharge the electrodes, and finally the system is regenerated by flushing a

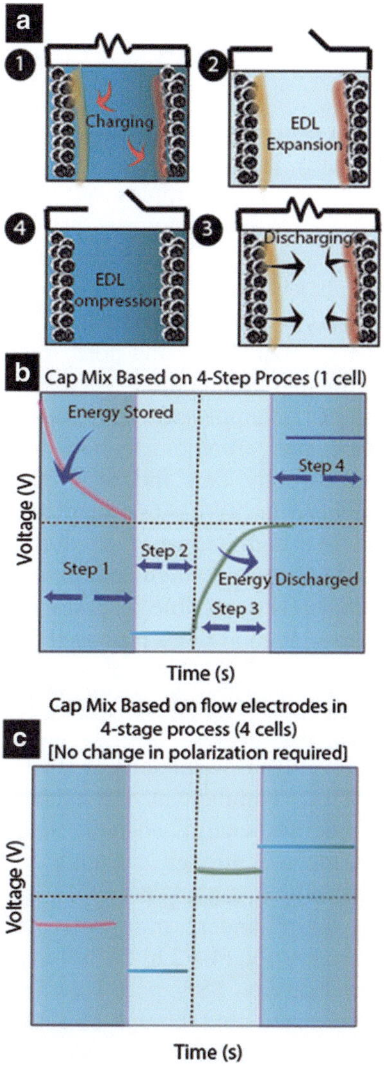

Fig. 11.6 Four-step capacitive mixing based on film electrodes (**a**) and associated voltage outputs with each step (**b**). Replacing film electrodes allows for continuous and stable voltage (power output) (**c**)

high-concentration solution through the cell and pumping in a high-concentration solution. Through this cycle energy can be extracted, in a similar process to a heat engine (which uses pressure and volume changes to extract energy). The efficiencies of current CapMix technologies are fundamentally hindered by the intermittent nature and complex architecture. Replacing the film electrodes allows for more constant voltage outputs (Fig. 11.6c) and thus constant power output. Flow electrodes may facilitate more scalable and versatile CapMix systems.

11.3 Carbon Materials as Active Materials in Suspension Electrodes

The world of carbon is vast with an array of different types of carbon allotropes currently being explored and used in electrochemical energy storage systems (Table 11.1). Examples of carbon materials [50, 51] include carbon onions [52, 53], carbon nanotubes, activated carbon (AC), carbide-derived carbon (CDC) [54, 55], graphene [56, 57], graphene oxide, and templated carbons. These carbons are characterized as either porous or nonporous materials. Examples of porous carbons are AC, CDC, and templated carbon and are characterized by high specific surface area (SSA) and pore volumes. These properties often lead to high gravimetric and volumetric capacitances in electrochemical capacitors (Table 11.1). In contrast, examples of nonporous carbons are carbon nanotubes, graphene, carbon black, and carbon onions have minimal/limited inner porosities. While nonporous carbons may lack the SSA of their porous counterparts (except single layer graphene or SWCNTs), they are usually characterized with high conductivities due to a high degree of graphitization. Carbon onions in particular show notable performance in high-power applications, because their external surface is completely accessible for fast ion adsorption and desorption. For large-scale applications, such as those proposed for capacitive suspension electrodes, cost, lifetime, and scalability are the primary guiding parameters for the material's development. Thus, activated carbons (Table 11.1) are the leading material candidate currently being examined for flow electrodes utilizing electrostatic mechanisms for either ion removal from water sources (FCDI) or for high-power grid energy storage (EFC).

Most commercially available ACs are derived from abundant organic byproducts such as coconut shells or bamboo and can be manufactured in bulk quantities through simple carbonization and activation procedures. Typically, organic byproducts can be directly carbonized at elevated temperatures ($> 600°C$_under nitrogen or CO/CO_2. Depending on the duration of the carbonization (4–12 h) the bio-mass derived carbon can take on different properties (conductivity, SSA, pore volume). After complete carbonization, the SSA can be further increased through an activation process. Activation procedures are considered either physical or chemical and either utilize a gas (CO_2) or a solid (potassium hydroxide) as a reactant with the carbon product. The typical reaction that is undergone at the carbon surface in physical and chemical activation procedures are:

$$C_{(s)} + CO_{2(g)} \longrightarrow 2CO_{(g)} \tag{11.4}$$

or

$$6KOH_{(s)} + 2C_{(s)} \longrightarrow 2K_{(s)} + 3H_2(g) + 2K_2CO_{3(s)} \tag{11.5}$$

These reactions create a series of micro- and mesoporous networks in the carbon substrates and increase the available SSA of the carbon material. Active materials can be characterized as primarily microporous, mesoporous, or macroporous

Table 11.1 Brief summary of carbon materials currently being examined for electrochemical energy storage and defining characteristics

Material	Carbon onions	Carbon nanotubes	Activated carbon	Graphene
Conductivity	High	High	Low	High
Volumetric Capacitance	Low	Low	High	Moderate
Gravimetric Capacitance	Low	Low	Moderate	Moderate
Cost	High	High	Low	Moderate
Scalability	Low	Moderate	High	Moderate
Structure		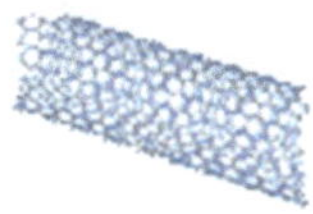	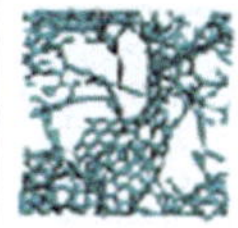	

depending on the distribution of pore sizes present in the sample. Microporous materials are characterized as materials with a majority of the pores with diameters <2 nm. Mesoporous materials typically have pores with pore diameters between 2 and 50 nm, and macroporous materials have average pore diameters >50 nm. It should be noted, highly microporous carbons with a great proportion of pores <1 nm may yield very high SSAs but low to moderate capacitances because ion transport within the pores may be limited by the small pore openings and/or hindered by more tortuous pore pathways.

For suspension-based flowable systems, such as the EFC and FCDI, several different types of activated carbons and active materials have been examined (Fig. 11.7). Figure 11.6a shows a TEM image of acetylene carbon black (CB), which is a common conductive additive used in suspension-based electrodes and film electrodes alike. CB is spherical and amorphous in nature and usually has a low specific surface area ($< 100\,m^2g^{-1}$) depending on the manufacturing process (Table 11.2). Due to its low SSA, it is usually included in incremental quantities (< 2 wt%/ < 1 vol% of the entire flow electrode) and dispersed well to ensure homogenous behavior. Aside from acetylene-based carbon black, Ketjen black is a popular additive used within semi-solid battery community, where conductivity of the active material (lithium titanate, lithium iron phosphate, etc.) is significantly lower than activated carbon [59]. Unlike acetylene carbon black, which is nonporous, Ketjen black has a significant surface area $\tilde{}800\,m^2g^{-1}$ [30, 59]. To achieve homogenous dispersion of the carbon black with the active material, ball milling and magnetic stirring are the typical techniques utilized.

Figure 11.7b shows carbon nanospheres synthesized through a facile hydrothermal route [58]. Glucose solutions (0.3–1 mol/L) readily turn into spherical carbon when put into an autoclave and heated at $200-600\,°C$ for 3–10 h. In addition to

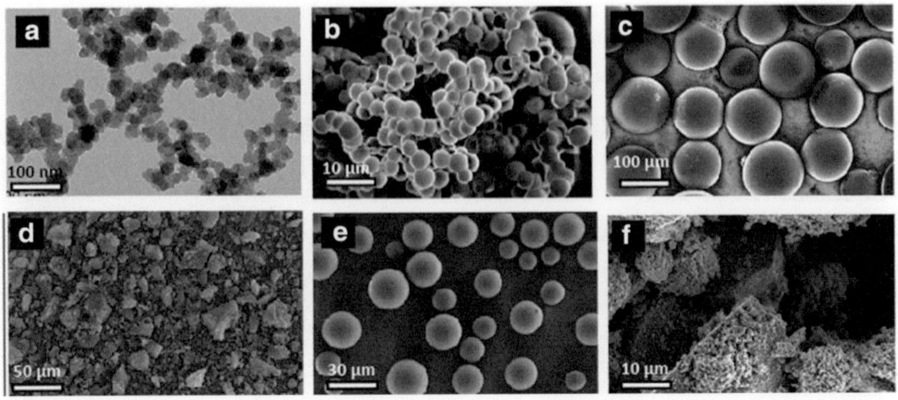

Fig. 11.7 TEM image of Acetylene carbon black (**a**) and SEM images of hydrothermal glucose-derived carbon nanospheres (**b**), phenolic-derived carbon spheres from MAST (**c**), high-surface area activated carbon (YP-50), MicroMast carbon spheres, and hydrothermally synthesized MnO_2 (**f**) (Reprinted with permission from [28, 29, 58]. Copyright 2014, American Chemical Society. Copyright 2012 and 2014 Elsevier)

glucose, there exists an array of other precursors that can be used to create tunable carbon materials [60, 61]. Depending on the synthesis conditions (time, temperature, and concentration), one can obtain carbon spheres with tunable sizes (diameters). Table 11.2 shows the surface characterization after physical activation in CO_2 for 4 h at 900 °C. This process is different from the two-step process described previously, as hydrothermal carbonization replaces carbonization with an inert environment. Following, hydrothermal carbonization the product is a spherical carbon with relatively low SSA (~600 m^2g^{-1}). To increase the material's SSA, the material can be activated with CO_2 as mentioned previously. The spherical geometry is advantageous for decreasing the viscosity and flow behavior of the suspension electrodes. Thus, in the EFC, several different spherical carbons (MAST Carbon, UK) have been examined including Fig. 11.7c, e. Figure 11.6c is MAST 125/250 and has an average particle size of 160 µm, whereas Fig. 11.7e (MAST micro beads) depicts much smaller spheres with an average particle size of 9 µm. These spherical carbon materials are derived from phenolic resins, rather than glucose as depicted in Fig. 11.7b.

Aside from spherical carbons, the most well-studied material in capacitive suspension electrodes is anisometric activated carbon derived from coconut shells. Figure 11.7d is a SEM micrograph of YP-50F activated carbon (AC) with particle sizes between 5 and 20 µm. This material is one of the most widely studied in conventional supercapacitors and is relatively inexpensive. The last material, Fig. 11.7f, is MnO_2 and will be discussed in detail in the section designated *Increasing the energy density of the EFC*.

Table 11.2 Porosity characteristics for various active materials for suspension electrodes, from N_2 gas sorption experiments

Material	BET specific surface area (m^2g^{-1})	Average pore size ϕ(nm)	Particle size ϕ(nm)
Acetylene carbon black	65	–	0.01–0.075
Nano-carbon spheres	2775	1.68	0.5
Mast 125/250	1576	4.7	160–190
YP-50 Activated Carbon	1472	1.3	5–20
Mast Microbeads	1127	7.8	9–20
Amorphous MnO_2	192	10	–

11.4 Electrochemical Properties of Suspension-Type Electrodes

Similar to conventional electrochemical energy storage devices, the EFC and SSFB are guided by traditional performance metrics such as power density and energy density. Unlike traditional storage devices, suspension (semisolid) flow systems are also concerned with the rheological properties of the suspension electrodes. The rheological (viscous) behavior of suspension electrodes is important because it determines the mechanical (flow) properties of the flow electrode. Suspensions take on different flow behaviors and can be characterized as liquid-like, gel-like and solid-like. These properties have implications for electron percolation, which will be described in more detail in the rheology section.

The electrochemical behavior of the suspension electrode is reliant on the electrolyte, the active material, and the underlying material organization (microstructure) that forms as result of individual material particles coming into and out of contact with each other. A typical film electrode is 100 μm or less, while suspension electrodes can be on the order of mm thicknesses. As electrodes become thicker (larger channel depth), the capacitance has been shown to decrease as material utilization is decreased.

The active material size, morphology, and compositional loading all affect the capacitance and available energy density of a suspension electrode. Figure 11.8 demonstrates the electrochemical properties of a suspension electrode composed of spherical activated carbon derived from phenolic resin and anisometric activated carbon. Each suspension is composed of 18 wt% active material (CS or AC), 2 wt% carbon black, and 80 wt% 1M sodium sulfate (1M $NaSO_4$) electrolyte. At 2 mV s^{-1}, the cyclic voltammogram of the CS-based suspension electrode appears rectangular, which is characteristic of an ideally polarizable material system (mostly electric double layer charge storage). At 0.75V the slope of the curve is nearly 0 indicating low resistance across the electrode. The activated carbon-based suspension electrode displays a more curved cyclic voltammogram. This usually occurs in more resistive or thicker electrodes, where efficient electron and percolation pathways cannot be achieved. Moreover, we see that the capacitance of the anisometric activated carbon is significantly less than that of the carbon spheres. This can be possibly attributed to the increase in SSA (1576 m^2 g^{-1} vs. 1472 m^2 g^{-1}). Nevertheless, the 5–8 % increase in SSA cannot attribute to the nearly 30 % increase in capacitance between the AC and CS. Thus, the decrease is most likely due to inefficient utilization of the material across the thickness of the electrode. In the case of the carbon spheres, connectivity is fairly uniform across the electrode structure; however, for activated carbon connectivity can be highly random due to agglomeration of the particles. This can lead to interstitial void regions. At 20 mV s^{-1} the same trend holds, between the AC and CS. The drop in capacitance is more significant for the AC, indicating that the connectivity of the flow-electrode active material plays a role in the rate performance and material utilization of the flow electrode.

The active material morphology also plays a role in the flow properties of a suspension electrode, as shown by the rheogram in Fig. 11.8e. Both suspensions show shear-thinning behavior, as there is a decrease in viscosity with increasing shear rate. Across all shear rates studied, the suspension based on spherical activated carbon is less viscous. The increased viscosity seen in the AC may be in part to the high agglomeration of the particles. In a suspension, van der Waal's forces of attraction are strong between small "nano-sized" particles and in high-concentration solutions. Moreover, these agglomerates formed will be of irregular shape, have a greater propensity to drag, and catch on edges of neighboring particles. From prior work, slurries with larger particles have shown lower viscosities than smaller particles with the same solid fraction. In addition, suspension with more uniform size distributions and greater sphericity have shown lower viscosities than those with wide size distributions and anisometric particles which agrees with the previous findings [62].

11.4.1 Oxygen-Enriched Carbons as Active Materials

The active material's morphology, compositional loading, electrolyte, and flow behavior all dictate electronic and ionic transport properties in a suspension electrode [28, 29, 63–66]. Since both types of suspension electrodes (EFC and FCDI) rely on the active material for charge storage (or ion removal), it is ideal to have a high active material mass loading. In desalination applications, which rely on the overall surface area of the active material for salt removal, a greater mass loading will lead to more efficient desalination. Furthermore, with high-mass loading, the

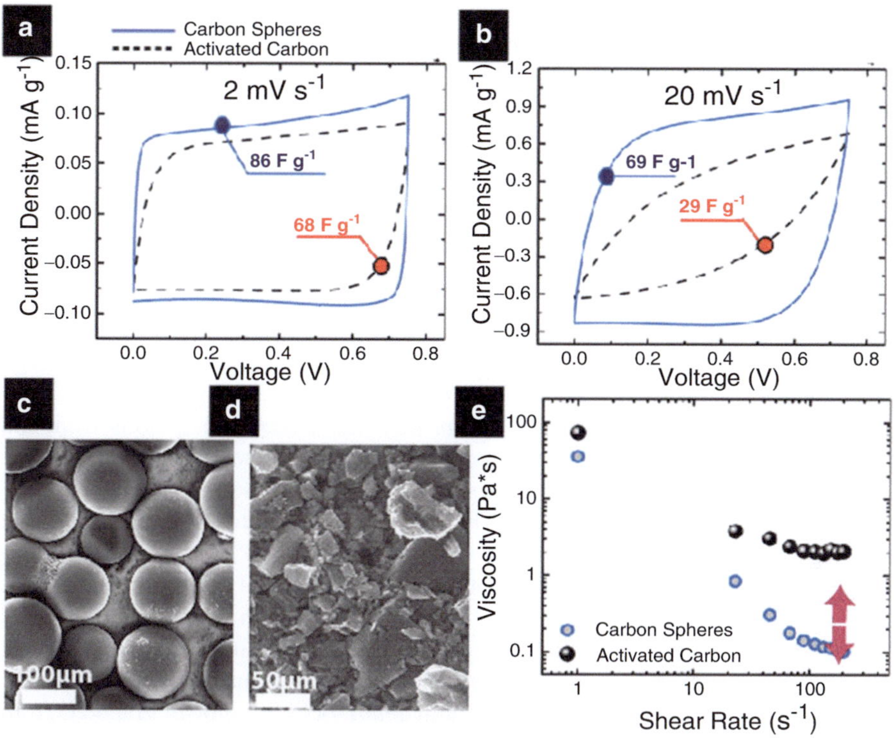

Fig. 11.8 Cyclic voltammetry at 2 mV s^{-1} for suspension electrodes composed of 20 wt% CS and 20 wt% AC (**a**) and 20 mV s^{-1} (**b**). SEM micrographs of the active material CS (**c**) and anisometric AC (**d**) and shear-controlled rheogram of the suspension electrodes composed of CS and AC (**e**) (Reprinted with permission from [28]. Copyright 2012 Elsevier)

electrode has a greater number of contact points, which enhances charge transport through the creation of percolation networks (Fig. 11.5a). Thus, for charge storage and transport properties, high-mass loading is advantageous with the limiting aspect being the flowability (rheological properties) of the suspension electrode. It has been shown that with increasing carbon content, the suspension electrode takes on different structural properties such as weak and strong gel characteristics. Moreover, strong gels demonstrate decreased resistances and in turn have been shown to require less electrical energy during operation [45]. The extent of gelation is directly related to the percolation threshold (the limit for which electrons freely flow through the thickness of the electrode) [59]. Needless to say, there are inherent tradeoffs between active material loading, in terms of viscosity, and key performance characteristics, such as energy and power density and energy efficiencies in suspension electrodes.

While surface chemistry has been widely studied within the supercapacitor community as a means to achieve additional charge storage through pseudocapacitive reactions, its role in flowable electrodes is slightly different as it affects both the viscosity of the slurry and wettability of the carbon [67–72]. Aqueous (water-based) electrolytes may not be able to fully penetrate hydrophobic activated carbon. The inability to completely "wet" the surface area limits full utilization of the material, as ions are limited to regions where the electrolyte can reach. Oxidizing carbon surfaces changes the surface from hydrophobic to hydrophilic, and thus oxidation has been shown to be an avenue toward increasing the wettability of the carbon and increasing surface accessibility for ions (and increase material capacitance) [70] in a flow electrode or film electrode.

Unique to suspension electrodes (flow electrodes) is the dynamic nature in which the suspended material interacts with each other. When a particle is submerged in an aqueous solution, it develops a surface electrical charge. When placed in a solution between the pH range of 5–9, the particle usually obtains a negative charge [73]. There are two primary mechanisms that cause surface charge to exist: (1) dissociation of ionogenic groups and (2) preferential adsorption of ions from the solution. The pH of the solution directly affects the surface charge (at low pH most particles exhibit a positive charge; as the pH is raised the surface charge switches from negative to positive). The point where it switches between positive and negative charge is known as the isoelectric point (IEP) or the point of zero charge [74]. As a result of this surface charge, a layer of counterions (ions of opposite charge) surround the particle. This layer is known as the diffuse double layer. Particles in the same solution tend to gain the same surface charge and thus naturally repel each other leading to a stable suspension of material. Most recently, stability of suspensions has been looked at in depth for exfoliation of graphene from graphite [75] and for processing nanomaterials like carbon nanotubes [76]. Nevertheless, for suspension electrodes isolated aggregation may have positive effects such as contributing to paths for enhanced transport properties.

These interactions between particles can be described by Derjaguin, Landau, Verwey, and Overbeek (DLVO) theory for colloidal stability. DLVO theory describes the interactions between particles as the summation of the forces of attraction (van der Waals) and repulsion (electrostatic) (Fig. 11.9a, b). In the case described above, when a material is placed in water, repulsion forces may dominate as the particles generate like charges. For suspension electrodes, our active material is going to be mixed with a high concentrated salt solution and thus strong aggregation effects are going to occur naturally. When the active material in a suspension is oxidized, the surface charge increases, in turn increasing the repulsion force contribution and suspension stability.

Figure 11.10a shows x-ray photoelectron spectra for AC and AC that has been lightly oxidized in nitric acid. Only a slight decrease in the BET specific surface area was observed after oxidation of the samples (~1322 $m^2\ g^{-1}$–1270 $m^2\ g^{-1}$). Oxidation can result in collapsing of micropores and decrease in SSA. Figure 11.10b shows the rheological properties of the suspension electrodes based on AC and oxidized AC. The suspension electrode based on 28 wt% oxidized AC demonstrated similar flow behavior to a suspension electrode composed of 20 wt% AC. Furthermore, the oxidized sample at 20 wt% was shown to be less viscous in nature than the unoxidized 20 wt%. These results clearly indicate the presence of

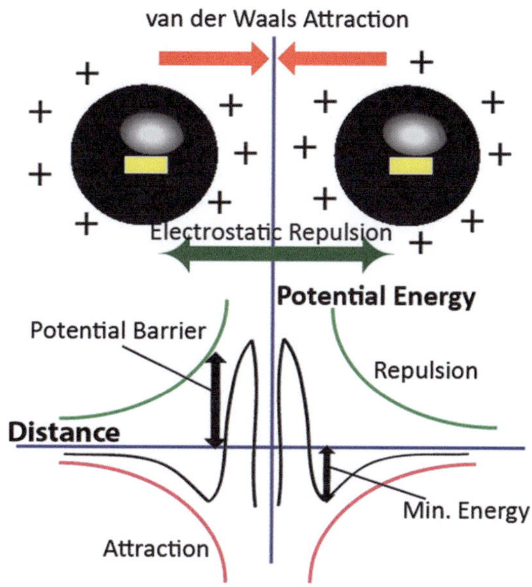

Fig. 11.9 Force interactions between active materials in a suspension electrode (**a**) and total DLVO profile as a function of the van der Waals and double layer repulsion forces (**b**)

oxygen functional groups positively enhances the flowability of suspension electrodes. The increased flowability may be in part due to a decrease in hydrophobic interactions between carbon particles as the surface becomes more acidic and hydrophilic in nature [77] and increased repulsion forces between particles for greater suspension stability (flowability).

Oxidizing the active material not only changes the flow behavior of the suspension electrode but also changes the electrochemical and system level operational properties. Figure 11.10c shows the normalized effluent concentration from a FCDI cell similar to Fig. 11.5. As a constant current of 30 A m^{-2} is applied across the flow cell, the concentration at the effluent decreases linearly over time, with no real apparent difference between the flow electrode composed of AC and oxidized AC. However, when you look at the subsequent deionization voltage during the same period, the oxidized sample demonstrates an increased (and increasing) voltage with time. The increased deionization voltage seen in the suspension electrode based on oxidized material is due to an increased resistance in the material as a result of the oxidation. Thus, although there is an observed decrease in viscosity in the oxidized sample due to greater suspension stability, it comes at the cost of higher deionized voltage (electrical energy). Thus, in the design of active materials for suspension electrodes, there are many parameters to consider, including viscosity, energy density, power density, and material resistances.

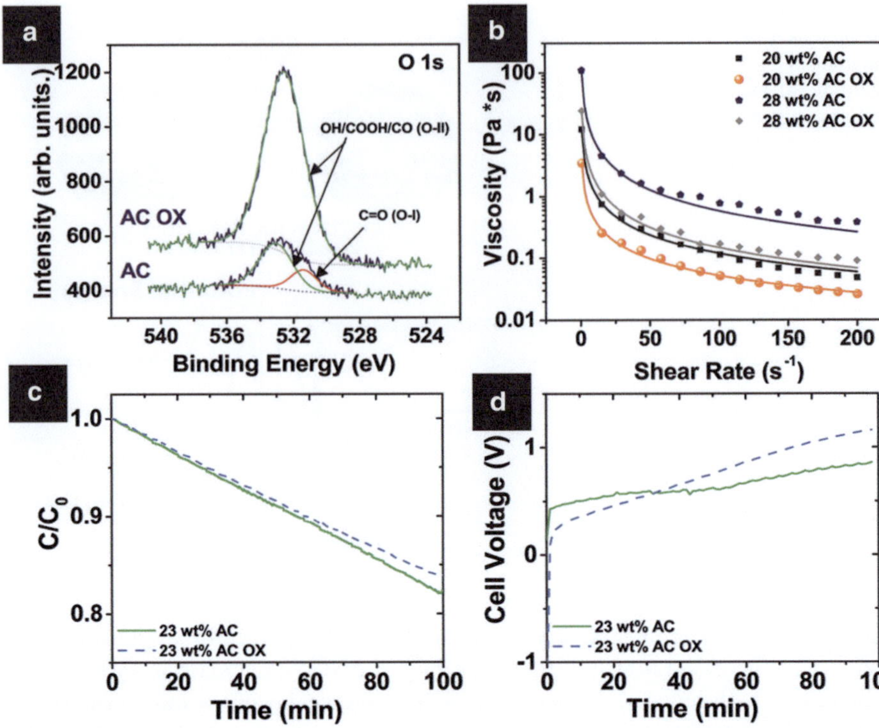

Fig. 11.10 X-ray photoelectron spectroscopy characterization of AC and oxidized AC and resulting rheological properties for different suspension (**b**). Normalized effluent concentration from FCDI cell operating in constant current (30 A m^{-2}) operation (**c**) and resulting deionization voltage for suspension electrodes based on oxidized material and non-oxidized material (**d**) (Reprinted with permission from [45]. Copyright 2015 American Chemical Society)

11.5 Increasing the Energy Density in Aqueous-Based Suspension Electrodes

Although suspension-type electrodes exhibit high-power density, they are limited in terms of their energy density. In order to compete in applications such as grid-level energy storage or in desalination applications, its energy density needs to be improved. The energy released by a suspension electrode, E, is

$$E = 0.5\,\mathrm{CV}^2 \tag{11.6}$$

where C is the capacitance (in F), and V is the operational voltage window (in volts). In order to increase the energy density, one can either increase the voltage window or the active material's capacitance. For aqueous electrolytes, the voltage window is thermodynamically limited to approximately 1.0 V because water decomposition occurs above 1.23 V [78]. Thus, a fundamental challenge resides in developing a suspension media that can withstand a large voltage window and/or has a high capacitance. Batteries which utilize faradic charge storage mechanisms are characteristically energy dense (low power), while electric double layer capacitors (EDLC) offer high power, but low energy. One way to increase the energy density in EDLCs is to make them more "battery-like." Pseudocapacitors (or redox electrochemical capacitors) and hybrid systems are two avenues toward increasing the energy density in electrochemical capacitors. They both display battery and EDLC characteristics as they store charge via electric double layer and faradic processes, with no or little diffusion limitations. In this section we focus on three different methods of achieving increased energy density is carbon-based suspension systems: (1) soluble carbonyl-based organic molecules with a carbon suspension (pseudocapacitor), (2) asymmetric device AC/metal oxide (hybrid) systems, and (3) soluble metal-redox couples combined with carbon suspensions (pseudocapacitor) or a "suspension redox flow battery."

11.5.1 Carbonyl-Based Organic Molecules as Redox Mediators

Carbonyl-based organic molecules (C-bOMs) can be used to increase the capacitance in aqueous carbon-based energy storage devices through the addition of fast and reversible redox reactions at the electrode|electrolyte interface (Fig. 11.11a). These C-bOMs can either be grafted on the surface of the electrode [80, 81] or dissolved in the electrolyte (redox-active electrolyte) [63, 82–84]. The majority component in suspension electrodes in the EFC/SSFB is the electrolyte, thus making the electrolyte a source for additional capacitance (faradic reactions), which is one way to achieve increased energy density.

Hydroquinone [82, 83], indigo carmine [84], p-phenylenediamine (PPD) [63, 85–88], and m-phenylenediamine [89] are just a few examples of organic compounds explored as C-bOMs for redox-active electrolytes in electrochemical capacitors. Figure 11.11a demonstrates cyclic voltammetry for a 23 wt% carbon sphere suspension with different concentrations of PPD dissolved in potassium hydroxide (KOH). In general, soluble C-bOMs undergo proton-coupled electron transfer (PCET) reactions at the electrode interface and are only stable within a specified pH range. Aromatic diamines such as PPD are examples of a C-bOMs that are stable in basic (proton deficient) solutions.

The CV of the suspension without any PPD is rectangular indicating mostly EDL charge storage (hashed region in Fig. 11.11a). However, as PPD (redox mediator) is added into the working electrolyte, there are noticeable broad peaks in the CV around 0.4 V [63]. These peaks can be attributed to faradic (electrochemical reactions) occurring along with electric double layer charge storage. Specifically these peaks can be ascribed to oxidation and reduction of p-phenylenediamine to p-phenylenediamine through the following two-proton, two-electron process:

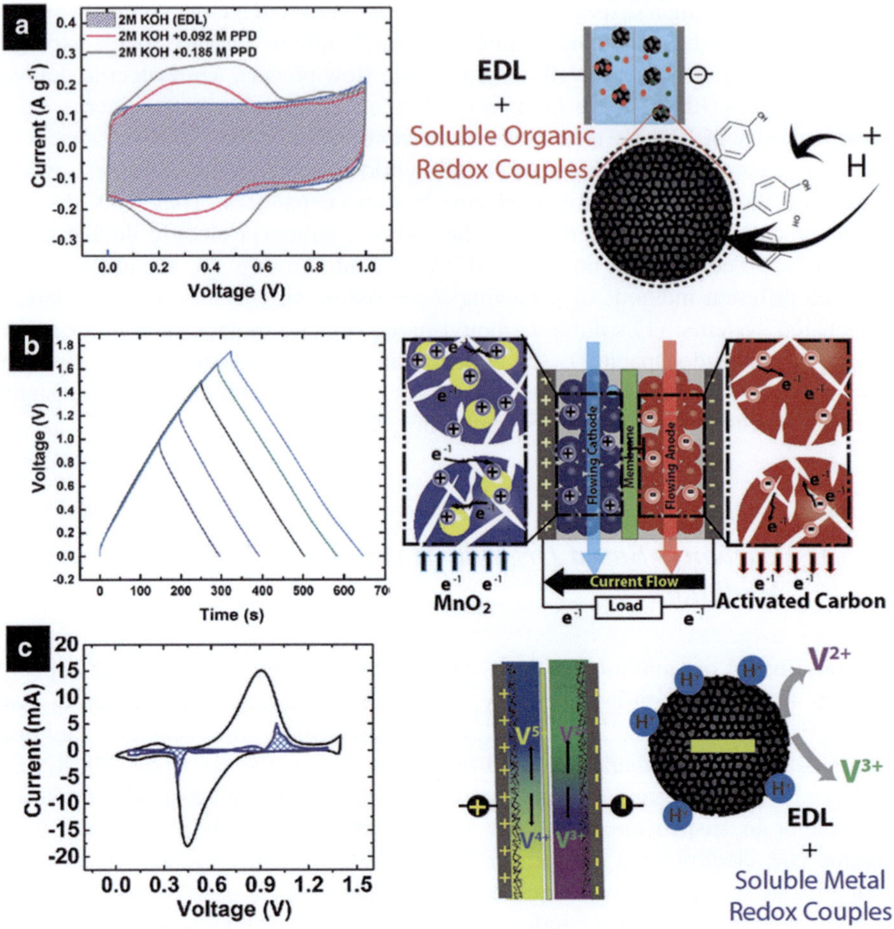

Fig. 11.11 Hybrid approaches for increasing the energy density in the EFC. Carbonyl-based organic molecules can be used as a redox mediator in an electrolyte for additional charge storage (**a**); an asymmetric configuration can be utilized for increasing the voltage window (**b**) and soluble metal-redox couples (VO_2^+/VO^{2+}) in sulfuric acid (**c**). Reprinted with permission from [29, 63, 79] (Copyright 2013, Elsevier, Copyright 2015, American Chemical Society, Copyright 2015, The Electrochemical Society)

$$C_6H_8N_2 \leftrightarrow C_6H_6N_2 + 2H^+ + 2e^- \tag{11.7}$$

The capacitance from a cyclic voltammetry is

$$\mathrm{C} = \frac{2}{\Delta E} x \frac{\int i\mathrm{dV}}{\nu \mathrm{m}} \tag{11.8}$$

where ΔE is the device's voltage window, I is the discharge current, V is the voltage, ν is the sweep rate, and m is the mass of one electrode. The capacitance is proportional to the area between the curves, and the presence of the peaks yields an increase in capacitance in the suspension electrodes. Moreover, increasing the amount of PPD dissolved in the electrolyte results in more charge storage (capacitance) and is represented by increased peak heights. The CVs in Fig. 11.11a are at a rate of $2\ \mathrm{mV\ s^{-1}}$, and the associated capacitances for each curve are $118\ \mathrm{F\ g^{-1}}$ (2M KOH), $175\ \mathrm{F\ g^{-1}}$ (2M KOH + 0.092 M PPD), and $225\ \mathrm{F\ g^{-1}}$ (2M KOH + 0.185 M PPD). Soluble C-bOMs in aqueous carbon-based suspensions represent a way to make significant impact in terms of energy density.

11.5.2 Asymmetric Activated Carbon/MnO_2 Suspension Systems

Asymmetric configuration is a common avenue currently being explored as a way to overcome voltage limitations in aqueous environments. In an asymmetric device, the anode and cathode are made of dissimilar materials. Typically, the anode is AC, while the cathode is a metal oxide such as MnO_2. The voltage window is expanded based on fact that you are able to take advantage of different potential windows in each electrode. Specifically, when MnO_2 is polarized positively, potential values greater than the thermodynamic potential for O_2 evolution can be achieved because of the faradic reaction at the electrode interface leading to the reduction of Mn (IV) to Mn(III). Moreover, AC can be polarized negatively and pushed to higher overpotential values for H_2 evolution because nascent hydrogen can be adsorbed in carbon's micropores and released during anodic oxidation [71, 90, 91] Thus, this method allows the voltage window to be expanded in an aqueous electrolyte (Fig. 11.11b). In symmetric systems, the onset of hydrogen evolution and oxidation limits the potential window [92]. Thus, by optimizing electrode design and by balancing the mass of the electrodes with respect to their associated capacitance, extended voltage windows and thus higher energy density can be achieved.

Figure 11.11b shows how the voltage window can be increased for an asymmetric suspension system, with AC as the anode and amorphous MnO_2 as the cathode [29]. It should be noted that metal oxides, common to battery systems, are typically less conductive than activated carbon, so metal oxides in suspension electrodes usually need a greater contribution of conductive additive (carbon black $\sim$4 wt%). This will have implications for not only the rate performance but also the rheological properties.

11.5.3 Soluble Metal-Redox Couples

Finally, soluble metal-redox couples can be utilized for additional charge storage in a similar way to C-bOMs. Redox-flow batteries (RFBs) utilize soluble ionic species for reversible energy storage and all charge storage is completed in a liquid (solution) phase [79]. In contrast to suspension electrode, the electrodes utilized for initiating the chemical reactions are stationary and usually a carbon felt or carbon fiber mat. RFBs are fundamentally limited by the solubility of the redox species, and sluggish reactions kinetics in the bulk of the electrolyte. The latter challenge is especially true for Vanadium-based systems which are fundamentally rate-limited. To improve reaction kinetics of each half-reaction (at the anode and cathode), the RFB community has concentrated a significant focus on developing the electrodelelectrolyte interface for decreased kinetic, ohmic, and transport losses. This has involved a significant amount of research on developing an electrode architecture for efficient convection processes. Examples of electrode architectures that have been examined include flow-through, flow-by, and interdigitated flow field configurations [93, 94]. Flowable (suspension) electrodes offer a fourth architecture that can be considered, as they provide a greater region (volume) of percolation beyond the surface of the electrode and decreased diffusion lengths for the ionic species to react. Figure 11.10c shows a carbon nanotube (CNT) flow electrode submersed in a redox-active electrolyte (1M $VOSO_4$/4 M H_2SO_4). The overall composition is 10 wt% CNT—90 % 1M $VOSO_4$/4 M H_2SO_4 [79]. CNTs alone are low-surface area active material and usually demonstrate low-energy densities $\left(< 0.05 \text{ Wh l}^{-1}\right)$; however, the energy density nears 2 Wh l^{-1} in the redox-active electrolyte containing the VO^{2+}/VO_2^{+} redox couple. The hatched region in Fig. 11.10c is an estimate of the electric double layer charge storage, while the remainder of the cyclic voltammogram is faradic charge storage. Combining typical metal-ion redox couples with carbon suspensions may be an avenue toward synergistically hybridizing the EFC with RFB technologies. Moreover, in Fig. 11.11c only one redox couple is utilized. In typical RFBs usually the anode and cathode contain different redox couples, which offers greater capacitance and higher voltage windows. These are open areas for research and study within suspension electrodes.

11.6 Experimental Testing of Suspension Electrodes

As suspension electrodes are inherently different from traditional film electrodes used in batteries and supercapacitors, there are slightly different test methods (protocols) employed. Much of the analysis and interpretation of the electrochemical data is similar to traditional supercapacitors and batteries, so in this section we focus on the methods rather than data analysis.

11.6.1 Electrochemical Characterization of Static Flow Electrodes

The most common experimental tests for suspension electrodes are static electrochemistry experiments. Static or stationary tests are usually done in a symmetric two-electrode configuration as shown in Fig. 11.12a. The EFC/FCDI utilizes suspension electrodes that are aqueous and can be assembled outside a glove box and thus this configuration works well. For carbon-based suspensions, all of the components (electrolyte, activated carbon, and carbon black) are measured to their desired ratio and mixed together for 1–2 h with a magnetic stirrer. Activated carbon is hydrophobic, so mixing can be challenging without the addition of heat. In the case of heating a suspension, it is customary to add an incremental (~200–500 mg) of deionized water, and then bring the suspension to a boil, and evaporate the excess water. Heating the suspension allows for greater wettability of the carbon, which allows for more stable and well-dispersed suspensions (Fig. 11.12b).

The two-electrode configuration in Fig. 11.12a is composed of two rectangular stainless steel blocks, with a latex gasket placed at the surface. The latex gasket has a defined thickness and serves as a template for isolating a suspension electrode. Gravimetrically equal amounts of suspensions are placed in the anode and cathode compartment and then the cell is assembled similar to Fig. 11.12c, with a separator between the anode and cathode compartment. It is important to note that contact resistance at the current collector is a challenge with suspension electrodes. In the block cells shown in Fig. 11.12, the cell is tightened with six screws to ensure good contact. In nonaqueous suspension electrodes (such as those seen in the semisolid battery), modified Swagelok systems have also been utilized. In these setups there is a compartment in the current collector for the cathodic suspension, and the other half of the cell is set aside for a lithium metal or another counter electrode. These cells usually have a spring closure to provide good contact between the current collector and the electrodes.

11.6.2 Intermittent Flow Tests

Intermittent flow tests are often performed on the optimized flowable electrode to better assess realistic flow conditions [95–97]. In this test uncharged suspensions are simultaneously pumped into the electroactive region via controlled syringes (Fig. 11.12d). The samples are then charged at 0.8 V and then pumped out the cell (Fig. 11.12e). After a set period of time, the suspension is reversed back into the cell and discharged at 0V (Fig. 11.12f). The resulting chronoamperometric profile can be examined to calculate the electrochemical performance of the flowable electrode under intermittent flow mode. In general, the discharge curve is used in extracting the gravimetric capacitance using the following equation:

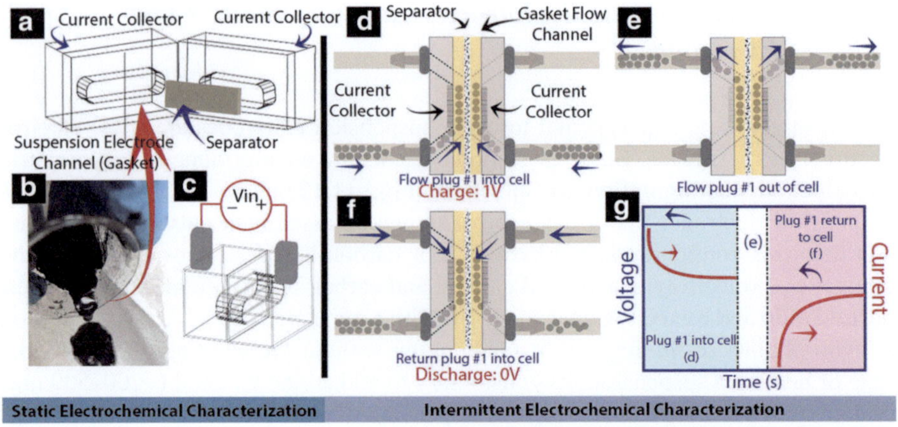

Fig. 11.12 Static electrochemical test setup in an exploded view (**a**) for suspension electrodes (**b**) and assembled for electrochemical characterization (**c**). The experimental procedure for assessing the material systems under flowing operation (**d–f**). In practice, slurry is pumped into the cell and charged at 1V (**d**) and then the suspension if pumped out of the cell (**e**) and finally returned into the cell for discharge (0V) (**f**). An example of the chronoamperometric data from the operation of the intermittent flow operation is shown in (**g**)

$$C_{\mathrm{gr}} = \frac{\frac{2}{\Delta E}\int i\mathrm{d}t}{m} \tag{11.9}$$

where ΔE is the discharge voltage window and m is the mass of carbon in one electrode. The volumetric capacitance, $C_{\mathrm{vol.}}$ was calculated as:

$$C_{\mathrm{vol}} = C_{\mathrm{gr}} \cdot \rho_{\mathrm{slurry}}\ , \tag{11.10}$$

where ρ_{slurry} is the density of the active slurry.

11.6.3 Electrical Conductivity of Suspension Electrodes

The electrical conductivity of suspension-based electrodes is an important parameter which describes the connectivity of active material and percolation pathways established. Suspension electrodes can be considered composite materials composed of both insulating and conducting material. Typically suspension electrodes contain three primary components: (1) an ion-conducting electrolyte, (2) solid active materials, and (3) a conductive additive (usually carbon based). In some cases the solid active material is conducting and enables efficient percolation pathways, while in other cases, such as that seen in polysulfide battery, the active material is either insulating or of low conductivity. The conductivity of a suspension provides an understanding of the degree of aggregation between particles and the electrical resistances that occur because of loss of contact points. Moreover, conductivity is especially important for suspension electrodes because under a shear environment, the material particles may come out of contact and decrease the conductivity during flow conditions. Thus, for comparison between samples and across literature, conductivity is an important property to be reported. Conductivity measurements of suspension electrodes can be done under alternative current (AC) or direct current (DC) methods [98]. Below we have outlined the two methods and procedures typically used:

11.6.3.1 Direct Current Conductivity Measurements

Measuring electrical conductivity of suspensions is a common practice within the field of nanofluids (commonly explored in heat transfer applications) and for mixed conducting composite materials for flexible and electrically conducting material systems (Fig. 11.13). Suspension electrodes for energy storage or desalination applications are very similar and have adopted similar protocols for reporting electrical conductivity values. Suspension electrodes are similar to the two previously described examples, but differ in two primary ways: (1) flow electrodes typically are significantly above the percolation threshold and (2) suspension electrodes usually contain an ionic solution and thus exhibit both ionic and electronic conductivities. This element of ionic conductivity often distorts the reported value for suspension electrodes. It is preferable to report the electrical conductivity of suspension electrodes with and without the salt in the solvent.

DC measurements can be obtained either through chronoamperometry (Fig. 11.12a) or cyclic voltammetry experiments (Fig. 11.12b). These tests can be completed in a similar methodology as the static electrochemical experiments described above, but with only one blocking electrode used as demonstrated in Fig. 11.12c. Using the chronoamperometric method, the voltage hold period should be 5–10 min (depending on the volume of suspension) used in the experiments. Once a stationary current is obtained, that value can be related to electrical resistivity and electrical conductivity:

$$\rho = \frac{\Delta V}{\Delta I_{\mathrm{DC}}} * \frac{\mathrm{SA}}{t} \tag{11.11}$$

$$\sigma = \frac{1}{\rho} \tag{11.12}$$

where SA is the surface area of the suspension electrode in contact with the current collector, t is the electrode thickness, and V and I are the applied voltage and steady-state current. In cyclic voltammetry the ratio of $\left(\frac{\Delta V}{\Delta I_{\mathrm{DC}}}\right)$ is acquired from the slope of the cyclic voltammetry. All suspension electrode electrical conductivities should be normalized to the test cell and electrical conductivity of the electrolyte. This can be

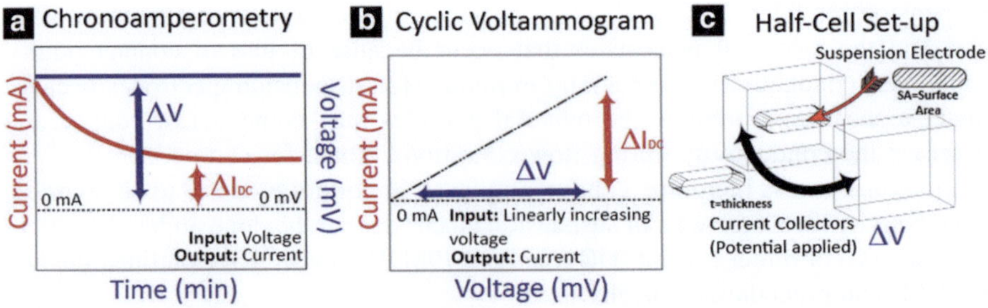

Fig. 11.13 Summary of direct current methods for measuring conductivity. Constant current chronoamperometry (**a**) and cyclic voltammetry (**b**) methods can be used in a half-cell (blocking electrode) setup (**c**) to extract conductivity measurements

accomplished by filling the cell with the electrolyte or solution, recording the I-V relations, and then normalizing the resistivity as:

$$\rho = \left[\left(\frac{\Delta V}{\Delta I_{\mathrm{DC}}} \right)_{\text{suspension}} - \left(\frac{\Delta V}{\Delta I_{\mathrm{DC}}} \right)_{\text{solution}} \right] * \frac{\mathrm{SA}}{t} \tag{11.13}$$

The electrical conductivity in the solution is not negligible in suspension electrodes and can result in overestimation of the electrical conductivity.

11.6.3.2 Alternating Current Electrical Conductivity Measurements

Electrical conductivity can also be extracted from an equivalent circuit and electrochemical impedance spectroscopy. The equivalent circuit superimposed on the Nyquist plot in Fig. 11.14 is an example of one equivalent circuit that is commonly used to extract the electronic resistance. In this circuit, R_i represents the migration of ions, while R_e represents the migration of electrons. The high-frequency intercept of the real axis is associated with the ionic and electronic resistance placed in parallel, while the high-frequency intercept is the electronic resistance. The electronic resistance can be used in a similar way as described in the DC section. Specifically, $R_e = V/I_{\mathrm{AC}}$, and then the resistivity and conductivity can be calculated accordingly. The semicircle is characteristic of a mixed ionic and electronic conductor. A sample that is purely electronically conducting will exhibit a very small semicircle and an inductive line, while a highly insulating system will have a semicircle characterized by a large diameter. Table 11.3 provides a summary of the different materials studied and their reported conductivities. Battery materials are typically much more insulating in comparison to activated carbon. In the case of references 34 and 64, samples were measured in an ionic solution. Moreover, in reference 64, the high-frequency intercept was utilized as the electronic resistance instead of the low frequency intercept, which leads to an overestimation of the electronic conductivity.

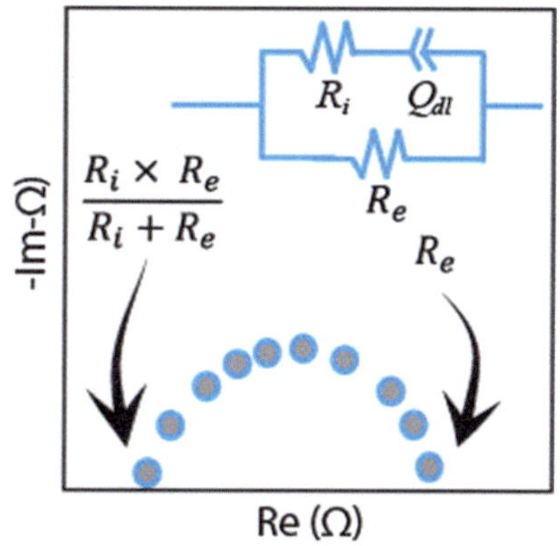

Fig. 11.14 Equivalent circuit superimposed on Nyquist plot for measuring electronic resistance (and conductivity) of a suspension electrode using AC methods

11.6.4 Rheological Properties of Suspension Electrodes

Suspension electrodes are composed of both solid and liquid constituents and display both liquid-like and solid-like properties. A purely elastic material, when perturbed from its ground state by an applied stress, will stretch, but ultimately recoil because all the input energy is stored and recovered [99]. In contrast, viscous materials resist shear flow and exhibit strain behavior in response to an applied stress. Suspension electrodes demonstrate both elastic and viscous material properties and can be considered a viscoelastic material [59, 66, 100]. Rheology, is the study of deformation and flow of materials and material systems and a new and emerging area is the rheology of suspension electrodes. Specifically, there is a need for a better understanding electron and ion transport kinetics are altered when a suspension electrode is exposed to shear dynamics [66]. Moreover, there is a growing need to develop the best methods for flowing suspension electrodes for energy-efficient charge storage or ion removal processes, which will require a good understanding of their rheological and electrorheological properties.

Rheology can be used to characterize key material characteristics such as viscosity. Viscosity describes a suspension electrodes' willingness to flow and is an important parameter to report on suspension electrodes. In this section we briefly describe the key qualitative and quantitative rheological methods used to characterize suspension electrodes and describe open areas for exploration that will lend to a better understanding of the mechanical properties of suspension electrodes.

11.6.4.1 Newtonian Versus Non-Newtonian Suspension Electrodes

Newtonian materials are fluids or suspension where the shear stress (τ) is linearly related to the shear rate ($\dot{\gamma}$):

Table 11.3 Reported suspension electrode electrical conductivities, including active material, electrolytes, and additives

Active material	Conductive additive	Electrolyte	Conductivity ($mS\,cm^{-1}$)	Method	Suspension or film	Reference
$LiFePO_4$	None	None	$3.7^{*}10^{-6}$	EIS	Film	[98]
AC	None	None	20	–	Film	–
$Li_4Ti_5O_{12}$ (22.4 v%)	KB (0.6 v %)	1 M $LiPF_6$ Dioxolane (70 v%)	0.01	EIS	Suspension	[24]
$LiCoO_2$ (22.4 v%)	KB (0.6 v %)	1 M $LiPF_6$ in alkyl carbonate (70 v %)	0.06	EIS		
–	KB (0.0007-0.034 v%)	1 M LFTSI in Propylene Carbonate	0.001-0.1	EIS	Suspension	[59]
–	C45 (0.0007-0.034 v%)		0.001-0.01			
$Li_4Ti_5O_{12}$ (15 wt%)	KB (3 wt%)	1 M LFTSI in Propylene Carbonate (82 w%)	0.01	EIS	Suspension	[66]
$Li_4Ti_5O_{12}$ (20 wt%)	KB (3 wt%)	1 M LFTSI in Propylene Carbonate (77 w%)	0.01			
$Li_4Ti_5O_{12}$	KB (3 wt%)	1 M LFTSI in Proplyene Carbonate (72 w%)	0.00001			
2.5 mol-S/L (Li_2S_8/Li_2S)	KB (1 v%)	0.5 M LiTFSI in TEGDME	2	EIS	Suspension	[34]
	KB (1.5 v %)		9			
	KB (2 v%)		18			
AC (14 wt %)	CB (2 wt%)	0.5 M $NaSO_4$	60	EIS	Suspension	[64]
AC (14 wt %)	CB (2 wt%)	1.0 M $NaSO_4$	70			
AC (14 wt %)	CB (2 wt%)	1.5 M $NaSO_4$	80			
AC (5 wt %)	–	17 mM NaCl	0.025	Direct Current	Suspension	[42]
AC (10 wt %)			0.06			
AC (20 wt %)			0.125			

KB Ketjen Black, *C-45* CNERGY™ Super C45 Carbon, *CB* carbon black, 100 % compressed, Alfa Aesar, *v%* volume %, *wt%* weight %

$$\tau = \eta\dot{\gamma} \tag{11.14}$$

where η is the (shear) viscosity. When this relationship does not hold, the material system being studied is considered non-Newtonian and displays complex flow behaviors. Thus, non-Newtonian materials' viscosities are independent of shear rate. Viscosity is an important rheological property of suspension electrodes because it is related to the pumping power (and thus, is related to energy consumption) of a system. Unlike Newtonian fluids, in non-Newtonian materials the rate of shear is a function of the shear stress:

$$\dot{\gamma} = f(\tau) \tag{11.15}$$

Depending on the nature of the function $f(\tau)$, these material systems can be characterized as Bingham plastic, shear-thinning, or shear-thickening suspension electrodes. Figure 11.15a demonstrates the relationship between shear stress and shearing rates for each of these types of systems. In general flow electrodes display shear-thinning or Bingham behavior [29, 34]. Shear thinning materials demonstrate a decrease in viscosity with increasing shearing rate (Fig. 11.15a). A common example of a shear-thinning (also referred to as a pseudoplastic) material is ketchup. Ketchup is a challenging material to flow, but when vigorously shaken (sheared), flows very easily. The power-law model is a commonly used to describe this behavior and is referred to as the Ostwald-deWaele model:

$$\tau = K\dot{\gamma}^{n-1}\dot{\gamma}. \tag{11.18}$$

11.6.4.2 Experimental Details

i. <u>Steady Shear Viscometry Tests</u> The most straightforward method for understanding stress–strain relationships of materials is to impose a shear strain and measure the resulting shear stress (Fig. 11.16a). In these tests macroscopic measurements (torque and rotational velocity) are extracted and related to material properties such as shear stress/shear rate via constitutive laws (Fig. 11.16a) which depend on the apparatus geometry [99, 101, 102]. Figure 11.15b demonstrates rotational geometries that can be employed for steady-shear viscometry tests. Parallel plates and the cone and plate setup have the benefit of demonstrating almost uniform strain behavior through the sample. In contrast, the concentric cylinder arrangement demonstrates nonhomogeneous strains and rates of strain across the thickness as the velocity profile diminishes with distance from in the interior cylinder boundary (Fig. 11.16a). Thus, in general, concentric cylinders are limited to low viscosity samples, whereas cone and plate and torsional parallel plates are used for more viscous samples.

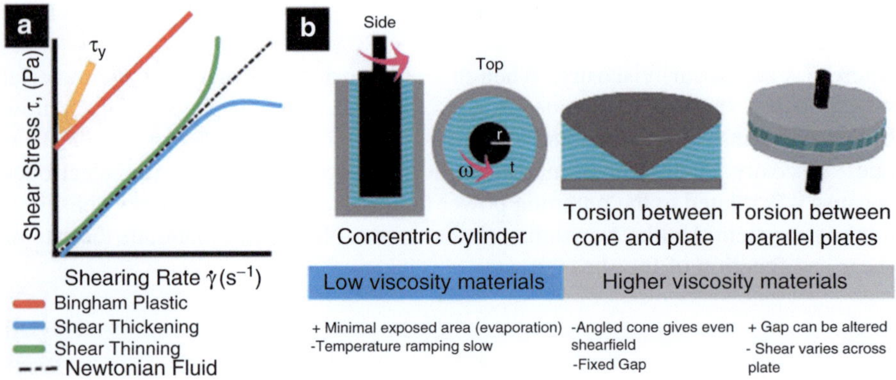

Fig. 11.15 Shear stress vs. shearing rate relationships for different types of fluids (**a**) and experimental geometries for characterizing rheological properties (**b**)

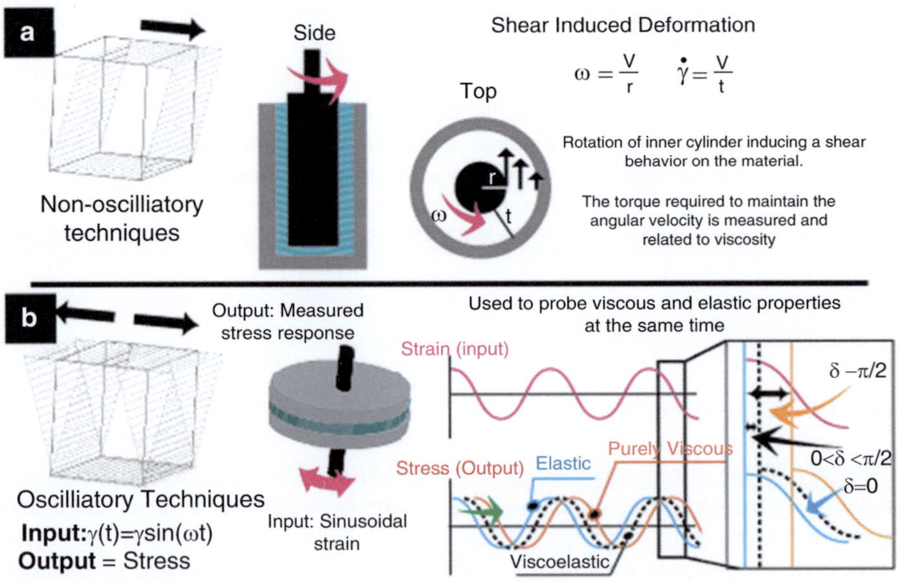

Fig. 11.16 Steady (**a**) and oscillatory (**b**) rheological techniques

Currently, an emerging concern for proper material characterization is understanding shear-banding effects. Shear banding is a phenomena seen in compositionally dense suspensions, where some of the fluid behaves as a liquid (sheared), while other parts of the sample act as a solid (unsheared) [103]. If shear banding occurs, the bulk properties derived from local constitutive laws based on continuum-derived flow profiles will be inaccurate. To address this challenge, different test protocols have been recommended when dealing with "yield stress fluids" (Bingham plastic materials). Typically, samples need to be pre-sheared prior to measurements to eliminate shear history effects of the material. High-rate pre-shears (100 s^{-1}) have been employed on carbon suspension electrodes [59] in order to erase the mechanical history. Next, it is recommended that suspension electrodes are tested at the highest shear rate and then progressively decrease the shear rate over time without passing by a resting period. This procedure helps to avoid shear banding and obtain homogenous flow behaviors. More work needs to be done on characterizing the bulk rheological properties of suspension electrodes to understand their flow behavior. Nevertheless, from steady-state experiments, key material properties such as shear modulus $G \equiv \tau/\gamma)$ and viscous stress $(G \equiv \tau/\dot{\gamma})$ can be obtained. Moreover, samples can be characterized as Newtonian/non-Newtonian and as shear-thinning, Bingham, or shear thickening.

ii. Unsteady Shear Methods—Small Amplitude Oscillatory Frequency Sweeps For complex fluids and material systems, oscillatory or non-steady-state (dynamic) techniques are used to probe mechanical properties (Fig. 11.16b). Oscillatory techniques enable characterization of structural and dynamic properties of suspension electrodes at different time scales, which can lead to a deeper understanding of how the microstructure (arrangement of active material) changes with flow rate. Traditionally oscillatory techniques use similar geometries as seen in the description of steady-shear experiments (Fig. 11.15b). The material (suspension) electrode is either placed between concentric cylinders or between parallel plates and coquette flow is induced via the movement of one of the plates (or by the movement of a cylinder). For oscillatory techniques, the rotating device is set to simple harmonic motion with shear displacements:

$$\gamma(t) = \gamma^0 \sin(\omega t) \tag{11.19}$$

where γ^0 is the amplitude, and ω is the frequency, and γ is the shear. The first derivative of the shear gives us the imposed shear rate:

$$\dot{\gamma}(t) = \omega \gamma^0 \cos(\omega t) \tag{11.20}$$

on suspension electrodes. In this technique, the measured value is the output stress (time dependent) which is also oscillatory in nature (Fig. 11.16b). The relationship between the oscillatory strain and stress measurements (in terms of phase) can lead to qualitative descriptions of the material system.

11.6.4.3 Electrorheology

Electrorheology is an experimental practice used to characterize suspensions and electrical and rheological properties simultaneously. Typically, this encompasses combining a potentiostat (capable of performing electrochemical impedance spectroscopy) and a shear-controlled rheometer. At different shear rates, the microstructure (and thus connectivity) can be altered [59]. This has implications for either increasing or decreasing the conductivity (electron percolation) throughout the suspension electrode. Thus, electrorheology is a powerful technique for characterizing how the microstructure and suspension properties change as a function of shear rate. Future work, in which these microstructural changes can be observed, will lend to an enhanced understanding of the effects of changing connectivity on material system properties. Small-angle neutron scattering is a technique that may enable this study.

11.7 Conclusions

Suspension (or semi-solid or flow) electrodes have been shown to be an avenue toward scalable electrochemical technologies for water deionization, energy generation, and energy storage. Previously, only two secondary battery systems were considered for GES, the redox flow battery and traditional solid state batteries based on film electrodes. The practice of suspending an active material in an electrolytic medium enables a host of new battery technologies to be considered for large-scale grid energy storage and opens the door for a range of cross- and interdisciplinary research topics.

Suspension electrodes are material systems and thus need to be qualitatively and quantitatively examined on a range of levels. The study of suspension electrodes requires not only an understanding of the key material properties but requires an understanding on how these material properties behave in different ionic solutions. In a suspension electrode the material morphology (spherical vs. anisometric) has a bearing on the conductivity and connectivity of the suspension electrode. The connectivity of the material plays a significant role on the overall material utilization and rate performance. Moreover, it was shown that active material surface heteroatoms have an impact on material resistance, suspension stability, and the rheological properties of a suspension electrode. These characteristics have major implications toward scale-up, energy efficiency, and pumping costs. Thus, while it is important to understand key material properties, we emphasize the importance of understanding how these nanoscale properties translate to system and material system level performance in suspension electrodes.

For flow electrode technologies, we are concerned with energy density, power density, and rheological (viscous) properties. It is important to strike an even balance within each of these areas. Carbon-based suspension electrodes (especially when based in aqueous solutions) are going to be fundamentally limited by energy density. Thus, there are several strategies to improve the energy density in suspension electrodes, such as hybridizing the EFC with soluble organic molecules and metal-based redox couples and utilizing asymmetric architectures with pseudocapacitive metal oxide cathodes. Each of these areas has their benefits and disadvantages; however, the versatility and open frontier of organic molecules appears to be a promising route toward high energy density systems based on carbon conducting networks.

Flowable electrode systems based on low-cost, environmentally benign, and abundant materials (e.g., carbon) have the greatest chance of making lasting impacts in the fields of energy storage and water desalination. The subject presented here represents a new and emerging field, which bridges electrochemistry, material science, and aspect of colloidal and rheological systems to describe the fundamental kinetic, mechanical, and electrochemical properties of suspension electrodes. A synergistic understanding and marriage of all of these fields will yield high-performing, scalable, and energy-efficient flowable electrode systems.

Acknowledgment K.B. Hatzell acknowledges support from the NSF Graduate Research Fellowship (Grant # 1002809).

References

1. Dunn B, Kamath H, Tarascon J-M (2011) Electrical energy storage for the grid: a battery of choices. Science 334:928–935
2. Yang Z, Zhang J, Kintner-Meyer MC, Lu X, Choi D, Lemmon JP, Liu J (2011) Electrochemical energy storage for a green grid. Chem Rev 111:3577
3. Liu J (2013) Addressing the grand challenges in energy storage. Adv Funct Mater 23:924–928
4. Darling RM, Gallagher KG, Kowalski JA, Ha S, Brushett FR (2014) Pathways to low-cost electrochemical energy storage: a comparison of aqueous and nonaqueous flow batteries. Energy Environ Sci 7:3459–3477
5. Hostick D, Belzer D, Hadley S, Markel T, Marnay C, Kintner-Meyer M (2012) Renewable electricity futures study. Volume 3: End-use electricity demand (report). National Renewable Energy Laboratory (NREL), Golden, CO
6. Gyuk I, Johnson M, Vetrano J, Lynn K, Parks W, Handa R, Kannberg L, Hearne S, Waldrip K, Braccio R (2013) Grid energy storage (Technical report). Department of Energy (DOE), Washington, DC
7. Potential Reliability Impacts of Emerging Flexible Resources (Technical Report) (2010) North American Electric Reliability Corporation (NERC), Princeton, NJ
8. Hittinger E, Whitacre J, Apt J (2012) What properties of grid energy storage are most valuable? J Power Sources 206:436–449
9. Whitacre J, Wiley T, Shanbhag S, Wenzhuo Y, Mohamed A, Chun S, Weber E, Blackwood D, Lynch-Bell E, Gulakowski J (2012) An aqueous electrolyte, sodium Ion functional, large format energy storage device for stationary applications. J Power Sources 213:255–264
10. Barnhart CJ, Benson SM (2013) On the importance of reducing the energetic and material demands of electrical energy storage. Energy Environ Sci 6:1083–1092
11. Whitacre J, Tevar A, Sharma S (2010) Na $_4$ Mn_9o_{18} as a positive electrode material for an aqueous electrolyte sodium-ion energy storage device. Electrochem Commun 12:463–466
12. Whitacre J, Shanbhag S, Mohamed A, Polonsky A, Carlisle K, Gulakowski J, Wu W, Smith C, Cooney L, Blackwood D (2014) A polyionic, large-format energy storage device using an aqueous electrolyte and thick-format composite $Nati_2(Po_4)_3$/activated carbon negative electrodes. Energy Technol 3:20–31
13. Skyllas-Kazacos M, Chakrabarti M, Hajimolana S, Mjalli F, Saleem M (2011) Progress in flow battery research and development. J Electrochem Soc 158:R55–R79
14. Skyllas-Kazacos M, Rychcik M, Robins RG, Fane A, Green M (1986) New all-vanadium redox flow cell. J Electrochem Soc 133:1057–1058

15. Weber AZ, Mench MM, Meyers JP, Ross PN, Gostick JT, Liu Q (2011) Redox flow batteries: a review. J Appl Electrochem 41:1137–1164
16. Huskinson B, Marshak MP, Suh C, Er S, Gerhardt MR, Galvin CJ, Chen X, Aspuru-Guzik A, Gordon RG, Aziz MJ (2014) A metal-free organic–inorganic aqueous flow battery. Nature 505:195–198
17. Yang B, Hoober-Burkhardt L, Wang F, Prakash GS, Narayanan S (2014) An inexpensive aqueous flow battery for large-scale electrical energy storage based on water-soluble organic redox couples. J Electrochem Soc 161:A1371–A1380
18. Brushett FR, Vaughey JT, Jansen AN (2012) An all-organic non-aqueous lithium-ion redox flow battery. Adv Energy Mater 2:1390–1396
19. Hernández-Burgos K, Burkhardt SE, Rodríguez-Calero GG, Hennig RG, Abruña HD (2014) Theoretical studies of carbonyl-based organic molecules for energy storage applications: the heteroatom and substituent effect. J Phys Chem C 118:6046–6051
20. Burkhardt SE, Lowe MA, Conte S, Zhou W, Qian H, Rodríguez-Calero GG, Gao J, Hennig RG, Abruña HD (2012) Tailored redox functionality of small organics for pseudocapacitive electrodes. Energy Environ Sci 5:7176–7187
21. Glicksman R, Morehouse C (1959) Investigation of the electrochemical characteristics of organic compounds Iv. Quinone compounds. J Electrochem Soc 106:741–745
22. Hernandez-Burgos K, Rodríguez-Calero GG, Zhou W, Burkhardt SE, Abruña HCD (2013) Increasing the gravimetric energy density of organic based secondary battery cathodes using small radius cations (Li+ and Mg2+). J Am Chem Soc 135:14532–14535
23. Nagarjuna G, Hui J, Cheng KJ, Lichtenstein T, Shen M, Moore JS, Rodríguez-López J (2014) Impact of redox-active polymer molecular weight on the electrochemical properties and transport across porous separators in nonaqueous solvents. J Am Chem Soc 136:16309–16316
24. Duduta M, Ho B, Wood VC, Limthongkul P, Brunini VE, Carter WC, Chiang Y-M (2011) Semi-solid lithium rechargeable flow battery. Adv Energy Mater 1:511–516
25. Presser V, Dennison CR, Campos J, Knehr KW, Kumbur EC, Gogotsi Y (2012) The electrochemical flow capacitor: a new concept for rapid energy storage and recovery. Adv Energy Mater 2:895–902
26. Jeon S-I, Park H-R, Yeo J-G, Yang S, Cho CH, Han MH, Kim D-K (2013) Desalination via a new membrane capacitive deionization process utilizing flow electrodes. Energy Environ Sci 6:1471–1475
27. Hatzell M, Hatzell KB, Logan BE (2014) Using flow electrodes in multiple reactors in series for continuous energy generation from capacitive mixing. Environ Sci Technol Lett 12:474–478
28. Campos J, Beidaghi M, Hatzell KB, Dennison C, Presser V, Kumbur EC, Gogotsi Y (2013) Investigation of carbon materials for use as a flowable electrode in electrochemical flow capacitors. Electrochim Acta 98:123–130
29. Hatzell KB, Fan L, Beidaghi M, Boota M, Pomerantseva E, Kumbur EC, Gogotsi Y (2014) Composite manganese oxide percolating networks as a suspension electrode for an asymmetric flow capacitor. ACS Appl Mater Interfaces 6:8886–8893
30. Hamelet S, Tzedakis T, Leriche J-B, Sailler S, Larcher D, Taberna P-L, Simon P, Tarascon J-M (2012) Non-aqueous Li-based redox flow batteries. J Electrochem Soc 159:A1360–A1367
31. Ventosa E, Buchholz D, Klink S, Flox C, Chagas LG, Vaalma C, Schuhmann W, Passerini S, Morante JR (2015) Non-aqueous semi-solid flow battery based on Na-Ion chemistry. P2-Type $Na_x ni_{0.22} co_{0.11}$ $Mn_{0.66}$ O2–$Nati_2$ $(Po_4)_3$. Chem Commun. doi:10.1039/C4CC09597A
32. Zhao Y, Si S, Wang L, Liao C, Tang P, Cao H (2014) Electrochemical study on polypyrrole microparticle suspension as flowing anode for manganese dioxide rechargeable flow battery. J Power Sources 248:962–968

33. Zhao Y, Si S, Wang L, Tang P, Cao H (2014) Electrochemical behavior of polyaniline microparticle suspension as flowing anode for rechargeable lead dioxide flow battery. J Electrochem Soc 161:A330–A335
34. Fan F, Woodford W, Li Z, Baram N, Smith KC, Helal A, McKinley GH, Carter WC, Chiang Y-M (2014) Polysulfide flow batteries enabled by percolating nanoscale conductor networks. Nano Lett 14:2210–2218
35. Chen H, Zou Q, Liang Z, Liu H, Li Q, Lu Y-C (2015) Sulphur-impregnated flow cathode to enable high-energy-density lithium flow batteries. Nat Commun 6. doi:10.1038/ncomms6877
36. Porada S, Lee J, Weingarth D, Presser V (2014) Continuous operation of an electrochemical flow capacitor. Electrochem Commun 178–181
37. Oncsik T, Trefalt G, Csendes Z, Szilagyi I, Borkovec M (2014) Aggregation of negatively charged colloidal particles in the presence of multivalent cations. Langmuir 30:733–741
38. Sinha P, Szilagyi I, Montes Ruiz-Cabello FJ, Maroni P, Borkovec M (2013) Attractive forces between charged colloidal particles induced by multivalent ions revealed by confronting aggregation and direct force measurements. J Phys Chem Lett 4:648–652
39. Elimelech M, Phillip WA (2011) The future of seawater desalination: energy, technology, and the environment. Science 333:712–717
40. Shannon MA, Bohn PW, Elimelech M, Georgiadis JG, Marinas BJ, Mayes AM (2008) Science and technology for water purification in the coming decades. Nature 452:301–310
41. Biesheuvel PM, Porada S, Levi M, Bazant MZ (2014) Attractive forces in microporous carbon electrodes for capacitive deionization. J Solid State Electrochem 1–12. doi:10.1007/s10008-014-2383-5
42. Porada S, Weingarth D, Hamelers HV, Bryjak M, Presser V, Biesheuvel M (2014) Carbon flow electrodes for continuous operation of capacitive deionization and capacitive mixing energy generation. J Mater Chem A 2:9313–9321
43. Biesheuvel PM, Porada S, van der Wal A, Presser V (2013) Water desalination by capacitive deionization. In: Gogotsi Y, Presser V (eds) Carbon nanomaterials, 2nd edn. CRC Press, Boca Raton, p 419
44. Porada S, Borchardt L, Oschatz M, Bryjak M, Atchison J, Keesman K, Kaskel S, Biesheuvel P, Presser V (2013) Direct prediction of the desalination performance of porous carbon electrodes for capacitive deionization. Energy Environ Sci 6:3700–3712
45. Hatzell KB, Hatzell MC, Cook KM, Boota M, Housel GM, McBride A, Kumbur EC, Gogotsi Y (2015) The effect of oxidation of carbon material on suspension electrodes for flow-electrode capacitive deionization. Environ Sci Technol. doi:10.1021/es5055989
46. Jeon S-I, Park J-S, Yeo J-G, Yang S, Choi J, Kim DK (2014) Ion storage and energy recovery of flow-electrode capacitive deionization process. J Mater Chem A 2:6378–6383
47. Gendel Y, Rommerskirchen AKE, David O, Wessling M (2014) Batch mode and continuous desalination of water using flowing carbon deionization (FCDI) technology. Electrochem Commun 46:152–156
48. Biesheuvel P, Van der Wal A (2010) Membrane capacitive deionization. J Membr Sci 346:256–262
49. Hatzell KB, Iwama E, Ferris A, Daffos B, Urita K, Tzedakis T, Chauvet F, Taberna P-L, Gogotsi Y, Simon P (2014) Capacitive deionization concept based on suspension electrodes without Ion exchange membranes. Electrochem Commun 43:18–21
50. Simon P, Gogotsi Y (2012) Capacitive energy storage in nanostructured carbon–electrolyte systems. Acc Chem Res 50:1094–1103
51. Simon P, Gogotsi Y (2008) Materials for electrochemical capacitors. Nat Mater 7:845–854
52. Portet C, Yushin G, Gogotsi Y (2007) Electrochemical performance of carbon onions, nanodiamonds, carbon black and multiwalled nanotubes in electrical double layer capacitors. Carbon 45:2511–2518
53. McDonough JK, Frolov AI, Presser V, Niu J, Miller CH, Ubieto T, Fedorov MV, Gogotsi Y (2012) Influence of the structure of carbon onions on their electrochemical performance in supercapacitor electrodes. Carbon 50:3298–3309

54. Gogotsi Y, Nikitin A, Ye H, Zhou W, Fischer JE, Yi B, Foley HC, Barsoum MW (2003) Nanoporous carbide-derived carbon with tunable pore size. Nat Mater 2:591–594
55. Chmiola J, Yushin G, Gogotsi Y, Portet C, Simon P, Taberna P-L (2006) Anomalous increase in carbon capacitance at pore sizes less than 1 nanometer. Science 313:1760–1763
56. Novoselov K, Geim AK, Morozov S, Jiang D, Katsnelson M, Grigorieva I, Dubonos S, Firsov A (2005) Two-dimensional gas of massless Dirac fermions in graphene. Nature 438:197–200
57. Geim AK, Novoselov KS (2007) The rise of graphene. Nat Mater 6:183–191
58. Zhang C, Hatzell KB, Boota M, Dyatkin B, Beidaghi M, Long D, Qiao W, Kumbur EC, Gogotsi Y (2014) Highly porous carbon spheres for electrochemical capacitors and capacitive flowable suspension electrodes. Carbon 77:155–164. doi:10.1016/j.carbon.2014.05.017
59. Youssry M, Madec L, Soudan P, Cerbelaud M, Guyomard D, Lestriez B (2013) Nonaqueous carbon black suspensions for lithium-based redox flow batteries: rheology and simultaneous rheo-electrical behavior. Phys Chem Chem Phys 15:14476–14486
60. Gogotsi YG, Nickel KG, Bahloul-Hourlier D, Merle-Mejean T, Khomenko GE, Skjerlie KP (1996) Structure of carbon produced by hydrothermal treatment of B-Sic powder. J Mater Chem 6:595–604
61. Basavalingu B, Calderon Moreno JM, Byrappa K, Gogotsi YG, Yoshimura M (2001) Decomposition of silicon carbide in the presence of organic compounds under hydrothermal conditions. Carbon 39:1763–1766
62. Boylu F, Dinçer H, Ateşok G (2004) Effect of coal particle size distribution, volume fraction and rank on the rheology of coal–water slurries. Fuel Process Technol 85:241–250. doi:10.1016/s0378-3820(03)00198-x
63. Hatzell KB, Beidaghi M, Campos JW, Dennison CR, Kumbur EC, Gogotsi Y (2013) A high performance pseudocapacitive suspension electrode for the electrochemical flow capacitor. Electrochim Acta 111:888–897
64. Dennison C, Beidaghi M, Hatzell KB, Campos J, Gogotsi Y, Kumbur E (2014) Effects of flow cell design on charge percolation and storage in the carbon slurry electrodes of electrochemical flow capacitors. J Power Sources 247:489–496
65. Smith KC, Chiang Y-M, Carter WC (2014) Maximizing energetic efficiency in flow batteries utilizing non-Newtonian fluids. J Electrochem Soc 161:A486–A496
66. Madec L, Youssry M, Cerbelaud M, Soudan P, Guyomard D, Lestriez B (2014) Electronic vs ionic limitations to electrochemical performance in $Li_4ti_5o_{12}$-based organic suspensions for lithium-redox flow batteries. J Electrochem Soc 161:A693–A699
67. Seredych M, Hulicova-Jurcakova D, Lu GQ, Bandosz TJ (2008) Surface functional groups of carbons and the effects of their chemical character, density and accessibility to ions on electrochemical performance. Carbon 46:1475–1488
68. Bandosz TJ, Jagiello J, Schwarz JA, Krzyzanowski A (1996) Effect of surface chemistry on sorption of water and methanol on activated carbons. Langmuir 12:6480–6486
69. Conway B, Birss V, Wojtowicz J (1997) The role and utilization of pseudocapacitance for energy storage by supercapacitors. J Power Sources 66:1–14
70. Hulicova-Jurcakova D, Seredych M, Lu GQ, Bandosz TJ (2009) Combined effect of nitrogen-and oxygen-containing functional groups of microporous activated carbon on its electrochemical performance in supercapacitors. Adv Funct Mater 19:438–447
71. Frackowiak E, Beguin F (2001) Carbon materials for the electrochemical storage of energy in capacitors. Carbon 39:937–950
72. Khomenko V, Raymundo-Piñero E, Béguin F (2010) A new type of high energy asymmetric capacitor with nanoporous carbon electrodes in aqueous electrolyte. J Power Sources 195:4234–4241
73. Kim J, Lawler DF (2005) Characteristics of zeta potential distribution in silica particles. Bull Kor Chem Soc 26:1083
74. Morrison ID, Ross S (2002) Colloidal dispersions: suspensions, emulsions, and foams. Wiley-Interscience, New York

75. Li D, Mueller MB, Gilje S, Kaner RB, Wallace GG (2008) Processable aqueous dispersions of graphene nanosheets. Nat Nanotechnol 3:101–105
76. Sun Z, Nicolosi V, Rickard D, Bergin SD, Aherne D, Coleman JN (2008) Quantitative evaluation of surfactant-stabilized single-walled carbon nanotubes: dispersion quality and its correlation with zeta potential. J Phys Chem C 112:10692–10699
77. Sugiura M, Esumi K, Meguro K, Honda H (1985) Surface treatment of meso-carbon microbeads by oxygen plasma. Bull Chem Soc Jpn 58:2638–2640
78. Fic K, Lota G, Meller M, Frackowiak E (2012) Novel insight into neutral medium as electrolyte for high-voltage supercapacitors. Energy Environ Sci 5:5842–5850
79. Hatzell K, Boota M, Kumbur E, Gogotsi Y (2015) Flowable conducting particle networks in redox-active electrolytes for grid energy storage. J Electrochem Soc 162:A5007–A5012
80. Anjos DM, McDonough JK, Perre E, Brown GM, Overbury SH, Gogotsi Y, Presser V (2013) Pseudocapacitance and performance stability of quinone-coated carbon onions. Nano Energy 2:702–712
81. Pognon G, Brousse T, Demarconnay L, Bélanger D (2011) Performance and stability of electrochemical capacitor based on anthraquinone modified activated carbon. J Power Sources 196:4117–4122
82. Boota M, Hatzell K, Kumbur E, Gogotsi Y (2015) Towards high-energy-density pseudocapacitive flowable electrodes by the incorporation of hydroquinone. ChemSusChem. doi:10.1002/cssc.201402985
83. Roldán S, Blanco C, Granda M, Menéndez R, Santamaría R (2011) Towards a further generation of high energy carbon based capacitors by using redox-active electrolytes. Angew Chem Int Ed 50:1699–1701
84. Roldán S, González Z, Blanco C, Granda M, Menéndez R, Santamaría R (2011) Redox-active electrolyte for carbon nanotube-based electric double layer capacitors. Electrochim Acta 56:3401–3405
85. Yu H, Wu J, Lin J, Fan L, Huang M, Lin Y, Li Y, Yu F, Qiu Z (2013) A reversible redox strategy for Swcnt-based supercapacitors using a high-performance electrolyte. ChemPhysChem 14:394–399. doi:10.1002/cphc.201200816
86. Kuang Y, Liu Z, Zhou H, Huang Z, Wang W, Zeng F (2013) Graphene covalenty functionalized with poly (P-phenylenediamine) as high performance electrode material for supercapacitor. J Mater Chem A 1:3454–3462
87. Liu Z, Zhou H, Huang Z, Wang W, Zeng F, Kuang Y (2013) Graphene covalently functionalized with poly (P-phenylenediamine) as high performance electrode material for supercapacitors. J Mater Chem A 1:3454–3462
88. Yu H, Wu J, Fan L, Luo GG, Lin J, Huang M (2012) A simple and high-effective electrolyte mediated with P-phenylenediamine for supercapacitor. J Mater Chem 22:19025–19030
89. Yu H, Fan L, Wu J, Lin Y, Lin J, Huang M, Lan Z (2012) Redox-active alkaline electrolyte for carbon-based supercapacitor with pseudocapacitive performance and excellent cyclability. RSC Adv 2(17):6736–6740
90. Khomenko V, Raymundo-Pinero E, Frackowiak E, Beguin F (2006) High-voltage asymmetric supercapacitors operating in aqueous electrolyte. Appl Phys A 82:567–573
91. Raymundo-Pinero E, Khomenko V, Frackowiak E, Beguin F (2005) Performance of manganese oxide/Cnts composites as electrode materials for electrochemical capacitors. J Electrochem Soc 152:A229–A235
92. Bichat M, Raymundo-Piñero E, Béguin F (2010) High voltage supercapacitor built with seaweed carbons in neutral aqueous electrolyte. Carbon 48:4351–4361
93. Aaron D, Liu Q, Tang Z, Grim G, Papandrew A, Turhan A, Zawodzinski T, Mench M (2012) Dramatic performance gains in vanadium redox flow batteries through modified cell architecture. J Power Sources 206:450–453
94. Darling RM, Perry ML (2014) The influence of electrode and channel configurations on flow battery performance. J Electrochem Soc 161:A1381–A1387

95. Li Z, Smith KC, Dong Y, Baram N, Fan FY, Xie J, Limthongkul P, Carter WC, Chiang Y-M (2013) Aqueous semi-solid flow cell: demonstration and analysis. Phys Chem Chem Phys 15:15833–15839
96. Smith KC, Brunini VE, Dong Y, Chiang Y-M, Carter WC (2014) Electroactive-zone extension in flow-battery stacks. Electrochim Acta 147:460–469
97. Boota M, Hatzell K, Beidaghi M, Dennison C, Kumbur E, Gogotsi Y (2014) Activated carbon spheres as a flowable electrode in electrochemical flow capacitors. J Electrochem Soc 161: A1078–A1083
98. Wang C, Hong J (2007) Ionic/electronic conducting characteristics of Lifepo4 cathode materials the determining factors for high rate performance. Electrochem Solid-State Lett 10:A65–A69
99. Barnes HA, Hutton JF, Walters K (1989) An introduction to rheology. Elsevier, Amsterdam/New York
100. Madec L, Youssry M, Cerbelaud M, Soudan P, Guyomard D, Lestriez B (2015) Surfactant for enhanced rheological, electrical, and electrochemical performance of suspensions for semi-solid redox flow batteries and supercapacitors. Chem Plus Chem, 80(2):396–401
101. Chhabra RP, Richardson JF (2011) Non-Newtonian flow and applied rheology: engineering applications. Butterworth-Heinemann, Amsterdam/Boston
102. Tanner RI (2000) Engineering rheology. Oxford University Press, Oxford
103. Ovarlez G, Cohen-Addad S, Krishan K, Goyon J, Coussot P (2013) On the existence of a simple yield stress fluid behavior. J Non-Newtonian Fluid Mech 193:68–79

Chapter 12
Membrane Separators for Electrochemical Energy Storage Technologies

Patrick T. Nonjola, Nomthandazo Mutangwa, and Hongze Luo

12.1 Introduction

Given the increase in energy consumption as the world's population grows, the shortage of traditional energy supplies (i.e., petroleum, oil, and gas), and the environmental impact caused by conventional power generation systems, it has become imperative to utilize unconventional energy sources and renewables and to redesign traditional processes to make them more energy efficient. Within this context, electrochemical energy storage and conversion play a significant role [1]. In most electrochemical energy technologies, the electrode and electrolyte materials must possess the required ionic and electronic transport properties. A great deal of research is still ongoing at a fundamental level to study and optimize the electrochemistry of candidate materials, composites, and assemblies (such as catalyst and interface designs) [2, 3]. The focus will be on two dominant electrochemical energy storage devices, batteries, and supercapacitors. These devices are usually composed of a positive (cathode) and a negative electrode (anode), separated by a separator and electrolyte solution [4]. Beside active research activities in developing high surface area electrodes and high efficient electrolytes, there has also been growing research activities in developing membrane separator materials utilized in most batteries and supercapacitors technologies [2, 5, 6]. Membrane separator is the critical part of cell component to ensure cell safety and also for converting the chemical energy to electrical energy. The main functions of a membrane separator are (I) to isolate the anode and the cathode so that no electrons can flow between them within the cell and (II) to allow the ions that are contained in

P.T. Nonjola (✉) • N. Mutangwa • H. Luo
Material Science and Manufacturing, Council for Scientific and Industrial Research, PO Box 395, Pretoria +27(0)001, South Africa
e-mail: snonjola@csir.co.za

K.I. Ozoemena, S. Chen (eds.), *Nanomaterials in Advanced Batteries and Supercapacitors*, Nanostructure Science and Technology,
DOI 10.1007/978-3-319-26082-2_12

the liquid electrolyte unrestricted passage between anode and cathode. These ions may be lithium ions (Li^+) in lithium–ion batteries (LIBs) or non-electrochemically active anions or cations in the case of redox flow batteries (RFBs) and electron isolating of counter-charged electrodes in supercapacitors [7]. This chapter will briefly summarize (with no intention to cover all aspects) the current understanding and development prospects associated with well-established membranes and also consolidate the basis for the modification of existing membrane types utilized as separators in electrochemical energy storage technologies. This includes principles of operation (performance and cost targets) as well as the status and challenges in materials (mechanical robustness and porosity–transport properties), chemistries, and technologies.

12.2 Battery Technology

This section will give a brief explanation of the working principle of primary and secondary batteries and clarify the differences between secondary batteries and redox flow batteries. Batteries are by far the most common form of storing electrical energy and have achieved significant commercial success especially in the automotive market with their high energy density and proven record in the consumer electronics market [8]. However, battery comprises of several technologies applying different operation principles and materials and can be demonstrated by two different categories: firstly, **electrochemical batteries**, which consist of (a) primary batteries and (b) secondary batteries. Although primary batteries (non-rechargeable) have their uses, secondary batteries (rechargeable) appear to be more dominant. In primary battery the electrochemical reaction is not reversible, are expensive, and environmental unfriendly due to its disposable character [9]. Secondary batteries allow the reversible electrochemical reaction, where the electrical energy can convert back to chemical energy. These cyclability and reusability are desirable for most energy storage applications. Among the types of secondary (rechargeable) batteries, lithium–ion batteries (LIBs) have gained particular interest because of their higher power densities, high capacity, slow self-discharge rate, low weight, high open circuit voltage, and negligible memory effect [10]. Figure 12.1 depicts the basic elements of a battery [2]. Secondly, the **flow batteries (FBs)** (rechargeable, flow through) are classified into redox flow batteries (RFBs) and hybrid flow batteries. RFB is considered the best option for medium- to large-scale storage owing to an excellent combination of energy efficiency, capital cost, and life cycle costs without specific site requirement [11]. Of all the RFBs developed to date, only the all-vanadium redox flow battery (V-RFB) developed at the University of New South Wales has received more attention due to its high energy efficiency of over 80 % in large installations and a long cycle life. The development of V-RFB has been assisted by tremendous improvements in electrode [12] and membrane materials [13].

As schematically shown in Fig. 12.2, RFB stores electrical energy in two soluble redox couples contained in external electrolyte tanks, sized in accordance with

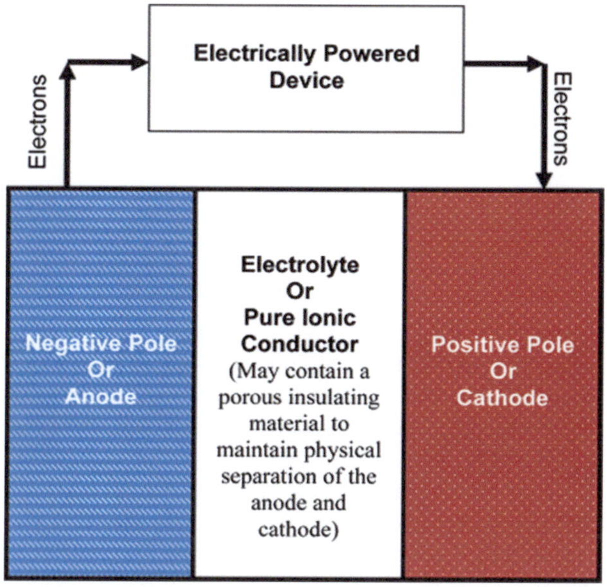

Fig. 12.1 Diagram representations of a battery powering a device (Reproduced with permission from Ref. [2], copyright 2004, American Chemical Society)

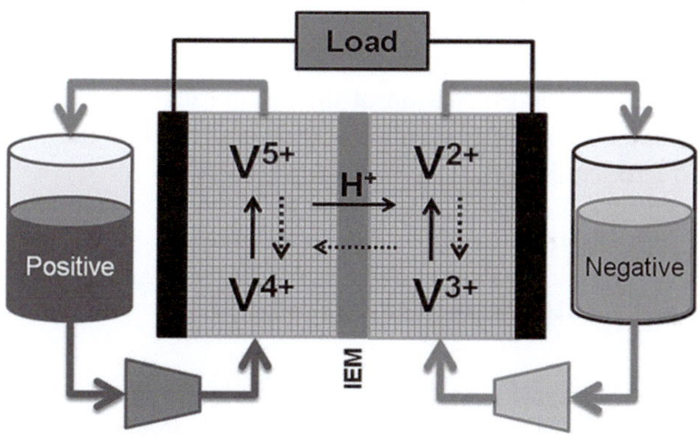

Fig. 12.2 Overview of a redox flow battery, showing two storage containers for the redox couples, one electrochemical cell in which charge and discharge reactions occur and a power source and load, respectively (Reproduced with permission from Ref. [14a], copyright 2004, Elsevier B.V)

application requirements [14]. Liquid electrolytes are pumped from storage tanks to flow-through electrodes where chemical energy is converted to electrical energy (discharge) or vice versa (charge).

The electrolytes flowing through the cathode and anode are often different and are referred to as anolyte and catholyte, respectively. Between the anode and cathode compartments is an ion-exchange membrane separator that allows diffusion of ions across the membrane while preventing the cross-mixing of the electrolyte solutions from these two reservoirs.

Unlike other electrochemical energy storage systems, flow batteries store energy in the form of reduced and oxidized electroactive species in the electrolyte, while batteries store energy within the electrode structure. The chemical reactions

Table 12.1 General comparison of electrochemical battery and redox flow battery

Category	Site of reactants	Electrolyte conditions	Separator
Electrochemical battery	Active electrode material	Static and held within cell	Microporous polymer separator
Redox flow battery	Aqueous electrolyte in tanks	Electrolyte recycles through the cell	Ion-exchange membrane (cationic/ionic)
Fuel cell	Gaseous or liquid fuel(anode) plus air (cathode)	Solid polymer or ceramic acts as solid electrolyte within cell	Ion-exchange membrane polymer or ceramic

involved in RFBs are reversible during charge and discharge. As a result, the electrolyte of the flow cell should always remain within the system, while in a fuel cell (FC) the reactants are stored externally to the cell and are consumed to release energy. The differences between batteries, redox flow cells, and fuel cells are shown in Table 12.1 [15].

Membrane separators play a key role in all battery systems mentioned above in converting chemical energy to electrical energy. A good overview of separators is provided by Arora and Zhang [16]. Various types of membrane separators used in batteries must possess certain chemical, mechanical, and electrochemical properties based on their applications, with particular emphasis on separators for lithium–ion batteries (LIBs). For the RFB, the separators usually allow the reactive ions to transport through them during the charging process. After the charge process, the membrane should block the transport of reactive chemicals as redox flow battery [11]. Detailed discussion of the membrane separator requirements, properties, and characterization techniques will be described in the following sections with respect to LIBs and RFBs.

12.3 Supercapacitor Technology

Supercapacitor is also an important electrochemical energy storage device that has attracted increasing attentions due to its advantages such as the high-rate capability in both charge and discharge processes and long cycle life as high as 10^6 cycles over traditional electrochemical energy storage devices [17]. A simple capacitor consists of two conductive plates (known as supercapacitors) separated by a nonconductive dielectric as illustrated in Fig. 12.3 [18]. There are three types of supercapacitors and various subclasses within these three classifications [19]. The three types of supercapacitors include **(1) electrochemical double-layer capacitors** (EDLCs), which are usually constructed from two carbon-based electrodes, an electrolyte, and a separator [20]. The electrodes do not allow recombination of ions, producing a double layer of charge at each electrode [21].

(2) Pseudo-capacitors use two different types of electrode material to store charge: conducting polymers and metal oxides. **(3) Hybrid capacitors** can be achieved by employing the battery electrode materials and the electrochemical

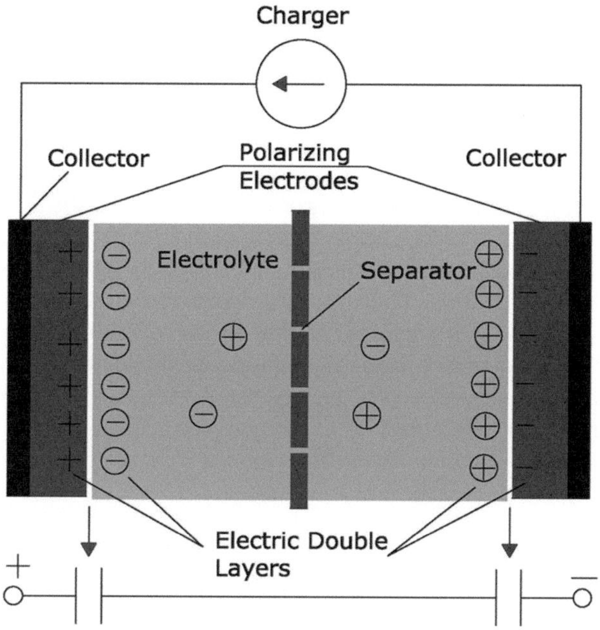

Fig. 12.3 Supercapacitor illustrating the energy storage in the electric double layers at the electrode–electrolyte interfaces (Reproduced with permission from Ref. [18c], copyright 2008, ESG)

capacitor materials within one device [22]. Like the other types of capacitors, hybrid capacitors are distinguished by their electrode configuration: composite, asymmetric, or battery type [23, 24].

Each of the elements that a supercapacitor consists of contributes to its final characteristics, which specify the ability to store and release an electric charge. Apart from the electrode–electrolyte system, the separator constitutes another significant component [25], which fulfills several important tasks, such as first of all electron isolating of counter-charged electrodes, with simultaneous assurance of high electric conductivity of electrolyte contained in its porosity [26]. However, research on the choice of the separator material for electrochemical capacitors is not carried out very often, and mostly researchers focus on the materials whose task is to store charges [27].

12.4 Membrane Separators and Their Applications in Lithium–Ion Batteries (LIBs)

12.4.1 Separator Requirements and Characterization

The main purpose of a membrane separator in lithium–ion battery (LIB) is to prevent contact between the anode and the cathode, while facilitating the diffusion of ions between the two electrodes [28]. Separator serves as an inactive fundamental and critical component and plays an important role in determining the device performance and ensures safety [16]. Lithium ions are able to flow between the two electrodes via an electrolyte medium through small pores in the separator. The structural and functional properties by which the separator material is characterized are very demanding and often inconsistent with each other for a particular application. The structural properties include chemical (molecular) and microcrystalline nature, thickness, pore size, pore size distribution, porosity, and various chemical and physical properties such as chemical stability and electrolyte uptake. The functional properties of interest are electrical resistivity, permeability, and transport number. It is useful to characterize separator materials in terms of their structural and functional properties and to establish a correlation of these properties with their performance in batteries. Separator characteristics are measured using a number of American Society for Testing and Materials (ASTM), standard test methods developed for characterizing plastic sheets and films, as well as industry-specific methods developed by research laboratories and separator manufacturers [16, 28].

12.4.1.1 Mechanical Strength

This is characterized in terms of the tensile strength along the machine direction (MD) and the transverse direction (TD), the tear resistance, and the puncture strength. All these parameters are described by Young's modulus. The separator must be mechanically strong, especially in the MD, and enough to withstand the tension of the winding operation during battery assembly. The puncture strength is defined as the maximum load required for a given needle to puncture a given separator, and it can be measured with a tensile tester. The minimum requirement for the mechanical and puncture strengths is 1000 kg cm^{-2} and 300 g, respectively, for a 25 μm membrane in LIB.

12.4.1.2 Chemical Stability

The separator material must be chemically stable against the electrolyte and electrode materials, especially under the strongly reductive and oxidative environments when the battery is fully charged. It should not degrade or loose mechanical strength or produce impurities, which can interfere with the function of the battery. An easy method to verify chemical stability is by calendar life testing.

12.4.1.3 Thermal Shrinkage

When the temperature rises to the softening temperature, the membrane tends to shrink, even if the porosity is very low, because of the difference in the density between the crystalline and amorphous phases of polyolefin materials. For example, the PE can shrink as much as 10 % when exposed to a temperature of 120 °C for only 10 min. The thermal shrinkage should be minimized. For the Li–ion battery, the shrinkage is required to be not more than 5 % after 60 min at 90 °C.

12.4.1.4 Permeability

Separator should not limit the electrical performance of the battery under normal conditions. Typically the presence of a separator increases the effective resistance of the electrolyte by a factor of 6–7. The ratio of the resistance of the separator filled with electrolyte divided by the resistance of the electrolyte alone is called MacMullin number. MacMullin numbers as high as eight have been used in high-power Li–ion batteries. For batteries used in hybrid electric vehicles (HEV) and in power tools, the MacMullin number should be lower for the purpose of safety and a long cycle life. Air permeability can be used indirectly to estimate the MacMullin value, which is defined as the time required for a specific amount of air to pass through a specific area of the separator under a specific pressure (Gurley value). When the porosity and thickness of the separators are fixed, the Gurley value reflects the tortuosity of the pores. The separator with uniform permeability is essential for the long cycle life of a battery. Variations in permeability will result in uneven current density distribution, which has been verified as the main reason for the formation of dendrite Li on the negative electrode.

12.4.1.5 Wettability

The separator should wet easily in the electrolyte and retain the electrolyte permanently. The former facilitates the process of electrolyte filling in battery assembly and the latter increases cycle life of the battery. There is no generally accepted test for separator wettability. The filling speed depends on the type of the materials, porosity, and pore size of the separators.

12.4.1.6 Thickness

In the current consumer applications, <25 μm seems to be the standard thickness for consumer rechargeable batteries. In addition, a uniform thickness (~40 μm) is critical for long cycle life of the batteries. A low thickness is required for high energy and power densities. However, this adversely affects the mechanical strength and safety. The thickness can be measured using the T411om-83 method developed under the auspices of the Technical Association of the Pulp and Paper Industry.

12.4.1.7 Porosity

Control of porosity is very important for battery separators. An appropriate porosity is necessary to hold sufficient liquid electrolyte for the ionic conductivity between the electrodes. Typically, the Li–ion battery separators have a porosity of 40 %. However, too high porosity will adversely impact the shutdown performance because in this case, the pores cannot be closed effectively and the membrane tends to shrink as it melts or softens. The porosity can be measured using liquid or gas absorption methods. In addition to the actual porosity, the uniformity of porosity across the separator is critically important. Nonuniform porosity leads to nonuniform current density and reduces the battery performance because the active materials coming into contact with the high-porosity regions may have to work harder compared with those coming into contact with the low-porosity regions. Typically, the LIB separators have a porosity of 40 %.

12.4.1.8 Pore Size

The pore size must be smaller than the particle size of the electrode components, including the electrode active materials and the conducting additives. In practical cases, membranes with submicron pore sizes have proven adequate to block the penetration of particles since the tortuous structure of the pores assists in blocking the particles from reaching the opposite electrode. Uniform distribution and a tortuous structure of the pores are both highly desirable since the former ensures a uniform current distribution throughout the separator and the latter suppresses the growth of dendritic lithium. The distribution and structure of pores can be analyzed using a Capillary Flow Porometer (Porous Materials Inc.) or scanning electron microscopy. Typically, a submicron pore size (less than 1 μm) is desirable for separators used in lithium–ion batteries.

12.4.1.9 Dimensional Stability

The separator should lay flat and not curl when it is laid out and soaked with liquid electrolyte. The separator should remain stable in dimensions over a wide temperature range.

12.4.1.10 Shutdown

In a Li–ion battery, the separator is required to be capable of battery shutdown at the temperature below that at which thermal runaway occurs, and the shutdown should not result in loss of mechanical integrity. Otherwise, the electrodes could come into direct contact and the resulting chemical reactions cause thermal runaway. The shutdown characteristics can be examined by differential scanning calorimetry or by observing the resistance change of the electrolyte-soaked membrane with temperature increase. For the polyethylene–polypropylene (PE–PP) bilayer separators used currently in Li–ion batteries, they have approximately 130 °C shutdown temperatures and approximately 165 °C melting temperature.

12.4.1.11 Cost

According to the current material prices and Li–ion technology, the cost of the separator can be over 20 % of the total cost of a Li–ion battery. The majority of the separator cost is in the manufacturing process. Therefore, a cost-effective process is necessary for the reduction of separator cost. Many of the properties above are associated with each other and may be in a trade-off relationship. For example, reducing the separator thickness increases battery energy and power densities, but it inevitably lowers the mechanical strength of the separator. In practical applications, the requirements above should be appropriately weighed among the performance, safety, and cost. The above list presents a broad spectrum of requirements for separators in LIBs. The order of importance of the various criteria varies, depending on the device applications.

12.4.2 Types of Separators

Separators for LIBs can be divided into different types, depending on their physical and chemical characteristics for a particular application. Although separators are effective in preventing electrical shorts between anode and cathode, their presence between the two electrodes decreases the effective conductivity of the electrolyte, raising cell impedance. Typical separators are between 20 and 30 μm thick [26]. In addition, the separator should be able to shut the battery down when overheating occurs for battery safety. However, as stated in previous section, practical separators cannot simultaneously possess all these ideal properties, and it becomes extremely important to balance different separator properties to achieve high-performance batteries. For the purpose of this chapter, based on structure and composition, LIBs membrane separators can be divided into five major types as described by Zhang et al. [29]: (1) microporous membranes, (2) modified microporous membranes, (3) nonwoven mats, (4) composite membranes, and (5) electrolyte membranes. *Microporous membrane separators* are characterized by pore sizes in the range of micrometers. They can be classified into monolayer and multilayer microporous membranes, depending on the number of layers. General specifications of the microporous separators utilized in LIBs provided by the US Advanced Battery Consortium (USABC) are summarized in Table 12.2 [28]. *Modified microporous membrane separators* are the membranes modified from conventional microporous membranes via surface modification like grafting methods using

Table 12.2 Key requirements of microporous separators for LIB

Parameters	Target
Thickness (μm)	<25
Porosity (%)	~40
Pore size (μm)	<1
Gurley value[a] (s)/100 cm^3	~25
Tensile strength ($kgcm^{-2}$)	<2 %
MacMullin number	<8
Puncture strength (g/ml)	>300
Chemical stability	Long enough time
Thermal stability	<5 % shrinkage
Melting point (°C)	>150
Skew (mm/m)	<2

plasma and irradiation or coating with a different polymer. *Nonwoven mat separators* have web structures bonded together with entangling fibers prepared by melt blown, wet laid, and electrospinning techniques. Since they have small fiber diameters, the nonwoven mat separators show higher porosities than other types of separators. *Composite membrane separators* are prepared by coating or filling inorganic materials to microporous membranes or nonwoven mats. Thus, they have outstanding thermal stability and exceptional wettability that other types cannot achieve.

Electrolyte membranes include a solid ceramic electrolyte, solid polymer electrolyte, gel polymer electrolyte, and composite electrolyte. They work as both separator and electrolyte and offer high battery safety. Each separator type, respectively, possesses intrinsic features to satisfy the requirements discussed in the previous section, including thickness, porosity, thermal property, wettability, mechanical, and chemical properties.

12.4.2.1 Microporous Membrane Separators

Most of microporous membrane separators are produced by a number of manufacturers and generally made from a polyolefin, mainly polyethylene (PE), polypropylene (PP), and their blends or multilayer forms such as PE-PP and PP-PE-PP [26, 29, 30]. The manufacturing process of these thin film microporous membranes can vary leading to large changes in material properties. The two main manufacturing processes can be classified into dry process and wet process [16, 28]. In the dry process, the olefin is extruded above its melting point in order to bring the separator to its designated thickness and then annealed. Additional stretching of the separator induces small micropores that are aligned in a linear fashion. Due to the organization of these micropores, the separator's mechanical properties are anisotropic and show the greatest strength in its direction of stretch (machine direction) [30]. In the wet process, a polyolefin resin is mixed with a hydrocarbon then heated until the mixture melts. The melted mixture is then extruded as a sheet and the liquid is

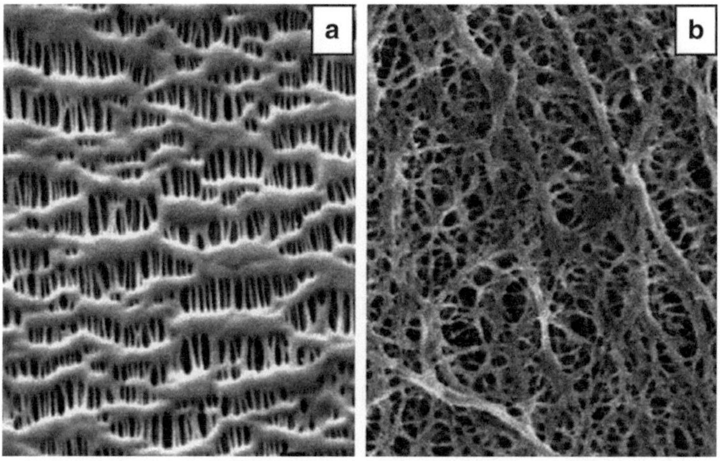

Fig. 12.4 Microstructure of the microporous polyolefin membranes made by (**a**) dry process and (**b**) wet process, respectively (permission required from ACS) (Reproduced with permission from Ref. [16], copyright 2004, American Chemical Society)

Table 12.3 Commercial separator properties[a]

Manufacturer	Celgard	Entek	Exxon	Degussa
Membrane	2325	Teklon	Tonen-1	Separion
Process	Dry	Wet	Wet	Wet
Composition	PP-PE-PP	PE	PE	Ceramic-PET-ceramic
Thickness (μm)	25	25	25	25
Porosity (%)	41	38	36	41
Mp (°C)	134/166	135/163	135	220
References	Data sheet	Data sheet	Data sheet	Data sheet

[a]Separator specification is found on data sheet for each product

extracted with a volatile liquid leaving behind micropores. The mechanical properties of these microporous separators are generally more isotropic since the spaces are not introduced mechanically. Microporous membranes prepared by the dry process have slit-like pores, while those from the wet process show interconnected and elliptical pores. Figure 12.4 compares the microstructures of microporous membranes prepared by the dry process and wet process [16]. The membranes formed by the dry process are more appropriate for high-power density batteries because of their open and straight porous structure. On the other hand, the membranes prepared by the wet process are more suitable in long cycle life batteries because interconnected pores and the tortuous structure are beneficial in preventing the growth of dendrites during charging and discharging.

Table 12.3 lists the major separator manufacturers, showing that most of the commercial separators are made of microporous polyolefin membranes [28, 31].

Celgard and Degussa separators are trilayer structures manufactured via dry and wet processes, respectively, while Entek and Exxon separators are single layer of PE manufactured via the wet process. Polyethylene terephthalate (PET) and polyvinylidene fluoride (PVDF) have also been used in commercial separators but are far less commonplace than the polyolefin films. Commercially available LIB microporous polyolefin separators have been found to be compatible with the cell chemistry and also be cycled for several hundred cycles without significant degradation in chemical or physical properties. However, these separators provide relatively low thermal stability, poor wettability, and unsatisfactory electrolyte retention. Accordingly, the physical properties of polyolefin microporous separators need to be improved in order to be applied as the separators for high-performance and safe LIB.

A variety of polymers such poly(vinylidene fluoride) (PVDF), polyacrylonitrile (PAN), and poly(methyl methacrylate) (PMMA) are utilized to overcome these drawbacks [32]. These microporous membranes can be prepared by using different methods, including dip coating, laminating, slot die gravure casting phase inversion, and so forth, as indicated by Huang et al. [26]. *Microporous PVDF membranes* are physicochemically and electrochemically stable in lithium–ion batteries and also mechanically strong with great wettability due to their good affinity to liquid electrolyte solutions [33]. *Microporous PAN membranes* exhibit good processability, high thermal stability, desirable morphology for high electrolyte uptake, excellent electrochemical stability, and good compatibility with electrodes. In addition, PAN can act as a polymer matrix to help maintain the liquid electrolyte and participate in lithium–ion transport due to the interaction between the lithium–ion and the $C\equiv N$ groups of PAN. PAN membranes can also minimize the formation of dendrites during the charging–discharging process of LIBs [34]. However, PAN membranes often suffer from electrolyte leakage during long-term storage [35]. *PMMA membranes* show high equivalent conductivity and good thermal and electrochemical stabilities and also exhibit high ionic conductivity and good adhesion to the electrodes due to the formation of gel phases with liquid electrolyte [36]. However, PMMA has poor mechanical strength due to its amorphous structure [37]. Table 12.4 lists the structural characteristics and properties of various monolayer membranes, including materials, processing methods, electrolyte uptake, ionic conductivities, and electrochemical performance.

Microporous membranes consisting of multilayers of PP and PE have been designed as thermal shutdown separators with a safety advantage. Membranes prepared from a PE melt at 135 °C and have no mechanical integrity above melting point, and those made by laminating layers of PP and PE have mechanical integrity up to 165 °C (the melting point of PP) as illustrated in Fig. 12.5. Separator shutdown is a useful mechanism for limiting temperature and preventing emitting in short-circuited cells. Shutdown usually takes place close to the melting temperature of the polymer when the pores collapse, turning the porous ionically conductive polymer film into a nonporous insulating layer between the electrodes. Like most batteries, LIB will generate heat if overcharged, and above the threshold temperature, the battery starts to "self-heat" because of exothermic reactions within

Table 12.4 Summary of monolayer membranes

Materials	Solvents	Processing methods	Membrane thickness	Liquid electrolyte[a]	Electrolyte uptake	Ionic conductivity	Performance
PP, PE [30]		Dry and wet processes	25–50 μm	$LiAsF_6$-cyclic ester/ether			Low shutdown temperature
PE [38]	Paraffin oil	Wet process	25 μm	LiPF6-PC			High mechanical strength
PVDF [39]	Acetone/ethanol	Phase inversion	35-50 μm	$LiPF_6$-EC/DMC/DEC		3.0×10^{-3} to 5.0×10^{-3} S cm-1	Stable discharge curves and low temperature performances
PVDF [36d]	NMP	Phase inversion	35-70 μm	$LiPF_6$-EC/DMC		0.7×10^{-3} to 2.1×10^{-3} S cm^{-1}	Good cycling ability and high capacity at high charge rate
PAN [40]	DMF	Phase inversion	60–70 μm	$LiClO_4$-EC/DMC, $LiPF_6$-EC/DMC, and $LiBF_4$-EC/DMC		2.5×10^{-3} to 2.8×10^{-3} S cm^{-1}	Good electrochemical stability
PAN, PMMA [41]	EC/PC	Casting method	50–100 μm	$LiN(SO_2CF_3)_2$-EC/DMC or $LiClO_4$-EC/PC		0.1×10^{-3} to 1.9×10^{-3} S cm^{-1}	Good electrochemical stability
PMMA [42]	PC	Casting method		$LiClO_4$-PC		1.0×10^{-4} S cm^{-1}	Low activation energy of conduction
PVDF, PVDF-HFP [43]		Phase inversion	120–220 μm	LiPF6-EC/DEC	Enhanced	7.2×10^{-3} S cm^{-1}	Better ion transport
PVDF-HFP/PAN blend [44]	DMF	Solvent evaporation	150–200 μm	$LiPF_6$-EC/DEC	Enhanced	1.5×10^{-3} to 2.0S cm^{-1}	Good electrochemical stability and high specific cyclic stability
PVDF/ PMMA blend [45]	THF	Solvent evaporation		LiClO4-EC/PC		0.4×10^{-4} to 4.0×10^{-4} S cm^{-1}	Good mechanical strength and high ion diffusion coefficient
PEO-EC blend PVDF-HFP/ PVAc blend [46]	THF	Solvent evaporation	140 μm	LiClO4-EC/PC		2.3×10^{-3} S cm^{-1}	Good thermal property and high electrochemical stability

[a]Liquid electrolyte was used during the measurements of electrolyte uptake, ionic conductivity, and electrochemical performance

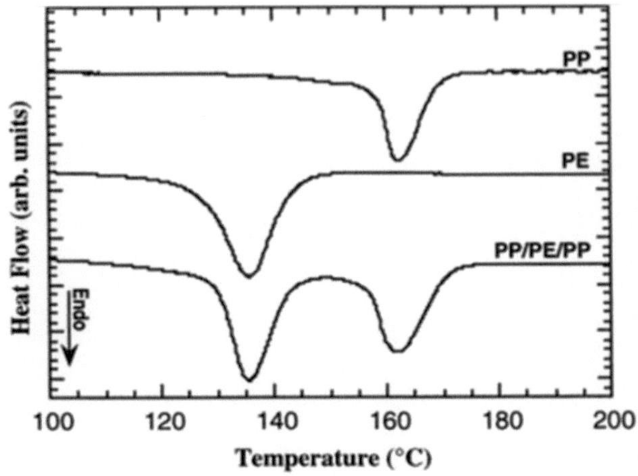

Fig. 12.5 DSC thermograms of representative polyolefin separators (PP), (PE) and (PP/PE/PP) (Reproduced with permission from Ref. [30], copyright 1999, Elsevier Science S.A)

the components of the cell [47]. An extensive study on the structure–property relations of the microporous polyolefin separators can be found in the work by Venugopal et al. [30].

Huai et al. [48] and Park et al. [49] prepared multilayer membranes that have the unique porous structure of PAN and PMMA nanoparticle coatings by conventional emulsion polymerization and applied on the surface of polyolefin nonwoven fabrics as shown in Fig. 12.6. The polymer nanoparticles formed close-packed coating layers on the polyolefin nonwoven fabric surface, forming a unique porous structure. These membranes showed good wettability and exhibited high ionic conductivity.

Microporous multilayer membranes combine the advantages of each component layers and the issues established from the drawbacks of monolayer membranes. Even though several microporous multilayer membranes, such as Celgard® trilayer separators, have been used in commercial LIBs, improvements in terms of thickness, wettability, and ionic conductivity are still needed.

12.4.2.2 Modified Microporous Membranes

Although most studies focus on modifying the structure and properties of microporous polyolefin membranes, efforts have been undertaken on the modification of microporous membranes made of other polymers to overcome poor wettability and electrolyte retention against electrolyte solutions containing polar solvents [50]. The most effective method of modifying microporous membranes is by treating them with a wetting agent or grafting hydrophilic functional groups onto the surface and pores of the separators [51]. Various techniques such as plasma, UV irradiation, gamma ray, electron beam irradiation, and polymer coating have been used to modify the structure of microporous polyolefin membrane separators [29]. For example, gamma and UV irradiations were used to modify PVDF

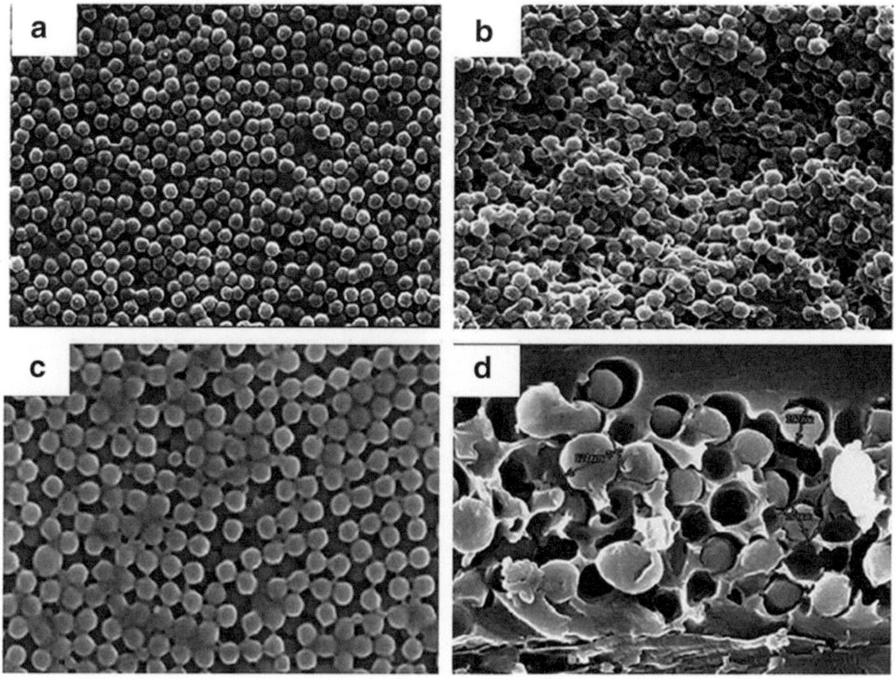

Fig. 12.6 SEM images of (**a**) surface and (**b**) cross section of PAN nanoparticle coated (Reproduced with permission from Ref. [48], copyright 2010, Springer-Verlag). (**c**) surface and (**d**) cross section of PMMA nanoparticle-coated multilayer membrane (Reproduced with permission from Ref. [49], copyright 2010, Elsevier B.V)

homopolymers [52, 53] and copolymers [54] membranes. These modified membranes showed improved mechanical and electrochemical properties compared to unmodified membranes. Besides the grafting methods, the interfacial adhesion between separators and electrodes is critical in determining the long-term performance of LIB. For improved interfacial adhesion, a thin polymer layer such as poly (ethylene oxide) (PEO) and poly(vinylidene fluoride) (PVDF) are gelled by liquid electrolytes to coat the surfaces of polyolefin membranes [55]. In practical processes, the polymer coating can be applied to the microporous membrane by a means of dipping or spraying. A heating step is required to promote the in situ formation of the gel polymer electrolyte when the polymer-coated membranes are used as the separator. The negative effect of the approaches above is that the polymer layer made by the solution method is structurally dense, which inevitably blocks the penetration of liquid electrolyte into the pores of the microporous membrane. Therefore, the polymer layer with a porous structure is highly desirable so as to facilitate the electrolyte filling process [56].

As a comparison, Fig. 12.7 shows the surface images of a microporous PE membrane before and after the polymer coating by different methods [57]. The

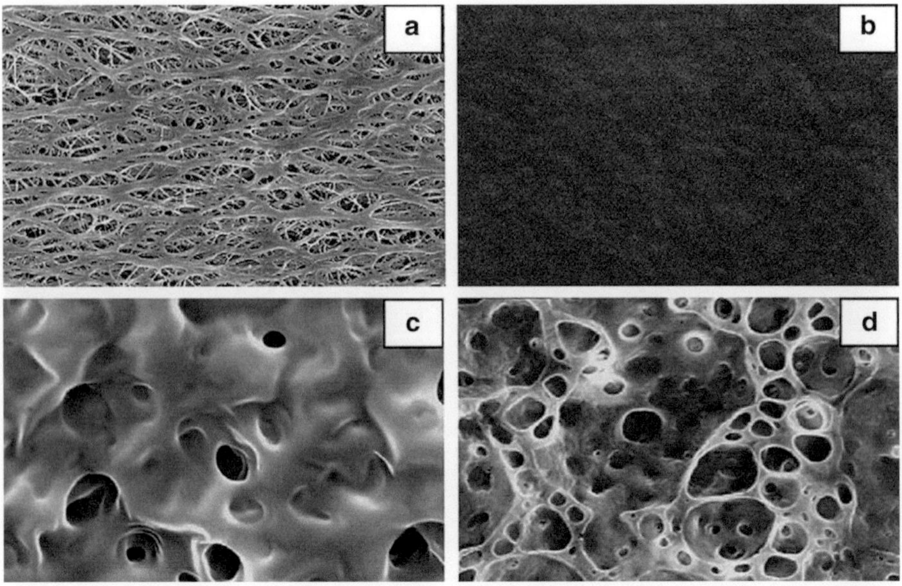

Fig. 12.7 Micrographs of the surface structure of a microporous PE membrane before and after polymer coating treatment. (**a**) Bare membrane, (**b**) with polymer coating by solution method, (**c**) with polymer coating by phase inversion, and (**d**) with polymer–silica coating by phase inversion (Reproduced with permission from Ref. [57], copyright 2006, Elsevier Ltd)

differences in the surface morphology of the polymer layers coated by the solution and phase inversion methods are clearly defined.

In summary, the modification of microporous membrane separators has led to improvements in mechanical properties, wettability, and ionic conductivity. In addition, the electrochemical performance, especially high-rate capability, of lithium–ion batteries can also be improved by using modified microporous membrane separators as indicated in the review article by Zhang et al. [29].

12.4.2.3 Nonwoven Material Separator

Although nonwoven mats have long been used as separators for different batteries, they have very limited applications in LIBs [29]. A nonwoven film is a textile product prepared directly from fibers that are bonded together [26]. The fibrous structure of nonwoven materials offers a high porosity, which is necessary for high electrolyte absorbance and low ionic resistance, and results in good charge–discharge cycles of the battery. In contrast to woven materials, the fibers in nonwovens are randomly distributed, whereby an orientated microstructure does not occur. Compared with microporous polyolefin membranes, a nonwoven film generally possesses much higher porosity and lower weight. Table 12.5 provides the techniques of nonwoven film fabrication, related synthetic materials, and film

Table 12.5 Fabrication of nonwoven films

Technique	Material	Film properties
Papermaking process	Polyolefin, PA, PTFE, PVDF, PVC, polyester, etc.	High porosity (60–80 %)
Melt-blowing method		Large pore size (20–50 μm)
Electrospinning		

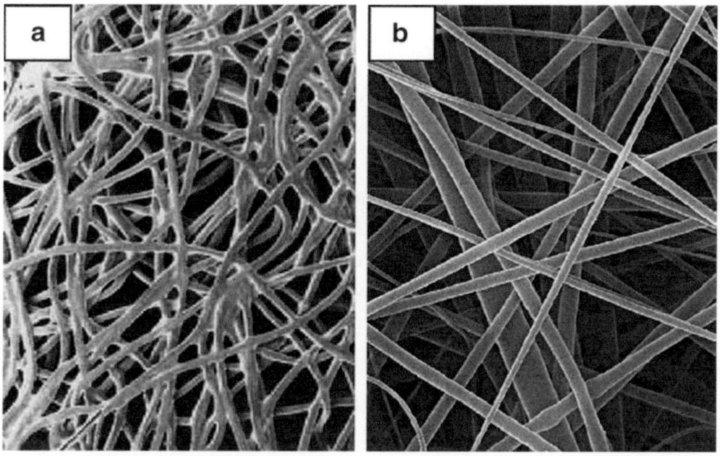

Fig. 12.8 Shows the microstructure of typical nonwoven mats made by different process and used in exploratory research for a Li–ion battery separator. (**a**) Wet and (**b**) electrospinning methods, respectively (Reproduced with permission from Ref. [28], copyright 2007, Elsevier B.V)

properties [28]. The arrangement of the fibers is the main advantage of a nonwoven material compared with woven structures for the battery separator applications. Figure 12.8 shows the microstructure of typical nonwoven mats made by different process and used in exploratory research for a LIB separator [28, 58]. Their structures are featured by the labyrinth-like pores, which are very helpful in preventing the growth of dendritic Li–in rechargeable Li and Li–ion batteries.

These separators generally have a thickness of 100–200 μm and a base density of 9–30 g cm^{-2}. In an effort to reduce the thickness, Kritzer [59] developed a technology that made it possible to produce nonwoven mats with a thickness of 20 μm or less.

However, the nonwoven separators have some drawbacks, such as large pore size and thicker nature. Due to the disadvantage mentioned above, there are two ways to employ the nonwoven films: (1) make gel polymer electrolyte as the supporting framework and (2) form composite separator by coating a layer of oxide particles on each side. These processes will be highlighted in the following sections. Most electrospun nonwoven mats have a desirable structure and properties for use as LIB separators. However, compared with microporous membranes and modified microporous membranes, electrospun nonwoven mats are relatively weak, and their mechanical properties must be improved to survive the winding process used in the commercial battery assembly. In addition, electrospinning is a relatively slow process, which limits the production rate of electrospun nonwoven mats. As a result, electrospun nanofiber mat separators are relatively expensive and can only be used in LIBs for special applications.

12.4.2.4 Polymer Electrolyte Separator

Polymer electrolyte separators have attracted considerable attention due to their vast applications in the development of ionic devices in solid state. These materials have useful chemical and mechanical properties that allow ease of fabrication in thin film form at desired sizes. They are ionically conductive and also can act as both separator and electrolyte between the two electrodes [60]. Basically, polymer electrolyte can be classified into the following categories: a solid ceramic electrolyte, solid polymer electrolyte, gel polymer electrolyte, and composite electrolyte. Detailed descriptions are discussed below:

(1) *Solid polymer electrolytes* (*SPEs*)

In SPEs, the polymer host itself is used as a solid solvent along with Li salt and does not contain any organic liquids. The SPEs suffer from poor ionic conductivities (10^{-5} S cm^{-1} at 20 °C) but are safer due to the absence of any organic solvent which can cause environmental hazards. SPE films can be prepared either using the solution cast technique or the hot press technique [61].

(2) *Solid ceramic electrolyte*

Solid ceramic electrolytes are classified into crystalline electrolytes, including the lithium super ionic conductor (LISICON)-type, thio-LISICON-type, garnet-type, perovskite-type, NASICON-type, and glass–ceramic electrolytes [62].

(3) *Gel polymer electrolytes* (*GPEs*)

GPEs are also known as hybrid polymer electrolytes, formed by trapping liquid electrolytes in a polymer matrix. The advantage of this type of electrolyte is that higher ionic conductivity can be achieved as compared with SPEs. The ion conductivity in these electrolytes takes place through the liquid electrolyte, with the host polymer mostly providing the structural support. Unfortunately, the loss of mechanical properties is observed in GPEs, and liquid electrolytes tend to provoke undesired interfacial reactions of the GPEs with electrodes during cell operation [63]. This may induce vigorous proliferation of a resistive layer on the electrodes, which hinders ionic transport at the interface of the GPE and the liquid electrolyte.

(4) *Composite polymer electrolytes* (*CPEs*)

More attention has been focused on CPEs, wherein nanosized fillers, such as zeolites, super-acid sulfated zirconia, alumina (Al_2O_3), silica (SiO_2), and titania (TiO_2), not only improved the transport properties but also the mechanical and electrochemical properties of the SPEs [64].

Comparing to solid electrolytes and liquid electrolytes, gel polymer electrolytes (GPEs) have several advantages, including no shape restrictions, faster charging–discharging, and higher power density. Currently, there are four major polymer host materials for GPEs: polyethylene oxide (PEO), polyacrylonitrile (PAN), polymethyl methacrylate (PMMA), and polyvinylidene fluoride (PVDF) [65]. In the past two decades, PEO-based electrolytes were the major polymer host matrix used in batteries, and significant research efforts have been placed on their development and improvement. The conduction of PEO-based gel polymer electrolytes is mainly though the complexes between lithium–ion and ether oxygen atom [66]. PEO-based electrolyte is the first studied system, which could be easily casted as thin membranes. However, due to PEO's high degree of crystallinity, the ionic conductivity of PEO-based electrolytes is low and varies from 10^{-8} S cm^{-1} to 10^{-4} S cm^{-1} at temperature between 40 °C and 100 °C [67]. Ito et al. observed that the ionic conductivity increases as the plasticizer increases, but the interfacial properties become worse due to the presence of hydroxyl end groups [68]. In the later studies, PAN was used as the electrolyte host matrix material because of its small thermal resistance and flame-retardant property. According to Feuiliade et al., the ionic conductivity of PAN-based gel electrolyte is between 10^{-5} S cm^{-1} and 10^{-3} S cm^{-1}, and the amount of transferred lithium–ion is larger than PEO-based gel polymer electrolyte [69]. Studies by Choi and coworkers showed that combining PAN and PEO together, which is 10PEO-40PAN-12$LiClO_4$-38EC/BL, could improve the mechanical flexibility, ionic conductivity, and interfacial properties of GPE [70]. PMMA-based GPEs are also used due to their enhanced interface stability and lower cost, due to rich raw materials and simple synthesis process, among other host matrix materials. However, its poor mechanical flexibility narrowed down the applications. Copolymerization between PMMA and other polymer provides better performance to lithium–ion polymer battery [71].

The most common polymer used in today's lithium–ion polymer batteries is PVDF, which supports high concentration of charge carriers [72]. Due to the semicrystalline structure of PVDF, lithium ions are drafted into the PVDF membrane when they stay on the surface [73]. Thus, a GPE membrane with fully interconnected open microspores, i.e., higher interfacial surface area, enhances ion storage and mobility [74]. Although the mobility of PVDF-based GPEs is greater than other polymer host matrix materials, the ionic conductivity of GPEs is still lower than liquid electrolytes. Introducing ionic liquids or nanosize fillers in GPE system could help to improve the performance of lithium–ion polymer batteries.

Ionic liquids (IL) are nonvolatile, nonflammable molten salts with low melting points. They exhibit high ionic conductivity, high thermal and chemical stability, a wide electrochemical window, and low toxicity [75]. Fernicaola et al. incorporated ionic liquids in an organic electrolyte solution to increase the ionic conductivity and stabilize the lithium ions carried on the surface of PVDF-base membrane [76]. Among aprotic ILs, protic ILs, and zwitter ILs, the aprotic class of ILs with high mobility and ion concentration stands out for advanced electrochemical systems, which consist of large irregular cations and small anions [77]. Sakaebe et al. studied and compared a few room temperature ionic liquid containing quaternary ammonium cation and imide anion and concluded that quaternary ammonium cations could stabilize the reduction on the lithium metal in LIB [78]. Overall, room temperature ionic liquids could improve the performance of lithium–ion polymer battery [79].

PG Bruce et al. reported nanomaterials could increase the rate of lithium insertion/removal, enhance the electron transport within the particles, and change the electrode potentials, but nanomaterials might cause more side reactions and increase the energy density [80]. Yang et al. demonstrated the ceramic fillers in PVDF-based gel polymer electrolytes improves interfacial stability between the electrode and the electrolyte, especially the cells with Al_2O_3 fillers that capacity remained 95 % of the initial capacity after 100 cycles at a C/2 rate [81].

12.4.2.5 Inorganic Composite Membrane Separators

Polymer separators made from microporous polyolefins and polymer matrices in the gel electrolytes have been widely used in the commercial lithium–ion batteries as indicated in previous sections. However, these organic materials usually have low melting points and poor mechanical strengths and could undergo obvious dimensional changes at elevated temperatures or be easily punctured by lithium dendrite commonly formed at high rates or long-term cycling, leading to some potential safety issues. Inorganic materials have been widely used to modify the polymer separator due to their excellent thermal stability and electrolyte wettability [82]. These polymer separators modified by inorganic materials exhibit improved thermal properties and cell performances [83]. It has been reported that addition of nanosize fillers, such as zeolites, alumina (Al_2O_3), silica (SiO_2), zirconia (ZrO_2), titania (TiO_2), or their mixture, not only improve the transport properties but also the mechanical and electrochemical properties of these polymer separators [84]. When compared with the commercial polymer separator, the inorganic separator has the advantages of "absolutely" thermal stability, strong electrolyte absorption, and no dendrite puncturing problems. The safety characteristic of the inorganic separator makes it possible to constitute a large-sized single cell, which has advantages over the battery pack consisting of many small cells. Although the composite separators described above offer excellent wettability and extremely thermal stability, they are not mechanically strong enough to withstand handling in cell winding and assembly [85]. To solve this drawback, Degussa commercialized a ceramic separator, Separion (trade name), as illustrated in Fig. 12.9 by combining the characteristics of polymeric nonwoven and ceramic materials [86].

The Separion separator is a flexible perforated nonwoven mat coated with a porous ceramic layer on its each side [86]. Typical physical properties of the Separion and Celgard membranes are summarized in Table 12.6, which clearly shows that the Separion separators have excellent wettability, high permeability (i.e., low Gurley value), a high meltdown temperature, and negligible shrinkage at high temperatures.

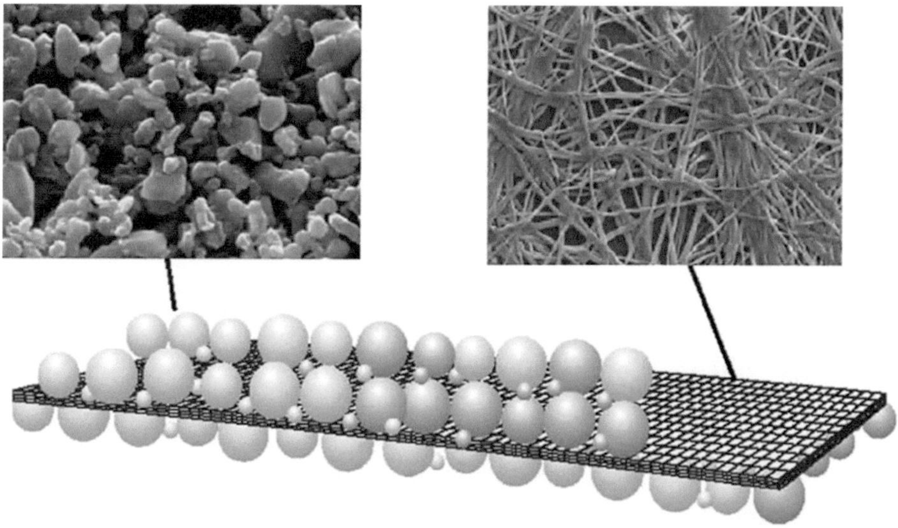

Fig. 12.9 Schematic and SEMs of a Degussa composite separator (Reproduced with permission from Ref. [86], copyright 2007, Elsevier B.V)

Table 12.6 Physical properties of Separion and Celgard membranes

Trade name	Separion	Separion	Celgard	Celgard
Separator brand	S240-P25	S240-P35	Celgard 2340	Celgard 2500
Composition	Al_2O_3/SiO_2	Al_2O_3/SiO_2	PP-PE-PP	PP
Support matrix	PET nonwoven	PET nonwoven	N/A	N/A
Thickness (μm)	25 ± 3	25 ± 3	38	25
Average pore size (μm)	0.24	0.45	0.038×0.90	0.209×0.054
Gurley value[b] (s)	10–20	5–10	31	9
Porosity (%)	>40	>45	45	55
Temperature stability (°C)	210	210	135/163	163
Thermal shrinkage[c] (%)	<1	<1	5	3
Tensile strength (MD)[d]	>3 N cm^{-1}	>3 N cm^{-1}	2100 kg cm^{-1}	1200 kg cm^{-1}
Tensile strength (TD)			100 kg cm^{-1}	115 kg cm^{-1}
Wettability for PC, EC, GBL	Excellent	Excellent	Not wet	Good

[a]The data were cited directly from the product brochures

[b]Gurley value was expressed as the time in seconds required to pass 100 ml of air through 6.45 cm^2 (1 $in.^2$) of membrane under a pressure of 31.0 cm (12.2 in.) of water

[c]Separion shows a value measured at 200 °C for 24 h, while Celgard a value measured at 90 °C for a fixed time

[d]Different units were used to express tensile strength by two companies

As discussed above, the incorporation of inorganic particles is regarded as promising approach to obtain high-performance LIB separators. The inorganic filler contribution is to enhance mechanical and electrochemical properties of the membrane separators used in lithium–ion batteries. Although the potential applications of purely inorganic separators are extensive, many challenges limit their applications in the lithium–ion batteries [87]. Besides the poor flexibility mentioned above, another obstacle is preparing the thin separator with high porosity and good mechanical strength.

12.5 Membrane Separators for Redox Flow Batteries (RFBs)

12.5.1 Properties and Requirements of Vanadium RFB Membranes

Since the discovery of the all-vanadium redox flow battery (V-RFB) by Maria Skyllas-Kazacos, the V-RFB has been going a long way and also reached the state of commercialization [88]. However, there is still research on improving the system in terms of stability and performance. One main research area beside the electrode and the electrolyte is dedicated to the membrane separator research. Membranes used as separators in the V-RFB have to separate efficiently the half cells and be anion or proton conductive to equilibrate the ionic charge during charge and discharge procedures. Beside ion conductivity and separation properties, the membrane has also to withstand the harsh conditions inside the V-RFB since the vanadium electrolyte solvent will be 3–6 M sulfuric acid, and the VO^{+2} produced during charging has very strong oxidative properties. In V-RFB systems, an ideal membrane should offer the following characteristics [89]:

- Good chemical stability under acidic conditions
- Resistance to the highly oxidizing environment of the positive half-cell electrolyte, low electrical resistance
- Low permeability to the vanadium or polyhalide ions
- High permeability to the charge-carrying hydrogen ions
- Good mechanical properties
- Low cost
- High counter ion conductivity

In addition to the abovementioned characteristics, an important property that an ideal membrane must possess is the ability to prevent the preferential transfer of water from one half-cell to the other, as this results in flooding of one half-cell while diluting the other [90]. Ion-exchange membranes (IEMs) have been investigated for V-RFB applications. In addition to the traditional dense membrane, microporous membranes (or separators) were also explored for VRB chemistries [16, 90]. However, a primary goal in V-RFB membrane research is to limit the vanadium cation diffusion through the membrane, while maintaining high oxidative stability in the presence of VO_2^+ (where vanadium is in the +5 oxidation state) and related species [91].

12.5.2 Types of Membrane Separators in V-RFB

There has been an intensive effort to apply high conductivity, low crossover ion-exchange membranes (IEMs) in different RFB systems such as iron–chromium, hydrogen–bromine, vanadium–bromine, nonaqueous [92], and other types of RFB chemistries [93]. IEMs of cation and anion selectivity are usually used as the membrane separators for V-RFB. These membranes are generally fabricated from polymeric materials containing pores with diameters of less than 20 Å. There are different methods of classifying ion-exchange membranes: based on their function, materials constituting the membranes, etc. Classification based on function is that ion-exchange membranes have an electrical charge, which is positive or negative. The function of ion-exchange membranes is determined from the species of the charge (positive or negative) of the ion-exchange groups fixed in the membranes and their distribution: (*1*) *cation exchange membranes* (*CEMs*), in which cation exchange groups (negatively charged) such as $-SO_3^-$, $-COO^-$, $-PO_3^{2-}$, $-PO_3H^-$, and $-C_6H_4O^-$ and other groups provide a negative fixed charge in aqueous or mixed water and organic solvent solutions [94], and (*2*) *anion exchange membranes* (*AEMs*), in which anion exchange groups (positively charged) such as $-NH_3^+$, $-NRH_2^+$, $-NR_2H^+$, $-NR_3^+$, and $-SR_2^+$ thus allow the passage of anions. Composite membranes can be produced to contain a cation selective layer (with negative fixed ionic groups) and an anionic layer (with positive ionic groups). Apart from polymeric ion-exchange membranes, ion-exchange membranes are also prepared from inorganic material like zeolites, bentonite, or phosphate salts. However, they have certain disadvantages because of their high cost and relatively bad electrochemical properties [95]. Table 12.7 highlights the difference between CEMs and AEMs.

In general, membranes can be classified based on a lot of factors like nature of the material, morphology, pore size, the driving force, and their configuration. When identified by morphology, membranes are often classified by their structures, being homogenous, heterogeneous, inter-polymeric, symmetric, or asymmetric [96].

Table 12.7 Difference between CEM and AEM

Cation exchange membranes	Anion exchange membranes
Low selectivity	High selectivity
High conductivity	Medium conductivity
High electrolyte transfer	Low electrolyte transfer
High oxidation stability	Medium to good oxidation stability
High cost for PFSA	Medium to low cost

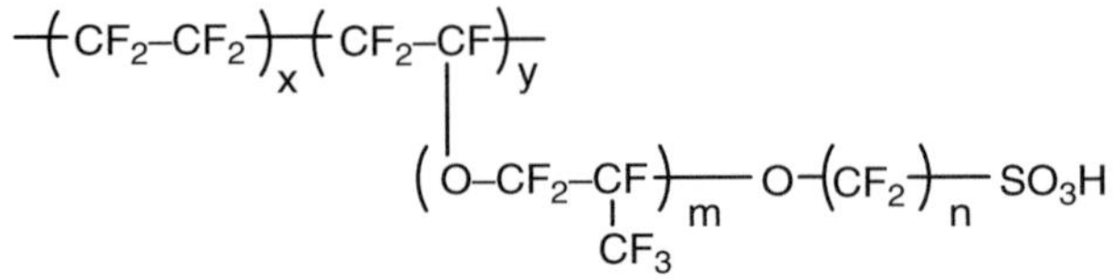

Fig. 12.10 Chemical structures of commercially available perfluorosulfonic acids cation exchange membranes (Reproduced with permission from Ref. [97], copyright 2000, Elsevier Science Ltd)

12.5.2.1 Cation Exchange Membranes (CEMs)

CEM is the most reported membrane for the V-RFB application in the scientific literature due to the high transport rates for protons. Several commercially available perfluorosulfonic acids (PFSAs), CEMs from Asahi Glass (Flemion), Dow Chemicals (Dow), Asahi Chemicals (Aciplex), and Du Pont (Nafion), that have been applied and tested in RFBs are illustrated in Fig. 12.10 [97]. Two of the abovementioned membranes (Flemion and Nafion) were identified as suitable membranes for the V-RFB. Apart from the membranes mentioned above, properties and cycling performance of other commercial membranes compared to Nafion [98] have been evaluated by Vafiais and Skyllas-Kazacos for the V-RFBs [99].

Nafion has superior chemical resistance when compared to other hydrocarbon CEMs, as these (hydrocarbons CEM) degrade in the vanadium electrolyte, while Nafion membrane does not [100]. Despite this, vanadium active species permeate through the Nafion membrane and decrease the Coulombic efficiency. The diffusion coefficient of vanadium ions across the Nafion membrane has been determined to be in the order of $V^{2+} > VO^{2+} > VO_2^+ > V^{3+}$ by Sun et al. [101]. The two major disadvantages of Nafion for V-RFB applications are the costs and the vanadium crossover. The high costs are mainly due to the complex production method and low production volumes to date [102]. Vanadium permeation is due to the well-interconnected hydrophilic and broad channels [103]. To potentially lower the cost of these systems compared to PFSA benchmarks, aromatic CEMs have been explored to good effect in a range of studies by Zhang et al. [104]. Figure 12.11 depicts the difference of a swollen Nafion membrane compared with an aromatic CEM (sulfonated poly(ether ether ketone) (SPEEK)) [105]. The conception expresses the highly permeable phase separation of Nafion and the dead end and narrow channels of SPEEK which do not contribute as effectively to the ion conductivity. This leads to the two main issues of Nafion in V-RFB:

- Self-discharge and low-energy efficiency due to vanadium permeation (crossover)
- Dilution and concentrating of the electrolyte solutions, due to high levels of water transfer

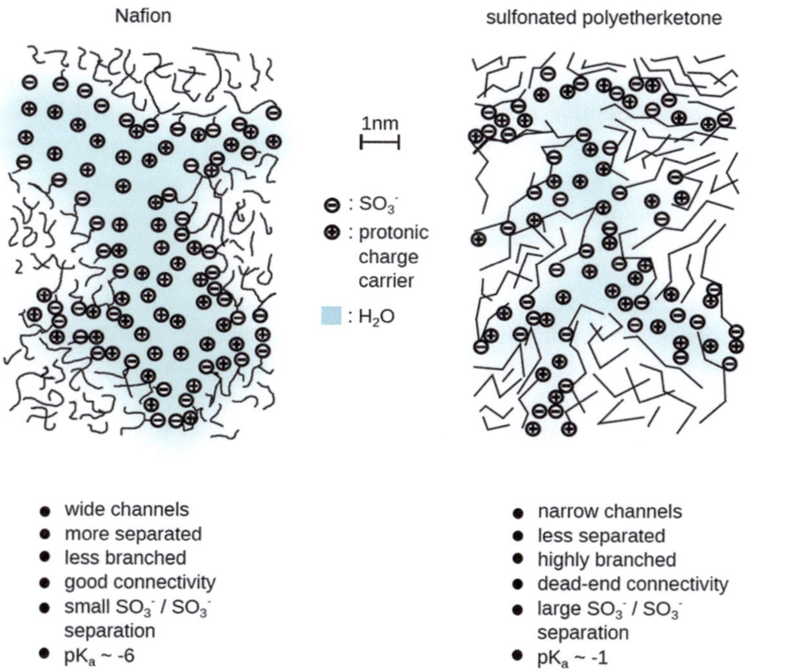

Fig. 12.11 Redrawn schematics of the microstructures of Nafion and a sulfonated poly ether ketone (Reproduced with permission from Ref. [105], copyright 2001, Elsevier Science B.V)

Fig. 12.12 Low capacity fade observed in a V-RFB using a low crossover sulfonated and selectively fluorinated aromatic CEM (*filled symbols*) compared to Nafion® NRE-212 (*open symbols*) (Reproduced with permission from Ref. [106], copyright 2012, Royal Society of Chemistry)

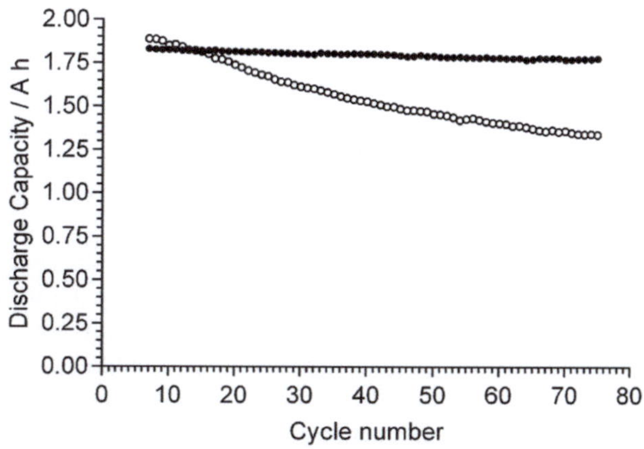

Figure 12.12 shows a V-RFB performance comparison of a sulfonated aromatic CEM versus a Nafion membrane [106]. The aromatic backbone was selectively fluorinated to prevent oxidative degradation of the aromatic CEM structure that has been reported for aromatic CEMs exposed to high oxidation state vanadium [107]. Due to the noticeable different ionic domain morphology in the sulfonated aromatic CEM, there is much less vanadium ion crossover in aromatic CEMs compared to the Nafion [105].

The lower vanadium crossover eases the capacity loss with cycling therefore helping to maintain reasonable ionic conductivity. Despite their excellent in situ V-RFB performance compared to PFSA membranes, the drawback of aromatic CEMs is their oxidative lifetime stability which is lower than that of Nafion for long-term grid storage applications.

12.5.2.2 Modified Composite Membranes

On the searching for alternatives to Nafion, considerable work has been done on modification methods of CEMs. Composite ion-exchange membranes with inorganic materials such as silica [108], zirconia [109], silica titanium oxide [110], zeolites [111], and silicon–aluminum oxide [112] have been in the focus of the research in the field of direct methanol fuel cells in order to reduce the methanol crossover. The crossover phenomena of methanol are caused by two different effects: (1) the protonic drag of methanol, where the methanol is easily transported together with protons through the membrane, and (2) the diffusion through the water-filled channels. Although Nafion membrane has a high ionic conductivity and good chemical stability in an oxidizing electrolyte, the high permeability of active species across the membrane has been a major technical problem for RFBs. The following methods were applied to CEMs in order to produce low vanadium permeable CEMs: (1) electrolyte soaking, (2) oxidation polymerization, (3) electrodeposition, (4) in situ sol gel, and (5) impregnation method [113].

To improve the ion selectivity or chemical properties, Nafion membrane has been modified as hybrid–composite membrane using organic materials such as poly (4-vinyl pyridine) [114], pyrroles [115], polyaniline [116], polyethylenimine (PEI) [117], poly(diallyldimethylammonium chloride-polyanion poly(sodium styrene sulfonate) (PDDA-PSS) [118], organically modified silicate (ORMOSIL) [119], and zirconium phosphate [120] hybrid membranes. As shown in Fig. 12.13, the sulfate V-FRBs with the Nafion–ORMOSIL membrane exhibited the highest discharge capacity and voltage, while the one with the unmodified Nafion membrane was the lowest [121].

Some developments have focused on the modified–composite membranes based on non-Nafion or other non-perfluorinated separators for V-RFB as indicated in Table 12.8. Early approaches include modification of low cost separators, such as Daramic [94, 122], and low-density polyethylene (LDPE) grafting and sulfonation process [123]. Although the resulting membranes have high IEC and conductivity, the permeability of active species across the membrane is still significant. Therefore, polyvinylidene (PVDF) (as poly(vinylidene fluoride)-graft-poly (styrenesulfonic acid) (PVDF-*g*-PSSA-PMAc) has been suggested to be grafted into matrix membranes. PVDF-*g*-PSSA-PMAc was reported to have good chemical stability and low ion permeability in vanadium electrolyte. Similar to PVDF, other matrix materials such as sulfonated poly(arylene thioether ketone) (SPTK), sulfonated poly(tetrafluoroethylene) (SPTKK) [124], and poly(tetrafluoroethylene) (PTFE) [125] have also been investigated.

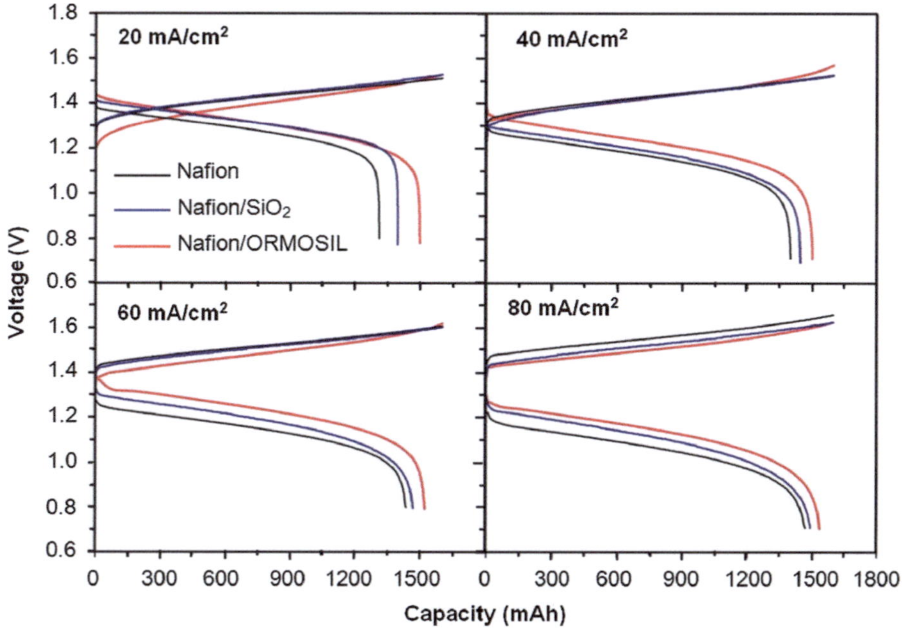

Fig. 12.13 Charge–discharge curves of VRB with Nafion, Nafion–SiO_2, and Nafion–ORMOSIL hybrid membranes at different current densities. The charge capacity was controlled to be 1600 mAh, corresponding to a redox couples utilization of 75 %. Mixtures of 40 mL of 2 M V3þ/V4þ and 2.5 M H_2SO_4 solutions were used as the starting anolyte and catholyte (Reproduced with permission from Ref. [121], copyright 2007, Elsevier B.V)

Table 12.8 Summary and comparison of modified–composite membranes used in all-vanadium redox flow battery

Membrane	Thickness (μm)	IEC (mmol/g)	Permeability (ppm)			Ionic conductivity (mS/cm)	Liquid uptake (wt %)
			V^{3+}	VO^{2+}	VO_2^+		
Nafion–SiO_2 [126]	204	0.96	<134	<82	<17.8	56.2	21.5
Modified Nafion117 [127]	201	0.88		34.6	–	5	–
SPEEK [103]	88	1.76	–	12.4	–	5	28.6
PVDF-*g*-PSSA-co-PMAc [128]	70	1.95	11.2	0.73	1.1	100	22
AIEM [129]	42	0.97	–	22	–	31	25.4
SPEEK [130]	151	1.59	–	94	–	2.2	36
SFPEK [131]	–	1.29	–	12	–	10.5	11.9
SPTKK [131]	–	1.91	–	31	–	13.6	19.3
Nafion117 [132]	178	0.94	<600	<550	<120	58.7	26.0
Nafion115 [103]	127		0.91	–	79.5	13.4	26

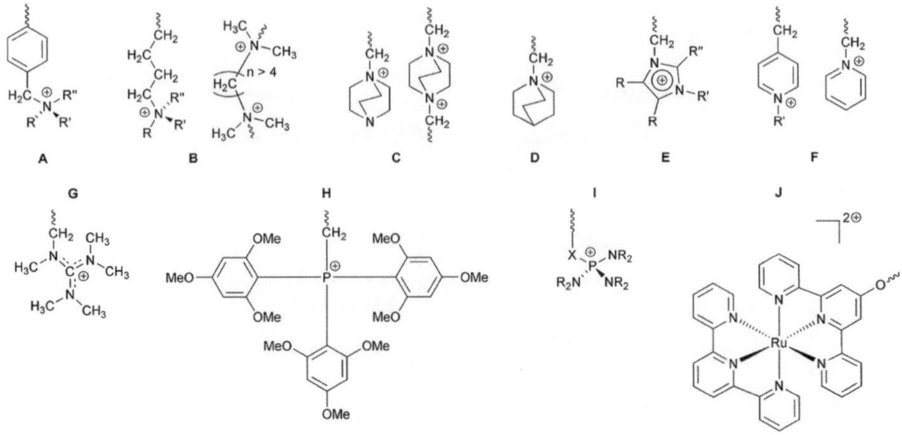

Fig. 12.14 Commonly encountered AEM cationic head-groups (those containing N–H and P–H bonds are omitted): A = benzyltrialkylammonium (the benchmark benzyltrimethylammonium is where R, R′, and R″ are methyl groups), B = alkyl-side-chain (benzene-free) quaternary ammonium (QA) and cross-linking diammonium groups (where the link chain is >C4 in length [preferably >C6 in length]), C = DABCO-based QA groups (more stable when only 1N atom is quaternized [cross-linked systems where both Ns are quaternized are also of interest but are less stable in alkali]), D ¼ quinuclidinium-based QA groups, E = imidazolium groups (where R = Me or H and R′, R″ = alkyl or aryl groups [not H]), F = pyridinium groups, G = pentamethylguanidinium groups, H = alkali stabilized quaternary phosphonium groups, I = P–N systems (where X = –SO_2– or –NR'– groups and where R = alkyl, aryl, or unsaturated cyclic systems), and J is an exemplar metal containing cationic group [133]

12.5.2.3 Anion Exchange Membranes (AEMs)

Next to cation exchange membranes, anion exchange membranes were also investigated for applications in VRB systems. The usage of AEMs is possible since only a charge transfer by ions is necessary to equilibrate the overall charge of the system. In AEMs, for example, $SO_2{}^{-4}$ will travel through the membrane during charge and discharge in order to equilibrate the charge. However, AEMs are membranes which have positively charged functional groups as shown in Fig. 12.14, which allow the transport of anions, while rejecting cations as indicated in a recent review that summarizes the properties of AEMs for the use in alkaline fuel cells [133]. Early studies of anion exchange membranes (AEMs) used in V-FRB were reported by Mohammadi and Skyllas-Kazacos [134]. AEMs with fixed cationic groups are attractive alternatives to CEMs due to the positively charged electroactive vanadium species in V-RFBs and have also been investigated as ultra-low vanadium crossover and water crossover IEMs.

Even though the AEM rejects cations, protons will diffuse through the membrane since their exclusion by the anion exchange membrane is less effective. To reduce the permeability of the vanadium ions and water, the AEMs are modified by cross-linking or incorporating cation exchange groups, which will result in improved performance and energy efficiency of V-FRBs (some results are

Table 12.9 Anion exchange membranes for vanadium RFB application comparing the proton conductivity and the performance of the developed membranes and Nafion of which both performance values appear in the same work

Membrane material	Conductivity (mS/cm)	Vanadium RFB performance (developed membrane/Nafion)		
		CE%	EE%	VE%
Poly(methacryloxyethyl dimethyl ammonium chloride) [98]	3.04	–	–	–
Chloromethylated poly(phthalazione ether sulfone) [138]	–	–	–	–
Cross-linked new Selemion [135]	–	–	82/–	87/–
Quaternized poly(phthalazinone ether sulfone) [137]	~10	98.7/95.9	83.4/86	84.5/89.7

CE coulomb efficiency, *EE* energy/efficiency, and *VE* voltage efficiency

summarized in Table 12.9) [135]. Other anion membranes explored include the quaternized poly(phthalazinone ether sulfone ketone) (QPPESK) [136] and poly (phthalazinone ether sulfone ketone) (QPPES) [137]. The QPPESK membrane demonstrated a good chemical stability in the VO_2^+ solutions for 20 days. The results of VRB single-cell tests showed a higher energy efficiency (88.3 %) from the cell with the QPPESK membrane than that with the Nafion 117 membrane (82.9 %) [136]. Qiu et al. [98] prepared an ETFE-based anion exchange membrane (AEM) to reduce the permeability of vanadium ions in vanadium redox flow batteries. Experimental results indicated a high ion-exchange capacity, lower area resistance, and much lower vanadium ion permeability of the ETFE-based AEM membranes compared to the Nafion 117 membranes.

To date studies have focused on basic descriptions of AEM performance in RFBs in order to classify the structure–property relationships for these materials in the unique environment of a redox flow cell. More advanced studies into the optimization of the polymer-tethered cationic groups and membranes with engineered physical structures (to decrease thickness and increase mechanical strength) are ongoing to continue to boost the performance of the membrane in these systems [139].

12.6 Membranes for Supercapacitors

12.6.1 Properties of Membrane Separators for Supercapacitors

The experimental study and modeling of the electrochemical characteristics of the separator material properties and influence of membrane parameters on the completed supercapacitor characteristics is an important problem taking into account the development of supercapacitors of high specific performances. The functional

Table 12.10 Supercapacitor competitive performance parameters using organic electrolyte

Parameters	Target
Active layer thickness	100 μm
Volumetric capacitance	100 F/cm^3
Electrolyte	0.05 S/cm
Cell voltage	2.5 V
Current collector thickness	25 μm
Separator porosity	50 %
Separator thickness	25 μm
Distributed resistance in pores	0.5 S/cm

components of a supercapacitor crucial to its operation are the electrodes, electrolyte, and separator as described in Sect. 12.4 [25]. In particular, the utilization of high surface area electrodes and the high efficient electrolytes in supercapacitors have stimulated intense research on separators. One of the key queries about designing high-performance supercapacitor is the choice of separators [2]. The separators used in supercapacitors prevent the occurrence of electrical contact between the two electrodes, but ion permeability, allowing ionic charge transfer, still takes place [26]. The basic principles to obtain a competitive supercapacitor performance is that the separator should have a high ionic electrolyte conductance, high ionic separator conductance, high electronic separator resistance, high electrode electronic conductance, large electrode surface, low separator, and electrodes thickness as listed in Table 12.10 [20]. Up to now, few research studies on the application of separators for supercapacitors have been performed.

12.6.2 Types of Membrane Separators for Supercapacitors

In general, the separators used between electrodes in supercapacitor have been constructed of porous polymeric membrane, nonwoven mat, and composite membrane separator, but prior known separators have had a tendency to dry out or collapse over a period of time or exhibit poor ionic conductivity [26]. Most commercial available separators are mainly designed for battery use; hence an accurate evaluation of the separator is essential to achieve the exceptional performance of supercapacitors. Polymer (typically polypropylene) or paper separators are used with organic electrolytes and ceramic, or glass fiber separators are often used with aqueous electrolytes. Cameron and Fitzsimmons [140], while comparing a paper separator with a proton-exchange membrane on the base of PVDF, conclude that a correctly selected separator should not contribute to the internal resistance that depends only on the electrolyte resistance and also not influence the electric capacity of a capacitor. The aforementioned authors also concluded that the separator material, its dimensions, and texture should not have any influence on the energy density and power density of the supercapacitor, respectively. Apart from the abovementioned PVDF membrane, microporous separators such as

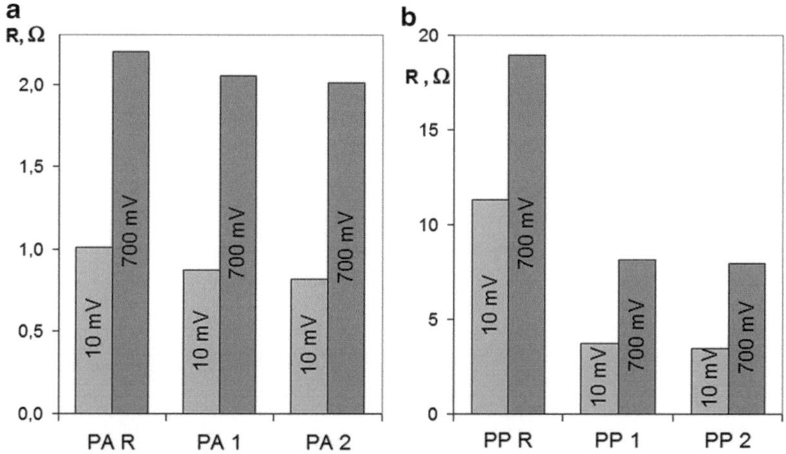

Fig. 12.15 Resistance of the electrolyte contained in the separators: (**a**) PA separators and electrolyte 1 M H_2SO_4 in water, (**b**) PP separators and electrolyte 1 M $TEABF_4$ in polyethylene carbonate (PC) (Reproduced with permission from Ref. [143], copyright 2014, Springer-Verlag, Berlin Heidelberg)

polypropylene (PP) and polyethylene (PE) made of this material are also used [141]. Winternheimer et al. [142] expanded a technological process of producing microporous biaxially oriented PP foil that produced excellent resistance to breakdowns and resistance to tensile strain of the separator, connected with long-term stability of work in both batteries and electrochemical capacitors. A considerable limitation in the use of these polymers is connected with their wettability in various electrolytes used in capacitors, as well as problems in removing moisture, in particular with PP materials. A possible solution was suggested by Szubzda et al. [143] where low-energy plasma technologies were used for modifying polymer PA (polyamide) and PP polymer nonwoven fabrics surfaces. The tests on these polymers confirmed the worth of this modification for cleaning surfaces and improving the wettability of the separators by the electrolyte, which in turn resulted in lowering of the internal resistance of the supercapacitor and significant improvements of the device power as indicated in Fig. 12.15.

As reported by Karabelli et al. [144], macroporous separator based on VDF homopolymer was prepared by a phase inversion technique using a solvent–non-solvent mixture. Their comparison demonstrated that, among the fluorinated polymers, the homopolymer PVDF is the most suitable grade to be implemented in supercapacitors. This separator provided higher conductivities than the commercial separator cellulose and above all Celgard as indicated in Table 12.11 [144]. The authors also reported that attempts to improve the mechanical strength of the separator and liquid electrolyte pair by decreasing separator porosity showed that the target could be reached without sacrificing conductivity.

Wu et al. [145] showed that macroporous eggshell membrane (ESM) as the separator of supercapacitor has favorable porous structures, high decomposition

Table 12.11 Conductivity and porosity values of soaked PVDF-based membranes and other commercial membranes at 25 °C

Sample	Porosity (%)	Conductivity ($mScm^{-1}$)	N_m	τ
AN-$TEABF_4$(1 mol^{-1})	–	55	–	–
M1	75 ± 5	18	3.1	1.6
M2	68 ± 8	15	3.6	1.6
M4	50 ± 1	4	11	2.4
Cellulose	65	10	5.5	1.9
Celgard™ 2500	53	4	14	2.7

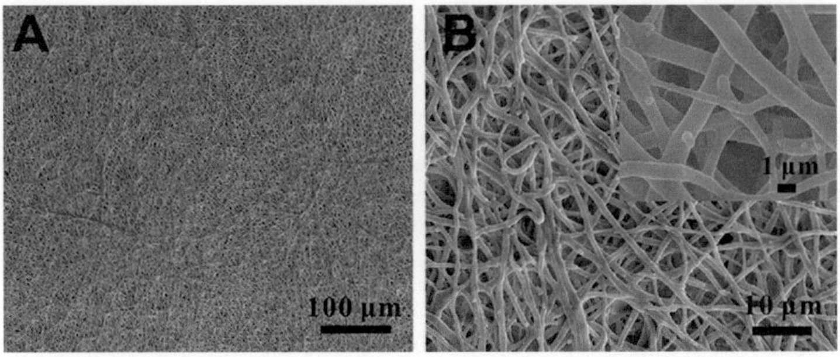

Fig. 12.16 SEM photographs of ESM at low (**a**) and high (**b**) magnifications (Reproduced with permission from Ref. [145], copyright 2012, Elsevier B.V)

temperature (over 200 °C), low swelling degree, and good mechanical strength. The typical SEM images of the natural ESM are shown in Fig. 12.16. It is shown that the membrane is a macroporous network composed of interwoven and coalescing shell membrane fibers ranging in diameter from 0.5 to 1 μm. The presence of macropores with pore sizes of 1–3 μm is evident, providing a superhighway for diffusing ions with a low resistance so as to enhance the performance of supercapacitors.

Furthermore, it can be concluded as indicated in Fig. 12.17 that ESM could be the best candidate for separator from the excellent electrochemical performances of supercapacitor, such as low resistance, quick charge–discharge ability (τ is 4.76 s), and outstanding cycling stability (92 % retention after 10,000 cycles) [145]. In a report by Morin [146], different nonwoven separators (Dreamweaver International (DWI) and two suppliers, referred as competitor) were tested at different microns (40, 30, 25) in supercapacitors assembled of commercial electrodes in a pouch cell. Separator properties, scanning electron micrographs (SEM), and capacitor performance were all measured for each material. Scanning electron micrographs (SEMs) as shown in Fig. 12.18 revealed that the separator marked (a) (competitor) is composed primarily of fibrillated cellulose microfibers with diameters in the range from 1 to 4 μm. Separator marked (b) (DWI) has similar composition, with a much higher population of fibers with diameters in the range of 0.2–0.4 μm.

An interesting comparison was made between the competitor's 30 μm material and the DWI 30 μm material (Silver AR30) as illustrated in Fig. 12.19. In those

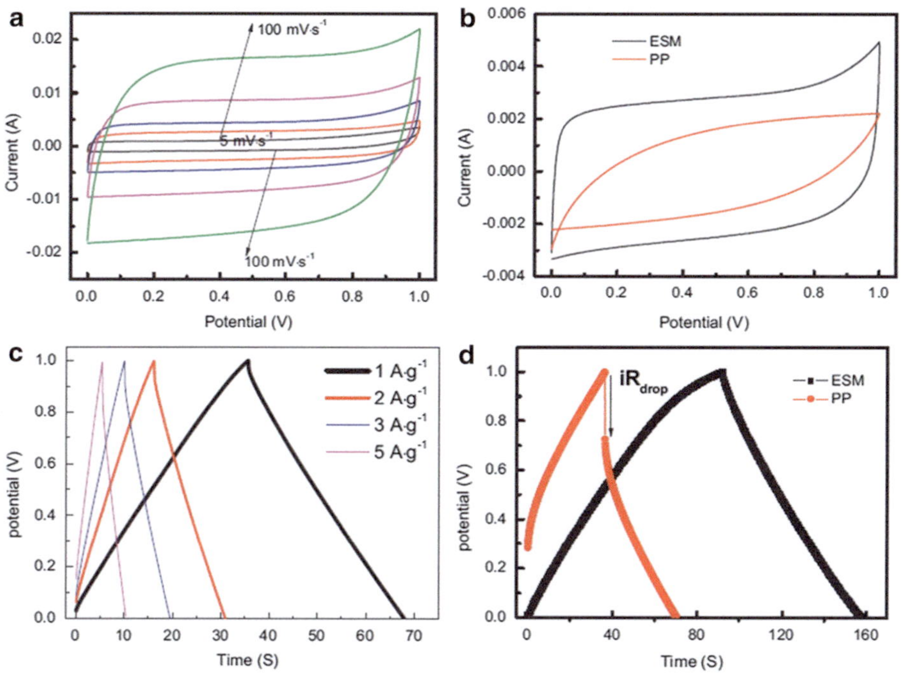

Fig. 12.17 (**a**) Rate-dependent cyclic voltammograms of supercapacitor with ESM in 1 mol L^{-1} Na_2SO_4 aqueous solution at various scan rates (5–100 mV s^{-1}). (**b**) Cyclic voltammograms of supercapacitors with ESM and PP separators in 1 mol L^{-1} Na_2SO_4 aqueous solution at the scan rates of 15 mV s^{-1}. (**c**) GCD curves of AC electrodes in 1 mol L^{-1} Na_2SO_4 aqueous solution at the charge–discharge current density of 0.5, 1, 2, 3, and 5 A g^{-1}. (**d**) GCD curves of supercapacitors with ESM and PP separators in 1 mol L^{-1} Na_2SO_4 aqueous solution at the charge–discharge current density of 0.5 A g^{-1} (Reproduced with permission from Ref. [145], copyright 2012, Elsevier B.V)

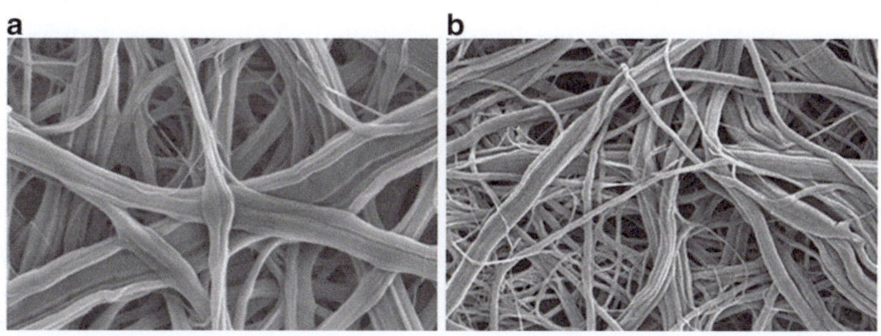

Fig. 12.18 Representative high magnification scanning electron microscope images (**a**) competitor separator and (**b**) DWI separator (Reproduced with permission from Ref. [146], copyright 2014, http://www.dreamweaverintl.com. Accessed 02 Dec 2014)

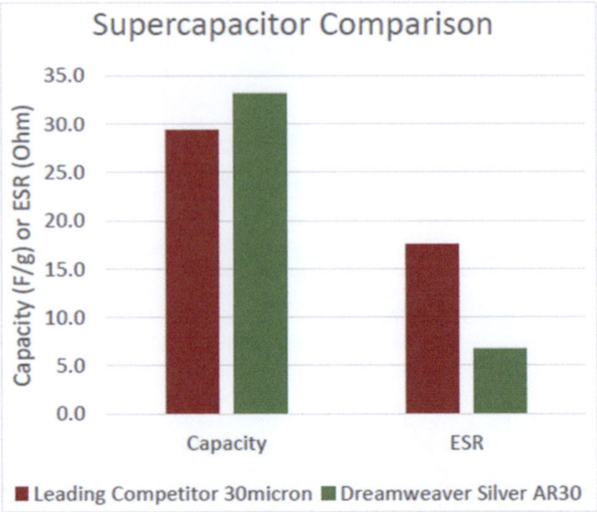

Fig. 12.19 Capacitor performance of the materials showed a similar 24 h self-discharge (Reproduced with permission from Ref. [146], copyright 2014, http://www.dreamweaverintl.com. Accessed 02 Dec 2014)

cells, the DWI separator provided 13 % higher capacitance, 61 % lower ESR, and equivalent self-discharge. At 40 and at 25 μm, the product from the two companies' performance is more similar.

Nafion membrane has been a common separator used in supercapacitors and had a setback due to its expensiveness and most importantly limited raw material. While eggshell membrane abovementioned has also been tipped to be one of the promising materials, its extravagant nature of usage and costly nature need to be taken into consideration. Mastragostino and Soavi considered polyvinyl alcohol (PVA) and lauroyl chitosan to be of wonderful materials, with the former being very good in mechanical strength and cost effective and later having the high levels of mechanical strength and ionic liquid retention required in EDLC cells [147]. Sa'adu et al. [148] explored the nobility of PVA and phosphoric acid (H_3PO_4) with cellulose filter paper and another separate solution of the combination of polymethyl(methacrylate) and lauroyl chitosan (PLC) and commercially prepared carbon nanotube (CPMWCNT) used as an electrolyte and electrode, respectively, for the fabrication of supercapacitor. The results suggested that the electrolyte performance was perfect due to high conductivity recorded when tested for the electrochemical impedance spectroscopy (EIS). The commercial carbon nanotube was shown to have thermal stability capable of being used as an electrode material for the purpose of this experiment. The overall formation gave balanced results for the supercapacitor as far as its efficiency, specific capacitance, and energy and power densities of the capacitance. High capacity of the supercapacitor can be achieved by enhancing the quality of the polymer electrolyte and reducing the thickness of the electrode.

12.7 Conclusions

The separator is a crucial component of electrochemical energy storage technologies especially in batteries and supercapacitors. Its main function is to prevent physical contact of the electrodes while permitting ions to flow freely. The separator itself does not participate in any cell reactions; however, its properties significantly determine the performance and safety of the batteries. For high energy and power densities, the separator is required to be very thin and highly porous, while it adversely affects the safety and cycle life of the battery as a result of the reduced mechanical strength.

Separators in most lithium–ion batteries are made of either microporous polymeric films or nonwoven fabrics. The nonwoven fabrics consist of a single polyolefin or a combination of polyolefins. The safety requirement is a top priority for rechargeable LIBs, especially those used in hybrid electric vehicles and power tools. However, none of polyolefin separators can withstand temperatures above the melting point (near 165 °C) of PP. The inorganic composite separator, which combines the characteristics of flexible polymers and hydrophilic inorganic materials, exhibits exceptional thermal stability and excellent wettability. Therefore, the inorganic composite separator exhibits overwhelming advantages in terms of safety and electrolyte wettability.

Vanadium RFB relies on ion-exchange (cation/anion)-based membranes, and the reflection on their properties in terms of conductivity, chemical stability, and crossover properties have been extensively reported. Although many membranes provide promising results, meeting all the requirements for the commercialization of an economically viable system is still a challenge that requires further development in order to achieve the required cost structure for large-scale grid-connected applications. The high reactivity of V^{5+} as a strong oxidant makes it challenging to select materials in terms of the long-term durability in V-RFBs. Although Nafion continues to be one of the widely studied membranes in V-RFB systems, however, its high cost, high levels of water transfer, and high diffusivity values for the vanadium ions limit its use in commercial applications.

One of the key difficulties about designing high-performance supercapacitor is the choice of separators. Separators used in supercapacitor have been constructed of rubber, plastic, aqua gel, resorcinol formaldehyde polymers, polyolefin films, etc., to prevent the conduction of electrons between the electrodes, but such prior known separators have had a tendency to dry out or collapse over a period of time or exhibit poor ionic conductivity. Therefore, a need exists for supercapacitor separators, which are made of highly porous materials that provide minimal resistance for electrolyte ion's movement and that at the same time, have electronic insulator properties between opposing electrodes.

Future development of separators will be made by balancing the performance against safety and cost. The high cost of separators is mainly due to their production process; therefore, developing a more cost-effective process is very important for the reduction of separator cost.

References

1. (a) Inagaki M, Konno H, Tanaike O (2010) Carbon materials for electrochemical capacitors. J Power Sources 195(24):7880–7903; (b) Yang Z, Zhang J, Kintner-Meyer MCW, Lu X, Choi D, Lemmon JP, Liu J (2011) Electrochemical energy storage for green grid. Chem Rev 111(5):3577–3613. doi:10.1021/cr100290v; (c) Nyholm L, Nyström G, Mihranyan A, Strømme M (2011) Toward flexible polymer and paper-based energy storage devices. Adv Mater 23(33):3751–3769. doi:10.1002/adma.201004134
2. Winter M, Brodd RJ (2004) What are batteries, fuel cells, and supercapacitors? Chem Rev 104:4245–4269. doi:10.1021/cr020730k
3. (a) Roberts B, McDowall J (2005) Commercial successes in power storage: advances in power electronics and battery application yield new opportunities. IEEE Power Energy Mag 3(2), March; (b) Shibata A, Sato K (1999) Development of a vanadium redox flow battery for electricity storage. IEEE Power Eng J 13(3):130–135; (c) Koshimizu G, Numata T (2005) Applications of energy storage for stabilization of wind power in power systems. Annual meeting presentation, Subaru Project, ESA, May 24
4. Yang Z, Zhang J, Kintner-Meyer MCW, Lu X, Choi D, Lemmon JP, Liu J (2010) Electrochemical energy storage for green grid. Chem Rev doi:10.1021/cr100290v
5. Nor NSM, Deraman M, Omar, Taer E, Awitdrus, Farma R, Basri RH, Dolah BNM (2013) Nanoporous separators for supercapacitor using activated carbon monolith electrode from oil palm empty fruit bunches. Paper presented at the AIP conference proceedings 1586:68–73. doi:10.1063/1.4866732
6. Chatzivasileiadi A, Ampatzi E, Knight I (2013) Characteristics of electrical energy storage technologies and their applications in buildings. Renew Sustain Energy Rev 25:814–830. doi:10.1016/j.rser.2013.05.023
7. Kreuer KD (2014) Ion conducting membranes for fuel cells and other electrochemical devices. Chem Mater 26:361–380. doi:10.1021/cm402742u
8. Hunt GL (1988) The great battery search. Spectr IEEE 35(11):21–28
9. Whittingham MS (2004) Lithium batteries and cathode materials. Chem Rev 104:4271–4302. doi:10.1021/cr020731c
10. Price A, Bartley S, Male S, Cooley G (1999) A novel approach to utility scale energy storage. J Power Eng 13(3):122–129
11. Prifti H, Parasuraman A, Winardi S, Lim TM, Skyllas-Kazacos M (2010) Membranes for redox flow battery applications. Membranes 2:275–306. doi:10.3390/2020275
12. (a) Rychcik M, Skyllas-Kazacos M (1987) Evaluation of electrode materials for vanadium redox cell. J Power Sources 19(1):45–54. doi:10.1016/0378-7753(87)80006-X; (b) Hagg CM, Skyllas-Kazacos M (2002) Novel bipolar electrodes for battery applications. J Appl Electrochem 32(10):1063–1069. doi:10.1023/A:1021228304148
13. (a) Qiu Y, Li MY, Ni JF, Zhai ML, Peng L, Xu L, Zhou HH, Li JQ, Wei GS (2007) Preparation of ETFE-based anion exchange membrane to reduce permeability of vanadium ions in vanadium redox battery. J Membr Sci 297(1–2):174–180. doi:10.1016/j.memsci.2007.03.042; (b) Vafiadis H, Skyllas-Kazacos M (2006) Evaluation of membranes for the novel vanadium bromine redox flow cell. J Membr Sci 279(1–2):394–402. doi:10.1016/j.memsci.2005.12.028
14. (a) Ponce de León C, Frías-Ferrer A, González-García J, Szánto DA, Walsh FC (2006) Redox flow cells for energy conversion. J Power Sources 160(1):716–732. doi:10.1016/j.jpowsour.2006.02.095; (b) Skyllas-Kazacos M, Chakrabarti MH, Hajimolana SA, Mjalli FS, Saleem M (2011) Progress in flow battery research and development. J Electrochem Soc 158:R55–R79. doi:10.1149/1.3599565; (c) Nourai A (2002) Large-scale electricity storage technologies for energy management. Power Eng Soc Summer Meet 1:310–315, July 2002
15. Butler PC, Miller DW, Verardo AE (1982) 17th intersociety energy conversion engineering conference, vol 2, p 653, Los Angeles

16. Arora P, Zhang ZJ (2004) Battery separators. Chem Rev 104(10):4419–4462. doi:10.1021/cr020738u
17. (a) Miller JR, Simon P (2008) Electrochemical capacitors for energy management. Science 321:651–652. doi:10.1126/science.1158736; (b) Yang X, Cheng C, Wang Y, Qiu L, Li D (2013) Liquid-mediated dense integration of graphene materials for compact capacitive energy storage. Science 341:534–537. doi:10.1126/science.1239089
18. (a) Halper MS, Ellenbogen JC (2006) Supercapacitors: a brief overview, The MITRE Corporation, McLean, Virginia, Technical report 06-0667, Mar 2006; (b) Rufer A, Hotellier D, Barrade P (2004) A supercapacitor-based energy storage substation for voltage compensation in weak transportation networks. IEEE Trans Power Deliv 19(2):629–636; (c) Jayalakshmi M, Balasubramani K (2008) Simple capacitors to supercapacitors – an overview. Int J Electrochem Sci 3(11):1196–1217
19. Béguin F, Frackowiak E (eds) (2013) Supercapacitors-materials, systems, and applications. Wiley-VCH, Weinheim
20. Schneuwly A, Gallay R (2000) Properties and applications of supercapacitors from the state-of-the-art to future trends. Paper presented at the proceeding PCIM, Boston, 3–5 Oct 2000
21. (a) Yu A, Chabot V, Zhang J (eds) (2013) Electrochemical supercapacitors for energy storage delivery – fundamentals and applications. CRC Press, Taylor and Francis Group; (b) Conway BE(1st ed) (1999) Electrochemical supercapacitors: scientific fundamentals and technological applications. Kluwer Academic/Plenum Publishers
22. Sun XZ, Zhang X, Huang B, Zhang HT, Zhang DC, Ma YW (2013) (LiNi0.5Co0.2Mn0.3O2 + AC)/graphite hybrid energy storage device with high specific energy and high rate capability. J Power Sources 243:361–368. doi:10.1016/j.jpowsour.2013.06.038
23. Davies A, Yu AP (2011) Material advancements in supercapacitors: from activated carbon to carbon nanotube and graphene. Can J Chem Eng 89(6):1342–1357. doi:10.1002/cjce.20586
24. (a) Amatucci GG, Badway F, Du Pasquier A, Zheng T (2001) An asymmetric hybrid nonaqueous energy storage cell. J Electrochem Soc 148(8):A930–A939. doi:10.1149/1.1383553; (b) Cericola D, Kötz R (2012) Hybridization of rechargeable batteries and electro chemical capacitors: principles and limits. Electrochim Acta 72:1–17. doi:10.1016/j.electacta.2012.03.151
25. Cheng SC, Liang YZ, Zhao JM, Zhang CH, Sun SY, Zhou NT, Qiu YP (2013) Zhang XW (2013) heat treatment of electrospun Polyvinylidene fluoride fibrous membrane separators for rechargeable lithium-ion batteries. J Power Sources 240:204–211. doi:10.1016/j.jpowsour.2013.04.019
26. Huang XS (2011) Separator technologies for lithium-ion batteries. J Solid State Electrochem 15:649–662. doi:10.1007/s10008-010-1264-9
27. Balducci A, Soavi F, Mastragostino M (2006) The use of ionic liquids as solvent-free green electrolytes for hybrid supercapacitors. Appl Phys A 82:627–632. doi:10.1007/s00339-014-8674-y
28. Zhang SS (2007) A review on the separators of liquid electrolyte Li-ion batteries. J Power Sources 164:351–364. doi:10.1016/j.jpowsour.2006.10.065
29. Lee H, Yanilmaz M, Toprakci O, Fu K, Zhang X (2014) A review of recent developments in membrane separators for rechargeable lithium-ion batteries. Energy Environ Sci 7:3857–3886. doi:10.1039/c4ee01432d
30. Venugopal G, Moore J, Howard J, Pendalwar S (1999) Characterization of microporous separators for lithium-ion batteries. J Power Sources 77:34–41. PII: S0378-7753-98.00168-2
31. Kirchhöfer M, von Zamory J, Paillard E, Passerini S (2014) Separators for Li-Ion and Li-metal battery including ionic liquid based electrolytes based on the TFSI− and FSI− Anions. Int J Mol Sci 15:14868–14890. doi:10.3390/ijms150814868
32. (a) Abraham KM, Alamgir M (1994) Room temperature polymer electrolytes and batteries based on them. Solid State Ionics 70–71(1):20–26. doi:10.1016/0167-2738(94)90283-6; (b) Shiao HC, Chua D, Lin HP, Slane S, Salomon M (2000) Low temperature electrolytes

for Li-ion PVDF cells. J Power Sources 87:167–173. PII: S0378- 7753-99.00470-X; (c) Panero S, Scrosati B (2000) Gelification of liquid–polymer systems: a valid approach for the development of various types of polymer electrolyte membranes. J Power Sources 90:13–19. PII: S0378- 7753-99.00470-X

33. (a) Bottino A, Camera-Roda G, Capannelli G, Munari S (1991) The formation of microporous polyvinylidene difluoride membranes by phase separation. J Membr Sci 57(1):1–20. doi:10.1016/S0376-7388(00)81159-X; (b) Boudin F, Andrieu X, Jehoulet C, Olsen II (1999) Microporous PVdF gel for lithium-ion batteries. J Power Sources 81–82:804–807. PII: S0378-7753-99.00154-8; (c) Saunier J, Alloin F, Sanchez J, Caillon G (2003) Thin and flexible lithium-ion batteries: investigation of polymer electrolytes. J Power Sources 119–121:454–459. doi:10.1016/S0378-7753(03)00197-6; (d) Magistris A, Mustarelli P, Parazzoli F, Quartarone E, Piaggio P, Bottino A (2001) Porosity and conductivity of PVDF films for polymer electrolytes. J Power Sources 97–98:657–660. PII: S0378-7753(01)00644-9; (e) Djian D, Alloin F, Martinet S, Lignier H (2009) Macroporous poly(vinylidene fluoride) membrane as a separator for lithium-ion batteries with high charge rate capacity. J Power Sources 187(2):575–580. doi:10.1016/j.jpowsour.2008.11.027; (f) Jiang Z, Carroll B, Abraham KM (1997) Studies of some poly(vinylidine fluoride) electrolytes Electrochim. Acta 42 (17):2667–2677. PII: s00134686(97)000054; (g) Magistris A, Quartarone E, Mustarelli P, Saito Y, Kataoka H (2002) PVDF-based porous polymer electrolytes for lithium batteries. Solid State Ion 152–153:347–354. PII: S0167-2738(02)00335-1
34. (a) Min HS, Ko JM, Kim DW (2003) Preparation and characterization of porous polyacrylonitrile membranes for lithium ion polymer batteries. J Power Sources 119–121:469–472. doi:10.1016/S0378-7753(03)00206-4; (b) Jung B, Yoon JK, Kim B, Rhee HW (2005) Effect of crystallization and annealing of polyacrylonitrile membranes for ultrafiltration. J Membr Sci 246(1):67–76. doi:10.1016/j.memsci.2004.08.012
35. Gopalan A, Santhosh P, Manesh K, Nho J, Kim S, Hwang C, Lee K (2008) Development of electrospun PVDF–PAN membrane-based polymer electrolytes for lithium batteries. J Membr Sci 325:683–690. doi:10.1016/j.memsci.2008.08.047
36. Lee MH, Kim HJ, Kim E, Rhee SB, Moon MJ (1996) Effect of phase separation on ionic conductivity of poly(methy1 methacrylate)-based solid polymer electrolyte. Solid State Ion 85(1–4):91–98. PII: SO167-2738(96)00046-X
37. Manuel Stephan A, Nahm KS (2006) Review on composite polymer electrolytes for lithium batteries. Polymer 47:5952–5964. doi:10.1016/j.polymer.2006.05.069
38. Ihm DW, Noh JG, Kim JY (2002) Effect of polymer blending and drawing conditions on properties of polyethylene separator prepared for Li-ion secondary battery. J Power Sources 109:388–393. PII: S0378-7753(02)00097-6
39. Saunier J, Alloin F, Sanchez J, Caillon G (2003) Thin and flexible lithium-ion batteries: investigation of polymer electrolytes. J Power Sources 119–121:454–459. doi:10.1016/S0378-7753(03)00197-6
40. Min HS, Ko JM, Kim DW (2003) Preparation and characterization of porous polyacrylonitrile membranes for lithium ion polymer batteries. J Power Sources 119–121:469–472. doi:10.1016/S0378-7753(03)00206-4
41. Appetecchi GB, Croce F, Scrosati B (1997) High performance electrolyte membranes for plastic lithium batteries. J Power Sources 66(1–2):77–82. PII:SO378-7753(96)02484-6
42. Bohnke O, Frand G, Rezrazi M, Rousselot C, Truche C (1993) Fast ion transport in new lithium electrolytes gelled with PMMA. 1. Influence of polymer concentration. Solid State Ion 66(1–2):97–104. doi:10.1016/0167-2738(93)90032-X
43. Magistris A, Quartarone E, Mustarelli P, Saito Y, Kataoka H (2002) PVDF-based porous polymer electrolytes for lithium batteries. Solid State Ion 152–153:347–354. PII: S0167-2738 (02)00335-1
44. Subramania A, Sundaram NTK, Kumar GV (2006) Structural and electrochemical properties of micro-porous polymer blend electrolytes based on PVdF-co-HFP-PAN for Li-ion battery applications. J Power Sources 153(1):177–182. doi:10.1016/j.jpowsour.2004.12.009

45. Nicotera I, Coppola L, Oliviero C, Castriota M, Cazzanelli E (2006) Investigation of ionic conduction and mechanical properties of PMMA–PVdF blend-based polymer electrolyte. Solid State Ion 177(5–6):581–588. doi:10.1016/j.ssi.2005.12.028
46. Choi NS, Lee YG, Park JK, Ko JM (2001) Preparation and electrochemcial characteristics of plasticized polymer electrolytes based upon a P(VdF-co-HFP)/PVAc blend. Electrochim Acta 46(10–11):1581–1586. doi:10.1016/s0013-4686(00)00756-8
47. Dahn JR, Fuller EW, Obrovac M, von Sacken U (1994) Thermal stability of LixCoO2, LixNiO2 and λ-MnO2 and consequences for the safety of Li-ion cells. Solid State Ion 69 (3–4):265–270. doi:10.1016/0167-2738(94)90415-4
48. Huai Y, Gao J, Deng Z, Suo J (2010) Preparation and characterization of a special structural poly(acrylonitrile)-based microporous membrane for lithium-ion batteries. Ionics 16:603–611. doi:10.1007/s11581-010-0431-4
49. Park JH, Park W, Kim JH, Ryoo D, Kim HS, Jeong YU, Kim DW, Lee SY (2011) Close-packed poly(methyl methacrylate) nanoparticle arrays-coated polyethylene separators for high-power lithium-ion polymer batteries. J Power Sources 196:7035–7038. doi:10.1016/j.jpowsour.2010.09.102
50. Wang HP, Huang H, Wunder SL (2000) Novel microporous poly(vinylidene fluoride) blend electrolytes for lithium-ion batteries. J Electrochem Soc 147:2853–2861. doi:10.1149/1.1393616
51. (a) Taskier (1982) HT, Hydrophilic polymer coated microporous membranes capable of use as a battery separator. US Patent 4,359,510, 3 Sept 1982; (b) Gineste JL, Pourcelly G (1995) Polypropylene separator grafted with hydrophilic monomers for lithium batteries. J Membr Sci 107:155–164. SSD10376-7388(95)001 12–3; (c) Senyarich S, Viaud P (2000) Method of forming a separator for alkaline electrolyte secondary electric cell. US Patent 6,042,970, 28 Mar 2000; (d) Gao K, Hu XG, Yi TF, Dai CS (2006) PE-g-MMA polymer electrolyte membrane for lithium polymer battery. Electrochim Acta 52:443–449. doi:10.1016/j.electacta.2006.05.049
52. Jarvis CR, Macklin WJ, Macklin AJ, Mattingley NJ, Kron E (2001) Use of grafted PVdF based polymers in lithium batteries. J Power Sources 97:664–666. PII: S0378-7753(01)00696-6
53. Jeong KU, Chae HD, Lim CH, Lee HK, Ahn JH, Nah C (2009) Fabrication and characterization of electrolyte membranes based on organoclay/tripropylene glycol diacrylate/poly (vinylidene fluoride) electrospun nanofiber composites. Polym Int 59:249–255. doi:10.1002/pi.2716
54. Song MK, Kim YT, Kim YT, Cho BW, Popov BN, Rhee HW (2003) Thermally stable gel polymer electrolytes. J Electrochem Soc 150(4):A439–A444. doi:10.1149/1.1556592
55. (a) Kim DW, Ko JM, Chun JH, Kim SH, Park JK (2001) Electrochemical performances of lithium-ion cells prepared with polyethylene oxide-coated separators. Electrochem Commun 3:535–538. PII:S1388-2481(01)00214-4; (b) Kim DW, Noh KA, Chun JH, Kim SH, Ko JM (2001) Highly conductive polymer electrolytes supported by microporous membrane. Solid State Ion 144:329–337. PII: S0167-2738Ž01.00977-8; (c) Eschbach FO, Oliver M (1997) Gel electrolyte bonded rechargeable electrochemical cell and method of making same. US Patent 5,681,357, 28 Oct 1997; (d) Hamano K, Shiota H, Shiraga S, Aihara S, Yoshida Y, Murai M, Inuzuka T (1999) Lithium-ion secondary battery and method of fabricating thereof. US Patent 5,981,107, 9 Nov 1999
56. Jeong YB, Kim DW (2004) Effect of thickness of coating layer on polymer-coated separator on cycling performance of lithium-ion polymer cells. J Power Sources 128:256–262. doi:10.1016/j.jpowsour.2003.09.073
57. (a) Oh JS, Kang YK, Kim DW (2006) Lithium polymer batteries using the highly porous membrane filled with solvent-free polymer electrolyte. Electrochim Acta 52:1567–1570. doi:10.1016/j.electacta.2006.02.062; (b) Lee YM, Choi NS, Lee JA, Seol WH, Cho KY, Jung YJ, Kim JW, Park JK (2005) Electrochemical effect of coating layer on the separator based on PVdF and PE non-woven matrix. J Power Sources 146:431–435. doi:10.1016/j.

jpowsour.2005.03.047; (c) Kim JY, Kim SK, Lee SJ, Lee SY, Lee HM, Ahn S (2004) Preparation of micro-porous gel polymer for lithium ion polymer battery. Electrochim Acta 50:363–366. doi:10.1016/j.electacta.2004.01.131

58. Song MK, Kim YT, Cho YJ, Cho BW, Popov BN, Rhee HW (2004) Composite polymer electrolytes reinforced by non-woven fabrics. J Power Sources 125:10–16. doi:10.1016/S0378-7753(03)00826-7
59. Kritzer P (2006) Nonwoven support material for improved separators in Li–polymer batteries. J Power Sources 161:1335–1340. doi:10.1016/j.jpowsour.2006.04.142
60. Cheita JR, Dutta A, Dass NN (2002) Characterization and conductivity studies of poly (2-dimethylamino ethylmethacrylate) and its hydrochloride salt in solid state. In: Chowdari BVR (ed) Proceedings of the 8th Asian conference, Langkawi
61. (a) Nookala M, Kumar B, Rodrigues S (2002) Ionic conductivity and ambient temperature Li electrode reaction in composite polymer electrolytes containing nanosize alumina. J Power Sources 111(1):165–172. doi:10.1016/S0378-7753(02)00303-8; (b) Cui ZY, Xu YY, Zhu LP, Wang JY, Xi ZY, Zhu BK (2008) Preparation of PVDF/PEO-PPO-PEO blend microporous membranes for lithium ion batteries via thermally induced phase separation process. J Membr Sci 325(2):957–963. doi:10.1016/j.memsci.2008.09.022
62. Cao C, Li ZB, Wang XL, Zhao XB, Han WQ (2014) Recent advances in inorganic solid electrolytes for lithium batteries. Front Energy Res 2(25):1–10. doi:10.3389/fenrg.2014.00025
63. (a) Wang Y, Ma X, Zhang Q, Tian N (2010) Synthesis and properties of gel polymer electrolyte membranes based on novel comb-like methyl methacrylate copolymers. J Membr Sci 349(1–2):279–286. doi:10.1016/j.memsci.2009.11.060; (b) Lee KH, Park JK, Kim WJ (2000) Electrochemical characteristics of PAN ionomer based polymer electrolytes. Electrochim Acta 45(8–9):1301–1306. doi:10.1016/S0013-4686(99)00336-9
64. (a) Kim KM, Ryu KS, Kang SG, Chang SH, Chung IJ (2001) The effect of silica addition on the properties of poly((vinylidene fluoride)-*co*-hexafluoropropylene)-based polymer electro lytes. Macromol Chem Phys 202:866–872. doi:10.1002/1521-3935(20010301)202; (b) Kim KM, Park NG, Ryu KS, Chang SH (2002) Characterization of poly(vinylidenefluoride-*co*-hexafluoropropylene)-based polymer electrolyte filled with TiO_2 nanoparticles. Polymer 43 (11):3951–3957. doi:10.1016/S0032-3861(02)00215-X
65. (a) Armand M, Tarascon JM (2008) Building better batteries. Nature 451:652–657. doi:10.1038/451652a; (b) Manuel Stephan A (2006) Review on gel polymer electrolytes for lithium batteries. Eur Polym J 42:21–42. doi:10.1016/j.eurpolymj.2005.09.017
66. Bruce PG (1995) Structure and electrochemistry of polymer electrolytes. Electrochim Acta 40(13–14):2077–2085. doi:10.1016/0013-4686(95)00144-4
67. Song J, Wang Y, Wan C (1999) Review of gel-type polymer electrolytes for lithium-ion batteries. J Power Sources 77(2):183–197. PII: S0378-7753(98)00193-1
68. Ito Y, Kanehori K, Miyauchi K, Kudo T (1987) Ionic conductivity of electrolytes formed from PEO-LiCF3SO3 complex low molecular weight poly (ethylene glycol). J Mater Sci 22 (5):1845–1849
69. Feuillade G, Perche P (1975) Ion-conductive macromolecular gels and membranes for solid lithium cells. J Appl Electrochem 5:63–69
70. Choi B, Kim Y, Shin H (2000) Ionic conduction in PEO–PAN blend polymer electrolytes. Electrochim Acta 45:1371–1374. PII: S0013-4686(99)00345-X
71. (a) Ramesh S, Leen KH, Kumutha K, Arof A (2007) FTIR studies of PVC/PMMA blend based polymer electrolytes. Spectrochim Acta A Mol Biomol Spectrosc 66:1237–1242. doi:10.1016/j.saa.2006.06.012; (b) Lee H, Yoo JK, Park JH, Kim JH, Kang K, Jung YS (2012) A stretchable polymer–carbon nanotube composite electrode for flexible lithium-ion batteries: porosity engineering by controlled phase separation. Adv Energy Mater 2:976–982. doi:10.1002/aenm.201100725
72. (a) Esterly DM (2002) Manufacturing of poly (vinylidene fluoride) and evaluation of its mechanical properties. A Master's thesis submitted on the 9th of August 2002, University of

Blackburn, Virginia; (b) Choe H, Giaccai J, Alamgir M, Abraham K (1995) Preparation and characterization of poly (vinyl sulfone)-and poly (vinylidene fluoride)-based electrolytes. Electrochim Acta 40(13–14):2289–2293.

73. (a) Gentili V, Panero S, Reale P, Scrosati B (2007) Composite gel-type polymer electrolytes for advanced, rechargeable lithium batteries. J Power Sources 170:185–190. doi:10.1016/j.jpowsour.2007.04.008; (b) Salimi A, Yousefi AA (2003) Analysis Method: FTIR studies of &-phase crystal formation in stretched PVDF films. Polym Test 22:699–704. doi:10.1016/S0142-9418(03)00003-5; (c) Kim JR, Choi SW, Jo SM, Lee WS, Kim BC (2004) Electrospun PVdF based fibrous polymer electrolytes for lithium ion polymer batteries. Electrochim Acta 50:69–75; (d) Zhang HP, Zhang P, Li ZH, Sun M, Wu YP, Wu HQ (2007) A novel sandwiched membrane as polymer electrolyte for lithium ion battery. Electrochem Commun 9:1700–1703. doi:10.1016/j.elecom.2007.03.021; (e) Ji GL, Zhu BK, Cui ZY, Zhang CF, Xu YY (2007) PVDF porous matrix with controlled microstructure prepared by TIPS process as polymer electrolyte for lithium ion battery. Polymer 48:6415–6425. doi:10.1016/j.polymer.2007.08.049
74. (a) Boudin F, Andrieu X, Jehoulet C, Olsen II (1999) Microporous PVdF gel for lithium-ion batteries. J Power Sources 81–82:804–807. PII: S0378- 7753 99 00154–8; (b) Choi SW, Jo SM, Lee WS, Kim YR (2003) An electrospun poly(vinylidene fluoride) nanofibrous membrane and its battery applications advanced materials 15:2027–2032. doi:10.1002/adma.200304617; (c) Montazami R, Liu S, Liu Y, Wang D, Zhang Q, Heflin JR (2011) Thickness dependence of curvature, strain, and response time in ionic electroactive polymer actuators fabricated via layer-by-layer assembly. J App Phys 109:104301–104305. doi:10.1063/1.3590166; (d) Montazami R, Wang D, Heflin JR (2012) Influence of conductive network composite structure on the electromechanical performance of ionic electroactive polymer actuators. Inter J Smart Nano Mater 3:204–213. doi:10.1080/19475411.2011.650232
75. Cheruvally G, Kim JK, Choi JW, Ahn JH, Shin YJ, Manuela J, Raghavan P, Kim KW, Ahn HJ, Choi DS, Song CE (2007) Electrospun polymer membrane activated with room temperature ionic liquid: Novel polymer electrolytes for lithium batteries. J Power Sources 172:863–869. doi:10.1016/j.jpowsour.2007.07.057
76. Fernicola A, Scrosati A, Ohno H (2006) Potentialities of ionic liquids as new electrolyte media in advanced electrochemical devices. Ionics 12:95–102. doi:10.1007/s11581-006-0023-5
77. (a) Matsumoto H, Sakaebe H, Tatsumi K, Kikuta M, Ishiko E, Kono M (2006) Fast cycling of $Li/LiCoO_2$ cell with low-viscosity ionic liquids based on bis (fluorosulfonyl) imide $[FSI]^-$. J Power Sources 160:1308–1313. doi:10.1016/j.jpowsour.2006.02.018; (b) Balducci A, Soavi F, Mastragostino M (2006) The use of ionic liquids as solvent-free green electrolytes for hybrid supercapacitors. Appl Phys A 82:627–632. doi:10.1007/s00339-005-3402-2
78. Sakaebe H, Matsumoto H (2003) N – Methyl-N – propylpiperidinium bis (trifluoromethanesulfonyl) imide (PP13–TFSI)–novel electrolyte base for Li battery. Electrochem Commun 5:594–598. doi:10.1016/S1388-2481(03)00137-1
79. (a) Fuller J, Carlin RT, Osteryoung RA (1997) The room temperature ionic liquid 1-ethyl-3-methylimidazolium tetrafluoroborate: electrochemical couples and physical properties. J Electrochem Soc 144:3881–3886. doi:10.1149/1.1838106; (b) Fung Y, Zhou R (1999) Room temperature molten salt as medium for lithium battery. J Power Sources 81:891–895. doi:10.1016/S0378-7753(99)00127-5
80. (a) Bruce PG, Scrosati B, Tarascon JM (2008) Nanomaterials for rechargeable lithium batteries. Angew Chem Int Ed 47:2930–2946. doi:10.1002/anie.200702505; (b) Subramania A, Kalyana Sundaram NT, Sathiya Priya AR, Vijaya Kumar G (2007) Preparation of a novel composite micro-porous polymer electrolyte membrane for high performance Li-ion battery. J Membr Sci 294:8–15. doi:10.1016/j.memsci.2007.01.025; (c) Balaya P, Bhattacharyya AJ, Jamnik J, Zhukovskii YF, Kotomin EA, Maier J (2006) Nano-ionics in the context of lithium batteries. J Power Sources 159:171–178. doi:10.1016/j.jpowsour.2006.04.115

81. Yang CM, Kim HS, Na BK, Kum KS, Cho BW (2006) Gel-type polymer electrolytes with different types of ceramic fillers and lithium salts for lithium-ion polymer batteries. J Power Sources 156:574–580. doi:10.1016/j.jpowsour.2005.06.018
82. (a) Wu CG, Lu MI, Tsai CC, Chuang HJ (2006) PVdF-HFP/metal oxide nanocomposites: the matrices for high-conducting, low-leakage porous polymer electrolytes. J Power Sources 159:295–300. doi:10.1016/j.jpowsour.2006.04.108; (b) Jeong HS, Kim DW, Jeong YU, Lee SY (2010) Effect of phase inversion on microporous structure development of Al_2O_3/poly (vinylidene fluoride-hexafluoropropylene)-based ceramic composite separators for lithium-ion batteries. J Power Sources 195:6116–6121. doi:10.1016/j.jpowsour.2009.10.085; (c) Kim M, Han GY, Yoon KJ, Park JH (2010) Preparation of a trilayer separator and its application to lithium-ion batteries. J Power Sources 195:8302–8305. doi:10.1016/j.jpowsour.2010.07.016; (d) Liao YH, Rao MM, Li WS, Yang LT, Zhu BK, Xu R, Fu CH (2010) Fumed silica-doped poly(butyl methacrylate-styrene)-based gel polymer electrolyte for lithium ion battery. J Membr Sci 352:95–99. doi:10.1016/j.memsci.2010.01.064
83. (a) Zhang SS, Xu K, Jow TR (2004) An inorganic composite membrane as the separator of Li-ion batteries. J Power Sources 140:361–364. doi:10.1016/j.jpowsour.2004.07.034; (b) Takemura D, Aihara S, Hamano K, Kise M, Nishimura T, Urushibata H, Yoshiyasu H (2005) A powder particle size effect on ceramic powder based separator for lithium rechargeable battery. J Power Sources 146:779–783. doi:10.1016/j.jpowsour.2005.03.159; (c) Goodenough JB, Kim Y (2010) Challenges for rechargeable Li batteries. Chem Mater 22:587–603. doi:10.1021/cm901452z
84. Croce F, Curini R, Martinelli A, Persi L, Ronci F, Scrosati B, Caminiti R (1999) Physical and chemical properties of nanocomposite polymer electrolytes. J Phys Chem B 103:10632–10638. doi:10.1021/jp992307u
85. Fergus JW (2010) Ceramic and polymeric solid electrolytes for lithium-ion batteries. J Power Sources 195:4554–4569. doi:10.1016/j.jpowsour.2010.01.076
86. (a) Augustin S, Hennige VD, Horpel G, Hying C (2002) Ceramic but flexible: new ceramic membrane foils for fuel cells and batteries. Desalination 146:23–28. PII: S0011-9164(02) 00465-4; (b) Hennige V, Hying C, Horpel G, Novak P, Vetter J (2006) Separator provided with asymmetrical pore structures for an electrochemical cell. US Patent Appl. 20060078791-A1. 13 Apr 2006
87. Li J, Daniel C, Wood D (2011) Materials processing for lithium-ion batteries. J Power Sources 196:2452–2460. doi:10.1016/j.jpowsour.2010.11.001
88. Chieng SC, Kazacos M, Skyllas-Kazacos M (1992) Preparation and evaluation of composite membrane for vanadium redox battery applications. J Power Sources 39:11–18. doi:10.1016/0378-7753(92)85002-R
89. Skyllas-Kazacos M, Milne NA, KazacosGC (2007) Membrane properties and behaviour in the generation 2 Vanadium Bromide Redox Flow batteries. Paper presented at the 16th international federation for heat treatment and surface engineering Congress, Brisbane, 30 Oct–2 Nov 2007
90. Sukkar T, Skyllas Kazacos M (2003) Modification of membranes using polyelectrolytes to improve water transfer properties in the vanadium redox battery. J Membr Sci 222:249–264
91. Kim S, Tighe TB, Schwenzer B, Yan J, Zhang J, Liu Z, Yang M, Hickner A (2011) Chemical and mechanical degradation of sulfonated poly(sulfone) membranes in vanadium redox flow batteries. J Appl Electrochem 41:1201–1213. doi:10.1007/s10800-011-0313-0
92. (a) Kim DH, Seo SJ, Lee MJ, Park JS, Moon SH, Kang YS, Choi YW, Kang MS (2014) Pore-filled anion-exchange membranes for non-aqueous redox flow batteries with dual-metal-complex redox shuttles. J Membr Sci 454:44–50. doi:10.1016/j.memsci.2013.11.051; (b) Maurya S, Shin SH, Sung KW, Moon SH (2014) Anion exchange membrane prepared from simultaneous polymerization and quaternization of 4-vinyl pyridine for non-aqueous vanadium redox flow battery applications. J Power Sources 255:325–334. doi:10.1016/j.jpowsour.2014.01.047; Shinkle AA, Sleightholme AES, Griffith LD, Thompson LT, Monroe

CW (2012) Degradation mechanisms in the non-aqueous vanadium acetylacetonate redox flow battery. J Power Sources 206:490–496. doi:10.1016/j.jpowsour.2010.12.096
93. Alotto P, Guarnieri M, Moro F (2014) Redox flow batteries for the storage of renewable energy: a review. Renew Sustain Energy Rev 29:325–335. doi:10.1016/j.rser.2013.08.001
94. Mohammadi T, Skyllas-Kazacos M (1995) Preparation of sulfonated composite membrane for vanadium redox flow battery applications. J Membr Sci 107:35–45. SSD10376-7388 (95) 00096–8
95. Xu T (2005) Ion exchange membranes: state of their development and perspective. J Membr Sci 263:1–29. doi:10.1016/j.memsci.2005.05.002
96. Strathmann H (1983) Synthetic membranes and their preparation. In: Proceedings of the NATO advanced study institute on synthetic membranes: science, engineering, and applications alcabideche, Portugal, 26 June–8 July 1983
97. Rikukawa M, Sanui K (2000) Proton-conducting polymer electrolyte membranes based on hydrocarbon polymers. Prog Polym Sci 25:1463–1502. PII: S0079-6700(00)00032-0
98. Qiu Y, Li MY, Ni JF, Zhai ML, Peng J, Xu L, Zhou HH, Li JQ, Wei GS (2007) Preparation of ETFE-based anion exchange membrane to reduce permeability of vanadium ions in vanadium redox battery. J Membr Sci 297:174–180. doi:10.1016/j.memsci.2007.03.042
99. Vafiadis H, Skyllas-Kazacos M (2006) Evaluation of membranes for the novel vanadium bromine redox flow cell. J Membr Sci 279:394–402. doi:10.1016/j.memsci.2005.12.028
100. (a) Skyllas-Kazacos M, Kasherman D, Hong DR, Kazacos M (1991) Characteristics and performance of 1kW UNSW vanadium redox battery J Power Sources 35(4):399–404; (b) Mohammadi T, Skyllas-Kazacos M (1997) Evaluation of the chemical stability of some membranes in vanadium solution. J Appl Electrochem 27(2):153–160
101. Sun CX, Chen J, Zhang HM, Han X, Luo QT (2010) Investigations on transfer of water and vanadium ions across Nafion membrane in an operating vanadium redox flow battery. J Power Sources 195:890–897. doi:10.1016/j.jpowsour.2009.08.041
102. Gubler L, Gürsel SA, Scherer GG (2005) Radiation grafted membranes for polymer electrolyte fuel cells. Fuel Cells 5(3):317–335. doi:10.1002/fuce.200400078
103. Jia C, Liu J, Yan C (2010) A significantly improved membrane for vanadium redox flow battery. J Power Sources 195(13):4380–4383. doi:10.1016/j.jpowsour.2010.02.008
104. Li X, Zhang H, Mai Z, Zhang H, Vankelecom I (2011) Ion exchange membranes for vanadium redox flow battery (VRB) applications. Energy Environ Sci 4:1147–1160. doi:10.1039/C0EE00770F
105. Kreuer KD (2001) On the development of proton conducting polymer membranes for hydrogen and methanol fuel cells. J Membr Sci 185(1):29–39. PII: S0376-7388(00)00632-3
106. Chen D, Kim S, Li L, Yang G, Hickner M (2012) Stable fluorinated sulfonated poly(arylene ether) membranes for vanadium redox flow batteries. RSC Adv 2:8087–8094. doi:10.1039/c2ra20834b
107. (a) Kim S, Tighe T, Schwenzer B, Yan J, Zhang J, Liu J, Yang Z, Hickner M (2011) A chemical and mechanical degradation of sulfonated poly(sulfone) membranes in vanadium redox flow batteries. J Appl Electrochem 41:1201–1213. doi:10.1007/s10800-011-0313-0; (b) Jung MSJ, Parrondo J, Arges CG, Ramani V (2013) Polysulfone-based anion exchange membranes demonstrate excellent chemical stability and performance for the all-vanadium redox flow battery. J Mater Chem A 1:10458–10464. doi:10.1039/C3TA11459G
108. (a) Jiang R, Kunz HR, Fenton JM (2006) Composite silica/Nafion® membranes prepared by tetraethylorthosilicate sol–gel reaction and solution casting for direct methanol fuel cells. J Membr Sci 272(1–2):116–124. doi:10.1016/j.memsci.2005.07.026; (b) Jung DH, Cho SY, Peck DH, Shin DR, Kim JS (2002) Performance evaluation of a Nafion/silicon oxide hybrid membrane for direct methanol fuel cell. J Power Sources 106(1–2):173–177. PII: S0378-7753 (01)01053-9; (c) Antonucci PL, Arico AS, Creti P, Ramunni E, Antonucci V (1999) Investigation of a direct methanol fuel cell based on a composite Nafion-silica electrolyte for high temperature operation. Solid State Ion 125(1–4):431–437. PII: S0167-2738(99)00206-4

109. (a) Bauer F, Willert-Porada M (2004) Microstructural characterization of Zr-phosphate–Nafion® membranes for direct methanol fuel cell (DMFC) applications. J Membr Sci 233 (1–2):141–149. doi:10.1016/j.memsci.2004.01.010; (b) Arbizzani C, Donnadio A, Pica M, Sganappa M, Varzi A, Casciola M, Mastragostino M (2010) Methanol permeability and performance of Nafion–zirconium phosphate composite membranes in active and passive direct methanol fuel cells. J Power Sources 195(23):7751–7756. doi:10.1016/j.jpowsour.2009.07.034
110. Daiko Y, Klein LC, Kasuga T, Nogami M (2006) Hygroscopic-oxides/Nafion® hybrid electrolyte for direct methanol fuel cells. J Membr Sci 281(1–2):619–625. doi:10.1016/j.memsci.2006.04.033
111. Yildirim MH, Curoś AR, Motuzas J, Julbe A, Stamatialis DF, Wessling M (2009) Nafion®/H-ZSM-5 composite membranes with superior performance for direct methanol fuel cells. J Membr Sci 338(1–2):75–83. doi:10.1016/j.memsci.2009.04.009
112. Ismail A, Othman N, Mustafa A (2009) Sulfonated polyether ether ketone composite membrane using tungstosilicic acid supported on silica–aluminium oxide for direct methanol fuel cell (DMFC). J Membr Sci 329(1–2):18–29. doi:10.1016/j.memsci.2008.11.052
113. (a) Mohammadi T, Skyllas-Kazacos M (1995) Characterisation of novel composite membrane for redox flow battery applications. J Membr Sci 98(1–2):77–87. SSD10376-7388 (94) 00 178–2; (b) Chieng S, Kazacos M, Skyllas-Kazacos M (1992) Modification of Daramic, microporous separator, for redox flow battery applications. J Membr Sci 75(1–2):81–91. doi:10.1016/0376-7388(92)80008-8; (c) Tian B, Yan CW, Wang F (2004) Proton conducting composite membrane from Daramic/Nafion for vanadium redox flow battery. J Membr Sci 234(1–2):51–54. doi:10.1016/j.memsci.2004.01.012; (c) Mohammadi T, Skyllas-Kazacos M (1995) Use of polyelectrolyte for incorporation of ion-exchange groups in composite membranes for vanadium redox flow battery applications. J Power Sources 56(1):91–96. SSDIO378-7753(95)02222-3
114. Woong JC, Venkataramani S, Kim SC (2006) Modification of Nafion membrane using poly (4-vinyl pyridine) for direct methanol fuel cell. Polym Int 55:491–499. doi:10.1002/pi.1986
115. Zeng J, Jiang CP, Wang YH, Chen JW, Zhu SF, Zhao BJ, Wang RL (2007) Studies on polypyrrole modified nafion membrane for vanadium redox flow battery. Electrochem Commun 10:372–375. doi:10.1016/j.elecom.2007.12.025
116. Tan S, Bélanger D (2005) Characterization and transport properties of Nafion/polyaniline composite membrane. J Phys Chem B 109:23480–23490. doi:10.1021/jp054724e
117. Luo QT, Zhang HM, Chen J, Qian P, Zhai YF (2007) Modification of Nafion membrane using interfacial polymerization for vanadium redox flow battery applications. J Membr Sci 311:98–103. doi:10.1016/j.memsci.2007.11.055
118. Xi JY, Wu ZH, Teng XG, Zhao YT, Chen LQ, Qiu XP (2008) Self-assembled polyelectrolyte multilayer modified Nafion membrane with suppressed vanadium ion crossover for vanadium redox flow batteries. J Mater Chem 18:1232–1238. doi:10.1039/B718526J
119. Schulte D, Drillkens J, Schulte B, Sauer DU (2010) Nafion hybrid membranes for use in redox flow batteries. J Electrochem Soc 157:A989–A992. doi:10.1149/1.3456625
120. Hwang GJ, Ohya H (1996) Preparation of cation exchange membrane as a separator for the all-vanadium redox flow battery. J Membr Sci 120:55–67. Pll S0376-7388(96)00135-4
121. (a) Xi J, Wu Z, Qiu X, Chen L (2007) Nafion/SiO2 hybrid membrane for vanadium redox flow battery. J Power Sources 166:531–536. doi:10.1016/j.jpowsour.2007.01.069; (b) Teng X, Zhao Y, Xi J, Wu Z, Qiu X, Chen L (2009) Nafion/organically modified silicate hybrids membrane for vanadium redox flow battery. J Power Sources 189:1240–1246. doi:10.1016/j.jpowsour.2008.12.040
122. Zhao P, Zhang HM, Zhou HT, Yi BL (2005) Nickel foam and carbon felt applications for sodium polysulfide/bromine redox flow battery electrodes. Electrochim Acta 51:1091–1098. doi:10.1016/j.electacta.2005.06.008
123. (a) Mohammadi T, Skyllas-Kazacos M (1995) Use of polyelectrolyte for incorporation of ion-exchange groups in composite membranes for vanadium redox flow battery applications.

J Power Sources 56(1):91–96; (b) Hwang GJ, Ohya H (1996) Preparation of cation exchange membrane as a separator for the all-vanadium redox flow battery. J Membr Sci 120:55–67. PII S0376-7388(96)00135-4
124. Chen DY, Wang SJ, Xiao M, Meng YH (2010) Synthesis and characterization of novel sulfonated poly(arylene thioether) ionomers for vanadium redox flow battery applications. Energy Environ Sci 3:622–628. doi:10.1039/B917117G
125. Qiu JY, Ni JF, Zhai ML, Peng J, Zhou HH, Li JQ, Wei GS (2007) Radiation grafting of styrene and maleic anhydride onto PTFE membranes and sequent sulfonation for applications of vanadium redox battery. Radiat Phys Chem 76:1703–1707. doi:10.1016/j.radphyschem.2007.01.012
126. Xi J, Wu Z, Qiu X, Chen L (2007) Nafion/SiO_2 hybrid membrane for vanadium redox flow battery. J Power Sources 166:531–536. doi:10.1016/j.jpowsour.2007.01.069
127. Teng X, Zhao Y, Xi J, Wu Z, Qiu X, Chen L (2009) Nafion/organically modified silicate hybrids membrane for vanadium redox flow battery. J Power Sources 189:1240–1246. doi:10.1016/j.jpowsour.2008.12.040
128. Qiu J, Zhai M, Chen J, Wang Y, Peng J, Xu L, Li J, Wei G (2009) Performance of vanadium redox flow battery with a novel amphoteric ion exchange membrane synthesized by two-step grafting method. J Membr Sci 342:215–220. doi:10.1016/j.memsci.2009.06.043
129. Qiu J, Zhao L, Zhai M, Ni J, Zhou H, Peng J, Li J, Wei G (2008) Pre-irradiation grafting of styrene and maleic anhydride onto PVDF membrane and subsequent sulfonation for application in vanadium redox batteries. J Power Sources 177:617–623. doi:10.1016/j.jpowsour.2007.11.089
130. Chen D, Wang S, Xiao M, Meng Y (2010) Synthesis and properties of novel sulfonated poly (arylene ether sulfone) ionomers for vanadium redox flow battery. Energy Convers Manag 51:2816–2824. doi:10.1016/j.enconman.2010.06.019
131. Chen D, Wang S, Xiao M, Meng Y (2009) Synthesis and characterization of novel sulfonated poly(arylene thioether) ionomers for vanadium redox flow battery application. Energy Environ Sci 3:622–628. doi:10.1039/B917117G
132. Teng X, Zhao Y, Xi J, Wu Z, Qiu X, Chen L (2009) Nafion/organic silica modified TiO_2 composite membrane for vanadium redox flow battery via in situ sol–gel reactions. J Membr Sci 341:149–154. doi:10.1016/j.memsci.2009.05.051
133. Varcoe JR, Atanassov P, Dekel DR, Herring AM, Hickner MA, Kohl PA, Kucernak AR, Mustain WE, Nijmeijer K, Scott K, Xu T, Zhuang L (2014) Anion-exchange membranes in electrochemical energy systems. Energy Environ Sci 7:3135–3191. doi:10.1039/c4ee01303d
134. Mohammadi T, Skyllas Kazacos M (1997) Evaluation of the chemical stability of some membranes in vanadium solution. J Appl Electrochem 27:153–160. doi:10.1023/A:1018495722379
135. Hwang GJ, Ohya H (1997) Crosslinking of anion exchange membrane by accelerated electron radiation as a separator for the all-vanadium redox flow battery. J Membr Sci 132 (1):55–61. PII S0376-73 88(97)00040-9
136. Jian XG, Yan C, Zhang HM, Zhang SH, Liu C, Zhao P (2007) Synthesis and characterization of quaternized poly(phthalazinone ether sulfone ketone) for anion-exchange membrane. Chin Chem Lett 18:1269–1272. doi:10.1016/j.cclet.2007.08.022
137. Xing D, Zhang S, Yin C, Zhang B, Jian X (2010) Effect of amination agent on the properties of quaternized poly(phthalazinone ether sulfone) anion exchange membrane for vanadium redox flow battery application. J Membr Sci 354(1–2):68–73. doi:10.1016/j.memsci.2010.02.064
138. Xing D, Zhang S, Yin C, Yan C, Jian X (2009) Preparation and characterization of chloromethylated/quaternized poly(phthalazinone ether sulfone) anion exchange membrane. Mater Sci Eng B 157(1–3):1–5. doi:10.1016/j.mseb.2008.11.019
139. (a) Maurya S, Shin S H, Kim M K, Yun S H, Moon S H (2013) Stability of composite anion exchange membranes with various functional groups and their performance for energy conversion. J Membr Sci 443:28–35. doi:10.1016/j.memsci.2013.04.035; (b) Maurya S,

Shin SH, Sung KW, Moon SH (2014) Anion exchange membrane prepared from simultaneous polymerization and quaternization of 4-vinyl pyridine for non-aqueous vanadium redox flow battery applications. J Power Sources 255:325–334. doi.org/10.1016/j.jpowsour.2014.01.047; (c) Fang J, Xu H, Wei X, Guo M, Lu X, Lan C, Zhang Y, Liu Y, Peng T (2013) Preparation and characterization of quaternized poly (2,2,2-trifluoroethyl methacrylate-co-Nvinylimidazole) membrane for vanadium redox flow battery. Polym Adv Technol 24:168–173. doi:10.1002/pat.3066; (d) Wang Y, Qiu J, Peng J, Xu L, Li J, Zhai M (2011) Study on the chemical stability of the anion exchange membrane of grafting dimethylaminoethyl methacrylate. J Membr Sci 376:70–77. doi:10.1016/j.memsci.2011.04.002

140. Cameron CG, Fitzsimmons SM (2008) Supercapacitor separators and polypyrrole composites, defence R&D Canada—Atlantic, Technical memorandum, DRDC Atlantic TM 2008-219
141. Kim BC, Too CO, Kwon JS, Ko JM, Wallace GG (2011) A flexible capacitor based on conducting polymer electrodes. Synth Met 161:1130–1132
142. Fischle H, Busch D, Schmitz B, Herrmann H, Winternheimer S (2008) New separator material for supercapacitors. Paper presented at the 3rd European symposium on supercapacitors and applications. ESSCAP '08 Rome
143. Szubzda B, Szmaja A, Ozimek M, Mazurkiewicz S (2014) Polymer membranes as separators for supercapacitors. Appl Phys A 117:1801–1809. doi:10.1007/s00339-014-8674-y
144. Karabelli D, Leprêtre JC, Alloin F, Sanchez JY (2011) Poly(vinylidene fluoride)-based macroporous separators for supercapacitors. Electrochim Acta 57:98–103. doi:10.1016/j.electacta.2011.03.033
145. Yu H, Tang Q, Wu J, Lin Y, Fan L, Huang M, Lin J, Li Y, Yu F (2012) Using eggshell membrane as a separator in supercapacitor. J Power Source 206:463–468. doi:10.1016/j.jpowsour.2012.01.116
146. Morin B (2014) A comparison of nonwoven separators for supercapacitors. http://www.dreamweaverintl.com. Accessed 02 Dec 2014
147. Mastragostino M, Soavi F (2007) Strategies for high-performance supercapacitors for HEV. J Power Source 174:89–93. doi:10.1016/j.jpowsour.2007.06.009
148. Hashim MA, Sa'adu L, Baharuddin M, Dasuki KA (2014) Using PVA, methacrylate and Lauroyl Chitosan as separator in supercapacitors. J Mater Sci Res 3(1):25–29. doi:10.5539/jmsr.v3n1p25

Chapter 13
Nanocomposite Polymer Electrolytes in Electrochemical Energy Storage Systems

Kumar Raju and Samuel A. Suthanthiraraj

13.1 Introduction

During the twentieth century, most synthetic polymers had been traditionally used as structural materials or electric insulators. Today, they have become very useful as electronic and ionic conductors. For example, when combined with appropriate salts, their ionic conductivity can be enhanced and used as electrolytes in electrochemical devices. Polymer electrolyte complexes were firstly developed in 1975 by Wright et al. [1] who showed that polyethylene oxide (PEO) could serve as a viable host material for sodium and potassium salts, thus generating a solid electrical-conducting polymer–salt complex. Motivated by this interesting development, Michel Armand, who had previously suggested the possible application of graphite as intercalation compounds for electrodes, quickly realized that lithium/PEO complexes could serve as excellent solid electrolytes for intercalation electrodes. According to Armand et al. [2], a lithium salt with large soft anions (such as $LiClO_4$ or $LiN(CF_3SO_3)_2$) could be dissolved in a solvating high molecular weight polymer matrix containing electron-donating groups (such as PEO) through direct coordination of the cation and electron pairs to form solid electrolytes. The solid electrolyte complex obtained by this favorable competition between the solvation energy and the lattice energy of the salt becomes a good conductor at $60 - 80°C$.

K. Raju (✉)
Energy Materials, Materials Science and Manufacturing, Council for Scientific and Industrial Research (CSIR), Pretoria 0001, South Africa

Department of Energy, University of Madras, Guindy Campus, Chennai 600 025, India
e-mail: Kraju@csir.co.za

S.A. Suthanthiraraj
Department of Energy, University of Madras, Guindy Campus, Chennai 600 025, India

K.I. Ozoemena, S. Chen (eds.), *Nanomaterials in Advanced Batteries and Supercapacitors*, Nanostructure Science and Technology,
DOI 10.1007/978-3-319-26082-2_13

Indeed, Armand's work sparked a lot of interest as it brought new perspectives in the international solid-state ionics community.

Solvation mechanisms in polymer electrolytes involve: (i) *the lithium ions which are more likely to be wrapped by polymer chain* and (ii) *the possibility to coordinate with heteroatoms which separate them from the anions*. In the beginning years, the conductivity of the polymer electrolytes was believed to take place mainly in the crystalline phase of the PEO helices. Later, it was established that the conductivity mainly depends on the number of the ionic carriers and on their mobility which require local relaxation and segmental mobility of polymer chain. However, with an increase in the ionic carriers as a result of the increased concentration of the salt, the mobility of ions is decreased by the occurrence of ion–ion association [3–5]. Indeed, the dynamics of ion in an electrolyte are well dependent on the flexibility of the polymer chain, i.e., amorphous state which allows ions to move faster than the crystalline counterparts [6, 7].

Generally, solid polymer electrolytes suffer from poor conductivity, mechanical stability, and low solubility of lithium salts which lead to a significant challenge to achieve superior electrochemical performance. However, various types of polymer electrolytes have been developed like (i) solid polymer electrolyte (SPE), (ii) gel polymer electrolyte (GPE), and (iii) composite polymer electrolyte (CPE) to increase the ion mobility in the solid electrolyte [8–10]. Despite having adequate conductivity with gel polymer electrolytes by adding organic solvents like ethylene carbonate (EC) and propylene carbonate (PC) to the solid electrolyte, the low mechanical ruggedness and loss of compatibility have limited the choice for practical applications. One of the numerous efforts to rectify the challenge has been the use of inorganic filler particles into the polymer system [11]. Since the last decade, nanomaterials have been used to improve the performance of polymer electrolytes for application in energy storage systems, notably lithium-ion batteries and supercapacitors. This review provides an overview of the development in nanocomposite polymer electrolytes and their physicochemical properties.

13.2 Nanocomposite Polymer Electrolytes

The emergence of nanomaterials in the late 1990s provided a new lease on life and benefits into the electrochemical devices due to their attractive virtues of viable physicochemical and mechanical properties [12, 13]. It has generally been established that the addition of ceramic nanoparticulate ($\tilde{}$ 20 nm) fillers boosts the segmental mobility of polymer hosts and their interfacial properties in contact with electrodes [14, 15]. This increase in mobility is explained by the enhanced degree of amorphocity of the polymer matrices and to maintain the liquid-like behavior at the microscopic level. On the other hand, the addition of filler nanoparticles improves the mobile ion concentration due to the more dissociation of salt by Lewis acid–base interactions [16–19]. According to the Lewis acid–base approach, the ion–ion and the polymer–filler–salt interactions result in the changes in the degree of dissociation and flexibility of polymer chain, thus influencing the

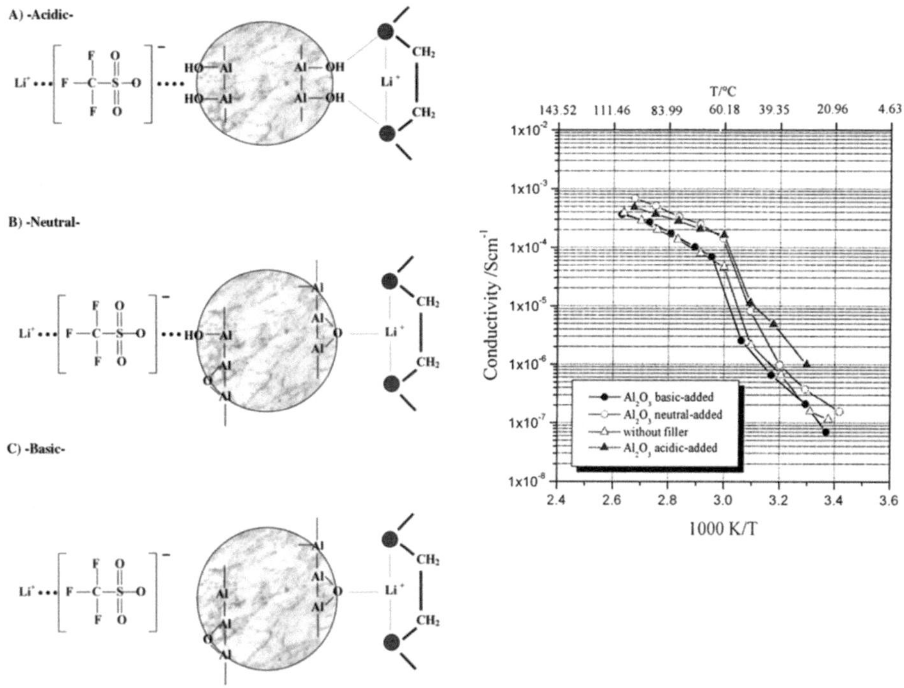

Scheme 13.1 Pictorial model of the surface interactions between three forms of dispersed nanosized Al_2O_3 ceramic and the PEO–$LiSO_3CF_3$ electrolyte complex. (**a**) Al_2O_3 acidic; (**b**) Al_2O_3 neutral; (**c**) Al_2O_3 basic. Inset represents the conductivity Arrhenius plots of $P(EO)_{20}LiSO_3CF_3$ + 10 w/o Al_2O_3 nanocomposite samples differing only by the type of surface states of the selected ceramic filler (Reprinted from Ref. [20]. Copyright 2001, with permission from Elsevier)

ion transport. As explained in Scheme 13.1, Scrosati et al. [20] proposed a model to account for the role of the ceramic fillers in enhancing the transport properties of PEO-based composite polymer electrolytes. To test their model, they selected the $P(EO)_{20}LiSO_3CF_3$ polymer matrix with the addition of 10 w/o Al_2O_3, this ceramic being available in three forms, i.e., acidic, basic, and neutral. The expected surface interactions of these three different ceramic forms are pictorially summarized in Scheme 13.1. In summary, the effect of the ceramics increases in the sequence: Al_2O_3 acidic > Al_2O_3 neutral > Al_2O_3 basic.

Nanocomposite polymer electrolytes provide several advantages over the conventional solid and liquid electrolytes in terms of producing thin films of conducting materials with good mechanical and electrochemical properties. Homogeneous dispersion of filler nanoparticles within the host polymers improves the ion transport and the physicochemical properties of electrolyte complex. For instance, high surface area nanoparticles such as Al_2O_3, ZrO_2, TiO_2, SiO_2, fumed silica, and ceramic powders were introduced into polymer–salt matrices to explore dimensionally stable polymer electrolytes for electrochemical devices [21]. It is foreseeable that smaller size particles for a similar volume fraction of the ceramic phase would impart an improved performance as compared to larger size particles because they cover more surface area [22] (Fig. 13.1).

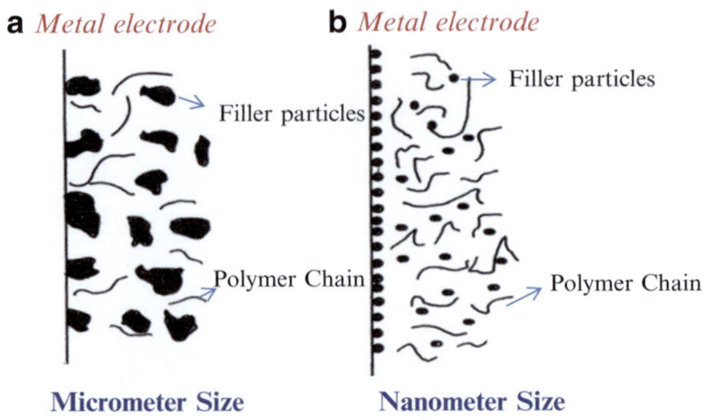

Fig. 13.1 Schematic representation of polymer host, nano- and micron-sized inorganic filler in the polymer host. (**a**) Micron-sized and(**b**) nanosized particles (Reprinted from Ref. [22]. Copyright 1994, with permission from Elsevier)

13.2.1 Preparation of Nanocomposite Polymer Electrolyte

A simple and viable method of preparing nanocomposite polymer electrolytes is the solution casting technique. Typically, the host polymers are first dissolved thoroughly in low-boiling solvents by continuous stirring followed by an addition of appropriate amount of salts to polymer solution with further stirring until a homogeneous solution is attained. Subsequently, nanoparticles of ceramic fillers are mixed thoroughly with the solution. The resulting slurry is then cast in Teflon-lined petri dish followed by vacuum drying, yielding mechanically stable thin film (Scheme 13.2). The resulting thin films are stored in moisture-free atmosphere for further electrochemical analysis.

13.2.2 Determining the Critical Properties of Nanocomposite Polymer Electrolytes

The understanding of the ion transport mechanism in nanoparticle-filled electrolyte complex is one of the primary research focus areas to achieve sufficient ionic conductivity. Indeed, the performance of composite electrolyte is mainly dependent on their morphology, segmental motion, and ionic interactions and the mobility, thermal behavior, and interfacial phenomena at the electrodes. Importantly, when the ion carriers increase by increasing the salt concentration, the mobility of ion pair is decreased. Vibrational spectroscopies (FTIR and Raman) are the perfect tools to investigate the ion–ion interactions, ion–polymer interactions, and surface interactions of filler nanoparticles in polymer electrolyte which are very sensitive to the local environment of vibrating molecules. An increasing or decreasing intensity of stretching vibration bands are used to quantify the fraction of free ions and ion pairs. The variation of the fraction of free ions and that of ion pairs has been calculated quantitatively by means of the band fitting of the SO_3 stretching mode. A Gaussian–Lorentzien or Voigt function has been used to fit the SO_3 stretching

Scheme 13.2 Preparation of mechanically stable nanocomposite polymer electrolyte thin film

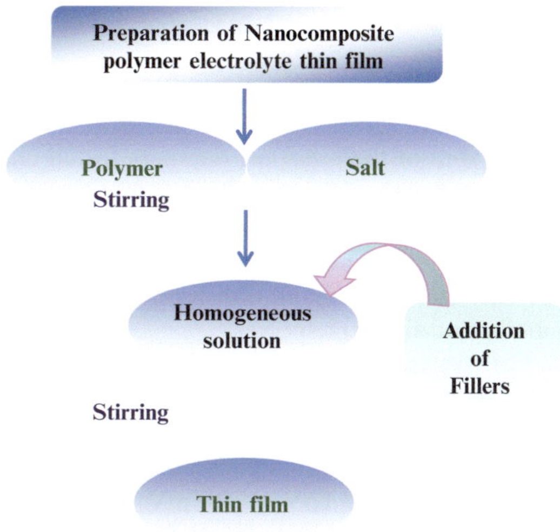

region, and all the deconvoluted spectra were best fitted by the above function with a linear background appearing to comprise two distinct contributions. These deconvoluted peaks appearing at lower and higher wave numbers are related to the presence of free anion and ion pairs, respectively [23–28]. It may be estimated using the peak profile analysis of integrated intensities of SO_3 peaks using the formulas:

$$Fraction\ of free\ ions = \frac{Area\ of\ free\ anion\ peak}{Total\ peak\ area} \tag{13.1}$$

$$Fraction\ of\ ion - pairs\ = \frac{Area\ of\ ion - pair\ peak}{Total\ peak\ area} \tag{13.2}$$

As we discussed in Scheme 13.1, Scrosati et al. [20] developed Lewis acid–base approach to explain the interactions between polymer chain, counter ions, and filler surface groups that result in greater dissociation of salt. The concentration of mobile charge carriers increases by the inclusion of filler nanoparticles; the dissociation of cation is evaluated by the *transference number* of cation, t_{M+}, which quantifies the fraction of cations contributing to the ionic conductivity [29]. It should be noted that the *transport number* and *transference number*, both of which are used in polymer electrolyte studies, are different terms. The transference number (t_i) is defined as the net number of faradays of charge carried across the reference plane by the cation constituent in the direction of the cathode during the passage of one faraday of charge across the plane. For an associated system, containing only M^+, X^-, MX, M_2X^+, and $MX_2{}^-$, the transference number of the X constituent, t_x, may be related to the individual transport numbers by

$$t_X = t_{X-} + 2\, t_{MX2-} - t_{M2X+} \tag{13.3}$$

A similar equation for the cation transference number, $t_{M,}$ can be given, $t_M + t_x = 1$.

The sum of the transport numbers for all charged species is unity, and it is the same in the case of the transference numbers for all salt constituents as well. When a salt is dissociated fully into two simple species, the transport numbers are equal to the transference numbers. Since ion association is a common phenomenon in polymer electrolytes, the transference number is predominantly used to study the mobility of different species.

Cationic transference numbers are usually obtained using the combination of DC polarization and AC impedance technique, originally developed by Bruce and Vincent for ideal solid electrolytes [30, 31]. A standard symmetric electrochemical cell of the form metal/single ion conductor/metal was polarized by application of a small DC potential ($\Delta V = {}^{\sim}10$ mV), and the resulting current transient was measured. The transference number was calculated using the following equation:

$$t_{M+} = \frac{I_s}{I_o}\left(\frac{\Delta V - I_o R_o}{\Delta V - I_s R_f}\right) \tag{13.4}$$

where I_o denotes the initial current, I_s the steady-state current, R_o the initial interfacial resistance, R_s the steady-state interfacial resistance, and $t_{Ag}{}^{+}$ the transference number of silver ions at room temperature. The initial current, I_o, is considered to be due to migration of both cations and anions. Because of the cell polarization, the current decreases over time to a steady-state value, I_s, which is considered to be due to the migration of the cations only. Also, the interfacial resistances of the passivating layers (solid electrochemical interface between electrodes and solid polymer electrolyte or SEI) before and after DC polarization are determined by electrochemical impedance spectroscopy (EIS) as the initial (R_o) and steady-state (R_f) resistances. Table 13.1 exemplifies some transference number of different polymer electrolytes with and without fillers.

Despite having quantified amount of ion carriers, the ion mobility certainly associated with segmental motion of polymer chain, i.e., flexibility which requires amorphous phase to speed up the ion mobility. An elegant way to predict the flexibility of polymer electrolyte is measuring the glass transition of temperature, T_g using differential scanning calorimetry (DSC) [33–35]. The resulting decrease in T_g results in an increase in the amorphous phase which would enhance polymer flexibility. Another way to evaluate the segmental motion of polymer chain is quasielastic neutron scattering technique which infers slowing segmental mobility due to the interactions between the cation and polymer backbone [36, 37].

EIS is the conventional and most critical technique for determining the conductivity of polymer electrolytes. EIS technique is well described in the literature and several books. The point of intersection of the impedance plot on the real axis at the high frequency region, within the frequency range under investigation,

Table 13.1 Transference number of different polymer electrolytes with and without fillers

Electrolyte complex	Filler	Temperature (° C)	Transference number t_+
$PEO - LiClO_4$		100	0.18–0.19
$P(EO)_8 - LiClO_4$		90	0.19–0.20
$P(EO)_8 - LiClO_4$	TiO_2	90	0.50–0.60
$P(EO)_8 - LiClO_4$	Al_2O_3	90	0.31–0.33
$P(EO)_{30} - LiClO_4$	SiO_2	100	0.22–0.23

Reprinted from Ref. [32] with permission from Elsevier

predominantly gives the bulk resistance (R_b) of the sample, thus eliminating other effects such as electrode polarization, grain boundaries, etc. The bulk conductivity of a given sample may be expressed as

$$\sigma = \frac{t}{R_b A} \tag{13.5}$$

where "t" is the thickness of the sample, "A" is the area of cross section, and "R_b" is the bulk resistance of the sample.

13.3 Ion Transport Mechanism in Polymer Electrolytes

Ionic transport in polymer electrolyte materials is enhanced through defects or disorder, and these ions move through the lattice by jumping process. Two different types of ionic hopping (intra-chain and interchain) processes occurring in the polymeric system are presented pictorially in Fig. 13.2.

Generally, both cations and anions are in mobile state in the case of polymer electrolytes. In particular, the ion transport is closely associated with the segmental motion of the polymer chain and ion mobility, the ion mobility confined to the amorphous domains than the crystalline region, and that local segmental motion plays a vital role in the conduction mechanism.

Armand et al. [39] have reported that the conductivity (σ), collected over a wide temperature range of amorphous polymer–salt complexes, is more accurately represented by a Vogel–Tamman–Fulcher (VTF) [40–42] equation than a simple Arrhenius function. If in case log σ vs. 1/T plot shows a curvature instead of a linear behavior as described by Arrhenius, the ion transport is not only due to hopping process of ions but also due to some other mechanisms occurring within the amorphous polymer electrolytes. In order to identify the exact mechanism of high ionic conductivity and ion mobility in polymer–salt complexes, many quantitative theoretical models have been proposed.

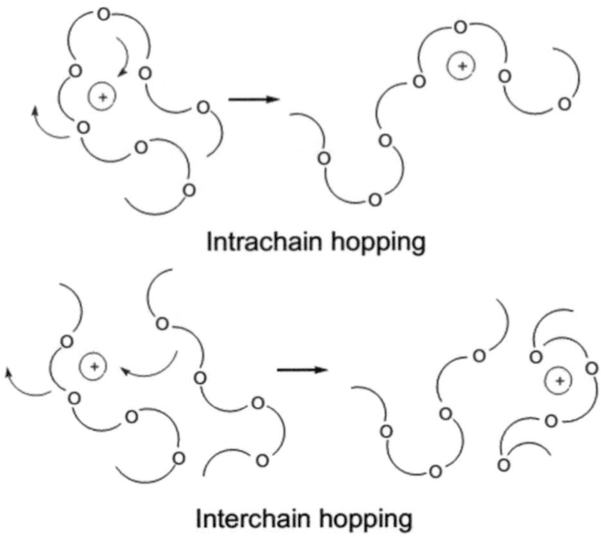

Fig. 13.2 Ionic conduction mechanism in solid polymer electrolytes [5, 38] (Reprinted with permission from Ref. [5]. Copyright 1991. Wiley VCH)

13.3.1 Arrhenius Theory

The ionic conduction in solid electrolyte increases with increase in temperature. A variation of ionic conduction with temperature may be expressed by Arrhenius equation given as

$$\sigma = \frac{\sigma_o}{T} e^{\left(-\frac{Ea}{kT}\right)} \tag{13.6}$$

where σ_o is the pre-exponential factor, E_a is the activation energy, k is the Boltzmann constant, and T is the absolute temperature [43–45]. Normally, Arrhenius plot (log σ vs. 1000/T) would show a linear behavior. Ion transport in these materials is enhanced through defects or disorder, and the mobile ions must move through the lattice by hopping process. Ion hopping may take place in two ways, i.e., interchain and intra-chain hopping. Depending upon the ionic radii, the movement of the counter ion may be hindered. In polymer electrolytes, when the temperature increases, the amorphous region is increased, and thus the nonlinear behavior is observed in the (log σ vs. 1/T) plot. This nonlinear behavior of polymer electrolytes may be explained by the Vogel–Tamman–Fulcher theory.

13.3.2 Free-Volume Model (VTF Model)

Temperature-dependent conductivity of a polymer electrolyte can be described accurately by the VTF equation. According to the free-volume theory [46], the ionic transport mechanism in a polymer electrolyte is mainly due to the segmental motion. The segmental motion has promoted ion movement through the formation and destruction of a coordination sphere of the solvated ion, thereby creating a space (free-volume) into which the ions could diffuse under the influence of the electrical field. The Vogel–Tamman–Fulcher (VTF) equation is given by

$$\sigma = \frac{\sigma_o}{T^{1/2}}\, e^{\left(\frac{-B}{k(T_o - T)}\right)} \tag{13.7}$$

where B is the pseudo activation energy of the ion and T_o is the thermodynamic glass transition temperature or the ideal glass transition temperature, which is usually 50 K lower than the glass transition temperature (T_g). At temperatures below the ideal glass transition temperature (T_o), the free volume has been assumed to vanish.

13.3.3 Williams–Landel–Ferry (WLF) Model

The Williams–Landel–Ferry (WLF) model [47] is an empirical model to describe the temperature dependence of a number of relaxation and transport processes in the vicinity of the glass transition temperature. The WLF model expresses a characteristic property, e.g., reciprocal dielectric relaxation time and magnetic resonance relaxation rate, in terms of shift factor, a_T, which is the ratio of any mechanical relaxation process at temperature T to its value at a reference temperature T_g, and is defined by

$$\log a_T + \text{const} = -\left[\frac{C_1(T - T_g)}{C_2 + (T - T_g)}\right] \tag{13.8}$$

where T_g is the reference temperature, and C_1 and C_2 are constants. The shift factor a_T is universal for all polymers, and the universality of the shift factor is due to the dependence of relaxation rates on free volume of the polymer electrolyte. This shift factor is correlated with the ionic conductivity suggesting that ionic motion is indeed promoted by local polymer segmental motion. By coupling the WLF equation with the Stokes–Einstein equation (Eq. 13.7) and by assuming the polymer electrolyte as a strong electrolyte, the temperature dependence of conductivity may be written in the WLF form (Eq. 13.8),

$$\log \frac{\sigma(T)}{\sigma(T_g)} = \left[\frac{C_1(T - T_g)}{C_2 + (T - T_g)}\right] \tag{13.9}$$

where σ (T) and σ (T_g) are the ionic conductivities at temperatures, T and T_g, respectively, and C1 and C2 are constants. The WLF equation, which is an extension of VTF empirical equation, holds good for a number of polymer electrolyte systems.

13.4 Electrochemical Stability

The electrochemical stability of an electrolyte specimen is one of the essential parameters when the electrochemical power sources are concerned. Potential window, the potential range under which the electrolyte is neither oxidized nor reduced, is often used to characterize the electrochemical stability [21, 22, 48]. This value must cover potentials from the anodic limit E_{ox} and cathodic limit E_{Red}. The potential window is generally measured by cyclic voltammetry (CV) or linear sweep voltammetry (LSV). Therefore, this value depends on the experimental conditions such as scan rate and current density. The electrolyte must be also chemically stable with the anode and cathode.

The stable electrochemical properties such as low interfacial resistance and high polymer electrolyte decomposition potential are very important for an electrolyte to be suitable for high-performance polymer electrolyte-based electrochemical power devices. Performing linear sweep voltammetry (LSV) on a cell with one blocking electrode is the most common way to evaluate the electrochemical stability window. The blocking electrode is polarized anodically, and the electrochemical stability window is identified by the rapid increase of the anodic current observed when the decomposition of the polymer electrolyte.

13.5 Applications of Nanocomposite Polymer Electrolytes

13.5.1 Lithium-Ion Batteries

A lithium-ion battery is an electrochemical device that converts chemical energy into electrical energy by redox reaction. High-energy density rechargeable lithium-ion batteries are considered as more efficient energy storage devices for use from portable consumer electronics to large-scale applications such as electric vehicles and hybrid electric vehicles which are competitive to the present internal combustion engine. The maximum electrical energy, which critically depends on the materials, could be achieved by choosing perfect electrode and electrolyte materials. Indeed, the nanocomposite polymer electrolytes in lithium-ion batteries are recognized as the pioneering technology to achieve good energy and power density without sacrificing the compatibility over commercial organic liquid electrolytes. The low elastic moduli of polymers with ceramic fillers offer mechanically stable

Table 13.2 Different types of nanocomposite polymer electrolytes with their ionic conductivity at room temperature

Polymer–salt complex	Fillers	Conductivity ($S\ cm^{-1}$)	Reference
PEO – $LiBF_4$	Nanosized Al_2O_3	~10^{-4}	[61]
PEO – $LiClO_4$	SiC	~10^{-7}	[62]
PEO – $LiClO_4$	$\alpha -$ Al_2O_3	~10^{-5}	[63]
PEO – $LiClO_4$	$AlCl_3$	~10^{-5}	[23]
PEO-PEG-LiI	Al_2O_3	~10^{-6}	[64]
PEG – $LiCF_3SO_3$	SiO_2	~10^{-5}	[65]
PEO – $LiClO_4$	$BaTiO_3$, CeO_2, Fe_2O_3, MoS_2	~$10^{-7} - 10^{-6}$	[66]
PEO-PMMA-LiI	MgO, SiO_2	~10^{-4}	[67]
PVA	Fumed silica	~10^{-4}	[68]
PVDF-HFP	Cellulose		[69]
PEO – $LiClO_4$	ZnO		[70]
PVDF-HFP	$MgAl_2O_4$		[71]
PEG 400	Fluorinated – Fe_2O_3		[72]
PAN – $LiClO_4$	$Li_{0.33}La_{0.557}TiO_2$ nanowire		[73]

with flexible thin film of conducting medium to simplify the cell design with high durability and safety.

Host polymers used for nanocomposite polymer electrolyte preparation include poly(ethylene oxide) (PEO), poly(vinylidene fluoride) (PVDF), poly(vinylidene fluoride-co-hexafluoropropylene) (PVDF-HFP), poly(methyl methacrylate) (PMMA), poly(vinyl alcohol) (PVA), and poly(acrylonitrile) (PAN) with the different lithium salts like $LiClO_4$, $LiN(CF_3SO_2)_2$, $LiCF_3SO_3$, LIBOB, $LiPF_6$, and $LiBF_4$, and most of them are formed by dispersing inorganic fillers like Al_2O_3, SiO_2, TiO_2, ZnO, Fe_2O_3, SnO_2, $BaTiO_3$, $PbTiO_3$, MgO, CeO_2, and $LiNbO_3$ [48–54]. Some of the nanoparticulate fillers for the electrolyte complexes are tabulated in Table 13.2 with their ionic conductivities.

Addition of active and passive (i.e., conducting and nonconducting) fillers like γ-$LiAlO_2$, Li_3N [55, 56] and SiO_2, TiO_2, and MgO [57–59] into the polymer matrix enhances the ionic conductivity which may be attributed to an increase in volume fraction of the amorphous phase and thus improves the electrode–electrolyte interface which is more stable and boosts the efficient cycling performance of energy storage devices in comparison to the filler-free electrolytes. Substantially, Kumar and Scanlon [60] suggested that both the nitride fillers like AIN and BN demonstrated an enhanced ion transport and electrode–electrolyte interfacial stability than the ceramic fillers such as Al_2O_3 and SiO_2 owing to the formation of Li_3N near the electrode thus facilitating the ionic conductivity. Table 13.3 shows some of the lithium-ion batteries fabricated with polymer electrolytes and their electrochemical performances.

Table 13.3 Some of the polymer electrolytes used in lithium-ion battery and their performance

Cell configuration	Specific capacity ($mAhg^{-1}$)	Reference
$Li_4Ti_5O_{12}$/ GPE/$LiNi_{0.5}Mn_{1.5}O_4$	125	[74]
$Li_4Ti_5O_{12}$/GPE/$LiMn_2O_4$	146	[74]
Li/PLTB@PVDF-HFP/$LiMn_2O_4$	103	[75]
SnC/GPE/$LiNi_{0.5}Mn_{1.5}O_4$	120	[76]
PMA/PEG – SiO_2, lithium organic battery	418	[77]

13.5.2 Supercapacitor

Supercapitors or ultra-capacitors are important electrochemical energy storage devices with high power densities compared to conventional electrochemical energy storage systems such as batteries and fuel cells [78, 79]. In general, the electric double-layer capacitors (EDLC) store charge at the interface by physical adsorption without involving faradaic process, whereas the fast intercalation processes are involved in the pseudocapacitors [80, 81]. An extraordinary storing of electrical energy with exceptional power has been projected to increase the awareness and development of important technologies such as hybrid electric vehicles, portable electronics, and power saving units. Recently, the flexible and wearable electronics has attracted intensively due to their reliability and easy handling and possesses the great promises to be used as flexible energy storage devices [82]. EDLC is durable and more stable for capacitive storage especially for the portable electronics, but their energy density is relatively low when compared to the pseudocapacitors. Therefore, pseudocapacitors have become the most successful and substantial storage device for the flexible and wearable electronics. Although, the pseudocapacitor has more charge storage capability, their rate capability and conductivity is relatively low.

The most developed and advantageous supercapacitors are based on liquid electrolytes. The use of liquid electrolytes may lead to some critical problems with regard to practical applications, including leakage, combustion in the case of organic electrolytes, corrosion, poisoning and pollution to the environment, and fabrication, especially packaging difficulties. Much effort has been taken toward the replacement of liquid electrolytes with polymer electrolytes with high mechanical strength, free from leakage as well as corrosion and easy to package for transportation [83]. For instance, Osaka et al. [84] introduced a poly(vinylidene fluoride) gel electrolytes with composite electrodes and achieved a specific capacitance of 123 F/g followed by Lantham et al. [85] who reported that a poly (urethane) gel electrolyte-based supercapacitor had a specific capacity up to 35 F/g. Composites of PEO with PAN and poly(methyl methacrylate) (PMMA) were also attempted, and specific capacitances of the order of 0.6 F/cm^2, energy density of 0.85 Wh/kg, and a power density of 300 W/kg were obtained.

13.6 Summary and Future Perspectives

Solid polymer electrolytes/nanocomposite polymer electrolytes based all solid-state flexible energy storage devices are attaining massive success in the portable electronics markets in the recent years. However, the incarnation of ceramic fillers incorporated polymer electrolytes that offer both mechanically stable and electrochemically viable materials. In this chapter, we have discussed the breakthroughs in the last decades and some of the key parameters which are critical to the polymer electrolytes in order to bring this material into portable electronics industries. None of the available electrolytes is found to be optimal in all aspects, i.e., high electrical conductivity, stable, high transference number, low electrochemical polarization, and inherently safer. Composite polymer electrolytes suffer from low transference number, in which more work is certainly needed to address this persisting challenge to increase the cationic transference number close to the unity and the room temperature ionic conductivity $\geq 10^{-3}$S cm^{-1}. It can be improved by extending the investigation further by the way of substituting an electrochemically stable polymer with different side chains and triggering the sensitive properties of fillers, i.e., volume fraction, filler size, and surface nature.

Acknowledgments We thank Prof KI Ozoemena (CSIR, South Africa) for the insightful comments and discussions on this book chapter.

References

1. Wright PV (1975) Electrical conductivity in ionic complexes of poly (ethylene oxide). Br Polym J 7:319–327
2. Armand MB, Chabagno JM, Duclot M (1978) In: Second international meeting on solid electrolytes. St. Andrews, Scotland
3. Bruce DW, O'Hare D, Walton RI (2011) Energy materials. John Wiley & Sons Ltd, Chichester, UK
4. MacCallum JR, Vincent CA (1987) Polymer electrolytes reviews-I. Elsevier, London
5. Gray FM (1991) Solid polymer electrolytes: fundamentals and technological applications. VCH, New York
6. Berthier C, Gorecki W, Minier M, Armand MB, Chabagno JM, Rigaud P (1983) Microscopic investigation of ionic conductivity in alkali metal salts-poly(ethylene oxide) adducts. Solid State Ion 11:91–95
7. Shriver DF, Farrington GC (1985) Solid ionic conductors. Chem Eng News 63:42–44
8. Frech R, Huang W (1994) Polymer conformation and ionic association in complexes of lithium, sodium and potassium triflate with poly (ethylene oxide) oligomer. Solid State Ion 72:103–108
9. Abraham KM, Alamgir M (1990) Li + –conductive solid polymer electrolytes with liquid-Like conductivity. J Eletrochem Soc 137:1657–1658
10. Capuano F, Croce F, Scrosati B (1991) Composite polymer electrolytes. J Eletrochem Soc 138:1918–1922
11. Quartarone E, Mustarelli P, Magistris A (1998) PEO-based composite polymer electrolytes. Solid State Ion 110:1–14

12. Armand M, Tarascon JM (2008) Building better batteries. Nature 451:652–657
13. Croce F, Appetecchi GB, Persi L, Scrosati B (1998) Nanocomposite polymer electrolytes for lithium batteries. Nature 394:456–458
14. Best AS, Adebahr J, Jacobsson P, MacFarlane DR, Forsyth M (2001) Microscopic interactions in nanocomposite electrolytes. Macromolecules 34:4549–4555
15. Castro WA, Zapata VH, Vargas RA, Mellander B-E (2007) Electrical conductivity relaxation in PVOH-$LiClO_4$-Al_2O_3. Electrochim Acta 53:1422–1426
16. Jayathilaka PARD, Dissanayake MAKL, Albinsson I, Mellander B-E (2002) Effect of nano-porous Al_2O_3 on thermal, dielectric and transport properties of the $(PEO)_9$ LiTFSI polymer electrolyte system. Electrochim Acta 47:3257–3268
17. Das SK, Bhattacharyya AJ (2009) Oxide particle surface chemistry and ion transport in Soggy sand electrolytes. J Phys Chem C 113:6699–7705
18. Money BK, Hariharan K, Swenson J (2012) Glass transition and relaxation processes of nanocomposite polymer electrolytes. J Phys Chem B 116:7762–7770
19. Do NST, Schaetzl DM, Dey B, Seabaugh AC, Fullerton-Shirey S (2012) Influence of Fe_2O_3 nanofiller shape on the conductivity and thermal properties of solid polymer electrolytes: nanorods versus nanospheres. J Phys Chem C 116:21216–21223
20. Croce F, Persi L, Scrosati B, Serraino-Fiory F, Plichta E, Hendrickson MA (2001) Role of the ceramic fillers in enhancing the transport properties of composite polymer electrolytes. Electrochim Acta 46:2457–2461
21. Manuel Stephan A, Nahm KS (2006) Review on composite polymer electrolytes for lithium batteries. Polymer 47:5952–5964
22. Kumar B, Scanlon LG (1994) Polymer-ceramic composite electrolytes. J Power Sources 52:261–268
23. Wieczorek W, Florjanczyk Z, Stevens JR (1995) Composite polyether based solid electrolytes. Electrochim Acta 40:2251–2258
24. Grams 8 software (1994) Galactic Industries Crop
25. Bernson A, Lindgren J (1993) Free ions and ion pairing/clustering in the system $LiCF_3SO_3$ -PPOn. Solid State Ion 60:37–41
26. Bernson A, Lindgren J (1993) Ion aggregation and morphology for poly (ethylene oxide)-based polymer electrolytes containing rare earth metal salts. Solid State Ion 60:31–36
27. Suthanthiraraj SA, Kumar R, Paul BJ (2010) Vibrational spectroscopic and electrochemical characteristic of poly (propylene glycol)–silver triflate polymer electrolyte system. Ionics 16:145–151
28. Kumar R, Suthanthiraraj SA (2014) Ion dynamics and segmental relaxation of CeO_2 nanoparticles loaded soft-matter like gel polymer electrolyte. J Non-Cryst Solids 405:76–82
29. Bruce PG, Hardgrave MT, Vincent CA (1992) The determination of transference numbers in solid polymer electrolytes using the Hittorf method. Solid State Ion 53-56:1087–1094
30. Evans J, Vincent CA, Bruce PG (1987) Electrochemical measurement of transference numbers in polymer electrolytes. Polymer 28:2324–2328
31. Watanabe M, Nagano S, Sanui K, Ogata N (1988) Estimation of Li + transport number in polymer electrolytes by the combination of complex impedance and potentiostatic polarization measurements. Solid State Ion 28–30:911–917
32. Chung SH, Wang Y, Persi L (2001) Enhancement of ion transport in polymer electrolytes by addition of nanoscale inorganic oxides. J Power Sources 97–98:644–648
33. Chen HW, Chang FC (2001) The novel polymer electrolyte nanocomposite composed of poly (ethylene oxide), lithium triflate and mineral clay. Polymer 42:9763–9769
34. Wieczorek W, Raducha D, Zalewska A, Stevens JR (1998) Effect of salt concentration on the conductivity of PEO-based composite polymeric electrolytes. J Phys Chem B 102:8725–8731
35. Kumar R, Suthanthiraraj SA (2014) Segmental mobility and relaxation processes of Fe_2O_3 nanoparticle-loaded fast ionic transport nanocomposite gel polymer electrolyte. J Solid State Electrochem 18:1647–1656

36. Mao G, Perea RF, Howells WS, Price DL, Saboungi ML (2000) Relaxation in polymer electrolytes on the nanosecond timescale. Nature 35:415–419
37. Shirey SKF, Maranas JK (2009) Effect of $LiClO_4$ on the structure and mobility of PEO-based solid polymer electrolytes. Macromolecules 42:2142–2156
38. Gray FM (1997) Polymer electrolytes. The Royal Society of Chemistry, Cambridge, UK
39. Armand B, Chabagno JM, Duclot MJ (1979) Fast ion transport in solids. Elsevier, Amsterdam
40. Vogel H (1921) The law of the relation between the viscosity of liquids and the temperature. Phys Z 22:645–646
41. Tammann G, Hesse W (1926) Die abhängigkeit der viscosität von der temperatur bie unterkühlten flüssigkeiten. Z anorg allg Chem 156:245–257
42. Fulcher GS (1925) Analysis of recent measurements of the viscosity of glasses. J Am Ceram Soc 8:339–355
43. Olsen II, Koksbang R (1996) A temperature study of the ionic conductivity of a hybrid polymer electrolyte. J Electrochem Soc 143:570–574
44. Shi J, Puhu C, Chen R, Ying SK (1990) Recent advances in fast Ion conducting materials and devices. World Scientific, Singapore, p 267
45. Zahurak S, Kaplan M, Rietman E, Murphy D, Cava R (1988) Phase relationships and conductivity of the polymer electrolytes poly (ethylene oxide)/lithium tetrafluoroborate and poly (ethylene oxide)/lithium trifluoromethanesulfonate. Macromolecules 21:654–660
46. Cohen MH, Turnbull D (1959) Molecular transport in liquids and glasses. J Chem Phys 31 (5):1164–1169
47. Watanabe M, Itoh MS, Sanui K, Ogata N (1987) Carrier transport and generation processes in polymer electrolytes based on poly(ethylene oxide) networks. Macromolecules 20:569–573
48. Croce F, Curini R, Martinelli A (1999) Physical and chemical properties of nanocomposite polymer electrolytes. J Phys Chem B 103:10632–10638
49. Sun H, Takeda Y, Imanishi N, Yamamoto O, Sohn H (2000) Ferroelectric materials as a ceramic filler in solid composite polyethylene Oxide-Based electrolytes. J Electrochem Soc 147:2462–2467
50. Tominaga Y, Asai S, Sumita M, Panero S, Scrosati B (2005) A novel composite polymer electrolyte: effect of mesoporous SiO_2 on ionic conduction in poly (ethylene oxide)–$LiCF_3SO_3$ complex. J Power Sources 146:402–406
51. Dissanayake M, Jayathilaka P, Bokalawala R, Albinsson I, Mellander B (2003) Effect of concentration and grain size of alumina filler on the ionic conductivity enhancement of the $(PEO)_9$ $LiCF_3SO_3$: Al_2O_3 composite polymer electrolyte. J Power Sources 119:409–414
52. Xiong H, Zhao X, Chen J (2001) New polymer-inorganic nanocomposites: PEO-ZnO and PEO–ZnO–$LiClO_4$ films. J Phys Chem B 105:10169–10174
53. Sun H, Sohn H, Yamamoto O, Takeda Y, Imanishi N (1999) Enhanced lithium-ion transport in PEO-based composite polymer electrolytes with ferroelectric $BaTiO_3$. J Electrochem Soc 146:1672–1676
54. Appetecchi G, Scaccia S, Passerini S (2000) Investigation on the stability of the lithium-polymer electrolyte interface. J Electrochem Soc 147:4448–4452
55. Appetecchi GB, Croce F, Persi L, Ronci F, Scrosati B (2000) Transport and interfacial properties of composite polymer electrolytes. Electrochim Acta 45:1481–1490
56. Fan J, Fedkiw PS (1997) Composite electrolytes prepared from fumed silica, polyethylene oxide oligomers, and lithium salts. J Electrochem Soc 144:399–408
57. Scrosati B, Croce F, Persi L (2000) Impedance spectroscopy study of PEO-based nanocomposite polymer electrolytes. J Electrochem Soc 147:1718–1721
58. Kumar J, Rodrigues SJ, Kumar B (2010) Interface-mediated electrochemical effects in lithium/polymer-ceramic cells. J Power Sources 195:327–334
59. Kumar B, Scanlon LG (1999) Polymer–ceramic composite electrolytes: conductivity and thermal history effects. Solid State Ion 124:239–254
60. Kumar B, Scanlon LG, Spry RJ (2001) On the origin of conductivity enhancement in polymer-ceramic composite electrolytes. J Power Sources 96:337–342

61. Krawiec W, Scanlon LG, Fellner JP, Vaia RA, Vasudevan S, Giannelis EP (1995) Polymer nanocomposites: a new strategy for synthesizing solid electrolytes for rechargeable lithium batteries. J Power Sources 54:310–315
62. Choi B, Shin K (1996) Effects of SiC fillers on the electrical and mechanical properties of $(PEO)_{16}$ $LiClO_4$ electrolytes. Solid State Ion 86:303–306
63. Nairn K, Forsyth M, Every H, Greville M, MacFarlane D (1996) Polymer-ceramic ion-conducting composites. Solid State Ion 86:589–593
64. Wieczorek W, Stevens JR, Florjanczyk Z (1996) Composite polyether based solid electrolytes. The Lewis acid–base approach. Solid State Ion 85:67–72
65. Golodnitsky D, Ardel G, Peled E (1996) Effect of plasticizers on the CPE conductivity and on the Li-CPE interface. Solid State Ion 85:231–238
66. Peled E, Golodnitsky D, Ardel G, Eshkenazy V (1995) The sei model—application to lithium-polymer electrolyte batteries. Electrochim Acta 40:2197–2204
67. Slane S, Salomon M (1995) Composite gel electrolyte for rechargeable lithium batteries. J Power Sources 55:7–10
68. Hu XL, Hou GM, Zhang MQ, Rong MZ, Ruan WH, Giannelis EP (2012) A new nanocomposite polymer electrolyte based on poly(vinyl alcohol) incorporating hypergrafted nano-silica. J Mater Chem 22:18961–18967
69. Kelley J, Simonsen J, Ding J (2013) Poly(vinylidene fluoride-co-hexafluoropropylene) nanocomposites incorporating cellulose nanocrystals with potential applications in lithium ion batteries. J Appl Polym Sci 127:487–493
70. Patil SU, Yawale SS, Yawale SP (2014) Conductivity study of PEO–$LiClO_4$ polymer electrolyte doped with ZnO nanocomposite ceramic filler. Bull Mater Sci 37:1403–1409
71. Padmaraja O, Rao BN, Jena P, Venkateswarlu M, Satyanarayanaa N (2014) Electrochemical studies of electrospun organic/inorganic hybrid nanocomposite fibrous polymer electrolyte for lithium battery. Polymer 55:1136–1142
72. Bertasia F, Negro E, Vezzùc K, Nawna G, Pagota G, Noto VD (2015) Single-ion-conducting nanocomposite polymer electrolytes for lithium batteries based on lithiated-fluorinated-iron oxide and poly(ethylene glycol) 400. Electrochim Acta 175:113–123. doi:10.1016/j.electacta.2015.03.149
73. Liu W, Liu N, Sun J, Hsu PS, Li Y, Lee WH, Cui Y (2015) Ionic conductivity enhancement of polymer electrolytes with ceramic nanowire fillers. Nano Lett 15:2740–2745
74. Reale P, Panero S, Scrosati B (2005) Sustainable high-voltage lithium ion polymer batteries. J Electrochem Soc 152(10):A1949–A1954
75. Qina B, Liua Z, Dinga G, Duana Y, Zhanga C, Cui Z (2014) A single-ion gel polymer electrolyte system for improving cycle performance of LiMn2O4 battery at elevated temperatures. Electrochim Acta 141:167–172
76. Hassoun J, Panero S, Reale P, Scrosati B (2009) A new, safe, high-rate and high-energy polymer lithium –ion battery. Adv Mater 21:4807–4810
77. Zhu Z, Hong M, Guo D, Shi J, Tao Z, Chen J (2014) All-solid-state lithium organic battery with composite polymer electrolyte and Pillar[5]quinone cathode. J Am Chem Soc 136:16461–16464
78. Zhang LL, Zhao X (2009) Carbon-based materials as supercapacitor electrodes. Chem Soc Rev 38:2520–2531
79. Chmiola J, Yushin G, Gogotsi Y, Portet C, Simon P, Taberna PL (2006) Anomalous increase in carbon capacitance at pore sizes less than 1 nanometer. Science 313:1760–1763
80. Zhai Y, Dou Y, Zhao D, Fulvio PF, Mayes RT, Dai S (2011) Carbon materials for chemical capacitive energy storage. Adv Mater 23:4828–4850
81. Wang G, Zhang L, Zhang J (2012) A review of electrode materials for electrochemical supercapacitors. Chem Soc Rev 41:797–828
82. Kötz R, Carlen M (2000) Principles and applications of electrochemical capacitors. Electrochim Acta 45:2483–2498

83. Panero S, Clemente A, Spila E (1996) Solid state supercapacitors using gel membranes as electrolytes. Solid State Ion 86:1285–1289
84. Osaka T, Liu X, Nojima M, Momma T (1999) An electrochemical double layer capacitor using an activated carbon electrode with gel electrolyte binder. J Electrochem Soc 146:1724–1729
85. Latham RJ, Rowlands SE, Schlindwein WS (2002) Supercapacitors using polymer electrolytes based on poly (urethane). Solid State Ion 147:243–248

Chapter 14
Computational Modelling as a Value Add in Energy Storage Materials

R. Regina Maphanga and Phuti E. Ngoepe

14.1 Introduction

Energy storage is undoubtedly one of the greatest challenges in the twenty-first century Arico et al. [1]. It has become a growing global concern as a result of increased energy demand, combined with drastic increases in the price of refined fossil fuels and the environmental consequences of their use. Furthermore, energy storage for portable electronic devices, electric vehicles, and renewable energy sources is a pressing technological challenge for which rechargeable lithium batteries are key candidate systems. Despite the fact that rechargeable lithium batteries have revolutionized portable electronic devices, it has proved impossible to scale up first-generation batteries for renewable energy storage or transport applications. To facilitate viable electric vehicles and renewable sources, energy density and discharge rates of batteries need be increased. One avenue of exploration in achieving such goals is the use of nanomaterials. Furthermore, with a continuously increasing power of modern high-performance computers, sophisticated simulations and calculations are exploited for innovative materials, including energy storage materials.

Among the numerous metal oxides that have the required properties for use in energy storage devices, manganese dioxide (MnO_2) is a promising candidate. In particular, new applications ranging from mobile units to electric vehicles have accelerated developments of alkaline, lithium-ion, and lithium-air batteries and supercapacitors where MnO_2 plays a significant role as an electrode material. The advantages of MnO_2 are low cost, natural abundance, and environmental

R.R. Maphanga (✉) • P.E. Ngoepe
Materials Modelling Centre, University of Limpopo, Private bag x 1106, Sovenga, 0727 Polokwane, South Africa
e-mail: rapela.maphanga@ul.ac.za; phuti.ngoepe@ul.ac.za

K.I. Ozoemena, S. Chen (eds.), *Nanomaterials in Advanced Batteries and Supercapacitors*, Nanostructure Science and Technology,
DOI 10.1007/978-3-319-26082-2_14

compatibility as well as high capacity. It has many different crystal structures, such as the $\alpha-, \beta-, \gamma-$ and $\delta-$ types, due to corner- and/or edge-sharing arrangements of octahedral units. These structural varieties provide the possibility to control the thermal and electrochemical properties of MnO_2, as well as the recent ability to synthesize different nano-architectures of MnO_2, such as nanowires, nanorods, nanobelts, nanoflowers, etc. Debart et al. [2], Jiao and Bruce [3], Zhou et al. [4], Matsumoto et al. [5], provides an opportunity of tuning device performance for various electrochemical applications as compared to the bulk form.

Materials on the nanometer scale display different chemical and physical properties from bulk systems. Recently, special attention has been focused on the fabrication of the hierarchical superstructures based on low-dimensional nanocrystals Wang et al. [6], which are expected to play a crucial role in fabricating the next generation of nanodevices because they can be used as both building units and interconnections, and will offer opportunities to explore their novel collective optical, mechanical, magnetic, and electronic properties. There has been an increased interest in manganese dioxide nanostructures for their potential applications in catalysis, rechargeable batteries, ion-sieves and supercapacitors. Several experimental methods (such as liquid phase, thermal decomposition, electrochemical deposition, sol-gel, and molten salt) have been developed for the synthesis of MnO_2 nanostructures, including nanorods, nanowires, nanobelts, nanoneedles and nanosheets. However, the experimental organization of MnO_2 nanosheet clusters into hierarchical superstructures still remains a challenge. The preparation of the hierarchical MnO_2 sheets via electrochemical deposition were explored in Feng et al. [7] MnO_2. Research interest in the fundamental and functional properties of solid oxide electrodes provides information that can be used to improve the existing technology for further performance enhancement of electroactive materials. However, the fundamental basis for the contrasting intercalation properties of nanostructured and bulk materials is not fully understood. One of the major demands for developing this material is to modify and strengthen the structural stability in order to prevent the rapid capacity from fading during the process of charge/discharge cycling Fernandes et al. [8], Ghaemi et al. [9], Susana-Pilla et al. [10], Kucza [11].

Recently, the low-temperature hydrothermal method was used to synthesize uniform nanoclusters of γ–MnO_2. Lithiumions were treated with MnO_2 nanoclusters without any intercalation to investigate the effect of lithium ions on the morphology, particle size, and charge/discharge behaviors of the synthesized MnO_2 Karami et al. [12]. It has been found that the solid-state treatment can change the morphology and particle size of γ–MnO_2 nanoclusters, without considerable change in the phase composition. In addition, low-dimensional structures of MnO_2 have been synthesized as nanorods and nanosheets Jiao and Bruce [3] and modelled as nanoparticle Sayle et al. [13] and varying nano-architectures, that is, nanosheet, nanorod and nanoporous Sayle et al. [14]. The control of morphologies, sizes, architectures, and patterns of nanocrystals has become a dominant theme in materials science because these parameters are the key elements that determine physical properties. It is well known that battery performance is highly dependent on MnO_2 morphologies and crystallographic forms.

It is beyond the scope of this chapter to give an exhaustive review of all research activities on energy storage materials using both computer simulation methods and experiments; excellent reviews and articles can be found elsewhere Arico et al. [1], Walsh et al. [15], Islam and Fisher [16], Bruce et al. [17], Whittingham [18], Armand and Tarascon [19], Fergus [20], Thackeray et al. [21], Masquelier and Croguennec [22], Bruce et al. [23], Scrosati et al. [24], Scrosati and Garche [25], Park et al. [26], Etacheri et al. [27], Takada [28], Goodenough and Park [29], Goodenough and Kim [30], Ellis and Nazar [31], Ellis et al. [32], Zaghib et al. [33], Patil et al. [34], Tompsett et al. [35], Tompsett et al. [36]. This chapter gives a brief description of the role that computer simulation methods play in understanding fundamental properties of battery materials. In particular, the use of computer modelling methods in simulating manganese dioxide as an electrode material for lithium-ion batteries is discussed. Novel evolutionary methods such as global minimization Woodley [37] and amorphization and recrystallization method Sayle and Johnston [38] play a significant role and will be reviewed on identified energy materials. Atomistic simulation of various nanostructures is demanding and challenging, and so the simulation strategies for generating full atomistic models for nanomaterials and spanning all spatial dimensions including nanoparticle, nanorod, nanosheet and nanoporous will be demonstrated; such nanostructures have all been synthesized experimentally Jiao and Bruce [3], Debart et al. [2], Matsumoto et al. [5], Li et al. [39], Zhou et al. [4].

14.2 Computer Simulation Methods

The description of computer simulation methods will be brief as more comprehensive reviews are available elsewhere Catlow [40], Gale [41], Harding [42], Koch and Holthausen [43], Smith and Forester [44], Walsh et al. [15]. Four main classes of computer simulation technique have been used in the study of nanostructured MnO_2 as a cathode material. The techniques are interatomic potential-based, electronic structure, global minimization, and amorphization and recrystallization methods.

14.2.1 Atomistic Simulation Methods

The atomistic simulation technique based on the Born model of solids is briefly described. The basis of this method is to calculate the total energy of the system under study. These rest upon the specification of an appropriate potential model, which expresses the total energy of the system as a function of the nuclear coordinates. The interaction between ions is constituted of two parts:

(1) short-range interactions that involve the attractions and repulsions of the ions, and (2) long-range interactions that describe the Coulombic interaction. The Born model of solids assumes that the energy of the system is given by the sum of all pair-wise interactions between the atoms (see Born and Huang [45]). The Buckingham potential is often used for the short-range term and was used in this study, expressed as:

$$Ur_{ij} = \left[\frac{q_i q_j}{r_{ij}} + A_{ij} e^{\left(-r_{ij}/\rho_{ij}\right)} - \frac{C_{ij}}{r_{ij}^6}\right] \tag{14.1}$$

where U_{ij} is the potential energy between ions, *i* and *j*, separated by distance *r*. Often the shell model is added as a computationally inexpensive means of reproducing electronic polarizability effects, and has proven to be effective in simulating dielectric and lattice dynamical properties. The simplest form of potential is one in which electronic polarizability is neglected, and is called a "rigid ion model." Good quality interatomic potentials are available for a wide range of transition metal oxides. In molecular modelling, the interest is specifically in the minimum point on the energy surface. Minimum energy arrangements of the atoms correspond to the stable states of the system; any movement away from a minimum gives a configuration with high energy. The minimum with the very lowest energy is known as the "global energy minimum."

A minimization algorithm is used to identify the geometries of the system that correspond to the minimum points on the energy surface. The highest point on the pathway between minima is of special interest, and is known as the "saddle point," with the arrangement of atoms being the transition structure. In order for the potential model to be useful in predicting perfect lattice properties, it must be combined with an energy minimization technique to bring the system to a state of mechanical equilibrium. Energy minimization provides the lowest energy configuration of a given structure through a search of the potential energy surface using efficient minimization routines. There are two approaches that are common for minimizing the lattice energy – either at constant volume or constant pressure. For bulk systems, three-dimensional periodic boundary conditions are normally applied to the simulation box, in effect rendering the structure infinite in extent. The most widely used code for energy minimization using effective potentials is GULP Gale and Rohl [46]. Atomistic modelling of nanostructures, surface structures and energies uses similar methodology and interatomic potentials to bulk calculations.

The molecular dynamics (MD) technique comprises numerically solving Newton's equations of motion for a large ensemble of particles at a finite temperature and pressure. Repetition of the integration algorithm over many time steps yields a detailed picture of the evolution of ion positions and velocities as a function of time. MD is a deterministic method; thus, the state of the system at any future time can be predicted from its current state. Two properties MD simulations are eminently good at providing within the current context of lithium-ion conducting materials are (1) diffusion coefficients, and (2) ion transport mechanisms as functions of temperature and/or composition Islam and Fisher [16]. DL_POLY is a widely used molecular dynamics code Smith and Forester [44]. The advantage of interatomic potential methods is that they are computationally inexpensive, and with modern high performance computing facilities, it is possible to efficiently simulate large ensembles of atoms – from thousands to millions. The major disadvantage of these methods is that explicit electronic structure information about the system is not included.

14.2.2 Electronic Structure Methods

Density functional theory (DFT) has since advanced to become a very important approach for the description of the ground state properties by determining the energy of many-electron, and therefore many-atom systems. DFT is the most widely used theoretical framework for accurate calculations of a system's electronic ground state. The minimum value of the total energy functional is the ground state energy of the system, and the density that yields this minimum value is the exact single particle ground state density. Thus, DFT expresses the energy as a functional of the electron density, which in turn depends on the positions of the atoms; the total energy is expressed as:

$$E = E[\rho(r), R_\alpha] \tag{14.2}$$

where the electron density, ρ and the total energy, E depend on the type and arrangement of the atomic nuclei; R_α denotes the positions of the nuclei in the system. This equation is the key to the atomic-scale understanding of structural, electronic, and magnetic properties of matter. Then the total energy is decomposed into the three terms of the Schrödinger equation: the kinetic energy, Coulomb energy, and the exchange-correlation energy due to classical electrostatics interactions among all charged particles in the system. DFT relates the ground-state energy to the electron density, r, and in the Kohn–Sham formalism Kohn and Sham [47]. The total energy, E, is given by

$$\left(-\frac{\hbar^2}{2m}\nabla^2 + V_H[\rho(r)] + V_N(r) + V_{XC}[\rho(r)]\right)\psi_i(r) = E_i\psi_i(r) \tag{14.3}$$

where the first term is the kinetic energy, V_H is the Hartree term representing the electrostatic energy of an electron moving independently in the mean electrostatic field due to all other electrons, V_N is the energy due to all nuclei, and V_{XC} is the exchange-correlation energy, accounting for Pauli exclusion and electron correlation effects. Unlike the first three terms, the exchange-correlation term cannot be explicitly obtained because the exact exchange-correlation energy is not known. The only way to solve the problem is by way of approximate methods. One of the first approximations to be developed is the local density approximation (LDA), which estimates the exchange-correlation potential as if the electrons were surrounded by a uniform electron gas at each point. The basic idea in LDA is that for any atomic arrangement – such as crystal surface or molecule – there is a certain electron density ρ(r) at each point in space. This is an excellent approximation for metallic systems, but represents a severe simplification in systems with strong varying electron density. When materials contain localized electrons, such as most oxides or salts, LDA reveals systematic shortcomings, leading to overestimation of binding energies Freeman and Wimmer [48], with the associated underestimation of crystal lattice constants, for example. Despite the remarkable success of the LDA, it also has limitations. For systems where the density varies slowly, the LDA tends to perform well and chemical trends are well reproduced. In strong correlated systems where independent particle pictures break down, the LDA is very inaccurate. To overcome this, the generalized gradient approximation (GGA) was developed, in which a dependence on the gradient of the electron density is introduced. To avoid electron self-interaction errors that occur in standard LDA or GGA for strongly correlated electronic systems, DFT + U methods are often used for open-shell transition metal compounds, where U is an on-site Coulomb interaction parameter. The U parameter is derived either by fitting to experimental data (e.g., cell voltages or band gaps), or by self-consistent calculations. There is also an increasing use of hybrid functionals, which include a fraction of the Hartree-Fock exchange, and can also correct self-interaction present in the Hartree term, although the amount of exact exchange required can also be system-

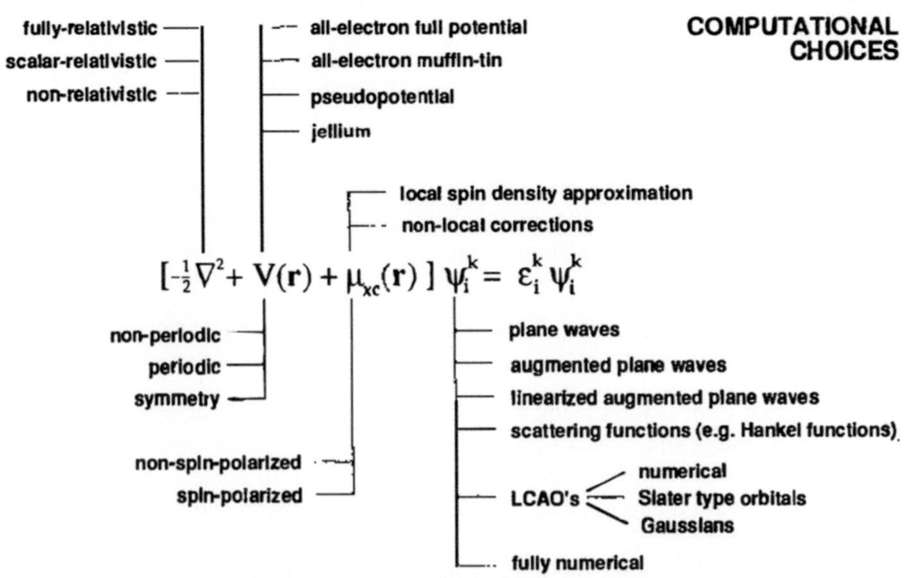

Fig. 14.1 Overview of theoretical and computational choices for the solution of the density functional one-particle equations Wimmer [49]

dependent, introducing another empirical term. An overview of the major choices between different electronic structure methods is summarized in Fig. 14.1.

In the past two decades there has been an explosion in the number of studies carried out on condensed matter systems, including lithium-ion and lithium-air battery materials, using density functional theory methods. There are a number of density functional theory codes available, such as CASTEP Payne et al. [50], VASP Kresse and Furthmuller [51], CRYSTAL Civalleri et al. [52], etc. Most importantly, the development of computational materials science has been assisted by the continuing rapid growth in computer power as well as advances in theory and more efficient software.

14.2.3 Evolutionary Algorithm Methods

Computer simulation methods such as DFT are well established and have proved invaluable as complements to experimental methods. However, new materials with ever-increasing complex structures are synthesized, including nanostructures, and it may soon prove to be impossible to generate models that are sufficiently realistic and adequately described using such methods. Hence, it could be possible to generate such models by simulating the synthetic process itself. The two different approaches are considered in this chapter. These evolutionary simulation methods allow the simulation itself to direct the atomistic configuration, where the simulation is based on some evolutionary procedure that may include nucleation and growth, or generic algorithms.

14.2.3.1 Amorphization and Recrystallization Method

An evolutionary simulation technique that can be used to generate complex models is called "amorphization and recrystallization" (A&R) Sayle and Johnston [38]. The method allows a spontaneous evolution of all structural features that would exist within a real system. The procedure Sayle [53], attempts to avoid the inherent limitations of molecular dynamics when applied to crystalline solids which, owing to the barrier heights for ionic mobility, cannot explore configurational space for the system within the timescales accessible. The very short time scales accessible to atomistic dynamical simulation is typically 1 ns, which is a major limitation. Consequently, for high crystalline materials, in which ionic migration is slow, dynamical simulation is not an appropriate technique for exploring the energy barriers for migration because no migration would be observed within the time scales accessible. However, owing to high ion mobility in an amorphous system, they can migrate more quickly (within the time scale accessible) to low-energy positions compared with performing similar simulations without the amorphous transition. Therefore, A&R strategy provides a mechanism for overcoming, in part, the considerable (time) limitation associated with dynamical

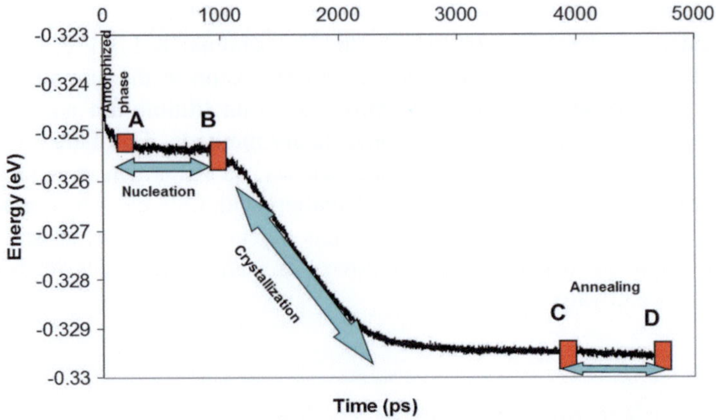

Fig. 14.2 Calculated energy as a function of time of the MnO_2 nanoparticle depicting various stages of amorphization and recrystallization technique. Adapted with permission from Sayle et al. [13]. Copyright 2005 American Chemical Society

simulations. The technique involves allowing the system to undergo, under dynamical simulation, a spontaneous amorphization prior to recrystallizing. The A&R procedure starts by generating an amorphous configuration that is then allowed to nucleate and recrystallize.

Figure 14.2 shows the variation of the configuration energy with time. The figure depicts the nucleation and crystal growth of the system using the amorphization and recrystallization method; in this case, an example of an MnO_2 nanoparticle is shown. In the first 1000 ps (between points A and B), the energy remains constant, and it corresponds with the evolution of a small crystalline region with the system. This is a nucleation seed that initiates crystallization of the surrounding regions. Beyond point B, the configuration energy reduces up to C, which is related to the crystallization process. The configuration energy plateaus from C up to D, which corresponds to fully crystallized structure after a duration of 4000 ps.

During the amorphization and recrystallization process, the ions are able to move into low-energy configurations within the timescales (typically nanoseconds) accessible to the simulation. Central to this method is the dynamical simulation, as applied to an amorphous system that allows the ions to evolve or assemble into an appropriate (low-energy) configuration owing to the much-higher ionic mobility of species within an amorphous structure, compared with a crystalline one. Prolonged dynamical simulation results in recrystallization of the amorphous system, with the evolution of the structural features.

The conditions under which the system is constrained are critical in generating the desired high-energy amorphous transition. If the conditions are not appropriate, the system may not go amorphous and therefore no structural modifications will evolve. Alternatively, the system may undergo amorphous transition, although it may have insufficient energy to allow complete exploration of the potential surface. To induce amorphization, the system can be constrained under compressive or tensile stress. The subsequent application of high- temperature dynamical simulation to the system then results in its amorphization. The material, as it recrystallizes, will develop microstructural features observed experimentally including, for example, grain boundaries, dislocations, point defects, surface structures, faceting, and morphology Sayle and Parker [54]. A&R is, therefore, an appropriate simulation tool to explore systems with complex microstructures, such as MnO_2, because it does not require one to propose and generate a realistic structure that includes all the microstructural features prior to simulating with dynamical or static methods.

14.2.3.2 Global Minimization Method

There are various search algorithms implemented in global minimization methods, namely Hybrid Simulated Annealing-Monte Carlo Basin Hopping (SA-MCBH); Genetic Algorithm (GA); and Genetic Algorithm Hybrid. Their description can be found elsewhere Catlow et al. [55].

Energies of nanoclusters can be calculated using methods based either on interatomic potentials or electronic structure techniques. The interatomic potential (IP) method is computationally inexpensive and allows large numbers and sizes of clusters to be explored. However, they have a number of limitations that may result in giving inaccurate results. Combining the two approaches (interatomic potential and density functional theory) proved to be insightful, where IP methods are being used in the first stage to explore the range of cluster sizes and structures, while DFT methods are used in the second stage to refine the energy ordering and structures of a selected subset of clusters. Thus, those found to be low in energy are from the first stage. In addition, the DFT methods can be used to test the reliability of the first stage calculations by probing cluster structures, which are likely to be sensitive to the parameterization of interatomic potential used. Interatomic potentials are commonly parameterized by reference to bulk crystalline properties. This may cause difficulties when they are applied to clusters where the coordination number and bonding may differ. It is vital that the potential-based calculations be validated by DFT methods. Low-energy configurations for $(MnO_2)_n$ clusters are predicted using a multistage procedure (see Fig. 14.3). For each cluster, in a first stage low-energy, stationary points are found on three different energy hypersurfaces that are defined by interatomic potentials.

All IP calculations are based on the Born model of ionic solids, in which manganese and oxygen ions interact via long-range Coulombic interactions and short-range parameterized interactions. The potential parameters were previously developed by Matsui Akoagi and Matsui [56] for isostructural TiO_2 and were refined and used to model bulk, surfaces and structures of MnO_2 Sayle et al. [13], Sayle et al. [14], Maphanga et al. [57]. For the shell model, potentials developed by Ammundsen et al. [58] were employed and fitted to reproduce the lattice parameters of MnO_2 bulk structures. The potentials gave good agreement with the crystal structures of both pyrolusite and ramsdellite polymorphs of MnO_2. The evolutionary algorithm implemented within the GULP Gale and Rohl [46] code was used to

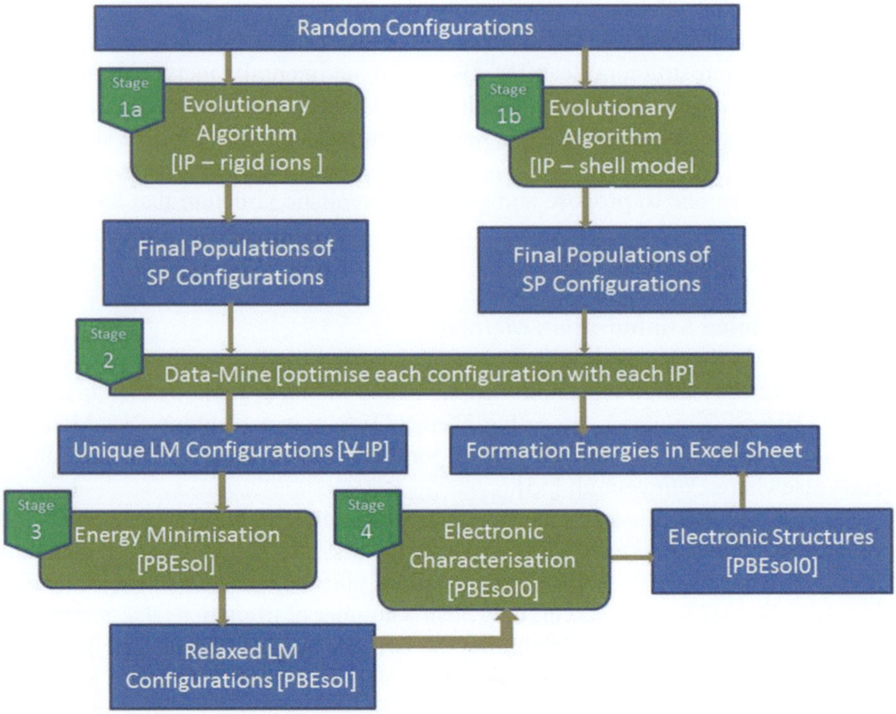

Fig. 14.3 Schematic representation of the many stages used to investigate $(MnO_2)_n$ cluster. The *rectangles* represent an input/output for the codes used: GULP in Stages 1, a script that calls GULP in Stage 2, and FHI-AIMS in Stages 3 and 4

perform IP calculations, which are based on two potential models, namely, the partially charged rigid ion model and the fully charged shell model. The energy is given by the sum of the two-body interactions, and its analytical expression is:

$$V(r_{ij}) = \frac{q_i q_j}{r_{ij}} + B_0 + B_1 r + B_2 r^2 + B_3 r^3 + B_4 r^4 + B_5 r^5 \tag{14.4}$$

where the first term is the Coulomb contribution to the energy between two point charges q_i and q_j at a distance of r_{ij} apart. The analytical expression for the Buckingham potential used in this work is given in Eq. 14.1.

An evolutionary algorithm (EA) method for global optimization in Stage 1 was used to search for lower local minima (LM) on the energy hypersurface to predict stable and low-energy metastable atomic configurations. Various relaxed configurations, initially random stationary points on the energy hypersurface, from a population in which there is competition to survive and procreate is simulated. The probability of success of any current configuration is based on its energy relative to the other configurations in the population. In the current study, only the 30 best unique structures are kept at the end of each cycle. Within a population, if the difference in the energy of formation between the two configurations is within E_d, then the higher energy configuration is removed from the population. The EA method was applied at least twice to each hypersurface, when $E_d = 0.01$ eV and $E_d = 0.004$ eV. The atoms of each isolated cluster are constrained to be within a spherical container with a radius of 8 Å. The constraint prevents any initial fragmentation of a cluster and helps to speed up convergence of the local optimizer. Each LM corresponds to a configuration of atoms of the cluster, referred to as "plausible structure."

14.2.4 Generating Nanostructures

Nanomaterials offer the ability to change the properties of the material or proffer new properties by simply adjusting the size. It is not surprising that unprecedented efforts have focused on the synthesis, Pinna and Niederberger [59] structure, Thomas and Midgley [60]; Möbus and Inkson [61] properties, Lai et al. [62]; Hahn [63] and applications Feng et al. [64]; Berry [65]; Tarnuzzer et al. [66] of such nanomaterials. However, structural features that exist on the nanoscale are difficult to characterize experimentally, and therefore atomistic computer simulation offers a unique window of exploration into such materials Sayle et al. [67]. A general and systematic simulation strategy for creating models (atom coordinates) of nano materials has been developed Sayle et al. [67]. In particular, amorphized nanoparticles are assembled into periodic arrays using crystallographic rules, and they facilitate their aggregation and crystallization into framework nano-architectures. Thus, an atomistic model for nanostructure can be generated by constructing a simulation cell with lattice parameter in accordance with the experiment and positioning nanoparticles at base positions to facilitate the symmetry and periodic boundary conditions and replicate the material infinitely in three (bulk) or two (surface) dimensions. Under molecular dynamics simulation, the amorphous nanoparticle will agglomerate with its periodic neighbors to formulate the walls of the nanomaterial Tarnuzzer et al. [66]. The size of the nanoparticle can be adjusted to enable the wall thicknesses of the model to agree with the wall thickness of the real material. A schematic representation of this process is shown in Fig. 14.4.

To facilitate the variety of different nanostructures, the simulation cell size, as seen in Table 14.1, was adjusted to enable neighboring nanoparticles to agglomerate. For example, for a nanoparticle (0D), neighboring nanoparticles are not supposed to agglomerate. Accordingly, the size of the simulation cell was maintained high enough (in all dimensions) to prevent neighboring nanoparticles to first attract one another and then agglomerate upon contact. To facilitate a nanorod, the simulation cell was reduced in one dimension to enable neighboring nanoparticles to agglomerate in one direction, thus facilitating a rod.

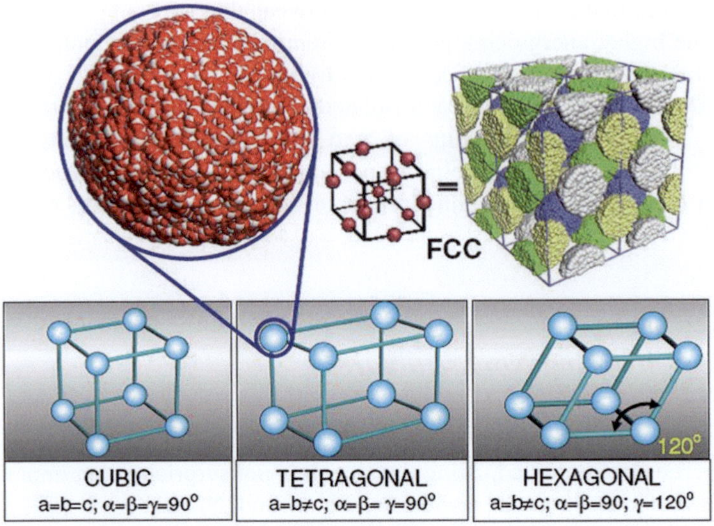

Fig. 14.4 Schematic illustrating the systematic construction of a nanostructure. Possible simulation cells, for example, *cubic*, *tetragonal*, and *hexagonal* are shown below, and a nanoparticle is positioned at base positions to conform to a particular space group. *Top right* shows nanoparticles positioned in a face-centred cubic array. Reprinted with permission from Sayle et al. [14]. Copyright 2009 American Chemical Society

Table 14.1 Dimensionality, simulation duration, cell sizes and wall thicknesses pertaining to each of the nanostructures

Dimension	Name	Simulation time (ns)	Simulated cell size (nm)	Wall thickness (nm)
0-D	Particle	25	10.0 × 10.0 × 10.0	8
1-D	Rod	21	10.0 × 8.5 × 10.0	6
2-D	Sheet	25	8.5 × 8.5 × 8.5	3–6
3-D	Porous	27	7.4 × 7.4 × 7.4	4–7
	Bulk	25	6.3 × 6.3 × 6.3	–

A schematic representation is shown in Fig. 14.5. Agglomeration of neighboring nanoparticles in two and three dimensions facilitates nanosheets and nanoporous architectures, respectively. To facilitate a model for the bulk material, the simulation cell size was reduced so that the atoms, comprising the amorphous nanoparticle, filled the simulation cell with no empty space. This was achieved by applying constant pressure MD simulation with a 50 kbar pressure; similar to a liquid, the amorphous MnO_2 material under MD simulation, fills the simulation cell completely. Finally, each system was crystallized by performing constant volume MD simulation at 2000 K for a duration sufficient to ensure that the nanostructures were no longer changing structurally or energetically.

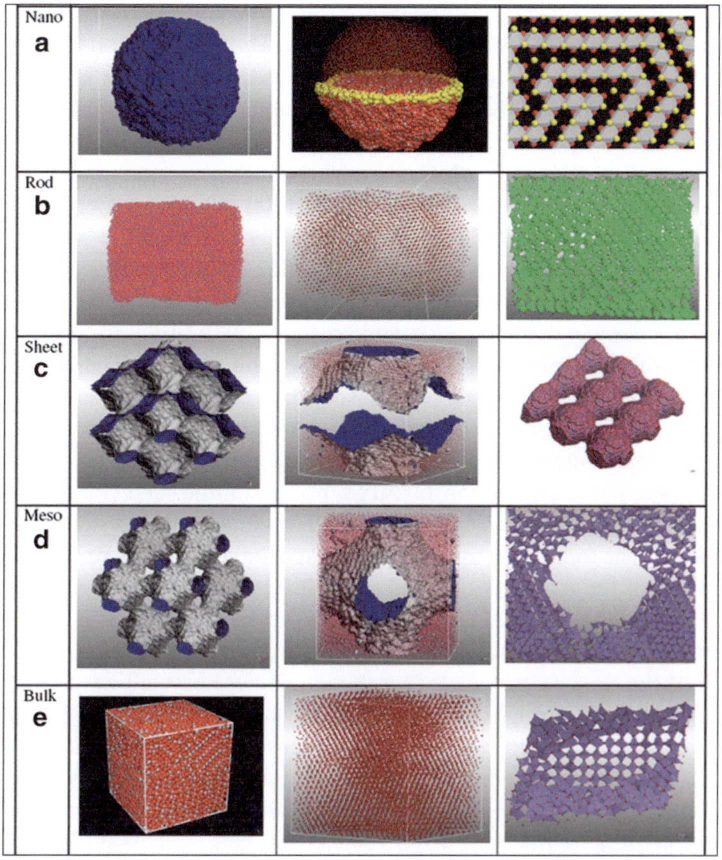

Fig. 14.5 Various structures of MnO_2 (**a**) surface-rendered model for the nanoparticle revealing its spherical morphology; (**b**) morphology of the nanorod; (**c**) nanosheet; (**d**) nanoporous; and (**e**) bulk system. Adapted with permission from Sayle et al. [13] and Sayle et al. [14]. Copyright 2005 and 2009 American Chemical Society

14.3 Nanostructures

14.3.1 MnO_2 Nanoparticle

Sayle et al. [13] generated models of MnO_2 nanoparticles, with full atomistic detail, using a simulated amorphization and recrystallization strategy. In particular, a 24696-atom "cube" of MnO_2 was amorphized by increasing the lattice parameter by 36 % (tension-induced). To amorphize and recrystallize the system, long-duration MD simulations were carried out with a time-step of 0.005 ps, at 2000 K. This was performed under NVE and NPT ensembles for amorphization and recrystallization respectively, for a total duration of 4000 ps. Finally, after the system underwent a full recrystallization process, it was cooled gradually; when MD was run for 500 ps at 1500 K, 250 ps at 1000 K, 250 ps at 500 K and 500 ps at 0 K, the latter acted effectively as an energy minimization. Closer inspection of the nanoparticle deduced that the nucleation seed initiates the crystallization of the entire system. At 1750 ps, the system reveals an octahedral pattern conforming to the pyrolusite structure of MnO_2, grain boundaries, and initial stages of microtwinning. A final structure at 0 K (after 4000 ps) shows a crystalline nanoparticle that adopts the pyrolusite structure; this comprises a wealth of twin boundaries, rather than existing as a single crystal. Twin boundaries are evidenced by observing where the parallel lines of manganese ions change the direction.

Figure 14.6 depicts the nucleation and crystal growth of the MnO_2 nanoparticle system using the amorphization and recrystallization method. The first 1000 ps is related to the nucleation process. A corresponding snapshot of a movie at 500 ps, derived from the trajectories of an MnO_2 nanoparticle, depicts an amorphous structure. A closer inspection depicts the evolution of a small crystalline region within the system at 750 ps. Prior to 750 ps, no crystalline seed could be identified. This is regarded as a nucleating seed, which nucleates the crystallization of the entire system. The reduction in energy marked beyond 1000 ps corresponds to crystallization of the nanoparticle, and crystalline patterns emerge. After 1750 ps, the crystalline region is much larger and is shown in Fig. 14.6. It is apparent from inspection of this figure that microtwinning has occurred. In particular, close inspection of the snapshots reveals that the manganese crystallizes into octahedral holes not only in a straight/parallel fashion, which propagates the pyrolusite structure, but also at specific angles, which results in the formation of the twin boundaries. Inspection also reveals that the crystallization proceeds from a *single* crystalline seed; it is conceivable that two or more seeds could have evolved with crystallization fronts emanating from both seeds. The system is fully crystalline after 4000 ps and this part of the simulation is regarded as simulated annealing.

The nanoparticle plays a significant role in amorphization and recrystallization methods since it is the basis for the production of other nanoartchitectures such as nanoporous, nanorods, and bulk, where simulation commences with amorphized nanoparticles. In the next section, the generation and characterization of the nanoporous and bulk structures will be explored, together with their lithiation for lithium-ion and lithium-air battery applications.

14.3.2 Other MnO_2 Nanostructures

The important chemistry facilitating charge/discharge occurs deep within the framework architecture of the MnO_2 as an electrode material – a region notoriously difficult to probe experimentally Thomas and Midgley [60]. The simulation of the various nanostructural architectures is challenging. Consequently, the simulation strategies for generating full atomistic models for nanomaterials, spanning all spatial dimensions including nanoparticle (0D), nanorod (1D), nanosheet (2D),

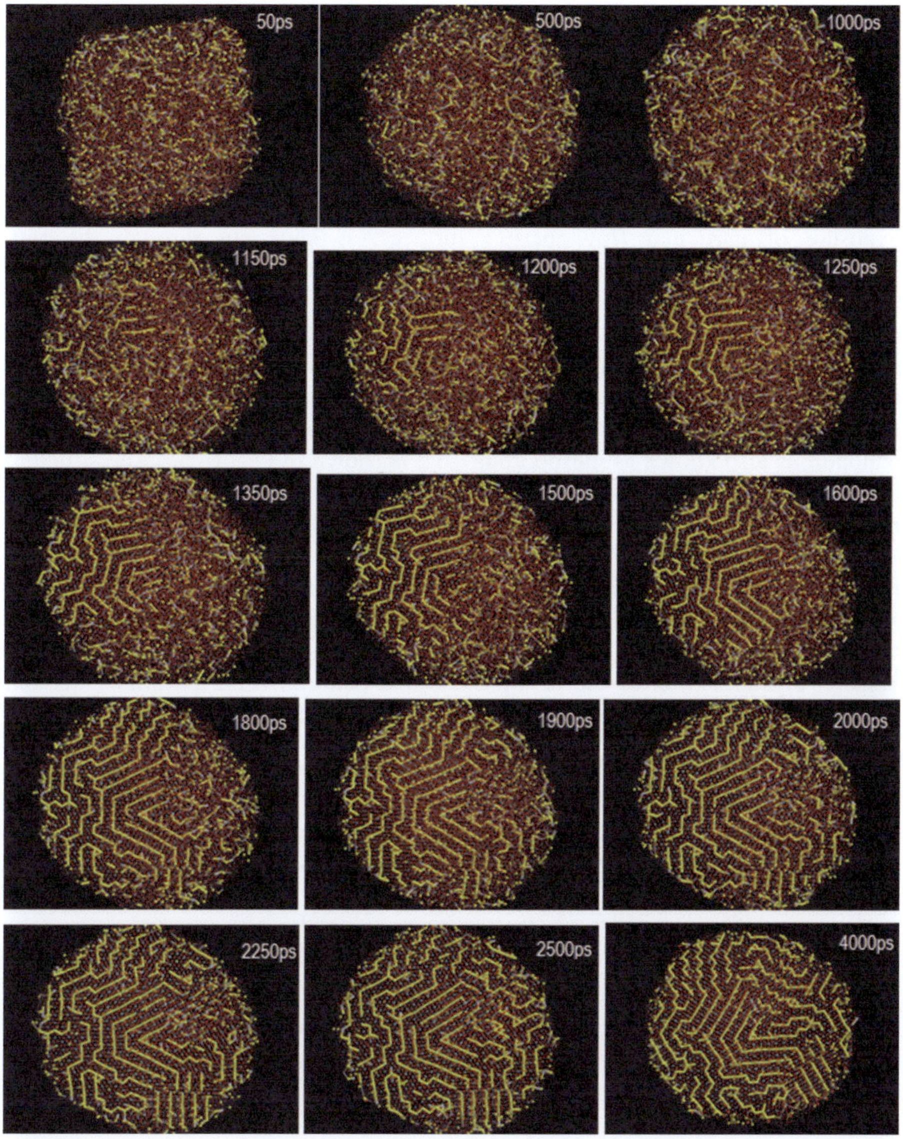

Fig. 14.6 Snapshots of the atom positions comprising the MnO_2 nanoparticle taken as a function of time, illustrating the crystallization process. Species colored *yellow* correspond to manganese and are bonded to each other to illustrate more clearly the crystallinity, connectivity, and evolution of twinning. The figures were constructed to ensure clarity of the crystallization process and therefore most of the ions comprising the nanoparticle are shown as *dots*. The time, shown in ps, is usefully correlated with the energy trace in Fig. 14.2. Manganese ions are colored *yellow*, oxygen ions are *red*. Reprinted with permission from Sayle et al. [13]. Copyright 2005 American Chemical Society

and nanoporous (3D) are demonstrated; such nanostructures have all been synthesized experimentally Debart et al. [2], Jiao and Bruce [3], Zhou et al. [4], Matsumoto et al. [5], Li et al. [39]. We illustrate the approach by generating nanostructural models of MnO_2, which is an important material with respect to energy storage.

The atomistic structure of the MnO_2 nanorod (1D) is shown in Fig. 14.5b. Closer inspection of the surface of the nanorod does not reveal any obvious faceting or exposure of particular (low-energy) surfaces. A slice cut through the rod, Fig. 14.5b, reveals considerable microtwinning of the lattice and vacancies. The atomistic structure of the MnO_2 nanosheet (2D) is shown in Fig. 14.5c. The model reveals that the nanostructure exposes both concave and convex surfaces with some evidence of faceting, including surface steps, edges and corners. It is difficult to assign a particular index to the surface because of the small area and high (positive and negative) curvature of the nanosheet. The atomistic structure of the MnO_2 nanoporous also known as mesoporous (3D) is shown in Fig. 14.5d. The structure clearly reveals that the architecture comprises positive curvature (convex) of the material, derived from the spherical building blocks. However, under MD simulation, the nanostructure to agglomerate and the ions move in an attempt to minimize the surface energy. The configurational energy of each system, calculated as a function of time, is shown in Fig. 14.7, which reveals complete crystallization for "bulk," "nanosheet" and "nanoporous" architectures after about 1–2 ns. Conversely, crystallization of the nanoparticle and nanorod occurred after 3 ns and 8 ns respectively.

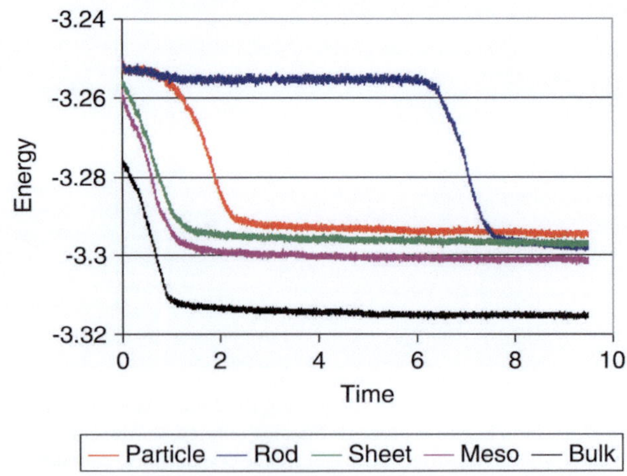

Fig. 14.7 Configuration energy (10^5eV) calculated as a function of time (ns) for each of the nanostructures. The abrupt change in energy reflects a phase change and is associated with the amorphous-crystalline transition; the difference in energy between the starting (amorphous) configuration and the final (crystalline state) loosely reflects the heat of crystallization. It takes more time to crystallize the nanorod system. This was attributed to the reamorphization of the nucleating seed and the time required to (spontaneously) evolve a new nucleating seed. Reprinted with permission from Sayle et al. [14]. Copyright 2009 American Chemical Society

Atomistic models are of little value if they simply confirm experimental observations; rather, the power and value of computer simulations are in their predictive capability. The close analogy between experiment and simulated models at three levels of higherarchical structural complexity – crystal structure, microstructure (microtwinning and point defects) – and nanostructure (particle, rod, sheet, porous) enables us to have confidence in using the models to predict properties related to energy storage. A few examples based on simulation of MnO_2 as an electrode material are discussed in the next section.

14.4 Electrochemical Predictions

When used as a rechargeable battery material, the host MnO_2 must undergo charge/discharge cycling, where Li^+ species are respectively deinserted/inserted into the MnO_2 lattice without structural modification. Clearly, to glean insight into this process and help understand how the host lattice can maintain structural integrity under cycling, it is important to understand how the Li^+ species enter the host. Equipped with an atomistic structural model, it is possible to explore this process computationally. Conversely, it is very difficult to extract such insight, deep within the nanostructure, using experimental methods. In Sects. 4.1, 4.2, 4.3 and 4.4, amorphization and recrystallization methods, together with graphical techniques, are used to explore the MnO_2 nanostructures and the bulk system.

14.4.1 Nanostructures

The atomistic structure of the MnO_2 nanoparticle was discussed in the previous section, which reveals a spherical morphology. The nanoparticle MnO_2 crystallizes into the pyrolusite structure in accordance with the experiment. The polymorph was not imposed on the structure; rather, the model structure was allowed to crystallize into any polymorph it desires – driven by the interatomic potential model describing the ionic interactions rather than any (perhaps erroneous) decision to impose any particular polymorph. Similar to the nanoparticle, all the other nanostructures crystallized into the pyrolusite polymorph of MnO_2.

Nanoporous MnO_2, when used as a host lattice in a rechargeable battery, can undergo many charge/discharge cycles while maintaining its structural integrity Jiao and Bruce [3], Bruce et al. [17]. Conversely, the parent bulk material does not proffer such properties; rather, the structure of the bulk material deforms upon lithium insertion Jiao and Bruce [3]. This behavior was attributed to the ability of the nanoporous architecture to "expand into the pores." A systematic strategy – analogous to methods widely and routinely used to model crystal structure – was used to generate MnO_2 nanostructures Sayle et al. [14]. In particular, the symmetry associated with the nanostructure, coupled with basis *nanoparticles,* was used to prescribe full atomistic models of nanoparticles (0D), nanorods (1D), nanosheets (2D), and nanoporous (3D) architectures as discussed in Sect. 3.2. Nanoporous MnO_2 can act as a host lattice for the insertion and deinsertion of lithium with application in rechargeable lithium batteries. The models reveal that the periodic framework architecture, together with microtwinning, enables insertion of lithium anywhere on the (internal) surface and facilitates lithium transport in all three spatial directions within the host lattice. To understand how nanoporous MnO_2 can facilitate lithium cycling and maintain structural integrity, the internal surfaces of the material are explored.

Figure 14.8a shows a perspective view inside a cavity within the nanoporous MnO_2 showing cavity and interconnecting channels. The internal surface of the system is explored. Two lithium ions, colored yellow, indicate a path that the lithium might take once inside the MnO_2. The lithium ion changes direction at the microtwin boundary. At the end (direction of the arrow) it appears that the tunnel is blocked by a manganese ion. However, further analysis using molecular graphics revealed that the tunnel was not blocked; rather, the "blockage" is actually another twin boundary that is not in the plane of the page. Accordingly, a closer inspection of the structure reveals that the lithium ion is free to insert at many positions on the internal surface and can move in all three spatial directions once inside the nanoporous material (see Fig. 14.8b). The results predict that the considerable twinning observed in the structural models enables lithium to insert anywhere on the internal surface and move spatially in all three directions. This will facilitate an even expansion of the framework upon insertion/deinsertion and thus enable the material to expand elastically and protect against plastic deformation upon cycling. However, an even expansion necessitates that the porous material be periodic. When lithium is introduced into the surface of the parent bulk MnO_2, would change the lattice parameter of the uppermost surface layer from $4.40 - 5.01$Å (6.5 %). This strain can likely be accommodated for a single "layer" – the energy associated with maintaining favorable (cation-anion) interactions far outweighing the energy required to strain the surface layer. We propose that the nanoporous host architecture is able to deform elastically, rather than plastically, during cycling: Jiao and Bruce found that the lattice parameter of the porous $\beta-MnO_2$ changes from $a = 4.40, c = 2.88$Å to $a = 5.01, c = 2.81$Å upon insertion Jiao and Bruce [3].

We predict that to maximize its electrochemical properties, the MnO_2 host should be symmetrically porous, heavily twinned, and the wall thickness minimized. Thus there exists a critical thickness pertaining to the elastic-plastic deformation for host lattices, above which the strain associated with lithium insertion is accommodated via plastic, rather than elastic; deformation of the host lattice leads to property fading upon cycling. Further supporting evidence comes from Debart et al. [2], who found that the energy storage capacities of nanorods of $\beta-MnO_2$, 100 nm in diameter, fade very rapidly upon cycling, which indicates a plastic rather than elastic deformation upon cycling. Accordingly, based upon this experimental data, we predict that the critical thickness lies between 10 and 100 nm. We note that

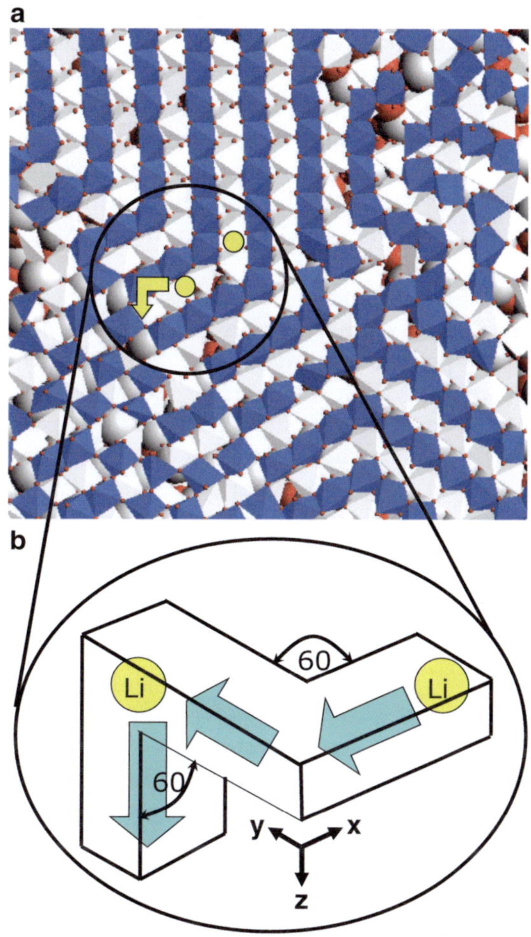

Fig. 14.8 Atomistic model and schematic illustrating the 3D mobility of Li in nanoporous MnO_2; Li atoms are schematic. (**a**) Atomistic model showing the 1 × 1 tunnels; (**b**) schematic showing how the Li can move in all three spatial directions in the porous MnO_2 Reprinted with permission from Sayle et al. [14]. Copyright 2009 American Chemical Society

the bulk crystalline $\beta-MnO_2$ converts to $LiMn_2O_4$ spinel on chemical intercalation Jiao and Bruce [3]. In addition, Tang and co-workers find that the "lithium insertion behavior in the first discharge process depends strongly upon the size of the $\beta-MnO_2$ nanocrystals" Tang et al. [68]. This prediction may help explain why certain nanoforms of MnO_2 are electrochemically active, while others are not.

14.4.2 Lithiated Nanostructures

Central to porous nanomaterials, with applications spanning from catalysts to fuel cells, is their "fragile" structure, which must remain structurally intact during the application's lifespan. In this section, atomistic simulation methods are used to characterize the structural durability of MnO_2 nanoporous materials by exploring its mechanical strength. Rechargeable lithuim-ion battery materials suffer volume changes during charge/discharge cycles as lithium ions are repeatedly inserted and extracted from the host, causing failure as a result of localized stress. However, nanoporous MnO_2 does not suffer structural collapse during cycling. Previous studies showed that mesoporous and needle-like nanostructured pyrolusite MnO_2 allow good lithium ions intercalation Jiao and Bruce [3], Luo et al. [69], Cheng et al. [70], Tang et al. [68]. Both pore size and wall thickness of the mesoporous structures have been demonstrated to affect the rate capability Ren et al. [71]. The mesoporous MnO_2 cell has a capacity of 284 mAh/g and good cycling stability Jiao and Bruce [3].

Analysis of the simulated crystallization, using molecular graphics, reveals the spontaneous evolution of a pseudo-crystalline seed, conforming to a pyrolusite MnO_2 polymorph, in accord with the experiment. The nucleating seed initiated the crystallization of the surrounding amorphous sea of ions. It is noted that the host MnO_2 was able to crystallize around the lithium species, enabling the lithium to be located within the 1×1 tunnels of the pyrolsite MnO_2 as observed experimentally Luo et al. [69]. Inspection of the final, low temperature structural model, using graphical techniques reveals that the simulated crystallization has facilitated the evolution of a variety of important microstructural features observed experimentally Chabre and Pannetier [72], including grain-boundaries, twin-boundaries, microtwinning and point defects. The calculated density and surface area for $Li_{0.16}MnO_2$ are 2.8g/cm^3 and 168m^2/g respectively, with a 4–8 nm wall thickness in accord with experiment Luo et al. [69]. We predict that during lithium intercalation, associated with charge/discharge cycles, the host lattice contracts/expands by 16 %, respectively. In particular, when lithium is introduced into the MnO_2 host (discharging), the volume increases (per MnO_2 unit) from 28.4 Å^3 (pure MnO_2) to 29.2 Å^3($Li_{0.16}MnO_2$) – a 2.8 % increase. Assuming that the lattice expands linearly with the introduction of lithium, then the expansion associated with lithium intercalation into MnO_2 up to $Li_{0.92}MnO_2$ is predicted to be about 16 %. Calculated stress-strain curves for MnO_2 and $Li_{0.16}MnO_2$ reveal that the material expands elastically up to 5 % (uniaxial) and 10 % (hydrostatic) strain with plastic deformation occurring above these levels. In summary, when used as an electrode in a rechargeable lithium-ion battery, symmetrically nanoporous MnO_2 can host flooding of all pores by electrolytes because of pore accessibility. Accordingly, during charge/discharge cycles, lithium ions can be inserted at all positions on the internal surfaces, enabling the entire symmetrically porous MnO_2 architecture to expand homogeneously; any slight inhomogeneity in either the structure or insertion process, can be accommodated by elastic deformation (up to 10 % volume change).

14.4.3 Bulk

In this section, the A&R simulation technique is used to characterize and explore microstructural complexities of $\gamma-MnO_2$ to help the experiment understand the material. The structural characterization of $\gamma-MnO_2$ materials by diffraction techniques is complex, and characterization at the atomistic level is difficult experimentally. At present there is no efficient method for characterizing industrial samples of $\gamma-MnO_2$, and therefore no means of relating their atomistic scale arrangement with their behavior in batteries. Many theoretical methods that are used to investigate the structure and energetics of a system proceed by proposing an atomistic model, which is then simulated using static or dynamical methods. However, for systems that comprise complex microstructures, such as $\gamma-MnO_2$, the starting structural models can be difficult to generate. For example, the model needs to include within a *single* simulation cell: (1) ramsdellite/pyrolusite intergrowths, the interfacial configuration of these intergrowths and the relative sizes of the domains – this is certainly a challenging undertaking in itself. It is important to ensure that "some" of the domains are (2) twinned, and finally (3) point defects need to be introduced. Moreover, all microstructural features (1), (2), and (3) are associated with combinations and permutations associated with location, arrangement, and concentration.

A "cube" of MnO_2, comprising about 25,000 ions, was cut from the parent material. The coordinates of all the component manganese and oxygen ions were then changed to facilitate a 36 % increase in the lattice parameter, which resulted in a highly tensioned nanocube (Fig. 14.9a). Constant volume molecular dynamics simulation, applied to this system, caused the Mn and O ions to accelerate and move in order to reduce the lattice parameter (Fig. 14.9b); the system can perhaps be likened to an elastic band under tension, which is then released. The ions, as they accelerated, collided with other ions, which changed their trajectories. The summation of these collisions and trajectories caused the system to amorphize (Fig. 14.9c). To generate models for bulk MnO_2, the MD simulation was changed from constant volume to constant pressure. In particular, the 50 kbar pressure that was imposed on the system reduced the size of the simulation cell (Fig. 14.9d). And, as the simulation cell volume reduced, the ions comprising the nanoparticle moved to fill the simulation cell. The simulation cell continued to shrink until the MnO_2 completely filled the available volume with no void space (Fig. 14.9e). The resulting system was periodic in three dimensions and, therefore, representative of the bulk material. Finally, the pressure was relieved and the system cooled, under MD, to 0 K (Fig. 14.9f). To validate the simulated models, XRD patterns were calculated and compared with experimental results. Calculated XRD patterns for the microstructural models for MnO_2 are very broad compared with the XRD patterns for the parent materials (pyrolusite and ramsdellite) and correlate with those for $\gamma-MnO_2$. Central to simulated A&R is its ability to facilitate the evolution of a wealth of highly complex microstructural features into a single simulation cell.

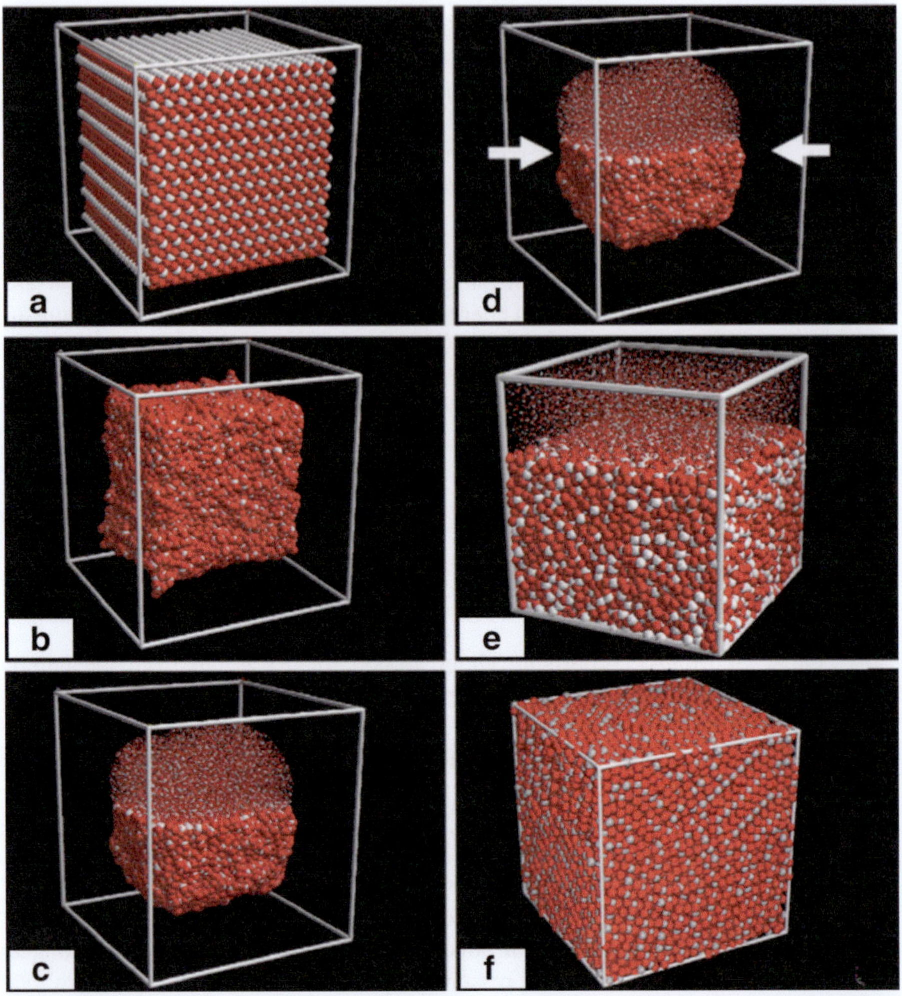

Fig. 14.9 MnO_2 structure depicting the amorphization (**a–c**) and recrystallization (**d–f**) processes; (**a**) starting configuration; (**b**) configuration after 5 ps of MD; (**c**) after 100 ps of MD; (**d**) as (**c**), but with 50 kbar pressure imposed upon the system; (**e**) configuration after imposing 50 kbar pressure and MD for 125 ps; (**f**) final, zero pressure, 0 K, structure. Cell dimensions for (**a–c**) are fixed (NVE). Manganese is colored *gray* and oxygen is *red*. Reprinted from Sayle et al. [73], Copyright 2006, with permission from Elsevier

14.4.4 Lithiated Bulk MnO_2

Building on previous studies on the characterization of the $\gamma-MnO_2$ bulk system, we use the technique to explore and characterize the microstructure of lithium-inserted bulk MnO_2. The simulated amorphization and recrystallization method has been used to generate atomistic models of Li_xMnO_2 (with $x = 0.06 - 0.33$) by

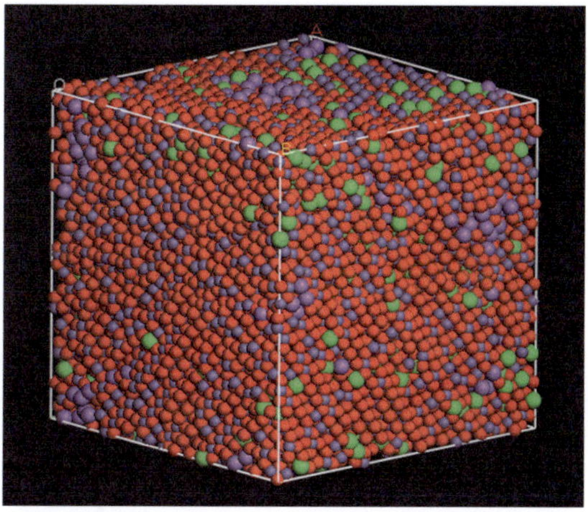

Fig. 14.10 Snapshot of the final configuration for $Li_{0.12}MnO_2$ (1000 Li ions) Mn^{4+} is colored *purple* (small), Mn^{3+} is *purple* (big), oxygen is *red*, and lithium *green*

simulating crystallization, starting from an amorphous precursor, comprising about 25,000 atoms. Thus, lithium ions were inserted into an amorphous MnO_2 host, and crystallized at a temperature of 2000 K and a pressure of 50 kbar. To facilitate charge neutrality, an equivalent number of Mn^{4+} species were reduced to Mn^{3+}.

Figure 14.10 gives a graphic representation of the final crystallized configuration for $Li_{0.12}MnO_2$ (1000 Li ions) at 0 K. Calculated XRD patterns for systems with various lithium concentrations show slight shift peak positions and subtle differences with respect to split peaks. The calculated XRD patterns are in accord with experiment. The calculated XRD patterns for the microstructural models for Li $-MnO_2$ are very broad compared to experimental XRD patterns, which we attribute to the microtwinning. Simulated X-ray diffraction patterns of Li $-$ MnO_2, with microtwinning and a high concentration of lithium, agree better with experimental XRD patterns. Hence, one can conclude that the structure is strongly modified, probably with an important decrease of pyrolusite value in the system and appearance of other kinds of defects. A closer examination of XRD patterns for the microstructures generated has been carried out and were compared with literature results (Fig. 14.11) by Bowden et al. [74], whereby lithiated manganese dioxide (LiMD) structures were studied using XRD, TEM and NMR. The structures ranged from perfect (defect-free) ramsdellite as a function of increasing De Wolff defects to the limiting case of perfect pyrolusite.

Microstructural details of a slice cut through the system perpendicular to [001], for $Li_{0.06}MnO_2$ is shown in Fig. 14.12. Polyhedron representation is used to illustrate the structure and the wealth twin boundaries more clearly. The figure reveals microstructural features observed within the system. They include pyrolusite – square, pyrolusite single twin; triangle, pyrolusite heavily twinned; green rectangle, stacking fault; thin rectangle, double chains formation; thick rectangle, grain boundary; yellow oval, defect cluster; green oval and Mn^{4+} vacancies – circle. The red polyhedrons are the Mn^{3+} ions and the yellow are the lithium ions.

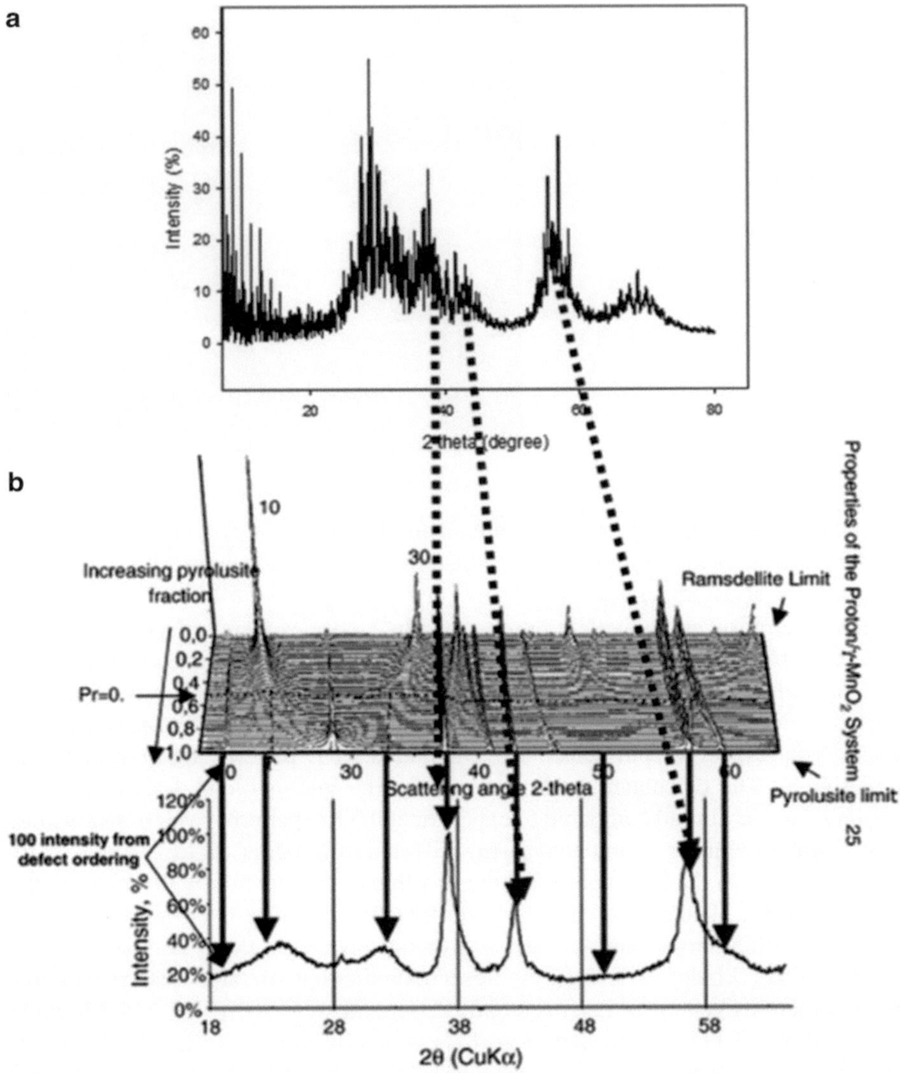

Fig. 14.11 Comparison between calculated and literature XRD patterns. Figure (**a**) is this study and (**b**) is taken from Ref. [23] obtained with simulated DIFFAX Program. *Dashed arrows* show common peaks between this study and literature XRD patterns Maphanga et al. [75]

The system tends to show a complex defect cluster – green oval where the lithium ions are located. The cluster is comprised of several Mn^{4+} and oxygen vacancies. On changing the concentration to $Li_{0.33}MnO_2$ microstructural details are more complex. Furthermore, the structure comprises dislocations, isolated and associated point defects such as vacancies, interstitials, grain boundaries, and wealth of twin boundaries. Observation of increasing concentration on stabilization of ramsdellite

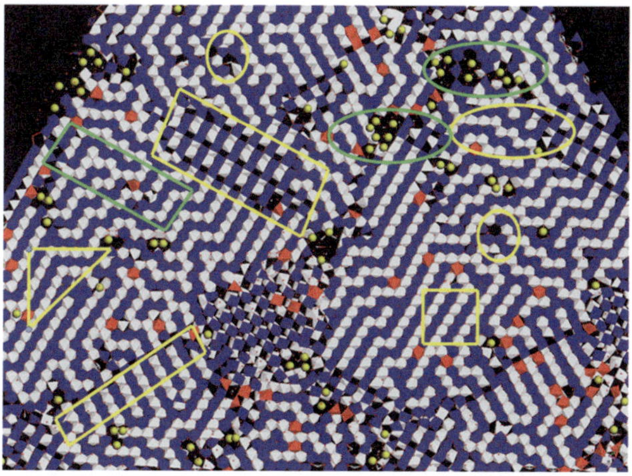

Fig. 14.12 Graphical representation of the final configuration for $Li_{0.06}MnO_2$ (520 Li ions) at 0 K. The *blue* and *white* octahedrons represent the upper and lower layers of Mn^{4+} ions. The *red* octahedrons represent the Mn^{3+} ions. The Li^+ ions are *yellow* Maphanga et al. [75]

structure as compared to that in the pure bulk MnO_2. The results support the possibility of intrinsic defects facilitating the stabilization of ramsdellite domains (2×1 channels) within the system in accordance with proposals based on experimental data Chabre and Pannetier [72]. In particular, to accommodate lithium ions in the host MnO_2, the material evolves a 2×1 (ramsdellite) channel compared to (1×1) channels of the parent (pyrolusite).

14.5 MnO_2 Nanoclusters

Low-lying local minima on the energy of formation landscape defined by interatomic potentials (IPs) are found by using one of a number of global optimization algorithms Woodley and Catlow [76], Woodley [77]. The use of IPs approaches reduces the computational cost of locating plausible structures, which can be refined at a later stage. Searching landscapes based on IPs has been successfully applied to finding a wide range of bulk metal oxide phases Putz et al. [78], Woodley et al. [79], Woodley [80], Woodley et al. [81], Woodley and Catlow [82], Schön [83], Schön and Jansen [84] and therefore such techniques have been applied to clusters Catlow et al. [55], such as, MgO Roberts and Johnston [85], ZnO Al-Sunaidi et al. [86], ZnS Woodley et al. [87], Hamad et al. [88], TiO_2 Hamad et al. [82], SiO_2 Flikkema and Bromley [89], ZrO_2 Woodley et al. [90], Al_2O_3 Woodley [80] and In_2O_3 Walsh and Woodley [91]. The IPs are typically fitted so as to reproduce the structures and properties of the bulk phases, which may be one of the causes of a change in the ranking in terms of a stability order for the clusters.

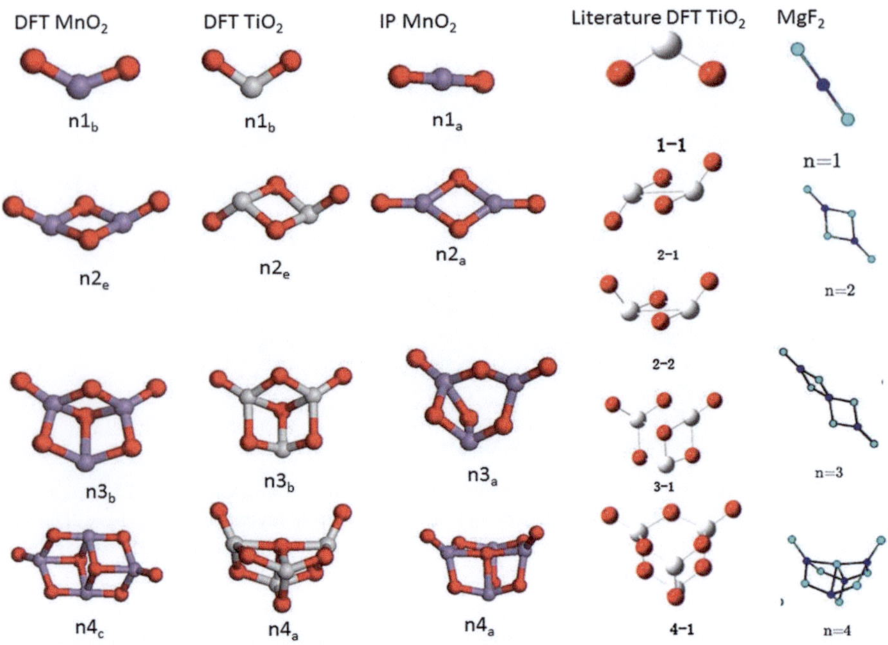

Fig. 14.13 Global minima for $(MnO_2)_n$, $n = 1$ to 4. The *gray* color represents manganese atoms and *red* the oxygen atoms. Lowest energy configurations for $(MnO_2)_n$, $n = 1 - 4$ from (**a**) IP and (**b**) DFT calculations and (**c**) DFT $(TiO_2)_n$ nanoclusters. *Purple* balls represent manganese atoms, *gray* are titanium atoms, and *red* balls oxygen atoms

When the plausible structures are found, they are refined using a DFT approach. Previous studies reported stable structures of $(ZrO_2)_n$, where $n = 1 - 8$ nanoclusters, which were obtained by data mining from plausible $(TiO_2)_n$ structures found using global optimization techniques.

Global minima configurations for MnO_2, obtained using the evolutionary algorithm methods, are discussed in this section. The lowest energy local minima configurations for $(MnO_2)_n$, clusters ($n = 1 - 4$) are presented in Fig. 14.13 together with the order of stability as measured by IP and DFT PBEsol0 Walsh et al. [15]. The saddle point configurations for MnO_2 in comparison with GM for TiO_2 and MgF_2 clusters are reported. This is in agreement with previous studies on MgF_2 clusters Francisco et al. [92] and TiO_2. For $n = 1$, at the initial stage, interatomic potential predicts a linear configuration with an angle of 180°. This is in agreement with previous studies on MgF_2 clusters Francisco et al. [92]. The structure bends into a boomerang (n1a) during relaxation in Stages 3 and 4 when the DFT methods are employed. Table 14.1 gives the geometric parameters (bond lengths and angles) for the global minima $n = 1$ clusters. Previous studies reported that d-orbitals have influence on the nanoclusters. When the high spins are included in the configurations, the bond angles decrease from 180° to 129°, hence the polarizability of the ions within the system.

For $n = 2$, configuration n2a is the global minimum for the IP method. When the DFT method is used to relax plausible structures, n2b configuration is obtained as the global minima for MnO_2 clusters. Both systems predict a configuration with two-terminal Mn-O bond as the global minima in agreement with previous studies on TiO_2 DFT nanoclusters simulations Hamad et al. [93]. The first and second local minima are predicted to be the configurations with two terminal Mn-O and one terminal and three bridging O atoms, respectively. Both systems predict a configuration with two-terminal Mn-O bond as the global minima in agreement with previous studies on TiO_2 DFT nanoclusters calculations.

Cluster	Bond lengths (Å)			Bond angles (°)		
		IP	DFT		IP	DFT
$n = 1$	Mn-O	1.683	1.579	O-Mn-O	180.000	129.040
$n = 2$	Mn-O1	1.687	1.575	O1-Mn-Mn	141.315	121.433
	Mn-O2	1.822	1.778	O2-Mn-O2	102.629	91.780
				Mn-O2-Mn	77.371	88.180

The cluster n3a is predicted to be the global minimum for IP calculations. The lowest energy configuration, $n3_b$ for $(MnO_2)_3$ is composed of two terminal oxygen atoms and one threefold coordinated oxygen atom in agreement with previous DFT calculations. DFT calculations for the nanocluster $n = 3$, predicts the configuration of the lowest GM to be identical to those found for TiO_2, ZrO_2 and HfO_2 in previous studies. Thus, the DFT method predicts that the GM configuration is composed of three tetragonal sides of a cuboid with two singly coordinated oxygen atoms. The GM configuration for DFT-PBEsol0 can be constructed from smaller clusters, in particular, $n1_b$ and $n2_b$. Thus, the smallest lowest energy configurations can be used as building blocks for the larger clusters. The configuration n4c is found to be the global minima for $(MnO_2)_4$, which is composed of two terminal oxygen atoms and six bridging oxygen atoms. The GM configurations found for $n = 4$ can be constructed from smaller clusters. Numerous DFT studies on $(TiO_2)_n$ showed that DFT calculations predict the lowest energy global minima better than the other methods, such as Hatree-Fock (HF) Hamad et al. [93] and pair-potential Yu and Freas [94]. The latter two methods neglect the partial covalent nature of Ti-O bonds.

The clusters are ranked according to PBEsol0 functional because it is one of the least biased among complementary functionals. The small clusters are isostructural to those predicted for titania, silica, and hafnia Catlow et al. [55] and characterized by manganese ions concentrating in the core region of the cluster decorated by dangling, or singly coordinated, oxygen ions at the periphery. However, when the energy of formation is calculated using a DFT approach, after relaxation, the dangling oxygen atoms for MnO_2 do not remain in the plane containing the nearest Mn_2O_2 tetragon.

14.6 Conclusions

The properties of a nanomaterial can be fine-tuned or new properties proffered compared to the bulk parent material by changing the size of one or more dimensions in the nanometer range. However, populating nanostructure-property correlation tables, which will prove invaluable in the intelligent design of new nanomaterials, necessitates characterization of the nanostructure with atomistic detail. This is difficult experimentally, and therefore atomistic computer simulation offers a unique window of exploration into such materials. Conversely, while approaches to generating atomistic models for crystal structures are well developed and have been widely used for over 60 years, Mott and Littleton [95], Lidiardi [96], analogous systematic approaches for generating atomistic models for nanostructures are not yet available. In this chapter, systematic strategies were developed and used to generate full atomistic models for MnO_2 nanomaterials, spanning all spatial dimensions including nanoparticle (0D), nanorod (1D), nanosheet (2D), and nanoporous (3D). Specifically, it was shown how amorphous model nanobuilding blocks, playing the role of atoms in meta-crystallography, can be easily positioned into any location to attain desired or prescribed architectures. The nanoparticles can then be encouraged to agglomerate, under MD simulation, to formulate the walls of the nanoporous framework architecture.

Generated models for the MnO_2 nanoparticle are crystalline, accommodate very complex microstructures, and the associated atomistic relaxation and rumpling are considerable. Is it suggested that the complex microstructure observed within the model, together with the ionic relaxation and rumpling, might help explain the absence of long-range order associated with the experimentally fabricated nanoparticles. Also, from the simulations, the crystallization speed could be estimated. In particular, the diameter of the nanoparticle is 8 nm, and this takes about 1500 ps to crystallize. Therefore, the crystallization front, at 2000 K, can be estimated to move at 8 nm/1500 ps or about 5 m/s.

Various complex microstructures for different materials and nanoarchitectures are generated and provide very valuable insights. Mechanical load simulations were carried out to help understand why the nanoporous structure possesses exemplary electrochemical properties. Symmetrically porous β–MnO_2 can host the storage and transport of lithium (charge carriers) up to a composition of $Li_{0.92}MnO_2$, commensurate with 284 $mAhg^{-1}$, yet retain its structural integrity despite the high lattice expansion and contraction associated with charge/discharge cycles. The simulations reveal that for MnO_2 to be a viable host within rechargeable battery systems, the material must be symmetrically porous to enable the material to expand linearly with lithium intercalation. Non-symmetrically porous MnO_2 will suffer plastic deformation. We propose that other (nanoporous) intercalation hosts will follow similar mechanisms.

To understand the topological formation of MnO_2 small clusters at nanoscale, a combination of global search techniques and density functional theory method are employed to elucidate the energy landscapes of MnO_2 clusters. The two methods were used successfully to predict the global and local minima configurations of small $(MnO_2)_n$ nanoclusters. The IP method allows a large number of configurations to be explored, while DFT-PBEsol0 takes into consideration the electronic configuration of the system. The results are of great importance in understanding the formation and nucleation of MnO_2 nanostructures. The analysis of MnO_2 nanoclusters obtained by the interatomic potential method reveals that the clusters are formed by three basic structural motifs: (1) a planar or quasi-planar MnO_2 rhombus, (2) a planar or quasi-planar MnO_3 unit with a threefold-coordinated Mn atom, (3) and a distorted MnO_4 tetrahedron with a tetra-coordinated Mn atom. Owing to the complexity of the MnO_2, experimental investigations on the nucleation and small cluster have not been reported. In summary, atomistic computer simulation methods are well positioned to help provide insights and help experiment at the atomistic level, pertaining to nanostructuring of energy storage materials. Continued advances in the development of nanostructured electrode materials could result in energy storage devices with high-energy density and high-power density. Therefore continuous efforts on developing methods of synthesizing nanostructured materials are essential for solving scientific challenges related to energy. Properties such as nucleation, transport, voltage profiles, ragonne plots, and mictrostructure of various energy materials can be predicted by various modelling methods.

Acknowledgements We acknowledge the financial support from the National Research Foundation and Centre for High Performance Computing in South Africa for computing resources.

References

1. Arico AS, Bruce PG, Scrosat B et al (2005) Nanostructured materials for advanced energy conversion and storage devices. Nat Mater 4:366–377
2. Debart A, Paterson AJ, Bao J et al (2008) Alpha- nanowires: a catalyst for the O_2 electrode in rechargeable lithium batteries. Angew Chem Int Ed 47:4521–4524
3. Jiao F, Bruce PG (2007) Mesoporous crystalline – a reversible positive electrode for rechargeable lithium batteries. Adv Mater 19:657–660
4. Zhou Q, Li X, Li YG et al (2006) Synthesis and electrochemical properties of semicrystalline gyroidal mesoporous. Chin J Chem 24:835–839
5. Matsumoto Y, Ida S, Inoue T (2008) Photodeposition of metal oxide at the TiO_x nanosheet to observe the photocatalytic active site. J Phys Chem C 112:11614–11616
6. Wang D, Luo H, Kou R et al (2004) A general route to macroscopic hierarchical 3D nanowire networks. Angew Chem Int Ed 43:6169–6173
7. Feng ZP, Li GR, Zhong JH et al (2009) Multilayer nanosheet clusters evolved from monolayer nanosheets and their predominant electrochemical properties. Electrochem Commun 11:706–710
8. Fernandes JB, Desai BD, Dalal VNK (1984) Studies on chemically precipitated Mn (IV) oxides. Electrochim Acta 29:181–185
9. Ghaemi M, Ghavami RK, Khosravi-Fard L et al (2004) Electrolytic via non-isothermal electrode heating: a promising approach for optimizing performances of electroactive materials. J Power Sources 125:256–266

10. Susana-Pilla A, Duarte MME, Mayer CE (2004) Manganese dioxide electrodeposition in sulphate electrolytes: the influence of ferrous ions. J Electroanal Chem 569:7
11. Kucza W (2002) Structural stability of polymorphs and their reactivity vs. lithium. Electrochem Commun 4:669–673
12. Karami H, Ramandi-Ghamoshi M, Moeini S et al (2010) Low temperature hydrothermal synthesis of nanoclusters as positive material of RAM batteries. J Clust Sci 21:21–34
13. Sayle TXT, Catlow CRA, Maphanga RR et al (2005) Generating nanoparticles using simulated amorphization and recrystallization. J Am Chem Soc 127:12828–12837
14. Sayle TXT, Maphanga RR, Ngoepe PE et al (2009) Predicting the electrochemical properties of nanomaterials used in rechargeable Li batteries: simulating nanostructure at the atomistic level. J Am Chem Soc 131:6161–6173
15. Walsh A, Sokol AA, Catlow CRA (2013) Computational methods for energy materials. Wiley, Chichester
16. Islam MS, Fisher CAJ (2014) Lithium and sodium battery cathode materials: computational insights into voltage, diffusion and nanotrutural properties. Chem Soc Rev 43:185–204
17. Bruce PG, Scrosati B, Tarascon JM (2008) Nanomaterials for rechargeable lithium batteries. Angew Chem Int 47:2930–2946
18. Whittingham MS (2004) Lithium batteries and cathode materials. Chem Rev 104:4271–4302
19. Armand M, Tarascon JM (2008) Building better batteries. Nature 451:652–657
20. Fergus JW (2010) Recent developments in cathode materials for lithium-ion batteries. J Power Sources 195:939–954
21. Thackeray MM, Wolverton C, Isaacs ED (2012) Electrical energy storage for transportation-approaching the limits of, and going beyond, lithium-ion batteries. Energy Environ Sci 4:7854–7863
22. Masquelier C, Croguennec L (2013) In-depth safety-focused analysis of solvents used in electrolytes for large scale lithium ion batteries. Chem Rev 113:6552–6591
23. Bruce PG, Freunberger SA, Hardwick LJ et al (2012) Li - O_2 and Li-S batteries with high energy storage. Nat Mater 11:19–29
24. Scrosati B, Hassoun J, Sun YK (2011) Lithium-ion batteries. A look into the future. Energy Environ Sci 4:3287–3295
25. Scrosati B, Garche J (2010) Lithium batteries: status, prospects and future. J Power Sources 195:2419–2430
26. Park M, Zhang X, Chunh M et al (2010) A review of conduction phenomena in Li-ion batteries. J Power Sources 195:7904–7929
27. Etacheri V, Marom R, Elazari R et al (2011) Challenges in the development of advanced Li-ion batteries: a review. Energy Environ Sci 4:3243–3262
28. Takada K (2013) Progress and prospective of solid-state lithium batteries. Acta Mater 61:759–770
29. Goodenough JB, Park KS (2013) The Li-ion rechargeable battery: a perspective. J Am Chem Soc 135:1167–1176
30. Goodenough JB, Kim Y (2010) Challenges for rechargeable Li batteries. Chem Mater 22:587–603
31. Ellis BL, Nazar LF (2012) Sodium and sodium-ion energy storage batteries. Curr Opin Solid State Mater Sci 16:168–177
32. Ellis BL, Lee KT, Nazar LF (2010) Positive electrode materials for Li-ion and Li-batteries. Chem Mater 22:691–714
33. Zaghib K, Guerfi A, Hovington P et al (2013) Review and analysis of nanostructured olivine-based lithium recheargeable batteries: status and trends. J Power Sources 232:357–369
34. Patil A, Patil V, Shin DW et al (2008) Issue and challenges facing rechargeable thin film lithium batteries. Mater Res Bull 43:1913–1942
35. Tompsett DA, Parker SC, Islam MS (2014) Rutile -MnO_2 surfaces and vacancy formation for high electrochemical and catalytic performance. J Am Chem Soc 136:1418–1426

36. Tompsett DA, Parker SC, Bruce PG et al (2013) Nanostructuring of: the important role of surface to bulk ion migration. Chem Mater 25:536–541
37. Woodley SM (2004) Applications of evolutionary computation in chemistry. Springer-Verlag, Berlin
38. Sayle DC, Johnston RL (2003) Evolutionary techniques in atomistic simulation: thin films and nanoparticles. Curr Opin Solid State Mater Sci 7:3–12
39. Li GR, Qu DL, Yu XL et al (2008) Microstructural evolution of from porous structures to clusters of nanosheet arrays assisted by gas bubbles via electrodeposition. Langmuir 24:4254–4259
40. Catlow CRA (1997) Computer modelling of in inorganic crystallography. Academic, San Diego
41. Gale JD (1997) GULP: a computer program for the symmetry-adapted simulation solids. J Chem Soc Faraday Trans 93:629–637
42. Harding JH (1990) Computer simulation of defects in ionic solids. Rep Prog Phys 53:1403–1466
43. Koch W, Holthausen MC (2001) A chemist's guide to density functional theory. Wiley-VCH, Weinheim
44. Smith W, Forester TR (1994) DL_POLY_2.0: a general-purpose parallel molecular dynamics simulation package. J Mol Graph 14:136–141
45. Born M, Huang K (1954) Dynamical theory of crystal lattices. Oxford University Press, Oxford
46. Gale JD, Rohl AL (2003) The general utility lattice program (GULP). Mol Simul 29:291–341
47. Kohn W, Sham LJ (1965) Self-consistent equations including exchange and correlation effects. Phys Rev 140:A1133–A1138
48. Freeman AJ, Wimmer E (1995) Density functional theory as a major tool in computational materials science. Annu Rev Mater Sci 25:7–36
49. Wimmer E (1991) Density functional methods in chemistry. Springer-Verlag New York Inc., New York
50. Payne MC, Teter MP, Allan DC et al (1992) Iterative minimization techniques for ab initio total-energy calculations: molecular dynamics and conjugate gradients. Rev Mod Phys 64:1045–1097
51. Kresse G, Furthmuller J (1996) Efficient iterative schemes for ab initio total-energy calculations using a plane-wave basis set. Phys Rev B 54:11169–11186
52. Civalleri B, D'Arco P, Orlando R et al (2001) Hartree-Fock geometry optimisation of periodic systems with the crystal code. Chem Phys Lett 348:131–138
53. Sayle DC (1999) The predicted 3-D atomistic structure of an interfacial screw-edge dislocation. J Mater Chem 9:2961–2964
54. Sayle DC, Parker SC (2003) Encapsulated oxide nanoparticles: the influence of the microstructure on associated impurities within a material. J Am Chem Soc 125:8581–8588
55. Catlow CRA, Bromley ST, Hamad S et al (2010) Modelling nano-clusters and nucleation. Phys Chem Chem Phys 12:786–811
56. Akoagi M, Matsui M (1991) Molecular dynamics simulation of the structural and physical properties of the four polymorphs of TiO_2. Mol Simul 6:239–247
57. Maphanga RR, Parker SC, Ngoepe PE (2009) Atomistic simulation of the surface structure of electrolytic manganese dioxide. Surf Sci 603:3184–3190
58. Ammundsen B, Islam MS, Jones DJ et al (1999) Local structure and defect chemistry of substituted lithium manganate spinels: X-ray absorption and computer simulation studies. J Power Sources 81–82:500–504
59. Pinna N, Niederberger M (2008) Surfactant-free nonaqueous synthesis of metal oxide nanostructures. Angew Chem Int Ed 47:5292–5304
60. Thomas JM, Midgley PA (2004) High-resolution transmission electron microscopy. The ultimate nanoanalytical technique. Chem Commun 11:1253–1267

61. Möbus G, Inkson BJ (2007) Nanoscale tomography in materials science. Mater Today 10:18–25
62. Lai M, Kulak AN, Law D et al (2007) Profiting from nature: macroporous copper with superior mechanical properties. Chem Commun 34:3547–3549
63. Hahn H (2003) Unique features and properties of nanostructured materials. Adv Eng Mater 5:277–284
64. Feng XD, Sayle DC, Wang ZL et al (2006) Converting ceria polyhedral nanoparticles into single-crystal nanospheres. Science 312:1504–1508
65. Berry CC (2005) Possible exploitation of magnetic nanoparticle-cell interaction for biomedical applications. J Mater Chem 15:543–547
66. Tarnuzzer RW, Colon J, Patil S et al (2005) Vacancy engineered ceria nanostructures for protection from radiation-induced cellular damage. Nano Lett 5:2573–2577
67. Sayle DC, Seal S, Wang Z et al (2008) Mapping nanostructure: a systematic enumeration of nanomaterials by assembling nanobuilding blocks at crystallographic positions. ACS Nano 2:1237–1251
68. Tang WP, Yang XJ, Liu ZH et al (2003) Preparation of -MnO_2 nanocrystal/acetylene black composites for lithium batteries. J Mater Chem 13:2989–2995
69. Luo JY, Zhang JJ, Xia YY (2006) Highly electrochemical reaction of lithium in the ordered mesoporosus -MnO_2. Chem Mater 18:5618–5623
70. Cheng FY, Zhao JZ, Song W et al (2006) Facile controlled synthesis of MnO_2 nanostructures of novel shapes and their application in batteries. Inorg Chem 45:2038–2044
71. Ren Y, Hardwick LJ, Bruce PG (2010) Lithium intercalation into mesoporous anatase with an ordered 3D pore structure. Angew Chem Int Ed 49:2570–2574
72. Chabre Y, Pannetier J (1995) Structural and electrochemical properties of the proton gamma-MnO_2 system. Prog Solid State Chem 23:1–130
73. Sayle TXT, Catlow CRA, Maphanga RR et al (2006) Evolving microstructure in MnO_2 using amorphization and recrystallization. J Cryst Growth 294:118–129
74. Bowden W, Grey CP, Hackney S et al (2006) Lithiation of ramsdellite-pyrolusite MnO_2: NMR, XRD, TEM and electrochemical investigation of the discharge mechanism. J Power Sources 153:265–273
75. Maphanga RR, Sayle DC, Sayle TXT et al (2011) Amorphization and recrystallization study of lithium insertion into manganese dioxide. Phys Chem Chem Phys 13:1307–1313
76. Woodley SM, Catlow CRA (2008) Crystal structure prediction from first principles. Nat Mater 7:937–946
77. Woodley SM (2004) Prediction of inorganic crystal framework structures – Part 2 – Using a genetic algorithm and a direct approach to exclusion zones. Phys Chem Chem Phys 6:1823–1829
78. Putz H, Schon JC, Jansen M (1998) Investigation of the energy landscape of Mg_2OF_2. Comput Mater Sci 11:309–322
79. Woodley SM, Battle PD, Gale JD et al (1999) The prediction of inorganic crystal structures using a genetic algorithm and energy minimisation. Phys Chem Chem Phys 1:2535–2542
80. Woodley SM (2007) Engineering microporous architectures: combining evolutionary algorithms with predefined exclusion zones. Phys Chem Chem Phys 9:1070–1077
81. Woodley SM, Battle PD, Gale JD et al (2004) Prediction of inorganic crystal framework structures. Phys Chem Chem Phys 6:1815–1822
82. Woodley SM, Catlow CRA (2009) Structure prediction of titania phases: implementation of Darwinian versus Lamarckian concepts in an evolutionary algorithm. Comput Mater Sci 45:84–95
83. Schön JC (2004) Enthalpy landscapes of the earth alkaline metal oxides. Z Anorg Allg Chem 630:2354–2366. doi:10.1002/zaac.200400289
84. Schön JC, Jansen M (1995) Determination of candidate structures for simple ionic compounds through cell optimisation. Comput Mater Sci 4:43–58

85. Roberts C, Johnston RL (2001) Investigation of the structures of MgO clusters using a genetic algorithm. Phys Chem Chem Phys 3:5024–5034
86. Al-Sunaidi AA, Sokol AA, Catlow CRA et al (2008) Structures of zinc oxide nanoclusters: as found by revolutionary algorithm techniques. J Phys Chem C 112:18860–18875
87. Woodley SM, Sokol AA, Catlow CRA (2004) Structure prediction of inorganic nanoparticles with predefined architecture using a genetic algorithm. Z Anorg Allg Chem 630:2343–2353
88. Hamad S, Woodley SM, Catlow CRA (2009) Experimental and computational studies of ZnS nanostructures. Mol Simul 35:1015–1032
89. Flikkema E, Bromley ST (2004) Dedicated global optimization search for ground state silica nanoclusters: $(SiO_2)_N$ ($N = 6-12$). J Phys Chem B 108:9638–9645
90. Woodley SM, Hamad S, Catlow CRA (2010) Exploration of multiple energy landscapes for zirconia nanoclusters. Phys Chem Chem Phys 12:8454–8465
91. Walsh A, Woodley SM (2010) Evolutionary structure prediction and electronic properties of indium oxide nanoclusters. Phys Chem Chem Phys 12:8446–8453
92. Francisco EA, Pendás AM, Blanco MA (2005) Global optimization of ionic $MgnF_2n$ ($n = 1–30$) clusters. J Chem Phys 123:234305–234308
93. Hamad S, Catlow CRA, Woodley SM et al (2005) Structure and stability of small TiO_2 nanoparticles. J Phys Chem B 109:15741–15748
94. Yu W, Freas RB (1990) Formation and fragmentation of gas-phase titanium/oxygen cluster positive ions. J Am Chem Soc 112:7126–7133
95. Mott NF, Littleton MJ (1938) Conduction in polar crystals. I. Electrolytic conduction in solid salts. Trans Faraday Soc 34:485–499
96. Lidiardi AB (1989) The Mott-Littleton method: an introductory survey. J Chem Soc Faraday Trans 85:341–349

Chapter 15
Mathematical Modelling and Simulation of Supercapacitors

Innocent S. Ike and Sunny Iyuke

List of Symbols

C	is the specific capacitance of an EDLC, F/cm^3 or F/kg
A	is the specific surface area of the electrode, cm^2
d	is the effective thickness of the electric double layer, cm
n	is the mean number of the electrons transferred in the redox reaction
F	is the Faraday constant
M	is the molar mass of the metal oxide
m	is the mass of the capacitor, kg
V	is the operating voltage window, V
R	is the series resistance of all components/layers in the capacitor, Ω
E	is the energy density of a capacitor, Wh/kg

I.S. Ike (✉)
Materials for Energy Research Group (MERG), University of the Witwatersrand, Private Bag 3, Johannesburg 2050, South Africa

Department of Science and Technology/National Research Foundation, DST/NRF Centre of Excellence in Strong Materials (CoE-SM), University of the Witwatersrand, Private Bag 3, Johannesburg 2050, South Africa

School of Chemical and Metallurgical Engineering, University of the Witwatersrand, Private Bag 3, Johannesburg 2050, South Africa

Department of Chemical Engineering, Federal University of Technology, Owerri, Imo State, Nigeria
e-mail: innocent.ike@students.wits.ac.za

S. Iyuke
School of Chemical and Metallurgical Engineering, University of the Witwatersrand, Private Bag 3, Johannesburg 2050, South Africa

K.I. Ozoemena, S. Chen (eds.), *Nanomaterials in Advanced Batteries and Supercapacitors*, Nanostructure Science and Technology,
DOI 10.1007/978-3-319-26082-2_15

P_{MAX}	is the optimum power density of a capacitor, kW/kg
R_{esr}	is the equivalent series resistance, Ω
I_{cell}	is the constant current for charging or discharging the capacitor, A
V_{cell}	is the capacitor cell voltage, V
C_{dl}	is the double-layer capacitance, F
V_{sc}	is the voltage across the C_{dl}, V
i_{sc}	is the current used to charge or discharge, A
$V_{cell}^{max}$is	the maximum cell voltage, V
V_{SC}^{O}	is the voltage across the double-layer capacitance at the end of charging, V
R_{lk}	is the parallel leakage resistance, Ω
i_{lk}	is the parallel leakage current, A
i_F	is the current of solvent electrochemical decomposition, A
t	is the time, s
x	is the position along a specific direction, cm
a	is electrode/electrolyte interface surface, cm^2/cm^3
k_x	is a factor accounting for electrode roughness
K_v	is a coefficient that is dependent on applied technology
C_{diff}	is the differential capacitance of an electrochemical capacitor, F
C_V	is the specific capacitance of the electrode with double layer, F/cm^3
L_D	is the thickness of the electrode with EDL, cm
L_R	is the thickness of the redox couple electrode, cm
C	is the concentration, mol/cm^3
D	is the coefficient of diffusivity, cm^2/s
C^0	is the initial concentration of mobile ion in the electrode, mol/cm^3
i_0	is the exchange current density, A/cm^2
U_{eq}	is the equilibrium potential of the capacitor, V
A	is the supercapacitor cross-sectional area, cm^2
L_S	is the separator's thickness, cm
t_C	is the cut-off time, s
m	is the mass of either capacitor, kg
m_S and m_H	are the mass of the symmetric and hybrid asymmetric supercapacitor, respectively
$\dot{q}$	is the heat generation rate
$\dot{q}_{irr}$ and $\dot{q}_{rev}$	are the irreversible and reversible heat production rates
$\dot{q}_{E,j}$	is the heat production rate due to the effects from Joule heating
$\dot{q}_{E,d}$	is the heat production rate due to the effects of ion diffusion
$\dot{q}_{E,s}$	is the heat production rate due to the effects of steric
$\dot{q}_{S,c}$	is the heat production rate due to the effects mixing heat with concentration gradient
$\dot{q}_{S,T}$	is the heat production rate due to the effects mixing heat with temperature gradient

Greek

η_D	is the overpotential in the double-layer electrode, V
ρ_D	is the density of the electrodes with EDL, g/cm^3
ρ_S	is the density of separator, g/cm^3
ρ_e	is the density of electrode, g/cm^3
ε_0	is the porosity of electrode
ε_S	is the porosity of separator
ρ_R	is the density of the redox electrode, g/cm^3
α_a and α_c	are the charge transfer coefficients of positive and negative electrodes, respectively
η_R	is the overpotential of the redox couple electrode, V
θ_s	is the surface state of charge
θ^0	is the initial state of charge
θ	is the state of charge
ϕ_1	is the potential in the solid matrix of the electrode, V
ϕ_2	is the potential in the electrolyte, V
ϕ_{21}	is the electrolyte potential at the interface of electrode/electrolyte, V
σ	is the conductivity of the electrode, $\Omega^{-1}cm^{-1}$
ε_r	is the relative permittivity of the medium in the electric double layer
ε^0	is the permittivity of vacuum
κ_s	is the conductivity of the ions in electrolyte, $\Omega^{-1}cm^{-1}$
κ	is the conductivity of the electrolyte, $\Omega^{-1}cm^{-1}$
κ_0	is the bulk conductivity of the electrolyte, $\Omega^{-1}cm^{-1}$

15.1 Introduction

Electrochemical capacitors (ECs) that store charges both electrostatically using reversible adsorptions of ions of the electrolyte onto active electrode materials (electric double layer) and through a faradaic reaction have greater energy density than traditional electrostatic capacitors. ECs have recently been the focus of serous researches because of their high prospects as energy stocking systems. Electric double-layer capacitors (EDLCs) have high accessible specific surface area SSA and stockpile electric charges using reversible electric double layer at the boundary of electrode/electrolyte [1–7]. EDLCs have high power rates, long life cycle, high efficiency, environmental friendliness and safety which are a merit over others. EDLC has higher energy density than convectional dielectric capacitors while still maintaining a reasonably high power density like the convectional capacitors [8].

EC has been given much attention due to the practical prospective in applications with growing power demands. Due to the reversibility of the electrochemical energy storage, supercapacitors are charged, discharged and recharged very fast with good long cycle life. Energy stockpiling is through electrostatic charge and not by electrochemical reactions like in the battery [9].

There are two basic operating mechanisms through which ECs store the energy: (i) EDLC which is as a result of the electric double layer (EDL) gathering around the surface of electrode at the boundary between the electrode and electrolyte [7, 10]. The specific capacitance of an EDLC, C, is determined by an expression

$$C = \frac{\varepsilon_r \varepsilon^0 A}{d} \tag{15.1}$$

where ε_r is the relative permittivity of medium in EDL, ε^0 is the permittivity of vacuum, A is the specific surface area of electrode and d is effective thickness of the EDL. (ii) Pseudo-capacitance is the result of the redox reaction between the electrode materials and the electrolyte [11–13]. This mechanism of charge accumulation at the electrode surface is by redox reaction. The theoretical pseudo-capacitance of metal oxide can be expressed as

$$C = \frac{nF}{MV} \tag{15.2}$$

where n is the mean number of the electrons transferred in the redox reaction, F is the Faraday constant, M is the molar mass of the metal oxide and V is the operating voltage window. The energy density (E) of a capacitor is estimated by the formula given below:

$$E = \frac{CV^2}{2m} \tag{15.3}$$

The optimum power density of a capacitor is estimated by the expression given below:

$$P_{MAX} = \frac{V^2}{4mR} \tag{15.4}$$

where R is the series resistance of all components/layers in the capacitor. These capacitances can simultaneously be created in one capacitor to produce what is known as hybrid capacitors. By achieving faradaic and electrostatic mechanisms of charge storage simultaneously in a capacitor, the hybrid capacitor can produce higher energy and power densities without compromising good cycle stability. Battery-type and composite hybrids are other kinds of hybrid capacitors that are obtainable. Different types of capacitors that merge different electrode materials as a composite material so as to improve the capacitor's performance have been developed.

The main performance parameters of capacitors are mass-/volume-specific capacitance, energy and power density, rate capability and cycling stability

[14]. It is desirable to reduce the equivalent series resistance (R) and increase the capacitance (C) and the operating voltage window (V) in order to increase the energy and power densities of a capacitor. The maximum operating voltage window (Vm) attainable for capacitor is solely dependent upon the type of electrolyte used, and this voltage is limited by the thermodynamic stability of the electrolyte. The maximum operating voltage window Vm for the electrochemical capacitors using aqueous electrolyte is 1 V. Some recent directions of investigation in capacitors are to introduce nonaqueous electrolytes with high optimum operating voltage window Vm. A thin, high surface-area EDL with high surface area (about 200m^2/g) and very little charge separation ($\delta \sim 10$Å) [11] is a necessity for a larger double-layer capacitance to be achieved.

Conway [11] indicated that double-layer capacitance is generated from the potential-dependence of the surface density of charges stored electrostatically at electrodes interfaces. Schematic diagram of the main composition and design of an EDLC and the movement of cations and anions in the device is shown in Fig. 15.1. Electric double-layer capacitors must utilise two layers, one working contrary the other during charge or discharge, as presented by Conway [11] and depicted in Fig. 15.2a. Contrary to the EDLCs, the pseudo-capacitors store energy via faradaic reaction that involves fast and reversible redox reactions among electrolyte and electrode materials at electrode surface [7], as shown in Fig. 15.2b. Pseudocapacitors can attain much higher capacitance compared with the EDL capacitance. Notwithstanding, small power density that results from the low electrical conductivity which mitigates quick movement of electron and the readily destroyed structure of the materials during electrolysis contradicts more realistic of these active materials to pseudo-capacitors. Hybrid capacitors achieve higher specific energy and power without compromising good life cycle by using both charge storage mechanisms simultaneously to store electric charges as shown in Fig. 15.2c.

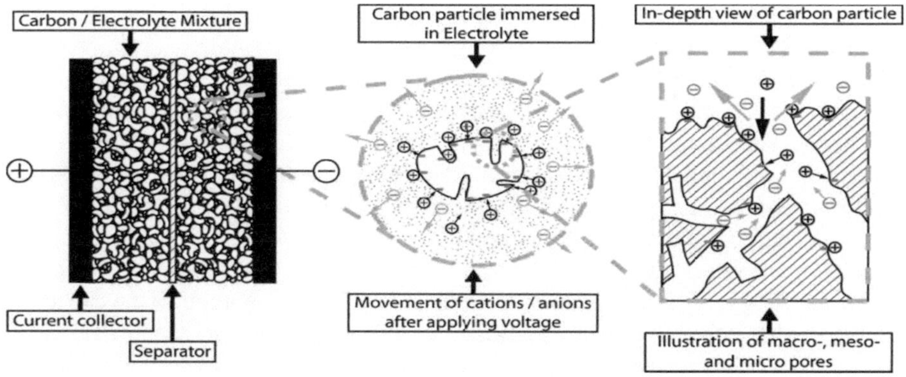

Fig. 15.1 Illustration of the basic components and design of an EDLC. *Arrows* indicate the direction of movement for the cations and anions [15] (Reprinted from Electrochimica Acta, Vol 55, F. Favier, Julia Kowal, Dirk Uwe Sauer, Modelling the effects of charge redistribution during self-discharge of supercapacitors, 7516–7523, Copyright (2010), with permission from Elsevier)

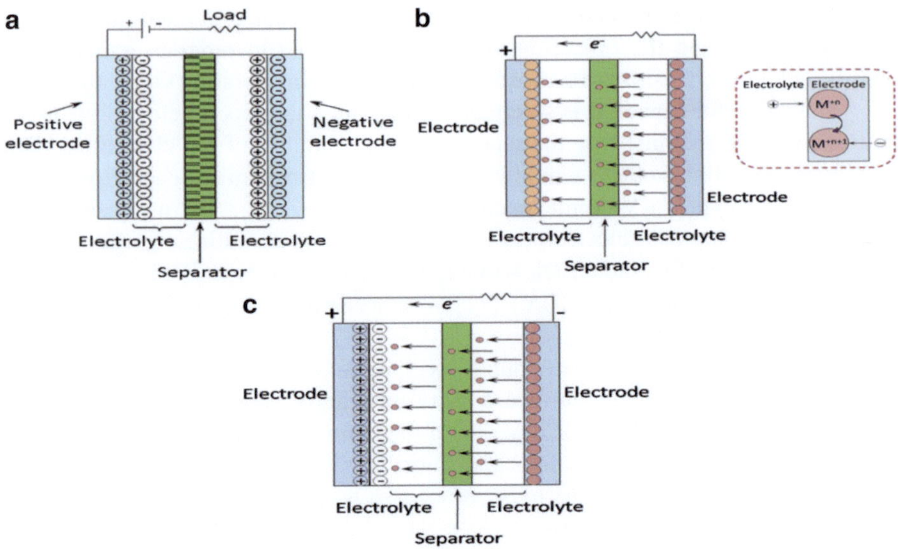

Fig. 15.2 Schematic representation of electrochemical capacitor's types: (**a**) EDLC type, (**b**) pseudo-capacitor type, (**c**) hybrid capacitor type (the figure is not to scale) [16] (Reprinted (adapted) from (J. Energy Eng. 139 (2) (2013) 72). Copyright (2011), with permission from American Society of Civil Engineers)

Electrochemical capacitors consist of two carbon material electrodes immersed into an electrolyte. The electrodes are in-between a membrane separator that is a nonelectric conductor and allows the movement of the charged ions but disallows electric contact as shown in Fig. 15.1. A good separator is electrochemically stable and of high porosity and high thermal and chemical inertia. Furthermore, the thickness of a separator needs to be as thin as possible and relatively inexpensive in terms of cost since a separator is not active in a capacitor. Its thinness in size is nevertheless constrained by the following conditions: (1) the electrical short circuit risk owing to free carbon particles that will likely provide a contact between the electrodes [17] and (2) the mechanical strength so as to permit the winding processes. The design components which can be examined more precisely using microstructural models like the one reported Verbrugge and Liub [18] are electrode thickness and its inner area, porosities and particle sizes, and the electrolyte and electrode materials employed. ECs experience self-heating which increases with an increase in the interfacial electric field and temperature during charge and discharge processes. The self-heating phenomenon increases the temperature of the capacitor by increasing irreversibly side reactions that in turn reduces the electrolyte ionic conductivity. This reduction in the electrolyte ionic conductivity results in further heat production that might result to heat runaway with time, if every other things are the same.

The features and parameters of processes transpiring in ECs are much interconnected and are conditional on the physical, electrical, electrochemical and crystallographic parameters of electrodes and electrolyte, and the entire cell and its components. Therefore, model of capacitor to estimate its energy, capacity, and other functional features is inconceivable without due consideration of device component variables. A great number of researchers ignore the physical, electrical, and structural features of electrode materials and its components during modelling and prediction of capacitor's parameters. This resulted to the discrepancy between theoretical results and experimental measurements of actual capacitors parameters [2]. Most of the available models were developed on the assumption that the energy storage mechanism is solely by electric double layer and overlooks the side reactions at the boundary [19]. Even the models which accounted for these chemical side reactions [20–22] were oversimplified by not incorporating the temperature consideration into the models. Furthermore, the effect of pores and its different sizes has not been handled qualitatively, and no specific model has accounted for all the necessary factors in a critical way. Thus, it is significant to recognise the setbacks in the available models and then derive the models based on first principles view to compensate for the discrepancies in the modelling.

A model developed from first principles view is able to forecast the internal temperature and interfacial electric field in real life so as to promote more realistic lifetime predictions, to promote the process of designing the ECs as compared to other electrical energy storage devices when aiming at maintaining a bus voltage around specific extents for arbitrary loading distributions and to aid in the development of porous structures that permits quicker rates of releasing stored charge and energy [3]. Available models for ECs and other energy storage devices are presently incapable of precisely predicting the internal temperature rise inside the capacitor, interfacial field and temporal voltage reaction for arbitrary charge–discharge current distributions.

In this overview, we examined up to date progress in the field of ECs modelling and by considering the two main types of electrochemical capacitors and their hybrid capacitors. We analysed their mechanisms with the intention to utilize their successful manners to acquire enhanced device performance. The available models of ECs were assessed and the merits and challenges of every type examined with the aim to utilize the comparative merits and compensate the challenges to achieve synergic and more practical ECs models with thermal model considering temperature distribution within cylindrical and rectangular shaped device.

The basic purposes are to arrive at mathematical description of features of symmetric and asymmetric ECs and parameters of the processes therein subject to all parameters of electrode materials, electrolyte, and the design of components and the entire cell; to assemble theoretical ground for determination and enhancement of parameters of different kinds and designs of symmetric and asymmetric ECs while considering the features of electrode materials, designs, and the electrodes and separator spatial structures; to produce recommendations for design and the optimization approach for ECs cell with a wide range of applications via modelling; to obtain optimal parameters of ECs subject to the type and value of all electrodes materials parameters, electrolyte parameters, separator parameters and the amount of applied charge and discharge currents.

15.2 Current Status of Supercapacitor Modelling and Simulation

Supercapacitors theoretical models begin from the original Helmholtz model to mean-field continuum models, the surface curvature-based post-Helmholtz models and the current atomistic simulations. Practical models of ECs can be assembled by use of high stage of progress in classical and quantum molecular dynamics techniques along with the parallel high performance computing. These classes of ECs have been proposed by authors/researchers: empirical models, dissipative transmission line models, continuum models (Poisson-Nernst-Planck equations), atomistic models (Monte Carlo, molecular dynamics), Quantum models (ab initio quantum chemistry and Density Functional Theory, DFT) and simplified analytical models. Each type of model emerged for different objectives and has different merits and challenges associated with it.

15.2.1 Empirical Models

This equivalent circuit model is frequently called a three-branch model and assumes that the long-term branch capacitance only considers the difference in charge for a maximum time of 30 min, while the immediate branch depends on the voltage. The earliest observed models which considered changes of capacitance over time were produced by Zubieta and Bonert [23]. Empirical models are better in estimating accurately the EC's electrical behaviour than ordinary *RC* circuit, and also it allows whosoever uses it to readily integrate the model into his/her system simulations. The empirical models could also be utilised in characterisation of self-discharge and/or leakage current parameters [24].

Ban et al [8] reported a straightforward mathematical models with a voltage-independent parallel leakage process and electrochemical disintegration to describe ECs conducts during charging and discharging in order to present a basic understanding of supercapacitors behaviors during charging and discharging. Experiment data were simulated to acquire the suitable parameters values such as specific capacitance and equivalent series resistance from where specific energy and power can be estimated via models. The acquired simulated parameter values when parallel leakage and solvent disintegration were considered could be employed in prediction of supercapacitor's self-discharge, shelf-life, performance/efficiency and the practicable cell voltage maximum. The models are thus effective in evaluation and identification of supercapacitor cells and comprehending supercapacitors charging or discharging behavior. Also, the models could be effective in acquiring supercapacitors important parameters like equivalent series resistance and capacitance based on the documented outlines/plots of charging–discharging.

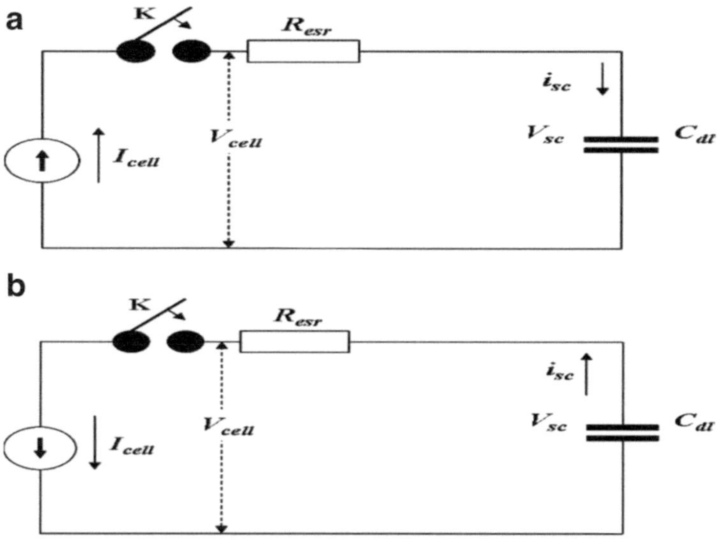

Fig. 15.3 Proposed equivalent circuits of a double-layer capacitor at a constant current charging (**a**) and discharging (**b**) in the absence of parallel leakage phenomena and chemical disintegration of the solvent. K, electric switcher; R_{esr}, equivalent series resistance; I_{cell}, constant current for charging or discharging the capacitor; V_{cell}, capacitor cell voltage; C_{dl}, double-layer capacitance; V_{sc}, voltage across the C_{dl}; and i_{sc}, current used to charge or discharge respectively [8] (Reprinted from Electrochimica Acta, Vol. 90, Shuai Ban, Jiujun Zhang, Lei Zhang, Ken Tsay, Datong Song, Xinfu Zou, Charging and discharging electrochemical supercapacitors in the presence of both parallel leakage process and electrochemical decomposition of solvent, 542–549, Copyright (2013), with permission from Elsevier)

A schematic diagram of electrical equivalent circuit of double layer capacitors during charging/discharging at a constant cell current in without parallel leakage and chemical disintegration of solvent is shown in Fig. 15.3. The current used in charging or discharging the double-layer capacitance (i_{sc}) is equivalent to a constant charging current (Icell) determined by equation 5 assuming that the voltage across the capacitor (Vsc) is zero before charging process commence:

$$I_{cell} = i_{SC} \tag{15.5}$$

The voltage across the capacitor when charging process has commenced by putting on the switch is given as Eq. 15.6 below:

$$V_{SC} = \frac{1}{C_{dl}} \int_{t=0}^{t=t} i_{SC} \, dt = \frac{1}{C_{dl}} \int_{t=0}^{t=t} I_{cell} \, dt = \frac{I_{cell}}{C_{dl}} \, t \tag{15.6}$$

The supercapacitor cell voltage during charging process is therefore given as Eq. 15.7 below:

$$V_{cell} = I_{cell}\ R_{esr} + V_{SC} = I_{cell}\ R_{esr} + \frac{I_{cell}}{C_{dl}}\ t \tag{15.7}$$

When the ECs are charged to a designed cell voltage of $V_{cell}^{\max}$, a constant current (Icell) discharging process commenced right away, as depicted in Fig. 15.3b. Note that at this desired cell voltage (or maximum cell voltage) of $V_{cell}^{\max}$, the voltage across the double-layer capacitance (C_{dl}), V_{SC}^{O}, should be given by Eq. 15.8, and in the instance of discharging, this V_{SC}^{O} will be the load to discharge the capacitors, and the cell voltage (V_{cell}) during discharging process can be expressed as Eq. 15.9 shown below:

$$V_{SC}^{O} = V_{cell}^{\max} - I_{cell}\ R_{esr} \tag{15.8}$$

$$V_{cell} = V_{SC}^{O} - I_{cell}\ R_{esr} - \frac{I_{cell}}{C_{dl}}\ t \tag{15.9}$$

Parameters like double-layer capacitance (C_{dl}), voltage across the double-layer capacitance at the end of charging (V_{SC}^{O}) and equivalent series resistance (R_{esr}) are simulated from Eqs. 15.8 and 15.9, respectively, using both the charge and discharge curves.

Figure 15.4 shows the schematic diagram of electrical equivalent circuit of double-layer capacitor for constant current charging–discharging with involvement of parallel leakage phenomena and chemical disintegration of the solvent. A constant charging current (I_{cell}) integration equation for the charge and discharge processes is given as Eqs. 15.10 and 15.11, respectively:

$$i_{SC} + \exp\left(\frac{-\Delta V^{O}}{b_p + b_n}\right)\ \exp\left(\frac{1}{(b_p + b_n)C_{dl}}\int_{t=0}^{t=t} i_{SC}\ dt\right) + \frac{1}{R_{lk}C_{dl}}\int_{t=0}^{t=t} i_{SC}\ dt\ -\ I_{cell} = 0 \tag{15.10}$$

$$i_{SC} - \exp\left(\frac{V_{SC}^{O} - \Delta V^{O}}{b_p + b_n}\right)\ \exp\left(-\frac{1}{(b_p + b_n)C_{dl}}\int_{t=0}^{t=t} i_{SC}\ dt\right) - \frac{1}{R_{lk}C_{dl}}\int_{t=0}^{t=t} i_{SC}\ dt\ -\ I_{cell} = 0 \tag{15.11}$$

The cell voltages (V_{cell}) for the charge and discharge processes in Fig. 15.4 are expressed, respectively, as Eqs. 15.12 and 15.13 shown below:

$$V_{cell} = I_{cell}\ R_{esr} + V_{SC} = I_{cell}\ R_{esr} + \frac{1}{C_{dl}}\int_{t=0}^{t=t} i_{SC}\ dt \tag{15.12}$$

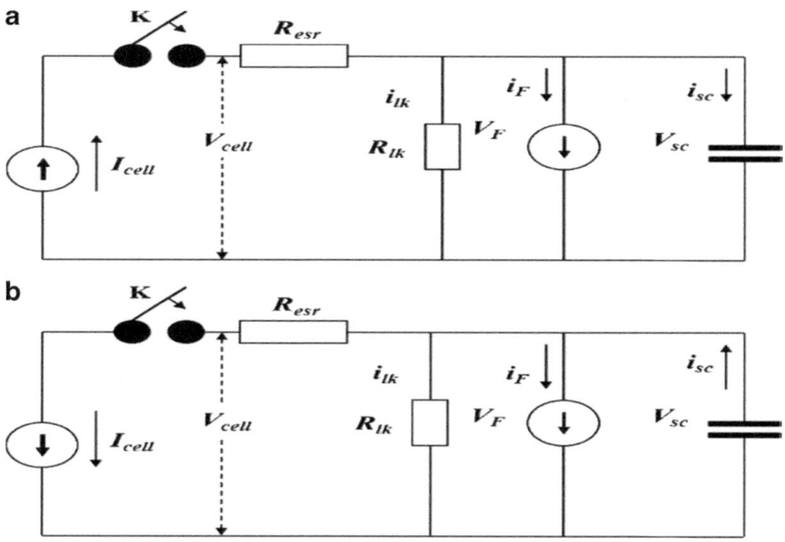

Fig. 15.4 Proposed equivalent circuits for a double-layer capacitor at a constant current charging (**a**) and discharging (**b**) in the presence of electrochemical decomposition of the solvent. K, electric switcher; R_{esr}, equivalent series resistance; R_{lk}, parallel leakage resistance; i_{lk}, parallel leakage current; I_{cell}, constant current for charging or discharging; V_{cell}, capacitor cell voltage; i_F, current of solvent electrochemical decomposition; C_{dl}, double-layer capacitance; i_{sc}, current used to charge or discharge the double-layer capacitance; V_F, electrode potential of the solvent electrochemical decomposition; and V_{sc} (= V_F), voltage across the double-layer capacitance, respectively [8] (Reprinted from Electrochimica Acta, Vol. 90, Shuai Ban, Jiujun Zhang, Lei Zhang, Ken Tsay, Datong Song, Xinfu Zou, Charging and discharging electrochemical supercapacitors in the presence of both parallel leakage process and electrochemical decomposition of solvent, 542–549, Copyright (2013), with permission from Elsevier)

$$V_{cell} = -I_{cell}\ R_{esr} + V_{SC} = V_{SC}^{O} - I_{cell}\ R_{esr} - \frac{1}{C_{dl}} \int_{t=0}^{t=t} i_{SC}\ dt \tag{15.13}$$

The models developed by Ban et al [8] could be used to explain charge and discharge characteristics of capacitors involving voltage-independent parallel leakage process and solvent disintegration as proposed by its capability to fit data from experiment to good extent. However, little disagreement which might have emanated from experimental minor factors like gaseous evolution influence, decomposition of current collectors, crystallization of electrolyte inside separator, assembly standard at the discharge plots in particular.

The major challenges of these models are their small quality for the conditions of operation which are quite different from those utilised in obtaining the parameters [25] and to predict long time ranges when heat produced within the system alters the capacitor electrical behaviours. Since these models were not developed from the capacitor basic physics, they are able to estimate temperature growth within the system and can only be correct if the temperature rise is negligible. On this understanding, it is unrealistic to utilise the models in accurate prediction of a real-life electrochemical capacitor.

15.2.2 Dissipation Transmission Line Models

The electrode pore structure mitigates the ionic motions to regions situated deep into a pore and is the main cause of nonlinear growth of terminal voltage. It is therefore beneficial to get a model which explains the movement and steady-state reaction of electrolyte portion since its mass portion is presented to first order by a simple resistor. The earliest porous electrode model that handled EDLC with a porous electrode wall as a transmission line, a shared electric double-layer capacitor and electrolyte resistance was developed by de Levie [26].

Influence of ions reduction on porous electrodes charging rate requires additional careful examination for various reasons:

(1) It will enhance understanding of EDLC's maximum performance in relation with the device parameters. For instance, ions reduction might happen in electrodes and the separator, and tuning geometry may enhance power and linearity.
(2) The investigation of this process presents basic questions concerning the interactions between the adsorption and ions transfer in nanoporous conductors, and the effect of varying ion mobilities on the effects.

The de Levie model treats the pore as a distributed *RC* circuit (Fig. 15.5) and states that voltage and current vary with respect to time *t* and position *x* along the pore through the following differential Eqs. 15.14 and 15.15:

$$\frac{dV}{dx} = -IR \tag{15.14}$$

$$\frac{dI}{dx} = -C\frac{dV}{dt} \tag{15.15}$$

where *C* and *R* are capacitance per unit length and resistance per unit length, respectively. De Levie assumed that the concentration (and therefore solution resistance) throughout the entire pore is constant, but the different solution environment outside of the pore must also be considered. So as to estimate the voltage outside the pore, the bulk solution between the pore entrance mark and the midpoint

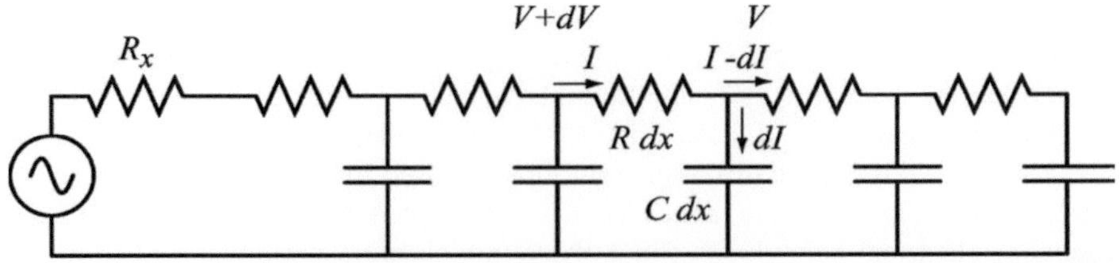

Fig. 15.5 De Levie's transmission line equivalent circuit for an electrolyte-filled conducting pore [27] (Reprinted from Journal of The Electrochem. Soc., 157 (8) A912-A918 (2010). Reproduced, by permission of The Electrochemical Society)

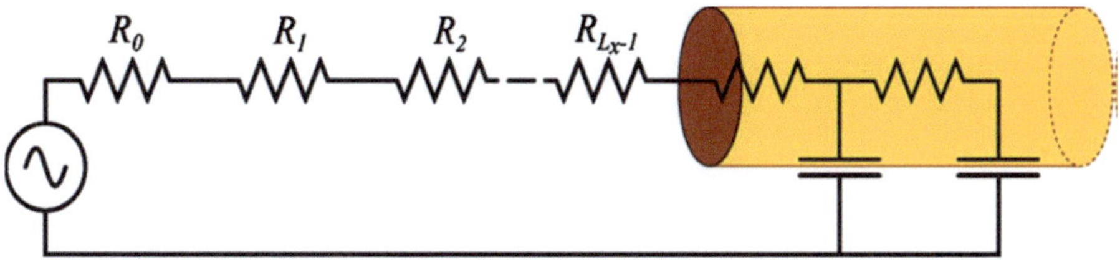

Fig. 15.6 (*Colour online*) Extension of circuit model outside pore [27] (Reprinted with permission from Journal of The Electrochem. Soc., 157 (8) A912-A918 (2010). Copyright 2010, The Electrochemical Society)

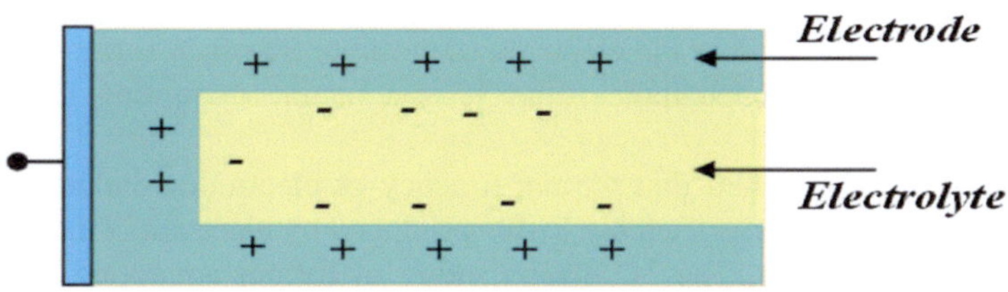

Fig. 15.7 One-dimensional pore [28]. ©(2010) IEEE (Reprinted, with permission, from [N. Bertrand, J. Sabatier, O. Briat, J.-M. Vinassa, IEEE Trans. Ind. Electron. 57 (12) (2010) 3991])

between the working and counter electrodes' mark ought to be handled as a series of discrete resistors with each resistance dependent on the salt concentration at that specific position (Fig. 15.6).

The administering expressions for electrochemical capacitor shown in this section consider an electric double-layer capacitor, a rough portion of the electrode/electrolyte interface and without voltage and temperature variations of the parameters. The capacitor is explained by a one-dimensional pore, 1D pore, as shown in Fig. 15.7.

At the boundary of electrode and electrolyte, the current densities in electrode and electrolyte are given by i_s and i_e, respectively, as given by Eqs. 15.16 and 15.17 shown below:

$$i_s = -\sigma \ \frac{\partial \phi_1}{\partial x} \tag{15.16}$$

$$i_e = -k \ \frac{\partial \phi_2}{\partial x} \tag{15.17}$$

$$\frac{\partial i_e}{\partial x} = - \ a \ . \ C_{dl} \frac{\partial (\phi_1 - \phi_{21})}{\partial t} - k_x . \ (\phi_{21} - \phi_2). \tag{15.18}$$

In the equations above, ϕ_1 and ϕ_2 are potential in solid matrix of the electrode and electrolyte, respectively; ϕ_{21} is the electrolyte potential at the boundary between the electrode and electrolyte; σ and k represent the conductivities of electrode and

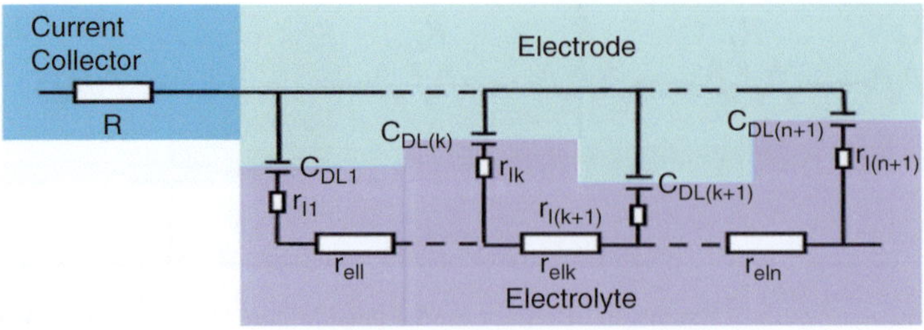

Fig. 15.8 Half porous electrode model discretisation [28]. ©(2010) IEEE (Reprinted, with permission, from [N. Bertrand, J. Sabatier, O. Briat, J.-M. Vinassa, IEEE Trans. Ind. Electron. 57 (12) (2010) 3991])

electrolyte, respectively; a is the electrode/electrolyte interface surface; C_{dl} represents the double-layer capacitance; and k_x is a factor accounting for electrode roughness.

This model has revealed that the dependency of electrode conductivity on its functioning is small, and as such the losses produced by electrode's internal ohmic resistance are negligible. The 1D discretisation of former equations, in n layers, resulted to an electrical circuit. Wires and separator effect which was neglected in the past has now been considered by an RS resistance in series with the pore model, as shown in Fig. 15.8. The double-layer capacitor $C_{\mathrm{DL}}(k)$ is instigated by the potential at the boundary electrode and electrolyte, while the resistive components $r_{\mathrm{el}}(k)$ resulted from the conductivity of electrolyte. The rough portion of electrode/electrolyte interface produces extra electrolyte resistive components $r_I(k)$ which is in series with DLC.

The EDLC is a physical component which has a demanded capacitance, a parasitic inductance as a result of the geometry, resistances owing to internal resistances of electrons and ions and resistance as a result of leakages current between its electrodes.

The equivalent series resistance ESR which is the summation of the series and parallel and resistances R_s and R_p, respectively, is accountable for the electrical losses that produce the supercapacitor heatings.

The main transmission line model utilised in description of frequency characteristic of the capacitance and series resistance, which ignored voltage and temperature variations subject to capacitance and series resistance, has previously developed by de Levie [26]. A simple model that accounts for extra linear variation of the capacitance with tension has also been suggested by Zubieta et al. [23]. Dougal et al. [29] and Belhachemi et al. [30] have used comparable models whose capacitance consists of a steady portion C_o and a linear voltage variant one $C_V = K_V \ . \ U$, where $K_v$is a coefficient that is dependent on applied technology. The overall capacitance at the voltage U is given by the expression below:

$$C = C_0 + C_V \qquad (15.19)$$

The relation interaction between current and voltage has to be developed from already established expression relating the current to charge:

$$i = \frac{dQ}{dt} \tag{15.20}$$

Substituting for charge equation subject to U and C and accounting for indirect dependence of C on time, it is not difficult to prove that equation for current is as shown below:

$$i(t) = (C_0 \ + 2 \ . \ K_V \ . \ U) \ . \ \frac{dU}{dt} \tag{15.21}$$

Differential capacitance of an electrochemical capacitor can be defined as the below given expression compared to classical expression:

$$C_{diff} = C_0 + 2 \ . \ K_V \ . \ U \tag{15.22}$$

The stored energy is as presented in the expression given below:

$$E(U) = \left(C_0 \ + \ \frac{4}{3} K \ |U|\right) \ * \ \frac{U^2}{2} \tag{15.23}$$

Conclusively, the current and energy for a particular voltage are quite more than the expectations from classical equations in the instance of steady capacitance.

Values of capacitance and resistance vary with the frequency spectrum, and the performance obviously can be estimated with an impedance spectrum analyser [31]. In order to consider the voltage, temperature and frequency reliabilities, a simple equivalent circuit was derived by Rafik et al. (Fig. 15.9) by combining the de

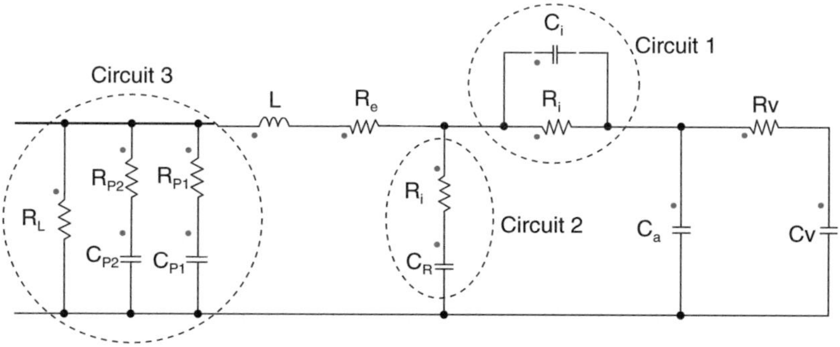

Fig. 15.9 Equivalent circuit for capacitance and series resistance dependences as a function of frequency, voltage and temperature [24] (Reprinted from Journal of Power Sources, Vol. 165, Issue 2, Bor Yann Liaw, H. Gualous, R. Gallay, A. Crausaz, A. Berthon, Frequency, thermal and voltage supercapacitor characterization and modelling, Pages 928–934, Copyright (2007), with permission from Elsevier)

Levie frequency model and Zubieta voltage model and inclusion of more function to account temperature dependence.

The equivalent scheme consists of a fixed capacitance C_o in parallel with a voltage linearly growing one C_v. The series resistance and capacitance–voltage variations are sensitive in the small frequency domain that is considered with a simple R_vC_v circuit. To obtain the variation of frequency, resistance R_T and capacitance C_R which behaviour is one of the low-pass filters with cut-off frequency $\tau = R_TC_R$ have been introduced in circuit 2. "Circuit 1", which involves two RC parallel branches with various time constants each and also a parallel R_F resistance that produces the long-time leakage current, has been considered for the heat variation, while circuit 3 is required to complete the modelling with respect to leakage current and charge redistribution. Obviously, the available capacitance is at its maximum for small frequency, as shown by more time allowed for electrolyte ions to get to the surface deep in pores of the electrodes. Only apparent electrode surface is available for ions at higher frequency, hence a very low capacitance.

The series resistance consists of electronic part which is due to internal ohmic resistance in the electrodes and the ionic which emerges from the mobility of ions in electrolyte.

The major merit of the dissipative transmission line model is the capacity to produce direct well-acceptable access among pore structures and response time in the field of ECs. It also gives a first-order estimation of the exponential growth/reduction in voltage level while considering a steady current of charge and discharge cycle and is built upon physical structure of the interface instead of merely matching the experimental measurements by integrating passive circuit element. Nevertheless, this approach accounts for small area of interfacial dynamics and is basically utilised in specific aims like improvement of electrode synthesis [32], study of self-discharge characteristics [33] and prediction of electrode surface impedance [34].

15.2.3 Continuum Models (Poisson–Nernst–Planck Equations)

The very precise modelling method presently available utilised the Poisson–Nernst–Planck (PNP) electrodiffusion theory to be the boundary of electrode and electrolyte, while the physicochemical parameters became dependent on the local surroundings contrary to being constant because the value of the electric field at the boundary is quite big. The PNP theory can estimate electrolyte properties correctly only if the physicochemical parameters are independent of the local surroundings. Figure 15.11 represents the schematic model of EDL structures while considering an electrode that is positively charged for an instance. Helmholtz's [35] initial proposal of an EDL model that consists of a simple separation of charges at electrode/electrolyte interface like ordinary parallel capacitor to estimate the

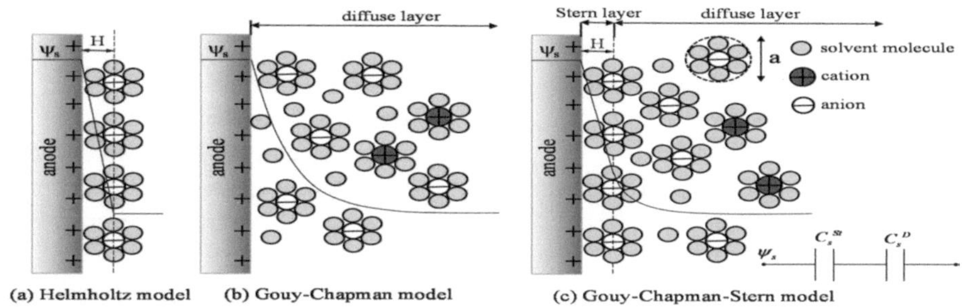

Fig. 15.10 Schematics of the electric double-layer structure showing the arrangement of solvated ions near the electrode/electrolyte interface in the Stern layer and the diffuse layer. (**a**) Helmholtz model, (**b**) Gouy–Chapman model and (c) Gouy–Chapman–Stern model [38] (Reprinted (adapted) with permission from H. Wang, L. Pilon, J. Phys. Chem. C 115 (2011) 16711. Copyright 2011 American Chemical Society)

capacitance ordinary parallel capacitor to estimate the capacitance without considering the surface potential and electrolyte concentration [36] was as represented in Fig. 15.10a. The Helmholtz model structure is like the structure of conventional capacitors using two planar electrodes separated by a dielectric.

Gouy and Chapman [36] modified the Helmholtz model to Gouy–Chapman model in 1910 and 1913, respectively, by considering the issue that concentration of ions has to be continuous in electrolyte solution. Ions with an opposed to the sign of electrons on the electrode are distributed in the area of thickness higher than Helmholtz layer H rather than being packed close at the electrode surface as demonstrated in Fig. 15.10b. This model considered the impacts of the electrode potential and ionic concentration on the local electrolyte concentration and potential field by utilising Boltzmann distribution function [35]. Chapman developed and solved the Poisson–Boltzmann equation steady state wisely in order to estimate electric potential in the diffuse region. This theory estimates wrongly huge ionic concentrations of potentials of minute portion of 1 V even with very dilute solutions and as such can hardly be utilised to model real ECs with potential of 1V at the surface and electrolyte concentration of 1 mol/L, since the ions were taken as point charges where their sizes are specific in real life [37].

Stern in 1924 integrated the Helmholtz and Gouy–Chapman models to obtain the Gouy–Chapman–Stern EDL theory that is generally utilised in the evolution of EDLCs; this model clearly explains the ion concentration in the Stern and diffuse layer as could be clearly seen in Fig. 15.10c. Grahame [39] modified the Stern model by noting that ions undergo adsorption at the surface of the electrode results to double layer of different thicknesses in the Stern layer region [11]. This model derived by Stern and Grahame is often called the Stern model [40]. Hainan Wang and Laurent Pilon produced three-dimensional (3D) model from continuum theory for characterising EDLCs with organised electrode structures with meso-pores and critically considering the three-dimensional electrode morphology, definite size of ions, Stern and diffuse region and variation of the dielectric permittivity of electrolyte with local electric field [41]. Pairs of boundary conditions to estimate the Stern region without characterising it within electrolyte environment were developed, and the model was utilised in estimating the electrode configuration of CP204-S15 meso-pores EDLC that was previously synthesised and described by Woo et al. [41]. This created the possibility of simulating EDLCs with three-dimensional organised electrode structures and correctly considering the Stern and diffuse layers, definite size of ions and the variation of electrolyte permittivity on the local electric field [42].

15.2.4 Atomistic Models (Monte Carlo Molecular Dynamics)

Molecular modelling is very significant for optimal design of EDLCs, and important merits of this method of modelling are its capability to estimate processes that is impossible to be noticed straight away experimentally. For instance, the orientation of ions in electrolyte [43] and alteration of electrode form as a result of polarisation [11] are essential conditions that affect electrode capacitance and are hardly determined from experiment. The interwoven effect of these variables can be comprehensive and utilised in design optimisation of new EDLCs through exact simulations and physical modelling. Because of the complex interactions among electrode and electrolyte properties, together with its effect on the energy and power densities at large, active optimisation of EDLC design remains a serious difficulty. The electrolyte to be used simultaneously with specific pore geometries and sizes is an illustration of design considerations [37, 39, 40]. When the ions have various kinetic sizes during undercharge and discharge, then models need to be derived for these specific electrolytes, and optimum size of the pores in each electrode shall be different [41, 42, 44]. Electric conductivity of room temperature ionic liquid (RTIL) electrolytes can be increased by introducing organic solvents while using optimal concentration of electrolyte [45, 46], but another issue on optimisation here becomes the influence of pore size on power density [47]. The energy density of EDLCs can be improved by using pores of diameter that is approximately equal to the diameters of the ions [7, 37]; power density will be restricted by high diffusion resistance created by small pores.

Molecular modelling will obviously enhance the comprehension of the equilibrium and dynamic processes taking place in EDLC on atomic state, though its accuracy greatly depends on how authentic the force fields applied in explaining molecular behaviours in the fluid phase and force fields of the electrode modelled. Monte Carlo (MC) simulation that relies on statistical mechanics via significant choosing of phase space of a molecular system and molecular dynamics (MD) is the highly acceptable molecular simulation method [28, 48]. MC simulation is restricted to estimation of equilibrium properties since it is based on statistical mechanics, while MD simulations evaluate Newton's equations of motion for several molecular systems which properties are averaged over a small simulated time to estimate system's properties. Another merit of MD simulations is its ability to estimate dynamic properties like diffusion of ions.

Choosing the electrolyte and solvent models and the electrode model (polarisable electrodes) with steady potential when charges on electrode atoms vary during the prediction is an essential step that can strongly affect the correctness of EDLCs' molecular simulation results aimed at the improvement of molecular models. Introducing electrode polarisability which improves the surface charge within the regions through induced charges, which finally grows the field and capacitance [49], is cogent both for the numerical distinction generated in predicted results and correct representation of basic physical aspect affecting capacitance and dynamics that occur in EDLCs. Erroneous small capacitance value as compared to experimental results will obviously be obtained if polarisation of the electrode is ignored.

An important field of research in molecular modelling that attracts more attention recently and obviously be a guide in the characterisation of novel EDLCs with improved energy density is the simulation of surface functionalities and nanosurface tuning simultaneously with optimum choice of RTIL electrolyte on the basis of size.

Simulations of RTILs at planar surfaces have revealed important understanding of the EDL structure and assisted in describing the experimental difference noticed in differential capacitance, though this knowledge can hardly be utilised when designing EDLCs with high energy density because this type of electrodes has small surface areas and is not suitable. Porous electrodes are highly preferred for providing high surface area, but are certainly controlled by several physics from simulation approach.

Classical and quantum molecular dynamics simulations can produce the most exact explanation of the EDLs. The state of art of classical and quantum molecular dynamics and utilization of parallel high performance computations allow assembling of practical ECs models. The consistent classical examination of EDL at the interface of metal and ionic liquid was conducted in reference [50]. The molecular dynamics simulation of solid electrolytes employing carbon nanotubes (CNT) was performed by Chaban et al. [43]. Such classical model creates a platform for acceptable description of electrolyte ionic subsystem but failed to give information on electron subsystem of the material of the electron–hole conductor of the electrode.

Molecular dynamics (MD) simulation presents the most basic and modifiable avenue for analysing molecular behaviours and has been greatly utilised in modelling electro-osmotic flow [44–47, 51, 52] as well as treating higher charge densities which are of great significance in EDLCs [53–55]. However, this is capital intensive as regards computational cost and as such is unrealistic to handle time and length scales obtainable in several applications. Calculating long-range Coulombic relationship and using boundary conditions generate a great challenge, thus an alternative approach is highly required.

The maximum specific capacity of EDL on pure defect-free graphite electrode surface that accurately fitted with all available experimental data was determined by employing results of quantum molecular dynamics computations of Lankin et al. [56]. The explanation of the dense region of the EDL is not straightforward because potentials distributions in the EDL cannot be explained by the mean charge densities. In this instance, connections among the positions of charged particles have to be accounted for. On this understanding, the assemblage of a consistent model of the dense region of the EDL is quite strenuous.

Molecular dynamics findings of temperature effect on capacitance showed a positive capacitance effect on temperature for OLC-based electrochemical capacitors and low effect of capacitance on temperature for CNT-based electrochemical capacitors, in good agreement with empirical findings. Molecular revelations into RTIL-based electrochemical capacitors surveyed towards this direction will enhance and improve the design of present-day energy storage device and electrochemical capacitors in particular.

In consideration of mean-field theory and Monte Carlo (MC) simulations, Kondrat et al. [57, 58] credited the growth of capacitance to the "superionic state" of ions residing in the pores. Molecular dynamics (MD) simulations have generated exactly the same capacitance–voltage profile as those obtained by experiments with solvent electrolytes and RTILs considering the van der Waals relationships and complex ionic form [59, 60].

Also DFT investigation revealed a bell-shaped curve owing to the co-ion elimination and introduction of counterion without capacitance growth [61]. This observation in DFT computations is likely as a result of the simplified RTIL ions in the constrained local model as well as the chosen size of the pore. Huang et al. [62] presented the exohedral electrochemical capacitor model for spherical electrodes and cylindrical electrodes so as to give reasons for relating capacitance and the size of either OLC or CNT. The origin of the curvature effect that the EDL capacitance grows with a decay in the size of electrodes exhibited by the theoretical models is still not completely understood.

Limited investigations have been channelled towards the dependency of the differential capacitance variations on the potentials for electrochemical capacitors with OLC-/CNT-based spherical electrodes and RTIL electrolytes, while much attention has been given proving that different C-V profiles were noticed in capacitors with planar electrodes [63–67].

RTIL-based capacitor potentials operating in a large temperature window enhance its applicability during strong conditions owing to its excellent thermal stability [62, 68], and several research groups have utilised the capacitor's temperature dependence on the capacitance. Some theoretical and experimental research showed that capacitance grows with temperature [62, 69], some showed that capacitance decays with a growth in temperature [70, 71], and others showed even a complex capacitance–temperature dependency. Lin et al. [72] of recent demonstrated that the capacitance of electrochemical capacitors using OLC electrodes and RTIL electrolytes grows with an increase in temperature, while that of a vertically aligned CNT electrodes with RTIL electrolytes was approximately independent of the temperature. MD simulations were utilised in modelling electrochemical capacitors with RTIL electrolyte and OLC/CNT electrodes to be dependent on temperature in order to obtain molecular understanding into the temperature variation of capacitance [73, 74].

A Monte Carlo investigation revealed that a change in the position of the counterions on the charged surface and the packing of ions in particular demonstrate a strong responsibility on the temperature reliability on the EDL close to the charged pore wall [75].

MD simulations have a peculiar instrument to produce total molecular revelations into the behaviour of RTIL-based electrochemical capacitor capacitance using several kinds of carbon electrodes. The capacitance of porous capacitors was noticed to agree well with a particular form of RTILs, the general characters of pores and the applied potential.

Upon the improvement of capacitors with improved energy density, enhancing the power density too is another critical part of electrochemical capacitors [76] which is greatly controlled by electrolyte ion diffusion rate. Whereas numerous studies have been carried out on the properties of capacitors that are in equilibrium state, very few attempts have been channelled to finding the dynamic process of charge and discharge of capacitors using modelling approach. There is a great need for improved theoretical and experimental modelling since classical explanation of EDL formation for RTILs electrolytes totally failed, owing to very complex molecular form and unavailability of a solvent. Since how the differential capacitance of CNT-based capacitors depends on the potential is still confusing till now, MD simulation can offer much help in the development of plans for choosing optimum size of SWCNT electrodes for supercapacitors. Most of experimental research noticed capacitance growth with temperature rise [62, 69, 77], while some other research revealed capacitance decay with temperature rise [70].

Majority of the research forecast that the capacitance decreases with an increase in temperature or at least a complex behaviour between capacitance and temperature [68, 71, 75, 78] as was shown in MD simulations by Vatamanu et al. [71], while Monte Carlo simulations by Boda and Henderson [78] revealed a bell-shaped profile for capacitance–temperature relationships.

Recent research with vertically aligned CNT electrodes and RTIL electrolytes showed that capacitance was almost invariant with the temperature [72]. Thus, deeper studies are required in order to confirm a concrete explanation of how the temperature depends on capacitance and also the basic mechanisms that produce this wonderful process.

Guang Feng et al. [73] noticed that the capacitance of EDL close to the CNT electrode is almost independent of temperature within the range of 260 and 400 K, which agreed well with new research results by Lin et al. [72], though they noted that the capacitor power and charge and discharge rates that severely rely on ion mobility instead of packing cannot be evaluated from the obtained temperature reliance on capacitance.

Using EDLCs as excellent electrical energy storage system is gaining much attention, but finding out new electrolyte and electrode materials with high capacitive values and optimising the entire device are important areas that molecular modelling can greatly assist. Molecular modelling has recently helped in a great way in the explanation of the process of charge dynamics and storage in EDLCs as regards to results that are opposing long perceptions of the EDL. Molecular simulations held that structure of RTILs EDL at planar electrode surfaces is associated with big potential overshadow and several regions of ions with alternate charge [79]. Nano-pores have more charge storage capability compared to planar electrodes since the overall charge on electrode is equilibrated by a single portion of ions not being overshadowed [80]. Optimal energy density is dependent on pore diameter and the optimum pore diameter grows with voltage [81]. This electrode polarisation which improves surface charge through induced charges that leads to growth in electric field strength and capacitance is a crucial feature that must be integrated into molecular models aimed at correctly matching its results with experimental results and exactly portray real-life devices.

Comparatively, movement effects were modelled to small level, and opinions for utilising these influences to improve the EDLC performance were omitted. Extensive research in molecular modelling should be increased to accommodate more emphasis on ionic transport and charge and discharge kinetics to determine and enhance efficiency of dense RTIL electrolytes in carbon electrodes of nano-pores, since modelling of the ion transport effects has not given due attention in the past. Again, simulations that can accurately estimate kinetics during different charge and discharge rates will certainly improve the optimisation of energy and power densities. EDLC movement was derived to only learn the effects of polarisation resting, temperature, electrode shape and ion size. Introducing organic solvent to RTIL electrolytes results to improved electrolyte function by increasing ionic conductivity and relative permittivity and also reducing the viscosity and ESR but pays compromised with a decrease in density of ions that lowers the capacitance.

Simulations of very complex structures of electrodes like three-dimensional, such as hierarchical porous system that has not been accounted for, have to be introduced in molecular models so as to realise a high-level reality and as such obtain novel platform to further enhance EDLC performance [82, 83]. On the same manner, pseudo-capacitors and hybrid capacitors with battery-type and EDLC electrodes are fields of research that are drawing great attention now, though realistic models for the capacitors are presently constrained to little MD simulations [84, 85] and ab initio computations [86]. Developing active models in these directions will certainly be very important for accurate identification of situations and materials that could result to enormous advantages in EDLC performance.

15.2.5 Quantum Models (Ab Initio Quantum Chemistry and DFT) of Electrochemical Capacitors

Another available simulation method that solves quantum mechanics equations for many systems is relying on quantum density functional theory (q-DFT). In q-DFT, the overall particles participating in the prediction are much small compared to what was obtained from MC and MD predictions. This method of predictions can be utilised to supplement the MC and MD predictions by proving the force fields or other predictions like the position of electrolyte ion from the electrode surface [79], but very difficult to expand to a feasible description of a porous electrode owing to high cost of computations.

The non-atomistic simulation method that approximate the MC and MD simulations and can explain EDLCs with porous electrodes by reducing base potential energy and getting local density curves of solvent and electrolyte models, and from their estimate the characters of EDL [61] is classical density functional theory (c-DFT).

Most of the time, c-DFT is likened to MC simulations that use ancient models to learn different structures of EDL as functions of ion size, electrolyte valences and pore size [61], because it cannot consider many important processes like surface roughness, ambiguous electrode geometry and electrode surface polarisation.

These simulations are often compared with modified Poisson–Boltzmann theory and can demonstrate deviation from Gouy–Chapman–Stern behaviours [87], though its use in modelling real-life EDLC systems is restricted owing to simplification of the ionic structure. c-DFT permits a careful examination of the effects of important parameters like ion size, electrode geometry and electric potential with ease compared to experimental approach [42], that is, an excellent method for estimating capacitance and dynamics of EDLCs.

c-DFT models cannot be utilised in estimating capacitance in capacitors that use electrodes with nano-pores which are below 1 nm and/or well-packed electrolyte molecules like in RTILs [63], and these situations are now greatly common in EDLCs and so the use of molecular simulations is favoured owing to its atomistic approach.

Car-Pirandello method found on the electron density functional theory (DFT) [88] is among suitable methods for resolving difficulties with employing the ab initio molecular dynamics; it permits repetitive examination of the attributes of the ionic sub-system of electrolyte and electron-hole subsystem of electrode in a single computation.

This theory is built on the belief that the energy of electronic subsystem is dependent on the electron density E^{KS} $[n(r)]$ and that the energy at the base state (T = 0) corresponds to the limit of this functional. Lankin et al. [56] assembled a quantum mechanical model that give a platform to explain the features of electrolyte and electron-hole subsystem of electrode and also electrical charge at electrolyte and electrode material interface from the density functional theory.

The justification for recent research focus on ionic liquids come from the simulations unique behaviours of electrolyte with low vapour pressure, all round functionalities and electrochemical potential range. Kornyshev and coresearchers [57] described the unusual growth as image forces which exponentially block the repulsion of similar-charged counterions within small-pore model of charged ions in a metallic slit pore. De-en and Jianzhong [61] resolved the challenge of microscopic attitude of the interface of electrode and electrolyte and the capacitance dependence on the pore size using the classical density functional theory (c-DFT).

Although c-DFT was extensively utilised to investigate electric double-layer structure in aqueous systems, it was not to examine the differential capacitance at interface of electrode/nonaqueous electrolyte and pore size variation with EDLC. De-en and Jianzhong discovered the enormous ability of c-DFT approach in giving a microscopic insight with little molecular coverage and calculation cost and also the capability to handle a wider pore size window as compared with ion size at bigger scales. Integration of polar solvent into c-DFT computations so as to examine the desolvation theory and find out the opposition between the ionic liquids and organic electrolytes is quite easy.

Fundamental purposes and uses of c-DFT to electrolyte environments have been examined of recent in some reviews [89, 90], where it were presented that mathematical background of c-DFT is identical to the famous electronic DFT [91]. c-DFT for electrolytes presents the free energy to be dependent on ion local densities and solvent molecules rather than expressing the system energy to be dependent on the electron density.

The straightforwardness of this theoretical model also improves the examination of where the oscillation of capacitance in ionic liquid electrolyte came from as well as complete understanding of core parameters in large range of environments. c-DFT suggests a small gap that makes the profile of ion density to interfere with the EDLs between two charged surfaces and also strong vibration-like profiles in the density of ions close to a charged surface [92, 93]. c-DFT is expected to be capable of estimating very complex forms of capacitance–potential curves with its charging processes and the changes in the curvature shape with pore size, since the model for the ionic liquid replicates real-life situation [94].

A total image of capacitance during transit of the pore size from the microscopic range to mesoscopic scales ($> 2\,\text{nm}$) is provided by c-DFT simulations, and the capacitance virtually becomes independent of pore size if the pore size goes beyond a few nanometres. Modelling of EDL behaviour in the past relied on sharing the pore size frame into several regions, such as below 1, 1–2, 2–5 nm, etc. [95, 96], which becomes irrelevant as it produces a steady change of structure and capacitance of EDL, while the pore size varies on c-DFT application. Absolute agreement with the experimental results shows that c-DFT accounted for the most significant physics in the system.

c-DFT research has not addressed the issues on size of solvent molecules, solvent polarity, concentrations of ions, ionic charges, cations–anions size disparity and the shape of the electrode, which are important parameters that will improve the comprehension of organic electrolyte EDLCs [97, 98]. It is greatly important to consider the influence of these parameters while matching integral capacitance with the differential capacitance [99].

The c-DFT method that stands for the state-of-the-art manner by experimentalists to determine surface area and pore size distribution of real porous materials [100, 101] can handle the modelling difficulty on how to confirm an estimation of the interactions between the structure and capacitance in real porous material and a particular electrolyte. Consequently, c-DFT modelling can account for all phenomena of EDLC charge storage like calculation of porosity, capacitance calculation, as well as charge and discharge kinetics. Breakthrough is still not an insight for the c-DFT as fully accounting for the complex porous electrodes and estimating the perfect electrolyte and porous design to optimise the energy and power densities of real-life EDLCs are still a stiff challenge. A strong difficulty from the modelling viewpoint remains how to clearly prove the estimation of the interactions between the structure and capacitance in real porous material and a specific electrolyte.

15.2.6 Simplified Analytical Models

This kind of model is derived on the basis of fundamental of electrochemistry and physics of the ECs. Algebraic and differential equations were utilised in characterising the chemical and physical processes controlling the properties (charged species transport and electrochemical reaction rate) of the system [20, 22, 28, 48, 102, 103]. The ECs are categorised into two on the basis of chemical and physical processes that accounts for the total capacitance of the system: EDLC which stores charge electrostatically and pseudo-capacitor which stores charge by fast and quick reversible chemical reaction. This type of EC models is generally derived relying on the given fundamental assumptions: the system has a constant capacitance at the interface, the rates of the spontaneous generation (spontaneous charge) of non-equilibrium charge carriers are much smaller compared with the rates of the spontaneous recombination (self-discharge) of non-equilibrium charge carriers, the electrodes has no self-discharge effects, the ionic concentrations of electrolyte and conductivity of electrodes change in a small range, the type of conductivity of the electrode pore does not change during charging and discharging, the capacitance depends very little on the value of the potential, the temperature is constant and uniform and the electrolytes are binary without electrochemical reactions. Several areas of ECs like the effect of side reactions [48], electrode pore structure [28, 102] and estimation of specific energy and power [103] were investigated using these assumptions. Karthik et al. [2] presented a reduced EC models examined with scaling arguments, correlated and proved with experimental data without considering energy change expression and heat generation by ignoring temperature influence and side reactions.

Major efforts related to double-layer capacitors have been directed to understanding the influence of electrode thickness and pore structure on the energy density and device capacity. Efforts towards developing pseudo-capacitors have mostly focused on creating different electrode materials with most emphasis given to metal oxides [2, 6] and some effort going towards understanding charge–discharge behaviour of these redox couple electrodes [11]. Devices in which the two electrodes exhibit the same capacitive behaviour are known as symmetric capacitors [2]. If the capacitive behaviour of the two electrodes is different, the device is known as an asymmetric capacitor.

15.2.6.1 Governing Equations for the Double-Layer Electrodes

The density of current in the electrode and electrolyte phases is presented as follows [48]:

$$i_1 = -\ \sigma \frac{\partial \varphi_1}{\partial x} \tag{15.24}$$

$$i_2 = -\ \kappa \frac{\partial \varphi_2}{\partial x} \tag{15.25}$$

where i_1 and i_2 are current densities in electrode and the electrolyte, respectively; σ and k are matrix and solution phase conductivities, respectively; and φ_1 and φ_2 are the potentials in the electrode and electrolyte, respectively.

The solution phase conductivity is related to the porosity of double-layer electrode:

$$\kappa = \kappa_0 \varepsilon_0^{1.5} \tag{15.26}$$

where κ_0 is the conductivity of bulk ions in electrolyte and ε_0 is the electrode porosity.

The conservation of charge implies that:

$$i_{cell} = i_1 + i_2 \tag{15.27}$$

$$\frac{\partial i_1}{\partial x} = -\frac{\partial i_2}{\partial x} = a i_n \tag{15.28}$$

where i_{cell} is the total applied current density, a is the interfacial area over unit volume and i_n is the double-layer current per interfacial area.

It follows from the assumption that every faradaic phenomenon produces current–voltage curves similar to that of a supercapacitor, and so the interaction among potential difference and current transferred at the interface follows the expression given below:

$$i_n = -C_V \frac{\partial(\varphi_1\ -\ \varphi_2)}{\partial t} \tag{15.29}$$

where C_V is the specific capacitance of the electrode with double layer.

On combining Eqs. 15.24–15.29, the resultant equation is given as:

$$\frac{\partial \eta_D}{\partial t} = \frac{1}{\left(\frac{aC}{\sigma} + \frac{aC}{\kappa}\right)} \frac{\partial^2 \eta_D}{\partial x^2} \tag{15.30}$$

where the overpotential in the double-layer electrode η_D is given as Eq. 15.31 below:

$$\eta_D = \varphi_1 - \varphi_2 \tag{15.31}$$

A solution to Eq. 15.31 could be achieved by imposing the initial condition that the overpotential is zero everywhere at time $t = 0$ and using the boundary conditions below:

$$x = 0 \qquad \frac{\partial \eta_D}{\partial x} = \frac{-i_{cell}}{\sigma} \tag{15.32}$$

$$x = L_D \qquad \frac{\partial \eta_D}{\partial x} = \frac{-i_{cell}}{\kappa} \tag{15.33}$$

where L_D is the thickness of the electrode with EDL. In this configuration, $x = 0$ is defined as the interface between the double-layer electrode and current collector, and $x = {}_D L$ is defined as the interface between the double-layer electrode and separator. Analytical solutions for the electrode voltage as a function of discharge current have been developed previously by Srinivasan and Weidner [22, 104, 105].

15.2.6.2 Governing Equations for the Separator

The electronic conductivity of the separator is zero, and as such:

$$i_1 = 0 \tag{15.34}$$

$$i_2 = -\kappa_s \frac{\partial \varphi_2}{\partial x} \tag{15.35}$$

where κ_s is the conductivity of the ions in electrolyte. Thus, current density in the separator follows Ohm's law.

15.2.6.3 Governing Equations for the Redox Couple Electrodes

Discharge in the redox couple electrode is controlled by diffusion of a mobile ion through the film. Therefore, solving the time-dependent diffusion equation for ion concentration given below is necessary:

$$\frac{\partial C}{\partial t} = D \frac{\partial^2 C}{\partial x^2} \tag{15.36}$$

where C is the concentration and D is the coefficient of diffusivity.

The initial condition states that at time $t = 0$, the concentration of mobile ion is uniform throughout the electrode, C^0, and the boundary conditions could be derived as shown below:

$$x = 0 \qquad \frac{\partial C}{\partial x} = 0 \tag{15.37}$$

$$x = L_R \qquad \frac{\partial C}{\partial x} = \frac{i_{cell}}{DF} \tag{15.38}$$

where $x = 0$ is the interface between the redox couple electrode and current collector and $x = L_R$ is the interface between the redox couple electrode and separator. Mobile ion diffusion into the redox couple electrode sustains the reaction at the film/separator interface, at $x = L_R$. The kinetic expression at this interface has been given previously [102, 103] and was extended to the hybrid asymmetric supercapacitors by John A. et al. [104] as given in Eq. 15.39:

$$i_{cell} = i_0 \left[\theta_s \ \exp \frac{\alpha_a F}{RT} \eta_R - [1 - \theta_s] \exp \frac{\alpha_c F}{RT} \eta_R \right] \tag{15.39}$$

where i_0 is the exchange current density; θ_s is the surface state of charge; α_a and α_c are the charge transfer coefficients of the positive and negative electrodes, respectively; and η_R is the overpotential of the redox couple electrode relative to equilibrium potential U_{eq}:

$$\eta_R = \varphi_1 - \varphi_2 - U_{eq} \tag{15.40}$$

where U_{eq} varies with the state of charge and the redox electrode material. In the case of redox couple electrodes in hybrid asymmetric supercapacitors, the state of charge remains between approximately 0.4 and 0.6. The equilibrium potential of the redox couple electrode is presented as the Nernst equation [106]:

$$U_{eq} = U^0_{eq} + \frac{RT}{F} \ln \frac{\theta}{1 - \theta} \tag{15.41}$$

The concentration of mobile ion affects the redox electrode kinetics through Eq. 15.39. This model for the redox couple electrode is generic and could be applied to describe any electrode as long as parameters such as initial ion concentration and diffusivity are known. Diffusivity values that were previously reported by experimental methods [30] were utilised while making the assumption that diffusivity is constant.

15.2.6.4 Energy and Power Density of Supercapacitors

Different electrochemical energy storage devices can be compared using their respective energy and power densities, and any advantage that hybrid asymmetric

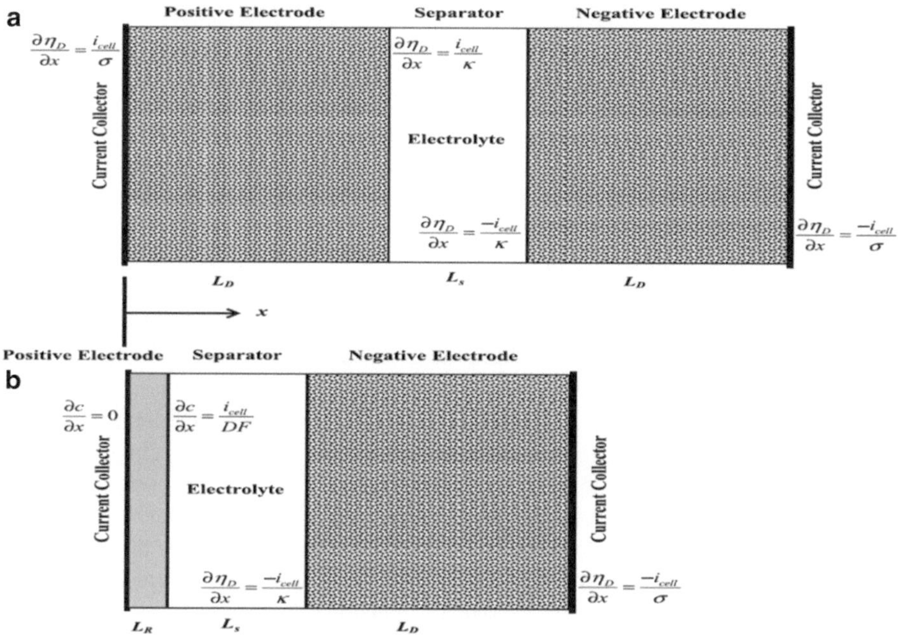

Fig. 15.11 Schematic representations of (**a**) the symmetric supercapacitor and (**b**) the hybrid asymmetric supercapacitor with the electrode dimensions not drawn to scale [104] (Reproduced by permission of The Electrochemical Society)

supercapacitors (Fig. 15.11a) possess over symmetric supercapacitors (Fig. 15.11b) can as well be demonstrated. The energy and power density are dependent on the capacitor mass. The mass of the symmetric supercapacitor is given as the mass of each porous electrode and electrolyte inside each electrode and inside separator:

$$m_S = 2L_D A\rho_D + L_S A\rho_S + 2\varepsilon_0 L_D \rho_e + \varepsilon_S L_S \rho_e \tag{15.42}$$

where A is the supercapacitor cross-sectional area, ρ_D is the density of the electrodes with EDL, L_S is the separator's thickness, ρ_S is the density of separator, ρ_e is the density of electrode, ε_0 is the porosity of electrode and ε_S is the porosity of separator. The difference between the mass of the hybrid asymmetric supercapacitor and symmetric supercapacitor is that the former only has one electrode with EDL, with a dense redox couple electrode as the positive electrode:

$$m_H = L_D A\rho_D + L_S A\rho_S + \varepsilon_0 L_D \rho_e + \varepsilon_S L_S \rho_e + L_R A\rho_R \tag{15.43}$$

where ρ_R is the density of the redox electrode.

The energy and power density are calculated by integrating cell potential over cut-off time, or the time required upon discharge for the two electrode potentials to equal one another, which varies with discharge current density i_{cell} as given below:

$$E = \frac{i_{cell}}{m} \int_0^{t_C} V_{cell} \; dt \tag{15.44}$$

$$P = \frac{i_{cell}}{mt} \int_0^{t_C} V_{cell} \; dt \tag{15.45}$$

where V_{cell} is the cell potential, t_C is the cut-off time and m is the mass of either capacitor, representing either m_S or m_H.

The moment the electrode parameters are known and the potential profile has been simulated, the energy and power density dependency of discharge rate could be determined. These models are generally applicable to any double-layer and redox couple electrode, so long as the properties of the materials are known.

The ability of this analytical model to describe the electrical functioning of the ECs utilising the partial differential equations which explains the electrochemical and physical processes within the capacitor is its main advantage. This method is more flexible to accommodating more parameters and sets of complementary equations and gives a very low degree of reliance on empirical evidence. However, this type of model is very difficult to estimate the ageing phenomena of battery and EC models which has the same analytical modelling approach as stated by Doyle et al. [103] in their article. This is basically because the non-homogeneity of the electrolyte at the interface and integration of thermal variables were not considered when deriving the analytical model. Kazaryan et al. [2] presented model for computation, control, and enhancement of the energy, power, and other parameters necessary for safe and long performance of different kinds and designs of asymmetric ECs while considering all electrode materials properties and spatial structures, designs, and separator properties. They ignored the effect of self-discharge on the negative electrode as well as the potential variation along the positive electrode during device charging and discharging processes.

Jin et al. [107] presented model for a straightforward planar EC and revealed that the polarisation density and electrolyte solution are very critical and cannot be ignored in the model derivation. Also they discovered that modelling of the surface contact area is of important concern since most of the charge is stored in this very little portion that its movement is dependent on several variables like electric and thermal fields, concentrations of species and impurities. Julian et al. [40] developed three-dimensional model for ECs using electrodes with cylindrical meso-pores and used it to study the influence of the following: pore radius, electrolyte permittivity, porosity, ions effective diameter and the characteristics of electrolyte on diffuse layer capacitance. They proved that trimming down of ion effectual diameter and radius of the pores produced a great growth in diffuse layer gravimetric capacitance. Ganesh and Sanjeev [108] derived from first concepts of ionic movement one-dimensional (1D) and two-dimensional (2D) models of EC without faradic reaction utilizing isotropic transfer properties to describe the regaining of potential throughout pause after discharge/charge and its dependence on current and concentration, as well as EC failure when charged at high currents with low electrolyte concentrations.

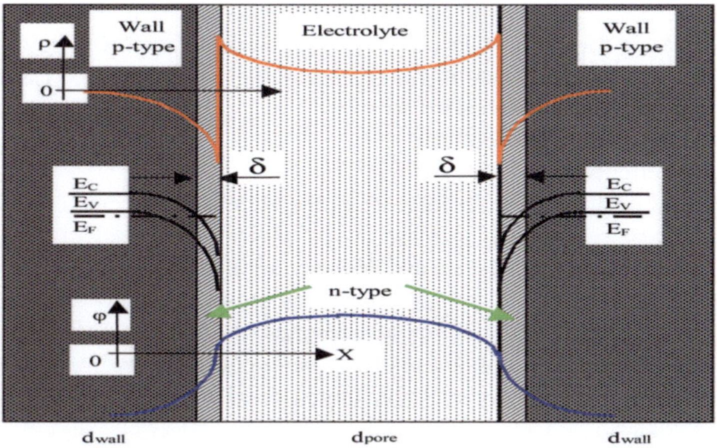

Fig. 15.12 Distributions of electric charge and potentials in the walls of *cylindrical*-shaped pores and in the electrolyte which is inside the pores of the negative electrode with p-type conductivity [2] (Reprinted with permission from J. Electrochem. Soc. 153 (2006) A1655. Copyright 2010, The Electrochemical Society)

Conductivity of electric double-layer electrode on charging and discharging may vary reasonably along with specific capacitance parameters, type of conductivity, and the electrophysical parameters of the electrode. This variation has significant influence on capacitance and other parameters of ECs. The variations in parameter must be account for in order to achieve more accurate results. Figure 15.12 depicts the dispersal of electric charge and potentials in the pore walls and in electrolyte inside the pores of negative carbon electrode with p-type conductivity. Change of the conductivity types that lead to appearance of physical p-n transition happens in the near-surface layer of walls of electrode pores during strong polarization. The width and dispersal of volume spatial charge of p-n transition relies on electrophysical parameters of electrode material, electrolyte, and the electrode potential.

EDL electrodes of supercapacitors are basically degenerate semiconductors of p-type conductivity in which the Fermi level (E_F) is in the valence band. The surface layers of wall pores plays crucial role in EDL parameters when strong polarization exists among the electrodes. When strong bend occurs on the bands in the near-surface layers of the wall pores by δ thickness, Fermi level E_F is above the conductivity band bottom (material in this area is a degenerate material with p-type conductivity) as could be seen in Fig. 15.12. Various nonlinear effects are anticipated in the near-surface layers of the walls and wall pores subject to electrode anisotropy as well as the pores various sizes and shapes. The capacitances of EDL from side of the electrolyte and from side of the solid body (electrode) are connected in series with each other, and parameters of wall pores [109] play crucial function in variation of EDL total capacitance.

Kazaryan et al. [2] developed analytical model of asymmetric ECs which presented the platform to compute the capacitors energy, capacity, power parameters, energy efficiency of charging and discharging cycles of subject to type and value of conductivity of electrodes with EDL, electrolyte conductivity, thickness and specific capacitance of electrodes as well as the amount of applied charging and discharging currents.

They equally noticed that the increase in the volumetric specific capacitance of the EDL electrode results in increase in the device specific energy parameters, along with increase in polarization and depolarization energy losses that appears as heat. The great part of the energy loss came from electrodes and electrolyte polarization resistance while some part resulted from the electrodes potentials depolarization. They also pointed out that the energy efficiency of ECs charge-discharge cycles depends reasonably on the electrode and electrolyte conductivity, electrode thickness, and amount of applied charge-discharge currents. An increase in electrode and electrolyte conductivity reduces nonlinearly the polarization and depolarization losses of energy during the device charging and discharging processes.

15.3 Heat Production and Heat Modelling of Electrochemical Capacitors

Electrochemical capacitors are usually charged and discharged at very high rate, which will obviously produce great amount of heat within the capacitor. This condition of increased temperature will certainly increase the capacitance and the self-discharge while the equivalent series resistance (ESR) will be decaying [24, 49, 108, 110, 111]. ECs encounter "self-heating" which is promoted with a growth in electric field and temperature during charge and discharge processes. The activity of self-heating grows the temperature of capacitor by enhancing nonreversible side reactions that reduce the electrolyte's ionic conductivity, which, when every all other things are same, generates more heat that over time creates thermal runaway. During EDLC charge and discharge processes, an amount of the electric charge is transformed to heat, and the heat production rate varies with condition of operation, material of construction and the capacitor's design [112]. Severe high temperatures result to the following: increased capacitor ageing rate [90–93, 97–101, 109, 112–117], increased self-discharge rates [113–116], increased cell pressure and possibly electrolyte evaporation [116]. ECs resistance increases while the capacitance decreases along with the device ageing, and this results in capacitor's temperature increase and voltage [117], which is capable of causing voltage instability and unfavourable overvoltage of each cell connected in series [112, 113]. Thermal modelling is very useful in estimating the temperature of operations, the thermal functioning of new EDLCs and development of thermal management plans for existing EDLC designs.

The lifetime expectancy and performance of EC decay in an irretrievable manner because of heat produced due to ohmic resistance losses of the capacitor's components and the electrolyte ionic resistance. The temperature within the electrochemical capacitors changes with time and position under consideration. Thermal energy is transported from high-temperature point to low-temperature point by electronic transit that transports electric charges in electrical conductors. The temperature at a specific point in the electrochemical capacitor is influenced by three modes of heat transfer that is related to operating condition and capacitor internal geometry: conduction, convection and radiation. An oxidative reaction of electrolyte is triggered off at elevated temperatures judging from the Arrhenius law and follows the same trend at higher voltages. Persistent generation of heat and subsequent temperature growth results in a reduction in capacitance and a rise internal ohmic resistance and self-discharge rate. It is very essential to know the thermal functioning of EDLCs owing to its great reliance on temperature.

The design and fabrication of modern-day capacitors and their cooling plans greatly rely on the success of thermal simulations models; thus, it is very essential to obtain mathematical description of the mechanisms that generates heat in the system.

Estimation of electrical heating functioning of capacitors could be very challenging due to huge series processes like ionic and electronic transport, heat and mass diffusion and the involvement of heterogeneous structures. Temperature distribution in supercapacitors depends on losses of power and energy within the devices.

The relation between the rate of heat generation during continuous charge–discharge processes and temperature $T\,(M, t)$ at specific position in a supercapacitor is mainly controlled by heat diffusion expression as has previously been presented by Guillemet et al. [111]:

$$\nabla \cdot (-k(M) \cdot \nabla T(M,t)) + C_p\rho \, \frac{\partial T(M,t)}{\partial t} = \dot{q}\cdot \tag{15.46}$$

The heat generation rate $\dot{q}\cdot$ is caused by the following:

- Transportation of electrolyte ions that are in electrodes and separator and transportation of electrons in current collectors and electrodes
- Reversible and irreversible reactions at the electrode/electrolyte interface of the porous structures
- Thermal contact, electrical ohmic resistances of the layers and polarisation and depolarisation resistances

Guillemet et al. [111] demonstrated from thermal examination carried out on ECs that highest temperature is obtained at the middle of the device as expected using the model they developed.

Anna d'Entremont and Laurent Pilon [118] derived a physical model that describes the diffusion of electrons, heat production and heat transports in EDLCs while charging and discharging. Their model is very significant because it estimates the spatial and temporal changes of the various heat production rates and temperature within EDLCs from basic principles and considered the irreversible Joule heating as well as three reversible heat production rates coming from ion diffusion, steric effects and changes in entropy of mixing. The reversible heat production rates are exothermic on charging the capacitor, endothermic while discharging it and localised in the electric double layers.

They developed this thermal model in order to understand electron diffusion, heat production and heat transport in electric double-layer capacitors at constant current charge and discharge cycles [118]. They achieved this by formulating irreversible Joule heating and reversible heat generation formula for electric double-layer capacitor and were able to reproduce experimental data earlier presented in literature hence validating their model. Their considerations were that the entropy of ions is reduced while ions form the EDL on charging, and the entropy of ions grows as they regain their uniform concentration on discharging [6], resulting to exothermic and endothermic processes which satisfied thermodynamics second law. The energy equation developed by Anna d'Entremont and Laurent Pilon [118] is expressed as Eq. 15.47 shown below:

$$\rho c_p \frac{\partial T}{\partial t} = \nabla \ . \ (k \ \nabla \ T) + \dot{q}_{irr} + \dot{q}_{rev} \tag{15.47}$$

where $\dot{q}_{irr}$ and $\dot{q}_{rev}$ are irreversible and reversible heat production rates; $\dot{q}_{irr} = \dot{q}_{E,j}$; and $\dot{q}_{rev} = \dot{q}_{E,d} + \dot{q}_{E,s} + \dot{q}_{S,c} + \dot{q}_{S,T}$. The heat production rates $\dot{q}_{E,j}$, $\dot{q}_{E,d}$, $\dot{q}_{E,s}$, $\dot{q}_{S,c}$ and $\dot{q}_{S,T}$ are the effects from Joule heating, ion diffusion, effects of steric, mixing heat with concentration gradient and mixing heat with temperature gradient, respectively.

They noticed that Joule heating term $\dot{q}_{E,j}$ is irreversible because the value remains positive, while $\dot{q}_{E,d}$ and $\dot{q}_{E,s}$ account for reversible heat production since their value can be positive or negative. Movement of ions when charging the capacitors is towards direction reducing electric potential energy to create the EDLs, and heat energy is released in the process. During capacitor discharge, transportation of ions is controlled by diffusion and steric effects since the EDLs are at rest and the fluxes then move towards the increase in electric potential and heat energy are then absorbed.

Heat energy is liberated when the potential energy of ions is reduced, and there is effective contribution to heat production rate because of chemical potential gradients, partial molar entropy and temperature. Heat is also liberated when solvent and/or ion molecules reduce the entropy as predicted by Schiffer et al. [49].

Irreversible Joule heating alone will obviously result to a linear temperature growth but further reversible heat production will make the temperature profiles to oscillate, since the reversible heat production is exothermic on charging and endothermic while discharging.

Furthermore, the increase in oscillation amplitude associated with ion diffusion was approximately twice that associated with either steric effects or heat of mixing with concentration gradient.

The temperature evolution they predicted remarkably looks like the one noticed experimentally by Guillemet et al. [111] and Gualous et al. [119], which showed that physical model pictured the physical processes controlling the heating functioning of EDLCs.

Burheim et al. [120] observed from experimental results that a smaller thermal conductivity of wet and dry electrodes made of OLC materials greatly showed that great effective pore volume is destructive to heat conduction and required to be considered while designing and fabricating electrodes from materials that has big pore volume or small packing density.

Modelling internal temperature distribution of electrochemical capacitors on a very high cycle and commercially sized unit showed that temperature slope of many degrees is dependent on internal ohmic resistance of the device and as such electrode materials are expected to be wetted in the long run.

15.4 Self-discharge and Charge Redistribution of Electrochemical Capacitors

The challenge of self-discharge in electrochemical capacitors is still to be completely summarized in overview articles up till date and this is vital since they are energy storage systems that undergo quick self-discharge phenomenon [15, 86, 120–136]. Self-discharge is basically regarded as the volume density of currents of recombination of charge carriers in an electrochemical capacitor. Thus, getting an understanding of mechanism or mechanisms of current leakage or self-discharge in electrochemical capacitors is the only way out of challenges of self-discharge in ECs. As a rule, in electrochemical capacitors, the rates of production of carriers are much lower to be compared with the rates of the recombination of carriers. Factors that affect the quantity of charge and energy stored at a specific temperature [130, 132] and the initial way of charge or charge origin [129] have been previously studied. Conway et al. [15, 122] suggested mathematical models that estimate the distribution of self-discharge on the basis of three likely mechanisms of self-discharge as a guide to discovery of self-discharge mechanism of electrochemical capacitors, which has posed been a crucial difficulty for the ECs applications.

ECs like other energy storage device experience self-discharge at completely charged state because they have greater free energy in the charged state than the discharged state [11]. A reduction in voltage of charged capacitor at resting state leads to reduction of power and energy densities of the ECs [11], which obviously create fluctuation in the charge storage device for critical use [6]. Self-discharge challenge has since been recognized and relatively large amount of research has been focused to the understanding of its mechanism. Self-discharge behaviour of EDLCs could be affected by the surface functionalities which add to charging process by electrolysis on active materials. It has been presented that level of retaining oxygen via either physical adsorption or acid functionalisation on electrode surface can affect the self-discharge rate in EDLCs [4, 11]. High concentration of complexes on electrode surface may grow the rate of self-discharge and as such predicting that the complexes catalyse the oxidation or reduction reactions of carbon electrodes [137]. A very smaller self-discharge current was recorded when the functional groups on the electrode surface were eliminated from electrode materials via peculiar heat treatment in non-reactive conditions [4].

In comparison, the fundamental of self-discharge procedures in pseudo-capacitive materials differs from those materials exhibiting double-layer capacitance such as carbon materials. This is because the charge transfer procedure in the materials is faradaic in nature as discussed earlier and involves different oxidation states which correlate with various different solid phases [125]. The potential reduction of the absolutely charged pseudo-capacitive electrode over time could be generated by a process that is diffusion controlled, leading to a rest in the oxidation state gradient that occurs from the outer surface into the stream of electrode materials [11]. This kind of self-discharge process would be accelerated where conducting materials like graphite with growing surface area are available [138]. Increased active prismatic sites as a result of grown surface area were thought to be catalysing self-discharge rate [138]. There was other method to investigate the self-discharge process and this was to move the Nernst potentials for water disintegration. It was asserted that decomposition of the impurity electrolyte could occur on surface of electrode as a result of variation in water decomposition potentials coming from dissolved H2 and O2 which might catalyse self-discharge of device capacitor at storage. Metallic impurity in electrolytes equally influences reasonably the self-discharge procedure [126, 127, 130]. Of late, effects of redistribution of charge on self-discharge distribution of porous electrode of EDLCs has been examined [15, 33, 129]. Employing high surface-area-activated carbon creates pores of various sizes ranging from macro to micro sizes. Therefore, charging/discharging of the electrode does not progress uniformly along the pore wall because of voltage difference that seems results in self-discharge effect in an open-circuit.

Self-discharge can result from ECs in fully charged condition but for some other reasons and mechanisms apart from what is obtainable in charged batteries, although the practical effect is similar (loss of charge and energy over time while at rest). Self-discharge creates great constraints and bottleneck in the applications and operations of ECs, especially for the device that will be used for 'standby' purposes.

It should be noted that perfect polarizable electrode not having any leakage, cannot exhibit self-discharge effect because self-discharge will only happen due to faradaic electron-transfer processes at and below the maximum potential achieved on charging or when electrodes were not perfectly sealed up.

Recently, many research groups have embarked on intense studies on self-discharge [24, 28, 48, 49, 87, 88, 102, 103, 107, 108, 110, 111, 121–131]; Ricketts and Ton-That [125] observed that this process of self-discharge includes fast diffusion process with a slow current leakage. Decrease in voltage resulting to loss of charge when the device is at rest could be due to side reactions owing to disintegration of electrolyte, redox reactions due to presence of impurities or obvious availability of functional groups on electrode surface. Faults emanating from the electrochemical capacitor production could lead to little short circuits between the electrodes [11] which results to noticeable self-discharge effects.

Maximilian et al. [139] from both simulations and experiments vividly demonstrated that most of the noticed reduction in voltage is not due to actual self-discharge since self-discharge is referred to processes whereby charge carriers are consumed in side reactions and are no longer presence for discharging, which accounts only for a very minute rate in the issue of the EDLCs. This noticeable reduction in voltage is mostly as a result of redistribution of charge carriers which is still retrievable from the capacitor under enough discharge time. These findings are of high value for supercapacitors that do not need to be recharged after a long time, and the outstanding energy is required for processes like engine cracking, since fully charged device exhibits a relatively slow rate of voltage and power loss. The charge carriers could be retrieved from the capacitors again during a very slow discharge.

Despite the increase of leakage current and self-discharge current and the decrease of energy and capacity parameters of the supercapacitors, many shuttle ions have negative effect on positive electrode of asymmetric supercapacitors. Impurity ions gradually penetrate the active materials crystal lattice during supercapacitor operation and result in a decrease of self-discharge currents and so decay the capacity parameters and life cycle of positive electrode and the supercapacitor at large. A significant degradation of stored energy which significantly increased at high voltage and temperature since self-discharge is greatly influenced by these parameters (voltage and temperature) mostly happens within the first hours in EDLCs with organic electrolytes. It is noticeable that modelling the whole self-discharge effect by an equivalent circuit gives efficient results and allows for estimation of self-discharge effects and subsequent losses of stored charge and energy for various electrochemical capacitors.

Tuning the self-discharge rate is therefore feasible for the SWNT supercapacitors by simply trimming the surface chemistry of the nanotubes as shown by Q. Zhang et al. [140].

The effects of surface chemistry on self-discharge by obstructing the electrostatic interaction among electrolytic ions and the SWNT surface are very important in wide circumstance of obtaining further understanding of self-discharge process and realising the self-discharge tuning through method of surface chemistry modification, that is of utmost significance in electrochemical capacitor's self-discharge study and will benefit possible implementation of ECs for energy storage.

According to Libin Chen et al. [141], electrochemical capacitor's quick self-discharge process can be effectively quenched by using an ion exchange membrane separator or $CuSO_4$ active electrolyte, which in turn gives understandings into the modern-day design of capacitors with high capacitance and enhanced energy sustainability. Ban et al. [8] emphasised that it is suitable not to charge electrochemical capacitor to thermodynamic voltage of solvent disintegration for safe operation of the capacitor and to avoid build of product gases such as O_2 and H_2 if using an aqueous electrolyte in a closed cell. This may obviously lead to mass transport limitations and also pressure building up, which will cause safety concerns and self-discharge.

It is suitable to create a recent widespread theory of self-discharge of present-day ECs in order to originate technology for manufacture of ECs of different systems, develop new ECs with optimal designs, improved energy capacity and operation parameters as well as calculation and control of parameters under various mechanisms self-discharge.

15.5 Modelling Challenges of Electrochemical Capacitor

Simple charging and discharging operation of electrochemical capacitor demands enough information on impact of the capacitor's geometry on its functioning, the effects of electrons and ions on electric potential and the temperature and electric field on chemical reactions at electrode surface. ECs modelling procedure includes representation of porous separator immersed in electrolyte, interface of electrodes and electrolyte, the electrodes and contacts at the boundary of electrode and current collector. Each stage/layer required to be modelled separately while handling the pores like heterogeneous structure with different interface and also the interface between electrode and electrolyte. The kinetics within the ECs is basically controlled by the temperature and electric field at the interface. It is always necessary to begin first with a plane system and gradually increase the models complexity so as to describe and specify such complex system.

Development of big signal models that are dependent on time assuming a constant temperature which is a significant route because it permits the physics of the interface at steady temperature to be proved in the beginning stage.

Effective optimisation of EDLC type is a great difficulty because of complex interactions among the electrode and electrolyte characteristics and their effects on entire capacitor's energy and power performance. The effects of particular adsorption of ions to surface of electrode, the kinetic of electrons, transport processes in solvated ions in pores of differing geometries and sizes and electrolytes to be used with pore of a given geometries and sizes are the factors to be considered in the design of EDLCs.

Prediction of surface tuning and its surface functionalisation along with optimum choice of RTIL electrolyte subject to size seems to be a significant field of study for molecular modelling which should be focused more on, as it might aid in design of new EDLCs with very improved energy densities. Predictions of RTILs with planar surfaces described the experimental difference in differential capacitance, but designing EDLCs with improved performance is hard to be utilised since porous electrodes are preferred due to their large surface area but are unfortunately controlled by great different physics.

Also, process of thermal runway is the main factor that is not included in all the available electrochemical capacitor models. The methods of all existing models neglected this heating process and also through oversimplification restrained incorporation of thermal variables. These models cannot be utilised to investigate the influence of capacitor design and operation of heat production during constant current charge and discharge cycles and so are difficult to develop a plan to control heat generation for the electrical energy storage device in order to keep the capacitor's functioning temperature constant at comfortable range. Models which modify the assumptions that the available analytical models were built upon by designating electrolyte solutions by employing every composing relations like heat generations, charge redistribution effects, self-discharge effects, polarisation and depolarisation losses of energy as well as taking due account of the capacitor component's key parameters, and other factors in a rigorous manner are required. This model will present the physical and electrochemical processes by accounting for charge conservation, typed and energy together with important composite equations for both EDL and faradic charging. This operation is to be performed on every layers and groups within the device cell at macro and micro scales.

However, charging and discharging of the capacitor requires understanding of the influence of cell capacitor geometry, contributions of electrons and ions to electric potential and chemical reactions subject to the electric field and temperature.

15.6 Conclusions

Modelling and simulation remain a key to positive result in developing a novel electrochemical capacitor with improved energy and power densities, thus mathematical theory that past equivalent circuits to combine charging to mechanics, energy dissipation and application of physics and chemistry of solvents and ions in nanoporous electrode materials with the electronic and ionic transfers in device electrodes is needed. Therefore, it is important to assemble from basic concepts of physics and electrochemistry the required mathematical model and theoretical ground for design, optimization and fabrication of ECs of coveted performance subject to more practical assumptions.

ECs are to be modelled subject to (i) electronic, atomic and solvent molecular polarisation; (ii) distortion of outer electron shells of ions and molecules of electrolyte in supercapacitor; (iii) variations in conductivity and density of electron levels on surface of the electrode subject to the potential; (iv) volume spatial charge emergence in the electrode's near-surface layers; and (v) great electrodes capacitance effect on potential in order to enhance the accuracy of theoretical computations of parameters and description of results from experiment for various device in wide voltage operating range. The growth of density of the quasi-free electrons (holes) and ions in the volume of spatial charge of the electrode and electrolyte, respectively, during the capacitors' charging creates great interactions that make capacitance value a function of electric charge, electrolyte features, and electrode materials. Thus, Eq. 15.1 gives a platform to measure highest expected amount of capacitance. In order to actualize an accurate capacitance value, it is suitable to consider the electric charge, electrolyte and electrode material properties in Eq. 15.1.

Results in literature commonly presented the fundamental relationship between ECs components, but none of the available model completely considered the ECs parameters and other factors such as internal temperature rise and interfacial electric field in a meticulous way, thus the difference in theoretical and experimental results of practical ECs parameters.

Although examination of temperature influences on performance of IL-based ECs with surface-curved electrodes presented by Guang Feng et al. [73] showed that capacitance of ECs of CNT electrodes is almost independent of the temperature because of small changes in the EDL formation, and the capacitance of OLC-based increases with temperature owing to reduction in EDL thickness with growth in temperature. This agreed completely with the recent experimental results presented by Lin et al. [72]. They equally noted that the real capacitor performance in terms of power and charging or discharging rates that is greatly dependent on the mobility of ions rather than ion packing and cannot be estimated from the capacitance behaviour with temperature. The mechanism at the beginning of the improved capacitance in nanoporous electrodes can be completely understood from MD simulations that include real-life carbon materials considering the effects of polarisation in proper manner [80], because introducing metallic walls polarization influences could permit absolute understanding of the emergence of a superionic state [57, 58].

Moreover, the expected lifetime and performance of EDLC decay irreversibly because of heat production via polarisation and depolarisation energy losses, electrode internal ohmic resistance and the resistance of ions in electrolyte. The self-heating processes increase the temperature of the capacitor and catalyse irreversibly side reactions which decay conductivity of ions in the electrolyte assuming all other factors are same and generate more heat that results to heat runaway with time.

Tuning the self-discharge rate is therefore achievable for SWNT ECs by simply trimming the surface chemistry of the nanotubes. The effects of surface chemistry on self-discharge by interference into electrostatic interaction among electrolytic ions and the SWNT surface are crucial in wide circumstances of getting further understanding of self-discharge process and realising the self-discharge tuning through method of surface chemistry modification, which is a step of utmost significance in electrochemical capacitor's self-discharge study and will benefit potential applications of ECs for energy storage. ECs' fast self-discharge process can be effectively reduced by utilising an ion exchange membrane separator or $CuSO_4$ active electrolyte, which gives insights into the present-day design of capacitors with high capacitance and improved energy sustainability.

It is suitable to charge the ECs below of solvent dis integration thermodynamic voltage for safe operation of capacitor and to avoid build-up of product gases such as O_2 and H_2 if using aqueous electrolyte in a closed cell. This may obviously lead to mass transport limitations and also pressure build-up that will cause safety concerns and self-discharge.

Thermal model is very crucial because it estimates the spatial and temporal changes in various heat production rates and the temperature within EDLCs from fundamental basis and accounts for the irreversible Joule heating and the three reversible heat production rates resulting ion diffusion, steric effects and changes in entropy of mixing.

Therefore, models that includes thermal model will give the required inner temperature distribution outline that will guide designing electrodes materials, estimating charging and discharging temperatures as well as assembling heat management/cooling plans for available and new EDLC designs with the functioning conditions. The integrated model will present the needed temperature profile of electrochemical capacitors that could be employed to find cell optimum loading while maintaining the quantity of ions via irreversible chemical reactions or generation of gases in electrolyte.

The efforts and models in view for enhancing the performance and functioning of electrochemical capacitor demand in-depth model of ECs based on formulation of much signal that integrates heat and electric fields along with the type and value of electrode conductivity, thickness, porosity and specific capacitance of electrodes, conductivity of electrolyte, thickness and porosity of separator and the amount of charging and discharging currents.

Acknowledgements The financial assistance of the Materials for Energy Research Group (MERG), Private Bag 3, Johannesburg 2050, South Africa and the National Research Foundation (NRF), South Africa, towards this research is hereby acknowledged. Opinions expressed and conclusions arrived at are those of the authors and are not necessarily to be attributed to the MERG and NRF.

The authors express their appreciation to Professor Kenneth I. Ozoemena for his well-purposeful comments and constructive suggestions, which have significantly contributed to the improvement of the manuscript.

References

1. Somasundaram K, Birgersson E, Mujumdar AS (2011) Analysis of a model for an electrochemical capacitor. J Electrochem Soc 158:A1220–A1230. doi:10.1149/2.062111jes
2. Kazaryan SA, Razumov SN, Litvinenko SV et al (2006) Mathematical model of heterogeneous electrochemical capacitors and calculation of their parameters. J Electrochem Soc 153: A1655–A1671. doi:10.1149/1.2212057
3. Pech D, Brunet M, Durou H et al (2010) Ultrahigh-power micrometre-sized supercapacitors based on onion-like carbon. Nat Nanotechnol 5:651–654. doi:10.1038/nnano.2010.162
4. Pandolfo AG, Hollenkamp AF (2006) Carbon properties and their role in supercapacitors. J Power Sources 157:11–27. doi:10.1016/j.jpowsour.2006.02.065

5. Frackowiak E (2007) Carbon materials for supercapacitor application. Phys Chem Chem Phys 9:1774–1785. doi:10.1039/B618139M
6. Simon P, Gogotsi Y (2008) Materials for electrochemical capacitors. Nat Mater 7:845–854. doi:10.1038/nmat2297
7. Zhang LL, Zhao XS (2009) Carbon-based materials as supercapacitor electrodes. Chem Soc Rev 38:2520–2531. doi:10.1039/B813846J
8. Ban S, Zhang J, Zhang L et al (2013) Charging and discharging electrochemical supercapacitors in the presence of both parallel leakage process and electrochemical decomposition of solvent. Electrochim Acta 90:542–549. doi:10.1016/j.electacta.2012.12.056
9. Petreus D, Moga D, Galatus R, Munteanu RA (2008) Modeling and sizing of supercapacitors. Adv Electr Comput Eng 8:15–22. doi:10.4316/aece.2008.02003
10. Bose S, Kuila T, Mishra AK et al (2011) Carbon-based nanostructured materials and their composites as supercapacitor electrodes. J Mater Chem 22:767–784. doi:10.1039/C1JM14468E
11. Conway BE (1999) Electrochemical supercapacitors: scientific fundamentals and technological applications. Kluwer Academic/Plenum, New York/London
12. Jiang H, Ma J, Li C (2012) Mesoporous carbon incorporated metal oxide nanomaterials as supercapacitor electrodes. Adv Mater 24:4197–4202. doi:10.1002/adma.201104942
13. Lokhande CD, Dubal DP, Joo O-S (2011) Metal oxide thin film based supercapacitors. Curr Appl Phys 11:255–270. doi:10.1016/j.cap.2010.12.001
14. Stoller MD, Ruoff RS (2010) Best practice methods for determining an electrode material's performance for ultracapacitors. Energy Environ Sci 3:1294–1301. doi:10.1039/C0EE00074D
15. Kaus M, Kowal J, Sauer DU (2010) Modelling the effects of charge redistribution during self-discharge of supercapacitors. Electrochim Acta 55:7516–7523. doi:10.1016/j.electacta.2010.01.002
16. Vangari M, Pryor T, Jiang L (2013) Supercapacitors: review of materials and fabrication methods. J Energy Eng 139:72–79. doi:10.1061/(ASCE)EY.1943-7897.0000102
17. Richner RP (2001) Entwicklung neuartig gebundener Kohlenstoff- materiaien fur electrische Dop-pelschichtkondensatorelektronden. Swiss Federal Institute of Technology Zurich
18. Verbrugge MW, Liu P (2005) Microstructural analysis and mathematical modeling of electric double-layer supercapacitors. J Electrochem Soc 152:D79–D87. doi:10.1149/1.1878052
19. Vazquez S, Lukic SM, Galvan E et al (2010) Energy storage systems for transport and grid applications. IEEE Trans Ind Electron 57:3881–3895. doi:10.1109/TIE.2010.2076414
20. Devan S, Subramanian VR, White RE (2004) Analytical solution for the impedance of a porous electrode. J Electrochem Soc 151:A905–A913. doi:10.1149/1.1739218
21. Kim H, Popov BN (2003) A mathematical model of oxide/carbon composite electrode for supercapacitors. J Electrochem Soc 150:A1153–A1160. doi:10.1149/1.1593039
22. Lin C, Popov BN, Ploehn HJ (2002) Modeling the effects of electrode composition and pore structure on the performance of electrochemical capacitors. J Electrochem Soc 149:A167–A175. doi:10.1149/1.1431575
23. Zubieta L, Bonert R (2000) Characterization of double-layer capacitors for power electronics applications. IEEE Trans Ind Appl 36:199–205. doi:10.1109/28.821816
24. Rafik F, Gualous H, Gallay R et al (2007) Frequency, thermal and voltage supercapacitor characterization and modeling. J Power Sources 165:928–934. doi:10.1016/j.jpowsour.2006.12.021
25. Diab Y, Venet P, Gualous H, Rojat G (2009) Self-discharge characterization and modeling of electrochemical capacitor used for power electronics applications. IEEE Trans Power Electron 24:510–517. doi:10.1109/TPEL.2008.2007116
26. De Levie R (1963) On porous electrodes in electrolyte solutions: I. Capacitance effects. Electrochim Acta 8:751–780. doi:10.1016/0013-4686(63)80042-0
27. Robinson DB, Wu C-AM, Jacobs BW (2010) Effect of salt depletion on charging dynamics in nanoporous electrodes. J Electrochem Soc 157:A912–A918. doi:10.1149/1.3416905

28. Bertrand N, Sabatier J, Briat O, Vinassa J-M (2010) Embedded fractional nonlinear supercapacitor model and its parametric estimation method. IEEE Trans Ind Electron 57:3991–4000. doi:10.1109/TIE.2010.2076307
29. Dougal RA, Gao L, Liu S (2004) Ultracapacitor model with automatic order selection and capacity scaling for dynamic system simulation. J Power Sources 126:250–257. doi:10.1016/j.jpowsour.2003.08.031
30. Belhachemi F, Rael S, Davat B (2000) A physical based model of power electric double-layer supercapacitors. In: Conference record of the 2000 I.E. industry applications conference, vol 5, pp 3069–3076
31. Kurzweil P, Fischle H-J (2004) A new monitoring method for electrochemical aggregates by impedance spectroscopy. J Power Sources 127:331–340. doi:10.1016/j.jpowsour.2003.09.030
32. Robinson DB (2010) Optimization of power and energy densities in supercapacitors. J Power Sources 195:3748–3756. doi:10.1016/j.jpowsour.2009.12.004
33. Black J, Andreas HA (2010) Prediction of the self-discharge profile of an electrochemical capacitor electrode in the presence of both activation-controlled discharge and charge redistribution. J Power Sources 195:929–935. doi:10.1016/j.jpowsour.2009.08.040
34. El Brouji EH, Briat O, Vinassa JM et al (2009) Impact of calendar life and cycling ageing on supercapacitor performance. IEEE Trans Veh Technol 58:3917–3929
35. Hunter RJ (1987) Foundations of colloid science. Clarendon, Oxford
36. Gouy G (1910) Sur la constitution de la charge electrique a la surface d'un electrolyte. J Phys Fr 9
37. Masliyah JH, Bhattacharjee S (2006) Electrokinetic and colloid transport phenomena. Wiley, Hoboken, NJ
38. Wang H, Pilon L (2011) Accurate simulations of electric double layer capacitance of ultramicroelectrodes. J Phys Chem C 115:16711–16719. doi:10.1021/jp204498e
39. Grahame DC (1947) The electrical double layer and the theory of electrocapillarity. Chem Rev 41:441–501. doi:10.1021/cr60130a002
40. Varghese J, Wang H, Pilon L (2011) Simulating electric double layer capacitance of mesoporous electrodes with cylindrical pores. J Electrochem Soc 158:A1106–A1114. doi:10.1149/1.3622342
41. Woo S-W, Dokko K, Nakano H, Kanamura K (2008) Preparation of three dimensionally ordered macroporous carbon with mesoporous walls for electric double-layer capacitors. J Mater Chem 18:1674–1680. doi:10.1039/B717996K
42. Wang H, Pilon L (2013) Mesoscale modeling of electric double layer capacitors with three-dimensional ordered structures. J Power Sources 221:252–260. doi:10.1016/j.jpowsour.2012.08.002
43. Chaban VV, Kalugin ON (2009) Structure and dynamics in methanol and its lithium ion solution confined by carbon nanotubes. J Mol Liq 145:145–151. doi:10.1016/j.molliq.2008.06.003
44. Freund JB (2002) Electro-osmosis in a nanometer-scale channel studied by atomistic simulation. J Chem Phys 116:2194–2200. doi:10.1063/1.1431543
45. Qiao R, Aluru NR (2003) Ion concentrations and velocity profiles in nanochannel electroosmotic flows. J Chem Phys 118:4692–4701
46. Thompson AP (2003) Nonequilibrium molecular dynamics simulation of electro-osmotic flow in a charged nanopore. J Chem Phys 119:7503–7511. doi:10.1063/1.1609194
47. Qiao R, Aluru NR (2004) Charge inversion and flow reversal in a nanochannel electro-osmotic flow. Phys Rev Lett 92:198301. doi:10.1103/PhysRevLett.92.198301
48. Johnson AM, Newman J (1971) Desalting by means of porous carbon electrodes. J Electrochem Soc 118:510–517. doi:10.1149/1.2408094
49. Schiffer J, Linzen D, Sauer DU (2006) Heat generation in double layer capacitors. J Power Sources 160:765–772. doi:10.1016/j.jpowsour.2005.12.070

50. Kislenko SA, Samoylov IS, Amirov RH (2009) Molecular dynamics simulation of the electrochemical interface between a graphite surface and the ionic liquid [BMIM][PF6]. Phys Chem Chem Phys 11:5584–5590. doi:10.1039/B823189C
51. Kim D, Darve E (2006) Molecular dynamics simulation of electro-osmotic flows in rough wall nanochannels. Phys Rev E 73:051203. doi:10.1103/PhysRevE.73.051203
52. Wu P, Qiao R (2011) Physical origins of apparently enhanced viscosity of interfacial fluids in electrokinetic transport. Phys Fluids 1994-Present 23:072005. doi:10.1063/1.3614534
53. Cui ST, Cochran HD (2002) Molecular dynamics simulation of interfacial electrolyte behaviors in nanoscale cylindrical pores. J Chem Phys 117:5850–5854. doi:10.1063/1.1501585
54. Feng G, Qiao R, Huang J et al (2010) Ion distribution in electrified micropores and its role in the anomalous enhancement of capacitance. ACS Nano 4:2382–2390. doi:10.1021/nn100126w
55. Wander MCF, Shuford KL (2010) Molecular dynamics study of interfacial confinement effects of aqueous NaCl brines in nanoporous carbon. J Phys Chem C 114:20539–20546. doi:10.1021/jp104972e
56. Lankin AV, Norman GE, Stegailov VV (2010) Atomistic simulation of the interaction of an electrolyte with graphite nanostructures in perspective supercapacitors. High Temp 48:837–845. doi:10.1134/S0018151X10060106
57. Kondrat S, Kornyshev A (2011) Superionic state in double-layer capacitors with nanoporous electrodes. J Phys Condens Matter 23:022201. doi:10.1088/0953-8984/23/2/022201
58. Kondrat S, Georgi N, Fedorov MV, Kornyshev AA (2011) A superionic state in nano-porous double-layer capacitors: insights from Monte Carlo simulations. Phys Chem Chem Phys 13:11359–11366. doi:10.1039/C1CP20798A
59. Yang L, Fishbine BH, Migliori A, Pratt LR (2009) Molecular simulation of electric double-layer capacitors based on carbon nanotube forests. J Am Chem Soc 131:12373–12376. doi:10.1021/ja9044554
60. Shim Y, Kim HJ (2010) Nanoporous carbon supercapacitors in an ionic liquid: a computer simulation study. ACS Nano 4:2345–2355. doi:10.1021/nn901916m
61. Jiang D, Wu J (2013) Microscopic insights into the electrochemical behavior of nonaqueous electrolytes in electric double-layer capacitors. J Phys Chem Lett 4:1260–1267. doi:10.1021/jz4002967
62. Silva F, Gomes C, Figueiredo M et al (2008) The electrical double layer at the [BMIM][PF6] ionic liquid/electrode interface—effect of temperature on the differential capacitance. J Electroanal Chem 622:153–160. doi:10.1016/j.jelechem.2008.05.014
63. Kornyshev AA (2007) Double-layer in ionic liquids: paradigm change? J Phys Chem B 111:5545–5557. doi:10.1021/jp067857o
64. Lockett V, Horne M, Sedev R et al (2010) Differential capacitance of the double layer at the electrode/ionic liquids interface. Phys Chem Chem Phys 12:12499–12512. doi:10.1039/C0CP00170H
65. Baldelli S (2008) Surface structure at the ionic liquid–electrified metal interface. Acc Chem Res 41:421–431. doi:10.1021/ar700185h
66. Trulsson M, Algotsson J, Forsman J, Woodward CE (2010) Differential capacitance of room temperature ionic liquids: the role of dispersion forces. J Phys Chem Lett 1:1191–1195. doi:10.1021/jz900412t
67. Feng G, Zhang JS, Qiao R (2009) Microstructure and capacitance of the electrical double layers at the interface of ionic liquids and planar electrodes. J Phys Chem C 113:4549–4559. doi:10.1021/jp809900w
68. Holovko M, Kapko V, Henderson D, Boda D (2001) On the influence of ionic association on the capacitance of an electrical double layer. Chem Phys Lett 341:363–368. doi:10.1016/S0009-2614(01)00505-X
69. Lockett V, Sedev R, Ralston J et al (2008) Differential capacitance of the electrical double layer in imidazolium-based ionic liquids: influence of potential, cation size, and temperature. J Phys Chem C 112:7486–7495. doi:10.1021/jp7100732

70. Alam MT, Islam MM, Okajima T, Ohsaka T (2007) Measurements of differential capacitance at mercury/room-temperature ionic liquids interfaces. J Phys Chem C 111:18326–18333. doi:10.1021/jp075808l
71. Vatamanu J, Borodin O, Smith GD (2010) Molecular insights into the potential and temperature dependences of the differential capacitance of a room-temperature ionic liquid at graphite electrodes. J Am Chem Soc 132:14825–14833. doi:10.1021/ja104273r
72. Lin R, Taberna P-L, Fantini S et al (2011) Capacitive energy storage from −50 to 100 °C using an ionic liquid electrolyte. J Phys Chem Lett 2:2396–2401. doi:10.1021/jz201065t
73. Feng G, Li S, Atchison JS et al (2013) Molecular insights into carbon nanotube supercapacitors: capacitance independent of voltage and temperature. J Phys Chem C 117:9178–9186. doi:10.1021/jp403547k
74. Li S, Feng G, Fulvio PF et al (2012) Molecular dynamics simulation study of the capacitive performance of a binary mixture of ionic liquids near an onion-like carbon electrode. J Phys Chem Lett 3:2465–2469. doi:10.1021/jz3009387
75. Boda D, Henderson D, Chan K-Y (1999) Monte Carlo study of the capacitance of the double layer in a model molten salt. J Chem Phys 110:5346–5350. doi:10.1063/1.478429
76. Cagle C, Feng G, Qiao R et al (2009) Structure and charging kinetics of electrical double layers at large electrode voltages. Microfluid Nanofluidics 8:703–708. doi:10.1007/s10404-009-0542-2
77. Masarapu C, Zeng HF, Hung KH, Wei B (2009) Effect of temperature on the capacitance of carbon nanotube supercapacitors. ACS Nano 3:2199–2206. doi:10.1021/nn900500n
78. Boda D, Henderson D (2000) The capacitance of the solvent primitive model double layer at low effective temperatures. J Chem Phys 112:8934–8938. doi:10.1063/1.481507
79. Feng G, Qiao R, Huang J et al (2010) The importance of ion size and electrode curvature on electrical double layers in ionic liquids. Phys Chem Chem Phys 13:1152–1161. doi:10.1039/C0CP02077J
80. Merlet C, Rotenberg B, Madden PA et al (2012) On the molecular origin of supercapacitance in nanoporous carbon electrodes. Nat Mater 11:306–310. doi:10.1038/nmat3260
81. Kondrat S, Pérez CR, Presser V et al (2012) Effect of pore size and its dispersity on the energy storage in nanoporous supercapacitors. Energy Environ Sci 5:6474. doi:10.1039/c2ee03092f
82. Feng G, Huang J, Sumpter BG et al (2011) A "counter-charge layer in generalized solvents" framework for electrical double layers in neat and hybrid ionic liquid electrolytes. Phys Chem Chem Phys 13:14723. doi:10.1039/c1cp21428d
83. Frackowiak E, Lota G, Pernak J (2005) Room-temperature phosphonium ionic liquids for supercapacitor application. Appl Phys Lett 86:164104. doi:10.1063/1.1906320
84. Xing L, Vatamanu J, Borodin O et al (2012) Electrode/electrolyte interface in sulfolane-based electrolytes for Li Ion batteries: a molecular dynamics simulation study. J Phys Chem C 116:23871–23881. doi:10.1021/jp3054179
85. Vatamanu J, Borodin O, Smith GD (2012) Molecular dynamics simulation studies of the structure of a mixed carbonate/LiPF6 electrolyte near graphite surface as a function of electrode potential. J Phys Chem C 116:1114–1121. doi:10.1021/jp2101539
86. Ozoliņš V, Zhou F, Asta M (2013) Ruthenia-based electrochemical supercapacitors: insights from first-principles calculations. Acc Chem Res 46:1084–1093. doi:10.1021/ar3002987
87. Hou C-H, Taboada-Serrano P, Yiacoumi S, Tsouris C (2008) Monte Carlo simulation of electrical double-layer formation from mixtures of electrolytes inside nanopores. J Chem Phys 128:044705. doi:10.1063/1.2824957
88. Pillay B, Newman J (1996) The influence of side reactions on the performance of electrochemical double-layer capacitors. J Electrochem Soc 143:1806–1814. doi:10.1149/1.1836908
89. Wu J (2006) Density functional theory for chemical engineering: from capillarity to soft materials. AIChE J 52:1169–1193. doi:10.1002/aic.10713
90. Wu J, Li Z (2007) Density-functional theory for complex fluids. Annu Rev Phys Chem 58:85–112. doi:10.1146/annurev.physchem.58.032806.104650

91. Hohenberg P, Kohn W (1964) Inhomogeneous electron gas. Phys Rev 136:B864–B871. doi:10.1103/PhysRev.136.B864
92. Jiang D, Meng D, Wu J (2011) Density functional theory for differential capacitance of planar electric double layers in ionic liquids. Chem Phys Lett 504:153–158. doi:10.1016/j.cplett.2011.01.072
93. Wu J, Jiang T, Jiang D et al (2011) A classical density functional theory for interfacial layering of ionic liquids. Soft Matter 7:11222–11231. doi:10.1039/C1SM06089A
94. Xing L, Vatamanu J, Borodin O, Bedrov D (2013) On the atomistic nature of capacitance enhancement generated by ionic liquid electrolyte confined in subnanometer pores. J Phys Chem Lett 4:132–140. doi:10.1021/jz301782f
95. Huang J, Sumpter BG, Meunier V (2008) A universal model for nanoporous carbon supercapacitors applicable to diverse pore regimes, carbon materials, and electrolytes. Chem Eur J 14:6614–6626. doi:10.1002/chem.200800639
96. Huang J, Sumpter BG, Meunier V (2008) Theoretical model for nanoporous carbon supercapacitors. Angew Chem Int Ed 47:520–524. doi:10.1002/anie.200703864
97. Feng G, Jiang D, Cummings PT (2012) Curvature effect on the capacitance of electric double layers at ionic liquid/onion-like carbon interfaces. J Chem Theory Comput 8:1058–1063. doi:10.1021/ct200914j
98. Nguyen PTM, Fan C, Do DD, Nicholson D (2013) On the cavitation-like pore blocking in Ink-bottle pore: evolution of hysteresis loop with neck size. J Phys Chem C 117:5475–5484. doi:10.1021/jp4002912
99. Pizio O, Sokołowski S, Sokołowska Z (2012) Electric double layer capacitance of restricted primitive model for an ionic fluid in slit-like nanopores: a density functional approach. J Chem Phys 137:234705. doi:10.1063/1.4771919
100. Ravikovitch PI, Neimark AV (2006) Density functional theory model of adsorption on amorphous and microporous silica materials. Langmuir 22:11171–11179. doi:10.1021/la0616146
101. Gor GY, Thommes M, Cychosz KA, Neimark AV (2012) Quenched solid density functional theory method for characterization of mesoporous carbons by nitrogen adsorption. Carbon 50:1583–1590. doi:10.1016/j.carbon.2011.11.037
102. Dunn D, Newman J (2000) Predictions of specific energies and specific powers of double-layer capacitors using a simplified model. J Electrochem Soc 147:820–830. doi:10.1149/1.1393278
103. Doyle M, Fuller TF, Newman J (1993) Modeling of galvanostatic charge and discharge of the lithium/polymer/insertion cell. J Electrochem Soc 140:1526–1533. doi:10.1149/1.2221597
104. Staser JA, Weidner JW (2014) Mathematical modeling of hybrid asymmetric electrochemical capacitors. J Electrochem Soc 161:E3267–E3275. doi:10.1149/2.031408jes
105. Srinivasan V, Weidner JW (1999) Mathematical modeling of electrochemical capacitors. J Electrochem Soc 146:1650–1658. doi:10.1149/1.1391821
106. Gomadam PM, Weidner JW, Dougal RA, White RE (2002) Mathematical modeling of lithium-ion and nickel battery systems. J Power Sources 110:267–284. doi:10.1016/S0378-7753(02)00190-8
107. Chang JH, Dawson FP, Lian KK (2011) A first principles approach to develop a dynamic model of electrochemical capacitors. IEEE Trans Power Electron 26:3472–3480. doi:10.1109/TPEL.2011.2161096
108. Gualous H, Bouquain D, Berthon A, Kauffmann JM (2003) Experimental study of supercapacitor serial resistance and capacitance variations with temperature. J Power Sources 123:86–93. doi:10.1016/S0378-7753(03)00527-5
109. Conway BE, Niu J, Pell WG (2003) Seminar on double layer capacitor and hybrid energy storage devices. In: Proceedings of the 13th international seminar on double layer capacity and hybrid energy storage devices
110. Kötz R, Hahn M, Gallay R (2006) Temperature behavior and impedance fundamentals of supercapacitors. J Power Sources 154:550–555. doi:10.1016/j.jpowsour.2005.10.048

111. Guillemet P, Scudeller Y, Brousse T (2006) Multi-level reduced-order thermal modeling of electrochemical capacitors. J Power Sources 157:630–640. doi:10.1016/j.jpowsour.2005.07.072
112. Henson W (2008) Optimal battery/ultracapacitor storage combination. J Power Sources 179:417–423. doi:10.1016/j.jpowsour.2007.12.083
113. Camara MB, Gualous H, Gustin F et al (2010) DC/DC converter design for supercapacitor and battery power management in hybrid vehicle applications #x2014;polynomial control strategy. IEEE Trans Ind Electron 57:587–597. doi:10.1109/TIE.2009.2025283
114. Greenwell W, Vahidi A (2010) Predictive control of voltage and current in a fuel cell #x2013; ultracapacitor hybrid. IEEE Trans Ind Electron 57:1954–1963. doi:10.1109/TIE.2009.2031663
115. Payman A, Pierfederici S, Meibody-Tabar F (2009) Energy management in a fuel cell/supercapacitor multisource/multiload electrical hybrid system. IEEE Trans Power Electron 24:2681–2691. doi:10.1109/TPEL.2009.2028426
116. Lemofouet S, Rufer A (2006) A hybrid energy storage system based on compressed air and supercapacitors with maximum efficiency point tracking (MEPT). IEEE Trans Ind Electron 53:1105–1115. doi:10.1109/TIE.2006.878323
117. Kurzweil P, Chwistek M (2008) Electrochemical stability of organic electrolytes in supercapacitors: spectroscopy and gas analysis of decomposition products. J Power Sources 176:555–567. doi:10.1016/j.jpowsour.2007.08.070
118. D' Entremont A, Pilon L (2014) First-principles thermal modeling of electric double layer capacitors under constant-current cycling. J Power Sources 246:887–898. doi:10.1016/j.jpowsour.2013.08.024
119. Gualous H, Louahlia H, Gallay R (2011) Supercapacitor characterization and thermal modelling with reversible and irreversible heat effect. IEEE Trans Power Electron 26:3402–3409. doi:10.1109/TPEL.2011.2145422
120. Burheim OS, Aslan M, Atchison JS, Presser V (2014) Thermal conductivity and temperature profiles in carbon electrodes for supercapacitors. J Power Sources 246:160–166. doi:10.1016/j.jpowsour.2013.06.164
121. Wang G, Zhang L, Zhang J (2012) A review of electrode materials for electrochemical supercapacitors. Chem Soc Rev 41:797–828. doi:10.1039/c1cs15060j
122. Conway BE, Pell WG, Liu T-C (1997) Diagnostic analyses for mechanisms of self-discharge of electrochemical capacitors and batteries. J Power Sources 65:53–59. doi:10.1016/S0378-7753(97)02468-3
123. Pell WG, Conway BE, Adams WA, de Oliveira J (1999) Electrochemical efficiency in multiple discharge/recharge cycling of supercapacitors in hybrid EV applications. J Power Sources 80:134–141. doi:10.1016/S0378-7753(98)00257-2
124. Ricketts BW, Ton-That C (2000) Self-discharge of carbon-based supercapacitors with organic electrolytes. J Power Sources 89:64–69. doi:10.1016/S0378-7753(00)00387-6
125. Niu J, Conway BE, Pell WG (2004) Comparative studies of self-discharge by potential decay and float-current measurements at C double-layer capacitor and battery electrodes. J Power Sources 135:332–343. doi:10.1016/j.jpowsour.2004.03.068
126. Kazaryan SA, Kharisov GG, Litvinenko SV, Kogan VI (2007) Self-Discharge Related to Iron Ions and its Effect on the Parameters of HES PbO2 | H2SO4 | C Systems. J Electrochem Soc 154:A751–A759. doi: 10.1149/1.2742335
127. Kazaryan SA, Litvinenko SV, Kharisov GG (2008) Self-discharge of heterogeneous electrochemical supercapacitor of PbO2 I H2SO4 I C related to manganese and titanium ions. J Electrochem Soc 155:A464–A473. doi:10.1149/1.2904456
128. Kim K-M, Hur J-W, Jung S-I, Kang A-S (2004) Electrochemical characteristics of activated carbon/Ppy electrode combined with P(VdF-co-HFP)/PVP for EDLC. Electrochim Acta 50:863–872. doi:10.1016/j.electacta.2004.02.059
129. Black J, Andreas HA (2009) Effects of charge redistribution on self-discharge of electrochemical capacitors. Electrochim Acta 54:3568–3574. doi:10.1016/j.electacta.2009.01.019

130. Andreas HA, Lussier K, Oickle AM (2009) Effect of Fe-contamination on rate of self-discharge in carbon-based aqueous electrochemical capacitors. J Power Sources 187:275–283. doi:10.1016/j.jpowsour.2008.10.096
131. Ishimoto S, Asakawa Y, Shinya M, Naoi K (2009) Degradation responses of activated-carbon-based EDLCs for higher voltage operation and their factors. J Electrochem Soc 156:A563–A571. doi:10.1149/1.3126423
132. Kowal J, Avaroglu E, Chamekh F et al (2011) Detailed analysis of the self-discharge of supercapacitors. J Power Sources 196:573–579. doi:10.1016/j.jpowsour.2009.12.028
133. Yang H, Zhang Y (2011) Self-discharge analysis and characterization of supercapacitors for environmentally powered wireless sensor network applications. J Power Sources 196:8866–8873. doi:10.1016/j.jpowsour.2011.06.042
134. Zhang Q, Rong J, Ma D, Wei B (2011) The governing self-discharge processes in activated carbon fabric-based supercapacitors with different organic electrolytes. Energy Environ Sci 4:2152–2159. doi:10.1039/C0EE00773K
135. Pell WG, Conway BE (2001) Voltammetry at a de Levie brush electrode as a model for electrochemical supercapacitor behaviour. J Electroanal Chem 500:121–133. doi:10.1016/S0022-0728(00)00423-X
136. Xu B, Wu F, Su Y et al (2008) Competitive effect of KOH activation on the electrochemical performances of carbon nanotubes for EDLC: balance between porosity and conductivity. Electrochim Acta 53:7730–7735. doi:10.1016/j.electacta.2008.05.033
137. Hsieh C-T, Teng H (2002) Influence of oxygen treatment on electric double-layer capacitance of activated carbon fabrics. Carbon 40:667–674. doi:10.1016/S0008-6223(01)00182-8
138. Skipworth E, Donne SW (2007) Role of graphite in self-discharge of nickel(III) oxyhydroxide. J Power Sources 174:186–190. doi:10.1016/j.jpowsour.2007.07.078
139. Niu J, Pell WG, Conway BE (2006) Requirements for performance characterization of C double-layer supercapacitors: applications to a high specific-area C-cloth material. J Power Sources 156:725–740. doi:10.1016/j.jpowsour.2005.06.002
140. Zhang Q, Cai C, Qin J, Wei B (2014) Tunable self-discharge process of carbon nanotube based supercapacitors. Nano Energy 4:14–22. doi:10.1016/j.nanoen.2013.12.005
141. Chen L, Bai H, Huang Z, Li L (2014) Mechanism investigation and suppression of self-discharge in active electrolyte enhanced supercapacitors. Energy Env Sci 7:1750–1759. doi:10.1039/C4EE00002A

Index

K.I. Ozoemena, S. Chen (eds.), *Nanomaterials in Advanced Batteries and Supercapacitors*, Nanostructure Science and Technology,
DOI 10.1007/978-3-319-26082-2

MIX
Papier aus verantwortungsvollen Quellen
Paper from responsible sources
FSC® C105338

If you have any concerns about our products,
you can contact us on
ProductSafety@springernature.com

In case Publisher is established outside the EU,
the EU authorized representative is:
Springer Nature Customer Service Center GmbH
Europaplatz 3, 69115 Heidelberg, Germany

Printed by Libri Plureos GmbH
in Hamburg, Germany